STATISTICAL PORTFOLIO ESTIMATION

STATISTICAL PORTFOLIO ESTIMATION

Masanobu Taniguchi, Hiroshi Shiraishi,
Junichi Hirukawa, Hiroko Kato Solvang,
and Takashi Yamashita

CRC Press
Taylor & Francis Group
Boca Raton London New York

CRC Press is an imprint of the
Taylor & Francis Group, an **informa** business
A CHAPMAN & HALL BOOK

CRC Press
Taylor & Francis Group
6000 Broken Sound Parkway NW, Suite 300
Boca Raton, FL 33487-2742

First issued in paperback 2021

Version Date: 20170720

ISBN 13: 978-1-03-209649-0 (pbk)
ISBN 13: 978-1-4665-0560-5 (hbk)

Publisher's Note
The publisher has gone to great lengths to ensure the quality of this reprint but points out that some imperfections in the original copies may be apparent.

Visit the Taylor & Francis Web site at
http://www.taylorandfrancis.com

and the CRC Press Web site at
http://www.crcpress.com

Contents

Preface

The field of financial engineering has developed as a huge integration of economics, probability theory, statistics, etc., for some decades. The composition of porfolios is one of the most fundamental and important methods of financial engineering to control the risk of investments. This book provides a comprehensive development of statistical inference for portfolios and its applications. Historically, Markowitz (1952) contributed to the advancement of modern portfolio theory by laying the foundation for the diversification of investment portfolios. His approach is called the mean-variance portfolio, which maximizes the mean of portfolio return with reducing its variance (risk of portfolio). Actually, the mean-variance portfolio coefficients are expressed as a function of the mean and variance matrix of the return process. Optimal portfolio coefficients based on the mean and variance matrix of return have been derived by various criteria. Assuming that the return process is i.i.d. Gaussian, Jobson and Korkie (1980) proposed a portfolio coefficient estimator of the optimal portfolio by making the sample version of the mean-variance portfolio. However, empirical studies show that observed stretches of financial return are often are non-Gaussian dependent. In this situation, it is shown that portfolio estimators of the mean-variance type are not asymptotically efficient generally even if the return process is Gaussian, which gives a strong warning for use of the usual portfolio estimators. We also provide a necessary and sufficient condition for the estimators to be asymptotically efficient in terms of the spectral density matrix of the return. This motivates the fundamental important issue of the book. Hence we will provide modern statistical techniques for the problems of portfolio estimation, grasping them as optimal statistical inference for various return processes. We will introduce a variety of stochastic processes, e.g., non-Gaussian stationary processes, non-linear processes, non-stationary processes, etc. For them we will develop a modern statistical inference by use of local asymptotic normality (LAN), which is due to LeCam. The approach is a unified and very general one. Based on this we address a lot of important problems for portfolios. It is well known that a Markowitz portfolio is to optimize the mean and variance of portfolio return. However, recent years have seen increasing development of new risk measures instead of the variance such as Value at Risk (VaR), Expected Shortfall (ES), Conditional Value at Risk (CVaR) and Tail Conditional Expectation (TCE). We show how to construct optimal portfolio estimators based on these risk measures. Thanks to the advancement of computer technology, a variety of financial transactions have became possible in a very short time in recent years. From this point of view, multiperiod problems are one of the most important issues in portfolio theory. We discuss this problems from three directions of perspective. In the investment of the optimal portfolio strategy, multivariate time series models are required. Moreover, the assumption that the innovation densities underlying those models are known seems quite unrealistic. If those densities remain unspecified, the model becomes a semiparametric one, and rank-based inference methods naturally come into the picture. Rank-based inference methods under very general conditions are known to achieve semiparametric efficiency bounds. However, defining ranks in the context of multivariate time series models is not obvious. Two distinct definitions can be proposed. The first one relies on the assumption that the innovation process is described by some unspecified independent component analysis model. The second one relies on the assumption that the innovation density is some unspecified elliptical density. Applications to portfolio management problems, rank statistics and mean-diversification efficient frontier are discussed. These examples give readers a practical hint for applying the introduced portfolio theories. This book contains applications ranging widely, from financial field to genome and medical science. These examples give readers a hint for applying the introduced portfolio theories.

This book is suitable for undergraduate and graduate students who specialize in statistics, mathematics, finance, econometrics, genomics, etc., as a textbook. Also it is appropriate for researchers in related fields as a reference book.

Throughout our writing of the book, our collaborative researchers have offered us advice, debate, inspiration and friendship. For this we wish to thank Marc Hallin, Holger Dette, Yury Kutoyants, Ngai Hang Chan, Liudas Giraitis, Murad S. Taqqu, Anna Clara Monti, Cathy W. Chen, David Stoffer, Sangyeol Lee, Ching-Kan Ing, Poul Embrechts, Roger Koenker, Claudia Klüppelberg, Thomas Mikosch, Richard A.Davis and Zudi Lu.

The research by the first author M.T. has been supported by the Research Institute for Science & Engineering, Waseda University, Japanese Government Pension Investment Fund and JSPS fundings: Kiban(A)(23244011), Kiban(A)(15H02061) and Houga(26540015). M.T. deeply thanks all of them for their generous support and kindness. The research by the second author H.S. has been supported by JSPS fundings: Wakate(B)(24730193), Kiban(C)(16K00036) and Core-to-Core Program (Foundation of a Global Research Cooperative Center in Mathematics focused on Number Theory and Geometry). The research by the third author J.H. has been supported by JSPS funding: Kiban(C)(16K00042).

Masanobu Taniguchi, Waseda, Tokyo,

Hiroshi Shiraishi, Keio, Kanagawa,

Junichi Hirukawa, Niigata,

Hiroko Kato Solvang, Institute of Marine Research, Bergen

Takashi Yamashita, Government Pension Investment Fund, Tokyo

Chapter 1

Introduction

This book provides a comprehensive integration of statistical inference for portfolios and its applications. Historically, Markowitz (1952) contributed to the advancement of modern portfolio theory by laying the foundation for the diversification of investment portfolios. His approach is called the mean-variance portfolio, which maximizes the mean of portfolio return with reducing its variance (risk of portfolio). Actually, if the return process of assets has the mean vector $\boldsymbol{\mu}$ and variance matrix Σ, then the mean-variance portfolio coefficients are expressed as a function of $\boldsymbol{\mu}$ and Σ.

Optimal portfolio coefficients based on $\boldsymbol{\mu}$ and Σ have been derived by various criteria. Assuming that the return process is i.i.d. Gaussian, Jobson and Korkie (1980) proposed a portfolio coefficient estimator of the optimal portfolio by substituting the sample mean vector $\hat{\boldsymbol{\mu}}$ and the sample variance matrix $\hat{\Sigma}$ into $\boldsymbol{\mu}$ and Σ, respectively. However, empirical studies show that observed stretches of financial return are often not i.i.d. and non-Gaussian. This leads us to the assumption that financial returns are non-Gaussian-dependent processes. From this point of view, Basak et al. (2002) showed the consistency of optimal portfolio estimators when portfolio returns are stationary processes. However, in the literature there has been no study on the asymptotic efficiency of estimators for optimal portfolios.

In this book, denoting optimal portfolios by a function $g = g(\boldsymbol{\mu}, \Sigma)$ of $\boldsymbol{\mu}$ and Σ, we discuss the asymptotic efficiency of estimators $\hat{g} = g(\hat{\boldsymbol{\mu}}, \hat{\Sigma})$ when the return is a vector-valued non-Gaussian stationary process. In Section 8.5 it is shown that $\hat{g}$ is not asymptotically efficient generally even if the return process is Gaussian, which gives a strong warning for use of the usual estimators $\hat{g}$. We also provide a necessary and sufficient condition for $\hat{g}$ to be asymptotically efficient in terms of the spectral density matrix of the return. This motivates the fundamental important issue of the book. Hence we will provide modern statistical techniques for the problems of portfolio estimation, grasping them as optimal statistical inference for various return processes. Because empirical financial phenomena show that return processes are non-Gaussian and dependent, we will introduce a variety of stochastic processes, e.g., non-Gaussian stationary processes, non-linear processes, non-stationary processes, etc. For them we will develop a modern statistical inference by use of local asymptotic normality (LAN), which is due to LeCam. The approach is a unified and very general one. We will deal with not only the mean-variance portfolio but also other portfolios such as the mean-VaR portfolio, the mean-CVaR portfolio and the pessimistic portfolio. These portfolios have recently wide attention. Moreover, we discuss multiperiod problems received which are much more realistic that (traditional) single period problems. In this framework, we need to construct a sequence or a continuous function of asset allocations. If the innovation densities are unspecified for the multivariate time series models in the analyses of optimal portfolios, semiparametric inference methods are required. To achieve semiparametric efficient inference, rank-based methods under very general conditions are widely used. As examples, we provide varous practical applications ranging from financial science to medical science.

This book is organized as follows. Chapter 2 explains the foundation of stochastic processes, because a great deal of data in economics, finance, engineering and the natural sciences occur in the form of time series where observations are dependent and where the nature of this dependence is of interest in itself. A model which describes the probability structure of the time series is called a stochastic process. We provide a modern introduction to stochastic processes. Because statistical analysis for stochastic processes largely relies on asymptotic theory, some useful limit theorems and central limit theorems are introduced. In Chapter 3, we introduce the modern portfolio theory of Markowitz (1952), the capital asset pricing model (CAPM) of Sharpe (1964) and the arbitrage pricing theory of Ross (1976). Then, we discuss other portfolios based on new risk measures such as VaR and CVaR. Moreover, a variety of statistical estimation methods for portfolios are introduced with some simulation studies. We discuss the multiperiod problem in Chapter 4. By using dynamic programming, a secence (or a continuous function) of portfolio weights is obtained in the discrete time case (or the continuous time case). In addition, the universal portfolio (UP) introduced by Cover (1991) is discussed. The idea of this portfolio comes from the information theory of Shannon (1945) and Kelly (1956).

The concepts of rank-based inference are simple and easy to understand even at an introductory level. However, at the same time, they provide asymptotically optimal semiparametric inference. We discuss portfolio estimation problems based on rank statistics in Chapter 5. Section 5.1 contains the introduction and history of rank statistics. The important concepts and methods for rank-based inference, e.g., maximal invariants, least favourable and U-statistics for stationary processes, are also introduced. In Section 5.2, semiparametrically efficient estimations in time series are addressed. As a introduction, we discuss the testing problem for randomness against the ARMA alternative, and testing for the ARMA model against other ARMA alternatives. The notion of tangent space, which is important for semiparametric optimal inference, is introduced, and the semiparametric asymptotic optimal theory is addressed. Semiparametrically efficient estimations in univariate time series, and the multivariate elliptical residuals case are discussed. In Section 5.3, we discuss the asymptotic theory of a class of rank order statistics for the two-sample problem based on the squared residuals from two classes of ARCH models. Since portfolio estimation includes inference for multivariate time series, independent component analysis (ICA) for multivariate time series is useful for rank-based inference, and is introduced in Section 5.4. First, we address the foregoing models for financial multivariate time series. Then, we discuss ICA modeling in both the time domain and the frequency domain. Finally, portfolio estimations based on ranks for an independent component and on ranks for elliptical residuals are discussed in Section 5.5.

The classical theory of portfolio estimation often assumes that asset returns are independent normal random variables. However, in view of empirical financial studies, we recognize that the financial returns are not Gaussian or independent. In Chapter 6, assuming that the return process is generated by a linear process with skew-normal innovations, we evaluate the influence of skewness on the asymptotics of mean-variance portfolios, and discuss a robustness with respect to skewness. We also address the problems of portfolio estimation (i) when the utility function depends on exogeneous variables and (ii) when the portfolio coefficients depend on causal variables. In the case of (ii) we apply the results to the Japanese pension investment portfolio by use of the canonical correlation method. Chapter 7 provides five applications: Japanese companies monthly log-returns analysis based on the theory of portfolio estimators, Japanese goverment pension investment fund analysis for the multiperiod portfolio optimization problem, microarray data analysis by rank-order statistics for an ARCH residual, and DNA sequence data analysis and a cortico muscular coupling analysis as applications for a mean-diversification portfolio. As for the statistical inference, Chapter 8 is the main chapter. To describe financial returns, we introduce a variety of stochastic processes, e.g., vector-valued non-Gaussian linear processes, semimartingales, ARCH(∞)-SM, CHARN process, locally stationary process, long memory process, etc. For them we introduce some fundamental statistics, and show their fundamental asymptotics. As the actual data analysis we have to choose a statistical model from a class of candidate processes. Section 8.4 introduces a generalized AIC(GAIC), and derives contiguous asymptotics of the selected order. We elucidate the philosophy of GAIC and the other criteria, which is helpful for use of these criteria. Section 8.5 shows that the generalized mean-variance portfolio estimator $\hat{g}$ is not asymptotically optimal in general even if the return process is a Gaussian process. Then we give a necessary and sufficient condition for $\hat{g}$ to be asymptotically optimal in terms of the spectra of the return process. Then we seek the optimal one. In this book we develop modern statistical inference for stochastic processes by use of LAN. Section 8.3 provides the foundation of the LAN approach, leading to a conclusion that the MLE is asymptotically optimal. Section 8.6 addresses the problem of shrinkage estimation for stochastic processes and time series regression models, which will be helpful when the return processes are of high dimension.

Chapter 2

Preliminaries

This chapter discusses the foundation of stochastic processes. Much of statistical analysis is concerned with models in which the observations are assumed to vary independently. However, a great deal of data in economics, finance, engineering, and the natural sciences occur in the form of time series where observations are dependent and where the nature of this dependence is of interest in itself. A model which describes the probability structure of a series of observations X_t, $t = 1, \ldots, n$, is called a stochastic process. An X_t might be the value of a stock price at time point t, the water level in a lake at time point t, and so on. The main purpose of this chapter is to provide a modern introduction to stochastic processes. Because statistical analysis for stochastic processes largely relies on asymptotic theory, we explain some useful limit theorems and central limit theorems.

2.1 Stochastic Processes and Limit Theorems

Suppose that X_t is the water level at a given place of a lake at time point t. We may describe its fluctuating evolution with respect to t as a family of random variables $\{X_1, X_2, \ldots\}$ indexed by the discrete time parameter $t \in \mathbb{N} \equiv \{1, 2, \ldots\}$. If X_t is the number of accidents at an intersection during the time interval $[0, t]$, this leads to a family of random variables $\{X_t : t \geq 0\}$ indexed by the continuous time parameter t. More generally,

Definition 2.1.1. *Given an index set T, a stochastic process indexed by T is a collection of random variables $\{X_t : t \in T\}$ on a probability space $(\Omega, \mathcal{F}, P)$ taking values in a set $\mathcal{S}$, which is called the state space of the process.*

If $T = \mathbb{Z} \equiv$ the set of all integers, we say that $\{X_t\}$ is a *discrete time* stochastic process, and it is often written as $\{X_t : t \in \mathbb{Z}\}$. If $T = [0, \infty)$, then $\{X_t\}$ is called a *continuous time* process, and it is often written as $\{X_t : t \in [0, \infty)\}$.

The state space $\mathcal{S}$ is the set in which the possible values of each X_t lie. If $\mathcal{S} = \{\ldots, -1, 0, 1, 2, \ldots\}$, we refer to the process as an integer-valued process. If $\mathcal{S} = \mathbb{R} = (-\infty, \infty)$, then we call $\{X_t\}$ a real-valued process. If $\mathcal{S}$ is the m-dimensional Euclidean space $\mathbb{R}^m$, then $\{X_t\}$ is said to be an *m-vector process*, and it is written as $\{\boldsymbol{X}_t\}$.

The distinguishing features of a stochastic process $\{X_t : t \in T\}$ are relationships among the random variables $X_t, t \in T$. These relationships are specified by the joint distribution function of every finite family $X_{t_1}, \ldots, X_{t_n}$ of the process.

In what follows we provide concrete and typical models of stochastic processes.

Example 2.1.2. *(AR(p)-, MA(q)-, and ARMA(p, q)-processes).*

Let $\{u_t : t \in \mathbb{Z}\}$ be a family of independent and identically distributed (i.i.d.) random variables with mean zero and variance σ^2 (for convenience, we write $\{u_t\} \sim$ i.i.d.$(0, \sigma^2)$). We define a

stochastic process $\{X_t : t \in \mathbb{Z}\}$ *by*

$$X_t = -(a_1 X_{t-1} + \cdots + a_p X_{t-p}) + u_t \tag{2.1.1}$$

where the a_j*'s are real constants* ($a_p \neq 0$). *The process is said to be an autoregressive process of order* p *(or AR(p)), which was introduced in Yule's analysis of sunspot numbers. This model has an intuitive appeal in view of the usual regression analysis. A more general class of stochastic model is*

$$X_t = -\sum_{j=1}^{p} a_j X_{t-j} + u_t + \sum_{j=1}^{q} b_j u_{t-j} \tag{2.1.2}$$

where the b_j*'s are real constants* ($b_q \neq 0$). *This process* $\{X_t : t \in \mathbb{Z}\}$ *is said to be an autoregressive moving average process (model) of order* (p, q), *and is denoted by ARMA*(p, q). *The special case of ARMA*$(0, q)$ *is referred to as a moving average process (model) of order* q, *denoted by MA*(q).

Example 2.1.3. *(Nonlinear processes).*

A stochastic process $\{X_t : t \in \mathbb{Z}\}$ *is said to follow a nonlinear autoregressive model of order* p *if there exists a measurable function* $f : \mathbb{R}^{p+1} \to \mathbb{R}$ *such that*

$$X_t = f(X_{t-1}, X_{t-2}, \ldots, X_{t-p}, u_t), \quad t \in \mathbb{Z} \tag{2.1.3}$$

where $\{u_t\} \sim$ *i.i.d.*$(0, \sigma^2)$. *Typical examples are as follows. Tong (1990) proposed a self-exciting threshold autoregressive model (SETAR*$(k; p, \ldots, p)$ *model) defined by*

$$X_t = a_0^{(j)} + a_1^{(j)} X_{t-1} + \cdots + a_p^{(j)} X_{t-p} + u_t, \tag{2.1.4}$$

if $X_{t-d} \in \Omega_j$, $j = 1, \ldots, k$, *where the* Ω_j *are disjoint intervals on* $\mathbb{R}$ *with* $\bigcup_{j=1}^{k} \Omega_j = \mathbb{R}$, *and* $d > 0$ *is called the "threshold lag." Here* $a_i^{(j)}$, $i = 0, \ldots, p$, $j = 1, \ldots, k$, *are real constants. In econometrics, it is not natural to assume a constant one-period forecast variance. As a plausible model, Engle (1982) introduced an autoregressive conditional heteroscedastic model (ARCH*(q)*), which is defined as*

$$X_t = u_t \sqrt{a_0 + \sum_{j=1}^{q} a_j X_{t-j}^2} \tag{2.1.5}$$

where $a_0 > 0$ *and* $a_j \geq 0$, $j = 1, \ldots, q$.

A continuous time stochastic process $\{X_t : t \in [0, \infty)\}$ *with values in the state space* $\mathcal{S} = \{0, 1, 2, \ldots\}$ *is called a counting process if* X_t *for any* t *represents the total number of events that have occurred during the time period* $[0, t]$.

Example 2.1.4. *(Poisson processes) A counting process* $\{X_t : t \in [0, \infty)\}$ *is said to be a homogeneous Poisson process with rate* $\lambda > 0$ *if*

(i) $X_0 = 0$,

(ii) *for all choices of* $t_1, \ldots, t_n \in [0, \infty)$ *satisfying* $t_1 < t_2 < \cdots < t_n$, *the increments*

$$X_{t_2} - X_{t_1}, X_{t_3} - X_{t_2}, \ldots, X_{t_n} - X_{t_{n-1}}$$

are independent,

(iii) *for all* $s, t \geq 0$,

$$P(X_{t+s} - X_s = x) = \frac{e^{-\lambda x}(\lambda t)^x}{x!}, x = 0, 1, 2, \ldots,$$

i.e., $X_{t+s} - X_s$ *is Poisson distributed with rate* λt.

The following example is one of the most fundamental and important continuous time stochastic processes.

Example 2.1.5. *(Brownian motion or Wiener process) A continuous time stochastic process* $\{X_t : t \in [0, \infty)\}$ *is said to be a Brownian motion or Wiener process if*

(i) $X_0 = 0$,

(ii) *for all choices of* $t_1, \ldots, t_n \in [0, \infty)$ *satisfying* $t_1 < t_2 < \ldots < t_n$, *the increments*

$$X_{t_2} - X_{t_1}, X_{t_3} - X_{t_2}, \ldots, X_{t_n} - X_{t_{n-1}}$$

are independent,

(iii) *for any* $t > s$,

$$P(X_t - X_s \leq x) = \frac{1}{\sqrt{2\pi c(t-s)}} \int_{-\infty}^{x} \exp\left\{-\frac{u^2}{2c(t-s)}\right\} du,$$

where c is a positive constant, i.e., $X_t - X_s \sim N(0, c(t-s))$.

The following process is often used to describe a variety of phenomena.

Example 2.1.6. *(Diffusion process) Suppose a particle is plunged into a nonhomogeneous and moving fluid. Let* X_t *be the position of the particle at time t and* $\mu(x, t)$ *the velocity of a small volume v of fluid located at time t. A particle within v will carry out Brownian motion with parameter* $\sigma(x, t)$. *Then the change in position of the particle in time interval* $[t, t + \Delta t]$ *can be written approximately in the form*

$$X_{t+\Delta t} - X_t \sim \mu(X_t, t)\Delta t + \sigma(X_t, t)(W_{t+\Delta t} - W_t),$$

where $\{W_t\}$ *is a Brownian motion process. If we replace the increments by differentials, we obtain the differential equation*

$$dX_t = \mu(X_t, t)dt + \sigma(X_t, t)dW_t, \tag{2.1.6}$$

where $\mu(X_t, t)$ *is called the drift and* $\sigma(X_t, t)$ *the diffusion. A rigorous discussion will be given later on. A stochastic process* $\{X_t : t \in [0, \infty)\}$ *is said to be a diffusion process if it satisfies (2.1.6). Recently this type of diffusion process has been used to model financial data.*

Since probability theory has its roots in games of chance, it is profitable to interpret results in terms of gambling.

Definition 2.1.7. *Let* $(\Omega, \mathcal{F}, P)$ *be a probability space,* $\{X_1, X_2, \ldots\}$ *a sequence of integrable random variables on* $(\Omega, \mathcal{F}, P)$, *and* $\mathcal{F}_1 \subset \mathcal{F}_2 \subset \cdots$ *an increasing sequence of sub* σ*-algebras of* $\mathcal{F}$, *where* X_t *is assumed to be* $\mathcal{F}_t$*-measurable. The sequence* $\{X_t\}$ *is said to be a martingale relative to the* $\mathcal{F}_t$ *(or we say that* $\{X_t, \mathcal{F}_t\}$ *is a martingale) if and only if for all* $t = 1, 2, \ldots$,

$$E(X_{t+1}|\mathcal{F}_t) = X_t \quad a.e. \tag{2.1.7}$$

Martingales may be understood as models for fair games in the sense that X_t *signifies the amount of money that a player has at time t. The martingale property states that the average amount a player will have at time* $(t+1)$, *given that he/she has amount* X_t *at time t, is equal to* X_t *regardless of what his/her past fortune has been.*

Nowadays the concept of a martingale is very fundamental in finance. Let B_t be the price (non-random) of a bank account at time t. Assume that B_t satisfies

$$B_t - B_{t-1} = \rho B_{t-1}, \qquad \text{for} \quad t \in \mathbb{N}.$$

Let X_t be the price of a stock at time t, and write the return as

$$r_t = (X_t - X_{t-1})/X_{t-1}.$$

If we assume

$$E\{r_t|\mathcal{F}_{t-1}\} = \rho \tag{2.1.8}$$

then it is shown that

$$E\left\{\frac{X_t}{B_t}\middle|\mathcal{F}_{t-1}\right\} = \frac{X_{t-1}}{B_{t-1}}, \tag{2.1.9}$$

(e.g., Shiryaev (1999)), hence, $\left\{\frac{X_t}{B_t}\right\}$ becomes a martingale. The assumption (2.1.8) seems natural in an economist's view. If (2.1.8) is violated, the investor can decide whether to invest in the stock or in the bank account.

In what follows we denote by $\mathcal{F}(X_1, \ldots, X_t)$ the smallest σ-algebra making random variables $X_1, \ldots, X_t$ measurable. If $\{X_t, \mathcal{F}_t\}$ is a martingale, it is automatically a martingale relative to $\mathcal{F}(X_1, \ldots, X_t)$. To see this, condition both sides of (2.1.7) with respect to $\mathcal{F}(X_1, \ldots, X_t)$. If we do not mention the $\mathcal{F}_t$ explicitly, we always mean $\mathcal{F}_t = \mathcal{F}(X_1, \ldots, X_t)$. In statistical asymptotic theory, it is known that one of the most fundamental quantities becomes a martingale under suitable conditions (see Problem 2.1).

In time series analysis, stationary processes and linear processes have been used to describe the data concerned (see Hannan (1970), Anderson (1971), Brillinger (2001b), Brockwell and Davis (2006), and Taniguchi and Kakizawa (2000)). Let us explain these processes in vector form.

Definition 2.1.8. *An m-vector process*

$$\{\boldsymbol{X}_t = (X_1(t), \ldots, X_m(t))' : t \in \mathbb{Z}\}$$

is called strictly stationary if, for all $n \in \mathbb{N}$, $t_1, \ldots, t_n, h \in \mathbb{Z}$, the distributions of $\boldsymbol{X}_{t_1}, \ldots, \boldsymbol{X}_{t_n}$ and $\boldsymbol{X}_{t_1+h}, \ldots, \boldsymbol{X}_{t_n+h}$ are the same.

The simplest example of a strictly stationary process is a sequence of i.i.d. random vectors $\boldsymbol{e}_t$, $t \in \mathbb{Z}$. Let

$$\boldsymbol{X}_t = \boldsymbol{f}(\boldsymbol{e}_t, \boldsymbol{e}_{t-1}, \boldsymbol{e}_{t-2}, \cdots), \quad t \in \mathbb{Z} \tag{2.1.10}$$

where $\boldsymbol{f}(\cdot)$ is an m-dimensional measurable function. Then it is seen that $\{\boldsymbol{X}_t\}$ is a strictly stationary process (see Problem 2.2).

If we deal with strictly stationary processes, we have to specify the joint distribution at arbitrary time points. In view of this we will introduce more convenient stationarity based on the first- and second-order moments. Let $\{\boldsymbol{X}_t : t \in \mathbb{Z}\}$ be an m-vector process whose components are possibly complex valued. The autocovariance function $\Gamma(\cdot, \cdot)$ of $\{\boldsymbol{X}_t\}$ is defined by

$$\Gamma(t, s) = Cov(\boldsymbol{X}_t, \boldsymbol{X}_s) \equiv E\{(\boldsymbol{X}_t - E\boldsymbol{X}_t)(\boldsymbol{X}_s - E\boldsymbol{X}_s)^*\}, \quad t, s \in \mathbb{Z}, \tag{2.1.11}$$

where $(\ )^*$ denotes the complex conjugate transpose of $(\ )$. If $\{\boldsymbol{X}_t\}$ is a real-valued process, then $(\ )^*$ implies the usual transpose $(\ )'$.

Definition 2.1.9. *An m-vector process* $\{\boldsymbol{X}_t : t \in \mathbb{Z}\}$ *is said to be second order stationary if*

(i) $E(\boldsymbol{X}_t^* \boldsymbol{X}_t) < \infty$ *for all* $t \in \mathbb{Z}$,

(ii) $E(\boldsymbol{X}_t) = \mathbf{c}$ *for all* $t \in \mathbb{Z}$, *where* $\mathbf{c}$ *is a constant vector,*

(iii) $\Gamma(t, s) = \Gamma(0, s - t)$ *for all* $s, t \in \mathbb{Z}$.

If $\{\boldsymbol{X}_t\}$ is second order stationary, (iii) is satisfied, hence, we redefine $\Gamma(0, s - t)$ as $\Gamma(s - t)$ for all $s, t \in \mathbb{Z}$. The function $\Gamma(h)$ is called the *autocovariance function (matrix) of* $\{\boldsymbol{X}_t\}$ *at lag* h. For fundamental properties of $\Gamma(h)$, see Problem 2.3.

Definition 2.1.10. *An m-vector process* $\{\boldsymbol{X}_t : t \in \mathbb{Z}\}$ *is said to be a Gaussian process if for each* $t_1, \ldots, t_n \in \mathbb{Z}$, $n \in \mathbb{N}$, *the distribution of* $\boldsymbol{X}_{t_1}, \ldots, \boldsymbol{X}_{t_n}$ *is multivariate normal.*

Gaussian processes are very important and fundamental, and for them it holds that second order stationarity is equivalent to strict stationarity (see Problem 2.5).

In what follows we shall state three theorems related to the spectral representation of second-order stationary processes. For proofs, see, e.g., Hannan (1970) and Taniguchi and Kakizawa (2000).

Theorem 2.1.11. *If* $\Gamma(\cdot)$ *is the autocovariance function of an m-vector second-order stationary process* $\{\boldsymbol{X}_t\}$, *then*

$$\Gamma(t) = \int_{-\pi}^{\pi} e^{it\lambda} dF(\lambda), \qquad t \in \mathbb{Z}, \tag{2.1.12}$$

where $F(\lambda)$ *is a matrix whose increments* $F(\lambda_2) - F(\lambda_1)$, $\lambda_2 \geq \lambda_1$, *are nonnegative definite. The function* $F(\lambda)$ *is uniquely defined if we require in addition that (i)* $F(-\pi) = \mathbf{0}$ *and (ii)* $F(\lambda)$ *is right continuous.*

The matrix $F(\lambda)$ *is called the spectral distribution matrix. If* $F(\lambda)$ *is absolutely continuous with respect to a Lebesgue measure on* $[-\pi, \pi]$ *so that*

$$F(\lambda) = \int_{-\pi}^{\lambda} f(\mu) d\mu, \tag{2.1.13}$$

where $f(\lambda)$ *is a matrix with entries* $f_{jk}(\lambda)$, $j, k = 1, \ldots, m$, *then* $f(\lambda)$ *is called the spectral density matrix. If*

$$\sum_{h=-\infty}^{\infty} ||\Gamma(h)|| < \infty, \tag{2.1.14}$$

where $||\Gamma(h)||$ *is the square root of the greatest eigenvalue of* $\Gamma(h)\Gamma(h)^*$, *the spectral density matrix is expressed as*

$$f(\lambda) = \frac{1}{2\pi} \sum_{h=-\infty}^{\infty} \Gamma(h) e^{-ih\lambda}. \tag{2.1.15}$$

Next we state the spectral representation of $\{\boldsymbol{X}_t\}$. For this the concept of an orthogonal increment process and the stochastic integral is needed. We say that $\{\boldsymbol{Z}(\lambda) : -\pi \leq \lambda \leq \pi\}$ is an m-vector-valued orthogonal increment process if

(i) $E\{\boldsymbol{Z}(\lambda)\} = \mathbf{0}, \quad -\pi \leq \lambda \leq \pi,$

(ii) the components of the matrix $E\{\boldsymbol{Z}(\lambda)\boldsymbol{Z}(\lambda)^*\}$ arc finite for all $\lambda \in [-\pi, \pi]$,

(iii) $E[\{\boldsymbol{Z}(\lambda_4) - \boldsymbol{Z}(\lambda_3)\}\{\boldsymbol{Z}(\lambda_2) - \boldsymbol{Z}(\lambda_1)\}^*] = \mathbf{0}$ if $(\lambda_1, \lambda_2] \cap (\lambda_3, \lambda_4] = \phi$,

(iv) $E[\{\boldsymbol{Z}(\lambda + \delta) - \boldsymbol{Z}(\lambda)\}\{\boldsymbol{Z}(\lambda + \delta) - \boldsymbol{Z}(\lambda)\}^*] \rightarrow \mathbf{0}$ as $\delta \rightarrow 0$.

If $\{Z(\lambda)\}$ satisfies (i)-(iv), then there exists a unique matrix distribution function G on $[-\pi,\pi]$ such that

$$G(\mu) - G(\lambda) = E[\{Z(\mu) - Z(\lambda)\}\{Z(\mu) - Z(\lambda)\}^*], \quad \lambda \le \mu.$$

Let $L_2(G)$ be the family of matrix-valued functions $M(\lambda)$ satisfying $tr\left\{\int_{-\pi}^{\pi} M(\lambda)dG(\lambda)M(\lambda)^*\right\} < \infty$. For $M(\lambda) \in L_2(G)$, there is a sequence of matrix-valued simple functions $M_n(\lambda)$ such that

$$tr\left[\int_{-\pi}^{\pi}\{M_n(\lambda) - M(\lambda)\}dG(\lambda)\{M_n(\lambda) - M(\lambda)\}^*\right] \to 0 \tag{2.1.16}$$

as $n \to \infty$ (e.g., Hewitt and Stromberg (1965)). Suppose that $M_n(\lambda)$ is written as

$$M_n(\lambda) = \sum_{j=0}^{n} M_j^{(n)}\chi_{(\lambda_j,\lambda_{j+1}]}(\lambda), \tag{2.1.17}$$

where $M_j^{(n)}$ are constant matrices and $-\pi = \lambda_0 < \lambda_1 < \cdots < \lambda_{n+1} = \pi$, and $\chi_A(\lambda)$ is the indicator function of A. For (2.1.17), we define the stochastic integral by

$$\int_{-\pi}^{\pi} M_n(\lambda)dZ(\lambda) \equiv \sum_{j=0}^{n} M_j^{(n)}\{Z(\lambda_{j+1}) - Z(\lambda_j)\} = I_n \quad (say).$$

Then it is shown that I_n converges to a random vector I in the sense that $||I_n - I|| \to 0$ as $n \to \infty$, where $||I_n||^2 = \mathrm{E}(I_n^* I_n)$. We write this as $I = l.i.m\ I_n$ (limit in the mean). For arbitrary $M(\lambda) \in L_2(G)$ we define the stochastic integral

$$\int_{-\pi}^{\pi} M(\lambda)dZ(\lambda) \tag{2.1.18}$$

to be the random vector I defined as above.

Theorem 2.1.12. *If $\{X_t : t \in \mathbb{Z}\}$ is an m-vector second-order stationary process with mean zero and spectral distribution matrix $F(\lambda)$, then there exists a right-continuous orthogonal increment process $\{Z(\lambda) : -\pi \le \lambda\pi\}$ such that*

(i) $E[\{Z(\lambda) - Z(-\pi)\}\{Z(\lambda) - Z(-\pi)\}^*] = F(\lambda), \quad -\pi \le \lambda \le \pi,$

(ii) $$X_t = \int_{-\pi}^{\pi} e^{-it\lambda}dZ(\lambda). \tag{2.1.19}$$

Although the substance of $\{Z(\lambda)\}$ is difficult to understand, a good substantial understanding of it is given by the relation

$$Z(\lambda_2) - Z(\lambda_1) = \underset{n\to\infty}{\mathrm{l.i.m}} \frac{1}{2\pi}\sum_{t=-n}^{n}{}' X_t \frac{e^{it\lambda_2} - e^{it\lambda_1}}{it}, \tag{2.1.20}$$

where $\sum'$ indicates that for $t = 0$ we take the summand to be $(\lambda_2 - \lambda_1)X_0$ (see Hannan (1970)).

Theorem 2.1.13. *Suppose that $\{X_t : t \in \mathbb{Z}\}$ is an m-vector second-order stationary process with mean zero, spectral distribution matrix $F(\cdot)$ and spectral representation (2.1.19), and that $\{A(j) : j = 0, 1, 2, \ldots\}$ is a sequence of $m \times m$ matrices. Further we set down*

$$Y_t = \sum_{j=0}^{\infty} A(j)X_{t-j}, \tag{2.1.21}$$

and

$$h(\lambda) = \sum_{j=0}^{\infty} A(j)e^{ij\lambda}.$$

Then the following statements hold true:

(i) *the necessary and sufficient condition that (2.1.21) exists as the l.i.m of partial sums is*

$$tr\left\{\int_{-\pi}^{\pi} h(\lambda)dF(\lambda)h(\lambda)^*\right\} < \infty, \tag{2.1.22}$$

(ii) *if (2.1.22) holds, then the process $\{\boldsymbol{Y}_t : t \in \mathbb{Z}\}$ is second order stationary with autocovariance function and spectral representation*

$$\Gamma_Y(t) = \int_{-\pi}^{\pi} e^{-it\lambda}h(\lambda)dF(\lambda)h(\lambda)^* \tag{2.1.23}$$

and

$$\boldsymbol{Y}_t = \int_{-\pi}^{\pi} e^{-it\lambda}h(\lambda)d\boldsymbol{Z}(\lambda), \tag{2.1.24}$$

respectively.

If $dF(\lambda) = (2\pi)^{-1}\sum d\lambda$, where all the eigenvalues of $\sum$ are bounded and bounded away from zero, the condition (2.1.22) is equivalent to

$$\sum_{j=0}^{\infty} ||A(j)||^2 < \infty. \tag{2.1.25}$$

Let $\{\boldsymbol{U}_t\} \sim$ i.i.d. $(\boldsymbol{0}, \sum)$. If $\{A(j)\}$ satisfies (2.1.25), then it follows from Theorem 2.1.13 that the process

$$\boldsymbol{X}_t = \sum_{j=0}^{\infty} A(j)\boldsymbol{U}_{t-j} \tag{2.1.26}$$

is defined as the *l.i.m* of partial sums, and that $\{\boldsymbol{X}_t\}$ has the spectral density matrix

$$f(\lambda) = \frac{1}{2\pi}\left\{\sum_{j=0}^{\infty} A(j)e^{ij\lambda}\right\}\sum\left\{\sum_{j=0}^{\infty} A(j)e^{ij\lambda}\right\}^*. \tag{2.1.27}$$

The process (2.1.26) is called the *m-vector linear process*, which includes m-vector AR(p), m-vector ARMA(p, q) written as VAR(p) and VARMA(p, q), respectively; hence, it is very general.

In the above the dimension of the process is assumed to be finite. However, if the number of cross-sectional time series data (e.g., returns on different assets, data disaggregated by sector or region) is typical large, VARMA models are not appropriate, because we need to estimate too many unknown parameters. For this case, Forni et al. (2000) introduced a class of dynamic factor models defined as follows:

$$\boldsymbol{X}_t = \sum_{j=0}^{\infty} B(j)\boldsymbol{u}_{t-j} + \boldsymbol{\xi}_t = \boldsymbol{Y}_t + \boldsymbol{\xi}_t, \quad (say) \tag{2.1.28}$$

where $\boldsymbol{X}_t = (X_1(t), \ldots, X_m(t))'$, $B(j) = \{B_{kl}(j) : k, l = 1, \ldots, m\}$ and $\boldsymbol{\xi}_t = (\xi_1(t), \ldots, \xi_m(t))'$. They impose the following assumption.

Assumption 2.1.14. **(1)** *The process* $\{\boldsymbol{u}_t\}$ *is uncorrelated with* $E\{\boldsymbol{u}_t\} = 0$ *and* $Var\{\boldsymbol{u}_t\} = \boldsymbol{I}_m$ *for all* $t \in \mathbb{Z}$.

(2) $\{\boldsymbol{\xi}_t\}$ *is a zero-mean vector process, and is mutually uncorrelated with* $\{\boldsymbol{u}_t\}$.

(3) $\sum_{j=0}^{\infty} \|B(j)\|^2 < \infty$.

(4) *Letting the spectral density matrix of* $\{X_t\}$ *be* $\boldsymbol{f}^X(\lambda) = \{f_{jl}^X(\lambda) : j, l = 1, \ldots, m\}$ *(which exists from the conditions (1)–(3)), there exists* $c^X > 0$ *such that* $f_{ij}^X(\lambda) \le c^X$ *for all* $\lambda \in [\pi, \pi]$ *and* $j = 1, \ldots, m$.

(5) *Let the spectral density matrices of* $\{\boldsymbol{Y}_t\}$ *and* $\boldsymbol{\xi}_t$ *be* $\boldsymbol{f}^Y(\lambda)$ *and* $\boldsymbol{f}^\xi(\lambda)$*, respectively, and let the jth eigenvalue of* $\boldsymbol{f}^Y(\lambda)$ *and* $\boldsymbol{f}^\xi(\lambda)$ *be* $p_{m,j}^Y(\lambda)$ *and* $p_{n,j}^\xi(\lambda)$*, respectively. Then it holds that*

(i) *there exists* $c^\xi > 0$ *such that* $\max_{1 \le j \le m} p_{n,j}^\xi(\lambda) \le c^\xi$ *for any* $\lambda \in [-\pi, \pi]$ *and any* $n \in \mathbb{N}$.

(ii) *The first q largest eigenvalues* $p_{n,j}^Y(\lambda)$ *of* $\boldsymbol{f}^Y(\lambda)$ *diverge almost everywhere in* $[-\pi, \pi]$.

Based on observations $\{\boldsymbol{X}_1, \ldots, \boldsymbol{X}_n\}$, Forni et al. (2000) proposed a consistent estimator $\hat{\boldsymbol{Y}}_t$ of $\boldsymbol{Y}_t$ under which m and n tend to infinity.

Example 2.1.15. *(spatiotemporal model) Recently the wide availability of data observed over time and space has developed many studies in a variety of fields such as environmental science, epidemiology, political science, economics and geography. Lu et al. (2009) proposed the following adaptive varying-coefficient spatiotemporal model for such data:*

$$X_t(\boldsymbol{s}) = a\{\boldsymbol{s}, \boldsymbol{\alpha}(\boldsymbol{s})'\boldsymbol{Y}_t(\boldsymbol{s})\} + \boldsymbol{b}\{\boldsymbol{s}, \boldsymbol{\alpha}(\boldsymbol{s})'\boldsymbol{Y}_t(\boldsymbol{s})\}'\boldsymbol{Y}_t(\boldsymbol{s}) + \epsilon_t(\boldsymbol{s}), \tag{2.1.29}$$

where $X_t(\boldsymbol{s})$ *is the spatiotemporal variable of interest, and t is time, and* $\boldsymbol{s} = (u, v) \in \mathcal{S} \subset \mathbb{R}^2$ *is a spatial location. Also,* $a(\boldsymbol{s}, z)$ *and* $\boldsymbol{b}(\boldsymbol{s}, z)$ *are unknown scalar and* $d \times 1$ *functions,* $\boldsymbol{\alpha}(\boldsymbol{s})$ *is an unknown* $d \times 1$ *index vector,* $\{\epsilon_t(\boldsymbol{s})\}$ *is a noise process which, for each fixed* $\boldsymbol{s}$*, forms a sequence of i.i.d. random variables over time and* $\mathbf{Y}_t(\boldsymbol{s}) = \{Y_{t1}(\boldsymbol{s}), \ldots, Y_{td}(\boldsymbol{s})\}'$ *consists of time-lagged values of* $X_t(\cdot)$ *in a neighbourhood of* $\boldsymbol{s}$ *and some exogenous variables. They introduced a two-step estimation procedure for* $\boldsymbol{\alpha}(\boldsymbol{s})$*,* $\boldsymbol{\alpha}(\boldsymbol{s}, *)$ *and* $\boldsymbol{b}(\boldsymbol{s}, *)$.

In Definition 2.1.7, we defined the martingale for sequences $\{X_n : n \in \mathbb{N}\}$. In what follows we generalize the notion to the case of an uncountable index set $\mathbb{R}^+$, i.e., $\{X_t : t \ge 0\}$. Let $(\Omega, \mathcal{F}, P)$ be a probability space. Let $\{\mathcal{F}_t : t \ge 0\}$ be a family (*filtration*) of sub-σ-algebras satisfying

$$\mathcal{F}_s \subseteq \mathcal{F}_t \subseteq \mathcal{F}, \quad s \le t,$$
$$\mathcal{F}_t = \bigcap_{s<t} \mathcal{F}_s,$$
$$\mathcal{F}_t = \mathcal{F}_t^P,$$

where $\mathcal{F}_t^P$ stands for completion of the σ-algebra $\mathcal{F}_t$ by the P-null sets from $\mathcal{F}$. Then, the quadruple $(\Omega, \mathcal{F}, \{\mathcal{F}_t : t \ge 0\}, P)$ is called a *stochastic basis*. Suppose that a stochastic process $\{X_t : t \ge 0\}$ is defined on $(\Omega, \mathcal{F}, \{\mathcal{F}_t : t \ge 0\}, P)$, and X_t's are $\mathcal{F}_t$-measurable. Then we say that they are *adapted with respect to* $\{\mathcal{F}_t : t \ge 0\}$, and often write $\{X_t, \mathcal{F}_t\}$.

Definition 2.1.16. *A stochastic process* $\{X_t, \mathcal{F}_t\}$ *is said to be a martingale if*

$$E|X_t| < \infty, \quad t \ge 0,$$
$$E\{X_t | \mathcal{F}_s\} = X_s \quad (P\text{-}a.s), \quad s \le t.$$

A random variable $\tau = \tau(\omega)$ *taking values in* $[0, \infty]$ *is called a Markov time if*

$$\{\omega : \tau(\omega) \le t\} \in \mathcal{F}_t, \quad t \ge 0.$$

The Markov times satisfying $P(\tau(\omega) < \infty) = 1$ are called stopping times.

Definition 2.1.17. *A process $\{X_t, \mathcal{F}_t\}$ is called a local martingale if there exists a sequence $\{\tau_n\}$ of stopping times such that $\tau_n(\omega) \le \tau_{n+1}(\omega)$, $\tau_n(\omega) \nearrow \infty$ (P-a.s.) as $n \to \infty$ and the processes $\{X_{\tau_n \wedge t}, \mathcal{F}_t\}$ are martingales.*

Henceforth we use the following notations:

$$\mathcal{D} \equiv \left[\{A_t, \mathcal{F}_t\} : \int_0^t |dA_s(\omega)| < \infty, t \ge 0, \omega \in \Omega, \textit{i.e., processes of bounded variation}\right]$$

and

$$\mathcal{M}_{loc} \equiv [\{M_t, \mathcal{F}_t\}; \textit{local martingales}].$$

Definition 2.1.18. *A stochastic process $\{X_t, \mathcal{F}_t\}$ is called a semimartingale if it admits the decomposition*

$$X_t = X_0 + A_t + M_t, \quad t \ge 0 \tag{2.1.30}$$

where $\{A_t, \mathcal{F}_t\} \in \mathcal{D}$ and $\{M_t, \mathcal{F}_t\} \in \mathcal{M}_{loc}$. The process $\{A_t\}$ is called a compensator of $\{X_t\}$.

Under ordinary circumstances the process of interest can be modeled in terms of a signal plus noise relationship:

$$\textit{process} = \textit{signal} + \textit{noise} \tag{2.1.31}$$

where the signal incorporates the predictable trend part of the model and the noise is the stochastic disturbance. Semimartingales are such. Sørensen (1991) discussed the process

$$dX_t = \theta X_t dt + dW_t + dN_t, \quad t \ge 0, \quad X_0 = x_0, \tag{2.1.32}$$

where $\{W_t\}$ is a standard Brownian motion and $\{N_t\}$ is a Poisson process with intensity λ. This process may be written in semimartingale form as

$$dX_t = \theta X_t dt + \lambda dt + dM_t, \tag{2.1.33}$$

where

$$M_t = W_t + N_t - \lambda t.$$

In Example 2.1.3, we introduced a few nonlinear processes. In what follows we explain more general ones.

Let $\{X_t\}$ and $\{u_t\}$ be defined on a stochastic basis $(\Omega, \mathcal{F}, \{\mathcal{F}_t\}, P)$, and X_t's are $\mathcal{F}_t$-measurable, u_t's are $\mathcal{F}_t$-measurable and independent of $\mathcal{F}_{t-1}$. Bollerslev (1986) introduced a generalized autoregressive conditional heteroscedastic model (GARCH(p,q)), which is defined by

$$\begin{cases} X_t = u_t \sqrt{h_t} \\ h_t = a_0 + \sum_{j=1}^{q} a_j X_{t-j}^2 + \sum_{j=1}^{p} b_j h_{t-j}, \end{cases} \tag{2.1.34}$$

where $a_0 > 0$, $a_j \ge 0$, $j = 1, \dots q$, and $b_j \ge 0$, $j = 1, \dots, p$.

Giraitis et al. (2000) introduced an ARCH(∞) model, which is defined by

$$\begin{cases} X_t = u_t \sqrt{h_t} \\ h_t = a_0 + \sum_{j=1}^{\infty} a_j X_{t-j}^2, \end{cases} \tag{2.1.35}$$

where $a_0 > 0$, $a_j \ge 0$, $j = 1, 2, \dots$. The class of ARCH(∞) models is larger than that of stationary GARCH(p,q) models.

Thc following example gives a very general and persuasive model which includes ARCH, SETAR, etc. as special cases.

Example 2.1.19. *A stochastic process* $\{\boldsymbol{X}_t = (X_{1,t}, \ldots, X_{m,t})' : t \in \mathbb{Z}\}$ *is said to follow a conditional heteroscedastic autoregressive nonlinear (CHARN) model if it satisfies*

$$\boldsymbol{X}_t = \boldsymbol{F}_{\boldsymbol{\theta}}(\boldsymbol{X}_{t-1}, \cdots, \boldsymbol{X}_{t-p}) + \boldsymbol{H}_{\boldsymbol{\theta}}(\boldsymbol{X}_{t-1}, \cdots, \boldsymbol{X}_{t-q})\boldsymbol{U}_t, \tag{2.1.36}$$

where $\boldsymbol{U}_t = (U_{1,t}, \ldots, U_{m,t})' \sim i.i.d.(\boldsymbol{0}, V)$, $\boldsymbol{F}_{\boldsymbol{\theta}} : \mathbb{R}^{mp} \to \mathbb{R}^m$, *and* $\boldsymbol{H}_{\boldsymbol{\theta}} : \mathbb{R}^{mq} \to \mathbb{R}^m \times \mathbb{R}^m$ *are measurable functions, and* $\boldsymbol{\theta} = (\theta_1, \ldots, \theta_r)' \in \Theta$ (an open set) $\subset \mathbb{R}^r$ *is an unknown parameter (Härdle et al. (1998)). This model is very general, and can be applied to analysis of brain and muscular waves as well as financial time series analysis (Kato et al. (2006)).*

When we analyze the nonlinear time series models, their stationarity is fundamental and important. We provide a sufficient condition for the CHARN model (2.1.36) to be stationary. Denote by $|A|$ *the sum of the absolute values of all the elements of a matrix or vector* A*. Let* $\boldsymbol{x} = (x_{11}, \ldots, x_{1m}, x_{21}, \ldots, x_{2m}, \ldots, x_{p1}, \ldots, x_{pm})' \in \mathbb{R}^{mp}$*. Without loss of generality we assume* $p = q$ *in (2.1.36).*

Theorem 2.1.20. *(Lu and Jiang (2001)) Suppose that* $\{\boldsymbol{X}_t\}$ *is generated by the CHARN model (2.1.36). Assume*

(i) $\boldsymbol{U}_t$ *has the probability density function* $p(\boldsymbol{u}) > 0$ *a.e.,* $\boldsymbol{u} \in \mathbb{R}^m$.

(ii) *There exist* $a_{ij} \geq 0$, $b_{ij} \geq 0$, $1 \leq i \leq m$, $1 \leq j \leq p$, *such that*

$$|\boldsymbol{F}_{\boldsymbol{\theta}}(\boldsymbol{x})| \leq \sum_{i=1}^{m}\sum_{j=1}^{p} a_{ij}|x_{ij}| + o(|\boldsymbol{x}|),$$

$$|\boldsymbol{H}_{\boldsymbol{\theta}}(\boldsymbol{x})| \leq \sum_{i=1}^{m}\sum_{j=1}^{p} b_{ij}|x_{ij}| + o(|\boldsymbol{x}|), \quad |\boldsymbol{x}| \to \infty.$$

(iii) $\boldsymbol{H}_\theta(\boldsymbol{x})$ *is a continuous and symmetric function with respect to* $\boldsymbol{x}$*, and there exists* $\lambda > 0$ *such that*

$$\{\text{the minimum eigenvalue of } \boldsymbol{H}_\theta(\boldsymbol{x})\} \geq \lambda$$

for all $\boldsymbol{x} \in \mathbb{R}^{mp}$.

(iv)

$$\max_{1 \leq i \leq m} \left\{ \sum_{j=1}^{p} a_{ij} + E|\boldsymbol{U}_1| \sum_{j=1}^{p} b_{ij} \right\} < 1.$$

Then, $\{\boldsymbol{X}_t\}$ *is strictly stationary.*

Suppose that we need to compute some statistical average of a strictly stationary process when we observe just a single realization of it. In such a situation, is it possible to determine the statistical average from an appropriate time average of a single realization. If the statistical (or ensemble) average of the process equals the time average, the process will be called ergodic. We mention some limit theorems related to the ergodic stationary process.

Let $\{\boldsymbol{X}_t = (X_1(t), \ldots, X_m(t))' : t \in \mathbb{Z}\}$ be a strictly stationary process defined on a probability space $(\Omega, \mathcal{F}, P)$. The σ-algebra $\mathcal{F}$ is generated by the family of all cylinder sets

$$\{\omega \,|(\boldsymbol{X}_{t_1}, \ldots, \boldsymbol{X}_{t_k}) \in B\},$$

where $B \in \mathcal{B}^{mk}$. For these cylinder sets we define the shift operator $A \to TA$ where, for a set A of the form

$$A = \{\omega \,|(\boldsymbol{X}_{t_1}, \ldots, \boldsymbol{X}_{t_k}) \in C\}, \quad C \in \mathcal{B}^{mk},$$

we have

$$TA = \{\omega | (\boldsymbol{X}_{t_1+1}, \ldots, \boldsymbol{X}_{t_k+1}) \in C\}.$$

This definition extends to all sets in $\mathcal{F}$. Since $\{\boldsymbol{X}_t\}$ is strictly stationary, A and $T^{-1}A$ have the same probability content. Then we say that T is *measure preserving*.

Definition 2.1.21. *Given a measure preserving transformation T, a measurable event A is said to be invariant if $T^{-1}A = A$.*

Denote the collection of invariant sets by $\mathcal{A}_I$.

Definition 2.1.22. *The process $\{\boldsymbol{X}_t : t \in Z\}$ is said to be ergodic if for all $A \in \mathcal{A}_I$, either $P(A) = 0$ or $P(A) = 1$.*

Now we state the following two theorems (see Stout (1974, p. 182) and Hannan (1970, p.204)).

Theorem 2.1.23. *Suppose that a vector process $\{\boldsymbol{X}_t : t \in Z\}$ is strictly stationary and ergodic, and that there is a measurable function $\phi : \mathbb{R}^\infty \to \mathbb{R}^k$. Let $\boldsymbol{Y}_t = \phi(\boldsymbol{X}_t, \boldsymbol{X}_{t-1}, \ldots)$. Then $\{\boldsymbol{Y}_t : t \in \mathbb{Z}\}$ is strictly stationary and ergordic.*

Since the i.i.d. sequences are strictly stationary and ergodic, we have

Theorem 2.1.24. *Suppose that $\{\boldsymbol{U}_t\}$ is a sequence of random vectors that are independent and identically distributed with zero mean and finite covariance matrix. Let*

$$\boldsymbol{X}_t = \sum_{j=0}^{\infty} A(j)\boldsymbol{U}_{t-j},$$

where $\sum_{j=0}^{\infty} \|A(j)\| < \infty$. Then $\{\boldsymbol{X}_t : t \in \mathbb{Z}\}$ is strictly stationary and ergodic.

The following is essentially the pointwise ergodic theorem.

Theorem 2.1.25. *If $\{\boldsymbol{X}_t : t \in \mathbb{Z}\}$ is strictly stationary and ergodic and $E\|\boldsymbol{X}_t\| < \infty$, then*

$$\frac{1}{n}\sum_{t=1}^{n} \boldsymbol{X}_t \overset{a.s.}{\to} E(X_1). \tag{2.1.37}$$

Also, if $E\|\boldsymbol{X}\|^2 < \infty$, then

$$\frac{1}{n}\sum_{t=1}^{n} \boldsymbol{X}_t\boldsymbol{X}'_{t+m} \overset{a.s.}{\to} E(\boldsymbol{X}_1\boldsymbol{X}'_{1+m}). \tag{2.1.38}$$

Proof. The assertion (2.1.37) is exactly the pointwise ergodic theorem (e.g., Theorem 3.5.7 of Stout (1974)). For (2.1.38), consider $\boldsymbol{X}_t\boldsymbol{X}'_{t+m}$. Fixing m and allowing t to vary, this constitutes a sequence of random matrices which again constitutes a vector strictly stationary process. Since $\{\boldsymbol{X}_t\}$ is ergodic, then so is $\{\boldsymbol{X}_t\boldsymbol{X}'_{t+m}\}$. Hence (2.1.38) follows from the pointwise ergodic theorem. □

For martingales, we mention the following two theorems on the strong law of large numbers (e.g., Hall and Heyde (1980), Heyde (1997)).

Theorem 2.1.26. *(Doob's martingale convergence theorem) If $\{X_n, \mathcal{F}_n, n \in \mathbb{N}\}$ is a martingale such that $\sup_{n\geq 1} E|S_n| < \infty$, then there exists a random variable X such that $E|X| < \infty$ and $X_n \overset{a.s.}{\to} X$.*

For locally square integrable martingale $M = \{M_t\}$ there exists a unique predictable process $\langle M, M\rangle_t$ for which

$$M_t^2 - \langle M, M\rangle_t \tag{2.1.39}$$

is a local martingale. The process $\langle M, M\rangle_t$ is called the *quadratic characteristic of M*. For an example of $\langle M, M\rangle_t$, see Problem 2.7. When $\{\boldsymbol{M}_t\}$ is a vector-valued local martingale, the quadratic characteristic $\langle \boldsymbol{M}, \boldsymbol{M}'\rangle_t$ is the one for which $\boldsymbol{M}\boldsymbol{M}_t' - \langle \boldsymbol{M}, \boldsymbol{M}'\rangle_t$ is a local martingale.

Theorem 2.1.27. *Suppose that $\{X_t, t \geq 0\}$ is a local square integrable martingale. Then,*

$$\frac{X_t}{A_t} \xrightarrow{a.s.} 0, \quad \text{as } t \to \infty, \tag{2.1.40}$$

for $A_t \equiv f(\langle X, X\rangle_t)$, $\lim_{t\to\infty} A_t = \infty$ and $\int_0^\infty \frac{dt}{f(t)^2} < \infty$.

We next state some central limit theorems which are useful to derive the asymptotic distribution of statistics for stochastic processes.

Theorem 2.1.28. *(Ibragimov (1963)) Let $\{X_t : t \in \mathbb{Z}\}$ be a strictly stationary ergodic process such that $E(X_t^2) = \sigma^2$, $0 < \sigma^2 < \infty$, $E\{X_t|\mathcal{F}_{t-1}\} = 0$, a.s., where $\mathcal{F}_t = \sigma(X_1, \ldots, X_t)$. Then*

$$\frac{1}{\sigma\sqrt{n}} \sum_{t=1}^{n} X_t \xrightarrow{d} N(0, 1).$$

Let $\{X_{n,t} : t = 1, \ldots, k_n\}$ be an array of random variables on a probability space $(\Omega, \mathcal{F}, P)$. Let $\{\mathcal{F}_{n,t} : 0 \leq t \leq k_n\}$ be any triangular array of sub σ-algebras of $\mathcal{F}$ such that for each n and $1 \leq t \leq k_n$, $X_{n,t}$ is $\mathcal{F}_{n,t}$-measurable and $\mathcal{F}_{n,t-1} \subset \mathcal{F}_{n,t}$. We write $S_n = \sum_{t=1}^{k_n} X_t$.

Theorem 2.1.29. *(Brown (1971)) Suppose that $\{X_{n,t}, \mathcal{F}_{n,t}, 1 \leq t \leq k_n\}$ is a martingale difference array satisfying*

(i)

$$\sum_{t=1}^{k_n} E\{X_{n,t}^2 \chi(|X_{n,t}| > \epsilon)\} \to 0 \quad \text{as} \quad n \to \infty \quad \text{for all } \epsilon > 0 \tag{2.1.41}$$

(Lindeberg condition).

(ii)

$$\sum_{t=1}^{k_n} E(X_{n,t}^2 | \mathcal{F}_{n,t-1}) \xrightarrow{p} 1. \tag{2.1.42}$$

Then,

$$S_n \xrightarrow{d} N(0, 1), \quad (n \to \infty).$$

Problems

2.1. Let $\boldsymbol{X}_n = (X_1, \ldots, X_n)'$ be a sequence of random variables forming a stochastic process, and having the probability density $p_\theta^n(\boldsymbol{x}_n)$, $\boldsymbol{x}_n = (x_1, \ldots, x_n)'$, where $\theta \in \Theta \subset \mathbb{R}^1$ (Θ is an open set). Assume that $p_\theta^n(\cdot)$ is differentiable with respect to θ. Write the log-likelihood function based on $\boldsymbol{X}_n$ as $L_n(\theta)$, and let $S_n = \partial/\partial\theta\{L_n(\theta)\}$ (score function). Assuming that $\partial/\partial\theta$ and E_θ are exchangeable, and the existence of moment of all the related quantities, show that $\{S_n, F_n\}$ is a martingale.

2.2. Let $\{\boldsymbol{X}_t\}$ be a process generated by (2.1.10). Then, show that it is strictly stationary.

2.3. Show that the autocovariance matrix $\Gamma(\cdot) = \{r_{ab}(\cdot) : a, b = 1, \ldots, m\}$ satisfies

(i) $\Gamma(h) = \Gamma(-h)'$ for all $h \in \mathbb{Z}$,

(ii) $|r_{ab}(h)| \leq \{r_{aa}(h)\}^{1/2}\{r_{bb}(h)\}^{1/2}$ for $a, b = 1, \ldots, m$ and $h \in \mathbb{Z}$,

(iii) $\sum_{a,b=1}^{n} \boldsymbol{v}_a^* \Gamma(b-a) \boldsymbol{v}_b \geq 0$ for all $n \in \mathbb{N}$ and $\boldsymbol{v}_1, \ldots, \boldsymbol{v}_n \in \mathbb{C}^m$.

2.4. Show that a strictly stationary process with finite second-order moments is second order stationary.

2.5. For Gaussian processes, show that second order stationarity is equivalent to strict stationarity.

2.6. Let $\{\boldsymbol{X}_t : t \in \mathbb{Z}\}$ be an m-vector process satisfying

$$\boldsymbol{X}_t + \Phi \boldsymbol{X}_{t-1} + \cdots + \Phi_p \boldsymbol{X}_{t-p} = \boldsymbol{U}_t + \Theta_1 \boldsymbol{U}_{t-1} + \cdots + \Theta_q \boldsymbol{U}_{t-q}, \tag{P.1}$$

where $\Phi_1, \ldots, \Phi_p, \Theta_1, \ldots, \Theta_q$ are real $m \times m$ matrices and $\{\boldsymbol{U}_t\} \sim$ i.i.d.$(\boldsymbol{0}, \sum)$. Defining $\Phi(z) = I + \Phi_1 z + \cdots + \Phi_p z^p$, where I is the $m \times m$ identity matrix, we assume

$$det\{\Phi(z)\} \neq 0 \quad \text{for all} \quad z \in \mathbb{C} \quad \text{such that} \quad |z| \leq 1. \tag{P.2}$$

The process $\{\boldsymbol{X}_t\}$ defined by (P.1) is called an *m-vector autoregressive moving average process of order (p, q)*, and is denoted by VARMA(p,q). Then, show that $\{\boldsymbol{X}_t\}$ is second order stationary, and has the spectral density matrix

$$f(\lambda) = \frac{1}{2\pi} \Phi(e^{i\lambda})^{-1} \Theta(e^{i\lambda}) \sum \Theta(e^{i\lambda})^* \Phi(e^{i\lambda})^{*\,-1}, \tag{P.3}$$

where $\Phi(z) = I + \Theta_1 z + \cdots + \Theta_q z^q$.

2.7. Let $a(t, \omega)$ be a square integrable predictable process, and $\{W_t\}$ is the standard Wiener process. Let $X = \{X_t, \mathcal{F}_t, t \geq 0\}$ be generated by

$$X_t = \int_0^t a(s, \omega) dW_s.$$

Then, show that

$$\langle X, X \rangle_t = \int_0^t a^2(s, \omega) ds.$$

2.8. Let $\{X_t\}$ be generated by the ARCH(∞) model defined as (2.1.35), and let $\mathcal{F}_t = \mathcal{F}(\ldots, X_1, X_2, \ldots, X_t)$. Then, show that $\{X_t, \mathcal{F}_t\}$ satisfies

$$E\{X_t | \mathcal{F}_{t-1}\} = 0, \quad a.e., \tag{P.4}$$

i.e., $\{X_t, \mathcal{F}_t\}$ is a *martingale difference sequence*. Hence, if $\{X_t\}$ is an ARCH(q) or GARCH(p, q) (see 2.1.34), then it becomes a martingale difference sequence.

2.9. Let $\{\boldsymbol{X}_t\}$ be generated by the CHARN(p, q) model defined by (2.1.36), and let $\mathcal{F}_t = \mathcal{F}(\boldsymbol{X}_1, ..., \boldsymbol{X}_t)$. Then, show that $\{\boldsymbol{X}_t, \mathcal{F}_t\}$ satisfies

$$E\{\boldsymbol{X}_t | \mathcal{F}_{t-1}\} = \boldsymbol{F}_\theta(\boldsymbol{X}_{t-1}, \ldots, \boldsymbol{X}_{t-p}), \quad a.e.$$

2.10. Simulate a series of $n = 100$ observations from the model (2.1.1) with $p = 1$ and $\{u_t\} \sim$ *i.i.d.* $N(0, 1)$. Then, plot their graphs for the case of $a_1 = 0.1, 0.3, 0.5, 0.7$ and 0.9.

2.11. Simulate a series of $n = 100$ observations from the model (2.1.5) with $q = 1$, $a_0 = 0.5$ and $\{u_t\} \sim$ *i.i.d.* $N(0, 1)$. Then, plot their graphs for the case of $a_1 = 0.1, 0.3, 0.5, 0.7$ and 0.9.

Chapter 3

Portfolio Theory for Dependent Return Processes

3.1 Introduction to Portfolio Theory

Modern portfolio theory introduced by Markowitz (1952) has become a broad theory for portfolio selection. He demonstrated how to reduce a standard deviation of return (portfolio risk) on a portfolio of assets. The Markowitz theory of portfolio management deals with individual assets in financial markets. It combines probability theory and optimization theory to model the behaviour of investors. The investors are assumed to decide a balance between maximizing the expected return and minimizing the risk of their investment decision. Portfolio return is characterized by the mean (i.e., expected return) and risk (defined by the standard deviation) of their portfolio of assets. These mathematical representations of mean and risk have allowed optimization tools to be applied to studies of portfolio management. The exact solution will depend on the level of the risk (in comparison with the rate of the mean) that the investors would bear. Even though many other models may treat different risks in place of the standard deviation, the trade-off between mean and risk has been the major issue which those theories try to solve.

If everybody uses the mean-variance approach to investing, and if everybody has the same estimates of the asset's expected returns, variances, and covariances, then everybody must invest in the same fund F of risky and risk-free assets. Because F is the same for everybody, it follows that in equilibrium, F must correspond to the market portfolio M, that is, a portfolio in which each asset is weighted by its proportion of total market capitalization. This observation is the basis for the **capital asset pricing model** (CAPM) proposed by Sharpe (1964). This model greatly simplifies the input for portfolio selection and makes the mean-variance methodology into a practical application. The CAPM result states that the expected return of each asset can be described as the function of "beta" which expresses the covariance with the market portfolio. Since a beta of a portfolio is equal to the weighted average of the betas of the individual assets that make up the portfolio, the covariance structure is greatly simplified by using the betas. Although the CAPM brings us a benefit for simplification, some researchers argued the validity of the CAPM assumptions such as the time independence, the uncorrelation and the Gaussianity. To overcome these problems, some extensions have been proposed so far.

In CAPM, the return of each asset is assumed as a regression model with a single explanatory variable (that is a market return), which implies that the market risk is the only source of risk besides the unique risk of each asset. However, there is some evidence that other common risk factors affect the market risk. Companies within the same country or the same industry appear to have common risks beyond the overall market risk. Also, research has suggested that companies with common characteristics such as a high book-to-market value have common risks, though this is controversial. A **factor model** that represents the connection between common risk factors and individual returns also leads to a simplified covariance structure, and provides important insight into the relationships among assets. The factor model framework leads to the **arbitrage pricing theory** (APT), introduced by Ross (1976), which is a pricing theory based on the principle of the absence of arbitrage.

The expected utility theory introduced by von Neumann and Morgenstern (1944) accounts for risk aversion in financial decision making, and provides a more general and more useful approach than the mean-variance approach. In view of the expected utility theory, the mean-variance approach has two pitfalls: First, the probability distribution of each asset return is characterized only by its first two moments. In the case of Gaussian distributions, the mean-variance model and utility theories are mainly compatible. The mean-variance-skewness model is introduced to relax the Gaussian assumption and to consider the skewness of the portfolio return. Second, the dependence structure is only described by the linear correlation coefficients of each pair of asset returns. In that case, serious losses are observed if extreme events are too underestimated.
Recent years have seen increasing development of new tools for risk management analysis. Many sources of risk have been identified, such as market risk, credit risk, counterparty default, liquidity risk, operational risk and so on. One of the main problems concerning the evaluation and optimization of risk exposure is the choice of **risk measures**. In portfolio theory, many risk measures instead of the variance have been introduced. According to Giacometti and Lozza (2004), we can distinguish two kinds of risk measures, namely, **dispersion measures** and **safety measures**. Dispersion measures include standard deviation, semivariance and mean absolute deviation. The "mean semivariance model" and "mean absolute deviation model" are proposed by using an alternative risk measure. The mean semivariance model deals with semivariance as the portfolio risk, which is closer to reality than the mean-variance model. In order to solve large-scale portfolio optimization problems, the mean absolute deviation model is considered, in which the portfolio risk is defined as the absolute deviation. Since investment managers frequently associate risk with the failure to attain a target return, the "mean target model" is introduced. This model optimizes distributions having below target returns. The safety measures describe the probability that the portfolio return falls under a given level. The typical model using this measure is "safety first" introduced by Roy (1952), Telser (1955) and Kataoka (1963). **Value at risk** (VaR) is the first attempt to take account of fat tailed and non-Gaussian returns; for example, when volatilities are random and possible jumps may occur, financial options are involved in the position, default risks are not negligible, or cross dependence between asset is complex, etc. Since VaR is a risk measure that only takes account of the probability of losses, and not of their size, other risk measures have been proposed. Among them, the expected shortfall (ES) as defined in Acerbi and Tasche (2001), also called **conditional value at risk** (CVaR) in Rockafellar and Uryasev (2000) or tail conditional expectation (TCE) in Artzner et al. (1999), is a class of risk measures which take account both of the probability of losses and of their size. Moreover, in Acerbi and Tasche (2001), **coherence** of these risk measures is proved. "Pessimistic portfolio" introduced by Bassett et al. (2004) minimizes **α-risks**, which include ES, CVaR and TCE. In order to explain the alternative calculation method for VaR or CVaR, we introduce the **copula dependence model**. Rockafellar and Uryasev (2000) discussed the procedure for calculating the mean-CVaR (or mean-VaR) portfolio based on the copula dependency structure. Thanks to the copula structure, the modeling of the univariate marginal and the dependence structure can be separated. We show that once a copula function and marginal distribution functions are determined, the mean-CVaR (or mean-VaR) portfolio is obtained.

3.1.1 *Mean-Variance Portfolio*

Suppose the existence of a finite number of assets indexed by $i, (i = 1, \ldots, m)$. Let $\boldsymbol{X}_t = (X_1(t), \ldots, X_m(t))'$ denote the random returns on m assets at time t. Assuming the stationarity of $\{\boldsymbol{X}_t\}$, write $E(\boldsymbol{X}_t) = \boldsymbol{\mu} = (\mu_1, \ldots, \mu_m)'$ and $Cov(\boldsymbol{X}_t, \boldsymbol{X}_t) = \Sigma = (\sigma_{ij})_{i,j=1,\ldots,m}$ (Σ is a nonsingular m by m matrix). Let $\boldsymbol{w} = (w_1, \ldots, w_m)'$ be the vector of portfolio weights. Then the return of the portfolio is $\boldsymbol{w}'\boldsymbol{X}_t$, and the expectation and variance are, respectively, given by $\mu(\boldsymbol{w}) = \boldsymbol{w}'\boldsymbol{\mu}$ and $\eta^2(\boldsymbol{w}) = \boldsymbol{w}'\Sigma\boldsymbol{w}$. Under the restriction $\sum_{i=1}^m w_i = 1$, we can plot all portfolios as points $(\sqrt{\eta^2(\boldsymbol{w})}, \mu(\boldsymbol{w}))$ on the mean-standard deviation diagram. The set of points that corresponds to portfolios is called the **feasible set** or **feasible region**, as shown in Figure 3.1. The feasible set[1] satisfies the following two important properties.

1. If $m \geq 3$, the feasible set will be a two-dimensional region.
2. The feasible region is convex to the left.

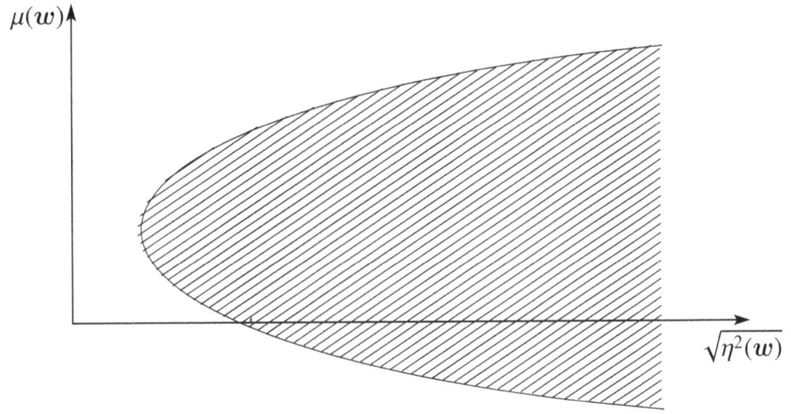

Figure 3.1: Feasible set (feasible region) with short selling allowed

The left boundary of a feasible set is called the **minimum-variance set**, because for a fixed mean rate of return, the feasible point with the smallest variance corresponds to the left boundary point. The minimum-variance set has a characteristic bullet shape. The upper portion of the minimum-variance set is called the **efficient frontier** of the feasible region, as shown in Figure 3.2. According

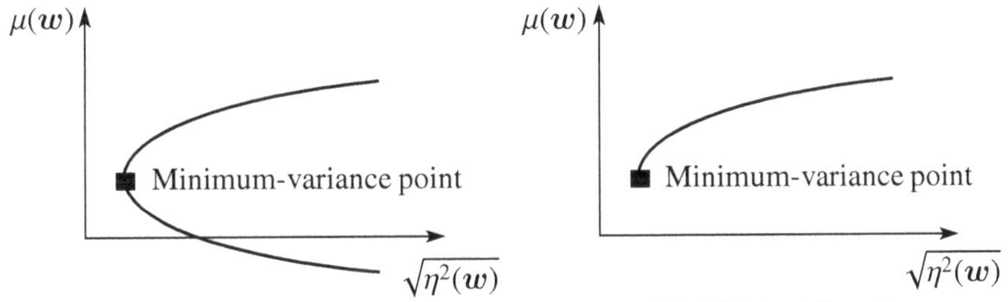

(a) **Minimum-variance set** (b) **Efficient frontier**

Figure 3.2: Minimum-variance set and efficient frontier

to Markowitz (1959), if a portfolio is represented by a point on the efficient frontier, the portfolio is called an **efficient portfolio** or, more precisely, the **mean variance efficient portfolio**. In other words, if a portfolio is not efficient, it is possible to find an efficient portfolio with either

1. greater expected return ($\mu(\boldsymbol{w})$) but no greater variance ($\eta^2(\boldsymbol{w})$), or
2. less variance (or standard deviation) but no less expected return.

Mean variance efficient portfolio weights have been proposed by various criteria. The following are typical ones.

Consider

$$\min_{\boldsymbol{w} \in \mathbb{R}^m} \eta^2(\boldsymbol{w}) \quad \text{subject to} \quad \mu(\boldsymbol{w}) = \mu_P, \ \sum_{i=1}^{m} w_i = 1 \tag{3.1.1}$$

[1]The sale of an asset that is not owned is called "short selling." If short selling is allowed, the portfolio weight w_i could be negative. Otherwise, additional constraints $w_i \geq 0$ for all i are added.

where μ_P is a given expected portfolio return by investors. This criterion indicates the variance minimizer of the set of portfolios for a given expected return as μ_P. The point of the portfolio (called the "minimum-variance portfolio") on the mean-standard deviation diagram corresponds to the point on the minimum-variance set. This portfolio is shown in Figure 3.3. Let $\boldsymbol{e} = (1, \ldots, 1)'$ (m-vector), $A = \boldsymbol{e}'\Sigma^{-1}\boldsymbol{e}$, $B = \boldsymbol{\mu}'\Sigma^{-1}\boldsymbol{e}$, $C = \boldsymbol{\mu}'\Sigma^{-1}\boldsymbol{\mu}$, and $D = AC - B^2$. The solution for $\boldsymbol{w}$ is given by

$$\boldsymbol{w}_0 = \frac{\mu_P A - B}{D}\Sigma^{-1}\boldsymbol{\mu} + \frac{C - \mu_P B}{D}\Sigma^{-1}\boldsymbol{e}. \tag{3.1.2}$$

In the case that the constraint $\mu(\boldsymbol{w}) = \mu_P$ is eliminated, the solution is given by

$$\boldsymbol{w}_0 = \frac{1}{A}\Sigma^{-1}\boldsymbol{e}. \tag{3.1.3}$$

This portfolio is called the "global minimum variance portfolio." The point of this portfolio on the mean-standard deviation diagram corresponds to the minimum-variance point in Figure 3.3. These solutions are obtained by solving the Lagrangian.
Next we consider

$$\max_{\boldsymbol{w}\in\mathbb{R}^m} \mu(\boldsymbol{w}) \quad \text{subject to} \quad \eta^2(\boldsymbol{w}) = \sigma_P^2, \ \sum_{i=1}^m w_i = 1 \tag{3.1.4}$$

where σ_P is a given standard deviation of portfolio return by investors. This criterion indicates the mean maximizer of the set of the portfolio for a given standard deviation of portfolio return as σ_P. This portfolio is also shown in Figure 3.3. If $J = B^2(1 - \sigma_P^2 A)^2 - A(1 - \sigma_P^2 A)(C - \sigma_P^2 B^2) \geq 0$ and $\sigma_P^2 A \geq 1$ are satisfied, then the solution for $\boldsymbol{w}$ is given by

$$\boldsymbol{w}_0 = \frac{1}{2\lambda_1}\Sigma^{-1}\boldsymbol{\mu} - \frac{\lambda_2}{2\lambda_1}\Sigma^{-1}\boldsymbol{e} \tag{3.1.5}$$

where

$$\lambda_1 = \frac{B - \lambda_2 A}{2}, \quad \lambda_2 = \frac{B(1 - \sigma_P^2 A) + \sqrt{J}}{A(1 - \sigma_P^2 A)}.$$

Note that if $\sigma_P^2 A = 1$, the solution describes the global minimum-variance portfolio.

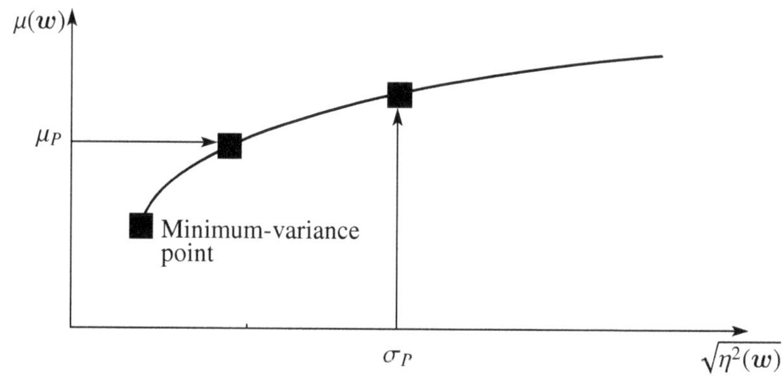

Figure 3.3: Mean variance efficient portfolio

Suppose there exists a risk-free asset with the return μ_f. Consider

$$\max_{\boldsymbol{w}\in\mathbb{R}^m} \frac{\mu(\boldsymbol{w}) - \mu_f}{\eta(\boldsymbol{w})} \quad \text{subject to} \quad \sum_{i=1}^m w_i = 1. \tag{3.1.6}$$

This efficient portfolio is called the "tangency portfolio." Given a point in the feasible region, we draw a line (l_f) between the risk-free asset and that point. When the line is a tangency line of the efficient frontier, the tangency point describes the tangency portfolio shown in Figure 3.4. To calculate the solution, we denote the angle between the line (l_f) for each point in the feasible region and the horizontal axis by θ. For any feasible portfolio, we have "$\tan\theta = \frac{\mu(\boldsymbol{w})-\mu_f}{\eta(\boldsymbol{w})}$." Since the tangency portfolio is the feasible point that maximizes θ or, equivalently, maximizes $\tan\theta$, the solution is given by

$$\boldsymbol{w}_0 = \frac{\Sigma^{-1}(\boldsymbol{\mu} - \mu_f \boldsymbol{e})}{\boldsymbol{e}'\Sigma^{-1}(\boldsymbol{\mu} - \mu_f \boldsymbol{e})}. \tag{3.1.7}$$

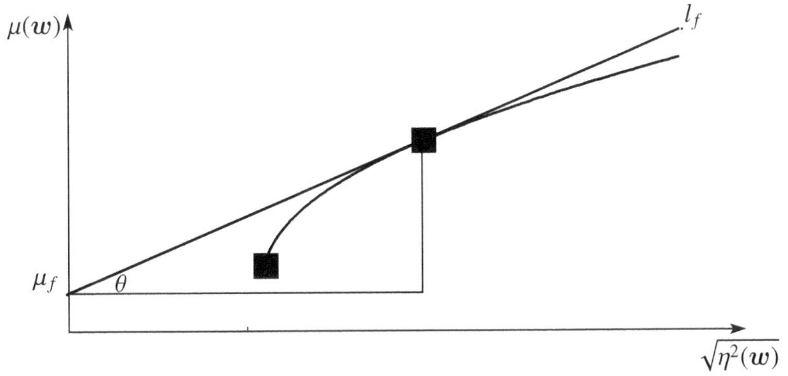

Figure 3.4: Tangency portfolio

The above portfolio selection problems such as (3.1.1), (3.1.4) and (3.1.6) consider cases where short sales for risky assets are allowed and lending or borrowing for a riskless asset is possible. If these assumptions are not satisfied, we need to add some additional constraints. In these cases, we cannot obtain explicit solutions, but we can get solutions by using "the quadratic programming method." The effect of these assumptions is explained by using the feasible region, as shown in Figure 3.5.

(a) Short sales for risky assets are not allowed and there is no risk-free asset: in this case, the feasible region is formed by all portfolios with nonnegative weights, which implies that additional constraints $w_i \geq 0$ for all i are added to the optimization problems.

(b) Short sales for risky assets are allowed and there is no risk-free asset: obviously the feasible region in this case contains that without short selling. However, the efficient frontiers of these two regions may partially coincide.

(c) Short sales for risky assets are allowed and only lending for a risk-free asset is allowed: in this case, the feasible region is surrounded by the finite line segments between the risk-free asset and the points in the feasible region, which implies that an additional constraint $w_f = 1 - \boldsymbol{w}'\boldsymbol{e} \geq 0$, that is, the nonnegative weight of the risk-free asset, is added to the optimization problems.

(d) Short sales for risky assets are allowed and both borrowing and lending for a risk-free asset are allowed: in this case, the feasible region is an infinite triangle whenever a risk-free asset is included in the universe of available assets.

3.1.2 Capital Asset Pricing Model

One of the important problems in the discipline of investment science is to determine the equibrium price of an asset. The capital asset pricing model (CAPM) developed primarily by Sharpe (1964),

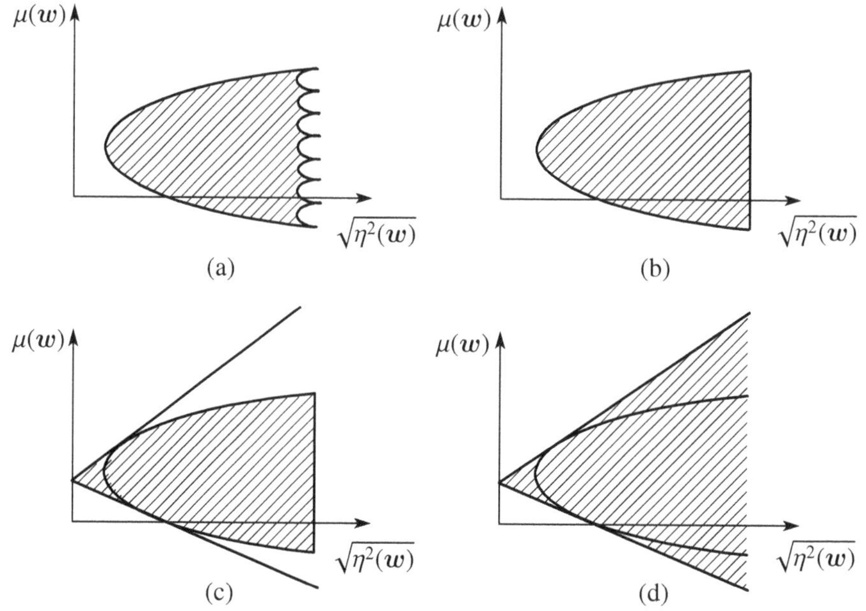

Figure 3.5: Feasible region

Lintner (1969) and Mossin (1966) answers this problem. The CAPM follows logically from the Markowitz mean-variance portfolio theory. The CAPM starts with the following assumptions.[2]

Assumption 3.1.1. *The assumptions of CAPM*

1. *The market prices are "in equilibrium." In particular, for each asset, supply equals demand.*
2. *Everyone has the same forecasts of expected returns and risks.*
3. *All investors choose portfolios optimally (efficiently) according to the principles of mean variance efficiency.*
4. *The market rewards for assuming unavoidable risk, but there is no reward for needless risks due to inefficient portfolio selection.*

Suppose that there exists a market portfolio with the expectation μ_M and the standard deviation σ_M, such as S&P500 and Nikkei 225. Then, we can plot the market portfolio on a $\mu - \sigma$ diagram. Given a risk-free asset with the return μ_f, we can draw a single straight line passing through the risk-free point and the market portfolio. This line is called the **capital market line (CML)**. (See Figure 3.6.)

The CML relates the expected rate of return of an efficient portfolio to its standard deviation, but it does not show how the expected rate of return of an individual asset relates to its individual risk. This relation is expressed by the CAPM. Under Assumption 3.1.1, if the market portfolio is mean variance efficient, the expected return $\mu_i = E[X_i(t)]$ of any asset i belongs with the market and can be written as

$$\mu_i - \mu_f = \beta_i(\mu_M - \mu_f) \tag{3.1.8}$$

where

$$\beta_i = \frac{\sigma_{iM}}{\sigma_M^2} \tag{3.1.9}$$

and σ_{iM} is the covariance between the returns on the asset i and the market portfolio. The value β_i is

[2]See, e.g., Luenberger (1997).

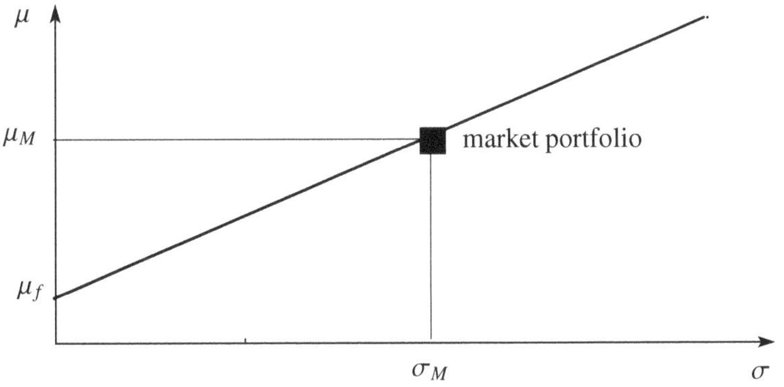

Figure 3.6: Capital market line

referred to as the **beta** of an asset. An asset's beta gives important information about the asset's risk characteristics by using the CAPM formula. The value $\mu_i - \mu_f$ is termed the **expected excess rate of return** (or the **risk premium**) of asset i; it is the amount by which the rate of return is expected to exceed the risk-free rate. It is easy to calculate the overall beta of a portfolio in terms of the betas of the individual assets in the portfolio. Suppose, for example, that a portfolio contains m assets with the weight vector $\boldsymbol{w} = (w_1, \ldots, w_m)'$. It follows immediately that

$$\mu(\boldsymbol{w}) - \mu_f = \beta(\boldsymbol{w})(\mu_M - \mu_f) \tag{3.1.10}$$

where $\beta(\boldsymbol{w}) = \boldsymbol{w}'\boldsymbol{\beta}$ and $\boldsymbol{\beta} = (\beta_1, \ldots, \beta_m)'$. In other words, the portfolio beta is just the weighted average of the betas of the individual assets in the portfolio, with the weights being identical to those that define the portfolio. The CAPM formula can be expressed in graphical form by regarding the formula as a linear relationship. This relationship is termed the **security market line (SML)**. (See Figure 3.7.) Under the equilibrium conditions assumed by the CAPM, any asset should fall on the SML.

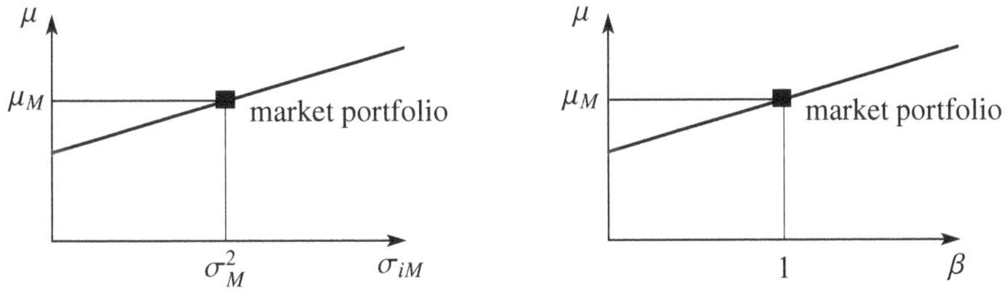

(a) **SML** on the $\mu - \sigma_{iM}$ diagram (b) **SML** on the $\mu - \beta$ diagram

Figure 3.7: Security market line

Let $X_i(t)$ be the return at time t on the ith asset. Similarly, let $X_M(t)$ and μ_f be the return on the market portfolio at time t and the risk-free asset. Consider the following regression model:

$$X_i(t) = \mu_f + \beta_i(X_M(t) - \mu_f) + \epsilon_i(t) \tag{3.1.11}$$

where $\epsilon_i(t)$ is a random variable with $E(\epsilon_i(t)) = 0$, $V(\epsilon_i(t)) = \sigma^2_{\epsilon i}$, $cov(\epsilon_i(t), X_M(t)) = 0$ and $cov(\epsilon_i(t), \epsilon_j(t)) = 0$ for $i \neq j$.[3] Taking expectations in (3.1.11), we get the form (3.1.8), which is

[3]This assumption is essentially the reason why we should hold a large portfolio of all stocks. But this assumption is sometimes unrealistic. In that case, correlation among the stocks in a market can be modeled using a "factor model" introduced in the next section.

the SML. In addition, we have

$$\sigma_i^2 = \beta_i^2\sigma_M^2 + \sigma_{\epsilon i}^2 \tag{3.1.12}$$

where $\sigma_i^2 = V(X_i(t))$ and $\sigma_M^2 = V(X_M(t))$. In the above equation, $\beta_i^2\sigma_M^2$ is called the **systematic risk**, which is the risk associated with the market as a whole. This risk cannot be reduced by diversification because every asset with nonzero beta contains this risk. On the other hand, $\sigma_{\epsilon i}^2$ is called **nonsystematic, idiosyncratic**, or **specific risk**, which is uncorrelated with the market and can be reduced by diversification. When you consider appropriate total asset number m of a portfolio in this model, it is worth it to reduce the nonsystematic risk for portfolios. Ruppert (2004, Example 7.2.) shows that the nonsystematic risk for a portfolio is decreasing as the asset number m tends to increase. Moreover, you can eliminate the nonsystematic risk by holding a diversified portfolio in which securities are held in the same relative proportions as in a broad market index. Suppose we have known β_i and $\sigma_{\epsilon i}^2$ for each asset in a portfolio and also known μ_M and σ_M^2 for the market and μ_f for a risk-free asset. Then, under the CAPM, we can compute the expectation and variance of the portfolio return $\boldsymbol{w}'\boldsymbol{X}_t$ as

$$\mu(\boldsymbol{w}) = \mu_f + \boldsymbol{w}'\boldsymbol{\beta}(\mu_M - \mu_f), \quad \eta^2(\boldsymbol{w}) = \boldsymbol{w}'\left(\boldsymbol{\beta}\boldsymbol{\beta}'\sigma_M^2 + \Sigma_\epsilon\right)\boldsymbol{w} \tag{3.1.13}$$

where $\boldsymbol{\beta} = (\beta_1, \ldots, \beta_m)'$ and $\Sigma_\epsilon = diag(\sigma_{\epsilon 1}^2, \ldots, \sigma_{\epsilon m}^2)$. By substituting (3.1.13) into (3.1.1), (3.1.4) or (3.1.6), we can obtain mean variance efficient portfolios in terms of $\boldsymbol{\beta}, \Sigma_\epsilon, \sigma_M^2, \mu_M$, and μ_f. Note that the CAPM leads to the reduction of the number of the parameter. If you assume the CAPM, the number of the parameter is $2 + 2m$ (i.e., $\beta_1, \ldots, \beta_m, \sigma_{\epsilon 1}^2, \ldots, \sigma_{\epsilon m}^2, \mu_M$ and σ_M^2). Otherwise, $2m + m(m-1)/2$ parameters (i.e., $\mu_1, \ldots, \mu_m, \sigma_{11}, \ldots, \sigma_{mm}$) are needed. That is the reason why the CAPM greatly simplifies the input for portfolio selection. However, there are the following problems on the validity of the CAPM assumptions.

1. Any or all of the quantities $\boldsymbol{\beta}, \Sigma_\epsilon, \sigma_M^2, \mu_M$, and μ_f could depend on time t. However, it is generally assumed that $\boldsymbol{\beta}$ and Σ_ϵ as well as σ_M^2 and μ_M of the market are independent of time t so that these parameters can be estimated assuming stationarity of the time series of returns.
2. The assumption that $\epsilon_i(t)$ is uncorrelated with $\epsilon_j(t)$ for $i \neq j$ is essential to the decomposition of the ith risk into the systematic risk and the nonsystematic risk. This assumption is sometimes unrealistic (see Ruppert (2004, Section 7.4.3)).
3. The CAPM is the basis of the mean variance portfolio and this portfolio fundamentally relies on the description of the probability distribution function (pdf) of asset returns in terms of Gaussian functions (see, e.g., Malevergne and Didier Sornette (2005)).

As for problem 1, Aue et al. (2012), Chochola et al. (2013, 2014) and others have constructed sequential monitoring procedures for the testing of the stability of portfolio betas. It is well known that if the model parameters β_i in the CAPM vary over time, the pricing of assets and predictions of risks may be incorrect and misleading. Therefore, some procedures for the detection of structural breaks in the CAPM have been proposed. On the other hand, Merton (1973) developed the intertemporal CAPM (ICAPM) in which investors act so as to maximize the expected utility of lifetime consumption and can trade continuously in time. In this model, unlike the single-period model, current demands are affected by the possibility of uncertain changes in future investment opportunities. Moreover, Breeden (1979) pointed out that Merton's intertemporal CAPM is quite important from a theoretical standpoint and introduced the consumption-based CAPM (CCAPM).
As for problem 2, the correlation among the stocks in a market sector can be modeled using the "factor model." For instance, Chamberlain (1983) and Chamberlain and Rothschild (1983) introduced an approximate factor structure. They claimed that the uncorrelated assumption for the traditional factor model included in CAPM is unnecessarily strong and can be relaxed.
As for problem 3, multi-moments CAPMs have been introduced. Since it is well known that the distributions of the actual asset returns have fat tails, Rubinstein (1973), Kraus and Litzenberger (1976),

Lim (1989) and Harvey and Siddique (2000) have underlined and tested the role of the skewness of the distribution of asset returns. In addition, Fang and Lai (1997) and Hwang and Satchell (1999) have introduced a four-moment CAPM to take into account the leptokurtic behaviour of the asset return distribution. Many other extensions have been presented, such as the VaR-CAPM by Alexander and Baptista (2002) or the distribution-CAPM by Polimenis (2002).

The CAPM is available not only for portfolio selection problem but also for the other investment problem such as investment implications, performance evaluation and project choice (see, e.g., Luenberger (1997)).

3.1.3 Arbitrage Pricing Theory

The information required by the mean-variance approach grows substantially as the number m of assets increases. There are m mean values, m variances, and $m(m-1)/2$ covariances: a total of $2m+m(m-1)/2$ parameters. When m is large, this is a very large set of required values. For example, if we consider a portfolio based on 1,000 stocks, 501,500 values are required to fully specify a mean-variance model. Clearly, it is a formidable task to obtain this information directly so that we need a simplified approach. Fortunately, the CAPM greatly simplifies the number of required values by using "beta" (in the case of the above example, 2,002 values are required). However, in the CAPM, the market risk factor is the only risk based on the correlation between asset returns. Indeed, there is some evidence of other common risk factors besides the market risk. **Factor models** generalize the CAPM by allowing more factors than simply the market risk and the unique risk of each asset. This model leads to a simplified structure for the covariance matrix, and provides important insight into the relationships among assets. The factors used to explain randomness must be chosen carefully and the proper choice depends on the universe, which might be population, employment rate, and school budgets. For common stocks listed on an exchange, the factors might be the stock market average, gross national product, employment rate, and so forth. Selection of factors in somewhat of an art, or a trial-and-error process, although formal analysis methods can also be helpful. The factor model framework leads to an alternative theory of asset pricing, termed **arbitrage pricing theory** (APT). This theory does not require the assumption that investors evaluate portfolios on the basis of means and variances; only that, when returns are certain, investors prefer greater return to lesser return. In this sense the theory is much more satisfying than the CAPM theory, which relies on both the mean-variance framework and a strong version of equilibrium, which assumes that everyone uses the mean-variance framework. However, the APT requires a special assumption, that is, the universe of assets being considered is large. In this theory, we must assume that there are an infinite number of securities, and that these securities differ from each other in nontrivial ways. For example, if you consider the universe of all publicly traded U.S. stocks, this assumption is satisfied well.

Time Series Factor Model Suppose there exist p kinds of risk factor $F_1(t), \dots, F_p(t)$ which depend on time t. In a "time series multifactor model,"[4] the random return on m assets at time t is written as

$$\boldsymbol{X}_t = \boldsymbol{a} + \boldsymbol{b}\boldsymbol{F}_t + \boldsymbol{\epsilon}_t \tag{3.1.14}$$

where $\boldsymbol{a} = (a_1, \dots, a_m)'$ is an m-dimensional constant vector, $\boldsymbol{b} = (b_{ij})_{i=1,\dots,m, j=1,\dots,p}$ is an $m \times p$ constant matrix, $\boldsymbol{F}_t = (F_1(t), \dots, F_p(t))'$ are p-dimensional random vectors with an m dimensional mean vector $\boldsymbol{\mu}_F$ and a $p\times p$ covariance matrix $\Sigma_F = (\sigma_{fij})_{i,j=1,\dots,m}$, and $\boldsymbol{\epsilon}_t$ are m-dimensional random vectors with a mean $\boldsymbol{0}$ and an $m \times m$ covariance matrix $\Sigma_\epsilon = diag(\sigma^2_{\epsilon 1}, \dots, \sigma^2_{\epsilon m})$. In addition, it is usually assumed that the errors $\boldsymbol{\epsilon}_t$ are uncorrelated with $\boldsymbol{F}_t$, that is, $Cov(\boldsymbol{\epsilon}_t, \boldsymbol{F}_s) = \boldsymbol{0}$. These are idealizing assumptions which may not actually be true, but are usually assumed to be true for the purpose of analysis. In this model, the a_i's are called **intercepts** because a_i is the intercept of the

[4] An alternative type of model is a cross-sectional factor model, which is a regression model using data from many assets but from only a single holding period (e.g., Ruppert (2004, Section 7.8.3)).

line for asset i with the vertical axis, and the b_i's are called **factor loadings** because they measure the sensitivity of the return to the factor.

If you consider the case of one factor (i.e., $p = 1$), it is called a single-factor model. In particular, if the factor $\boldsymbol{F}_t$ is chosen to be the excess rate of return on the market $X_M(t) - \mu_f$, then we can set $\boldsymbol{a} = \mu_f$ and $\boldsymbol{b} = \boldsymbol{\beta}$, and the single factor model is identical to the CAPM. A factor can be anything that can be measured and is thought to affect asset returns. According to Ruppert (2004), the following are examples of factors:

- Return on market portfolio (market model, e.g., CAPM)
- Growth rate of GDP
- Interest rate on short-term Treasury bills
- Inflation rate
- Interest rate spreads
- Return on some portfolio of stocks
- The difference between the returns on two portfolios.

Suppose we have known $\boldsymbol{a}, \boldsymbol{b}, \boldsymbol{\mu}_F, \Sigma_F$ and Σ_ϵ. Then, under the time series factor model, we can compute the expectation and variance of the portfolio return $\boldsymbol{w}'\boldsymbol{X}_t$ as

$$\mu(\boldsymbol{w}) = \boldsymbol{w}'(\boldsymbol{a} + \boldsymbol{b}\boldsymbol{\mu}_F), \quad \eta^2(\boldsymbol{w}) = \boldsymbol{w}'\left(\boldsymbol{b}\Sigma_F\boldsymbol{b}' + \Sigma_\epsilon\right)\boldsymbol{w}. \tag{3.1.15}$$

By substituting (3.1.15) into (3.1.1), (3.1.4) or (3.1.6), we can obtain mean variance efficient portfolios in terms of $\boldsymbol{a}, \boldsymbol{b}, \boldsymbol{\mu}_F, \Sigma_F$ and Σ_ϵ. In the usual representation of asset returns, a total of $2m + m(m-1)/2$ parameters are required to specify means, variances and covariances. In this model, a total of just $2m + mp + p + p(p-1)/2$ parameters are required . Similarly to the CAPM, under the time series multifactor model, the portfolio risk $\eta^2(\boldsymbol{w})$ can be divided into the systematic risk $\boldsymbol{w}'\boldsymbol{b}\Sigma_F\boldsymbol{b}'\boldsymbol{w}$ and the nonsystematic risk $\boldsymbol{w}'\Sigma_\epsilon\boldsymbol{w}$. The point of APT is that the values of $\boldsymbol{a}$ and $\boldsymbol{b}$ must be related if arbitrage opportunities[5] are to be excluded. This means that in the efficient portfolio based on $\mu(\boldsymbol{w})$ and the systematic risk, there is no arbitrage opportunity. On the other hand, the nonsystematic risk can be reduced by diversification because this term's contribution to overall risk is essentially zero in a well-diversified portfolio.

3.1.4 *Expected Utility Theory*

In order to model any decision problem under risk, it is necessary to introduce a functional representation of preferences which measures the degree of satisfaction of the decision maker. This is the purpose of the utility theory.

Suppose that there exists a set of possible outcomes which may have an impact on the consequences of the decisions: $\Omega = \{\omega_1, \ldots, \omega_k\}$. Let $p = \{p_1, \ldots, p_k\}$ be the probability of occurrence of Ω satisfied with $\forall i, 0 \le p_i \le 1$ and $\sum_{i=1}^k p_i = 1$. Then, a lottery L is defined by a vector $\{(\omega_1, p_1), \ldots, (\omega_k, p_k)\}$ and the set of all lotteries is defined by $\mathcal{L}$. Moreover, a compound lottery L_c can be defined as a lottery whose outcomes are also lotteries.[6]

The decision maker is assumed to be "rational" if one's preference relation (denoted by $\succeq$) over the set of lotteries $\mathcal{L}$ is a binary relation which satisfies the following axioms:

Axiom 3.1.2.

[5]"Arbitrage" means to earn money without investing anything. In other words, to get any return without any risk and cost (e.g., Luenberger (1997, Section 1.2.)).

[6]Assume that a compound lottery L_c has two outcomes of lottery $L^a = \{(\omega_1, p_1^a), \ldots, (\omega_k, p_k^a)\}$ and lottery $L^b = \{(\omega_1, p_1^b), \ldots, (\omega_k, p_k^b)\}$ where the respective probability that the L^a becomes the outcome is ϵ and the lottery L^b becomes the outcome is $1 - \epsilon$. Then, the probability that the outcome of L_c will be ω_i is given by $p_i^c = \epsilon p_i^a + (1-\epsilon)p_i^b$. Therefore, L_c has the same vector of probabilities as the convex combination $L_c = \epsilon L^a + (1-\epsilon)L^b$.

- *The relation $\succeq$ is complete, that is, all lotteries are always comparable by $\succeq$: $\forall L^a, \forall L^b \in \mathcal{L}$, $L^a \succeq L^b$ or $L^b \succeq L^a$.*
- *The indifference relation $\sim$ associated to the relation $\succeq$ is defined by $\forall L^a, \forall L^b \in \mathcal{L}$, $L^a \sim L^b \Leftrightarrow L^a \succeq L^b$ and $L^b \succeq L^a$.*
- *The relation $\succeq$ is reflexive: $\forall L \in \mathcal{L}$, $L \succeq L$.*
- *The relation $\succeq$ is transitive: $\forall L^a, \forall L^b, \forall L^c \in \mathcal{L}$, $L^a \succeq L^b$ and $L^b \succeq L^c \Leftrightarrow L^a \succeq L^c$.*

Another standard assumption is continuity: small changes on probabilities do not modify the ordering between two lotteries. This property is specified in the following axiom (**continuity**):

Axiom 3.1.3. *The preference relation $\succeq$ on the set $\mathcal{L}$ of lotteries is such that $\forall L^a, \forall L^b, \forall L^c \in \mathcal{L}$, if $L^a \succeq L^b \succeq L^c$ then there exists a scalar $\epsilon \in [0, 1]$ such that $L^b \sim \epsilon L^a + (1 - \epsilon)L^c$.*

This continuity axiom implies the existence of a functional $\mathcal{U} : \mathcal{L} \to \mathbb{R}$ such that

$$\forall L^a, \forall L^b, L^a \succeq L^b \Leftrightarrow \mathcal{U}(L^a) \geq \mathcal{U}(L^b).$$

To develop the analysis of economics under uncertainty, additional properties must be imposed on the preferences. One of the most important additional conditions is the **independence** axiom:

Axiom 3.1.4. *The preference relation $\succeq$ on the set $\mathcal{L}$ of lotteries is such that $\forall L^a, \forall L^b, \forall L^c \in \mathcal{L}$ and for all $\epsilon \in [0, 1]$, $L^a \succeq L^b \Leftrightarrow \epsilon L^a + (1 - \epsilon)L^c \succeq \epsilon L^b + (1 - \epsilon)L^c$.*

The independence axiom, implicitly introduced in von Neumann and Morgenstern (1944), characterizes the expected utility criterion: indeed, it implies that the preference functional $\mathcal{U}$ on the lotteries must be linear in the probabilities of the possible outcomes:

Lemma 3.1.5. *(existence of utility function) Assume that the preference relation $\succeq$ on the set $\mathcal{L}$ of lotteries satisfies the continuity and independence axioms. Then, the relation $\succeq$ can be represented by a preference function that is linear in probabilities: there exists a function u defined on the space of possible outcomes Ω and with values in $\mathbb{R}$ such that for any two lotteries $L^a = \{(p_1^a, \ldots, p_k^a)\}$ and $L^b = \{(p_1^b, \ldots, p_k^b)\}$, we have*

$$L^a \succeq L^b \Leftrightarrow \sum_{i=1}^{k} u(\omega_i)p_i^a \geq \sum_{i=1}^{k} u(\omega_i)p_i^b.$$

Proof: See Theorem 1.1 of Prigent (2007).

Using the properties of concavity/convexity of utility functions, we deduce a characterization of risk aversion:

Lemma 3.1.6. *(characterization of risk aversion) Let u be a utility function representing preferences over the set of outcomes. Assume that u is increasing. Then*

1) The function u is concave if and only if the investor is risk averse.

2) The function u is linear if and only if the investor is risk neutral.

3) The function u is convex if and only if the investor is risk loving.

Proof: See Theorem 1.2 of Prigent (2007).

Let X be a random variable which represents outcomes of a lottery L and suppose that a utility function u exists. Then another lottery $C[X]$ satisfied with $u(C[X]) = E[u(X)]$ is called the **certainty equivalent** of the lottery L. In addition, the difference $\pi[X] = E[X] - C[X]$ is called the **risk premium**, as introduced in Pratt (1964). By using the risk premium, we can measure the "degree" of risk aversion of an investor, which is equivalent to the **Arrow–Pratt measure** as below.

Lemma 3.1.7. *(degree of risk aversion) Let u and v be two utility functions representing preferences over wealth. Assume that they are continuous, monotonically increasing, and twice differentiable. Then the following properties are equivalent and characterize the "more risk aversion":*

1) The derivatives of both utility functions are such that $-\frac{\partial^2 u(x)}{(\partial x)^2} / \frac{\partial u(x)}{\partial x} > -\frac{\partial^2 v(x)}{(\partial x)^2} / \frac{\partial v(x)}{\partial x}$, *for every x in* $\mathbb{R}$.

2) There exists a concave function Ψ *such that* $u(x) = \Psi[v(x)]$, *for every* x *in* $\mathbb{R}$.

3) The risk premium satisfies $\pi_u(X) \geq \pi_v(X)$, *for all random real valued variables* X.

Proof: See Theorem 1.3 of Prigent (2007).

Definition 3.1.8. *The term* $A(x) = -\frac{\partial^2 u(x)}{(\partial x)^2} / \frac{\partial u(x)}{\partial x}$ *is called the Arrow Pratt measure of* ***absolute risk aversion (ARA)****. Another measure allows us to take account of the level of wealth: the ratio* $R(x) = xA(x)$, *which is called the Arrow Pratt measure of* ***relative risk aversion (RRA)****.*

From the above risk aversion measures, we can characterize some standard utility functions, and in particular the general class of HARA utilities, which are useful to get analytical results.

Definition 3.1.9. *A utility function u is said to have* ***harmonic absolute risk aversion (HARA)*** *if the inverse of its absolute risk aversion is linear in wealth.*[7]

Usually, three subclasses are distinguished:

Constant absolute risk aversion (CARA) In this class, the ARA $A(x)$ is constant ($= A$), the RRA $R(x)$ is increasing in wealth and the utility function u takes the form

$$u(x) = -\frac{\exp(-Ax)}{A}.$$

Constant relative risk aversion (CRRA) In this class, the RRA $R(x)$ is constant ($= R$), and the utility function u takes the form

$$u(x) = \begin{cases} x^{1-R}/(1-R) & if \quad R \neq 1 \\ ln[x] & if \quad R = 1 \end{cases}. \tag{3.1.16}$$

This kind of utility function exhibits a decreasing absolute risk aversion (DARA).

Quadratic utility function Quadratic utility function u takes the form

$$u(x) = a(b - x)^2.$$

Note that we have to restrict its domain as $(-\infty, b]$ on which u is increasing (if $a > 0$) and the ARA is increasing with wealth.

Other utility theories, such as weighted utility theory, rank-dependent expected utility theory (RDEU), anticipated utility theory, cumulative prospect theory, nonadditive expected utility and Regret theory, have been introduced (see Prigent (2007)).

As we mentioned above, under some specific assumptions, the investor's preference on the possible positions X can be represented by a utility function u. Once a utility function is defined, all alternative random wealth levels are ranked by evaluating their expected utility values $E[u(x)]$. Investors wish to form a portfolio to maximize the expected utility of final wealth. The investor's problem is

$$\max_{\boldsymbol{w} \in \mathbb{R}^m} E[U(\boldsymbol{w}'\boldsymbol{X}_t)] \quad \text{subject to} \quad \sum_{i=1}^{m} w_i = 1. \tag{3.1.17}$$

Note that when all returns are Gaussian random variables, the mean-variance criterion is also equivalent to the expected utility approach for any risk-averse utility function as follows:

Remark 3.1.10. *(Luenberger (1997), Ruppert (2004), Prigent (2007)) If the utility function U is quadratic,* $U(x) = x - \frac{k}{2}x^2$ *with* $k > 0$, *or exponential,* $U(x) = -\frac{\exp[-Ax]}{A}$ *with* $A > 0$ *and the returns on all portfolios are normally distributed, then the portfolio that maximizes the expected utility is on the efficient frontier.*

[7]HARA utility function u takes the form $u(x) = a(b + x/c)^{1-c}$ on the domain $b + x/c > 0$. The constant parameters a, b and c satisfy the condition $a(1 - c)/c > 0$ if $c \neq 0, 1$.

Suppose that all returns are non-Gaussian and a utility function U has the third derivative D^3U. Then, by Taylor expansion, the expected utility can be represented as

$$\begin{aligned} E[U(\boldsymbol{w}'\boldsymbol{X}_t)] &\approx U[E(\boldsymbol{w}'\boldsymbol{X}_t)] + \frac{1}{1!}DU[E(\boldsymbol{w}'\boldsymbol{X}_t)]E[\boldsymbol{w}'\boldsymbol{X}_t - E(\boldsymbol{w}'\boldsymbol{X}_t)] \\ &+ \frac{1}{2!}D^2U[E(\boldsymbol{w}'\boldsymbol{X}_t)]E[(\boldsymbol{w}'\boldsymbol{X}_t - E(\boldsymbol{w}'\boldsymbol{X}_t))^2] \\ &+ \frac{1}{3!}D^3U[E(\boldsymbol{w}'\boldsymbol{X}_t)]E[(\boldsymbol{w}'\boldsymbol{X}_t - E(\boldsymbol{w}'\boldsymbol{X}_t))^3] \\ &= U[\mu(\boldsymbol{w})] + \frac{1}{2}D^2U[\mu(\boldsymbol{w})]\eta^2(\boldsymbol{w}) + \frac{1}{6}D^3U[\mu(\boldsymbol{w})]c_3(\boldsymbol{w}) \end{aligned}$$

where $c_3(\boldsymbol{w}) = E[(\boldsymbol{w}'\boldsymbol{X}_t - E(\boldsymbol{w}'\boldsymbol{X}_t))^3]$ and $D^iU(x) = \frac{d^iU(x)}{(dx)^i}$ for $i = 1, 2, 3$. If the returns on all portfolios are not Gaussian, then we cannot ignore the higher-order moment of the above approximation. That is the motivation with which the mean-variance-skewness model is introduced.

Mean-variance-skewness model (MVS) The third moment of a return is called skewness, which measures the asymmetry of the probability distribution. A natural extension of the mean-variance model is to add skewness into consideration for portfolio management. There will be triple goals: maximizing the mean and the skewness, minimizing the variance. People interested in the use of skewness prefer a portfolio with a larger probability for large payoffs when the mean and variance remain the same. The mathematical programming formulation is as follows:

$$\max_{\boldsymbol{w}\in\mathbb{R}^m} E[(\boldsymbol{w}'\boldsymbol{X}_t - \mu(\boldsymbol{w}))^3] \quad \text{subject to} \quad \mu(\boldsymbol{w}) = \mu_P,\ \eta^2(\boldsymbol{w}) = \sigma_P^2,\ \sum_{i=1}^m w_i = 1. \tag{3.1.18}$$

The importance of higher-order moments in portfolio selection has been suggested by Samuelson (1970). But partially because of the difficulty of estimating the third-order moment for a large number of securities such as over a few hundred, and of solving nonconvex programs by using traditional computational methods, quantitative treatments of the third-order moment have been neglected for a long time.

When the utility functions are not observable, the theory of **stochastic dominance** is available to rank two random prospects X and Y. This theory uses the cumulative distribution functions F_X and F_Y of X and Y instead of $E[u(X)]$ and $E[u(Y)]$.

Definition 3.1.11. *Assume that the supports of all random variables are in an interval* $[a, b]$.

(1) X is said to dominate Y according to ***first-order stochastic dominance*** $(X \geq_1 Y)$ *if* $F_X(\omega) \leq F_Y(\omega)$ *for all* $\omega \in [a, b]$.

(2) X is said to dominate Y according to ***second-order stochastic dominance*** $(X \geq_2 Y)$ *if* $\int_a^\omega F_X(s)ds \leq \int_a^\omega F_Y(s)ds$ *for all* $\omega \in [a, b]$.

This definition is consistent with the expected utility, that is, $X \geq_1 Y$ if and only if $E[u(X)] \geq E[u(Y)]$ for any utility function u which is monotonically increasing, and $X \geq_2 Y$ if and only if $E[u(X)] \geq E[u(Y)]$ for any utility function u which is monotonically increasing and concave (see Propositions 1.3 and 1.4 of Prigent (2007)). Moreover, the second-order dominance property justifies the diversification investment strategy (see Remark 1.11 of Prigent (2007)).

The key property linked to the expected utility theory is the independence axiom, which may fail empirically and can yield some paradoxes, as shown by Allais (1953). Therefore, during the 1980s, the revision of the expected utility paradigm was intensely developed by slightly modifying or relaxing the original axioms. The following are typical ones.

Weighted utility theory One of the theories that was consistent with Allais, paradox was the "weighted utility theory," introduced by Chew and MacCrimmmon (1979). The idea is to apply a transformation on the initial probability. The basic result of Chew and MacCrimmon yields the following representation of preferences over lotteries $L = \{(x_1, p_1), \ldots, (x_k, p_k)\}$:

$$\mathcal{U}(L) = \sum_{i=1}^{k} u(x_i)\phi(p_i) \quad with \quad \phi(x_i) = p_i / \sum_{i=1}^{k} v(x_i)p_i$$

where u and v are two different elementary utility functions.

Prospect theory Another approach is the "prospect theory" introduced by Tversky and Kahneman (1979). The idea of the prospect theory is to represent the preferences by means of a function ϕ such that the utility of a lottery $L = \{(x_1, p_1), \ldots, (x_k, p_k)\}$ is given by

$$\mathcal{U}(L) = \sum_{i=1}^{k} u(x_i)\phi(p_i)$$

where ϕ is an increasing function defined on $[0, 1]$ with values in $[0, 1]$ and $\phi(0) = 0, \phi(1) = 1$. Using these transformations, the Allais paradox can be solved. However, these theories imply the violation of first-order stochastic dominance. To solve this problem, alternative approaches have been proposed.

Rank-dependent expected utility theory The rank dependent expected utility (RDEU) theory assumes that people consider cumulative distribution functions rather than probabilities themselves. For a lottery $L = \{(x_1, p_1), \ldots, (x_k, p_k)\}$ with $x_1 \leq x_2 \leq \cdots \leq x_k$, the utility is given by

$$\mathcal{U}(L) = \sum_{i=1}^{k} u(x_i) \left[\Phi(\sum_{j=1}^{i} p_j) - \Phi(\sum_{j=1}^{i-1} p_j) \right]$$

where the weights $\Phi(\sum_{j=1}^{i} p_j)$ depend on the ranking of the outcomes x_i. In this framework, it is possible to introduce preference representations that are compatible with first-order stochastic dominance. If Φ is not the identity but $u(x) = x$, the RDEU is the "dual theory" of Yaari (1987). Quiggin (1982) has proposed "anticipated utility theory" in which the "weak independence axiom" instead of the independence axiom is added. Then the utility is a class of the RDEU where Φ is nondecreasing on $[0, 1]$ to $[0, 1]$, and is concave on $[0, 1/2]$, $(\Phi(p_i) > p_i)$ and convex on $[1/2, 1]$, $(\Phi(p_i) < p_i)$ with $\Phi(1/2) = 1/2$ and $\Phi(1) = 1$. Tversky and Kahneman (1992) have introduced "cumulative prospect theory" in which utility functions are given for losses and gains. The utility is also a class of the RDEU where Φ is divided into a loss part Φ^- and a gain part Φ^+. The "Choquet expected utility" of Schmeidler (1989) is also a class of the RDEU. The model is based on the so-called "Choquet integral" (see Choquet (1953)). Consider two random variables X and Y associated respectively to lotteries L_X and L_Y, with the distribution functions F_X and F_Y. In the framework of the Choquet expected utility, X is preferred to Y if

$$E[u(X)] = \int_0^1 u(F_X^{-1}(t))dv(t) \geq \int_0^1 u(F_Y^{-1}(t))dv(t) = E[u(Y)]$$

where u is a utility function, v is a Choquet distortion measure and $F^{-1}(t)$ is tth quantile. The simplest Choquet distortion measure is given by the one-parameter family

$$v_\epsilon(t) = \min(t/\epsilon, 1) \quad for\ \epsilon \in (0, 1].$$

This family plays an important role in the portfolio application. Focusing for a moment on a single v_α, we have

$$E_{v_\alpha}[u(X)] = \frac{1}{\alpha} \int_0^\alpha u(F_X^{-1}(t))dt.$$

Based on this utility function, Bassett et al. (2004) introduced the "pessimistic portfolio" which we will discuss later.

3.1.5 Alternative Risk Measures

In portfolio theory, many risk measures alternative to the variance or the standard deviation have been introduced. As noted by Giacometti and Lozza (2004), we can distinguish two kinds of risk measures, that is, **dispersion measures** and **safety measures**.
In the class of the dispersion measures, the risks ρ are increasing, positive, and positively homogeneous, such as standard deviation, semivariance and mean absolute deviation. The following models are optimal portfolios to minimize dispersion measures alternative to the standard deviation.

Mean semivariance model Based on the observation that investors may only be concerned with the risk of return being lower than the mean (downside risk), the method of mean semivariance was proposed to model this fact (see, e.g., Markowitz (1959)). The portfolio optimization problem is defined as

$$\min_{\boldsymbol{w}\in\mathbb{R}^m} E[\min\{\boldsymbol{w}'\boldsymbol{X}_t - \mu(\boldsymbol{w}), 0\}^2] \quad \text{subject to } \mu(\boldsymbol{w}) = \mu_P, \ \sum_{i=1}^m w_i = 1. \tag{3.1.19}$$

Although the rationale behind introducing the model is intuitively closer to reality than the mean-variance model, it is not widely used (see Wang and Xia (2002)).

Mean absolute deviation model In order to solve large-scale portfolio optimization problems, Konno and Yamazaki (1991) considered the mean absolute deviation as the risk of portfolio investment. The portfolio optimization problem is defined as

$$\min_{\boldsymbol{w}\in\mathbb{R}^m} E[|\boldsymbol{w}'\boldsymbol{X}_t - \mu(\boldsymbol{w})|] \quad \text{subject to } \mu(\boldsymbol{w}) = \mu_P, \ \sum_{i=1}^m w_i = 1. \tag{3.1.20}$$

Using the historical data of the Tokyo Stock Exchange, Konno and Yamazaki compared the performance of the mean-variance model and that of the mean absolute deviation model, and found that the performance of those two models was very similar.

Mean target model Fishburn (1977) examined a class of mean risk dominance models in which risk equals the expected value of a function that is zero at or above a target return T, and is non-decreasing in deviations below T. Considering that investment managers frequently associate risk with the failure to attain a target return, the model is formulated as follows:

$$\min_{\boldsymbol{w}\in\mathbb{R}^m} \int_{-\infty}^{T} (T - x)^{\pi} dF_{\boldsymbol{w}'\boldsymbol{X}_t}(x) \quad \text{subject to } \sum_{i=1}^m w_i = 1 \tag{3.1.21}$$

where T is a specified target return, $\pi(> 0)$ is a parameter which can approximate a wide variety of attitudes towards the risk of falling below the target return, and $F_{\boldsymbol{w}'\boldsymbol{X}_t}(x)$ is the probability distribution function of $\boldsymbol{w}'\boldsymbol{X}_t$. This model avoids distributions having below-target returns. However, a major drawback of the computational complexity restricts this model's ample uses in practice.

It is known that dispersion measures are not consistent with first-order stochastic dominance. On the other hand, the safety measures introduced by Roy (1952), Telser (1955) and Kataoka (1963) are consistent with first-order stochastic dominance. The safety first rules are based on risk measures involving the probability that the portfolio return falls under a given level.

Safety first While Markowitz's mean-variance model measures a portfolio's variance as the risk, Roy proposed another criterion to minimize the probability that a portfolio reaches below a disaster level. Denoting by D the minimum acceptable level of return for the investment, Roy's criterion is

$$\min_{\boldsymbol{w}\in\mathbb{R}^m} P(\boldsymbol{w}'\boldsymbol{X}_t < D) \quad \text{subject to } \sum_{i=1}^m w_i = 1. \tag{3.1.22}$$

Another example is the Telser criterion:

$$\max_{\boldsymbol{w}\in\mathbb{R}^m} \mu(\boldsymbol{w}) \quad \text{subject to} \quad P(\boldsymbol{w}'\boldsymbol{X}_t < D) \leq \alpha, \ \sum_{i=1}^{m} w_i = 1. \tag{3.1.23}$$

With this different view of risk, the optimization problem can have several different formulations. One example is the Kataoka criterion:

$$\max_{\boldsymbol{w}\in\mathbb{R}^m} D(\boldsymbol{w}) \quad \text{subject to} \quad P(\boldsymbol{w}'\boldsymbol{X}_t < D(\boldsymbol{w})) \leq \alpha, \ \sum_{i=1}^{m} w_i = 1 \tag{3.1.24}$$

where α is an acceptable probability that the portfolio return may drop below the limit D.
The three previous criteria consist of a given probability threshold. Under Gaussian assumptions, Roy, Telser and Kataoka showed that all optimal solutions are obtained as a point on the mean variance efficient frontier. Obviously, if asset returns are not Gaussian due to having nonzero skewness or a fat tail, then optimal solutions may no longer be mean variance efficient. For general distribution, it is necessary to introduce the **value at risk (VaR)**.

Definition 3.1.12. *The α-VaR is given by*

$$\mathrm{VaR}_\alpha(X) := -\sup\{x|P(X \leq x) < \alpha\}. \tag{3.1.25}$$

Mean VaR Portfolio Consider

$$\min_{\boldsymbol{w}\in\mathbb{R}^m} \mathrm{VaR}_\alpha(\boldsymbol{w}'\boldsymbol{X}_t) \quad \text{subject to} \quad \mu(\boldsymbol{w}) = \mu_P, \ \sum_{i=1}^{m} w_i = 1. \tag{3.1.26}$$

The solution indicates minimizing the portfolio's VaR under a given expected value μ_P. It is easy to see that if the constraint $\mu(\boldsymbol{w}) = \mu_P$ is eliminated, the above problem corresponds to Kataoka's criterion.
As can be easily seen, VaR is a risk measure that only takes account of the probability of losses, and not of their size. Moreover, VaR is traditionally based on the assumption of normal asset returns and has to be carefully evaluated when there are extreme price fluctuations. Furthermore, VaR may not be convex for some probability distributions. To overcome these problems, other risk measures have been proposed such as **coherent risk measures** and **convex risk measures**. Regarding regulatory concerns in the financial sector, it has been interested in "how to measure portfolio risks." For this problem, Artzner et al. (1999) provided an axiomatic foundation for coherent risk measures.

Definition 3.1.13. *For real-valued random variables $X \in \mathcal{X}$ on $(\Omega, \mathcal{A})$, a mapping $\rho : \mathcal{X} \to \mathbb{R}$ is called a coherent risk measure if it is*

- *monotone: $X, Y \in \mathcal{X}$, with $X \leq Y \Rightarrow \rho(X) \geq \rho(Y)$.*
- *subadditive: $X, Y, X + Y \in \mathcal{X} \Rightarrow \rho(X + Y) \leq \rho(X) + \rho(Y)$.*
- *linearly homogeneous: for all $\lambda \geq 0$ and $X \in \mathcal{X}$, $\rho(\lambda X) = \lambda\rho(X)$.*[8]
- *translation invariant: for all $\lambda \in \mathbb{R}$ and $X \in \mathcal{X}$, $\rho(\lambda + X) = \rho(X) - \lambda$.*

These requirements rule out most risk measures traditionally used in finance. In particular, risk measures based on a variance or standard deviation are ruled out by the monotonicity requirement, and quantile-based risk measures such as VaR are ruled out by the subadditivity requirement.
α-risk is one of the important risk measures to have coherence properties. Let

$$\rho_{\nu_\alpha}(X) = -\int_0^1 F_X^{-1}(t)d\nu_\alpha(t) = -\frac{1}{\alpha}\int_0^\alpha F_X^{-1}(t)dt \tag{3.1.27}$$

[8]When the risk measure is not assumed to have variations in proportion to the risk variations themselves, linear homogeneity is no longer satisfied. A ρ is called "convex risk measure" if it is monotone, subadditive, translation invariant and convex, which means that for all $0 \leq \lambda \leq 1$ and $X, Y \in \mathcal{X}$, $\rho(\lambda X + (1 - \lambda)Y) \leq \lambda\rho(X) + (1 - \lambda)\rho(Y)$.

where F_X^{-1} is the generalized inverse of the cdf F_X, that is, $F_X^{-1}(p) = \sup\{x|F_X(x) \le p\}$, $\nu_\alpha(t) = \min\{t/\alpha, 1\}$ and $\alpha \in (0, 1]$. Variants of $\rho_{\nu_\alpha}(X)$ have been suggested under a variety of names, including expected shortfall (Acerbi and Tasche (2001)), conditional value at risk (Rockafellar and Uryasev (2000)) and tail conditional expectation (Artzner et al. (1999)). For the sake of brevity Bassett et al. (2004) called $\rho_{\nu_\alpha}(X)$ as α-risk of the random prospect X. Clearly, the α-risk is simply the negative Choquet ν_α expected return.

Remark 3.1.14. *The following risk measures correspond to the α-risk $\rho_{\nu_\alpha}(X)$.*

- *expected shortfall-1:* $\mathrm{ES}_\alpha(X) = -\dfrac{1}{\alpha}\displaystyle\int_0^\alpha F_X^{-1}(t)dt$
- *expected shortfall-2:* $\mathrm{ES}_\alpha(X) = -\dfrac{1}{\alpha}\Big(E[X1_{X\le q_\alpha(X)}] - q_\alpha(X)(P[X \le q_\alpha(X)] - \alpha)\Big)$
- *tail conditional expectation:* $\mathrm{TCE}_\alpha(X) = -E[X|X \le q_\alpha(X)]$[9]

where $q_\alpha(X)(= -\mathrm{VaR}_\alpha(X))$ is the quantile of X at the level $\alpha \in (0, 1]$.

Note that $\rho_{\nu_\alpha}(X)$ is usually equal and larger than $\mathrm{VaR}_\alpha(X)$ because $q_\alpha(X) \ge F_X^{-1}(t)$ for $t \in [0, \alpha]$. As for the portfolio selection theory, by using the α-risk, it is natural to consider the following criteria: $\rho_{\nu_\alpha}(X) - \lambda\mu(X)$ or, alternatively, $\mu(X) - \lambda\rho_{\nu_\alpha}(X)$. Minimizing the former criterion implies minimizing the α-risk subject to a constraint on mean return; maximizing the latter criterion implies maximizing the mean return subject to a constraint on the α-risk. Several authors have suggested these criteria as alternatives to the classical Markowitz criteria in which α-risk is replaced by the standard deviation of the random variable X. Since $\mu(X) = \int F_X^{-1}(t)dt = -\rho_1(X)$, these criteria are special cases of the following more general class.

Definition 3.1.15. *A risk measure ρ is called pessimistic if, for some probability measure ψ on $[0, 1]$,*

$$\rho(X) = -\int_0^1 \rho_{\nu_\alpha} d\psi(\alpha).$$

Pessimistic Portfolio Bassett et al. (2004) showed that decision making based on minimizing a "coherent risk measure" is equivalent to Choquet expected utility maximization using a linear form of the utility function and a pessimistic form of the Choquet distortion function.
Consider

$$\min_{\boldsymbol{w}\in\mathbb{R}^m} \rho_{\nu_\alpha}(\boldsymbol{w}'\boldsymbol{X}_t) \quad \text{subject to} \quad \mu(\boldsymbol{w}) = \mu_P, \ \sum_{i=1}^m w_i = 1. \tag{3.1.28}$$

A portfolio with the weight as the solution is called a **pessimistic portfolio**, which includes the mean-conditional value at risk (M-CVaR) portfolio, the mean-expected shortfall (M-ES) portfolio and the mean-tail conditional expectation (M-TCE) portfolio.

Although the α-risk provides a convenient one-parameter family of coherent risk measures, they are obviously rather simplistic. The α-risk ρ_{ν_α} can be extended to weighted averages of α-risk

$$\rho_\nu(X) = \sum_{k=1}^l \nu_k \rho_{\nu_{\alpha_k}}(X), \tag{3.1.29}$$

which is a way to approximate general pessimistic risk measures. Note that the weights $\nu_k : k = 1, \dots, l$ should be positive and sum to one.

[9]Note that $\mathrm{TCE}_\alpha(X)$ is not always equal to $\mathrm{ES}_\alpha(X)$. In that case, it may not be coherent (not subadditive), as shown in Acerbi (2004).

3.1.6 Copulas and Dependence

As we mentioned in 3.1.5, various risk measure alternatives to the variance or the standard deviation have been proposed so far in the portfolio optimization problem. VaR is widely used for financial risk measure (not only for portfolio optimization) because VaR is recommended as a standard by the Basel Committee. However, in recent years, VaR has been criticized as the risk measure since VaR is not coherent or convex. Due to nonconvexity, VaR may have many local extrema when optimizing a portfolio. As an alternative risk measure, CVaR is known to have better properties than VaR.

Definition 3.1.16. *The α-CVaR is given by*

$$\mathrm{CVaR}_\alpha(X) := E\{-X| - X \geq \mathrm{VaR}_\alpha(X)\} \tag{3.1.30}$$

where $\mathrm{VaR}_\alpha(X)$ *is the α-VaR defined by (3.1.25).*

Here β is a specified probability level in (0, 1) (in practice, $\beta = 0.90, 0.95$ or 0.99.). Pflug (2000) proved that CVaR is a coherent risk measure that is convex, monotonic, positively homogeneous and so on. Rockafellar and Uryasev (2000) proposed the mean-conditional value at risk (M-CVaR) portfolio, which minimizes the CVaR of a portfolio under the constraint for the portfolio mean.

Definition 3.1.17. *Let $\boldsymbol{w}'\boldsymbol{X} = \sum_{i=1}^m w_i X_i$ denote a portfolio return. Then, an M-CVaR portfolio is a solution of the following problem:*

$$\min_{\boldsymbol{w}\in\mathbb{R}^m} \mathrm{CVaR}_\beta(\boldsymbol{w}'\boldsymbol{X}) \quad \text{over} \quad \boldsymbol{w} \in \boldsymbol{W}(\mu_P) \tag{3.1.31}$$

where

$$\boldsymbol{W}(\mu_P) = \left\{\boldsymbol{w} = (w_1, \ldots, w_m)' | w_j \geq 0 \text{ for } j = 1, \ldots, m, \ \sum_{i=1}^m w_i = 1, \ \mu(\boldsymbol{w}) = E(\boldsymbol{w}'\boldsymbol{X}) \geq \mu_P\right\}.$$

According to Rockafellar and Uryasev (2000), this set $\boldsymbol{W}(\mu_P)$ is convex. Modeling for CVaR (including VaR) by use of copula dependency structure description is the subject of many studies, such as Palaro and Hotta (2006), Stulajter (2009), Kakouris and Rustem (2014) and so on. The advanced point of modeling with a copula is that for continuous multivariate distributions, the modeling of the univariate marginals and the multivariate or dependence structure can be separated, and the multivariate structure can be represented by a copula. The general representation of a multivariate distribution as a composition of a copula and its univariate margins is due to Sklar's theorem (e.g., Joe (2015)).

Theorem 3.1.18. *[Sklar's theorem] For an m-variate distribution $F \in \mathcal{F}(F_1, \ldots, F_m)$, with jth univariate marginal distribution F_j, the copula associated with F is a distribution function $C : [0, 1]^m \mapsto [0, 1]$ with $U(0, 1)$ margins that satisfies*

$$F(\boldsymbol{x}) = C(F_1(x_1), \ldots, F_m(x_m)), \quad \boldsymbol{x} \in \mathbb{R}^d.$$

(a) If F is a continuous m-variate distribution function with univariate margins $F_1, \ldots, F_m$, and quantile functions $F_1^{-1}, \ldots, F_m^{-1}$, then

$$C(\boldsymbol{u}) = F(F_1^{-1}(u_1), \ldots, F_m^{-1}(u_m)), \quad \boldsymbol{u} \in [0, 1]^m$$

is the unique choice.

(b) If F is an m-variate distribution of discrete random variables (more generally, partly continuous and partly discrete), then the copula is unique only on the set

$$\mathit{Range}(F_1) \times \cdots \times \mathit{Range}(F_m).$$

Proof: See Theorem 1.1 of Joe (2015).

Suppose that the m-variate distribution function F with a copula C is continuous with invariant marginal distribution functions $F_1, \ldots, F_m$. If $F_1, \ldots, F_m$ are absolutely continuous with respective densities $f_j(x) = \partial F_j(x)/\partial x$, and C has mixed derivative of order m, then, the joint density function f can be written as

$$f(\boldsymbol{x}) = f(x_1, \ldots, x_m) = \frac{\partial^m F(x_1, \ldots, x_m)}{\partial x_1 \cdots \partial x_m} = \frac{\partial^m C(F_1(x_1), \ldots, F_m(x_m))}{\partial x_1 \cdots \partial x_m}$$
$$= \frac{\partial^m C(u_1, \ldots, u_m)}{\partial u_1 \cdots \partial u_m} \prod_{i=1}^m \frac{\partial F_i(x_i)}{\partial x_i} = \frac{\partial^m C(u_1, \ldots, u_m)}{\partial u_1 \cdots \partial u_m} \prod_{i=1}^m f_i(x_i). \tag{3.1.32}$$

Definition 3.1.19. *If $C(\boldsymbol{u})$ is an absolutely continuous copula, then its density function*

$$c(\boldsymbol{u}) = c(u_1, \ldots, u_m) = \frac{\partial^m C(u_1, \ldots, u_m)}{\partial u_1 \cdots \partial u_m}, \quad \boldsymbol{u} \in (0,1)^m$$

is called the copula density (function).

It is easy to see that if there exists the copula density $c(\boldsymbol{u})$, it follows that

$$f(\boldsymbol{x}) = c(F_1(x_1), \ldots, F_m(x_m)) \prod_{i=1}^m f_i(x_i). \tag{3.1.33}$$

So far, varity copulas have been introduced. The following is the typical one.

Example 3.1.20. *[Gaussian copula] From the multivariate Gaussian distribution with zero means, unit variances and $m \times m$ correlation matrix Σ, we get*

$$C(\boldsymbol{u}; \Sigma) = \Phi_m\left(\Phi^{-1}(u_1), \ldots, \Phi^{-1}(u_m)\right), \quad \boldsymbol{u} \in [0,1]^m,$$

where $\Phi_m(\cdot; \Sigma)$ is the m-variate Gaussian cdf, Φ is the univariate Gaussian cdf, and Φ^{-1} is the univariate Gaussian inverse cdf or quantile function. The copula density is given by

$$c(\boldsymbol{u}; \Sigma) = \det(\Sigma)^{-1/2} \exp\left\{-\frac{1}{2}\boldsymbol{\xi}'(\Sigma^{-1} - I_m)\boldsymbol{\xi}\right\}, \tag{3.1.34}$$

where $\boldsymbol{\xi} = (\xi_1, \ldots, \xi_m)'$ and $\xi_i (i = 1, \ldots, m)$ is the u_i quantile of the standard Gaussian random variable X_i (i.e., $u_i = P(X_i < \xi_i)$, $X_i \sim N(0,1)$).

Suppose that a vector of random asset returns $\boldsymbol{X} = (X_1, \ldots, X_m)'$ follows a continuous m-variate distribution function $F_{\boldsymbol{X}}$ and the density function $f_{\boldsymbol{X}}$. Let $\boldsymbol{w} = (w_1, \ldots, w_m)'$ be the vector of portfolio weights. Then the return of the portfolio is $\boldsymbol{w}'\boldsymbol{X}$, and the distribution function can be written as

$$F_{\boldsymbol{w}'\boldsymbol{X}}(y) = P(\boldsymbol{w}'\boldsymbol{X} \le y) = \int_{\boldsymbol{w}'\boldsymbol{x} \le y} f_{\boldsymbol{X}}(\boldsymbol{x}) d\boldsymbol{x}.$$

If the density function $f_{\boldsymbol{X}}$ can be described by using the copula density and the marginal density $f_1(x), \ldots, f_m(x)$ as (3.1.33), then we can write

$$F_{\boldsymbol{w}'\boldsymbol{X}}(y) = \int_{\boldsymbol{w}'\boldsymbol{x} \le y} c(F_{\boldsymbol{X}}(\boldsymbol{x})) \prod_{i=1}^m f_i(x_i) d\boldsymbol{x}$$

where $F_{\boldsymbol{X}}(\boldsymbol{x}) = (F_1(x_1), \ldots, F_m(x_m))$. The variable transformation as $u_i = F_i(x_i)$ leads to

$$F_{\boldsymbol{w}'\boldsymbol{X}}(y) = \int_{\boldsymbol{w}'F_{\boldsymbol{X}}^{-1}(\boldsymbol{u}) \le y} c(\boldsymbol{u}) d\boldsymbol{u}$$

where $F_{\boldsymbol{X}}^{-1}(\boldsymbol{u}) = (F_1^{-1}(u_1), \ldots, F_m^{-1}(u_m))$ maps from $[0,1]^m$ to $\mathbb{R}^m$.

By using this result, VaR and CVaR are given as follows.

Proposition 3.1.21. *The α-VaR for $w'X$ is defined as*

$$\mathrm{VaR}_\alpha(w'X) = -\sup\left\{y \,\middle|\, \int_{w'F_X^{-1}(u)\le y} c(u)du < \alpha\right\}.$$

The α-CVaR for $w'X$ is defined as

$$\mathrm{CVaR}_\alpha(w'X) = -\frac{1}{\alpha}\int_{w'F_X^{-1}(u)\le -\mathrm{VaR}_\alpha(w'X)} w'F_X^{-1}(u)c(u)du.$$

Proof: It follows from the definition of α-VaR (see (3.1.25)) and α-CVaR (see (3.1.30)).

This result shows that once a copula density c and marginal distribution functions F_X are determined by simulation or estimation, we can calculate the portfolio $w'X$'s CVaR (or VaR), which implies that the optimal M-CVaR portfolio (or M-VaR portfolio) can be also obtained as the solution of (3.1.31) (or 3.1.26).

3.1.7 Bibliographic Notes

Markowitz (1952) was the original paper on portfolio theory and was expanded into a book Markowitz (1959). Elton et al. (2007), Luenberger (1997) Wang and Xia (2002), Ruppert (2004) and Malevergne and Didier Sornette (2005) provide an elementally introduction to portfolio selection theory such as mean variance portfolio, CAPM and APT. Prigent (2007) discusses the utility theory and risk measures in more detail. Koenker (2005) introduces the pessimistic portfolio. Wang and Xia (2002) discuss some other portfolios, taking care of the investment (transaction) cost. Joe (2015) and McNeil et al. (2015) provide some introduction of the general copula theory. Rockafellar and Uryasev (2000) and Kakouris and Rustem (2014) discuss a portfolio optimization problem by using copula. McNeil et al. (2015) provide some concepts and techniques of quantitative risk management (QRM) in which portfolio theory, finnancial time series modelling, copulas and extreme value theory are required.

3.1.8 Appendix

Proof of (3.1.2) Let $L(w)$ be

$$L(w) = \eta^2(w) - \lambda_1(\mu(w) - \mu_P) - \lambda_2(w'e - 1) = w'\Sigma w - \lambda_1(w'\mu - \mu_P) - \lambda_2(w'e - 1).$$

When w_0 is the minimizer of $L(w)$, it follows that

$$\frac{\partial}{\partial w}L(w)\Big|_{w=w_0} = \mathbf{0}, \quad \frac{\partial}{\partial \lambda_1}L(w)\Big|_{w=w_0} = 0, \quad \frac{\partial}{\partial \lambda_2}L(w)\Big|_{w=w_0} = 0.$$

Since

$$\frac{\partial}{\partial w}(w'\Sigma w) = 2\Sigma w, \quad \frac{\partial}{\partial w}(w'\mu) = \mu, \quad \frac{\partial}{\partial w}(w'e) = e,$$

we have

$$\frac{\partial}{\partial w}L(w) = 2\Sigma w - \lambda_1\mu - \lambda_2 e = \mathbf{0},$$

which implies that

$$w = \frac{\lambda_1\Sigma^{-1}\mu + \lambda_2\Sigma^{-1}e}{2}.$$

In addition, by substituting this result into

$$\frac{\partial}{\partial \lambda_1} L(\boldsymbol{w}) = \boldsymbol{w}'\boldsymbol{\mu} - \mu_P = 0, \quad \frac{\partial}{\partial \lambda_2} L(\boldsymbol{w}) = \boldsymbol{w}'\boldsymbol{e} - 1 = 0,$$

we have

$$\frac{\lambda_1 \boldsymbol{\mu}'\Sigma^{-1}\boldsymbol{\mu} + \lambda_2 \boldsymbol{e}'\Sigma^{-1}\boldsymbol{\mu}}{2} = \frac{\lambda_1 C + \lambda_2 B}{2} = \mu_P, \quad \frac{\lambda_1 \boldsymbol{e}'\Sigma^{-1}\boldsymbol{\mu} + \lambda_2 \boldsymbol{e}'\Sigma^{-1}\boldsymbol{e}}{2} = \frac{\lambda_1 B + \lambda_2 A}{2} = 1$$

Hence, it follows that

$$\lambda_1 = \frac{2\mu_P A - 2B}{AC - B^2} = \frac{2\mu_P A - 2B}{D}, \quad \lambda_2 = \frac{2C - 2\mu_P B}{AC - B^2} = \frac{2C - 2\mu_P B}{D}$$

and

$$\boldsymbol{w}_0 = \frac{\lambda_1}{2}\Sigma^{-1}\boldsymbol{\mu} + \frac{\lambda_2}{2}\Sigma^{-1}\boldsymbol{e} = \frac{\mu_P A - B}{D}\Sigma^{-1}\boldsymbol{\mu} + \frac{C - \mu_P B}{D}\Sigma^{-1}\boldsymbol{e}.$$

Proof of (3.1.3) Let $L(\boldsymbol{w})$ be

$$L(\boldsymbol{w}) = \eta^2(\boldsymbol{w}) - \lambda(\boldsymbol{w}'\boldsymbol{e} - 1) = \boldsymbol{w}'\Sigma\boldsymbol{w} - \lambda(\boldsymbol{w}'\boldsymbol{e} - 1).$$

When $\boldsymbol{w}_0$ is the minimizer of $L(\boldsymbol{w})$, it follows that

$$\frac{\partial}{\partial \boldsymbol{w}} L(\boldsymbol{w})\Big|_{\boldsymbol{w}=\boldsymbol{w}_0} = \boldsymbol{0}, \quad \frac{\partial}{\partial \lambda} L(\boldsymbol{w})\Big|_{\boldsymbol{w}=\boldsymbol{w}_0} = 0.$$

Hence we have

$$\frac{\partial}{\partial \boldsymbol{w}} L(\boldsymbol{w}) = 2\Sigma\boldsymbol{w} - \lambda\boldsymbol{e} = \boldsymbol{0},$$

which implies that

$$\boldsymbol{w} = \frac{\lambda\Sigma^{-1}\boldsymbol{e}}{2}.$$

In addition, by substituting this result into

$$\frac{\partial}{\partial \lambda} L(\boldsymbol{w}) = \boldsymbol{w}'\boldsymbol{e} - 1 = 0,$$

we have

$$\frac{\lambda \boldsymbol{e}'\Sigma^{-1}\boldsymbol{e}}{2} = \frac{\lambda A}{2} = 1$$

so that

$$\lambda = \frac{2}{A}$$

and

$$\boldsymbol{w}_0 = \frac{1}{A}\Sigma^{-1}\boldsymbol{e}.$$

Proof of (3.1.5) Let $L(\boldsymbol{w})$ be

$$L(\boldsymbol{w}) = \mu(\boldsymbol{w}) - \lambda_1(\eta^2(\boldsymbol{w}) - \sigma_P^2) - \lambda_2(\boldsymbol{w}'\boldsymbol{e} - 1) = \boldsymbol{w}'\boldsymbol{\mu} - \lambda_1(\boldsymbol{w}'\Sigma\boldsymbol{w} - \sigma_P^2) - \lambda_2(\boldsymbol{w}'\boldsymbol{e} - 1).$$

When $\boldsymbol{w}_0$ is the minimizer of $L(\boldsymbol{w})$, it follows that

$$\frac{\partial}{\partial \boldsymbol{w}} L(\boldsymbol{w})\Big|_{\boldsymbol{w}=\boldsymbol{w}_0} = \boldsymbol{0}, \quad \frac{\partial}{\partial \lambda_1} L(\boldsymbol{w})\Big|_{\boldsymbol{w}=\boldsymbol{w}_0} = 0, \quad \frac{\partial}{\partial \lambda_2} L(\boldsymbol{w})\Big|_{\boldsymbol{w}=\boldsymbol{w}_0} = 0.$$

Hence we have

$$\frac{\partial}{\partial \boldsymbol{w}} L(\boldsymbol{w}) = \boldsymbol{\mu} - 2\lambda_1\Sigma\boldsymbol{w} - \lambda_2\boldsymbol{e} = \boldsymbol{0},$$

which implies that

$$\boldsymbol{w} = \frac{\Sigma^{-1}\boldsymbol{\mu} - \lambda_2\Sigma^{-1}\boldsymbol{e}}{2\lambda_1}$$

for $\lambda_1 \neq 0$. In addition, by substituting this result into

$$\frac{\partial}{\partial \lambda_1} L(\boldsymbol{w}) = \boldsymbol{w}'\Sigma\boldsymbol{w} - \sigma_P^2 = 0, \quad \frac{\partial}{\partial \lambda_2} L(\boldsymbol{w}) = \boldsymbol{w}'\boldsymbol{e} - 1 = 0,$$

we have

$$\frac{\boldsymbol{\mu}'\Sigma^{-1}\boldsymbol{\mu} - 2\lambda_2\boldsymbol{\mu}'\Sigma^{-1}\boldsymbol{e} + \lambda_2^2\boldsymbol{e}'\Sigma^{-1}\boldsymbol{e}}{4\lambda_1^2} = \frac{C - 2\lambda_2 B + \lambda_2^2 A}{4\lambda_1^2} = \sigma_P^2,$$
$$\frac{\boldsymbol{e}'\Sigma^{-1}\boldsymbol{\mu} - \lambda_2\boldsymbol{e}'\Sigma^{-1}\boldsymbol{e}}{2\lambda_1} = \frac{B - \lambda_2 A}{2\lambda_1} = 1.$$

Hence, it follows that

$$A(1 - \sigma_P^2 A)\lambda_2^2 - 2B(1 - \sigma_P^2 A)\lambda_2 + (C - \sigma_P^2 B^2) = 0$$

and if $J = B^2(1 - \sigma_P^2 A)^2 - A(1 - \sigma_P^2 A)(C - \sigma_P^2 B^2) \geq 0$ is satisfied, we have

$$\lambda_2 = \frac{B(1 - \sigma_P^2 A) \pm \sqrt{J}}{A(1 - \sigma_P^2 A)}, \quad \lambda_1 = \frac{B - \lambda_2 A}{2}.$$

Let

$$\lambda_2^+ = \frac{B(1 - \sigma_P^2 A) + \sqrt{J}}{A(1 - \sigma_P^2 A)}, \quad \lambda_1^+ = \frac{B - \lambda_2^+ A}{2}, \quad \boldsymbol{w}^+ = \frac{\Sigma^{-1}\boldsymbol{\mu} - \lambda_2^+\Sigma^{-1}\boldsymbol{e}}{2\lambda_1^+}$$
$$\lambda_2^- = \frac{B(1 - \sigma_P^2 A) - \sqrt{J}}{A(1 - \sigma_P^2 A)}, \quad \lambda_1^- = \frac{B - \lambda_2^- A}{2}, \quad \boldsymbol{w}^- = \frac{\Sigma^{-1}\boldsymbol{\mu} - \lambda_2^-\Sigma^{-1}\boldsymbol{e}}{2\lambda_1^-}.$$

Then we have

$$(\boldsymbol{w}^+)'\boldsymbol{\mu} - (\boldsymbol{w}^-)'\boldsymbol{\mu} = \frac{\lambda_2^+ - \lambda_2^-}{\lambda_1^+\lambda_1^-}D = \frac{8(\sigma_P^2 A - 1)}{\sqrt{J}A}.$$

Therefore, if $\sigma_P^2 A \geq 1$, then $(\boldsymbol{w}^+)'\boldsymbol{\mu} \geq (\boldsymbol{w}^-)'\boldsymbol{\mu}$, which implies that

$$\boldsymbol{w}_0 = \frac{1}{2\lambda_1^+}\Sigma^{-1}\boldsymbol{\mu} - \frac{\lambda_2^+}{2\lambda_1^+}\Sigma^{-1}\boldsymbol{e}.$$

Proof of (3.1.7) Under $\boldsymbol{w}'\boldsymbol{e} = 1$, define $L(\boldsymbol{w})$ by

$$L(\boldsymbol{w}) = \tan\theta = \frac{\mu(\boldsymbol{w}) - \mu_f}{\eta(\boldsymbol{w})} = \frac{\boldsymbol{w}'(\boldsymbol{\mu} - \mu_f\boldsymbol{e})}{\sqrt{\boldsymbol{w}'\Sigma\boldsymbol{w}}}.$$

When $\boldsymbol{w}_0$ is the maximizer of $L(\boldsymbol{w})$, it follows that

$$\frac{\partial}{\partial \boldsymbol{w}} L(\boldsymbol{w})\bigg|_{\boldsymbol{w}=\boldsymbol{w}_0} = \boldsymbol{0}.$$

Hence we have

$$\frac{\partial}{\partial \boldsymbol{w}} L(\boldsymbol{w}) = \frac{(\boldsymbol{\mu} - \mu_f\boldsymbol{e})\sqrt{\boldsymbol{w}'\Sigma\boldsymbol{w}} - \boldsymbol{w}'(\boldsymbol{\mu} - \mu_f\boldsymbol{e})\frac{\Sigma\boldsymbol{w}}{\sqrt{\boldsymbol{w}'\Sigma\boldsymbol{w}}}}{\boldsymbol{w}'\Sigma\boldsymbol{w}} = \boldsymbol{0},$$

which implies that

$$\boldsymbol{w} = c_P\Sigma^{-1}(\boldsymbol{\mu} - \mu_f\boldsymbol{e})$$

where $c_P = \frac{\boldsymbol{w}'\Sigma\boldsymbol{w}}{\boldsymbol{w}'(\boldsymbol{\mu} - \mu_f\boldsymbol{e})}$. Because of $\boldsymbol{w}'\boldsymbol{e} = 1$, it follows that

$$c_P = \frac{1}{\boldsymbol{e}'\Sigma^{-1}(\boldsymbol{\mu} - \mu_f\boldsymbol{e})}.$$

3.2 Statistical Estimation for Portfolios

In this section, we discuss statistical estimation for the optimal portfolio weights introduced in Section 3.1. As we showed in Section 3.1, the class of mean-variance optimal portfolio weights can be written as a function $g = g(\boldsymbol{\mu}, \Sigma)$ of the mean vector $\boldsymbol{\mu}$ and covariance matrix Σ with respect to asset returns $\boldsymbol{X}(t)$. Several authors proposed estimators of g as functions of the sample mean vector $\hat{\boldsymbol{\mu}}$ and sample covariance matrix $\hat{\Sigma}$ (i.e., $\hat{g} = g(\hat{\boldsymbol{\mu}}, \hat{\Sigma})$) for independent returns of assets (e.g., Jobson et al. (1979), Jobson and Korkie (1981) and Lauprete et al. (2002)). However, empirical studies show that financial return processes are often dependent and non-Gaussian. From this point of view, Basak et al. (2002) showed the consistency of optimal portfolio estimators when the portfolio returns are stationary processes. In the literature there has been no study on the asymptotic efficiency of estimators for optimal portfolio weights. Shiraishi and Taniguchi (2008) considered the asymptotic efficiency of the estimators when the returns are non-Gaussian linear stationary processes. They give the asymptotic distribution of portfolio weight estimators $\hat{g}$ for the processes.

Next, we discuss the other statistical methods in portfolio optimization problems. First, we discuss an estimation method and its asymptotic property of "pessimistic portfolio," as we introduced in Section 3.1. According to Bassett et al. (2004), the asymptotic property is explained via quantile regression (QR) theory. Second, we briefly discuss shrinkage estimation of optimal portfolio weight discussed by Jorion (1986) and Ledoit and Wolf (2004), among others. Third, we briefly discuss Bayesian estimation theory in portfolio optimization problems. According to Aït-Sahalia and Hansen (2009), the econometrician's optimal portfolio weights are quite different from the above plug-in estimates, because the econometrician takes on the role of the investor by choosing portfolio weights that are optimal with respect to his or her subjective belief about the true return distribution. Next, we discuss factor analysis. As for the static factor model, three types of factor models are available for studying asset returns (see Connor (1995) and Campbell et al. (1996)), namely, macroeconomic factor model, fundamental factor model and statistical factor model. Here, we introduce static factor models and portfolio estimation methods. As for the dynamic factor model, we explain the weak points for the traditional (static) factor models and introduce the generalized dynamic factor model which provides the desirable properties, and construct a portfolio estimator. Finally, we discuss high-dimensional problems. Let n and m be the sample size and the sample dimension, respectively. We first show the asymptotic property for the estimators of some portfolios (the global minimum portfolio (MP), the tangency portfolio (TP) and the naive portfolio (NP)) under $n, m \to \infty$ and $m/n \to y \in (0, 1)$. Then, the asymptotic property for the estimator of MP is shown under $n, m \to \infty$ and $m/n \to y \in (1, \infty)$. Here the estimator is modified since the usual sample covariance matrix is singular under $m > n$.

3.2.1 *Traditional Mean-Variance Portfolio Estimators*

Suppose that a sequence of random returns $\{\boldsymbol{X}(t) = (X_1(t), \ldots, X_m(t))'; t \in \mathbb{Z}\}$ is a stationary process. Writing $\boldsymbol{\mu} = E\{\boldsymbol{X}(t)\}$ and $\Sigma = \text{Var}\{\boldsymbol{X}(t)\} = E[\{\boldsymbol{X}(t) - \boldsymbol{\mu}\}\{\boldsymbol{X}(t) - \boldsymbol{\mu}\}']$, we define $(m + r)$-vector parameter $\theta = (\theta_1, \ldots, \theta_{m+r})'$ by

$$\theta = (\boldsymbol{\mu}', vech(\Sigma)')'$$

where $r = m(m + 1)/2$. We estimate a general function $g(\theta)$ of θ, which expresses an optimal portfolio weight for the mean-variance model such as (3.1.2), (3.1.3), (3.1.5) and (3.1.7). Here it should be noted that the portfolio weight $\boldsymbol{w} = (w_1, \ldots, w_m)'$ satisfies the restriction $\boldsymbol{w}'\boldsymbol{e} = 1$. Then we have only to estimate the subvector $(w_1, \ldots, w_{m-1})'$. Hence we assume that the function $g(\cdot)$ is $(m - 1)$-dimensional, i.e.,

$$g : \theta \to \mathbb{R}^{m-1}. \tag{3.2.1}$$

In this setting we address the problem of statistical estimation for $g(\theta)$, which describes various mean variance optimal portfolio weights.
Let the return process $\{\boldsymbol{X}(t) = (X_1(t), \ldots, X_m(t))'; t \in \mathbb{Z}\}$ be an m-vector linear process

$$\boldsymbol{X}(t) = \sum_{j=0}^{\infty} A(j)\boldsymbol{U}(t - j) + \boldsymbol{\mu} \tag{3.2.2}$$

where $\{\boldsymbol{U}(t) = (u_1(t), \ldots, u_m(t))'\}$ is a sequence of independent and identically distributed (i.i.d.) m-vector random variables with $E\{\boldsymbol{U}(t)\} = \boldsymbol{0}$, $\text{Var}\{\boldsymbol{U}(t)\} = K$ (for short $\{\boldsymbol{U}(t) \sim i.i.d.(\boldsymbol{0}, K)\}$) and fourth-order cumulants. Here

$$A(j) = (A_{ab}(j))_{a,b=1,\ldots,m}, \ A(0) = I_m, \ \boldsymbol{\mu} = (\mu_a)_{a=1,\ldots,m}, \ K = (K_{ab})_{a,b=1,\ldots,m}.$$

We make the following assumption.
Assumption 3.2.1. *(i)* $\sum_{j=0}^{\infty} |j|^{1+\delta} \|A(j)\| < \infty$ *for some* $\delta > 0$
(ii) $\det\{\sum_{j=0}^{\infty} A(j)\boldsymbol{z}^j\} \neq 0$ *on* $\{\boldsymbol{z} \in \mathbb{C}; |\boldsymbol{z}| \leq 1\}$.

Assumption 3.2.1 (i) is the condition to guarantee finiteness for the fourth-moment of $\boldsymbol{X}(t)$. Assumption 3.2.1 (ii) is the "invertibility condition" to guarantee that $\boldsymbol{U}(t)$ is representable as the function of $\{\boldsymbol{X}(s)\}_{s \leq t}$.

The class of $\{\boldsymbol{X}(t)\}$ includes that of non-Gaussian vector-valued causal ARMA models. Hence it is sufficiently rich. The process $\{\boldsymbol{X}(t)\}$ is a second-order stationary process with spectral density matrix

$$f(\lambda) = (f_{ab}(\lambda))_{a,b=1,\ldots,m} = \frac{1}{2\pi} A(\lambda) K A(\lambda)^*$$

where $A(\lambda) = \sum_{j=0}^{\infty} A(j) e^{ij\lambda}$.
From the partial realization $\{\boldsymbol{X}(1), \ldots, \boldsymbol{X}(n)\}$, we introduce

$$\hat{\theta} = (\hat{\boldsymbol{\mu}}', vech(\hat{\Sigma})')', \quad \tilde{\theta} = (\hat{\boldsymbol{\mu}}', vech(\tilde{\Sigma})')'$$

where

$$\hat{\mu} = \frac{1}{n}\sum_{t=1}^{n} \boldsymbol{X}(t),\ \hat{\Sigma} = \frac{1}{n}\sum_{t=1}^{n}\{\boldsymbol{X}(t) - \hat{\boldsymbol{\mu}}\}\{\boldsymbol{X}(t) - \hat{\boldsymbol{\mu}}\}',\ \tilde{\Sigma} = \frac{1}{n}\sum_{t=1}^{n}\{\boldsymbol{X}(t) - \boldsymbol{\mu}\}\{\boldsymbol{X}(t) - \boldsymbol{\mu}\}'.$$

Denote the σ-field generated by $\{\boldsymbol{X}(s); s \leq t\}$ by $\mathcal{F}_t$. Also we introduce Ω_1 ($(m \times m)$-matrix), Ω_2^G ($(r \times r)$-matrix), Ω_2^{NG} ($(r \times r)$-matrix) and Ω_3 ($(m \times r)$-matrix) as follows

$$\begin{aligned}
\Omega_1 &= 2\pi f(0) \\
\Omega_2^G &= \left(2\pi \int_{-\pi}^{\pi} \{f_{a_1a_3}(\lambda)\overline{f_{a_2a_4}(\lambda)} + f_{a_1a_4}(\lambda)\overline{f_{a_2a_3}(\lambda)}\}d\lambda\right)_{a_1,a_2,a_3,a_4=1,\dots,m,a_1\geq a_2,a_3\geq a_4} \\
\Omega_2^{NG} &= \Omega_2^G + \left(\frac{1}{(2\pi)^2}\sum_{b_1,b_2,b_3,b_4=1}^{m} c^{\boldsymbol{U}}_{b_1,b_2,b_3,b_4}\int\int_{-\pi}^{\pi} A_{a_1b_1}(\lambda_1)A_{a_2b_2}(-\lambda_1)A_{a_3b_3}(\lambda_2)\right. \\
&\quad \left. \times A_{a_4b_4}(-\lambda_2)d\lambda_1 d\lambda_2\right)_{a_1,a_2,a_3,a_4=1,\dots,m,a_1\geq a_2,a_3\geq a_4} \\
\Omega_3 &= \left(\frac{1}{(2\pi)^2}\sum_{b_1,b_2,b_4=1}^{m} c^{\boldsymbol{U}}_{b_1,b_2,b_3}\int\int_{-\pi}^{\pi} A_{a_1b_1}(\lambda_1+\lambda_2)A_{a_2b_2}(-\lambda_1)A_{a_3b_3}(-\lambda_2)\right. \\
&\quad \left. d\lambda_1 d\lambda_2\right)_{a_1,a_2,a_3=1,\dots,m,a_2\geq a_3}
\end{aligned}$$

where $c^{\boldsymbol{U}}_{b_1,\dots,b_j}$'s are jth order cumulants of $u_{b_1}(t),\dots,u_{b_j}(t)$ $(j = 3, 4)$.

Remark 3.2.2. *Denoting*

$$\Omega = \begin{cases} \begin{pmatrix} \Omega_1 & \boldsymbol{0} \\ \boldsymbol{0} & \Omega_2^G \end{pmatrix} & \text{if } \{\boldsymbol{U}(t)\} \text{ is Gaussian} \\ \begin{pmatrix} \Omega_1 & \Omega_3 \\ \Omega_3' & \Omega_2^{NG} \end{pmatrix} & \text{if } \{\boldsymbol{U}(t)\} \text{ is non-Gaussian} \end{cases},$$

this Ω corresponds to the spectral representation of the asymptotic covariance matrix of $\sqrt{n}(\hat{\boldsymbol{\theta}} - \boldsymbol{\theta})$. For a simple example, we assume that $m = 1$ and $U(t) \overset{i.i.d.}{\sim} N(0, \sigma^2)$. Then, it follows that

$$\begin{aligned}
n\mathrm{Var}(\hat{\mu}) &= \frac{1}{n}\sum_{t_1,t_2=1}^{n} \mathrm{Cov}\{X(t_1), X(t_2)\} = \frac{1}{n}\sum_{t_1,t_2=1}^{n}\sum_{j_1,j_2=0}^{\infty} A(j_1)A(j_2)\mathrm{Cov}\{U(t_1-j_1), U(t_2-j_2)\} \\
&= \sum_{l=-n+1}^{n-1}(n-|l|)\sum_{j=0}^{\infty} A(j)A(j+l)\sigma^2 \to \sum_{l=-\infty}^{\infty}\sum_{j=0}^{\infty} A(j)A(j+l)\sigma^2 = \left\{\sum_{j=0}^{\infty} A(j)\right\}^2\sigma^2
\end{aligned}$$

under $A(j) = 0$ for $j < 0$. In the same way, we can write

$$\begin{aligned}
n\mathrm{Var}(\tilde{\Sigma}) &= \frac{1}{n}\sum_{t_1,t_2=1}^{n} \mathrm{Cov}\left[\{X(t_1)-\mu\}^2, \{X(t_2)-\mu\}^2\right] = \frac{2}{n}\sum_{t_1,t_2=1}^{n} \mathrm{Cov}\{X(t_1), X(t_2)\}^2 \\
&= 2\sum_{l=-n+1}^{n-1}(n-|l|)\left\{\sum_{j=0}^{\infty} A(j)A(j+l)\sigma^2\right\}^2 \to 2\sum_{l=-\infty}^{\infty}\left\{\sum_{j=0}^{\infty} A(j)A(j+l)\right\}^2\sigma^4 \\
&= 2\sum_{j_1,j_2,j_3\geq 0,\ j_1\leq j_2+j_3} A(j_1)A(j_2)A(j_3)A(j_2+j_3-j_1)\sigma^4.
\end{aligned}$$

On the other hand, Ω_1 and Ω_2^G can be written as

$$\Omega_1 = 2\pi f(0) = \sum_{j_1,j_2=0}^{\infty} A(j_1)A(j_2)\sigma^2 \int_{-\pi}^{\pi} e^{-i(j_1-j_2)\lambda} d\lambda = \left\{\sum_{j=0}^{\infty} A(j)\right\}^2 \sigma^2$$

$$\Omega_2^G = 4\pi \int_{-\pi}^{\pi} f(\lambda)\overline{f(\lambda)} d\lambda = \sum_{j_1,j_2,j_3,j_4=0}^{\infty} A(j_1)A(j_2)A(j_3)A(j_4) \int_{-\pi}^{\pi} e^{-i(j_1-j_2-j_3+j_4)\lambda} d\lambda$$

$$= 2 \sum_{j_1,j_2,j_3\geq 0,\ j_1\leq j_2+j_3} A(j_1)A(j_2)A(j_3)A(j_2+j_3-j_1)\sigma^4.$$

Theorem 3.2.3. *Under Assumption 3.2.1,*

$$\sqrt{n}(\hat{\theta}-\theta) \xrightarrow{\mathcal{L}} N(\mathbf{0}, \Omega).$$

Proof of Theorem 3.2.3. See Appendix.□

Here we assume that the function $g(\cdot)$ is $(m-1)$-dimensional, i.e.,

$$g : \theta \to \mathbb{R}^{m-1}. \tag{3.2.3}$$

For $g(\cdot)$ given by (3.2.3) we impose the following.

Assumption 3.2.4. *The function $g(\theta)$ is continuously differentiable.*

Under this assumption, the delta method is available and (3.1.2), (3.1.3), (3.1.5) and (3.1.7) satisfy this assumption. As a unified estimator for (mean-variance) optimal portfolio weights we introduce $g(\hat{\theta})$. For this we have the following result.

Theorem 3.2.5. *Under Assumptions 3.2.1 and 3.2.4,*

$$\sqrt{n}(g(\hat{\theta})-g(\theta)) \xrightarrow{\mathcal{L}} N\left(\mathbf{0}, \left(\frac{\partial g}{\partial \theta'}\right)\Omega\left(\frac{\partial g}{\partial \theta'}\right)'\right)$$

where $\partial g/\partial \theta'$ is the vector differential (i.e., $\partial g/\partial \theta' = (\partial g/\partial \theta_i)_{i=1,\ldots,m+r}$).

Proof of Theorem 3.2.5. The proof follows from Theorem 3.2.3 and the δ-method (e.g., Brockwell and Davis (2006)).□

3.2.2 Pessimistic Portfolio

In Section 3.1.5 we introduced the pessimistic Portfolio, which is a portfolio with the weight as the solution of

$$\min_{\boldsymbol{w}=(w_1,\ldots,w_m)'\in\mathbb{R}^m} \rho_{\nu_\alpha}(\boldsymbol{w}'\boldsymbol{X}) \quad \text{subject to } \mathrm{E(w'X)} = \mu_{\mathrm{P}}, \ \sum_{\mathrm{i}=1}^{\mathrm{m}} \mathrm{w_i} = 1 \tag{3.2.4}$$

where $\rho_{\nu_\alpha}(\boldsymbol{w}'\boldsymbol{X})$ is called the α-risk of the portfolio return $\boldsymbol{w}'\boldsymbol{X}$ and a pre-specified expected return μ_P. Here the α-risk $\rho_{\nu_\alpha}(\boldsymbol{w}'\boldsymbol{X})$ is defined as

$$\rho_{\nu_\alpha}(\boldsymbol{w}'\boldsymbol{X}) = -\int_0^1 F^{-1}_{\boldsymbol{w}'\boldsymbol{X}}(t)d\nu_\alpha(t) = -\frac{1}{\alpha}\int_0^\alpha F^{-1}_{\boldsymbol{w}'\boldsymbol{X}}(t)dt, \quad \alpha \in (0,1)$$

where $\nu_\alpha = \min\{t/\alpha, 1\}$ and $F^{-1}_{\boldsymbol{w}'\boldsymbol{X}}(t) = \sup\{x|F_{\boldsymbol{w}'\boldsymbol{X}}(x) \leq t\}$ denotes the quantile function of the portfolio return $\boldsymbol{w}'\boldsymbol{X}$ with distribution function $F_{\boldsymbol{w}'\boldsymbol{X}}$.

Bassett et al. (2004) showed that a portfolio with minimized α-risk can be constructed via the quantile regression (QR) methods of Koenker and Bassett (1978). QR is based on the fact that a quantile can be characterized as the minimizer of some expected asymmetric absolute loss function, namely,

$$\begin{aligned} F^{-1}_{\boldsymbol{w}'\boldsymbol{X}}(\alpha) &= \arg\min_{\theta} E\left[(\alpha\mathbb{I}\{\boldsymbol{w}'\boldsymbol{X} - \theta \geq 0\} + (1-\alpha)\mathbb{I}\{\boldsymbol{w}'\boldsymbol{X} - \theta < 0\})\,|\boldsymbol{w}'\boldsymbol{X} - \theta|\right] \\ &\equiv \arg\min_{\theta} E\left[\rho_\alpha(\boldsymbol{w}'\boldsymbol{X} - \theta)\right] \end{aligned}$$

where $\rho_\alpha(u) = u(\alpha - \mathbb{I}\{u < 0\})$, $u \in \mathbb{R}$ is called the check function (see Koenker (2005)) and $\mathbb{I}\{A\}$ is the indicator function defined by $\mathbb{I}\{A\} = \mathbb{I}_A(\omega) \equiv 1$ if $\omega \in A$, $\equiv= 0$ if $\omega \notin A$. To construct the optimal (i.e., α-risk minimized) portfolio, the following lemma is needed.

Lemma 3.2.6. *(Theorem 2 of Bassett et al. (2004)) Let $Y = \boldsymbol{w}'\boldsymbol{X}$ be a real-valued random variable with $E(Y) = E(\boldsymbol{w}'\boldsymbol{X}) = \mu_P < \infty$. Then,*

$$\min_{\theta\in\mathbb{R}} E\left[\rho_\alpha(Y - \theta)\right] = \alpha(\mu_P + \rho_{\nu_\alpha}(Y)).$$

Denoting $Y = \boldsymbol{w}'\boldsymbol{X}$ as a portfolio consisting of m different assets $\boldsymbol{X} = (X_1, \ldots, X_m)'$ with allocation weights $\boldsymbol{w} = (w_1, \ldots, w_m)'$ (subject to $\sum_{j=1}^m w_j = 1$), the optimization problem can be written as (3.2.4).

Suppose that we obtain a sequence of random returns $\{\boldsymbol{X}(t) = (X_1(t), \ldots, X_m(t))' : t = 1, \ldots, n, n \in \mathbb{N}\}$. Taniai and Shiohama (2012) introduced the sample or empirical version of (3.2.4) as

$$\min_{\boldsymbol{b}=(b_1,\ldots,b_m)'\in\mathbb{R}^m} \sum_{i=1}^{n+1} \rho_\alpha(\boldsymbol{Z}_i - \boldsymbol{W}_i'\boldsymbol{b}) \tag{3.2.5}$$

where $\bar{X}_i = n^{-1}\sum_{t=1}^{n} X_i(t)$ and

$$\begin{aligned} \boldsymbol{Z} &= (Z_1, \ldots, Z_n, Z_{n+1})' = (X_1(1), \ldots, X_1(n), \kappa(\bar{X}_1 - \mu_P))' \\ \boldsymbol{W} &= [\boldsymbol{W}_1, \ldots, \boldsymbol{W}_n | \boldsymbol{W}_{n+1}] \\ &= \begin{bmatrix} 1 & \cdots & 1 & 0 \\ X_1(1) - X_2(1) & \cdots & X_1(n) - X_2(n) & \kappa(\bar{X}_1 - \bar{X}_2) \\ \vdots & \ddots & \vdots & \vdots \\ X_1(1) - X_m(1) & \cdots & X_1(n) - X_m(n) & \kappa(\bar{X}_1 - \bar{X}_m) \end{bmatrix} \end{aligned}$$

with some $\kappa = O(n^{1/2})$. Denoting the minimizer of (3.2.5) as $\hat{\boldsymbol{\beta}}^{(n)}(\alpha) = (\hat{\beta}_1^{(n)}(\alpha), \ldots, \hat{\beta}_n^{(n)}(\alpha))'$, the optimal portfolio weights

$$\hat{\boldsymbol{w}}^{(n)}(\alpha) = \left(1 - \sum_{i=2}^{m} \hat{\beta}_i^{(n)}(\alpha), \hat{\beta}_2^{(n)}(\alpha), \ldots, \hat{\beta}_m^{(n)}(\alpha)\right)'$$

minimize α-risk. The large sample properties of $\hat{\boldsymbol{\beta}}^{(n)}(\alpha)$, especially its $\sqrt{n}$-consistency, can be implied from the standard arguments and assumptions in the QR context.

Let $\boldsymbol{\beta}(\alpha)$ be a QR coefficient of Z_i conditioning on $\boldsymbol{W}_i$ satisfied with

$$F_{Z_i}^{-1}(\alpha|\boldsymbol{W}_i) = \boldsymbol{W}_i'\boldsymbol{\beta}(\alpha) \tag{3.2.6}$$

where $F_Z^{-1}(\alpha|\boldsymbol{W}) = \inf\{x : P(Z \leq x|\boldsymbol{W}) \geq \alpha\}$. Note that here the QR model (3.2.6) has a random coefficient regression (RCR) interpretation of the form $Z_i = \boldsymbol{W}_i'\boldsymbol{\beta}(U_i)$ with component-wise monotone increasing function $\boldsymbol{\beta}$ and random variables U_i ~Uniform[0, 1]. Here a choice such that

$\beta(u) = (\beta_1(u), \beta_2(u), \ldots, \beta_m(u))' := (b_1 + F_\xi^{-1}(u), b_2, \ldots, b_m)'$ with F_ξ the distribution function of some independent and identically distributed (i.i.d.) n-tuple $(\xi_1, \ldots, \xi_n)$ yields

$$Z_i = \boldsymbol{W}_i'\beta(U_i) = \boldsymbol{W}_i' \begin{bmatrix} b_1 + \xi_i \\ b_2 \\ \vdots \\ b_m \end{bmatrix}.$$

Hence, recalling that the first component of $\boldsymbol{W}_i$ is 1 for $i = 1, \ldots, n$, it follows that, for any fixed $\alpha \in [0, 1]$, the QR coefficient $\beta(\alpha)$ can be characterized as the parameter $\boldsymbol{b} \in \mathbb{R}^m$ of a model such as

$$Z_i = \boldsymbol{W}_i'\boldsymbol{b} + \xi_i, \quad \xi_i \overset{iid}{\sim} G$$

where the density g of G is subject to

$$g \in \mathcal{F}^\alpha = \left\{ f : \int_{-\infty}^{0} f(x)dx = \alpha = 1 - \int_0^\infty f(x)dx \right\},$$

that is, $G^{-1}(\alpha) = 0$.
We impose the following assumptions.

Assumption 3.2.7. *The distribution functions $\{P(Z_i \le x|\boldsymbol{W}_i)\}$ are absolutely continuous, with continuous densities $f_i(\xi)$ uniformly bounded away from 0 and ∞ at the points $\xi_i(\alpha), i = 1, 2, \ldots$.*

Assumption 3.2.8. *There exist positive definite matrices D_0 such that*

(i) $\lim_{n\to\infty} \frac{1}{n+1} \sum_{i=1}^{n+1} \boldsymbol{W}_i\boldsymbol{W}_i' = D_0$

(ii) $\lim_{n\to\infty} \frac{1}{n+1} \sum_{i=1}^{n+1} f_i(\xi_i(\alpha))\boldsymbol{W}_i\boldsymbol{W}_i' = \lim_{n\to\infty} \frac{1}{n+1} \sum_{i=1}^{n+1} g(0)\boldsymbol{W}_i\boldsymbol{W}_i' = g(0)D_0$ *with* $g(0) \neq 0$

(iii) $\max_{i=1,\ldots,n+1} \|\boldsymbol{W}_i\|/\sqrt{n} \to 0.$

According to Koenker (2005), Assumption 3.2.8 (i) and (iii) are familiar throughout the literature on M-estimators for regression models; some variant of them is necessary to ensure that a Lindeberg condition is satisfied. Assumption 3.2.8 (ii) is really a matter of notational convenience and could be deduced from Assumption 3.2.8 (i) and a slightly strengthened version of Assumption 3.2.7.

Theorem 3.2.9. *(Theorem 4.1. of Koenker (2005)) Under Assumptions 3.2.7 and 3.2.8,*

$$\sqrt{n}\left\{\hat{\beta}^{(n)}(\alpha) - \beta(\alpha)\right\} \xrightarrow{d} N\left(\boldsymbol{0}, \frac{\alpha(1-\alpha)}{g(0)^2} D_0^{-1}\right).$$

Essentially, the result that the pessimistic portfolio obtained from $\hat{\beta}^{(n)}(\alpha)$ is minimizing the α-risk was of Bassett et al. (2004). Furthermore, Taniai and Shiohama (2012) showed that semiparametrically efficient inference of the optimal weights $\hat{\beta}^{(n)}(\alpha)$ is feasible.

3.2.3 Shrinkage Estimators

Let $\boldsymbol{X}(1), \ldots, \boldsymbol{X}(n)$ be a sequence of independent and identically distributed (m-dimensional) random vectors distributed as $N(\boldsymbol{\mu}, \Sigma)$, where $\boldsymbol{\mu}$ is an m-dimensional vector and Σ is the $m \times m$ positive definite matrix. The sample mean $\hat{\boldsymbol{\mu}} = 1/n \sum_{t=1}^{n} \boldsymbol{X}(t)$ seems the most fundamental and natural estimator of $\boldsymbol{\mu}$. However, if $m \ge 3$, Stein (1956) showed that $\hat{\boldsymbol{\mu}}$ is not admissible with respect to the mean squared error loss function. Furthermore, James and Stein (1961) proposed a shrinkage estimator $\hat{\mu}^S$ defined by

$$\hat{\mu}^S := \left(1 - \frac{m-2}{n\|\hat{\mu}\|^2}\right)\hat{\mu}$$

which improves $\hat{\mu}^S$ with respect to the mean squared error when $m \geq 3$.
For the portfolio choice problem, shrinkage estimation has been considered by Jobson et al. (1979), Jobson and Korkie (1981), Jorion (1986), Frost and Savarino (1986), Ledoit and Wolf (2003, 2004), among others. Traditional portfolio weight estimators are written as functions of the sample mean $\hat{\mu}$ and the sample covariance matrix $\hat{\Sigma}$. Jorion (1986) proposed a plug-in portfolio weight estimator constructed with shrinkage sample mean, that is,

$$\hat{\mu}^S := \delta\mu_0 e + (1-\delta)\hat{\mu}$$

where δ is a shrinkage factor satisfying $0 < \delta < 1$ and μ_0 is a common constant value. In addition, an optimal shrinkage factor δ^{opt} is given by

$$\delta^{opt} = \min\left\{1, \frac{(m-2)/n}{(\hat{\mu}-\mu_0 e)'\Sigma^{-1}(\hat{\mu}-\mu_0 e)}\right\}.$$

Furthermore, Frost and Savarino (1986) and Ledoit and Wolf (2003) and Ledoit and Wolf (2004) showed shrinkage estimation can be applied not only to mean, but also covariance matrices. They proposed a shrinkage covariance matrix estimator by using the usual sample covariance matrix $\hat{\Sigma}$ and a shrinkage target S (or its estimator $\hat{S}$), that is,

$$\hat{\Sigma}^S := \delta\hat{S} + (1-\delta)\hat{\Sigma}.$$

Ledoit and Wolf (2003) derive the following approximate expression for the optimal shrinkage factor:

$$\delta^{opt} \simeq \frac{1}{n}\frac{A-B}{C}$$

with

$$A = \sum_{i,j=1}^{m} asy\ var(\sqrt{n}\hat{\sigma}_{ij}),\ B = \sum_{i,j=1}^{m} asy\ cov(\sqrt{n}\hat{\sigma}_{ij}, \sqrt{n}\hat{s}_{ij}),\ C = \sum_{i,j=1}^{m}(\sigma_{ij} - s_{ij})^2.$$

In what follows, we briefly examine the effect of shrinkage estimators from a data generation process. We generate $K(= 3000)$ sequences of independent random vectors

$$\mathbb{X}_n^{(k)} := \left\{\boldsymbol{X}_t^{(k)} = (X_{1t}^{(k)}, \ldots, X_{qt}^{(k)})^\top\right\}_{t=1,\ldots,n}, \quad k = 1, \ldots, K$$

from a q-dimensional Gaussian distribution (i.e., $\boldsymbol{X}_t^{(k)} \overset{i.i.d.}{\sim} N_q(\boldsymbol{\mu}, \Sigma)$) where the components of the mean vector $\boldsymbol{\mu} = (\mu_1, \ldots, \mu_q)^\top$ and that of the covariance matrix $\Sigma = (\sigma_{ij})_{i,j=1,\ldots,q}$ are defined as follows:

$$\mu_i = 0.1\left(1 + \frac{i}{q}\right), \quad \sigma_{ii} = 1 + \frac{0.1i}{q} \quad \text{and} \quad \rho = \frac{\sigma_{ij}}{\sqrt{\sigma_{ij}\sigma_{ij}}} = 0.6\ (i \neq j).$$

Consider the following shrinkage estimators for the mean vector and the covariance matrix:

$$\hat{\mu}_k^S := \delta_1\mu_0 e + (1-\delta_1)\hat{\mu}^{(k)}, \quad \hat{\Sigma}_k^S := \delta_2\Sigma + (1-\delta_2)\hat{\Sigma}^{(k)}, \quad \tilde{\Sigma}_k^S := \delta_2 I + (1-\delta_2)\hat{\Sigma}^{(k)}$$

where $\delta_1, \delta_2 (= 0, 0.4, 0.8)$ are the shrinkage factors; $\mu_0 = 0.15$ is the shrinkage target mean; $\hat{\mu}^{(k)}$

and $\hat{\Sigma}^{(k)}$ are sample mean vector and sample covariance matrix based on $\mathbb{X}_n^{(k)}$, respectively. Note that $\hat{\Sigma}_k^S$ is the case that the shrinkage target covariance matrix corresponds to the true covariance matrix (i.e., $S = \Sigma$), and $\tilde{\Sigma}_k^S$ is the case that the shrinkage target does not equal the true covariance matrix (i.e., $S \neq \Sigma$).

Based on the above $\hat{\mu}_k^S$ and $\hat{\Sigma}_k^S$ (or $\tilde{\Sigma}_k^S$), we define an empirical version of the mean squared error (MSE) for the shrinkage portfolio estimator as follows;

$$\widehat{trV_S} := \frac{1}{K}\sum_{k=1}^{K} \| \sqrt{n}\left\{g(\hat{\mu}_k^S, \hat{\Sigma}_k^S) - g(\mu, \Sigma)\right\} \|^2, \quad \widetilde{trV_S} := \frac{1}{K}\sum_{k=1}^{K} \| \sqrt{n}\left\{g(\hat{\mu}_k^S, \tilde{\Sigma}_k^S) - g(\mu, \Sigma)\right\} \|^2$$

where $g(\cdot, \cdot)$ is a function of μ and Σ which corresponds to the portfolio weight defined by (3.1.2). Table 3.1 shows the empirical MSE $\widehat{trV_S}$ for $q = 3, 10, 30$ and $n = 10, 30, 100$ with $q < n$.

Table 3.1: Empirical version of the mean squared error ($\widehat{trV_S}$) for the shrinkage portfolio estimator when the asset size $q = 3, 10, 30$, the sample size $n = 10, 30, 100$ with $q < n$ and the shrinkage factors $\delta_1, \delta_2 = 0, 0.4, 0.8$.

	n	10			30			100		
q	$\delta_1 \setminus \delta_2$	0	0.4	0.8	0	0.4	0.8	0	0.4	0.8
3	0	9.954	2.994	1.945	3.489	2.333	1.861	2.685	2.097	1.807
	0.4	4.275	1.229	0.720	1.767	1.009	0.699	1.433	0.937	0.697
	0.8	1.439	0.351	0.114	0.918	0.362	0.139	0.882	0.432	0.221
10	0	-	-	-	33.448	12.221	8.590	12.485	9.546	8.396
	0.4	-	-	-	12.937	4.617	3.147	5.225	3.753	3.169
	0.8	-	-	-	2.807	0.913	0.516	1.920	1.160	0.848
30	0	-	-	-	-	-	-	83.558	36.967	27.585
	0.4	-	-	-	-	-	-	30.871	13.770	10.307
	0.8	-	-	-	-	-	-	5.832	3.177	2.582

The shrinkage effect for each n and q can be seen. Note that the empirical MSE for the usual sample portfolio estimator is shown in the case of $\delta_1 = \delta_2 = 0$. In view of the shrinkage effect for the mean, significant improvement (see the effect for the difference of δ_1) can be seen even if the shrinkage target mean vector $\mu_0 e$ does not correspond to the true mean vector μ. Table 3.2 shows the empirical MSE $\widetilde{trV_S}$ for $q = 3, 10, 30$ and $n = 10, 30, 100$ with $q < n$.

This result shows the risk that the analyst trusts the shrinkage target too much. Especially for $\delta_2 = 0.8$, the empirical MSEs of the shrinkage estimator are sometimes larger than that of the usual sample portfolio estimator (i.e., $\delta_1 = \delta_2 = 0$). However, if we set $\delta_1 = \delta_2 = 0.4$, then we can obtain improvement rather than the usual sample portfolio estimator in view of the empirical MSE. This phenomenon is one of the reasons the shrinkage estimator is useful for the portfolio selection problem.

3.2.4 *Bayesian Estimation*

From a Bayesian perspective, the effect of parameter uncertainty on the optimal portfolio problem is considered by expressing the investor's problem in terms of the predictive distribution of the future returns. Suppose that a sequence of random asset returns $\{\boldsymbol{X}(t) = (X_1(t), \ldots, X_m(t))'; t \in \mathbb{Z}\}$ is a stationary process with $E\{\boldsymbol{X}(t)\} = \mu$ and $\mathrm{Var}\{\boldsymbol{X}(t)\} = \Sigma$. Writing the unknown parameter, portfolio weight and investor's utility function as $\theta = (\mu, \Sigma) \in \mathbb{R}^m \times \mathbb{R}^{m+1}$, $w = (w_1, \ldots, w_m)' \in \mathbb{R}^m$ and $U : \mathbb{R} \to \mathbb{R}$, then the optimal portfolio weight w^{opt} is obtained by the solution of the following

Table 3.2: Empirical version of the mean squared error ($\widetilde{trV_S}$) for the shrinkage portfolio estimator when the asset size $q = 3, 10, 30$, the sample size $n = 10, 30, 100$ with $q < n$ and the shrinkage factors $\delta_1, \delta_2 = 0, 0.4, 0.8$.

	n	10			30			100		
q	$\delta_1 \setminus \delta_2$	0	0.4	0.8	0	0.4	0.8	0	0.4	0.8
3	0	9.954	7.126	8.144	3.489	4.435	7.187	2.685	3.732	6.815
	0.4	4.275	2.863	2.993	1.767	1.895	2.662	1.433	1.632	2.515
	0.8	1.439	0.740	0.436	0.918	0.649	0.455	0.882	0.699	0.587
10	0	-	-	-	33.448	28.874	35.504	12.485	17.974	32.207
	0.4	-	-	-	12.937	10.773	12.847	5.225	6.912	11.696
	0.8	-	-	-	2.807	1.902	1.790	1.920	1.864	2.281
30	0	-	-	-	-	-	-	83.558	83.934	112.359
	0.4	-	-	-	-	-	-	30.871	30.582	40.481
	0.8	-	-	-	-	-	-	5.832	5.683	7.248

optimization problem.

$$\max_{\boldsymbol{w}\in\mathbb{R}^m} E\{\boldsymbol{w}'\boldsymbol{X}(t)\} = \max_{\boldsymbol{w}} \int U\{\boldsymbol{w}'\boldsymbol{X}(t)\}p(\boldsymbol{X}(t)|\theta)d\boldsymbol{X}(t) \tag{3.2.7}$$

subject to $\boldsymbol{w}'\boldsymbol{e} = 1$, where $p(\boldsymbol{X}(t)|\theta)$ is the conditional density of asset returns under a given parameter θ. Suppose that $\boldsymbol{X} = \{\boldsymbol{X}(1), \ldots, \boldsymbol{X}(n)\}$ is observed from $\{\boldsymbol{X}(t)\}$. The empirical version of (3.2.7) can be written as

$$\max_{\boldsymbol{w}\in\mathbb{R}^m} E\{\boldsymbol{w}'\boldsymbol{X}(n+1)|\boldsymbol{X}\} = \max_{\boldsymbol{w}} \int U\{\boldsymbol{w}'\boldsymbol{X}(n+1)\}p(\boldsymbol{X}(n+1)|\boldsymbol{X})d\boldsymbol{X}(n+1) \tag{3.2.8}$$

subject to $\boldsymbol{w}'\boldsymbol{e} = 1$, where $\boldsymbol{X}(n+1)$ is the (yet unobserved) next period return, which implies that $p(\boldsymbol{X}(n+1)|\boldsymbol{X})$ means the predictive return density after observing $\boldsymbol{X}$. Denoting the conditional predictive return density under a given parameter θ and the joint posterior density after observing $\boldsymbol{X}$ by $p(\boldsymbol{X}(n+1)|\theta)$ and $p(\theta|\boldsymbol{X})$, respectively, the predictive return density $p(\boldsymbol{X}(n+1)|\boldsymbol{X})$ is given by

$$p(\boldsymbol{X}(n+1)|\boldsymbol{X}) = \int p(\boldsymbol{X}(n+1)|\theta)p(\theta|\boldsymbol{X})d\theta \tag{3.2.9}$$

(see, e.g., p. 35 of Rachev et al. (2008)). In addition, by using the continuous version of Bayes' theorem, the joint posterior density $p(\theta|\boldsymbol{X})$ can be written as

$$p(\theta|\boldsymbol{X}) = \frac{p(\boldsymbol{X}|\theta)p(\theta)}{p(\boldsymbol{X})} \propto L(\theta|\boldsymbol{X})p(\theta) \tag{3.2.10}$$

where $L(\theta|\boldsymbol{X})$ is the likelihood function for θ and $p(\theta)$ is a prior distribution of θ (called hyperparameters). Substituting (3.2.9) and (3.2.10) into (3.2.8), the empirical version of the optimization problem can be written as

$$\max_{\boldsymbol{w}\in\mathbb{R}^m} \int \left\{ \int U[\boldsymbol{w}'\boldsymbol{X}(n+1)]p(\boldsymbol{X}(n+1)|\theta)d\boldsymbol{X}(n+1) \right\} L(\theta|\boldsymbol{X})p(\theta)d\theta. \tag{3.2.11}$$

Comparing (3.2.7) and (3.2.11), we can describe the inference about the portfolio weight estimator.

i.i.d. Gaussian Return with Uninformative Prior When $\boldsymbol{X}(t)$'s are i.i.d. Gaussian random vectors with mean vector $\boldsymbol{\mu}$ and covariance matrix Σ, the likelihood function can be written as

$$L(\theta|\boldsymbol{X}) \propto |\Sigma|^{-n/2} \exp\left[-\frac{1}{2}\sum_{t=1}^{n}\{\boldsymbol{X}(t)-\boldsymbol{\mu}\}'\,\Sigma^{-1}\,\{\boldsymbol{X}(t)-\boldsymbol{\mu}\}\right]. \tag{3.2.12}$$

We consider the case when the investor is uncertain about the distribution of θ, and has no particular prior knowledge of θ. In that case, Jeffreys (1961) introduced the Jeffreys prior, that is,

$$p(\theta) \propto |\Sigma|^{-\frac{m+1}{2}}. \tag{3.2.13}$$

In general, the Jeffreys prior of a parameter θ is given by $p(\theta) = |I(\theta)|^{1/2}$, where $I(\theta)$ is the so-called Fisher's information matrix for θ, given by $I(\theta) = E\left\{\frac{\partial^2}{\partial\theta\partial\theta'}\log p(\boldsymbol{X}(t)|\theta)\right\}$. Then it can be shown that the distribution of $\boldsymbol{X}(n+1)|\boldsymbol{X}$ follows a multivariate Student's t-distribution with $n-m$ degrees of freedom. The predictive mean and covariance matrix of $\boldsymbol{X}(n+1)|\boldsymbol{X}$ are given by

$$\tilde{\boldsymbol{\mu}} = \hat{\boldsymbol{\mu}} \quad \text{and} \quad \tilde{\Sigma} = \frac{\mathrm{n}+1}{\mathrm{n}-\mathrm{m}-2}\hat{\Sigma} \tag{3.2.14}$$

where $\hat{\boldsymbol{\mu}} = \frac{1}{n}\sum_{t=1}^{n}\boldsymbol{X}(t)$ and $\hat{\Sigma} = \frac{1}{n}\sum_{t=1}^{n}\{\boldsymbol{X}(t)-\boldsymbol{\mu}\}\{\boldsymbol{X}(t)-\boldsymbol{\mu}\}'$. Hence, the Bayesian (mean-variance) optimal portfolio estimator is obtained by

$$g(\hat{\theta}^{bayes}) \quad \text{with} \quad \hat{\theta}^{\text{bayes}} = (\tilde{\boldsymbol{\mu}}, \tilde{\Sigma}) \tag{3.2.15}$$

where g is the optimal portfolio function defined in (3.2.3).

i.i.d. Gaussian Return with Informative Prior We consider that the investor has informative beliefs about the mean vector $\boldsymbol{\mu}$ and covariance matrix Σ. Suppose that the distributions of $\boldsymbol{\mu}|\Sigma$ and Σ are as follows:

$$\boldsymbol{\mu}|\Sigma \sim N\left(\eta, \frac{1}{\tau}\Sigma\right), \quad \Sigma \sim IW(\Omega, \nu), \tag{3.2.16}$$

where $IW(\Omega, \nu)$ is the inverse Wishart distribution with a parameter matrix Ω and a degree of freedom ν. Under the Gaussianity of $\boldsymbol{X}(t)$, (3.2.16) become the conjugate priors for $\boldsymbol{\mu}|\Sigma$ and Σ. Here the prior parameters τ and ν depend on the investor's confidence about η and Ω. In that case, it can be shown that the distribution of $\boldsymbol{X}(n+1)|\boldsymbol{X}$ follows a multivariate Student's t-distribution and the predictive mean and covariance matrix of $\boldsymbol{X}(n+1)|\boldsymbol{X}$ are given by

$$\tilde{\boldsymbol{\mu}} = \frac{\tau}{n+\tau}\eta + \frac{n}{n+\tau}\hat{\boldsymbol{\mu}} \tag{3.2.17}$$

$$\tilde{\Sigma} = \frac{n+1}{n(\nu+m-1)}\left\{\Omega + n\hat{\Sigma} + \frac{n\tau}{n+\tau}(\eta-\hat{\boldsymbol{\mu}})(\eta-\hat{\boldsymbol{\mu}})'\right\}. \tag{3.2.18}$$

In this case, the Bayesian (mean-variance) optimal portfolio estimator is obtained by substituting (3.2.17) and (3.2.18) for (3.2.14).

The Black–Litterman Model Black and Litterman (1992) applied the theoretical implications of an economic model to specifying a prior in the portfolio choice context. The model is called the Black–Litterman model which comes from the mixed estimation. According to Aït-Sahalia and Hansen (2009), the mixed estimation was first developed by Theil and Goldberger (1961) as a way to update the Bayesian inferences drawn from old data with the information contained in a set of new data. It applies more generally to the problem of combining information from two data sources into a single posterior distribution.

Let $\boldsymbol{X}(t)$ be returns of m-risky assets with the mean $\boldsymbol{\mu}$ and covariance matrix Σ. Assuming that Σ is known, the market capitalization weights $\boldsymbol{w}_{mkt}$ are observed and there exists a risk-free asset with the return $\mu_f = 0$, then the tangency portfolio weight is given by $\boldsymbol{w}_{mkt} = (1/\gamma)\Sigma^{-1}\boldsymbol{\mu}$ where γ is the aggregate risk aversion. Under the capital asset pricing model (CAPM), the mean vector $\boldsymbol{\mu}$ corresponds to the equilibrium risk premium $\boldsymbol{\mu}_{equil}$ therefore, we can write

$$\boldsymbol{\mu}_{equil} = \gamma\Sigma\boldsymbol{w}_{mkt}.$$

In the Black–Litterman model, the prior distribution of $\boldsymbol{\mu}$ is assumed by using the benchmark beliefs at this equilibrium risk premium, that is,

$$\boldsymbol{\mu} \sim N(\boldsymbol{\mu}_{equil}, \lambda\Sigma) \tag{3.2.19}$$

where λ measures the strength of the investor's belief in equilibrium.

On the other hand, the investor has a set of new views or forecasts $\boldsymbol{\eta} \in \mathbb{R}^k$ about a subset of $k \leq m$ linear combinations of returns $P\boldsymbol{X}(t)$, where P is a $k \times m$ matrix selecting and combining returns into portfolios for which the investor is able to express views. For instance, suppose that there exist four assets, A, B, C and D (i.e., $m = 4$), and the investor considers that

- asset A will give 1% monthly return,
- asset B will give 2% monthly return,
- asset C will outperform asset D by 0.5% (i.e., $k = 3$).

Then, P and $\boldsymbol{\eta}$ are written as

$$P = \begin{pmatrix} 1 & 0 & 0 & 0 \\ 0 & 1 & 0 & 0 \\ 0 & 0 & 1 & -1 \end{pmatrix}, \quad \boldsymbol{\eta} = \begin{pmatrix} 0.01 \\ 0.02 \\ 0.005 \end{pmatrix}.$$

The new views are assumed to be unbiased but imprecise, with distribution

$$\boldsymbol{\eta}|\boldsymbol{\mu} \sim N(P\boldsymbol{\mu}, \Omega) \tag{3.2.20}$$

where Ω expresses the new view's accuracy and correlations, and the investor needs to specify that. Combining (3.2.19) and (3.2.20) by use of Bayes' theorem, the joint posterior density $p(\boldsymbol{\mu}|\boldsymbol{\eta})$ can be written as

$$p(\boldsymbol{\mu}|\boldsymbol{\eta}) \propto p(\boldsymbol{\eta}|\boldsymbol{\mu})p(\boldsymbol{\mu}),$$

hence,

$$\boldsymbol{\mu}|\boldsymbol{\eta} \sim N(E(\boldsymbol{\mu}|\boldsymbol{\eta}), \mathrm{Var}(\boldsymbol{\mu}|\boldsymbol{\eta}))$$

where

$$\begin{aligned} E(\boldsymbol{\mu}|\boldsymbol{\eta}) &= \{(\lambda\Sigma)^{-1} + P'\Omega^{-1}P\}^{-1}\{(\lambda\Sigma)^{-1}\boldsymbol{\mu}_{equil} + P'\Omega^{-1}\boldsymbol{\eta}\} \\ &= \{(\lambda\Sigma)^{-1} + P'\Omega^{-1}P\}^{-1}\left\{\frac{\gamma}{\lambda}\boldsymbol{w}_{mkt} + P'\Omega^{-1}\boldsymbol{\eta}\right\} \\ \mathrm{Var}(\boldsymbol{\mu}|\boldsymbol{\eta}) &= \{(\lambda\Sigma)^{-1} + P'\Omega^{-1}P\}^{-1} \end{aligned}$$

In addition, the predictive mean and covariance matrix of $\boldsymbol{X}(n+1)|\boldsymbol{\eta}$ are given by

$$\tilde{\boldsymbol{\mu}} = E(\boldsymbol{\mu}|\boldsymbol{\eta}) \quad \text{and} \quad \tilde{\Sigma} = \{\Sigma^{-1} + \mathrm{Var}(\boldsymbol{\mu}|\boldsymbol{\eta})\}^{-1}. \tag{3.2.21}$$

In this case, the Bayesian (mean-variance) optimal portfolio estimator is obtained by substituting (3.2.21) for (3.2.14).

3.2.5 Factor Models

3.2.5.1 Static factor models

In practice, most financial portfolios consist of many assets, which implies that we need to consider high-dimensional statistical models. In that case, applying the traditional approach is often complicated and hard. To overcome this problem, we is commonly use static factor models to simplify portfolio analysis. In the literature, three types of static factor models are available for studying asset returns, namely, the macroeconomic factor model, fundamental factor model and statistical factor model. The macroeconomic factor model uses macroeconomic variables such as growth rate of GDP, interest rates, inflation rate, and unemployment rate to describe the common behavior of asset returns. The fundamental factor model uses firm or asset specific attributes such as firm size, book and market values, and industrial classification to construct common factors. The statistical factor model treats the common factors as unobservable to be estimated from the return series.

Macroeconomic Factor Model Suppose that there exist m assets and their returns at time t are expressed as $X_i(t)$ where $i = 1, \ldots, m$ and $t \in \mathbb{Z}$. Suppose also that there exist q ($q < m$) kinds of risk factor $F_1(t), \ldots, F_q(t)$ which depend on time t. A general form for the factor model is

$$\boldsymbol{X}(t) = \boldsymbol{a} + \boldsymbol{b}\boldsymbol{F}(t) + \boldsymbol{\epsilon}(t) \tag{3.2.22}$$

where $\boldsymbol{a} = (a_i)_{i=1,\ldots,m}$ is an m vector, $\boldsymbol{b} = (b_{ij})_{i=1,\ldots,m,j=1,\ldots,q}$ is an $m \times q$ matrix (b_{ij}'s are called factor loadings), $\boldsymbol{F}(t) = (F_1(t), \ldots, F_q(t))'$ is a vector of q common factors and $\boldsymbol{\epsilon}(t) = (\epsilon_1(t), \ldots, \epsilon_m(t))'$ is an m-dimensional error term. For asset returns, the factor $\boldsymbol{F}(t)$ is assumed to be an m-dimensional stationary process such that

$$E\{\boldsymbol{F}(t)\} = \boldsymbol{\mu}_F, \quad \mathrm{Var}\{\boldsymbol{F}(t)\} = \Sigma_F$$

and the error term $\boldsymbol{\epsilon}(t)$ is a white noise series and uncorrelated with the factor $\boldsymbol{F}(t)$, that is,

$$E\{\boldsymbol{\epsilon}(t)\} = \boldsymbol{0}, \quad \mathrm{Var}\{\boldsymbol{\epsilon}(t)\} = \Sigma_\epsilon = diag(\sigma_1^2, \ldots, \sigma_m^2), \quad \mathrm{Cov}\{\boldsymbol{F}(t), \boldsymbol{\epsilon}(t)\} = \boldsymbol{0}.$$

Then the mean vector and covariance matrix of the return $\boldsymbol{X}(t)$ is

$$E\{\boldsymbol{X}(t)\} = \boldsymbol{a} + \boldsymbol{b}\boldsymbol{\mu}_F, \quad \mathrm{Var}\{\boldsymbol{X}(t)\} = \boldsymbol{b}\Sigma_F\boldsymbol{b}' + \Sigma_\epsilon.$$

For macroeconomic factor models, the factors are observed and we can apply the least squares method to the multivariate linear regression (MLR) model ((3.2.22) is a special form of the MLR model). Observing $(\boldsymbol{X}(1), \boldsymbol{F}(1)), \ldots, (\boldsymbol{X}(n), \boldsymbol{F}(n))$, we have

$$\boldsymbol{X} := \begin{bmatrix} \boldsymbol{X}(1)' \\ \vdots \\ \boldsymbol{X}(n)' \end{bmatrix} = \begin{bmatrix} 1 & \boldsymbol{F}(1)' \\ \vdots & \vdots \\ 1 & \boldsymbol{F}(n)' \end{bmatrix} \begin{bmatrix} \boldsymbol{a}' \\ \boldsymbol{b}' \end{bmatrix} + \begin{bmatrix} \boldsymbol{\epsilon}(1)' \\ \vdots \\ \boldsymbol{\epsilon}(n)' \end{bmatrix} := \boldsymbol{G}\boldsymbol{A} + \boldsymbol{E}.$$

The ordinary least squares (OLS) estimator of $\boldsymbol{A}$ is obtained by

$$\hat{\boldsymbol{A}} := \begin{bmatrix} \hat{\boldsymbol{a}}' \\ \hat{\boldsymbol{b}}' \end{bmatrix} = (\boldsymbol{G}'\boldsymbol{G})^{-1}(\boldsymbol{G}'\boldsymbol{X}). \tag{3.2.23}$$

In addition, the estimator of Σ_ϵ is obtained by

$$\hat{\Sigma}_\epsilon = diag\left[\frac{1}{n-q-1}(\boldsymbol{X} - \boldsymbol{G}\hat{\boldsymbol{A}})'(\boldsymbol{X} - \boldsymbol{G}\hat{\boldsymbol{A}})\right]. \tag{3.2.24}$$

Based on (3.2.23), (3.2.24) and

$$\hat{\boldsymbol{\mu}}_F = \frac{1}{n}\sum_{t=1}^{n} \boldsymbol{F}(t), \quad \hat{\Sigma}_F = \frac{1}{n-1}\sum_{t=1}^{n}\{\boldsymbol{F}(t) - \hat{\boldsymbol{\mu}}_F\}\{\boldsymbol{F}(t) - \hat{\boldsymbol{\mu}}_F\}',$$

the mean-variance optimal portfolio estimator under the macroeconomic factor model is obtained by

$$g(\hat{\theta}^{MF}) = g(\hat{\mu}^{MF}, \hat{\Sigma}^{MF})$$

where

$$\hat{\mu}^{MF} = \hat{a} + \hat{b}\hat{\mu}_F, \quad \hat{\Sigma}^{MF} = \hat{b}\hat{\Sigma}_F\hat{b}' + \hat{\Sigma}_\epsilon$$

and g is the optimal portfolio function defined in (3.2.3).

Instead of OLS, there are some other methods such as ridge regression, LASSO and elastic-net. The criterion functions are as follows:

$$\begin{aligned}(\hat{a}, \hat{b}) = \arg\min_{(a,b)}\Bigg[&\sum_{t=1}^{n} \{X(t) - a - bF(t)\}\{X(t) - a - bF(t)\}' \\ &+ \lambda \sum_{i=1}^{q} \left\{\alpha\|b_i\|^2 + (1-\alpha)\|b_i\|\right\}\Bigg] \qquad (3.2.25)\end{aligned}$$

where $b = (b_1, \ldots, b_q)$. In (3.2.25), the value $\lambda = 0$ corresponds to the OLS method, $\alpha = 0$ corresponds to ridge regression, $\alpha = 1$ corresponds to (grouped) LASSO and otherwise the models show elastic-net. The differences between the three models and OLS are to impose penalties on the size of regression coefficients. These methods are called regularization and one of the most important purposes of regularization is to avoid overfitting.

Fundamental Factor Model In fundamental factor models, the factor loadings are assumed as observable asset-specific fundamentals such as industrial classification, market capitalization, book value, and style classification (growth or value) to construct common factors that explain the excess returns. In the literature, there are two approaches to this model: the BARRA approach proposed by Rosenberg (1974), founder of BARRA Inc, and the Fama–French approach proposed by Fama and French (1992). In what follows, we briefly discuss the BARRA approach. Suppose that the factor model

$$X(t) = bF(t) + \epsilon(t)$$

where $b = (b_{ij})_{i=1,\ldots,m, j=1,\ldots,q}$ is given $m \times q$ matrix, $F(t) = (F_1(t), \ldots, F_q(t))'$ is an unobserved factor and $\epsilon(t) = (\epsilon_1(t), \ldots, \epsilon_m(t))'$ is an error term. In this model, we have to estimate μ_F, Σ_F and Σ_ϵ. To do so, we use a two-step procedure as follows:

- In step one, the ordinary least squares (OLS) method is used at each time t to obtain preliminary estimates of $F(t)$, $\epsilon(t)$ and Σ_ϵ, namely,

$$\hat{F}_o(t) = (b'b)^{-1}b'X(t), \quad \hat{\epsilon}_o(t) = X(t) - b\hat{F}_o(t),$$

$$\hat{\Sigma}_{\epsilon,o} = diag\left\{\frac{1}{n-1}\sum_{t=1}^{n}\hat{\epsilon}_o(t)\hat{\epsilon}_o(t)'\right\}.$$

- In step two, by using the above $\hat{\Sigma}_{\epsilon,o}$, we use the generalized least squares (GLS) method to obtain a refined estimator of $F(t)$, $\epsilon(t)$ and Σ_ϵ, namely,

$$\hat{F}_g(t) = (b'\hat{\Sigma}_{\epsilon,o}^{-1}b)^{-1}b'\hat{\Sigma}_{\epsilon,o}^{-1}X(t), \quad \hat{\epsilon}_g(t) = X(t) - b\hat{F}_g(t),$$

$$\hat{\Sigma}_{\epsilon,g} = diag\left\{\frac{1}{n-1}\sum_{t=1}^{n}\hat{\epsilon}_g(t)\hat{\epsilon}_g(t)'\right\}.$$

The mean-variance optimal portfolio estimator under the BARRA factor model (fundamental factor model) is obtained by

$$g(\hat{\theta}^{FF}) = g(\hat{\mu}^{FF}, \hat{\Sigma}^{FF})$$

where

$$\hat{\mu}^{FF} = \boldsymbol{b}\hat{\mu}_F, \quad \hat{\Sigma}^{FF} = \boldsymbol{b}\hat{\Sigma}_F\boldsymbol{b}' + \hat{\Sigma}_{\epsilon,g},$$

and

$$\hat{\mu}_F = \frac{1}{n}\sum_{t=1}^{n}\hat{\boldsymbol{F}}_g(t), \quad \hat{\Sigma}_F = \frac{1}{n-1}\sum_{t=1}^{n}\{\hat{\boldsymbol{F}}_g(t) - \hat{\mu}_F\}\{\hat{\boldsymbol{F}}_g(t) - \hat{\mu}_F\}'.$$

Statistical Factor Model The statistical factor model aims to identify a few factors that can account for most of the variations in the covariance or correlation matrix obtained from the observed data. Consider the general factor model

$$\boldsymbol{X}(t) = \boldsymbol{a} + \boldsymbol{b}\boldsymbol{F}(t) + \epsilon(t)$$

where both $\boldsymbol{b}$ and $\boldsymbol{F}(t)$ are assumed to be unobservable. In addition, we assume that the factor $\boldsymbol{F}(t)$ and the error term $\epsilon(t)$ satisfy

$$E\{\boldsymbol{F}(t)\} = \boldsymbol{0}, \quad \mathrm{Var}\{\boldsymbol{F}(t)\} = I_q,$$
$$E\{\epsilon(t)\} = \boldsymbol{0}, \quad \mathrm{Var}\{\epsilon(t)\} = \Sigma_\epsilon, \quad \mathrm{Cov}\{\boldsymbol{F}(t), \epsilon(t)\} = \boldsymbol{0}$$

where I_q is the $q \times q$ identity matrix. In this model, it is well known that principal component analysis (PCA) is available to estimate the optimal portfolio. Denoting the covariance matrix of $\boldsymbol{X}(t)$ as Σ (nonnegative definite), it is easy to see that

$$\mathrm{Var}\{\boldsymbol{X}(t)\} = \Sigma = \boldsymbol{b}\boldsymbol{b}' + \Sigma_\epsilon.$$

Let $(\lambda_1, \boldsymbol{e}_1), \ldots, (\lambda_m, \boldsymbol{e}_m)$ be pairs of the eigenvalues and eigenvectors of Σ, where $\lambda_1 > \lambda_2 > \cdots > \lambda_m > 0$ and $\boldsymbol{e}_i = (e_{i1}, \ldots, e_{im})'$ satisfied with $\|\boldsymbol{e}_i\| = 1$ for $i = 1, \ldots, m$. Based on the first q pairs $(\lambda_i, \boldsymbol{e}_i)$ $(i = 1, \ldots, q)$, we have

$$\Sigma = [\boldsymbol{e}_1, \ldots, \boldsymbol{e}_q]\begin{bmatrix} \lambda_1 & & \\ & \ddots & \\ & & \lambda_q \end{bmatrix}\begin{bmatrix} \boldsymbol{e}_1' \\ \vdots \\ \boldsymbol{e}_q' \end{bmatrix} = PDP' = (PD^{1/2})(PD^{1/2})'.$$

If $\boldsymbol{X}(t)$ is perfectly fitted by $\boldsymbol{F}(t)$, the error term $\epsilon(t)$ must be vanish. In that case, we can write

$$\Sigma = \boldsymbol{b}\boldsymbol{b}' = (PD^{1/2})(PD^{1/2})'$$

which implies that $\boldsymbol{b} = PD^{1/2}$. In the same manner, we consider the empirical case. Again let $(\hat{\lambda}_1, \hat{\boldsymbol{e}}_1), \ldots, (\hat{\lambda}_m, \hat{\boldsymbol{e}}_m)$ be pairs of the eigenvalues and eigenvectors of sample covariance matrix

$$\hat{\Sigma} = \frac{1}{n-1}\sum_{t=1}^{n}\{\boldsymbol{X}(t) - \hat{\mu}\}\{\boldsymbol{X}(t) - \hat{\mu}\} \quad \text{with} \quad \hat{\mu} = \frac{1}{\mathrm{n}}\sum_{\mathrm{t}=1}^{\mathrm{n}}\mathrm{X(t)}$$

where $\hat{\lambda}_1 > \hat{\lambda}_2 > \cdots > \hat{\lambda}_m > 0$ and $\hat{\boldsymbol{e}}_i = (\hat{e}_{i1}, \ldots, \hat{e}_{im})'$ satisfied with $\|\hat{\boldsymbol{e}}_i\| = 1$ for $i = 1, \ldots, m$. Based on the first p pairs $(\hat{\lambda}_i, \hat{\boldsymbol{e}}_i)$ $(i = 1, \ldots, q)$, the estimator of factor loadings is given by

$$\hat{\boldsymbol{b}} = \hat{P}\hat{D}^{1/2} = [\hat{\boldsymbol{e}}_1, \ldots, \hat{\boldsymbol{e}}_q]\begin{bmatrix} \sqrt{\hat{\lambda}_1} & & \\ & \ddots & \\ & & \sqrt{\hat{\lambda}_q} \end{bmatrix}.$$

In addition, the estimator of Σ_ϵ is given by

$$\hat{\Sigma}_\epsilon = diag[\hat{\Sigma} - \hat{b}\hat{b}'].$$

The mean-variance optimal portfolio estimator under the statistical factor model is obtained by

$$g(\hat{\theta}^{SF}) = g(\hat{\mu}^{SF}, \hat{\Sigma}^{SF})$$

where

$$\hat{\mu}^{SF} = \hat{\mu}, \quad \hat{\Sigma}^{SF} = \hat{b}\hat{b}' + \hat{\Sigma}_\epsilon.$$

3.2.5.2 *Dynamic factor models*

Features of the traditional factor model defined by (3.2.22) are

- The number of risk factors (p) is finite,
- The risk factors $(F_1(t), \ldots, F_q(t))$ are uncorrelated with each other (i.e., $Cov(F_i(t), F_j(t)) = 0$ for $i \neq j$),
- The risk factor ($\boldsymbol{F}(t)$) and the error term ($\epsilon(t)$) are also uncorrelated (i.e., $Cov(\boldsymbol{F}(t), \epsilon(t)) = \boldsymbol{0}$).

Consider the representation of the MA(q) model given by

$$X(t) = a + \sum_{j=0}^{q} A(j)U(t-j) = a + (A(1), \ldots, A(q)) \begin{pmatrix} U(t-1) \\ \vdots \\ U(t-q) \end{pmatrix} + A(0)U(t),$$

as the classical factor model under the above restrictions, it is possible by the setting $\boldsymbol{b} = (A(1), \ldots, A(q))$, $\boldsymbol{F}(t) = (U(t-1), \ldots, U(t-q))'$ and $\epsilon(t) = A(0)U(t)$ we can assume that $U(t)$'s are uncorrelated with respect t. On the other hand, in the case of the AR(p) model given by

$$X(t) = \sum_{i=1}^{p} A(i)X(t-i) + U(t) = (A(1), \ldots, A(p)) \begin{pmatrix} X(t-1) \\ \vdots \\ X(t-p) \end{pmatrix} + U(t),$$

we need to consider another representation because of the uncorrelation within the risk factors. If $\{X(t)\}$ is the stationary process, we can write

$$X(t) = \sum_{j=0}^{\infty} \tilde{A}(j)U(t-j) = (\tilde{A}(1), \tilde{A}(2)^2, \ldots) \begin{pmatrix} U(t-1) \\ U(t-2) \\ \vdots \end{pmatrix} + \tilde{A}(0)U(t)$$

(e.g., Remark 4 of Brockwell and Davis (2006)), which implies that by setting $\boldsymbol{b} = (\tilde{A}(1), \tilde{A}(2), \ldots)$, $\boldsymbol{F}(t) = (U(t-1), U(t-2), \ldots)'$ and $\epsilon(t) = \tilde{A}(0)U(t)$, the uncorrelated conditions are satisfied but the number of risk factors is infinite.

To overcome this problem, Sargent and Sims (1977) and John (1977) proposed the "dynamic factor model" which is given by

$$\boldsymbol{X}(t) = \sum_{j=0}^{\infty} A(j)\boldsymbol{Z}(t-j) + \boldsymbol{U}(t) := \sum_{j=0}^{\infty} A(j)L^j \boldsymbol{Z}(t) + \boldsymbol{U}(t) := \underline{A}(L)\boldsymbol{Z}(t) + \boldsymbol{U}(t) \tag{3.2.26}$$

where $\boldsymbol{X}(t) = (X_1(t), \ldots, X_m(t))'$ is the m-vector of observed dependent variables, $\boldsymbol{U}(t) = (U_1(t), \ldots, U_m(t))'$ is the m-vector of residuals, $\boldsymbol{Z}(t) = (Z_1(t), \ldots, Z_q(t))'$ is the q-vector of factors with $q < m$, $A(j) = (A_{ab}(j))$ is the $m \times q$ matrix and L is the lag operator.

Although the dynamic factor model includes time series models such as VAR or VARMA models, the cross-correlation among idiosyncratic components is not allowed (i.e., $Cov(U_i(t), U_j(t)) = 0$ if $i \neq j$ is assumed).

On the other hand, Chamberlain (1983) and Chamberlain and Rothschild (1983) showed that the assumption that the idiosyncratic components are uncorrelated is unnecessarily strong and can be relaxed; they introduced approximate factor model which imposes a weaker restriction on the distribution of the asset returns.

Forni et al. (2000) argue "Factor models are an interesting alternative in that they can provide a much more parsimonious parameterization. To address properly all the economic issues cited above, however, a factor model must have two characteristics. First, it must be dynamic, since business cycle questions are typically dynamic questions. Second, it must allow for cross-correlation among idiosyncratic components, since orthogonality is an unrealistic assumption for most applications."

It is easy to see that dynamic factor model satisfies the first requirement but the second one is not satisfied (i.e., dynamic factor model has orthogonal idiosyncratic components), and approximate factor satisfies the second one but the first one is not satisfied (i.e., approximate factor model' is static).

The generalized dynamic factor model introduced by Forni et al. (2000) provides both features.

Definition 3.2.10. *[generalized dynamic factor model (Forni et al. (2000))] Suppose that all the stochastic variables taken into consideration belong to the Hilbert space $L_2(\Omega, \mathcal{F}, P)$, where $(\Omega, \mathcal{F}, P)$ is a given probability space; thus all first and second moments are finite. We will study a double sequence*

$$\{x_{it},\ i \in \mathbb{N}, t \in \mathbb{Z}\},$$

where

$$x_{it} = b_{i1}(L)u_{1t} + b_{i2}(L)u_{2t} + \cdots + b_{iq}(L)u_{qt} + \xi_{it}.$$

Remark 3.2.11. *For any $m \in \mathbb{N}$, a generalized dynamic model $\{X_i(t),\ i \in \mathbb{N}, t \in \mathbb{Z}\}$ with*

$$X_i(t) = b_{i1}F_1(t) + \cdots + b_{iq}F_q(t) + \epsilon_i(t)$$

can be described as

$$\boldsymbol{X}(t) := \begin{pmatrix} X_1(t) \\ \vdots \\ X_m(t) \end{pmatrix} = \begin{pmatrix} b_{11} & \cdots & b_{1q} \\ \vdots & \ddots & \vdots \\ b_{m1} & \cdots & b_{mq} \end{pmatrix} \begin{pmatrix} F_1(t) \\ \vdots \\ F_q(t) \end{pmatrix} + \begin{pmatrix} \epsilon_1(t) \\ \vdots \\ \epsilon_m(t) \end{pmatrix} := \boldsymbol{b}\boldsymbol{F}(t) + \epsilon(t),$$

which implies that the traditional (static) factor model defined by (3.2.22) with $\boldsymbol{a} = \boldsymbol{0}$ is a partial realization of the generalized factor model.

Forni et al. (2000) imposed the following assumptions (Assumptions 3.2.12–3.2.15) for the generalized dynamic factor model.

Assumption 3.2.12. *(i) The p-dimensional vector process $\{(u_{1t}, \ldots, u_{pt})', t \in \mathbb{Z}\}$ is orthonormal white noise, i.e., $E(u_{jt}) = 0$, $V(u_{jt}) = 1$ for any j, t, and $Cov(u_{j_1t_1}, u_{j_2t_2}) = 0$ for any j_1, j_2 and t_1, t_2 with $j_1 \neq j_2$ or $t_1 \neq t_2$.*

(ii) $\boldsymbol{\xi} = \{(\xi_{it}, i \in \mathbb{N}, t \in \mathbb{Z})\}$ is a double sequence such that, firstly,

$$\boldsymbol{\xi}_n = \{(\xi_{1t}, \ldots, \xi_{mt})', t \in \mathbb{Z})\}$$

is a zero-mean stationary vector process for any n, and, secondly, $Cov(\xi_{j_1t_1}, u_{j_2t_2}) = 0$ for any j_1, j_2, t_1, t_2.

(iii) the filters $b_{ij}(L)$ are one-sided in L (i.e., we can write $b_{ij}(L)u_{it} = \sum_{\ell=0}^{\infty} b_{ij}(\ell)L^{\ell}u_{it} = \sum_{\ell=0}^{\infty} b_{ij}(\ell)u_{it-\ell}$) and their coefficients are square summable (i.e., $\sum_{\ell=0}^{\infty} b_{ij}(\ell)^2 < \infty$).

This assumption implies that the sub-process $\boldsymbol{x}_m = \{\boldsymbol{x}_{mt} = (x_{1t}, \ldots, x_{mt})', t \in \mathbb{Z}\}$ is zero-mean and stationary for any m.

Let $\Sigma_m(\theta)$ denote the spectral density matrix of the vector process $\boldsymbol{x}_{mt} = (x_{1t}, \ldots, x_{mt})'$, that is,

$$\Sigma_m(\theta) = \frac{1}{2\pi} \sum_{k=-\infty}^{\infty} \Gamma_{mk} e^{-ik\theta}, \quad \Gamma_{mk} = Cov(\boldsymbol{x}_{mt}, \boldsymbol{x}_{mt-k}).$$

Let $\sigma_{ij}(\theta)$ be the (i, j)-th component (note that for $m_1 \neq m_2$, the (i, j)-th component of $\Sigma_{m_1}(\theta)$ corresponds to that of $\Sigma_{m_2}(\theta)$).

Assumption 3.2.13. *For any $i \in \mathbb{N}$, there exists a real $c_i > 0$ such that $\sigma_{ii}(\theta)$ for any $\theta \in [-\pi, \pi]$.*

This assumption expresses the boundedness for the spectral density matrix, but it is not uniformly bounded for i.

Let $\Sigma_m^{\xi}(\theta)$ and $\Sigma_m^{\chi}(\theta)$ denote the spectral density matrices of the vector process $\boldsymbol{\xi}_{mt} = (x_{1t}, \ldots, x_{mt})'$ and $\boldsymbol{\chi}_{mt} = \boldsymbol{x}_{mt} - \boldsymbol{\xi}_{mt}$ (the variable $\chi_{it} = x_{it} - \xi_{it}$ is the ith component of $\boldsymbol{\chi}_{mt}$ and is called the common component), respectively. Let $\lambda_{mj}^{\xi} \equiv \lambda_{mj}^{\xi}(\theta)$ and $\lambda_{mj}^{\chi} \equiv \lambda_{mj}^{\chi}(\theta)$ denote the dynamic eigenvalues of $\Sigma_m^{\xi}(\theta)$ and $\Sigma_m^{\chi}(\theta)$, respectively.[10] The λ_{mj}^{ξ} and λ_{mj}^{χ} are called idiosyncratic (dynamic) eigenvalue and common (dynamic) eigenvalue, respectively.

Assumption 3.2.14. *The first idiosyncratic dynamic eigenvalue λ_{m1}^{ξ} is uniformly bounded. That is, there exists a real Λ such that $\lambda_{m1}^{\xi}(\theta) \leq \Lambda$ for any $\theta \in [-\pi, \pi]$ and $m \in \mathbb{N}$.*

This assumption expresses the uniform boundedness for all idiosyncratic dynamic eigenvalues.

Assumption 3.2.15. *The first q common dynamic eigenvalues diverge almost everywhere in $[-\pi, \pi]$. That is, $\lim_{m\to\infty} \lambda_{mj}^{\chi}(\theta) = \infty$ for $j \leq q$, a.e. in $[-\pi, \pi]$.*

This assumption guarantees a minimum amount of cross-correlation between the common components (see Forni et al. (2000)).

Suppose that we observed an $m \times T$ data matrix

$$\boldsymbol{X}_m^T = (\boldsymbol{x}_{m1}, \ldots, \boldsymbol{x}_{mT}). \tag{3.2.27}$$

Then, the sample mean vector can be easily obtained by

$$\hat{\boldsymbol{\mu}}_m = \frac{1}{T} \sum_{t=1}^{T} \boldsymbol{x}_{mt}. \tag{3.2.28}$$

Let $\hat{\Sigma}_m(\theta)$ be a consistent periodogram-smoothing or lag-window estimator of the $m \times m$ spectral density matrix $\Sigma_m(\theta)$ of $\boldsymbol{x}_{mt}$, given by

$$\hat{\Sigma}_m(\theta) = \sum_{k=-M}^{M} \hat{\Gamma}_{mk} \omega_k e^{-ik\theta}, \quad \hat{\Gamma}_{mk} = \frac{1}{T-k} \sum_{t=1}^{T-k} (\boldsymbol{x}_{mt} - \hat{\boldsymbol{\mu}}_m)(\boldsymbol{x}_{m,t+k} - \hat{\boldsymbol{\mu}}_m)' \tag{3.2.29}$$

where $\omega_k = 1 - [|k|/(M+1)]$ are the weights corresponding to the Bartlett lag window for a fixed integer $M < T$. Let $\hat{\lambda}_j(\theta)$ and $\hat{\boldsymbol{p}}_j(\theta)$ denote the jth largest eigenvalue and the corresponding row eigenvector of $\hat{\Sigma}_m(\theta)$, respectively. Forni et al. (2005) propose the estimators of the auto-covariance matrices of $\boldsymbol{\chi}_{mt}$ and $\boldsymbol{\xi}_{mt}$ (i.e., $\Gamma_{mk}^{\chi} = Cov(\boldsymbol{\chi}_{mt}, \boldsymbol{\chi}_{m,t+k}), \Gamma_{mk}^{\xi} = Cov(\boldsymbol{\xi}_{mt}, \boldsymbol{\xi}_{m,t+k})$) as

$$\hat{\Gamma}_{mk}^{\chi} = \int_{-\pi}^{\pi} e^{ik\theta} \hat{\Sigma}_m^{\chi}(\theta) d\theta, \quad \hat{\Gamma}_{mk}^{\xi} = \int_{-\pi}^{\pi} e^{ik\theta} \hat{\Sigma}_m^{\xi}(\theta) d\theta \tag{3.2.30}$$

[10] The dynamic eigenvalue (λ_{mj}) of Σ_m is defined as the function associating with any $\theta \in [-\pi, \pi]$ the real nonnegative jth eigenvalue of $\Sigma_m(\theta)$ in descending order of magnitude.

where $\hat{\Sigma}^{\mathcal{X}}_m(\theta)$ and $\hat{\Sigma}^{\xi}_m(\theta)$ are estimators of the spectral density matrices $\Sigma^{\mathcal{X}}_m(\theta)$ and $\Sigma^{\xi}_m(\theta)$, given by

$$\hat{\Sigma}^{\mathcal{X}}_m(\theta) = \hat{\lambda}_1(\theta)\hat{\boldsymbol{p}}^*_1(\theta)\hat{\boldsymbol{p}}_1(\theta) + \cdots + \hat{\lambda}_q(\theta)\hat{\boldsymbol{p}}^*_q(\theta)\hat{\boldsymbol{p}}_q(\theta), \tag{3.2.31}$$

$$\hat{\Sigma}^{\xi}_m(\theta) = \hat{\lambda}_{q+1}(\theta)\hat{\boldsymbol{p}}^*_{q+1}(\theta)\hat{\boldsymbol{p}}_{q+1}(\theta) + \cdots + \hat{\lambda}_m(\theta)\hat{\boldsymbol{p}}^*_m(\theta)\hat{\boldsymbol{p}}_m(\theta). \tag{3.2.32}$$

Note that

- for the actual calculation, the integration is approximated as the sum for the discretized $\theta_h = \pi h/M$ with $h = -M, \ldots, M$,
- q is the number of dynamic factors Forni et al. (2000) suggest how to choose q,
- $\hat{\boldsymbol{p}}^*$ is the adjoint (transposed, complex conjugate) of $\hat{\boldsymbol{p}}$.

Since the (true) covariance matrix Γ_{m0} can be decomposed as the covariance matrices of $\Gamma^{\mathcal{X}}_{m0} = V(\boldsymbol{\chi}_{mt})$ and $\Gamma^{\xi}_{m0} = V(\boldsymbol{\xi}_{mt})$ from Assumption 3.2.12, we can obtain the covariance matrix estimator by

$$\tilde{\Gamma}_{m0} = \hat{\Gamma}^{\mathcal{X}}_{m0} + \hat{\Gamma}^{\xi}_{m0}. \tag{3.2.33}$$

From (3.2.28) and (3.2.33), we obtain the mean-variance optimal portfolio estimator under the dynamic factor model as

$$g(\hat{\boldsymbol{\mu}}_m, \tilde{\Gamma}_{m0}).$$

3.2.6 *High-Dimensional Problems*

It is known that many studies demonstrate that inference for the parameters arising in portfolio optimization often fails. The recent literature shows that this phenomenon is mainly due to a high-dimensional asset universe. Typically, such a universe refers to the asymptotics that the sample size (n) and the sample dimension (m) both go to infinity (see Glombek (2012)). In this section, we discuss asymptotic properties for estimators of the global minimum portfolio (MP), the tangency portfolio (TP) and the naive portfolio (NP) under $m, n \to \infty$. When m goes to infinity, we cannot analyze the portfolio weight in the same way as Theorem 3.2.5. However, it is possible to discuss the asymptotics for the portfolio mean, the portfolio variance, the Sharpe ratio and so on. Especially, in this section, we concentrate on the asymptotics for the portfolio mean and variance under $m, n \to \infty$. In the literature, there are some papers that treat this problem, but these are separated by (I) the case of $m/n \to y \in (0, 1)$ and (II) the case of $m/n \to y \in (1, \infty)$.

(I) is the case that $m < n$ is satisfied, which implies that it is natural to assume that there exists $\hat{\Sigma}^{-1}$ due to $\text{rank}(\hat{\Sigma}) = m$ where $\hat{\Sigma}$ is a sample covariance matrix based on $\boldsymbol{X}_1, \ldots, \boldsymbol{X}_n$. For case (I), we mainly follow Glombek (2012)). As we mentioned earlier, it is well known that the traditional estimator of mean-variance optimal portfolios ($g(\boldsymbol{\mu}, \Sigma)$) is given based on the sample mean vector $\hat{\boldsymbol{\mu}}$ and the sample covariance matrix $\hat{\Sigma}$ (i.e., $g(\hat{\boldsymbol{\mu}}, \hat{\Sigma})$). Under large n and large m setting, the spectral properties of $\hat{\Sigma}$ have been studied by many authors (e.g., Marčenko and Pastur (1967)). In particular, the eigenvalues of the sample covariance matrix differ substantially from the eigenvalues of the true covariance matrix for both n and m being large. That is why the optimal portfolio estimation involving the sample covariance matrix should take this difficulty into account. Bai et al. (2009) and El Karoui (2010) discussed this problem under (n, m)-asymptotics which correnspond to $n, m \to \infty$ and $m/n \to y \in (0, 1)$.

(II) is the case that there never exist $\hat{\Sigma}^{-1}$ because $\text{rank}(\hat{\Sigma}) = n < m$ in that case, we cannot construct estimators for portfolio quantities by use of the usual plug-in method. To overcome this problem, Bodnar et al. (2014) introduced use of a generalized inverse of the sample covariance matrix as the estimator for the covariance matrix Σ and proposed an estimator of the global minimum variance portfolio and showed its asymptotic property. For the case of (II), we will show Bodnar et al. (2014) result.

3.2.6.1 The case of $m/n \to y \in (0, 1)$

Suppose the existence of m assets and the random returns are given by $\boldsymbol{X}_t = (X_1(t), \ldots, X_m(t))'$ with $E(\boldsymbol{X}_t) = \boldsymbol{\mu}$ and $Cov(\boldsymbol{X}_t, \boldsymbol{X}_t) = \Sigma$ (assumed nonsingular matrix). Glombek (2012) analyze the inference for three types of portfolios as follows:

- MP: The global minimum portfolio of which the weight is defined in (3.1.3).
- TP: The tangency portfolio of which the weight is defined in (3.1.7) with $\mu_f = 0$.
- NP: The naive portfolio of which the weight is equal to each other.

Let $\boldsymbol{w}_{MP}, \boldsymbol{w}_{TP}, \boldsymbol{w}_{NP}$ denote the portfolio weights for MP, TP, NP, respectively. Then, we can write

$$\boldsymbol{w}_{MP} = \frac{1}{\boldsymbol{e}'\Sigma^{-1}\boldsymbol{e}}\Sigma^{-1}\boldsymbol{e}, \quad \boldsymbol{w}_{TP} = \frac{1}{\boldsymbol{e}'\Sigma^{-1}\boldsymbol{\mu}}\Sigma^{-1}\boldsymbol{\mu}, \quad \boldsymbol{w}_{NP} = \frac{1}{m}\boldsymbol{e}. \tag{3.2.34}$$

Let $\mu_{MP}, \mu_{TP}, \mu_{NP}$ denote the portfolio mean for MP, TP, NP, respectively. Since a portfolio mean is given by $\mu(\boldsymbol{w}) = \boldsymbol{w}'\boldsymbol{\mu}$, we can write

$$\mu_{MP} = \frac{\boldsymbol{\mu}'\Sigma^{-1}\boldsymbol{e}}{\boldsymbol{e}'\Sigma^{-1}\boldsymbol{e}}, \quad \mu_{TP} = \frac{\boldsymbol{\mu}'\Sigma^{-1}\boldsymbol{\mu}}{\boldsymbol{e}'\Sigma^{-1}\boldsymbol{\mu}}, \quad \mu_{NP} = \frac{\boldsymbol{\mu}'\boldsymbol{e}}{m}. \tag{3.2.35}$$

Furthermore, let $\sigma^2_{MP}, \sigma^2_{TP}, \sigma^2_{NP}$ denote the portfolio variance for MP, TP, NP, respectively. Since a portfolio variance is given by $\eta^2(\boldsymbol{w}) = \boldsymbol{w}'\Sigma\boldsymbol{w}$, we can write

$$\sigma^2_{MP} = \frac{1}{\boldsymbol{e}'\Sigma^{-1}\boldsymbol{e}}, \quad \sigma^2_{TP} = \frac{\boldsymbol{\mu}'\Sigma^{-1}\boldsymbol{\mu}}{(\boldsymbol{e}'\Sigma^{-1}\boldsymbol{\mu})^2}, \quad \sigma^2_{NP} = \frac{\boldsymbol{e}'\Sigma\boldsymbol{e}}{m^2}. \tag{3.2.36}$$

We suppose that $\boldsymbol{X}_1, \ldots, \boldsymbol{X}_n$ are observed; then the sample mean $\hat{\boldsymbol{\mu}}$ and the sample covariance matrix $\hat{\Sigma}$ (assumed nonsingular matrix) are obtained by

$$\hat{\boldsymbol{\mu}} = \frac{1}{n}\sum_{t=1}^{n}\boldsymbol{X}_t, \quad \hat{\Sigma} = \frac{1}{n}\sum_{t=1}^{n}(\boldsymbol{X}_t - \hat{\boldsymbol{\mu}})(\boldsymbol{X}_t - \hat{\boldsymbol{\mu}})', \tag{3.2.37}$$

which implies that the estimators for the above quantities are obtained by

$$\hat{\mu}_{MP} = \frac{\hat{\boldsymbol{\mu}}'\hat{\Sigma}^{-1}\boldsymbol{e}}{\boldsymbol{e}'\hat{\Sigma}^{-1}\boldsymbol{e}}, \quad \hat{\mu}_{TP} = \frac{\hat{\boldsymbol{\mu}}'\hat{\Sigma}^{-1}\hat{\boldsymbol{\mu}}}{\boldsymbol{e}'\hat{\Sigma}^{-1}\hat{\boldsymbol{\mu}}}, \quad \hat{\mu}_{NP} = \frac{\hat{\boldsymbol{\mu}}'\boldsymbol{e}}{m}, \tag{3.2.38}$$

$$\hat{\sigma}^2_{MP} = \frac{1}{\boldsymbol{e}'\hat{\Sigma}^{-1}\boldsymbol{e}}, \quad \hat{\sigma}^2_{TP} = \frac{\hat{\boldsymbol{\mu}}'\hat{\Sigma}^{-1}\hat{\boldsymbol{\mu}}}{(\boldsymbol{e}'\hat{\Sigma}^{-1}\hat{\boldsymbol{\mu}})^2}, \quad \hat{\sigma}^2_{NP} = \frac{\boldsymbol{e}'\hat{\Sigma}\boldsymbol{e}}{m^2}. \tag{3.2.39}$$

In what follows, we discuss the asymptotic property for these estimators under the following assumptions:

Assumption 3.2.16. *(1)* $\boldsymbol{X}_t \overset{i.i.d.}{\sim} N_m(\boldsymbol{\mu}, \Sigma)$.

(2) The (n, m)-asymptotics are given by $n, m \to \infty$ and $m/n \to y \in (0, 1)$.

(3) The (n, m)-limit

$$\mu_P \equiv \mu_P^{(m)} \to \mu_P^{(\infty)} \in \mathbb{R} \setminus \{0\}$$

exists for $P \in \{\mathrm{MP}, \mathrm{TP}, \mathrm{NP}\}$.

(4) We have either

a) $\sigma_P^2 \to 0$ or

b) $\sigma_P^2 \to (\sigma_P^{(\infty)})^2 \in (0, \infty)$, under the (n, m)-asymptotics, where $P \in \{\mathrm{MP}, \mathrm{TP}, \mathrm{NP}\}$.

Since the Gaussian assumption (1) is often criticized, Bai et al. (2009) discussed the (n, m)-asymptotics without the Gaussianity. As we mentioned, the (n, m)-asymptotics (2) guarantees that the sample covariance matrix remains nonsingular in the high-dimensional framework. According to Glombek (2012), Assumption (3) shows that the components of the vector $\boldsymbol{\mu}$ are summable with respect to the signed measure which is induced by the weight vector $\boldsymbol{w}_P$ where $P \in \{\text{MP}, \text{TP}, \text{NP}\}$ (i.e., $\mu_P^{(\infty)} = \sum_{i=1}^{\infty} w_{i,P}\mu_i < \infty$ for $\boldsymbol{\mu} = (\mu_i)_{i=1}^{\infty}, \boldsymbol{w}_P = (w_{i,P})_{i=1}^{\infty}$). Assumption (4a) is satisfied if $\lambda_{\max}(\Sigma) = o(d)$ where $\lambda_{\max}(\Sigma)$ denotes the largest eigenvalue of Σ. Economically, this condition may be interpreted as perfect diversification as the variance of the portfolio returns can be made arbitrarily small by increasing the number of considered assets (see Glombek (2012)). Assumption (4b) is a more "natural" case than (4a) because the positivity of this limit requires $\lambda_{\max}(\Sigma) = O(d)$ under the (n, m)-asymptotics.

For $\hat{\mu}_{MP}, \hat{\mu}_{TP}, \hat{\mu}_{NP}$, the consistencies under the (n, m)-asymptotics are given as follows:

Theorem 3.2.17 (Theorems 5.8, 5.17 and 5.25 of Glombek (2012)). *Under (1)–(3) and (4a) (only for $\hat{\mu}_{TP}^2$) of Assumption 3.2.16, we have*

$$\hat{\mu}_{MP} \xrightarrow{a.s.} \mu_{MP}^{(\infty)}, \quad \hat{\mu}_{TP} \xrightarrow{a.s.} \mu_{TP}^{(\infty)}, \quad \hat{\mu}_{NP} \xrightarrow{a.s.} \mu_{MP}^{(\infty)}.$$

Under (1)–(3) and (4b) of Assumption 3.2.16, we have

$$\hat{\mu}_{TP} \xrightarrow{p} \left(1 + \frac{y}{(SR_{TP}^{(\infty)})^2}\right)\mu_{TP}^{(\infty)},$$

where $(SR_{TP}^{(\infty)})^2 (< \infty)$ is given by $\lim_{m\to\infty} SR_{TP}$ with $SR_{TP}^2 = 1/(\boldsymbol{\mu}'\Sigma^{-1}\boldsymbol{\mu})$.

Proof: See Theorems 5.8, 5.17 and 5.25 of Glombek (2012), Corollary 2 and Lemma 3.1 of Bai et al. (2009) and Theorem 4.6 of El Karoui (2010). □

Note that there exists a bias for the estimator $\hat{\mu}_{TP}$ under the (n, m)-asymptotics in the natural case of Assumption 3.2.16 (4b). The next theorem gives the asymptotic normality for $\hat{\mu}_{MP}, \hat{\mu}_{TP}, \hat{\mu}_{NP}$ under the (n, m)-asymptotics.

Theorem 3.2.18. *[Theorems 5.9, 5.18 and 5.26 of Glombek (2012)] Under (1)–(3) and (4) of Assumption 3.2.16, we have*

$$\sqrt{n}\frac{\hat{\mu}_{MP} - \mu_{MP}}{\sqrt{\hat{\sigma}_{MP}^2(1 + \hat{\boldsymbol{\mu}}'\hat{\Sigma}^{-1}\hat{\boldsymbol{\mu}}) - \hat{\mu}_{MP}^2}} \xrightarrow{d} N\left(0, \frac{1}{1-y}\right),$$

$$\sqrt{n}\left\{\hat{\mu}_{TP} - \left(1 + \frac{m/n}{SR_{TP}^2}\right)\mu_{TP}\right\} \xrightarrow{d} N\left(0, \frac{V_{TP}^{\mu}(y)}{1-y}\right),$$

$$\sqrt{n}\frac{\hat{\mu}_{NP} - \mu_{NP}}{\hat{\sigma}_{NP}} \xrightarrow{d} N(0, 1),$$

where

$$V_{TP}^{\mu}(y) = (\sigma_{TP}^{(\infty)})^2 \frac{\mu_{TP}^{(\infty)}}{\mu_{MP}^{(\infty)}}\left\{1 + (SR_{TP}^{(\infty)})^2\right\} - (\mu_{TP}^{(\infty)})^2$$
$$+ (\sigma_{TP}^{(\infty)})^2 \frac{2y\left\{1 + (SR_{TP}^{(\infty)})^2\right\}\left\{(SR_{TP}^{(\infty)})^2 - (SR_{MP}^{(\infty)})^2\right\} + y^2\left\{(SR_{TP}^{(\infty)})^2 + (SR_{MP}^{(\infty)})^2 + 1\right\}}{(SR_{MP}^{(\infty)})^2(SR_{TP}^{(\infty)})^2},$$

and $(SR_{MP}^{(\infty)})^2 (< \infty)$ is given by $\lim_{m\to\infty} SR_{MP}$ with $SR_{MP}^2 = (\boldsymbol{e}'\Sigma^{-1}\boldsymbol{\mu})^2/(\boldsymbol{e}'\Sigma^{-1}\boldsymbol{e})$.

Proof: See Theorems 5.9, 5.18 and 5.26 of Glombek (2012) (see also Frahm (2010)).

Based on this result, it is possible to consider a (one- or two-sided) test for the hypothesis

$$H_0 : \mu_P = \mu_0 \text{ v.s. } H_1 : \mu_P \neq \mu_0$$

or to construct the confidence interval for μ_P (as in the case of μ_{TP}, we need an appropriate estimate of the asymptotic variance by using the bootstrap method or the shrinkage estimation method).

Similarly to the case of the mean, the consistency for $\hat{\sigma}^2_{MP}, \hat{\sigma}^2_{TP}, \hat{\sigma}^2_{NP}$ under the (n, m)-asymptotics is given as follows:

Theorem 3.2.19. *[Theorems 5.5, 5.15 and 5.23 of Glombek (2012)] Under (1)–(3) and (4a) (only for $\hat{\sigma}^2_{TP}$) of Assumption 3.2.16, we have*

$$\frac{\hat{\sigma}^2_{MP}}{\sigma^2_{MP}} \xrightarrow{a.s.} 1 - y, \quad \frac{\hat{\sigma}^2_{TP}}{\sigma^2_{TP}} \xrightarrow{a.s.} 1 - y, \quad \frac{\hat{\sigma}^2_{NP}}{\sigma^2_{NP}} \xrightarrow{a.s.} 1.$$

Under (1)–(3) and (4b) of Assumption 3.2.16, we have

$$\frac{\hat{\sigma}^2_{TP}}{\sigma^2_{TP}} \xrightarrow{p} (1 - y)\left(1 + \frac{y}{(SR^{(\infty)}_{TP})^2}\right).$$

Proof: See Theorems 5.5, 5.15 and 5.23 of Glombek (2012), Corollary 2 and Lemma 3.1 of Bai et al. (2009) and Theorem 4.6 of El Karoui (2010). □

This result shows that $\hat{\sigma}^2_{MP}, \hat{\sigma}^2_{TP}$ are inconsistent estimators for $\sigma^2_{MP}, \sigma^2_{TP}$, and $\hat{\sigma}^2_{NP}$ is a consistent estimator for σ^2_{NP} under (n, m)-asymptotics. The next theorem gives the asymptotic normality for $\hat{\sigma}^2_{MP}, \hat{\sigma}^2_{TP}, \hat{\sigma}^2_{NP}$ under (n, m)-asymptotics.

Theorem 3.2.20. *[Theorems 5.6, 5.16 and 5.24 of Glombek (2012)] Under (1)–(3) and (4) of Assumption 3.2.16, we have*

$$\sqrt{n}\left(\frac{\hat{\sigma}^2_{MP}}{\sigma^2_{MP}} - \left(1 - \frac{m}{n}\right)\right) \xrightarrow{d} N(0, 2(1 - y)),$$

$$\sqrt{n}\left\{\frac{\hat{\sigma}^2_{TP}}{\sigma^2_{TP}} - \left(1 - \frac{m/n}{SR^2_{TP}}\right)\right\} \xrightarrow{d} N(0, 2(1 - y)V^{\sigma}_{TP}(y)),$$

$$\sqrt{n}\left(\frac{\hat{\sigma}^2_{NP}}{\sigma^2_{NP}} - 1\right) \xrightarrow{d} N(0, 2),$$

where

$$V^{\sigma}_{TP}(y) = \frac{(SR^{(\infty)}_{TP})^4 + 4y(SR^{(\infty)}_{TP})^2 + y}{(SR^{(\infty)}_{MP})^2(SR^{(\infty)}_{TP})^2} + \frac{\left\{(SR^{(\infty)}_{TP})^2 - (SR^{(\infty)}_{MP})^2\right\}\left\{(SR^{(\infty)}_{TP})^4 + 2(SR^{(\infty)}_{TP})^2 + 3y\right\} + 2y^2\left\{1 + (SR^{(\infty)}_{TP})^2 + (SR^{(\infty)}_{MP})^2\right\}}{(SR^{(\infty)}_{MP})^2(SR^{(\infty)}_{TP})^2}.$$

Proof: See Theorems 5.6, 5.16 and 5.24 of Glombek (2012).

Based on this result, it is possible to consider a (one- or two-sided) test for the hypothesis

$$H_0 : \sigma^2_P = \sigma^2_0 \text{ v.s. } H_1 : \sigma^2_P \neq \sigma^2_0$$

or to construct the confidence interval for σ^2_P (as in the case of σ^2_{TP}, we need an appropriate estimate of the asymptotic variance by using the bootstrap method or the shrinkage estimation method).

3.2.6.2 The case of $m/n \to y \in (1, \infty)$

Suppose the existence of m assets and the random returns are given by $\boldsymbol{X}_t = (X_1(t), \ldots, X_m(t))'$ with $E(\boldsymbol{X}_t) = \boldsymbol{\mu}$ and $Cov(\boldsymbol{X}_t, \boldsymbol{X}_t) = \Sigma$ (assumed nonsingular matrix). Bodnar et al. (2014) analyze the inference for the global minimum portfolio (MP). Again, the portfolio weights ($\boldsymbol{w}_{MP}$) and the (oracle) portfolio variance ($\tilde{\sigma}^2_{MP}$) for MP are given as follows:

$$\boldsymbol{w}_{MP} = \frac{1}{\boldsymbol{e}'\Sigma^{-1}\boldsymbol{e}}\Sigma^{-1}\boldsymbol{e}, \quad \tilde{\sigma}^2_{MP} = \boldsymbol{w}'_{MP}\Sigma\boldsymbol{w}_{MP} = \frac{1}{\boldsymbol{e}'\Sigma^{-1}\boldsymbol{e}}. \tag{3.2.40}$$

Suppose that $\boldsymbol{X} \equiv (\boldsymbol{X}_1, \ldots, \boldsymbol{X}_n)$ are observed. In the case of (II), the sample covariance matrix $\hat{\Sigma}$ is singular and its inverse does not exist because of $n < m$. Thus, Bodnar et al. (2014) use the following generalized inverse of $\hat{\Sigma}$.

$$\hat{\Sigma}^* = \Sigma^{-1/2}(\boldsymbol{X}\boldsymbol{X}')^+\Sigma^{-1/2}, \tag{3.2.41}$$

where '+' denotes the Moore–Penrose inverse. It can be shown that $\hat{\Sigma}^*$ is the generalized inverse satisfying $\hat{\Sigma}^*\hat{\Sigma}\hat{\Sigma}^* = \hat{\Sigma}^*$ and $\hat{\Sigma}\hat{\Sigma}^*\hat{\Sigma} = \hat{\Sigma}$. Obviously, in the case of $y < 1$, the generalized inverse $\hat{\Sigma}^*$ coincides with the usual inverse $\hat{\Sigma}^{-1}$. Note that $\hat{\Sigma}^*$ cannot be determined in practice since it depends on the unknown matrix Σ. Pluging in this $\hat{\Sigma}^*$ instead of $\hat{\Sigma}^{-1}$ to the true portfolio quantities, the following portfolio estimators are obtained.

$$\boldsymbol{w}^*_{MP} = \frac{1}{\boldsymbol{e}'\hat{\Sigma}^*\boldsymbol{e}}\hat{\Sigma}^*\boldsymbol{e}, \quad (\tilde{\sigma}^*_{MP})^2 = (\boldsymbol{w}^*_{MP})'\Sigma(\boldsymbol{w}^*_{MP}). \tag{3.2.42}$$

We impose the following assumptions:

Assumption 3.2.21. *(1) The elements of the matrix $\boldsymbol{X} = (\boldsymbol{X}_1, \ldots, \boldsymbol{X}_n)$ have uniformly bounded $4 + \epsilon$ moments for some $\epsilon > 0$.*

(2) The (n, m)-asymptotics are given by $n, m \to \infty$ and $m/n \to y \in (1, \infty)$.

This assumption includes the Gaussian assumption such as (1) of Assumption 3.2.16, so that this is rather relaxed one. The consistency for $(\tilde{\sigma}^*_{MP})^2$ under (n, m)-asymptotics is given as follows:

Theorem 3.2.22. *[Theorem 2.2 of Bodnar et al. (2014)] Under Assumption 3.2.21, we have*

$$(\tilde{\sigma}^*_{MP})^2 \overset{a.s.}{\to} \frac{c^2}{c-1}\sigma^2_{MP}.$$

Proof: See Theorem 2.2 of Bodnar et al. (2014).

As in the case of $y < 1$, $(\tilde{\sigma}^*_{MP})^2$ is an inconsistent estimator for σ^2_{MP} under (n, m)-asymptotics.

Furthermore, they propose the oracle optimal shrinkage estimator of the portfolio weight for MP:

$$\boldsymbol{w}^{**}_{MP} = \alpha^+\boldsymbol{w}^*_{MP} + (1 - \alpha^+)\boldsymbol{b}, \quad \text{with} \quad \boldsymbol{b}'\boldsymbol{e} = 1 \tag{3.2.43}$$

where

$$\alpha^+ = \frac{(\boldsymbol{b} - \boldsymbol{w}^*_{MP})'\Sigma\boldsymbol{b}}{(\boldsymbol{b} - \boldsymbol{w}^*_{MP})'\Sigma(\boldsymbol{b} - \boldsymbol{w}^*_{MP})},$$

and the bona fide optimal shrinkage estimator of the portfolio weight for MP:

$$\boldsymbol{w}^{***}_{MP} = \alpha^{++}\frac{\hat{\Sigma}^+\boldsymbol{e}}{\boldsymbol{e}'\hat{\Sigma}^+\boldsymbol{e}} + (1 - \alpha^{++})\boldsymbol{b}, \quad \text{with} \quad \boldsymbol{b}'\boldsymbol{e} = 1 \tag{3.2.44}$$

where

$$\alpha^{++} = \frac{(m/n-1)\hat{R}_b}{(m/n-1)^2 + m/n + (m/n-1)\hat{R}_b},$$
$$\hat{R}_b = \frac{m}{n}\left(\frac{m}{n} - 1\right)(b'\hat{\Sigma}b)(e'\Sigma^+ e) - 1,$$

and $\hat{\Sigma}^+$ is the Moore–Penrose pseudo-inverse of the sample covariance matrix $\hat{\Sigma}$.

They investigate the difference between the oracle and the bona fide optimal shrinkage estimators for the MP as well as between the oracle and the bona fide traditional estimators. As a result, the proposed estimator shows significant improvement and it turns out to be robust to deviations from normality.

3.2.7 *Bibliographic Notes*

Koenker (2005) provides the asymptotic property of the pessimistic portfolio via quantile regression models. Aït-Sahalia and Hansen (2009) discuss briefly the shrinkage estimation in portfolio problems. Gouriéroux and Jasiak (2001), Scherer and Martin (2005), Rachev et al. (2008) and Aït-Sahalia and Hansen (2009) introduce Bayesian methods in portfolio problems. Tsay (2010) gives a general method to analyze factor models. Hastie et al. (2009) introduce estimation methods for linear regression models. Forni et al. (2000, 2005) discuss the generalized dynamic factor model. Glombek (2012) discusses asymptotics for estimators of the various portfolio quantities under $m, n \to \infty$ and $m/n \to y \in (0, 1)$. Bodnar et al. (2014) discusses asymptotics for that under $m, n \to \infty$ and $m/n \to y \in (1, \infty)$.

3.2.8 *Appendix*

Proof of Theorem 3.2.3 Denote the (a, b) component of $\hat{\Sigma}$ and $\tilde{\Sigma}$ by $\hat{\Sigma}_{ab}$ and $\tilde{\Sigma}_{ab}$, respectively. Then

$$\begin{aligned}
\sqrt{n}\left(\hat{\Sigma}_{ab} - \tilde{\Sigma}_{ab}\right) &= \frac{1}{\sqrt{n}}\left\{\sum_{t=1}^{n}(X_a(t) - \hat{\mu}_a)(X_b(t) - \hat{\mu}_b) - \sum_{t=1}^{n}(X_b(t) - \mu_b)(X_b(t) - \mu_b)\right\} \\
&= -\sqrt{n}(\hat{\mu}_a - \mu_a)(\hat{\mu}_b - \mu_b) \\
&= o_p(1).
\end{aligned}$$

Therefore

$$\begin{aligned}
\sqrt{n}(\hat{\theta} - \theta) &= \sqrt{n}\begin{pmatrix} \hat{\boldsymbol{\mu}} - \boldsymbol{\mu} \\ vech\{\tilde{\Sigma} - \Sigma\} \end{pmatrix} + o_p(1) \\
&= \frac{1}{\sqrt{n}}\sum_{t=1}^{n}\begin{pmatrix} \{X_1(t) - \mu_1\} \\ \vdots \\ \{X_m(t) - \mu_m\} \\ \{X_1(t) - \mu_1\}\{X_1(t) - \mu_1\} - \Sigma_{11} \\ \vdots \\ \{X_m(t) - \mu_m\}\{X_m(t) - \mu_m\} - \Sigma_{mm} \end{pmatrix} + o_p(1) \\
&\equiv \frac{1}{\sqrt{n}}\sum_{t=1}^{n}\boldsymbol{Y}(t) + o_p(1) \quad (say),
\end{aligned}$$

where $\{\boldsymbol{Y}(t)\} = \{(Y_1(t), \ldots, Y_{m+r}(t))'; t \in \mathbb{N}\}$. In order to prove the theorem we use the following lemma which is due to Hosoya and Taniguchi (1982).

Lemma 3.2.23. *(Theorem 2.2 of Hosoya and Taniguchi (1982)) Let a zero-mean vector-valued second-order stationary process* $\{\boldsymbol{Y}(t)\}$ *be such that*

(A-1) for a positive constant ϵ,

$$\mathrm{Var}\{E(Y_a(t+\tau)|\mathcal{F}_t)\} = O(\tau^{-2-\epsilon}) \text{ uniformly in } a = 1, \ldots, m+r,$$

(A-2) for a positive constant δ,

$$E|E\{Y_{a_1}(t+\tau_1)Y_{a_2}(t+\tau_2)|\mathcal{F}_t\} - E\{Y_{a_1}(t+\tau_1)Y_{a_2}(t+\tau_2)\}| = O[\{\min(\tau_1,\tau_2)\}^{-1-\delta}]$$

uniformly in t, *for* $a_1, a_2 = 1, \ldots, m+r$ *and* τ_1, τ_2 *both greater than* 0,

(A-3) $\{\boldsymbol{Y}(t)\}$ *has the spectral density matrix* $f^{\boldsymbol{Y}}(\lambda) = \{f^{\boldsymbol{Y}}_{ab}(\lambda); a, b = 1, \ldots, m+r\}$, *whose elements are continuous at the origin and* $f^{\boldsymbol{Y}}(0)$ *is nondegenerate,*

then $\frac{1}{\sqrt{n}}\sum_{t=1}^{n}\boldsymbol{Y}(t) \xrightarrow{\mathcal{L}} N(\boldsymbol{0}, 2\pi f^{\boldsymbol{Y}}(0))$, *where* $\mathcal{F}_t$ *is the* σ*-field generated by* $\{\boldsymbol{Y}(l); l \le t\}$.

Now returning to the proof of Theorem 3.2.3 we check the above (A-1)–(A-3).

(A-1). If $a \le m$, then

$$\begin{aligned} Y_a(t+\tau) &= X_a(t+\tau) - \mu_a \\ &= \sum_{k=1}^{m}\sum_{j=0}^{\infty} A_{ak}(j)u_k(t+\tau-j) \\ &= \sum_{k=1}^{m}\sum_{j=0}^{\tau-1} A_{ak}(j)u_k(t+\tau-j) + \sum_{k=1}^{m}\sum_{j=\tau}^{\infty} A_{ak}(j)u_k(t+\tau-j) \\ &\equiv A_a(t,\tau) + O^{(4)}_{\mathcal{F}_t}(\tau^{-1-\delta}), \quad (say). \end{aligned}$$

By use of a fundamental formula between $E(\cdot)$ and $cum(\cdot)$ (e.g., Brillinger (2001b)), we obtain

$$E|\{O^{(4)}_{\mathcal{F}_t}(\tau^{-1-\delta})\}^4| = E\left\{\sum_{k=1}^{m}\sum_{j=\tau}^{\infty} A_{ak}(j)u_k(t+\tau-j)\right\}^4 = O\left\{(\tau^{-1-\delta})^4\right\}$$

by Assumption 3.2.1. Similarly we can show

$$E|\{O^{(4)}_{\mathcal{F}_t}(\tau^{-1-\delta})\}^2| = O\left\{(\tau^{-1-\delta})^2\right\}.$$

Hence Schwarz's inequality yields

$$E|\{O^{(4)}_{\mathcal{F}_t}(\tau^{-1-\delta})\}^3| = O\{(\tau^{-1-\delta})^3\}, \quad E|\{O^{(4)}_{\mathcal{F}_t}(\tau^{-1-\delta})\}| = O\{(\tau^{-1-\delta})\}.$$

From (ii) of Assumption 3.2.1 it follows that $\mathcal{F}_t$ is equal to the σ-field generated by $\{\boldsymbol{U}(l); l \le t\}$. Therefore, noting that $A_a(t,\tau)$ is independent of $\mathcal{F}_t$, we obtain

$$E(Y_a(t+\tau)|\mathcal{F}_t) = E\{A_a(t,\tau) + O^{(4)}_{\mathcal{F}_t}(\tau^{-1-\delta})|\mathcal{F}_t\} = O^{(4)}_{\mathcal{F}_t}(\tau^{-1-\delta}).$$

Hence, we have

$$\mathrm{Var}\{E(Y_a(t+\tau)|\mathcal{F}_t)\} = \mathrm{Var}\{O^{(4)}_{\mathcal{F}_t}(\tau^{-1-\delta})\} = O(\tau^{-2-2\delta}), \ (for\ a \le m).$$

If $a > m$, then

$$\begin{aligned} &Y_a(t+\tau) \\ &= A_{a_1}(t,\tau)A_{a_2}(t,\tau) + O^{(4)}_{\mathcal{F}_t}(\tau^{-1-\delta})\{A_{a_1}(t,\tau) + A_{a_2}(t,\tau)\} + \{O^{(4)}_{\mathcal{F}_t}(\tau^{-1-\delta})\}^2 - \Sigma_{a_1a_2}. \end{aligned}$$

Similarly we obtain

$$E(Y_a(t+\tau)|\mathcal{F}_t) = E(A_{a_1}(t,\tau)A_{a_2}(t,\tau)) + \{O^{(4)}_{\mathcal{F}_t}(\tau^{-1-\delta})\}^2 - \Sigma_{a_1a_2}$$

which implies

$$\mathrm{Var}\{E(Y_a(t+\tau)|\mathcal{F}_t)\} = E|\{O^{(4)}_{\mathcal{F}_t}(\tau^{-1-\delta})\}^2 - O(\tau^{-2-2\delta})|^2 = O(\tau^{-4-4\delta}).$$

Thus we have checked (A-1).

(A-2). For $a_1, a_2 \le m$, noting that

$$Y_{a_1}(t+\tau_1)Y_{a_2}(t+\tau_2) = A_{a_1}(t,\tau_1)A_{a_2}(t,\tau_2) + O^{(4)}_{\mathcal{F}_t}(\tau_1^{-1-\delta})A_{a_2}(t,\tau_2) + O^{(4)}_{\mathcal{F}_t}(\tau_2^{-1-\delta})A_{a_1}(t,\tau_1) + O^{(4)}_{\mathcal{F}_t}(\tau_1^{-1-\delta})O^{(4)}_{\mathcal{F}_t}(\tau_2^{-1-\delta}),$$

we can see

$$E\{Y_{a_1}(t+\tau_1)Y_{a_2}(t+\tau_2)|\mathcal{F}_t\} = E\{A_{a_1}(t,\tau_1)A_{a_2}(t,\tau_2)\} + O^{(4)}_{\mathcal{F}_t}(\tau_1^{-1-\delta})O^{(4)}_{\mathcal{F}_t}(\tau_2^{-1-\delta}).$$

On the other hand, since

$$E\{Y_{a_1}(t+\tau_1)Y_{a_2}(t+\tau_2)\} = E\{A_{a_1}(t,\tau_1)A_{a_2}(t,\tau_2)\} + O\{(\min(\tau_1,\tau_2))^{-1-\delta}\},$$

we have

$$E|E\{Y_{a_1}(t+\tau_1)Y_{a_2}(t+\tau_2)|\mathcal{F}_t\} - E\{Y_{a_1}(t+\tau_1)Y_{a_2}(t+\tau_2)\}| = O\{(\min(\tau_1,\tau_2))^{-1-\delta}\}$$

for $a_1, a_2 \le m$. For cases of (i) $a_1 \le m, a_2 > m$, and (ii) $a_1, a_2 > m$, repeating the same arguments as in the above we can check (A-2).

(A-3). It follows from Assumption 3.2.1.

By Lemma 3.2.23 we have

$$\frac{1}{\sqrt{n}}\sum_{t=1}^{n} \boldsymbol{Y}(t) \xrightarrow{\mathcal{L}} N(\mathbf{0}, 2\pi f^{\boldsymbol{Y}}(0)).$$

Next we evaluate the asymptotic variance of $\frac{1}{\sqrt{n}}\sum_{t=1}^{n}\boldsymbol{Y}(t)$ concretely. Write $\boldsymbol{Y}_1(t) = (Y_1(t),\dots,Y_m(t))'$ and $\boldsymbol{Y}_2(t) = (Y_{m+1}(t),\dots,Y_{m+r}(t))'$. Then the asymptotic variance of $\frac{1}{\sqrt{n}}\sum_{t=1}^{n}\boldsymbol{Y}_1(t)$ and $\frac{1}{\sqrt{n}}\sum_{t=1}^{n}\boldsymbol{Y}_2(t)$ are evaluated by Hannan (1970) and Hosoya and Taniguchi (1982), respectively. What we have to evaluate is the asymptotic covariance $Cov\{\frac{1}{\sqrt{n}}\sum_{t=1}^{n}\boldsymbol{Y}_1(t), \frac{1}{\sqrt{n}}\sum_{t=1}^{n}\boldsymbol{Y}_2(t)\}$, whose typical component for $a_1 \le m$ and $a_2 > m$ is

$$\begin{aligned} &Cov\left\{\frac{1}{\sqrt{n}}\sum_{t=1}^{n} Y_{a_1}(t), \frac{1}{\sqrt{n}}\sum_{t=1}^{m} Y_{a_2}(t)\right\} \\ &= \frac{1}{n}\sum_{t=1}^{n}\sum_{l=1}^{\infty}\sum_{b_1,b_2,b_3=1}^{m} A_{a_1,b_1}(l)A_{a_2',b_2}(l)A_{a_3',b_3}(l)c^{\boldsymbol{U}}_{b_1,b_2,b_3} \\ &\to \frac{1}{(2\pi)^2}\sum_{b_1,b_2,b_3=1}^{m} c^{\boldsymbol{U}}_{b_1,b2,b_3}\int\int_{-\pi}^{\pi} A_{a_1b_1}(\lambda_1+\lambda_2)A_{a_2'b_2}(-\lambda_1)A_{a_3'b_3}(-\lambda_2)d\lambda_1 d\lambda_2 \end{aligned}$$

when $\{\boldsymbol{U}(t)\}$ is a non-Gaussian process. In the case of the Gaussian process of $\{\boldsymbol{U}(t)\}$, it is easy to see that $c^{\boldsymbol{U}}_{b_1,b_2,b_3}$ and $c^{\boldsymbol{U}}_{b_1,b_2,b_3,b_4}$ vanish. Hence we have the desired result. □

3.3 Simulation Results

This section provides various numerical studies to investigate the accuracy of the portfolio estimators discussed in Chapter 3. We provide simulation studies when the return processes are a vector autoregressive of order 1 (`VAR(1)`), a vector moving average of order 1 (`VMA(1)`) and a constant conditional correlation of order (1, 1) (`CCC(1,1)`) models, respectively. Then, the estimated efficient frontiers are introduced. In addition, we discuss the inference of portfolio estimators with respect to sample size, distribution, target return and coefficient.

Suppose that $\{\boldsymbol{X}(t) = (X_1(t), \ldots, X_m(t))'; t \in \mathbb{Z}\}$ follows a VAR(1) and a VMA(1) model, respectively, defined by

$$\text{VAR(1)} \quad : \quad \boldsymbol{X}(t) = \boldsymbol{\mu} + \boldsymbol{U}(t) + \Phi\{\boldsymbol{X}(t-1) - \boldsymbol{\mu}\} \tag{3.3.1}$$

$$\text{VMA(1)} \quad : \quad \boldsymbol{X}(t) = \boldsymbol{\mu} + \boldsymbol{U}(t) + \Theta\boldsymbol{U}(t-1) \tag{3.3.2}$$

where $\boldsymbol{\mu}$ is a constant m-vector; $\{\boldsymbol{U}(t)\}$ follows an i.i.d. multivariate normal distribution $N_m(\boldsymbol{0}, I_m)$ or a standardized multivariate Student-t distribution with 5 degrees of freedom satisfied with $E\{\boldsymbol{U}(t)\} = \boldsymbol{0}$ and $V\{\boldsymbol{U}(t)\} = I_m$; Φ is an $m \times m$ autoregressive coefficient matrix and Θ is an $m \times m$ moving average coefficient matrix. Then, under regularity conditions, the mean vector and the covariance matrix can be written as

$$\text{VAR(1)} \quad : \quad E\{\boldsymbol{X}(t)\} = \boldsymbol{\mu}, \quad V\{\boldsymbol{X}(t)\} = (I_m - \Phi\Phi')^{-1} \tag{3.3.3}$$

$$\text{VMA(1)} \quad : \quad E\{\boldsymbol{X}(t)\} = \boldsymbol{\mu}, \quad V\{\boldsymbol{X}(t)\} = I_m + \Theta\Theta'. \tag{3.3.4}$$

On the other hand, suppose that $\{\boldsymbol{X}(t) = (X_1(t), \ldots, X_m(t))'; t \in \mathbb{N}\}$ follows a CCC(1,1) model, defined by

$$\text{CCC(1,1)} \quad : \quad \boldsymbol{X}(t) = \boldsymbol{\mu} + D(t)\epsilon(t), \quad D(t) = diag\left(\sqrt{h_1(t)}, \ldots, \sqrt{h_m(t)}\right) \tag{3.3.5}$$

where $\boldsymbol{\mu}$ is a constant m-vector; $\{\epsilon(t) = (\epsilon_1(t), \ldots, \epsilon_m(t))'\}$ follows an i.i.d. multivariate normal distribution $N_m(\boldsymbol{0}, R)$ or a standardized multivariate Student-t distribution with 5 degrees of freedom satisfied with $E\{\epsilon(t)\} = \boldsymbol{0}$ and $V\{\epsilon(t)\} = R$; the diagonal elements of R equal 1 and the off-diagonal elements equal R_{ij}; the vector GARCH equation for $\boldsymbol{h}(t) = (h_1(t), \ldots, h_m(t))'$ satisfies

$$\boldsymbol{h}(t) = \begin{pmatrix} a_1 \\ \vdots \\ a_m \end{pmatrix} + \begin{pmatrix} A_1 & & 0 \\ & \ddots & \\ 0 & & A_m \end{pmatrix} \begin{pmatrix} h_1(t-1)\epsilon_1^2(t-1) \\ \vdots \\ h_m(t-1)\epsilon_m^2(t-1) \end{pmatrix} + \begin{pmatrix} B_1 & & 0 \\ & \ddots & \\ 0 & & B_m \end{pmatrix} \begin{pmatrix} h_1(t-1) \\ \vdots \\ h_m(t-1) \end{pmatrix}$$
$$:= \boldsymbol{a} + A\boldsymbol{r}^2(t-1) + B\boldsymbol{h}(t-1) \quad \text{(say)}.$$

Then, under regularity conditions, the mean vector and the covariance matrix can be written as

$$\text{CCC(1,1)} \quad : \quad E\{\boldsymbol{X}(t)\} = \boldsymbol{\mu}, \quad V\{\boldsymbol{X}(t)\} = \Omega(\boldsymbol{a}, A, B, R) = (\omega_{ij})_{i,j=1,\ldots,m} \tag{3.3.6}$$

where

$$\omega_{ij} = \begin{cases} \dfrac{a_i}{1 - A_i - B_i} & i = j \\ \dfrac{R_{ij}\sqrt{a_i a_j}}{\sqrt{(1 - A_i - B_i)(1 - A_j - B_j)}} & i \neq j \end{cases}.$$

3.3.1 Quasi-Maximum Likelihood Estimator

Given $\{\boldsymbol{X}(1), \ldots, \boldsymbol{X}(n); n \geq 2\}$, the mean vector and the covariance matrix estimators based on the quasi-maximum likelihood estimators are defined by

$$\text{VAR(1)} \quad : \quad \hat{\boldsymbol{\mu}}_{QML} = \hat{\boldsymbol{\mu}}, \quad \hat{\Sigma}_{QML} = (I_m - \hat{\Phi}\hat{\Phi}')^{-1} \tag{3.3.7}$$

$$\text{VMA(1)} \quad : \quad \hat{\boldsymbol{\mu}}_{QML} = \hat{\boldsymbol{\mu}}, \quad \hat{\Sigma}_{QML} = I_m + \hat{\Theta}\hat{\Theta}' \tag{3.3.8}$$

$$\text{CCC(1,1)} \quad : \quad \hat{\boldsymbol{\mu}}_{QML} = \hat{\boldsymbol{\mu}}, \quad \hat{\Sigma}_{QML} = \Omega(\hat{\boldsymbol{a}}, \hat{A}, \hat{B}, \hat{R}) \tag{3.3.9}$$

where $\hat{\boldsymbol{\mu}} = \frac{1}{n}\sum_{i=1}^{n} \boldsymbol{X}(t)$ and $\hat{\Phi}, \hat{\Theta}, \hat{\boldsymbol{a}}, \hat{A}, \hat{B}, \hat{R}$ are "(Gaussian) quasi-maximum (log)likelihood estimator (QMLE or GQMLE)" defined by $\hat{\eta} = \arg\max_{\eta \in E} L_n(\eta)$. Here the quasi-likelihood functions (QMLE) for the VAR model and the VMA model are given as follows:

$$\text{VAR(1)} \quad : \quad L_n^{(\text{var})}(\Phi) = \prod_{t=2}^{n} p\left\{\hat{\boldsymbol{Y}}(t) - \Phi\hat{\boldsymbol{Y}}(t-1)\right\} \tag{3.3.10}$$

$$\texttt{VMA(1)} \quad : \quad L_n^{(\texttt{vma})}(\Theta) = \prod_{t=1}^{n} p\left\{\hat{\boldsymbol{Y}}(t) + \sum_{j=1}^{t-1}(-1)^j\Theta^j\hat{\boldsymbol{Y}}(t-j)\right\} \tag{3.3.11}$$

where $p\{\cdot\}$ is the probability density of $\boldsymbol{U}(t)$ and $\hat{\boldsymbol{Y}}(t) = \boldsymbol{X}(t) - \hat{\boldsymbol{\mu}}$. The Gaussian quasi-maximum likelihood estimators (GQMLE) for CCC model are given as follows:

$$(\hat{\boldsymbol{a}}, \hat{A}, \hat{B}) = \arg\max_{(\boldsymbol{a},A,B)} \ell_V(\boldsymbol{a}, A, B), \quad \hat{R} = \arg\max_{R} \ell_C(\hat{\boldsymbol{a}}, \hat{A}, \hat{B}, R) \tag{3.3.12}$$

where $\ell_V^{(\texttt{ccc})}$ and $\ell_C^{(\texttt{ccc})}$ are a volatility part and a correlation part of the Gaussian log-likelihood defined by

$$\texttt{CCC(1,1)} \quad : \quad \ell_V^{(\texttt{ccc})}(\boldsymbol{a}, A, B) = -\frac{1}{2}\sum_{t=1}^{n}\left\{m\log(2\pi) + \log|\hat{D}_t|^2 + \hat{\boldsymbol{Y}}(t)'\hat{D}(t)^{-2}\hat{\boldsymbol{Y}}_t\right\} \tag{3.3.13}$$

$$\ell_C^{(\texttt{ccc})}(\boldsymbol{a}, A, B, R) = -\frac{1}{2}\sum_{t=1}^{n}\left\{\log|R| + \hat{\epsilon}(t)'R^{-1}\hat{\epsilon}(t)\right\} \tag{3.3.14}$$

where $\hat{\boldsymbol{Y}}(t) = \boldsymbol{X}(t) - \hat{\boldsymbol{\mu}}(t)$, $\hat{D}(t) = diag\left(\sqrt{\hat{h}_1(t)}, \ldots, \sqrt{\hat{h}_m(t)}\right)$, $\hat{\epsilon}(t) = \hat{D}(t)^{-1}\{\boldsymbol{X}(t) - \hat{\boldsymbol{\mu}}\}$ and

$$\hat{h}_i(t) = \begin{cases} a_i & t = 0 \\ a_i + A_i\{X_i(t-1) - \hat{\mu}_i\}^2 + B_i\hat{h}_i(t-1) & t = 1, \ldots, n \end{cases}.$$

Tables 3.3–3.5 provide $\hat{\boldsymbol{\mu}}$, $\hat{\Sigma}_{QML}$(or $\hat{\Sigma}_{GQML}$) and the (G)QMLEs for each model in the case of $m = 2$ and $n = 10, 100, 1000$. It is easy to see that in all cases the estimation accuracy tends to be better as the sample size increases.

Table 3.3: Estimated mean vector ($\hat{\boldsymbol{\mu}}$), covariance matrix ($\hat{\Sigma}_{QML}$) and VAR coefficient ($\hat{\Phi}$) (with the standard errors (SE)) for the VAR(1) model

	$\boldsymbol{\mu}$		Σ			Φ			
	μ_1	μ_2	σ_{11}	σ_{12}	σ_{22}	Φ_{11}	Φ_{12}	Φ_{21}	Φ_{22}
True	1.000	0.500	1.044	-0.015	1.012	0.200	-0.050	-0.050	0.100
				normal distribution					
$n = 10$	0.833	0.227	1.086	-0.049	1.768	0.235	0.274	0.151	-0.598
	(0.029)	(0.030)	(13.75)	(6.352)	(6.352)	(0.354)	(0.451)	(0.461)	(0.356)
$n = 100$	0.858	0.546	1.014	0.014	1.027	0.108	0.153	-0.052	0.049
	(0.033)	(0.031)	(0.051)	(0.030)	(0.030)	(0.100)	(0.103)	(0.102)	(0.103)
$n = 1000$	0.969	0.488	1.051	-0.014	1.006	0.215	-0.048	-0.048	0.065
	(0.031)	(0.031)	(0.014)	(0.007)	(0.007)	(0.003)	(0.029)	(0.032)	(0.031)
				t distribution with 5 degrees of freedom					
$n = 10$	1.223	0.552	-8.857	-5.238	-1.737	-0.460	-0.415	0.802	0.303
	(0.062)	(0.059)	(14.13)	(19.76)	(19.76)	(0.344)	(0.470)	(0.478)	(0.343)
$n = 100$	1.117	0.470	1.031	0.004	1.000	0.173	0.023	0.016	0.005
	(0.043)	(0.032)	(0.045)	(0.027)	(0.027)	(0.090)	(0.095)	(0.094)	(0.089)
$n = 1000$	0.913	0.507	1.047	-0.016	1.017	0.210	-0.065	-0.013	0.112
	(0.032)	(0.032)	(0.012)	(0.006)	(0.006)	(0.027)	(0.027)	(0.027)	(0.027)

Table 3.4: Estimated mean vector ($\hat{\mu}$), covariance matrix ($\hat{\Sigma}_{QML}$) and VMA coefficient ($\hat{\Theta}$) (with the standard errors (SE)) for the VMA(1) model

	μ		Σ			Θ			
	μ_1	μ_2	σ_{11}	σ_{12}	σ_{22}	Θ_{11}	Θ_{12}	Θ_{21}	Θ_{22}
True	1.000	0.500	1.012	-0.015	1.042	0.200	-0.050	-0.050	0.100
				normal distribution					
$n = 10$	1.285	0.818	4.643	-6.280	11.961	0.457	-0.432	-1.853	3.282
	(0.029)	(0.022)	(6.870)	(5.121)	(5.121)	(1.620)	(1.581)	(1.410)	(1.699)
$n = 100$	1.043	0.663	1.018	0.023	1.032	0.116	0.170	-0.071	-0.056
	(0.032)	(0.030)	(0.037)	(0.029)	(0.029)	(0.110)	(0.107)	(0.108)	(0.108)
$n = 1000$	1.037	0.444	1.005	-0.013	1.066	0.065	-0.100	-0.028	0.238
	(0.030)	(0.032)	(0.007)	(0.007)	(0.007)	(0.032)	(0.031)	(0.032)	(0.030)
				t distribution with 5 degrees of freedom					
$n = 10$	0.469	1.209	1.404	-0.106	1.033	0.496	-0.176	0.397	-0.048
	(0.099)	(0.038)	(7.853)	(5.057)	(5.057)	(1.641)	(1.453)	(1.486)	(1.640)
$n = 100$	1.019	0.362	1.010	-0.007	1.023	0.101	-0.047	-0.019	0.145
	(0.037)	(0.032)	(0.027)	(0.026)	(0.026)	(0.100)	(0.098)	(0.096)	(0.102)
$n = 1000$	0.982	0.429	1.008	-0.010	1.023	0.081	-0.060	-0.038	0.142
	(0.033)	(0.031)	(0.006)	(0.006)	(0.006)	(0.027)	(0.027)	(0.028)	(0.028)

Table 3.5: Estimated mean vector ($\hat{\mu}$), covariance matrix ($\hat{\Sigma}_{GQML}$) and CCC coefficient ($\hat{a}, \hat{A}, \hat{B}, \hat{R}$) (with the standard errors (SE)) for the CCC(1,0) model

	μ		Σ			$\hat{a}$		$\hat{A}$		$\hat{B}$		$\hat{R}$
	μ_1	μ_2	σ_{11}	σ_{12}	σ_{22}	a_1	a_2	A_1	A_2	B_1	B_2	R_{21}
True	1.000	0.500	0.375	0.075	0.166	0.300	0.150	0.200	0.100	0.000	0.000	0.300
						normal distribution						
$n = 10$	1.258	0.445	0.000	0.000	0.000	0.000	0.000	0.000	0.000	1.050	0.904	2.050
	(0.31)	(0.17)	(1.85)	(0.57)	(1.00)	(0.16)	(0.06)	(0.42)	(0.35)	(0.42)	(0.42)	(0.70)
$n = 100$	0.969	0.492	0.442	0.089	0.214	0.305	0.013	0.309	0.116	0.000	0.830	0.289
	(0.25)	(0.13)	(0.11)	(0.03)	(0.05)	(0.10)	(0.05)	(0.13)	(0.10)	(0.21)	(0.23)	(0.11)
$n = 1000$	0.967	0.491	0.372	0.069	0.170	0.298	0.156	0.197	0.085	0.000	0.000	0.275
	(0.24)	(0.12)	(0.09)	(0.03)	(0.04)	(0.07)	(0.04)	(0.06)	(0.04)	(0.09)	(0.19)	(0.15)
						t distribution with 5 degrees of freedom						
$n = 10$	0.901	0.753	0.117	0.102	0.302	0.116	0.222	0.000	0.265	0.004	0.000	0.545
	(0.35)	(0.20)	(18.5)	(0.86)	(2.08)	(0.46)	(0.14)	(0.50)	(0.36)	(0.41)	(0.42)	(0.61)
$n = 100$	1.062	0.435	0.819	0.183	0.417	0.458	0.382	0.439	0.083	0.000	0.000	0.314
	(0.26)	(0.13)	(6.80)	(0.48)	(1.60)	(0.20)	(0.09)	(0.27)	(0.23)	(0.19)	(0.22)	(0.11)
$n = 1000$	1.006	0.515	0.743	0.151	0.295	0.462	0.143	0.378	0.282	0.000	0.231	0.323
	(0.25)	(0.12)	(1.36)	(0.06)	(0.70)	(0.13)	(0.07)	(0.13)	(0.10)	(0.08)	(0.16)	(0.08)

3.3.2 *Efficient Frontier*

Let $\boldsymbol{X}(t) = (X_1(t), \ldots, X_m(t))'$ be the random returns on m assets at time t with $E\{\boldsymbol{X}(t)\} = \boldsymbol{\mu}$ and $V\{\boldsymbol{X}(t)\} = \Sigma$. Let $\boldsymbol{w} = (w_1, \ldots, w_m)'$ be the vector of portfolio weights. Then, the expected portfolio return is given by $\mu_P := \boldsymbol{w}'\boldsymbol{\mu}$, while its variance is $\sigma_P^2 := \boldsymbol{w}'\Sigma\boldsymbol{w}$. As already discussed in (3.1.2), the optimal portfolio weight for given expected return μ_P is expressed as

$$\boldsymbol{w}_0 = \frac{\mu_P\alpha - \beta}{\delta}\Sigma^{-1}\boldsymbol{\mu} + \frac{\gamma - \mu_P\beta}{\delta}\Sigma^{-1}\boldsymbol{e} \tag{3.3.15}$$

where $\alpha = e'\Sigma^{-1}e$, $\beta = \mu'\Sigma^{-1}e$, $\gamma = \mu'\Sigma^{-1}\mu$ and $\delta = \alpha\gamma - \beta^2$. By using this weight, the variance is expressed as

$$\sigma_P^2 = \frac{\mu_P^2\alpha - 2\mu_P\beta + \gamma}{\delta}. \tag{3.3.16}$$

Rewriting (3.3.16), we obtain

$$(\mu_P - \mu_{GMV})^2 = s(\sigma_P^2 - \sigma_{GMV}^2) \tag{3.3.17}$$

where

$$\mu_{GMV} = \frac{\beta}{\alpha}, \quad \sigma_{GMV}^2 = \frac{1}{\alpha} \quad and \quad s = \frac{\delta}{\alpha}.$$

Here μ_{GMV} and σ_{GMV}^2 are the expected return and the variance of the global minimum variance (GMV) portfolio (i.e., the portfolio with the smallest risk). The quantity s is called the slope coefficient of the efficient frontier.

By using the estimators of μ and Σ defined in (3.3.3)–(3.3.6) (or the sample covariance matrix), the estimated efficient frontier is given by

$$(\mu_P - \hat{\mu}_{GMV})^2 = \hat{s}(\sigma_P^2 - \hat{\sigma}_{GMV}^2) \tag{3.3.18}$$

where

$$\hat{\mu}_{GMV} = \frac{\hat{\beta}}{\hat{\alpha}}, \quad \hat{\sigma}_{GMV}^2 = \frac{1}{\hat{\alpha}} \quad and \quad \hat{s} = \frac{\hat{\delta}}{\hat{\alpha}}$$

with $\hat{\alpha} = e'\hat{\Sigma}^{-1}e$, $\hat{\beta} = \hat{\mu}'\hat{\Sigma}^{-1}e$, $\hat{\gamma} = \hat{\mu}'\hat{\Sigma}^{-1}\hat{\mu}$ and $\hat{\delta} = \hat{\alpha}\hat{\gamma} - \hat{\beta}^2$.

Figure 3.8 shows the true and the estimated efficient frontiers for each model. Here, the solid lines show the true efficient frontiers, ○ ($n = 10$), △ ($n = 100$) and + ($n = 1000$) plots show the estimated efficient frontiers. You can see that the estimated efficient frontier tends to come closer to the true efficient frontier as the sample size increases, because the slope coefficient ($\hat{s}$), the estimated mean ($\hat{\mu}_{GMV}$) and the estimated variance ($\hat{\sigma}_{GMV}^2$) of the GMV portfolio converge to the true values.

3.3.3 Difference between the True Point and the Estimated Point

Once μ_P is fixed in (3.3.16) or (3.3.17), a point (σ_P, μ_P) is determined on the σ-μ plane. Given $\hat{\mu}$, $\hat{\Sigma}$ and μ_P, the optimal portfolio weight estimator is obtained by

$$\hat{w}_{\mu_P} = \frac{\mu_P\hat{\alpha} - \hat{\beta}}{\hat{\delta}}\hat{\Sigma}^{-1}\hat{\mu} + \frac{\hat{\gamma} - \mu_P\hat{\beta}}{\hat{\delta}}\hat{\Sigma}^{-1}e \tag{3.3.19}$$

with $\hat{\alpha} = e'\hat{\Sigma}^{-1}e$, $\hat{\beta} = \hat{\mu}'\hat{\Sigma}^{-1}e$, $\hat{\gamma} = \hat{\mu}'\hat{\Sigma}^{-1}\hat{\mu}$ and $\hat{\delta} = \hat{\alpha}\hat{\gamma} - \hat{\beta}^2$. Then, in the same way as the true version, the estimated point $(\hat{\sigma}(\mu_P), \hat{\mu}(\mu_P))$ is determined on the σ-μ plane. Here $\hat{\mu}(\mu_P)$ and $\hat{\sigma}^2(\mu_P)$ are the optimal portfolio's conditional mean and variance, i.e.,

$$\hat{\mu}(\mu_P) := E\{\hat{w}'_{\mu_P}X(t)|\hat{w}_{\mu_P}\} = \hat{w}'_{\mu_P}\mu \tag{3.3.20}$$

$$\hat{\sigma}^2(\mu_P) := V\{\hat{w}'_{\mu_P}X(t)|\hat{w}_{\mu_P}\} = \hat{w}'_{\mu_P}\Sigma\hat{w}_{\mu_P}. \tag{3.3.21}$$

Figure 3.9 shows the degree of dispersion of the point $(\hat{\sigma}(\mu_P), \hat{\mu}(\mu_P))$ in the case of $\mu_P = 0.85$, $n = 10, 30, 100$ and 100 times generation of the VAR(1) model. Note that all estimated points are the right hand side of the (true) efficient frontiers. It can be seen that each point tends to come closer on the true efficient frontier as the sample size increases. In addition, we show the histogram of the following quantities in Figure 3.10:

$$\frac{\hat{\mu}(\mu_P)}{\mu_P}, \quad \frac{\hat{\sigma}(\mu_P)}{\sigma_P}, \quad \hat{d}(\mu_P) := \{\hat{\mu}(\mu_P) - \mu_P\}^2 + \hat{\sigma}^2(\mu_P) - \sigma_P^2.$$

For each histogram, the shape becomes sharp as the sample size increases. Regarding the ratio of σ_P, there exist points with $\frac{\hat{\sigma}(\mu_P)}{\sigma_P} < 1$ (i.e., $\hat{\sigma}(\mu_P) < \sigma_P$). These points must satisfy $\hat{\mu}(\mu_P) < \mu_P$ because all estimated points are the right hand side of the (true) efficient frontier.

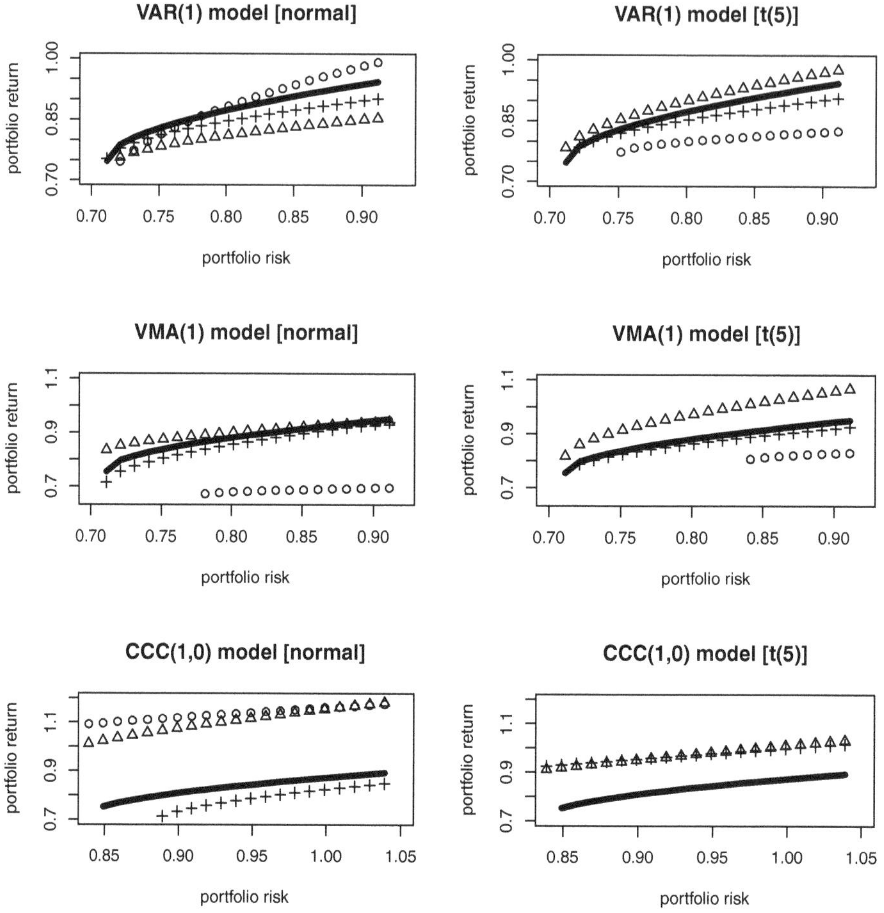

Figure 3.8: True and estimated efficient frontiers for each model. The solid lines show the true efficient frontiers, ○ (n = 10), △ (n = 100) and + (n = 1000) plots show the estimated efficient frontiers.

3.3.4 *Inference of* μ_P

Here we discuss the inference of estimation error when μ_P set down some values. In order to investigate the inference of estimation error, we introduce the following:

$$M_\mu := E\left\{\hat{\mu}(\mu_P) - \mu_P\right\} \tag{3.3.22}$$

$$M_\sigma := E\left\{\hat{\sigma}(\mu_P) - \sigma(\mu_P)\right\} \tag{3.3.23}$$

$$M_{(\mu,\sigma)} := E\left[\{\hat{\mu}(\mu_P) - \mu_P\}^2 + \left\{\hat{\sigma}^2(\mu_P) - \sigma^2(\mu_P)\right\}\right] = E\left\{\hat{d}(\mu_P)\right\} \tag{3.3.24}$$

Table 3.6 shows M_μ, M_σ and $M_{(\mu,\sigma)}$ in the case of $\mu_P = 0.748$ $(= \mu_{GMV})$, 0.850, 1.000, 1.500, 3.000, $n = 10, 30, 100$ and 100 times generation of the VAR(1) model. Note that we calculate the average of each quantity in place of the expectation. When μ_P is large, the estimation error is extremely large even if sample size increases.

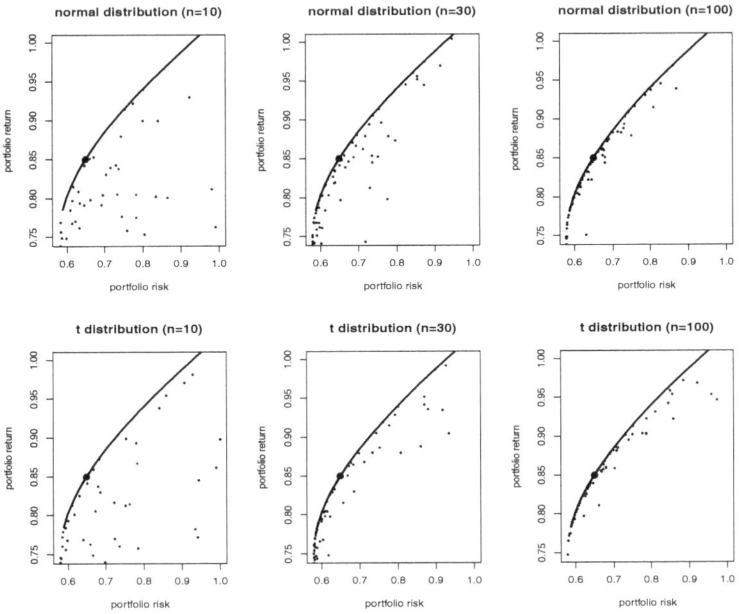

Figure 3.9: Results from 100 times generation of the VAR(1) model in the case of $n = 10, 30, 100$. For each generation, the optimal portfolio with a mean return of 0.85 is estimated. The estimated points are plotted as a small dot. The large dot is the true point and the solid line is the (true) efficient frontier.

Table 3.6: Inference of estimation error with respect to μ_P

	$n = 10$			$n = 30$			$n = 100$		
μ_P	M_μ	M_σ	$M_{(\mu,\sigma)}$	M_μ	M_σ	$M_{(\mu,\sigma)}$	M_μ	M_σ	$M_{(\mu,\sigma)}$
normal distribution									
0.748	0.017	0.892	4.812	-0.003	0.064	0.023	0.000	0.021	0.004
0.850	-0.020	0.965	6.313	-0.033	0.079	0.100	-0.003	0.035	0.036
1.000	-0.075	1.133	11.032	-0.078	0.088	0.401	-0.008	0.058	0.227
1.500	-0.261	1.971	48.987	-0.226	0.119	3.019	-0.024	0.138	2.106
3.000	-0.817	5.363	363.151	-0.670	0.265	25.516	-0.073	0.397	18.928
t distribution with 5 degrees of freedom									
0.748	0.000	0.708	5.105	-0.001	0.091	0.043	0.010	0.036	0.012
0.850	-0.092	0.701	4.491	-0.018	0.097	0.098	0.015	0.078	0.078
1.000	-0.229	0.758	7.307	-0.043	0.116	0.435	0.024	0.139	0.401
1.500	-0.684	1.456	48.368	-0.127	0.248	3.690	0.053	0.328	3.327
3.000	-2.052	4.048	465.153	-0.378	0.718	33.107	0.140	0.909	28.825

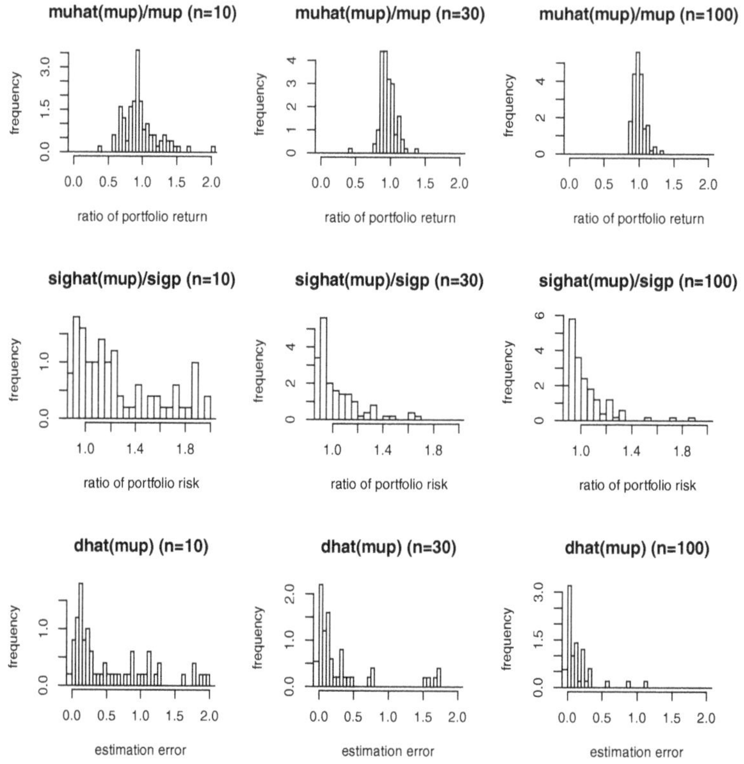

Figure 3.10: Histograms of $\hat{\mu}(\mu_P)/\mu_P$ (upper subplot), $\hat{\sigma}(\mu_P)/\sigma_P$ (middle subplot) and $\hat{d}(\mu_P)$ (lower subplot) with respect to the data of Figure 3.10 (normal distribution)

3.3.5 *Inference of Coefficient*

Here we discuss the inference of estimation error when VAR(1) coefficients set down some values. In place of the VAR(1) coefficient Φ in (3.3.1), we define

$$\Phi_1(\kappa) = I_3 - \kappa \begin{bmatrix} 0.8 & -0.05 & -0.04 \\ -0.05 & 0.85 & -0.03 \\ -0.04 & -0.03 & 0.9 \end{bmatrix}. \tag{3.3.25}$$

Table 3.7 shows M_μ, M_σ and $M_{(\mu,\sigma)}$ defined by (3.3.22)–(3.3.24) in the case of $\kappa = 1.0, 0.1, 0.01$, $n = 10, 100$, $\mu_P = 0.850$ and 100 times generation of the VAR(1) model. Note that when κ is close to 0, the model is near a unit root process. In the case of the near unit root process, it looks like the estimation errors are very large. In addition, in place of the VAR(1) coefficient Φ in (3.3.1), we define

$$\Phi_2(\kappa) = \kappa \begin{bmatrix} 0 & -0.05 & -0.04 \\ -0.05 & 0 & -0.03 \\ -0.04 & -0.03 & 0 \end{bmatrix} \tag{3.3.26}$$

Table 3.8 shows M_μ, M_σ and $M_{(\mu,\sigma)}$ defined by (3.3.22)–(3.3.24) in the case of $\kappa = -10.0, -1.0, 0, 1.0, 1.0, 10.0, n = 10, 100, \mu_P = 0.85$ and 100 times generation of the VAR(1) model. In the case of $\kappa < 0$, there exist negative correlations between the assets, and in the case of $\kappa > 0$, there exist positive correlations between the assets. If $\kappa = 0$, there is no correlation for all assets. In Tables 3.7 and 3.8, it looks like the estimation errors are unstable in the case of $n = 10$, which implies that the estimators do not converge. In the case of $n = 100$, the estimation error becomes large as the degree of correlation increases, especially for the positive correlation.

Table 3.7: Inference of estimation error with respect to $\Phi_1(\kappa)$

κ	$\det\Sigma$	$\sigma(\mu_P)$	$n=10$ M_μ	$n=10$ M_σ	$n=10$ $M_{(\mu,\sigma)}$	$n=100$ M_μ	$n=100$ M_σ	$n=100$ $M_{(\mu,\sigma)}$
normal distribution								
1.0	1.0	0.663	-0.269	2.962	47.739	0.022	0.720	0.924
0.1	234.3	1.687	0.096	1.877	381.533	0.000	-1.012	1.038
0.01	208390.1	5.244	-0.114	-3.167	40.837	0.000	-4.571	20.919
t distribution with 5 degrees of freedom								
1.0	1.0	0.663	0.271	4.098	166.713	0.073	0.754	0.946
0.1	234.3	1.687	-0.003	-0.053	4.144	0.003	-1.006	1.029
0.01	208390.1	5.244	0.059	-0.521	887.331	0.006	-4.569	20.901

Table 3.8: Inference of estimation error with respect to $\Phi_2(\kappa)$

κ	$\det\Sigma$	$\sigma(\mu_P)$	$n=10$ M_μ	$n=10$ M_σ	$n=10$ $M_{(\mu,\sigma)}$	$n=100$ M_μ	$n=100$ M_σ	$n=100$ $M_{(\mu,\sigma)}$
normal distribution								
-10.0	4.244	1.040	-0.269	2.585	45.649	0.022	0.344	0.523
-1.0	1.010	0.644	0.096	2.920	386.537	0.000	0.029	0.013
0	1.000	0.642	-0.114	1.433	32.856	0.000	0.030	0.020
1.0	1.010	0.644	-0.056	1.146	9.249	-0.001	0.024	0.016
10.0	4.244	1.040	-0.044	0.500	2.005	-0.012	0.045	0.044
t distribution with 5 degrees of freedom								
-10.0	4.244	1.040	0.271	3.721	163.767	0.073	0.377	0.520
-1.0	1.010	0.644	-0.003	0.989	5.119	0.003	0.036	0.017
0	1.000	0.642	0.059	4.080	903.705	0.006	0.032	0.020
1.0	1.010	0.644	-0.178	0.929	7.580	0.015	0.050	0.023
10.0	4.244	1.040	0.246	1.746	58.397	-0.012	0.081	0.072

3.3.6 Bibliographic Notes

Tsay (2010) provides the simulation technique for time series data. Ruppert (2004) discusses the application of resampling to the portfolio analysis.

Problems

3.1 Let X_t be random returns on $m(\geq 3)$ assets with $E(X_t) = \mu$ and $\text{Cov}(X_t) = \Sigma$ and without a risk-free asset. Then, show the feasible region (i.e., $W = \left\{(x,y) \in \mathbb{R}^2 | x = \sqrt{\eta^2(w)}, y = \mu(w), ||w|| = 1\right\}$) is convex to the left.

3.2 Prove Theorem 3.1.18.

3.3 Verify $\Omega_2^{NG} = \Omega_2^G$ if $\{U(t)\}$ is Gaussian in Remark 3.2.2 (Show fourth order cumulants $c^U_{b_1,b_2,b_3,b_4}$ vanish).

3.4 Suppose that $X_1, \ldots, X_n \overset{i.i.d.}{\sim} N_m(\mu, \Sigma)$ are obtained. Assuming $e'\Sigma^{-1}e \to \sigma^2 \in (0,\infty)$ under (n,m)-asymptotics (i.e., $n, m \to \infty$ and $m/n \to y \in (0,1)$), show

$$e'\hat{\Sigma}^{-1}e \to \frac{\sigma^2}{1-y}$$

where $\boldsymbol{e} = (1, \ldots, 1)'$($m$-vector) and $\hat{\Sigma}$ is the sample covariance matrix defined in (3.2.37).

3.5 Let $\{\boldsymbol{X}(t)\}$ be a process generated by (3.3.1). Then, show (3.3.3) provided

$$\det(I_m - z\Phi) \neq 0 \quad \text{for all } z \in \mathbb{C} \quad \text{such that } |z| \leq 1.$$

3.6 Let $\{\boldsymbol{X}(t)\}$ be a process generated by (3.3.5). Then, show (3.3.6) provided $|c_i| < 1$ for $i = 1, \ldots, m$ and $E\{h_{ij}(t)\} = \rho\sqrt{E\{h_{ii}(t)\}E\{h_{jj}(t)\}}$ for $i \neq j$.

3.7 Verify (3.3.16) and (3.3.17).

3.8 Suppose that $\{\boldsymbol{X}(t) = (X_1(t), X_2(t), X_3(t))', t \in \mathbb{Z}\}$ follows a VAR(1) defined by

$$\boldsymbol{X}(t) = \boldsymbol{\mu} + \boldsymbol{U}(t) + \Phi\{\boldsymbol{X}(t-1) - \boldsymbol{\mu}\} \tag{3.3.27}$$

where $\boldsymbol{\mu} = (0.01, 0.005, 0.003)'$; $\Phi = (\phi_{ij})_{i,j=1,2,3}$ with $\phi_{11} = 0.02^2, \phi_{22} = 0.015^2, \phi_{33} = 0.01^2$ and $\phi_{ij} = -0.5\sqrt{\phi_{ii}\phi_{jj}}$ for $i \neq j$; $\{\boldsymbol{U}(t)\} \overset{i.i.d.}{\sim} N_3(\boldsymbol{0}, I_3)$.

(i) Generate a sequence of 100($:= n$) observations $\boldsymbol{X}(1), \ldots, \boldsymbol{X}(100)$ from (3.3.27).

(ii) For a fixed 0.004($:= \mu_P$), compute

$$\tilde{\mu}(\mu_P) = \hat{\boldsymbol{w}}'_{\mu_P}\hat{\boldsymbol{\mu}}(= \mu_P), \quad \tilde{\sigma}(\mu_P) = \sqrt{\hat{\boldsymbol{w}}'_{\mu_P}\hat{\Sigma}\hat{\boldsymbol{w}}_{\mu_P}}.$$

(iii) Take 1000($:= B$) resamples $\{\boldsymbol{X}_b(1), \ldots, \boldsymbol{X}_b(n)\}_{b=1,\ldots,1000}$ from $\{\boldsymbol{X}(1), \ldots, \boldsymbol{X}(n)\}$, and for each $b = 1, \ldots, B$, compute

$$\tilde{\mu}_b(\mu_P) = \hat{\boldsymbol{w}}'_{\mu_P,b}\hat{\boldsymbol{\mu}}, \quad \tilde{\sigma}_b(\mu_P) = \sqrt{\hat{\boldsymbol{w}}'_{\mu_P,b}\hat{\Sigma}\hat{\boldsymbol{w}}_{\mu_P,b}}$$

where $\hat{\boldsymbol{w}}_{\mu_P,b}$ corresponds to $\hat{\boldsymbol{w}}_{\mu_P}$ based on $\{\boldsymbol{X}_b(1), \ldots, \boldsymbol{X}_b(n)\}$.

(iv) Get a confidence region

$$C_\alpha(\mu_P) = \left\{(\sigma, \mu) | L(\mu_P) = (\sigma - \tilde{\sigma}(\mu_P))^2 + (\mu - \tilde{\mu}(\mu_P))^2 \leq L_{\alpha,B}(\mu_P)\right\}$$

for a fixed 0.05($:= \alpha$) where $L_{\alpha,B}(\mu_P)$ is the $[(1-\alpha)B]$-th largest of $\{L_1(\mu_P), \ldots, L_B(\mu_P)\}$ with

$$L_b(\mu_P) = (\tilde{\sigma}_b(\mu_P) - \tilde{\sigma}(\mu_P))^2 + (\tilde{\mu}_b(\mu_P) - \tilde{\mu}(\mu_P))^2.$$

Chapter 4

Multiperiod Problem for Portfolio Theory

Portfolio optimization is said to be "myopic" when the investor does not know what will happen beyond the immediate next period. In this framework, basic results about single-period portfolio optimization (such as mean-variance analysis) are justified for short-term investments without portfolio rebalancing. Multiperiod problems are much more realistic than single-period ones. In this framework, we assume that an investor makes a sequence (in the discrete time case) or a continuous function (in the continuous time case) of decisions to maximize a utility function at each time. The fundamental method to solve this problem is **dynamic programming**. In this method, a **value function** which expresses the expected terminal wealth is introduced. The recursive equation with respect to the value function is the so-called **Bellman equation**. The **first-order conditions** (FOCs) to satisfy the Bellman equation are a key tool in solving the dynamic problem. The original literature on dynamic portfolio choice, pioneered by Samuelson (1969) and Fama (1970) in discrete time and by Merton (1969) in continuous time, produced many important insights into the properties of optimal portfolio policies.

Section 4.1 discusses multiperiod optimal portfolios in the discrete time. In this case, a sequence of optimal portfolio weights is obtained as the solutions of the FOCs. When the utility function is a constant relative risk aversion (CRRA) utility or a logarithmic utility, multiperiod optimal portfolios are derived as a sequence of maximizers of objective functions which are described as conditional expectations of the return processes. Since it is known that closed-form solutions are not available except for a few cases of utility functions and return processes, we consider a simulation approach for the derivation by using the bootstrap. By utilizing the technique, it is possible to estimate the multiperiod optimal portfolio even if the distribution of the return process is unknown.

In Section 4.2, we consider continuous time problems, which means that an investor is able to rebalance his/her portfolio continuously during the investment period. Especially, we assume that the price process of risky assets is a class of diffusion processes with jumps. Similarly to the discrete time case, multiperiod optimal portfolios are derived from Bellman equations for continuous time models (the so-called Hamilton–Jacobi–Bellman (HJB) equations). In the case of a CRRA utility function, it will be shown that multiperiod optimal portfolios correspond to a myopic portfolio. Then we discuss the estimation of multiperiod optimal portfolios for discretely observed asset prices. It would be worth considering the discrete observations from continuous time models in practice. In this setting, since an investor has to estimate multiperiod optimal portfolios based on observations, it is different from "stochastic control problems." For this problem, we introduce two types of methods: the generalized method of moments (GMM) and the threshold estimation method. In general, it is well known that maximum likelihood estimation is one of the best methods under the parametric model, but we cannot calculate the maximum likelihood estimation (MLE) for discrete observations. On the contrary, the above two methods are useful in this case.

Section 4.3 discussess universal portfolio (UP) introduced by Cover (1991). The idea of this portfolio comes from a different approach to traditional ones such as the mean variance approach. That is based on information theory introduced by Shannon (1948) (and many others), and the Kelly rule of betting the Kelly (1956).

4.1 Discrete Time Problem

In this section, we present some simulation-based methods for solving discrete time portfolio choice problems for multiperiod investments. In the same way as the single-period problem, multiperiod optimal portfolios are defined by a maximizer of an expectation of utility function. The fundamental method to solve this problem is dynamic programming. In this method, a value function is introduced as the conditional expectation of terminal wealth. Then, the optimal portfolio at each period is expressed as a maximizer of the value function.

4.1.1 Optimal Portfolio Weights

We now introduce the following notation.

- Trading time $t : t = 0, 1, \ldots, T$ where $t = 0$ and $t = T$ indicate the start point and the end point of the investments.
- Return process on risky assets $\{\boldsymbol{X}_t\}_{t=0}^{T-1} : \boldsymbol{X}_t = (X_{1,t}, \ldots, X_{m,t})'$ is a vector of random returns on m assets from time t to $t + 1$.
- Return process on a risk-free asset $\{X_{0,t}\}_{t=0}^{T-1} : X_{0,t}$ is a deterministic return from time t to $t + 1$.
- Information set process $\{\boldsymbol{Z}_t\}_{t=0}^{T-1} : \boldsymbol{Z}_t$ is a vector of state variables at the beginning of time t. The information set indicates the past asset returns and other economic indicators.
- Portfolio weight process[1] $\{\boldsymbol{w}_t\}_{t=0}^{T-1} : \boldsymbol{w}_t = (w_{1,t}, \ldots, w_{m,t})'$ is a portfolio weight at the beginning of time t and measurable (predictable) with respect to the past information $\boldsymbol{Z}_t$.
- Wealth process $\{W_t\}_{t=0}^{T}$:
  Under given W_t, the wealth at time $t + 1$ is defined by

$$W_{t+1} = W_t \{1 + \boldsymbol{w}_t' \boldsymbol{X}_t + (1 - \boldsymbol{w}_t' \boldsymbol{e}) X_{0,t}\} \qquad (4.1.1)$$

  for $t = 0, 1, \ldots, T$ where $\boldsymbol{e} = (1, \ldots, 1)'$ (m-vector). This constraint is called a "budget constraint" (Brandt et al. (2005)) or "self-financing" (Pliska (1997)) and describes the dynamics of wealth. The amount $1 + \boldsymbol{w}_t' \boldsymbol{X}_t + (1 - \boldsymbol{w}_t' \boldsymbol{e}) X_{0,t}$ specifies the total return of the portfolio from time t to $t + 1$. Especially, the amount $\boldsymbol{w}_t' \boldsymbol{X}_t$ arises from risky assets allocated by $\boldsymbol{w}_t$ and the amount $(1 - \boldsymbol{w}_t' \boldsymbol{e}) X_{0,t}$ arises from risk-free assets allocated by $1 - \boldsymbol{w}_t' \boldsymbol{e}$. Given a sequence of decisions $\{\boldsymbol{w}_\tau\}_{\tau=t}^{T-1}$, it is useful to observe that the terminal wealth W_T can be written as a function of current wealth W_t

$$W_T = W_t \prod_{\tau=t}^{T-1} \{1 + \boldsymbol{w}_\tau' \boldsymbol{X}_\tau + (1 - \boldsymbol{w}_\tau' \boldsymbol{e}) X_{0,\tau}\}.$$

- Utility function U: We assume that $W \mapsto U(W)$ is differentiable, concave, and strictly increasing for each $W \in \mathbb{R}$.

Under the above setting, we consider an investor's problem

$$\max_{\{\boldsymbol{w}_t\}_{t=0}^{T-1}} E[U(W_T)]$$

subject to the budget constraints (4.1.1). Following a formulation by dynamic programming (e.g., Bellman (2010)), it is convenient to express the expected terminal wealth in terms of a **value function** V which varies according to the current time, current wealth W_t and other state variables $\boldsymbol{Z}_t \in \mathbb{R}^K, K < \infty$:

$$V(t, W_t, \boldsymbol{Z}_t) = \max_{\{\boldsymbol{w}_u\}_{u=t}^{T-1}} E_t[U(W_T)]$$

[1] $\{\boldsymbol{w}_t\}$ is called "trading strategy" (Pliska (1997), Luenberger (1997)).

$$= \max_{\boldsymbol{w}_t} E_t \left[\max_{\{\boldsymbol{w}_u\}_{u=t+1}^{T-1}} E_{t+1}[U(W_T)] \right]$$
$$= \max_{\boldsymbol{w}_t} E_t \left[V\left(t+1, W_t \left\{1 + \boldsymbol{w}_t' \boldsymbol{X}_t + (1 - \boldsymbol{w}_t' \boldsymbol{e}) X_{0,t}\right\}, \boldsymbol{Z}_{t+1}\right)\right] \tag{4.1.2}$$

subject to the terminal condition

$$V(T, W_T, \boldsymbol{Z}_T) = U(W_T).$$

The expectations at time t (i.e., E_t) are taken with respect to the joint distribution of asset returns and next state, conditional on the information available at time t. The recursive equation (4.1.2) is the so-called **Bellman equation** and is the basis for any recursive solution of the dynamic portfolio choice problem. The **first-order conditions** (FOCs)[2] to obtain an optimal solution at each time t are

$$E_t \left[\partial_W V(t+1, W_t\{1 + \boldsymbol{w}_t' \boldsymbol{X}_t + (1 - \boldsymbol{w}_t' \boldsymbol{e}) X_{0,t}\}, \boldsymbol{Z}_{t+1})(\boldsymbol{X}_t - \boldsymbol{e} X_{0,t})\right] = \boldsymbol{0} \tag{4.1.3}$$

where ∂_W denotes the partial derivative with respect to the wealth W, that is,

$$\partial_W V(t+1, W_{t+1}, \boldsymbol{Z}_{t+1}) = \frac{\partial}{\partial W} V(t+1, W, \boldsymbol{Z}_{t+1}) \Big|_{W = W_{t+1}}.$$

These optimality conditions assume that the state variable $\boldsymbol{Z}_{t+1}$ is not impacted by the decision $\boldsymbol{w}_t$. These FOCs make up a system of nonlinear equations involving integrals that can in general be solved for $\boldsymbol{w}_t$ only numerically. Moreover, the second-order conditions are satisfied if the utility function is concave.

Example 4.1.1. *(CRRA Utility) In general, (4.1.3) can only be solved numerically. However, some analytic progress can be achieved in the case of the CRRA utility introduced in (3.1.16) with $R \neq 1$, that is, $U(W_T) = W_T^{1-\gamma}/(1-\gamma)$. Then, the Bellman equation simplifies to*

$$V(t, W_t, \boldsymbol{Z}_t) = \frac{(W_t)^{1-\gamma}}{1-\gamma} \psi(t, \boldsymbol{Z}_t)$$

where

$$\psi(t, \boldsymbol{Z}_t) = \max_{\boldsymbol{w}_t} E_t \left[\{1 + \boldsymbol{w}_t' \boldsymbol{X}_t + (1 - \boldsymbol{w}_t' \boldsymbol{e})\}^{1-\gamma} \psi(t+1, \boldsymbol{Z}_{t+1}) \right]$$

with $\psi(T, \boldsymbol{Z}_T) = 1$ (For details see Section 4.1.3). The corresponding FOCs are

$$E_t \left[\{1 + \boldsymbol{w}_t' \boldsymbol{X}_t + (1 - \boldsymbol{w}_t' \boldsymbol{e})\}^{-\gamma} \psi(t+1, \boldsymbol{Z}_{t+1})(\boldsymbol{X}_t - \boldsymbol{e} X_{0,t}) \right] = \boldsymbol{0}. \tag{4.1.4}$$

Although (4.1.4) is simpler than the general case, solutions can still only be obtained numerically. This example also helps to illustrate the difference between the dynamic and myopic (single-period) portfolio choice. If the excess returns $\boldsymbol{X}_t - \boldsymbol{e} X_{0,t}$ are independent of the state variables $\boldsymbol{Z}_{t+1}$, the $T-t$ period portfolio choice is identical to the single-period portfolio choice at time t. In that case, there is no difference between the dynamic and myopic portfolios. Let $\boldsymbol{w}_t^{dopt}$ and $\boldsymbol{w}_t^{mopt}$ be the optimal portfolio weights based on the dynamic and myopic (single-period) problems, respectively. If the above condition is satisfied, then it follows that

$$\boldsymbol{w}_t^{dopt} = \arg\max_{\boldsymbol{w}_t} E_t \left[\{1 + \boldsymbol{w}_t' \boldsymbol{X}_t + (1 - \boldsymbol{w}_t' \boldsymbol{e})\}^{1-\gamma} \right] \psi(t+1, \boldsymbol{Z}_{t+1})$$
$$= \arg\max_{\boldsymbol{w}_t} E_t \left[\{1 + \boldsymbol{w}_t' \boldsymbol{X}_t + (1 - \boldsymbol{w}_t' \boldsymbol{e})\}^{1-\gamma} \right]$$

[2]Here $\frac{\partial}{\partial \boldsymbol{w}_t} E_t[V] = E_t[\frac{\partial}{\partial \boldsymbol{w}_t} V]$ is assumed.

$$= \boldsymbol{w}_t^{mopt}. \tag{4.1.5}$$

In contrast, if the excess returns are not independent of the state variables, the dynamic portfolio choice may be substantially different from the single-period portfolio choice. The differences between the dynamic and myopic policies are called ***hedging demands*** *because the investor tries to hedge against changes in the investment opportunities due to deviations from the single-period portfolio choice. Let* $\boldsymbol{w}_t^{hopt}$ *be the portfolio weight concerned with hedging demands. Then, we can write*

$$\boldsymbol{w}_t^{dopt} = \boldsymbol{w}_t^{mopt} + \boldsymbol{w}_t^{hopt}$$

where

$$\boldsymbol{w}_t^{dopt} = \arg\max_{\boldsymbol{w}_t} E_t\left[\{1 + \boldsymbol{w}_t'\boldsymbol{X}_t + (1 - \boldsymbol{w}_t'\boldsymbol{e})\}^{1-\gamma}\,\psi(t+1, \boldsymbol{Z}_{t+1})\right]$$

$$\boldsymbol{w}_t^{mopt} = \arg\max_{\boldsymbol{w}_t} E_t\left[\{1 + \boldsymbol{w}_t'\boldsymbol{X}_t + (1 - \boldsymbol{w}_t'\boldsymbol{e})\}^{1-\gamma}\right].$$

Example 4.1.2. *(Logarithmic Utility) In case of the logarithmic utility, that is,* $U(W_T) = \ln(W_T)$*, the dynamic portfolio totally corresponds to the myopic portfolio. The Bellman equation is written as*

$$V(t, W_t, \boldsymbol{Z}_t) = \max_{\boldsymbol{w}_t} E_t\left[\ln\{1 + \boldsymbol{w}_t'\boldsymbol{X}_t + (1 - \boldsymbol{w}_t'\boldsymbol{e})\}\right]$$
$$+ \ln W_t + \sum_{\tau=t+1}^{T-1} E_t\left[\ln\{1 + \boldsymbol{w}_\tau'\boldsymbol{X}_\tau + (1 - \boldsymbol{w}_\tau'\boldsymbol{e})\}\right]. \tag{4.1.6}$$

Since the second and third terms of the right hand part of the above equation do not depend on the portfolio weights $\boldsymbol{w}_t$*, the optimal portfolio weight* $\boldsymbol{w}_t^{dopt}$ *is obtained by*

$$\boldsymbol{w}_t^{dopt} = \arg\max_{\boldsymbol{w}_t} E_t\left[\ln\{1 + \boldsymbol{w}_t'\boldsymbol{X}_t + (1 - \boldsymbol{w}_t'\boldsymbol{e})\}\right]$$

which corresponds to the single-period optimal portfolio weight $\boldsymbol{w}_t^{mopt}$*. This result leads to the "Kelly rule of betting" (Kelly (1956)).*

The conditions[3] for which the myopic portfolio choice corresponds to the dynamic one are summarized as follows:

- The investment opportunities are constant (e.g., in the case of i.i.d. returns).
- The investment opportunities vary with time, but are not able to hedge (e.g., in the case of (4.1.5)).
- The investor has a logarithmic utility on terminal wealth.

As for the other utility function, Li and Ng (2000) analyzed various formulations of the maximization of terminal quadratic utility under several hypotheses, provided explicit solutions in simplifying cases, and derived analytical expressions for the multiperiod mean variance efficient frontier. Leippold et al. (2002) provided an interpretation of the solution to the multiperiod mean variance problem in terms of an orthogonal set of basis strategies, each with a clear economic interpretation. They use the analysis to provide analytical solutions to portfolios consisting of both assets and liabilities. More recently, Cvitanic et al. (2008) connect this problem to a specific case of multiperiod Sharpe ratio maximization.

4.1.2 Consumption Investment

Intermediate consumption has traditionally been a part of the multiperiod optimal investment problem since Samuelson (1969) and Fama (1970). In this setting, the problem is formulated in

[3]For more detail, see Mossin (1968) or Aït-Sahalia and Hansen (2009)

terms of a sequence of consumptions and a terminal wealth. We also discuss the fundamental method to solve this problem. The future wealth may be influenced not only by the asset return but also by eventual transaction costs and other constraints such as the desire for intermediate consumption, minimization of tax impact or the influence of additional capital due to labour income. Suppose that there exists a consumption process $\{c_t\}_{t=0}^{T-1}$ which is a measurable (predictable) stochastic process.[4] Here c_t $(0 < c_t < 1)$ represents the weight of funds consumed by the investor at time t. A consumption investment plan consists of a pair $(\{\boldsymbol{w}_t\}_{t=0}^{T-1}, \{c_t\}_{t=0}^{T-1})$ where $\{c_t\}$ is a consumption process and $\{\boldsymbol{w}_t\}$ is a portfolio weight process. The investor seeks to choose the consumption investment plan that maximizes the expected utility over T periods. In particular, the investor faces a trade-off between consumption and investment. Consider

$$\max_{\{\boldsymbol{w}_t\}_{t=0}^{T-1}, \{c_t\}_{t=0}^{T-1}} E\left[\sum_{t=0}^{T-1} U_c(c_t) + U_W(W_T)\right]$$

subject to the budget constraint[5]

$$W_{t+1} = (1 - c_t)W_t \{1 + \boldsymbol{w}_t'\boldsymbol{X}_t + (1 - \boldsymbol{w}_t'\boldsymbol{e})X_{0,t}\}.$$

Here U_c denotes the utility of consumption c and U_W denotes the utility of wealth W. This model is called the "generalized consumption investment problem" (see Pliska (1997)). In this case, the value function $V(t, W_t, \boldsymbol{Z}_t)$ (i.e., the Bellman equation) is written as

$$\begin{aligned}
V(t, W_t, \boldsymbol{Z}_t) &= \max_{\{\boldsymbol{w}_u\}_{u=t}^{T-1}, \{c_u\}_{u=t}^{T-1}} E_t\left[\sum_{u=t}^{T-1} U_c(c_u) + U_W(W_T)\right] \\
&= \max_{\boldsymbol{w}_t, c_t} E_t\left[U_c(c_t) + \max_{\{\boldsymbol{w}_u\}_{u=t+1}^{T-1}, \{c_u\}_{u=t+1}^{T-1}} E_{t+1}\left[\sum_{u=t+1}^{T-1} U_c(c_u) + U_W(W_T)\right]\right] \\
&= \max_{\boldsymbol{w}_t, c_t} E_t\left[U_c(c_t) + V(t+1, (1 - c_t)W_t\{1 + \boldsymbol{w}_t'\boldsymbol{X}_t + (1 - \boldsymbol{w}_t'\boldsymbol{e})X_{0,t}\}, \boldsymbol{Z}_{t+1})\right] \\
&= \max_{\boldsymbol{w}_t, c_t} \{U_c(c_t) + E_t\left[V(t+1, (1 - c_t)W_t\{1 + \boldsymbol{w}_t'\boldsymbol{X}_t + (1 - \boldsymbol{w}_t'\boldsymbol{e})X_{0,t}\}, \boldsymbol{Z}_{t+1})\right]\}
\end{aligned}$$

subject to the terminal condition

$$V(T, W_T, \boldsymbol{Z}_T) = U_W(W_T).$$

The FOCs (which are also called **Euler conditions** (e.g., Gouriéroux and Jasiak (2001)) are given by

$$E_t\left[\partial_W V(t+1, W_{t+1}(\boldsymbol{w}_t, c_t), \boldsymbol{Z}_{t+1})(\boldsymbol{X}_t - \boldsymbol{e}X_{0,t})\right] = \boldsymbol{0} \tag{4.1.7}$$

$$\partial_c U_c(c_t) + E_t\Big[\partial_W V(t+1, W_{t+1}(\boldsymbol{w}_t, c_t), \boldsymbol{Z}_{t+1})(-W_t)\{1 + \boldsymbol{w}_t'\boldsymbol{X}_t + (1 - \boldsymbol{w}_t'\boldsymbol{e})X_{0,t}\}\Big] = 0 \tag{4.1.8}$$

where $W_{t+1}(\boldsymbol{w}, c) = (1 - c)W_t\{1 + \boldsymbol{w}'\boldsymbol{X}_t + (1 - \boldsymbol{w}'\boldsymbol{e})X_{0,t}\}$ and $\partial_c U_c(c_t) = \frac{\partial}{\partial c}U_c(c)|_{c=c_t}$. These are obtained by computing the first-order derivatives with respect to $\boldsymbol{w}_t$ and c_t. From (4.1.7) and (4.1.8), we have

$$\partial_c U_c(c_t) = E_t\left[\partial_W V(t+1, W_{t+1}(\boldsymbol{w}_t, c_t), \boldsymbol{Z}_{t+1})W_t(1 + X_{0,t})\right]. \tag{4.1.9}$$

This equation is called the "envelope relation" derived by Samuelson (1969) and based on (4.1.7) and (4.1.9), we can obtain the optimal solutions $\{\boldsymbol{w}_t\}_{t=0}^{T-1}$ and $\{c_t\}_{t=0}^{T-1}$ numerically.

The above portfolio problems are considered at the individual level. If the return process $\{\boldsymbol{X}_t\}$ consists of all assets on a market (i.e., $\boldsymbol{w}_t'\boldsymbol{X}_t$ represents a market portfolio) and c_t denotes the aggregate consumption of all investors, then it is known in the literature as the **consumption based capital asset pricing model** (CCAPM).

---

[4]We assume that $E_t(c_t) = c_t$ is satisfied for any t.

[5]Note that in this case the portfolio is not self-financed since the value of consumption spending does not necessarily balance the income pattern.

4.1.3 Simulation Approach for VAR(1) model

Since it is known that closed-form solutions are obtained only for a few special cases, the recent literature uses a variety of numerical and approximate solution methods to incorporate realistic features into the dynamic portfolio problem, such as those of Aït-Sahalia and Brandt (2001) and Brandt et al. (2005). In what follows, we introduce a procedure to construct dynamic portfolio weights based on the AR bootstrap following Shiraishi (2012). The simulation algorithm is as follows. First, we generate simulation sample paths of the vector random returns by using the AR bootstrap (Step1). Based on the bootstrapping samples, an optimal portfolio estimator, which is applied from time $T-1$ to the end of trading time T, is obtained under a constant relative risk aversion (CRRA) utility function (Step 2). Next we approximate the value function at time $T-1$ by linear functions of the bootstrapping samples (Step 3). This idea is similar to that of Aït-Sahalia and Brandt (2001) and Brandt et al. (2005). Then, an optimal portfolio weight estimator applied from time $T-2$ to time $T-1$ is obtained based on the approximated value function (Step 4). In the same manner as above, the optimal portfolio weight estimator at each trading time is obtained recursively (Steps 5-7).

Suppose the existence of a finite number of risky assets indexed by $i, (i = 1, \ldots, m)$. Let $\boldsymbol{X}_t = (X_{1,t}, \ldots, X_{m,t})'$ denote the random returns on m assets from time t to $t+1$.[6] Suppose too that there exists a risk-free asset with the return X_0.[7] Based on the process $\{\boldsymbol{X}_t\}_{t=0}^{T-1}$ and X_0, we consider an investment strategy from time 0 to time T where $T(\in \mathbb{N})$ denotes the end of the investment time. Let $\boldsymbol{w}_t = (w_{1,t}, \ldots, w_{m,t})'$ be vectors of portfolio weight for the risky assets at the beginning of time t. Here we assume that the portfolio weights $\boldsymbol{w}_t$ can be rebalanced at the beginning of time t and measurable (predictable) with respect to the past information $\mathcal{F}_t = \sigma(\boldsymbol{X}_{t-1}, \boldsymbol{X}_{t-2}, \ldots)$. Here we make the following assumption.

Assumption 4.1.3. *There exists an optimal portfolio weight* $\tilde{\boldsymbol{w}}_t \in \mathbb{R}^m$ *satisfied by* $|\tilde{\boldsymbol{w}}_t'\boldsymbol{X}_t + (1 - \tilde{\boldsymbol{w}}_t\boldsymbol{e})X_0| \ll 1$[8] *almost surely for each time* $t = 0, 1, \ldots, T-1$ *where* $\boldsymbol{e} = (1, \ldots, 1)'$.

Then the return of the portfolio from time t to $t+1$ is written as $1 + X_0 + \boldsymbol{w}_t'(\boldsymbol{X}_t - X_0\boldsymbol{e})$[9] and the total return from time 0 to time T (called terminal wealth) is written as

$$W_T := \prod_{t=0}^{T-1} \{1 + X_0 + \boldsymbol{w}_t'(\boldsymbol{X}_t - X_0\boldsymbol{e})\}.$$

Suppose that a utility function $U : x \mapsto U(x)$ is differentiable, concave and strictly increasing for each $x \in \mathbb{R}$. Consider an investor's problem

$$\max_{\{\boldsymbol{w}_t\}_{t=0}^{T-1}} E[U(W_T)].$$

In the same way as (4.1.2), the value function V_t can be written as

$$V_t \equiv \max_{\{\boldsymbol{w}_s\}_{s=t}^{T-1}} E[U(W_T)|\mathcal{F}_t] = \max_{\boldsymbol{w}_t} E\left[V_{t+1}|\mathcal{F}_t\right]$$

subject to the terminal condition $V_T = U(W_T)$. The first-order conditions (FOCs) are

$$E\left[\partial_W U(W_T)(\boldsymbol{X}_t - X_0\boldsymbol{e})\middle|\mathcal{F}_t\right] = \boldsymbol{0}$$

where $\partial_W U(W_T) = \frac{\partial}{\partial W} U(W)|_{W=W_T}$. These FOCs make up a system of nonlinear equations involving

[6] Suppose that $S_{i,t}$ is a value of asset i at time t. Then, the return is described as $1 + X_{i,t} = S_{i,t+1}/S_{i,t}$.

[7] Suppose that B_t is a value of risk-free asset at time t. Then, the return is described as $1 + X_0 = B_t/B_{t-1}$.

[8] We assume that the risky assets exclude ultra high risk and high return ones, for instance, the asset value $S_{i,t+1}$ may be larger than $2S_{i,t}$.

[9] Assuming that $\boldsymbol{S}_t := (S_{1,t}, \ldots, S_{m,t})' = B_t\boldsymbol{e}$, the portfolio return is written as $\{\boldsymbol{w}_t'\boldsymbol{S}_{t+1} + (1 - \boldsymbol{w}_t'\boldsymbol{e})B_{t+1})\}/\{\boldsymbol{w}_t'\boldsymbol{S}_t + (1 - \boldsymbol{w}_t\boldsymbol{e})B(t)\} = 1 + X_0 + \boldsymbol{w}_t'(\boldsymbol{X}_t - X_0\boldsymbol{e})$.

integrals that can in general be solved for $\boldsymbol{w}_t$ only numerically. According to the literature (e.g., Brandt et al. (2005)), we can simplify this problem in the case of a constant relative risk aversion (CRRA) utility function,[10] that is,

$$U(W) = \frac{W^{1-\gamma}}{1-\gamma}, \quad \gamma \neq 1$$

where γ denotes the coefficient of relative risk aversion. In this case, the Bellman equation simplifies to

$$\begin{aligned}
V_t &= \max_{\boldsymbol{w}_t} E\left[\max_{\{\boldsymbol{w}_s\}_{s=t+1}^{T-1}} E\left[\frac{1}{1-\gamma}(W_T)^{1-\gamma}|\mathcal{F}_{t+1}\right]\middle|\mathcal{F}_t\right] \\
&= \max_{\boldsymbol{w}_t} E\left[\max_{\{\boldsymbol{w}_s\}_{s=t+1}^{T-1}} E\left[\frac{1}{1-\gamma}\prod_{s=0}^{T-1}\{1 + X_0 + \boldsymbol{w}_s'(\boldsymbol{X}_s - X_0\boldsymbol{e})\}^{1-\gamma}|\mathcal{F}_{t+1}\right]\middle|\mathcal{F}_t\right] \\
&= \max_{\boldsymbol{w}_t} E\left[\frac{1}{1-\gamma}\prod_{s=0}^{t}\{1 + X_0 + \boldsymbol{w}_s'(\boldsymbol{X}_s - X_0\boldsymbol{e})\}^{1-\gamma}\right. \\
&\qquad \times \left.\max_{\{\boldsymbol{w}_s\}_{s=t+1}^{T-1}} E\left[\prod_{s=t+1}^{T-1}\{1 + X_0 + \boldsymbol{w}_s'(\boldsymbol{X}_s - X_0\boldsymbol{e})\}^{1-\gamma}|\mathcal{F}_{t+1}\right]\middle|\mathcal{F}_t\right] \\
&= \max_{\boldsymbol{w}_t} E\left[\frac{1}{1-\gamma}(W_{t+1})^{1-\gamma}\max_{\{\boldsymbol{w}_s\}_{s=t+1}^{T-1}} E\left[(W_{t+1}^T)^{1-\gamma}|\mathcal{F}_{t+1}\right]\middle|\mathcal{F}_t\right] \\
&= \max_{\boldsymbol{w}_t} E\left[U(W_{t+1})\Psi_{t+1}\middle|\mathcal{F}_t\right]
\end{aligned}$$

where $W_{t+1}^T = \prod_{s=t+1}^{T-1}\{1 + X_0 + \boldsymbol{w}_s'(\boldsymbol{X}_s - X_0\boldsymbol{e})\}$ and $\Psi_{t+1} = \max_{\{\boldsymbol{w}_s\}_{s=t+1}^{T-1}} E\left[(W_{t+1}^T)^{1-\gamma}|\mathcal{F}_{t+1}\right]$. From this, the value function V_t can be expressed as

$$V_t = U(W_t)\Psi_t$$

and Ψ_t also satisfies a Bellman equation

$$\Psi_t = \max_{\boldsymbol{w}_t} E\left[\{1 + X_0 + \boldsymbol{w}_t'(\boldsymbol{X}_t - X_0\boldsymbol{e})\}^{1-\gamma}\Psi_{t+1}|\mathcal{F}_t\right]$$

subject to the terminal condition $\Psi_T = 1$. The corresponding FOCs (in terms of Ψ_t) are

$$E\left[\{1 + X_0 + \boldsymbol{w}_t'(\boldsymbol{X}_t - X_0\boldsymbol{e})\}^{-\gamma}\Psi_{t+1}(\boldsymbol{X}_t - X_0\boldsymbol{e})\middle|\mathcal{F}_t\right] = \boldsymbol{0}. \tag{4.1.10}$$

It is easy to see that we cannot obtain the solution of (4.1.10) as an explicit form, so that we consider the (approximated) solution by AR bootstrap.

Suppose that $\{\boldsymbol{X}_t = (X_1(t), \ldots, X_m(t))'; t \in \mathbb{Z}\}$ is an m-vector AR(1) process defined by

$$\boldsymbol{X}_t = \boldsymbol{\mu} + A(\boldsymbol{X}_{t-1} - \boldsymbol{\mu}) + \boldsymbol{\epsilon}_t$$

where $\boldsymbol{\mu} = (\mu_1, \ldots, \mu_m)'$ is a constant m-dimensional vector, $\boldsymbol{\epsilon}_t = (\epsilon_1(t), \ldots, \epsilon_m(t))'$ are independent and identically distributed (i.i.d.) random m-dimensional vectors with $E[\boldsymbol{\epsilon}_t] = \boldsymbol{0}$ and $E[\boldsymbol{\epsilon}_t\boldsymbol{\epsilon}_t'] = \Gamma$ (Γ is a nonsingular m by m matrix), and A is a nonsingular m by m matrix. We make the following assumption.

[10] As we mentioned in Section 3.1.4, various types of utility functions have been proposed such as CARA, CRRA, DARA. Here, we deal with the CRRA utility function because it is easy to handle mathematically. If you consider other utility functions, the Bellman equation would be more complicated.

Assumption 4.1.4. $\det\{I_m - Az\} \neq 0$ *on* $\{z \in \mathbb{C}; |z| \leq 1\}$.

Given $\{\boldsymbol{X}_{-n+1}, \ldots, \boldsymbol{X}_0, \boldsymbol{X}_1, \ldots, \boldsymbol{X}_t\}$, the least-squares estimator $\hat{A}^{(t)}$ of A is obtained by solving

$$\hat{\Gamma}^{(t)}\hat{A}^{(t)} = \sum_{s=-n+2}^{t} \hat{\boldsymbol{Y}}_{s-1}^{(t)}(\hat{\boldsymbol{Y}}_s^{(t)})'$$

where $\hat{\boldsymbol{Y}}_s^{(t)} = \boldsymbol{X}_s - \hat{\boldsymbol{\mu}}^{(t)}$, $\hat{\Gamma}^{(t)} = \sum_{s=-n+1}^{t} \hat{\boldsymbol{Y}}_s^{(t)}(\hat{\boldsymbol{Y}}_s^{(t)})'$ and $\hat{\boldsymbol{\mu}}^{(t)} = \frac{1}{n+t}\sum_{s=-n+1}^{t} \boldsymbol{X}_s$. Then, the error $\hat{\boldsymbol{\epsilon}}_s^{(t)} = (\hat{\epsilon}_1^{(t)}(s), \ldots, \hat{\epsilon}_m^{(t)}(s))'$ is "recovered" by

$$\hat{\boldsymbol{\epsilon}}_s^{(t)} := \hat{\boldsymbol{Y}}_s^{(t)} - \hat{A}^{(t)}\hat{\boldsymbol{Y}}_{s-1}^{(t)} \quad s = -n+2, \ldots, t.$$

Let $F_n^{(t)}(\cdot)$ denote the distribution which puts mass $\frac{1}{n+t}$ at $\hat{\boldsymbol{\epsilon}}_s^{(t)}$. Let $\{\boldsymbol{\epsilon}_s^{(b,t)*}\}_{s=t+1}^{T}$ (for $b = 1, \ldots, B$) be i.i.d. bootstrapped observations from $F_n^{(t)}$, where $B \in \mathbb{N}$ is the number of bootstraps. Given $\{\boldsymbol{\epsilon}_s^{(b,t)*}\}$, define $\boldsymbol{Y}_s^{(b,t)*}$ and $\boldsymbol{X}_s^{(b_1,b_2,t)*}$ by

$$\begin{aligned} \boldsymbol{Y}_s^{(b,t)*} &= \left(\hat{A}^{(t)}\right)^{s-t}(\boldsymbol{X}_t - \hat{\boldsymbol{\mu}}^{(t)}) + \sum_{k=t+1}^{s}\left(\hat{A}^{(t)}\right)^{s-k}\boldsymbol{\epsilon}_k^{(b,t)*} \\ \boldsymbol{X}_s^{(b_1,b_2,t)*} &= \hat{\boldsymbol{\mu}}^{(t)} + \hat{A}^{(t)}\boldsymbol{Y}_{s-1}^{(b_1,t)*} + \boldsymbol{\epsilon}_s^{(b_2,t)*} \end{aligned} \tag{4.1.11}$$

for $s = t+1, \ldots, T$. Based on the above $\{\boldsymbol{X}_s^{(b_1,b_2,t)*}\}_{b_1,b_2=1,\ldots,B;s=t+1,\ldots,T}$ for each $t = 0, \ldots, T-1$, we construct an estimator of the optimal portfolio weight $\tilde{\boldsymbol{w}}_t$ as follows.

Step 1: First, we fix the current time t, which implies that the observed stretch $n+t$ is fixed. Then, we can generate $\{\boldsymbol{X}_s^{(b_1,b_2,t)*}\}$ by (4.1.11).

Step 2: Next, for each $b_0 = 1, \ldots, B$, we obtain $\hat{\boldsymbol{w}}_{T-1}^{(b_0,t)}$ as the maximizer of

$$E_{T-1}^*\left[\left\{1 + X_0 + \boldsymbol{w}'(\boldsymbol{X}_T^{(b_0,b,t)*} - X_0\boldsymbol{e})\right\}^{1-\gamma}\right] = \frac{1}{B}\sum_{b=1}^{B}\left\{1 + X_0 + \boldsymbol{w}'(\boldsymbol{X}_T^{(b_0,b,t)*} - X_0\boldsymbol{e})\right\}^{1-\gamma}$$

or the solution of

$$\begin{aligned} &E_{T-1}^*\left[\left\{1 + X_0 + \boldsymbol{w}'(\boldsymbol{X}_T^{(b_0,b,t)*} - X_0\boldsymbol{e})\right\}^{-\gamma}(\boldsymbol{X}_T^{(b_0,b,t)*} - X_0\boldsymbol{e})\right] \\ &= \frac{1}{B}\sum_{b=1}^{B}\left\{1 + X_0 + \boldsymbol{w}'(\boldsymbol{X}_T^{(b_0,b,t)*} - X_0\boldsymbol{e})\right\}^{-\gamma}(\boldsymbol{X}_T^{(b_0,b,t)*} - X_0\boldsymbol{e}) \\ &= 0 \end{aligned}$$

with respect to $\boldsymbol{w}$. Here we introduce a notation "$E_s^*[\cdot]$" as an estimator of conditional expectation $E[\cdot|\mathcal{F}_s]$, which is defined by $E_s^*[h(\boldsymbol{X}_{s+1}^{(b_0,b,t)*})] = \frac{1}{B}\sum_{b=1}^{B} h(\boldsymbol{X}_{s+1}^{(b_0,b,t)*})$ for any function h of $\boldsymbol{X}_{s+1}^{(b_0,b,t)*}$. This $\hat{\boldsymbol{w}}_{T-1}^{(b_0,t)}$ corresponds to the estimator of myopic (single-period) optimal portfolio weight.

Step 3: Next, we construct estimators of Ψ_{T-1}. Since it is difficult to express the explicit form of Ψ_{T-1}, we parameterize it as linear functions of $\boldsymbol{X}_{T-1}$ as follows:

$$\begin{aligned} \Psi^{(1)}(\boldsymbol{X}_{T-1}, \boldsymbol{\theta}_{T-1}) &:= [1, \boldsymbol{X}_{T-1}']\boldsymbol{\theta}_{T-1} \\ \Psi^{(2)}(\boldsymbol{X}_{T-1}, \boldsymbol{\theta}_{T-1}) &:= [1, \boldsymbol{X}_{T-1}', vech(\boldsymbol{X}_{T-1}\boldsymbol{X}_{T-1}')']\boldsymbol{\theta}_{T-1}. \end{aligned}$$

Note that the dimensions of $\boldsymbol{\theta}_{T-1}$ in $\Psi^{(1)}$ and $\Psi^{(2)}$ are $m+1$ and $m(m+1)/2 + m + 1$, respectively. The idea of $\Psi^{(1)}$ and $\Psi^{(2)}$ is inspired by the parameterization of the conditional expectations in Brandt et al. (2005).

In order to construct the estimators of $\Psi^{(i)}(i = 1, 2)$, we introduce the conditional least squares estimators of the parameter $\boldsymbol{\theta}_{T-1}^{(i)}$, that is,

$$\hat{\boldsymbol{\theta}}_{T-1}^{(i)} = \arg\min_{\boldsymbol{\theta}} Q_{T-1}^{(i)}(\boldsymbol{\theta})$$

where

$$\begin{aligned}Q_{T-1}^{(i)}(\boldsymbol{\theta}) &= \frac{1}{B}\sum_{b_0=1}^{B} E_{T-1}^{*}\left[(\Psi_{T-1} - \Psi^{(i)})^2\right] \\ &= \frac{1}{B}\sum_{b_0=1}^{B}\left[\frac{1}{B}\sum_{b=1}^{B}\left\{\Psi_{T-1}(\boldsymbol{X}_{T}^{(b_0,b,t)*}) - \Psi_{T-1}^{(i)}(\boldsymbol{X}_{T-1}^{(b_0,b_0,t)*}, \boldsymbol{\theta})\right\}^2\right]\end{aligned}$$

and

$$\Psi_{T-1}(\boldsymbol{X}_{T}^{(b_0,b,t)*}) = \left\{1 + X_0 + (\hat{\boldsymbol{w}}_{T-1}^{(b_0,t)})'(\boldsymbol{X}_{T}^{(b_0,b,t)*} - X_0\boldsymbol{e})\right\}^{1-\gamma}.$$

Then, by using $\hat{\boldsymbol{\theta}}_{T-1}^{(i)}$, we can compute $\Psi^{(i)}(\boldsymbol{X}_{T-1}^{(b_0,b,t)*}, \hat{\boldsymbol{\theta}}_{T-1}^{(i)})$.

Step 4: Based on the above $\Psi^{(i)}$, we obtain $\hat{\boldsymbol{w}}_{T-2}^{(b_0,t)}$ as the maximizer of

$$\begin{aligned}&E_{T-2}^{*}\left[\left\{1 + X_0 + \boldsymbol{w}'(\boldsymbol{X}_{T-1}^{(b_0,b,t)*} - X_0\boldsymbol{e})\right\}^{1-\gamma}\Psi^{(i)}(\boldsymbol{X}_{T-1}^{(b_0,b,t)*}, \hat{\boldsymbol{\theta}}_{T-1}^{(i)})\right] \\ &= \frac{1}{B}\sum_{b=1}^{B}\left\{1 + X_0 + \boldsymbol{w}'(\boldsymbol{X}_{T-1}^{(b_0,b,t)*} - X_0\boldsymbol{e})\right\}^{1-\gamma}\Psi^{(i)}(\boldsymbol{X}_{T-1}^{(b_0,b,t)*}, \hat{\boldsymbol{\theta}}_{T-1}^{(i)})\end{aligned}$$

or the solution of

$$\begin{aligned}&E_{T-2}^{*}\left[\left\{1 + X_0 + \boldsymbol{w}'(\boldsymbol{X}_{T-1}^{(b_0,b,t)*} - X_0\boldsymbol{e})\right\}^{-\gamma}(\boldsymbol{X}_{T-1}^{(b_0,b,t)*} - X_0\boldsymbol{e})\Psi^{(i)}(\boldsymbol{X}_{T-1}^{(b_0,b,t)*}, \hat{\boldsymbol{\theta}}_{T-1}^{(i)})\right] \\ &= \frac{1}{B}\sum_{b=1}^{B}\left\{1 + X_0 + \boldsymbol{w}'(\boldsymbol{X}_{T-1}^{(b_0,b,t)*} - X_0\boldsymbol{e})\right\}^{-\gamma}(\boldsymbol{X}_{T-1}^{(b_0,b,t)*} - X_0\boldsymbol{e})\Psi^{(i)}(\boldsymbol{X}_{T-1}^{(b_0,b,t)*}, \hat{\boldsymbol{\theta}}_{T-1}^{(i)}) \\ &= \boldsymbol{0}\end{aligned}$$

with respect to $\boldsymbol{w}$. This $\hat{\boldsymbol{w}}_{T-2}^{(b_0,t)}$ does not correspond to the estimator of myopic (single period) optimal portfolio weight due to the effect of $\Psi^{(i)}$.

Step 5: In the same manner of Steps 3–4, we can obtain $\hat{\boldsymbol{\theta}}_{s}^{(i)}$ and $\hat{\boldsymbol{w}}_{s}^{(b_0,t)}$, recursively, for $s = T-2, T-1, \ldots, t+1$.

Step 6: Then, we define an optimal portfolio weight estimator at time t as $\hat{\boldsymbol{w}}_{t}^{(t)} := \hat{\boldsymbol{w}}_{t}^{(b_0,t)}$ by Step 4. Note that $\hat{\boldsymbol{w}}_{t}^{(t)}$ is obtained as only one solution because $\boldsymbol{X}_{t+1}^{(b_0,b,t)*}(= \hat{\boldsymbol{\mu}}^{(t)} + \hat{A}^{(t)}(\boldsymbol{X}_t - \hat{\boldsymbol{\mu}}^{(t)}) + \boldsymbol{\epsilon}_{t+1}^{(b,t)*})$ is independent of b_0.

Step 7: For each time $t = 0, 1, \ldots, T-1$, we obtain $\hat{\boldsymbol{w}}_{t}^{(t)}$ by Steps 1–6. Finally, we can construct an optimal investment strategy as $\{\hat{\boldsymbol{w}}_{t}^{(t)}\}_{t=0}^{T-1}$.

Remark 4.1.5. *As we mentioned in Section 3.1.4, various types of utility functions have been proposed such as CARA, CRRA, DARA. Here, we deal with the CRRA utility function because it is easy to handle mathematically. If you consider other utility functions, the FOCs would be more complicated and can only solved numerically (e.g., generating sample paths).*

Regarding the validity of the approximation of the discrete time multiperiod optimal portfolio under the CRRA utility function, we need to evaluate the method in three view points.

The first view point is an existence of model misspecification. The quality of ARB approximation becomes poor when the model assumptions are violated. For instance, if the order of the autoregressive process is misspecified, the ARB method may be invalid; on the other hand, the moving block bootstrap (MBB) would still give a valid approximation.

The second view point is about the validity of the parametrization of Ψ in Step 3. Brandt et al. (2005) parameterize conditional expectations of the asset return processes as a vector of a polynomial based on the state variables, and estimate the parameters. The key feature of the parameterization is its linearity in the (nonlinear) functions of the state variables. Although the parameter space may not include the true conditional expectations, they conclude that it works well through some simulation results.

The third view point is about the length of the trading term T. The algorithm proceeds recursively backward. Due to the recursive calculations, the estimation errors propagate (and cumulate) from the end of investment time T to the current time.

4.1.4 Bibliographic Notes

Aït-Sahalia and Hansen (2009), Nicolas (2011) and Pliska (1997) give a general method to solve multiperiod portfolio problems. Gouriéroux and Jasiak (2001) introduce some econometric methods and models applicable to portfolio management, including CCAPM. Aït-Sahalia and Hansen (2009), Brandt et al. (2005) and Shiraishi (2012) investigate a simulation approach to solve dynamic portfolio problems. Shiraishi et al. (2012) discuss an optimal portfolio for the GPIF.

4.2 Continuous Time Problem

Merton (1969) considered a problem of maximizing the infinite-horizon expected utility of consumption by investing in a set of risky assets and a riskless asset. That is, the investor selects the amounts to be held in the m risky assets and the riskless asset at times $t \in [0, \infty)$, as well as the investor's consumption path. The available investment opportunities consist of a riskless asset with price $S_{0,t}$ and m risky assets with price $\boldsymbol{S}_t = (S_{1,t}, \ldots, S_{m,t})'$. In this section, following Aït-Sahalia et al. (2009), we will consider optimal consumption and portfolio rule when the asset price process $\{\boldsymbol{S}_t\}$ is a diffusion model with jump.

Before the discussion of the portfolio problem, we introduce three stochastic processes: the Lévy process, standard Brownian motion and the compound Poisson process.

Definition 4.2.1. *(Lévy process) A stochastic process $\{\boldsymbol{X}_t\}_{t\geq 0}$ on $\mathbb{R}^m$ defined on a probability space $(\Omega, \mathcal{F}, P)$ is a Lévy process if the following conditions are satisfied.*

(1) For any choice of $n \geq 1$ and $0 \leq t_0 < t_1 < \cdots < t_n$, random variables $\boldsymbol{X}_{t_0}, \boldsymbol{X}_{t_1} - \boldsymbol{X}_{t_0}, \ldots, \boldsymbol{X}_{t_n} - \boldsymbol{X}_{t_{n-1}}$ are independent (independent increments property).

(2) $\boldsymbol{X}_0 = \boldsymbol{0}$ a.s.

(3) The distribution of $\boldsymbol{X}_{s+t} - \boldsymbol{X}_t$ does not depend on s (temporal homogeneity or stationary increments property).

(4) It is stochastically continuous.

(5) There is $\Omega_0 \in \mathcal{F}$ with $P[\Omega_0] = 1$ such that, for every $\omega \in \Omega_0$, $\boldsymbol{X}_t(\omega)$ is right-continuous in $t \geq 0$ and has left limits in $t > 0$.

Definition 4.2.2. *(standard Brownian motion) A stochastic process $\{\boldsymbol{W}_t\}_{t\geq 0}$ on $\mathbb{R}^m$ defined on a probability space $(\Omega, \mathcal{F}, P)$ is a standard Brownian motion, it is a Lévy process and if,*

(1) for $t > 0$, $\boldsymbol{W}_t$ has a Gaussian distribution with mean vector $\boldsymbol{0}$ and covariance matrix tI_m (I_m is the identity matrix)

(2) there is $\Omega_0 \in \mathcal{F}$ with $P[\Omega_0] = 1$ such that, for every $\omega \in \Omega_0$, $\boldsymbol{W}_t(\omega)$ is continuous in t.

Definition 4.2.3. *(compound Poisson process) Let $\{N_t\}_{t\geq 0}$ be a Poisson process with constant intensity parameter $\lambda > 0$ and $\{\boldsymbol{Z}_n\}_{n=1,2,\ldots}$ be a sequence of independent and identically distributed* *(i.i.d.) random vectors (jump amplitudes) with the probability density* $f(\boldsymbol{z})$ *on* $\mathbb{R}^m \setminus \{0\}$. *Assume that* $\{N_t\}$ *and* $\{\boldsymbol{Z}_t\}$ *are independent. Define*

$$\boldsymbol{Y}_0 = \boldsymbol{0}, \quad \boldsymbol{Y}_t = \sum_{n=1}^{N_t} \boldsymbol{Z}_n.$$

Then the stochastic process $\{\boldsymbol{Y}_t\}_{t\geq 0}$ *on* $\mathbb{R}^m$ *is a compound Poisson process associated with* λ *and* f.

According to Sato (1999), *the most basic stochastic process modeled for continuous random motions is Brownian motion and that for jumping random motions is the Poisson process. These two belong to a class of Lévy processes. Lévy processes are, speaking only of essential points, stochastic processes with stationary independent increments.*

The following results guarantee that any Lévy process can be characterized by (A, ν, γ), which is called the generating triplet of the process.

Lemma 4.2.4. *(Theorems 7.10 and 8.1 of Sato (1999)) Let* $\{\boldsymbol{X}_t\}_{t\geq 0}$ *be a class of d-dimensional Lévy processes and denote its characteristic function by* $\mu(\boldsymbol{u}) := E[e^{i\boldsymbol{u}'\boldsymbol{X}_1}]$. *Then, we can write uniquely*

$$\mu(\boldsymbol{u}) = \exp\left[-\frac{1}{2}\boldsymbol{u}'A\boldsymbol{u} + i\gamma'\boldsymbol{u} + \int_{\mathbb{R}^d}\left(e^{i\boldsymbol{u}'\boldsymbol{x}} - 1 - i\boldsymbol{u}'\boldsymbol{x}\mathbb{I}_{\{\boldsymbol{x}:\|\boldsymbol{x}\|\leq 1\}}(\boldsymbol{x})\right)\nu(d\boldsymbol{x})\right]$$

where A (called the Gaussian covariance matrix) is a symmetric nonnegative-definite $d \times d$ *matrix,* ν *(called the Lévy measure of* $\{\boldsymbol{X}_t\}$*) is a measure on* $\mathbb{R}^d$ *satisfying*

$$\nu(\{0\}) = 0 \quad \text{and} \quad \int_{\mathbb{R}^d} (\|\boldsymbol{x}\| \wedge 1)\,\nu(d\boldsymbol{x}) < \infty,$$

and $\gamma \in \mathbb{R}^d$. *In addition, if* ν *satisfies* $\int_{\|\boldsymbol{x}\|\leq 1} \|\boldsymbol{x}\|\nu(d\boldsymbol{x}) < \infty$, *then we can rewrite*

$$\mu(\boldsymbol{u}) = \exp\left[-\frac{1}{2}\boldsymbol{u}'A\boldsymbol{u} + i\gamma_0'\boldsymbol{u} + \int_{\mathbb{R}^d}\left(e^{i\boldsymbol{u}'\boldsymbol{x}} - 1\right)\nu(d\boldsymbol{x})\right]$$

with $\gamma_0 \in \mathbb{R}^d$.[11]

Remark 4.2.5. *Let* $\{\boldsymbol{Y}_t\}$ *be a (d-dimensional) compound Poisson process with intensity parameter* $\lambda > 0$ *and probability density* $f(\boldsymbol{z})$ *of* $\boldsymbol{Z}_n$. *Then, the characteristic function can be written as*

$$\mu(\boldsymbol{u}) = E[e^{i\boldsymbol{u}'\boldsymbol{Y}_1}] = \exp\left[\lambda\left\{E(e^{i\boldsymbol{u}'\boldsymbol{Z}_n}) - 1\right\}\right] = \exp\left[\lambda\int_{\mathbb{R}^d}\left(e^{i\boldsymbol{u}'\boldsymbol{z}} - 1\right)f(\boldsymbol{z})d\boldsymbol{z}\right], \tag{4.2.1}$$

which implies that this process is a class of Lévy processes by setting

$$A \equiv \boldsymbol{0}, \quad \gamma \equiv \boldsymbol{0} \quad \text{and} \quad \nu(d\boldsymbol{z}) \equiv \lambda f(\boldsymbol{z})d\boldsymbol{z}.$$

4.2.1 Optimal Consumption and Portfolio Weights

Suppose that asset prices are the following exponential Lévy dynamics:

$$\frac{dS_{0,t}}{S_{0,t}} = rdt, \tag{4.2.2}$$

$$\frac{dS_{i,t}}{S_{i,t-}} = (r + R_i)dt + \sum_{j=1}^{m}\sigma_{ij}dW_{j,t} + dY_{i,t}, \quad i = 1, \ldots, m \tag{4.2.3}$$

with a constant rate of interest $r \geq 0$. The quantities R_i and σ_{ij} are constant parameters. We write $\boldsymbol{R} = (R_1, \ldots, R_m)'$ and $\Sigma = \boldsymbol{\sigma}\boldsymbol{\sigma}'$ where $\boldsymbol{\sigma} = (\sigma_{ij})_{i,j=1,\ldots,m}$. We assume that Σ is a nonsingular matrix,

[11] See p. 39 of Sato (1999).

$\boldsymbol{W}_t = (W_{1,t}, \dots, W_{m,t})'$ is an m-dimensional standard Brownian motion and $\boldsymbol{Y}_t = (Y_{1,t}, \dots, Y_{m,t})'$ is an m-dimensional compound Poisson process with intensity parameter $\lambda > 0$ and probability density $f(\boldsymbol{z})$ of $\boldsymbol{Z}_n$. Let $\mathcal{W}_t$ denote an investor's wealth at time t starting with the initial endowment $\mathcal{W}_0$. Writing $H_{0,t}$ and $H_{i,t}$ ($i = 1, \dots, m$) as the number of shares for the riskless and risky assets at time t, the wealth $\mathcal{W}_t$ can be written as

$$\mathcal{W}_t = H_{0,t}S_{0,t} + \sum_{i=1}^{m} H_{i,t}S_{i,t-} = H_{0,t}S_{0,t} + \boldsymbol{H}_t'\boldsymbol{S}_{t-} \tag{4.2.4}$$

where $\boldsymbol{H}_t = (H_{1,t}, \dots, H_{m,t})'$, $\boldsymbol{S}_t = (S_{1,t}, \dots, S_{m,t})'$ and $S_{0,t}, S_{1,t}, \dots, S_{m,t}$ are defined in (4.2.2) and (4.2.3). We assume that the investor consumes continuously at the rate C_t at time t. In the absence of any income derived outside his investments in these assets, we can write for a small period $[t-h, t)$

$$\begin{aligned}
-C_t h &\approx (H_{0,t} - H_{0,t-h})S_{0,t} + (\boldsymbol{H}_t - \boldsymbol{H}_{t-h})'\boldsymbol{S}_{t-} \\
&= (H_{0,t} - H_{0,t-h})(S_{0,t} - S_{0,t-h}) + (\boldsymbol{H}_t - \boldsymbol{H}_{t-h})'(\boldsymbol{S}_{t-} - \boldsymbol{S}_{t-h}) \\
&\quad + (H_{0,t} - H_{0,t-h})S_{0,t-h} + (\boldsymbol{H}_t - \boldsymbol{H}_{t-h})'\boldsymbol{S}_{t-}.
\end{aligned} \tag{4.2.5}$$

Taking the limits as $h \to 0$, we have

$$-C_t dt = dH_{0,t}dS_{0,t} + (d\boldsymbol{H}_t)'d\boldsymbol{S}_t + dH_{0,t}S_{0,t} + (d\boldsymbol{H}_t)'\boldsymbol{S}_{t-}. \tag{4.2.6}$$

Using Itô's lemma and plugging in (4.2.2), (4.2.3) and (4.2.6), the investor's wealth $\mathcal{W}_t \equiv X(S_{0,t}, \boldsymbol{S}_t, H_{0,t}, \boldsymbol{H}_t)$ follows the dynamics

$$\begin{aligned}
d\mathcal{W}_t &= dH_{0,t}S_{0,t} + (d\boldsymbol{H}_t)'\boldsymbol{S}_{t-} + H_{0,t}dS_{0,t} + \boldsymbol{H}_t'd\boldsymbol{S}_t + dH_{0,t}dS_{0,t} + (d\boldsymbol{H}_t)'d\boldsymbol{S}_t && (4.2.7)\\
&= -C_t dt + H_{0,t}dS_{0,t} + \boldsymbol{H}_t'd\boldsymbol{S}_{t-}. && (4.2.8)
\end{aligned}$$

Let $w_{0,t}$ and $\boldsymbol{w}_t = (w_{1,t}, \dots, w_{m,t})'$ denote the portfolio weights invested at time t in the riskless and the m risky assets. The portfolio weights can be written as

$$w_{0,t} = \frac{H_{0,t}S_{0,t}}{\mathcal{W}_t}, \quad w_{i,t} = \frac{H_{i,t}S_{i,t-}}{\mathcal{W}_t}, \; i = 1, \dots, m, \tag{4.2.9}$$

which implies that the portfolio weights satisfy

$$w_{0,t} + \sum_{i=1}^{m} w_{i,t} = \frac{1}{\mathcal{W}_t}(H_{0,t}S_{0,t} + \boldsymbol{H}_t'\boldsymbol{S}_{t-}) = 1, \tag{4.2.10}$$

by (4.2.4). By using $w_{0,t}$ and $\boldsymbol{w}_t$ in place of $H_{0,t}$ and $\boldsymbol{H}_t$, we can rewrite (4.2.8) as

$$\begin{aligned}
d\mathcal{W}_t &= -C_t dt + w_{0,t}\mathcal{W}_t \frac{dS_{0,t}}{S_{0,t}} + \sum_{i=1}^{m} \mathcal{W}_t \frac{dS_{i,t}}{S_{i,t-}} \\
&= -C_t dt + \mathcal{W}_t w_{0,t} r dt + \sum_{i=1}^{m} w_{i,t}\mathcal{W}_t \left\{ (r + R_i)dt + \sum_{j=1}^{m} \sigma_{i,j} dW_{j,t} + dY_{i,t} \right\} \\
&= (r\mathcal{W}_t + \boldsymbol{w}_t'\boldsymbol{R}\mathcal{W}_t - C_t)dt + \mathcal{W}_t \boldsymbol{w}_t'\boldsymbol{\sigma} d\boldsymbol{W}_t + \mathcal{W}_t \boldsymbol{w}_t' d\boldsymbol{Y}_t.
\end{aligned} \tag{4.2.11}$$

Here the second equation comes from (4.2.2) and (4.2.3).

The investor's problem at time t is then to pick the consumption and portfolio weight processes $\{C_s, \boldsymbol{w}_s\}_{t \le s \le \infty}$ which maximize the infinite horizon, discounted at rate β, expected utility of consumption (so-called value function)

$$V(\mathcal{W}_t, t) = \max_{\{C_s, \boldsymbol{w}_s\}_{t \le s \le \infty}} E_t \left[\int_t^{\infty} e^{-\beta s} U(C_s) ds \right] \tag{4.2.12}$$

subject to $\lim_{t\to\infty} E[V(\mathcal{W}_t, t)] = 0$ and (4.2.11) with given $\mathcal{W}_t$. Here $U(\cdot)$ denotes a utility function of consumption C_s and $E_t(\cdot)$ denotes the conditional expectation, that is, $E_t(\cdot) = E(\cdot|\mathcal{F}_t)$. For a small period $[t, t+h)$, this value function can be written as

$$\begin{aligned} V(\mathcal{W}_t, t) &= \max_{\{C_s, \boldsymbol{w}_s\}_{s\in[t,t+h)}} E_t\left[\max_{\{C_s, \boldsymbol{w}_s\}_{s\in[t+h,\infty]}} E_{t+h}\left\{\int_t^{\infty} e^{-\beta s} U(C_s)\right\} ds\right] \\ &= \max_{\{C_s, \boldsymbol{w}_s\}_{s\in[t,t+h)}} E_t\left[\int_t^{t+h} e^{-\beta s} U(C_s) ds \right. \\ &\qquad\qquad \left. + \max_{\{C_s, \boldsymbol{w}_s\}_{s\in[t+h,\infty]}} E_{t+h}\left\{\int_{t+h}^{\infty} e^{-\beta s} U(C_s)\right\} ds\right] \\ &= \max_{\{C_s, \boldsymbol{w}_s\}_{s\in[t,t+h)}} E_t\left[\int_t^{t+h} e^{-\beta s} U(C_s) ds + V(\mathcal{W}_{t+h}, t+h)\right]. \end{aligned}$$

Then, it follows that for a small period $[t, t+h)$

$$0 \approx \max_{\{C_s, \boldsymbol{w}_s\}_{s\in[t,t+h)}} E_t\left[e^{-\beta t} U(C_t) h + V(\mathcal{W}_{t+h}, t+h) - V(\mathcal{W}_t, t)\right].$$

Taking the limits as $h \to 0$, we have

$$0 = \max_{\{C_t, \boldsymbol{w}_t\}} \left[e^{-\beta t} U(C_t) + E_t\left\{\frac{dV(\mathcal{W}_t, t)}{dt}\right\}\right]. \tag{4.2.13}$$

Based on Itô's lemma for the diffusion process with jump (see e.g., Shreve (2004)), $dV(\mathcal{W}_t, t)$ can be written as

$$\begin{aligned} dV(\mathcal{W}_t, t) &= \frac{\partial V(\mathcal{W}_t, t)}{\partial t} dt + \frac{\partial V(\mathcal{W}_t, t)}{\partial \mathcal{W}_t} d\mathcal{W}_t^{(C)} + \frac{1}{2}\frac{\partial^2 V(\mathcal{W}_t, t)}{\partial \mathcal{W}_t^2} (d\mathcal{W}_t^{(C)})^2 \\ &\quad + 1_{\{\Delta N_t = 1\}} \{V(\mathcal{W}_t + \mathcal{W}_t \boldsymbol{w}_t' \boldsymbol{Z}_{N_t}, t) - V(\mathcal{W}_t, t)\} \end{aligned} \tag{4.2.14}$$

where $d\mathcal{W}_t^{(C)}$ denotes the continuous part of $d\mathcal{W}_t$ (i.e., $d\mathcal{W}_t^{(C)} = (r\mathcal{W}_t + \boldsymbol{w}_t' \boldsymbol{R} \mathcal{W}_t - C_t)dt + \mathcal{W}_t \boldsymbol{w}_t' \boldsymbol{\sigma} d\boldsymbol{W}_t$) and the fourth term of the right hand side describes the jump effect. Since the jump occurs discretely, the number of the jump from $t-$ to t is 0 or 1. If there is no jump in this period (i.e., $\Delta N_t = N_t - N_{t-} = 0$), the jump effect in dV is ignored, otherwise (i.e., $\Delta N_t = N_t - N_{t-} = 1$), the jump effect in dV is shown as $V(\mathcal{W}_t + \Delta \boldsymbol{Y}_t, t) - V(\mathcal{W}_t, t)$ where

$$\Delta \boldsymbol{Y}_t := \mathcal{W}_t \boldsymbol{w}_t' (\boldsymbol{Y}_t - \boldsymbol{Y}_{t-}) = \mathcal{W}_t \boldsymbol{w}_t' \left(\sum_{n=1}^{N_t} \boldsymbol{Z}_n - \sum_{n=1}^{N_{t-}} \boldsymbol{Z}_n\right) = \mathcal{W}_t \boldsymbol{w}_t' \boldsymbol{Z}_{N_t}$$

given $\mathcal{W}_t$, $\boldsymbol{w}_t$, Y_{t-} and $\Delta N_t = N_t - N_{t-} = 1$. From the definition of $\boldsymbol{W}_t$, the conditional expectations are given by

$$\begin{aligned} E_t\left(d\mathcal{W}_t^{(C)}\right) &= (r\mathcal{W}_t + \boldsymbol{w}' \boldsymbol{R} \mathcal{W}_t - C_t)dt + \mathcal{W}_t \boldsymbol{w}_t' \boldsymbol{\sigma} E(d\boldsymbol{W}_t) \\ &= (r\mathcal{W}_t + \boldsymbol{w}' \boldsymbol{R} \mathcal{W}_t - C_t)dt \end{aligned} \tag{4.2.15}$$

and

$$\begin{aligned} E_t\left\{(d\mathcal{W}_t^{(C)})^2\right\} &= (r\mathcal{W}_t + \boldsymbol{w}' \boldsymbol{R} \mathcal{W}_t - C_t)^2 dt^2 + \mathcal{W}_t^2 \boldsymbol{w}_t' \boldsymbol{\sigma} E(d\boldsymbol{W}_t d\boldsymbol{W}_t') \boldsymbol{\sigma}' \boldsymbol{w}_t \\ &\quad + (r\mathcal{W}_t + \boldsymbol{w}' \boldsymbol{R} \mathcal{W}_t - C_t)\mathcal{W}_t \boldsymbol{w}_t' \boldsymbol{\sigma} E(d\boldsymbol{W}_t) dt \\ &= \mathcal{W}_t^2 \boldsymbol{w}_t' \Sigma \boldsymbol{w}_t dt + o(dt). \end{aligned} \tag{4.2.16}$$

From the definition of N_t and $\boldsymbol{Z}_{N_t}$, the another conditional expectation is given by

$$E_t\left[1_{\{\Delta N_t = 1\}} \{V(\mathcal{W}_t + \mathcal{W}_t \boldsymbol{w}_t' \boldsymbol{Z}_{N_t}, t) - V(\mathcal{W}_t, t)\}\right]$$

$$= P\{\Delta N_t = 1|N_{t-}\} E_t\{V(\mathcal{W}_t + \mathcal{W}_t w_t' Z_{N_t}, t) - V(\mathcal{W}_t, t)\}$$
$$= \left(\lambda dt e^{-\lambda dt}\right) \int \{V(\mathcal{W}_t + \mathcal{W}_t w_t' z, t) - V(\mathcal{W}_t, t)\} d\nu(z)$$
$$= \lambda dt \int \{V(\mathcal{W}_t + \mathcal{W}_t w_t' z, t) - V(\mathcal{W}_t, t)\} d\nu(z) + o(dt) \qquad (4.2.17)$$

where $e^{-\lambda dt} = 1 - \lambda dt + o(dt)$ from Taylor expansion. Hence, from (4.2.14) to (4.2.17), we can write

$$E_t\left\{\frac{V(\mathcal{W}_t, t)}{dt}\right\} = \frac{\partial V(\mathcal{W}_t, t)}{\partial t} + \frac{\partial V(\mathcal{W}_t, t)}{\partial \mathcal{W}_t} E_t\left(\frac{d\mathcal{W}_t^{(C)}}{dt}\right) + \frac{1}{2}\frac{\partial^2 V(\mathcal{W}_t, t)}{\partial \mathcal{W}_t^2} E_t\left\{\frac{(d\mathcal{W}_t^{(C)})^2}{dt}\right\}$$
$$+ \frac{1}{dt} E_t\left[1_{\{\Delta N_t = 1\}}\{V(\mathcal{W}_t + \mathcal{W}_t w_t' Z_{N_t}, t) - V(\mathcal{W}_t, t)\}\right]$$
$$= \frac{\partial V(\mathcal{W}_t, t)}{\partial t} + \frac{\partial V(\mathcal{W}_t, t)}{\partial \mathcal{W}_t}(r\mathcal{W}_t + w' \boldsymbol{R} \mathcal{W}_t - C_t) + \frac{1}{2}\frac{\partial^2 V(\mathcal{W}_t, t)}{\partial \mathcal{W}_t^2} \mathcal{W}_t^2 w_t' \Sigma w_t$$
$$+ \lambda \int \{V(\mathcal{W}_t + \mathcal{W}_t w_t' z, t) - V(\mathcal{W}_t, t)\} \nu(dz) + o(1).$$

Summarizing the above argument, the Bellman equation (4.2.13) can be written as

$$0 = \max_{\{C_t, w_t\}} \{\phi(C_t, w_t; \mathcal{W}_t, t)\} \qquad (4.2.18)$$

where

$$\phi(C_t, w_t; \mathcal{W}_t, t) = e^{-\beta t} U(C_t) + \frac{\partial V(\mathcal{W}_t, t)}{\partial t} + \frac{\partial V(\mathcal{W}_t, t)}{\partial \mathcal{W}_t}(r\mathcal{W}_t + w' \boldsymbol{R} \mathcal{W}_t - C_t)$$
$$+ \frac{1}{2}\frac{\partial^2 V(\mathcal{W}_t, t)}{\partial \mathcal{W}_t^2} \mathcal{W}_t^2 w_t' \Sigma w_t + \lambda \int \{V(\mathcal{W}_t + \mathcal{W}_t w_t' z, t) - V(\mathcal{W}_t, t)\} \nu(dz)$$

subject to $\lim_{t\to\infty} E[V(\mathcal{W}_t, t)] = 0$.
Although the value function V is time dependent, it can be written from (4.2.12) for an infinite time horizon problem

$$e^{\beta t} V(\mathcal{W}_t, t) = \max_{\{C_s, w_s\}_{t \le s \le \infty}} E_t\left[\int_t^\infty e^{-\beta(s-t)} U(C_s) ds\right]$$
$$= \max_{\{C_{t+u}, w_{t+u}\}_{0 \le u \le \infty}} E_t\left[\int_0^\infty e^{-\beta u} U(C_{t+u}) du\right]$$
$$= \max_{\{C_u, w_u\}_{0 \le u \le \infty}} E_0\left[\int_0^\infty e^{-\beta u} U(C_u) du\right] := L(\mathcal{W}_t)$$

from the Markov property where $L(\mathcal{W}_t)$ is independent of time. Thus, (4.2.18) reduces to the following equation by using the time-homogeneous value function L:

$$0 = \max_{\{C_t, w_t\}} \{\phi(C_t, w_t; \mathcal{W}_t)\} \qquad (4.2.19)$$

where

$$\phi(C_t, w_t; \mathcal{W}_t) = U(C_t) - \beta L(\mathcal{W}_t) + \frac{\partial L(\mathcal{W}_t)}{\partial \mathcal{W}_t}(r\mathcal{W}_t + w_t' \boldsymbol{R} \mathcal{W}_t - C_t)$$
$$+ \frac{1}{2}\frac{\partial^2 L(\mathcal{W}_t)}{\partial \mathcal{W}_t^2} \mathcal{W}_t^2 w_t' \Sigma w_t + \lambda \int \{L(\mathcal{W}_t + \mathcal{W}_t w_t' z) - L(\mathcal{W}_t)\} \nu(dz) \qquad (4.2.20)$$

subject to $\lim_{t\to\infty} E[e^{-\beta t}L(\mathcal{W}_t)] = 0$.
In order to solve the maximization problem in (4.2.19), we calculate FOCs (first order conditions) for C_t and $\boldsymbol{w}_t$ separately. As for C_t, it follows that

$$\frac{\partial\phi(C_t, \boldsymbol{w}_t; \mathcal{W}_t)}{\partial C_t} = \frac{\partial U(C_t)}{\partial C_t} - \frac{\partial L(\mathcal{W}_t)}{\partial \mathcal{W}_t} = 0.$$

Given a wealth $\mathcal{W}_t$, the optimal consumption choice C_t^* is therefore

$$C_t^* = \left[\frac{\partial U}{\partial C}\right]^{-1} \frac{\partial L(\mathcal{W}_t)}{\partial \mathcal{W}_t}.$$

As for $\boldsymbol{w}_t$, the optimal portfolio weight $\boldsymbol{w}_t^*$ is obtained as the maximizer of the following objective function:

$$\begin{aligned}
&\frac{\partial L(\mathcal{W}_t)}{\partial \mathcal{W}_t}\boldsymbol{w}_t'\boldsymbol{R}\mathcal{W}_t + \frac{1}{2}\frac{\partial^2 V(\mathcal{W}_t)}{\partial \mathcal{W}_t^2}\mathcal{W}_t^2\boldsymbol{w}_t'\Sigma\boldsymbol{w}_t + \lambda\int \{L(\mathcal{W}_t + \mathcal{W}_t\boldsymbol{w}_t'\boldsymbol{z}) - L(\mathcal{W}_t)\}\,\nu(d\boldsymbol{z}) \\
&:= -\frac{\partial L(\mathcal{W}_t)}{\partial \mathcal{W}_t}\mathcal{W}_t\left\{-\boldsymbol{w}_t'\boldsymbol{R} + \frac{A}{2}\boldsymbol{w}_t'\Sigma\boldsymbol{w}_t + \lambda\psi(\boldsymbol{w}_t)\right\}
\end{aligned}$$

given a wealth $\mathcal{W}_t$, where

$$A = \left(-\frac{\partial L(\mathcal{W}_t)}{\partial \mathcal{W}_t}\right)^{-1}\frac{\partial^2 L(\mathcal{W}_t)}{\partial \mathcal{W}_t^2}\mathcal{W}_t \qquad (4.2.21)$$

$$\psi(\boldsymbol{w}_t) = \left(-\frac{\partial L(\mathcal{W}_t)}{\partial \mathcal{W}_t}\mathcal{W}_t\right)^{-1}\int \{L(\mathcal{W}_t + \mathcal{W}_t\boldsymbol{w}_t'\boldsymbol{z}) - L(\mathcal{W}_t)\}\,\nu(d\boldsymbol{z}). \qquad (4.2.22)$$

If L is a strictly convex function for $x > 0$, it satisfies that $-\frac{\partial L(\mathcal{W}_t)}{\partial \mathcal{W}_t}\mathcal{W}_t < 0$ (then, max becomes min) and A is a constant value, which implies that

$$\boldsymbol{w}_t^* := \arg\min_{\{\boldsymbol{w}_t\}} g(\boldsymbol{w}_t)$$

where

$$g(\boldsymbol{w}_t) = -\boldsymbol{w}_t'\boldsymbol{R} + \frac{A}{2}\boldsymbol{w}_t'\Sigma\boldsymbol{w}_t + \lambda\psi(\boldsymbol{w}_t).$$

Example 4.2.6. *(CRRA power utility) Suppose that the utility function U is a power utility,*[12] *that is,*

$$U(c) = \begin{cases} \frac{c^{1-\gamma}}{1-\gamma} & if\ c > 0 \\ -\infty & if\ c \le 0 \end{cases}$$

with CRRA coefficient $\gamma \in (0,1)\cup(1,\infty)$. Following Aït-Sahalia et al. (2009), we look for a solution to (4.2.19) in the form

$$L(x) = K^{-\gamma}\frac{x^{1-\gamma}}{1-\gamma}$$

where K is a constant value, then

$$\frac{\partial L(x)}{\partial x} = (1-\gamma)\frac{L(x)}{x}, \quad \frac{\partial^2 L(x)}{\partial x^2} = -\gamma(1-\gamma)\frac{L(x)}{x^2}.$$

[12] Aït-Sahalia et al. (2009) also considered the exponential and log utility cases, briefly.

Then (4.2.21) and (4.2.22) reduce to

$$A = \left\{-(1-\gamma)\frac{L(\mathcal{W}_t)}{\mathcal{W}_t}\right\}^{-1}\left\{-\gamma(1-\gamma)\frac{L(\mathcal{W}_t)}{\mathcal{W}_t^2}\mathcal{W}_t\right\} = \gamma \tag{4.2.23}$$

$$\begin{aligned}\psi(\boldsymbol{w}_t) &= \left\{-(1-\gamma)\frac{L(\mathcal{W}_t)}{\mathcal{W}_t}\mathcal{W}_t\right\}^{-1} L(\mathcal{W}_t)\int\left\{(1+\boldsymbol{w}_t'\boldsymbol{z})^{1-\gamma}-1\right\}\nu(d\boldsymbol{z})\\ &= -\frac{1}{1-\gamma}\int\left\{(1+\boldsymbol{w}_t'\boldsymbol{z})^{1-\gamma}-1\right\}\nu(d\boldsymbol{z}).\end{aligned} \tag{4.2.24}$$

In this case, $g(\boldsymbol{w}_t)$ is time independent under given constant $\boldsymbol{R}$, $A(=\gamma)$, Σ, λ and a function ν, so it is clear that any optimal solution will be time independent. Furthermore, the objective function is state independent, so any optimal solution will also be state independent. In addition, according to Aït-Sahalia et al. (2009), the constant K will be fully determined once we have solved for the optimal portfolio weights $\boldsymbol{w}_t^*$.

4.2.2 Estimation

As discussed previously, under the convexity of L, we can write optimal portfolio weights as

$$\boldsymbol{w}^* := \arg\min_{\boldsymbol{w}} g(\boldsymbol{w}) \tag{4.2.25}$$

where

$$g(\boldsymbol{w}) = -\boldsymbol{w}'\boldsymbol{R} + \frac{A}{2}\boldsymbol{w}'\Sigma\boldsymbol{w} + \lambda\psi(\boldsymbol{w}).$$

Note that $g(\boldsymbol{w})$ is time independent under some appropriate utility functions such as the CRRA power utility (see Example 4.2.6), which implies that the optimal portfolio $\boldsymbol{w}^*$ is also time independent. We now consider that the objective function g is parameterized by $\boldsymbol{\theta} \in \Theta \subset \mathbb{R}^q$, that is,

$$g(\boldsymbol{w}) \equiv g_{\boldsymbol{\theta}}(\boldsymbol{w}) = -\boldsymbol{w}'\boldsymbol{R}_{\boldsymbol{\theta}} + \frac{A}{2}\boldsymbol{w}'\Sigma_{\boldsymbol{\theta}}\boldsymbol{w} + \lambda_{\boldsymbol{\theta}}\psi_{\boldsymbol{\theta}}(\boldsymbol{w}).$$

Then, the optimal portfolio weights (4.2.25) can be written as

$$\boldsymbol{w}^*(\boldsymbol{\theta}) := \arg\min_{\boldsymbol{w}} g_{\boldsymbol{\theta}}(\boldsymbol{w}). \tag{4.2.26}$$

Therefore, once the parameter estimator $\hat{\boldsymbol{\theta}}$ of $\boldsymbol{\theta}$ is constructed based on discretely observed data $\{\boldsymbol{S}_t\}$, we can obtain the optimal portfolio estimator $\boldsymbol{w}^*(\hat{\boldsymbol{\theta}})$ by the plug-in method.

Suppose that the asset price processes $\{\boldsymbol{S}_t = (S_{1,t},\ldots,S_{m,t})'\}$ are defined by (4.2.3) with $r = 0$ and $S_{i,t} > 0$ for $i = 1,\ldots,m$ and $t \in [0,T]$. Write

$$X_{i,t} := \log S_{i,t} \quad \textit{and} \quad X_{i,0} = 0.$$

Based on Itô's lemma for the diffusion process with jump, it follows that

$$\begin{aligned}X_{i,t} &= \log S_{i,0} + \int_0^t \frac{\partial \log S_i}{\partial S_i} dS_{i,s}^{(C)} + \frac{1}{2}\int_0^t \frac{\partial^2 \log S_i}{\partial S_i^2}(dS_{i,t}^{(C)})^2\\ &\quad + \sum_{0<s\le t}(\log S_{i,s} - \log S_{i,s-})\\ &= \log S_{i,0} + \int_0^t \frac{dS_{i,s}^{(C)}}{S_{i,s-}} - \frac{1}{2}\int_0^t\left(\frac{dS_{i,s}^{(C)}}{S_{i,s-}}\right)^2 + \sum_{0<s\le t}(\log S_{i,s} - \log S_{i,s-})\end{aligned}$$

where

$$\log S_{i,0} = X_{i,0} = 0, \quad \frac{dS^{(C)}_{i,t}}{S_{i,t-}} = R_i dt + \sum_{j=1}^{m} \sigma_{ij} dW_{j,t}, \quad \left(\frac{dS^{(C)}_{i,t}}{S_{i,t-}}\right)^2 = \left(\sum_{j=1}^{m} \sigma_{ij}\right)^2 dt = \Sigma_{ii} dt$$

with $\boldsymbol{\sigma}\boldsymbol{\sigma}' = \left(\sum_{j_i,j_2=1}^{m} \sigma_{i_1 j_1}\sigma_{j_2 i_2}\right)_{i_1,i_2=1,\ldots,m} = (\Sigma_{i_1 i_2})_{i_1,i_2=1,\ldots,m} = \Sigma$.
Denote

$$\boldsymbol{\mu} := \boldsymbol{R} - \frac{1}{2}(\Sigma_{11}, \ldots, \Sigma_{mm})', \quad \boldsymbol{M}_t := (M_{1,t}, \ldots, M_{m,t})', \quad \boldsymbol{U}_r := (U_{1,r}, \ldots, U_{m,r})'$$

where

$$\begin{aligned}
\sum_{0<s\le t} (\log S_{i,s} - \log S_{i,s-}) &= \sum_{0<s\le t} \{\log(S_{i,s-}(1 + Y_{i,s} - Y_{i,s-})) - \log S_{i,s-}\} \\
&= \sum_{0<s\le t} \{\log(S_{i,s-}(1 + 1_{\{N_s - N_{s-}=1\}} Z_{i,N_s})) - \log S_{i,s-}\} \\
&= \sum_{0<s\le t} \log(1 + 1_{\{N_s - N_{s-}=1\}} Z_{i,N_s}) \\
&:= \sum_{0<s\le t} 1_{\{N_s - N_{s-}=1\}} U_{i,N_s} \quad (say) \\
&= \sum_{r=1}^{N_t} U_{i,r} \\
&= M_{i,t} \quad (say).
\end{aligned}$$

Then, we can write

$$\boldsymbol{X}_t := (X_{1,t}, \ldots, X_{m,t})' = \boldsymbol{\mu} t + \boldsymbol{\sigma} \boldsymbol{W}_t + \boldsymbol{M}_t, \quad \boldsymbol{M}_t := \sum_{r=1}^{N_t} \boldsymbol{U}_r. \tag{4.2.27}$$

In what follows, we will consider estimation theory of the parameter $\boldsymbol{\theta} := (\boldsymbol{\mu}', vec(\boldsymbol{\sigma})', \boldsymbol{\xi}')'$ (where $\boldsymbol{\xi}$ is a parameter which characterizes the intensity parameter $\lambda_{\boldsymbol{\xi}}$ of N_t and the density $f_{\boldsymbol{\xi}}(\boldsymbol{u})$ of $\boldsymbol{U}_r$) from discrete observations $\{\boldsymbol{X}_t = (X_{1,t}, \ldots, X_{m,t})' = (\log S_{1,t}, \ldots, \log S_{m,t})'; t = t_1, \ldots, t_n\}$. Especially, we will focus on two types of estimation method: the *generalized method of moments (GMM)* and the *threshold estimation method*.

4.2.2.1 Generalized method of moments (GMM)

Suppose that the log asset price process $\{\boldsymbol{X}_t = (X_{1,t}, \ldots, X_{m,t})'\}$ is defined by (4.2.27) and

$$\boldsymbol{U}_r \overset{i.i.d.}{\sim} N(\boldsymbol{\beta}, \Sigma_J).$$

In this case, we can treat the parameter vector as $\boldsymbol{\theta} := (\boldsymbol{\mu}', vech(\Sigma_C)', \lambda, \boldsymbol{\beta}', vech(\Sigma_J)')' \in \Theta \subset \mathbb{R}^q$ where $\Sigma_C = \boldsymbol{\sigma}\boldsymbol{\sigma}'$ and $q = 1 + 3m + m^2$.
We assume that observations are $\boldsymbol{X}_0, \boldsymbol{X}_1, \ldots, \boldsymbol{X}_n$ for $n \in \mathbb{N}$. Then, the conditional moments can be written as

$$\begin{aligned}
\boldsymbol{m}^{(1)}_t(\boldsymbol{\theta}) &:= \left(m^{(1)}_{i,t}(\boldsymbol{\theta})\right)_{i=1,\ldots,m} := E_{\boldsymbol{\theta}}(\boldsymbol{X}_t | \boldsymbol{X}_{t-1}) = \boldsymbol{X}_{t-1} + \boldsymbol{\mu} + \lambda\boldsymbol{\beta} \\
\boldsymbol{m}^{(2)}_t(\boldsymbol{\theta}) &:= \left(m^{(2)}_{ij,t}(\boldsymbol{\theta})\right)_{i,j=1,\ldots,m} := E_{\boldsymbol{\theta}}\left[\left\{\boldsymbol{X}_t - \boldsymbol{m}^{(1)}_t(\boldsymbol{\theta})\right\}\left\{\boldsymbol{X}_t - \boldsymbol{m}^{(1)}_t(\boldsymbol{\theta})\right\}' | \boldsymbol{X}_{t-1}\right] \\
&= \Sigma_C + \lambda\tilde{\Sigma}_J \\
\boldsymbol{m}^{(e)}_t(\boldsymbol{\theta}) &:= \left(m^{(e)}_{i,t}(\boldsymbol{\theta})\right)_{i=1,\ldots,m} := \left(E_{\boldsymbol{\theta}}(e^{X_{i,t}} | \boldsymbol{X}_{t-1})\right)_{i=1,\ldots,m} \\
&= \left(\exp\left\{X_{i,t-1} + \mu_i + \frac{1}{2}(\Sigma_C)_{ii} + \lambda\left(e^{\beta_i + \frac{1}{2}(\Sigma_J)_{ii}} - 1\right)\right\}\right)_{i=1,\ldots,m}
\end{aligned} \tag{4.2.28}$$

where $\tilde{\Sigma}_J = \Sigma_J + \beta\beta'$. By using these conditional moments, we introduce

$$h(\boldsymbol{X}_t, \boldsymbol{X}_{t-1}; \boldsymbol{\theta}) := \begin{pmatrix} h_1(\boldsymbol{X}_t, \boldsymbol{X}_{t-1}; \boldsymbol{\theta}) \\ h_2(\boldsymbol{X}_t, \boldsymbol{X}_{t-1}; \boldsymbol{\theta}) \\ h_e(\boldsymbol{X}_t, \boldsymbol{X}_{t-1}; \boldsymbol{\theta}) \end{pmatrix}$$

$$:= \begin{pmatrix} \boldsymbol{X}_t - \boldsymbol{m}_t^{(1)}(\boldsymbol{\theta}) \\ vech\left[\left\{\boldsymbol{X}_t - \boldsymbol{m}_t^{(1)}(\boldsymbol{\theta})\right\}\left\{\boldsymbol{X}_t - \boldsymbol{m}_t^{(1)}(\boldsymbol{\theta})\right\}' - \boldsymbol{m}_t^{(2)}(\boldsymbol{\theta})\right] \\ \left(e^{X_{i,t}} - m_{i,t}^{(e)}(\boldsymbol{\theta})\right)_{i=1,\dots,m} \end{pmatrix}. \tag{4.2.29}$$

Then, the $p(= 2m + m(m+1)/2)$-dimensional stochastic process $\{h(\boldsymbol{X}_t, \boldsymbol{X}_{t-1}; \boldsymbol{\theta}) : t = 1, 2, \dots\}$ satisfies

$$E_{\boldsymbol{\theta}}\{h(\boldsymbol{X}_t, \boldsymbol{X}_{t-1}; \boldsymbol{\theta})|\boldsymbol{X}_{t-1}\} = \boldsymbol{0} \quad \text{almost everywhere (a.e.)}$$

for all $t = 1, 2, \dots, n$. Moreover, we introduce a $q(= 1 + 3m + m^2)$-dimensional stochastic process $\{G_n(\boldsymbol{\theta}) : n = 1, 2, \dots\}$ by

$$G_n(\boldsymbol{\theta}) = \sum_{t=1}^{n} a(\boldsymbol{X}_{t-1}; \boldsymbol{\theta}) h(\boldsymbol{X}_t, \boldsymbol{X}_{t-1}; \boldsymbol{\theta})$$

where $a(\boldsymbol{X}_{t-1}; \boldsymbol{\theta})$ is a $q \times p$-dimensional function which depends only on $\boldsymbol{X}_{t-1}$ and $\boldsymbol{\theta}$, and is differentiable with respect to $\boldsymbol{\theta}$. Then, $G_n(\boldsymbol{\theta})$ is an unbiased martingale estimating function satisfying $E_{\boldsymbol{\theta}}\{G_n(\boldsymbol{\theta})\} = \boldsymbol{0}$ and

$$E_{\boldsymbol{\theta}}\{G_n(\boldsymbol{\theta})|\mathcal{F}_{n-1}\} = G_{n-1}(\boldsymbol{\theta}) \quad \text{a.e.}$$

for $n = 1, 2, \dots$, where $\mathcal{F}_{n-1}$ is the σ-field generated by the observations $\boldsymbol{X}_0, \boldsymbol{X}_1, \dots, \boldsymbol{X}_{n-1}$.

If $\|\mathrm{Var}_{\boldsymbol{\theta}}\{a(\boldsymbol{X}_{t-1}; \boldsymbol{\theta}) h(\boldsymbol{X}_t, \boldsymbol{X}_{t-1}; \boldsymbol{\theta})\}\| < \infty$ is satisfied, the martingale central limit theorem

$$\frac{1}{\sqrt{n}} G_n(\boldsymbol{\theta}) \xrightarrow{d} N(\boldsymbol{0}, V(\boldsymbol{\theta}))$$

holds where $V(\boldsymbol{\theta})$ is a $q \times q$-matrix given by

$$V(\boldsymbol{\theta}) = E_{\boldsymbol{\theta}}\{a(\boldsymbol{X}_{t-1}; \boldsymbol{\theta}) K_t(\boldsymbol{\theta}) a(\boldsymbol{X}_{t-1}; \boldsymbol{\theta})'\}$$

where $K_t(\boldsymbol{\theta}) = E_{\boldsymbol{\theta}}\{h(\boldsymbol{X}_t, \boldsymbol{X}_{t-1}; \boldsymbol{\theta}) h(\boldsymbol{X}_t, \boldsymbol{X}_{t-1}; \boldsymbol{\theta})' | \boldsymbol{X}_{t-1}\}$. Let $\hat{\boldsymbol{\theta}}_{GMM}$ be a solution of the estimating equation $G_n(\boldsymbol{\theta}) = \boldsymbol{0}$. By using Taylor expansion, we obtain

$$\boldsymbol{0} = G_n(\hat{\boldsymbol{\theta}}_{GMM}) = G_n(\boldsymbol{\theta}_0) + \dot{G}_n(\tilde{\boldsymbol{\theta}})(\hat{\boldsymbol{\theta}}_{GMM} - \boldsymbol{\theta}_0)$$

where $\dot{G}_n(\boldsymbol{\theta}) = \partial G_n(\boldsymbol{\theta})/\partial \boldsymbol{\theta}'$, $|\tilde{\boldsymbol{\theta}} - \boldsymbol{\theta}_0| \le |\hat{\boldsymbol{\theta}}_{GMM} - \boldsymbol{\theta}_0|$ and $\boldsymbol{\theta}_0$ is the true parameter. Then, assuming that

$$\frac{1}{n} \dot{G}_n(\tilde{\boldsymbol{\theta}}) \xrightarrow{p} E_{\boldsymbol{\theta}_0}\left\{a(\boldsymbol{X}_{t-1}; \boldsymbol{\theta}_0) \dot{h}(\boldsymbol{X}_t, \boldsymbol{X}_{t-1}; \boldsymbol{\theta}_0)\right\} := S(\boldsymbol{\theta}_0)$$

where $\dot{h}(\cdot, \cdot; \boldsymbol{\theta}) = \partial h(\cdot, \cdot; \boldsymbol{\theta})/\partial \boldsymbol{\theta}'$, we have

$$\sqrt{n}(\hat{\boldsymbol{\theta}}_{GMM} - \boldsymbol{\theta}_0) \xrightarrow{d} N\left(\boldsymbol{0}, S(\boldsymbol{\theta}_0)^{-1} V(\boldsymbol{\theta}_0) \left(S(\boldsymbol{\theta}_0)^{-1}\right)'\right) \tag{4.2.30}$$

as $n \to \infty$.

We now consider how to choose the best G_n in the class of unbiased martingale estimating functions. Let

$$\mathcal{G}_n := \left\{G_n(\boldsymbol{\theta}) : G_n(\boldsymbol{\theta}) = \sum_{t=1}^{n} a(\boldsymbol{X}_{t-1}; \boldsymbol{\theta}) h(\boldsymbol{X}_t, \boldsymbol{X}_{t-1}; \boldsymbol{\theta})\right\}$$

be a class of estimating functions satisfying (4.2.30). Here $\boldsymbol{a}$ is a function to be chosen and $\boldsymbol{h}$ is a fixed function defined in (4.2.29). For any $G_n(\boldsymbol{\theta})$, we define

$$I_{G_n}(\boldsymbol{\theta}) := \bar{G}_n(\boldsymbol{\theta})' \langle G(\boldsymbol{\theta})\rangle_n^{-1} \bar{G}_n(\boldsymbol{\theta})$$

where

$$\bar{G}_n(\boldsymbol{\theta}) = \sum_{t=1}^{n} \boldsymbol{a}(\boldsymbol{X}_{t-1};\boldsymbol{\theta}) E_{\boldsymbol{\theta}}\left\{\dot{\boldsymbol{h}}(\boldsymbol{X}_t, \boldsymbol{X}_{t-1};\boldsymbol{\theta})|\boldsymbol{X}_{t-1}\right\} \tag{4.2.31}$$

$$\langle G(\boldsymbol{\theta})\rangle_n = \langle G(\boldsymbol{\theta}), G(\boldsymbol{\theta})\rangle_n = \sum_{t=1}^{n} \boldsymbol{a}(\boldsymbol{X}_{t-1};\boldsymbol{\theta}) K_t(\boldsymbol{\theta}) \boldsymbol{a}(\boldsymbol{X}_{t-1};\boldsymbol{\theta})'. \tag{4.2.32}$$

Then, it follows that

$$\frac{1}{n} I_{G_n}(\boldsymbol{\theta}) \xrightarrow{p} S(\boldsymbol{\theta}) V(\boldsymbol{\theta})^{-1} S(\boldsymbol{\theta})',$$

which implies that the inverse of $I_{G_n}(\boldsymbol{\theta})$ estimates the covariance matrix of the asymptotic distribution of the estimator $\hat{\boldsymbol{\theta}}_{GMM}$. If an estimation function $G_n^* \in \mathcal{G}_n$ satisfies

$$I_{G_n^*}(\boldsymbol{\theta}) \geq I_{G_n}(\boldsymbol{\theta})$$

(i.e., $I_{G_n^*}(\boldsymbol{\theta}) - I_{G_n}(\boldsymbol{\theta})$ is nonnegative definite) for all $\boldsymbol{\theta} \in \Theta$, for all $G_n \in \mathcal{G}_n$ and for all $n \in \mathbb{N}$, G_n^* is called *optimal* in $\mathcal{G}_n$.

Lemma 4.2.7. *(Heyde (1988), Theorem 2.1 of Heyde (1997)) $G_n^* \in \mathcal{G}_n$ is an optimal estimating function within $\mathcal{G}_n$ if*

$$\bar{G}_n(\boldsymbol{\theta})^{-1} \langle G(\boldsymbol{\theta}), G^*(\boldsymbol{\theta})\rangle_n$$

is a constant matrix for all $G_n \in \mathcal{G}_n$.

Let $G_n, G_n^* \in \mathcal{G}_n$ be

$$G_n(\boldsymbol{\theta}) = \boldsymbol{a}(\boldsymbol{X}_{t-1};\boldsymbol{\theta})\boldsymbol{h}(\boldsymbol{X}_t, \boldsymbol{X}_{t-1};\boldsymbol{\theta}), \quad G_n^*(\boldsymbol{\theta}) = \boldsymbol{a}^*(\boldsymbol{X}_{t-1};\boldsymbol{\theta})\boldsymbol{h}(\boldsymbol{X}_t, \boldsymbol{X}_{t-1};\boldsymbol{\theta}),$$

respectively. Then, it follows that

$$\langle G(\boldsymbol{\theta}), G^*(\boldsymbol{\theta})\rangle_n = \sum_{t=1}^{n} \boldsymbol{a}(\boldsymbol{X}_{t-1};\boldsymbol{\theta}) K_t(\boldsymbol{\theta}) \boldsymbol{a}^*(\boldsymbol{X}_{t-1};\boldsymbol{\theta})'. \tag{4.2.33}$$

From (4.2.31) and (4.2.33), $\bar{G}_n(\boldsymbol{\theta})^{-1} \langle G(\boldsymbol{\theta}), G^*(\boldsymbol{\theta})\rangle_n$ is constant for all $G_n \in \mathcal{G}_n$ if

$$\boldsymbol{a}^*(\boldsymbol{X}_{t-1};\boldsymbol{\theta}) = -\dot{H}_t(\boldsymbol{\theta})' K_t(\boldsymbol{\theta})^{-1}$$

where $\dot{H}_t(\boldsymbol{\theta}) = E_{\boldsymbol{\theta}}\left\{\dot{\boldsymbol{h}}(\boldsymbol{X}_t, \boldsymbol{X}_{t-1};\boldsymbol{\theta})|\boldsymbol{X}_{t-1}\right\}$. An optimal estimating function is thus

$$G_n^*(\boldsymbol{\theta}) = -\dot{H}_t(\boldsymbol{\theta})' K_t(\boldsymbol{\theta})^{-1} \boldsymbol{h}(\boldsymbol{X}_t, \boldsymbol{X}_{t-1};\boldsymbol{\theta}).$$

From (4.2.29), $\dot{H}_t(\boldsymbol{\theta})$ is described as

$$\dot{H}_t(\boldsymbol{\theta}) = \begin{pmatrix} I_m & 0_{m\times m(m+1)/2} & \beta & \lambda I_m & 0_{m\times m(m+1)/2} \\ 0_{m(m+1)/2\times m} & I_{m(m+1)/2} & vech(\tilde{\Sigma}_J) & \dot{H}_t^{(2,\beta)}(\boldsymbol{\theta}) & \lambda I_{m(m+1)/2} \\ diag(\boldsymbol{m}_t^{(e)}(\boldsymbol{\theta})) & \dot{H}_t^{(e,\Sigma_C)}(\boldsymbol{\theta}) & \dot{H}_t^{(e,\lambda)}(\boldsymbol{\theta}) & \dot{H}_t^{(e,\beta)}(\boldsymbol{\theta}) & \dot{H}_t^{(e,\Sigma_J)}(\boldsymbol{\theta}) \end{pmatrix} \tag{4.2.34}$$

where

$$\dot{H}_t^{(2,\beta)}(\theta) = \left(\dot{H}_{ij,k,t}^{(2,\beta)}(\theta)\right)_{i,j,k=1,\ldots,m,j\leq k} = \left(\lambda\beta_i\delta(i,k) + \lambda\beta_j\delta(j,k)\right)_{i,j,k=1,\ldots,m,j\leq k}$$

$$\dot{H}_t^{(e,\Sigma_C)}(\theta) = \left(\dot{H}_{i,jk,t}^{(e,\Sigma_C)}(\theta)\right)_{i,j,k=1,\ldots,m,j\leq k} = \left(\frac{1}{2}m_{i,t}^{(e)}(\theta)\delta(i,j,k)\right)_{i,j,k=1,\ldots,m,j\leq k}$$

$$\dot{H}_t^{(e,\lambda)}(\theta) = \left(\dot{H}_{i,t}^{(e,\lambda)}(\theta)\right)_{i=1,\ldots,m} = \left(\left(\exp\{\beta_i + \frac{1}{2}(\Sigma_J)_{ii}\} - 1\right)m_{i,t}^{(e)}(\theta)\right)_{i=1,\ldots,m}$$

$$\dot{H}_t^{(e,\beta)}(\theta) = \left(\dot{H}_{i,j,t}^{(e,\beta)}(\theta)\right)_{i,j=1,\ldots,m} = \left(\lambda\exp\{\beta_i + \frac{1}{2}(\Sigma_J)_{ii}\}m_{i,t}^{(e)}(\theta)\delta(i,j)\right)_{i,j=1,\ldots,m}$$

$$\dot{H}_t^{(e,\Sigma_J)}(\theta) = \left(\dot{H}_{i,jk,t}^{(e,\Sigma_J)}(\theta)\right)_{i,j,k=1,\ldots,m,j\leq k} = \left(\frac{\lambda}{2}\exp\{\beta_i + \frac{1}{2}(\Sigma_J)_{ii}\}m_{i,t}^{(e)}(\theta)\delta(i,j,k)\right)_{i,j,k=1,\ldots,m,j\leq k}$$

for $\delta(i,j) = \begin{cases} 1 & if\ i = j \\ 0 & if\ i \neq j \end{cases}$ and $\delta(i,j,k) = \begin{cases} 1 & if\ i = j = k \\ 0 & if\ otherwise \end{cases}$.

Similarly, from (4.2.29), $K_t(\theta)$ is described as

$$K_t(\theta) = \begin{pmatrix} m_t^{(2)}(\theta) & K_t^{(1,2)}(\theta)' & K_t^{(1,e)}(\theta)' \\ K_t^{(1,2)}(\theta) & K_t^{(2,2)}(\theta) & K_t^{(2,e)}(\theta)' \\ K_t^{(1,e)}(\theta) & K_t^{(2,e)}(\theta) & K_t^{(e,e)}(\theta) \end{pmatrix} \quad (4.2.35)$$

where

$$\begin{aligned}
K_t^{(1,2)}(\theta) &= \left(K_{i,jk,t}^{(1,2)}(\theta)\right)_{i,j,k=1,\ldots,m,j\leq k} \\
&= \left(\lambda\left\{\beta_i(\Sigma_J)_{jk} + \beta_j(\Sigma_J)_{ik} + \beta_k(\Sigma_J)_{ij} + \beta_i\beta_j\beta_k\right\}\right)_{i,j,k=1,\ldots,m,j\leq k} \\
K_t^{(2,2)}(\theta) &= \left(K_{ij,kl,t}^{(2,2)}(\theta)\right)_{i,j,k,l=1,\ldots,m,i\leq j,k\leq l} \\
&= \Big(m_{ik,t}^{(2)}(\theta)m_{jl,t}^{(2)}(\theta) + m_{il,t}^{(2)}(\theta)m_{jk,t}^{(2)}(\theta) + \lambda\big\{(\tilde{\Sigma}_J)_{ij}(\tilde{\Sigma}_J)_{kl} + (\tilde{\Sigma}_J)_{ik}(\tilde{\Sigma}_J)_{jl} \\
&\qquad + (\tilde{\Sigma}_J)_{il}(\tilde{\Sigma}_J)_{jk} - 2\beta_i\beta_j\beta_k\beta_l\big\}\Big)_{i,j,k,l=1,\ldots,m,i\leq j,k\leq l} \\
K_t^{(1,e)}(\theta) &= \left(K_{i,j,t}^{(1,e)}(\theta)\right)_{i,j=1,\ldots,m} \\
&= \left(\left\{(\Sigma_C)_{ij} - \lambda\beta_i + \lambda(\beta_i + (\Sigma_J)_{ij})e^{\beta_j + \frac{1}{2}(\Sigma_J)_{jj}}\right\}m_{j,t}^{(e)}(\theta)\right)_{i,j=1,\ldots,m} \\
K_t^{(2,e)}(\theta) &= \left(K_{ij,k,t}^{(2,e)}(\theta)\right)_{i,j,k=1,\ldots,m,j\leq k} \\
&= \Bigg(\left[(\Sigma_C)_{ij} + \lambda\left\{(\Sigma_J)_{ij} + (\beta_i + (\Sigma_J)_{ik})(\beta_j + (\Sigma_J)_{jk})\right\}e^{\beta_k + \frac{1}{2}(\Sigma_J)_{kk}}\right]m_{k,t}^{(e)}(\theta) \\
&\qquad + \frac{K_{i,k,t}^{(1,e)}(\theta)K_{j,k,t}^{(1,e)}(\theta)}{m_{k,t}^{(e)}(\theta)} - m_{ij,t}^{(2)}(\theta)m_{k,t}^{(e)}(\theta)\Bigg)_{i,j,k=1,\ldots,m,j\leq k} \\
K_t^{(e,e)}(\theta) &= \left(K_{i,j,t}^{(1,2)}(\theta)\right)_{i,j=1,\ldots,m} \\
&= \left(\left[\exp\{(\Sigma_C)_{ij} + \lambda(e^{\beta_i + \frac{1}{2}(\Sigma_J)_{ii}} - 1)(e^{\beta_j + \frac{1}{2}(\Sigma_J)_{jj}} - 1)\} - 1\right]m_{i,t}^{(e)}(\theta)m_{j,t}^{(e)}(\theta)\right)_{i,j=1,\ldots,m}.
\end{aligned}$$

Let $\hat{\theta}_{GMM}^*$ be a solution of the estimating equation $G_n^*(\theta) = \mathbf{0}$. Then, we have

$$\sqrt{n}(\hat{\theta}_{GMM}^* - \theta_0) \xrightarrow{d} N\left(\mathbf{0}, V^*(\theta_0)^{-1}\right) \quad (4.2.36)$$

as $n \to \infty$, where

$$\frac{1}{n} I_{G_n^*}(\theta) = \frac{1}{n} \bar{G}_n^*(\theta)' \langle G^*(\theta) \rangle_n^{-1} \bar{G}_n^*(\theta) = \frac{1}{n} \sum_{t=1}^{n} \dot{H}_t(\theta)' K_t(\theta)^{-1} \dot{H}_t(\theta)$$
$$\xrightarrow{p} E_\theta\{\dot{H}_t(\theta)' K_t(\theta)^{-1} \dot{H}_t(\theta)\} = V^*(\theta).$$

Example 4.2.8. *(CRRA power utility) Consider the case of Example 4.2.6. Let* $\theta = (\mu', vech(\Sigma^{(C)})', \lambda, \beta', vech(\Sigma^{(J)})')'$. *Then, the optimal portfolio can be written by*

$$w^*(\theta) = \arg\min_{w \in \mathbb{R}^m} g_\theta(w)$$

where

$$g_\theta(w) = -w' \left\{ \mu + \frac{1}{2}(\Sigma_{11}^{(C)}, \ldots, \Sigma_{mm}^{(C)})' \right\} + \frac{\gamma}{2} w' \Sigma^{(C)} w + \lambda \psi_\theta(w)$$

and

$$\psi_\theta(w) = -\frac{1}{1-\gamma} \int_{z > -1} \left\{ (1 + w'z)^{1-\gamma} - 1 \right\} f_\theta(z) dz$$
$$f_\theta(z) = \phi_{(\beta, \Sigma^{(J)})}(\log(1+z_1), \ldots, \log(1+z_m)) \prod_{i=1}^{m} \frac{1}{1+z_i}.$$

Here $\phi_{(\beta,\Sigma^{(J)})}$ *denotes the p.d.f. of* $N(\beta, \Sigma^{(J)})$. *Then, from (4.2.36) and the* δ*-method, we have*

$$\sqrt{n}(w^*(\hat{\theta}_{GMM}^*) - w^*(\theta_0)) \xrightarrow{d} N\left(0, \dot{w}^*(\theta_0) V^*(\theta_0)^{-1} (\dot{w}^*(\theta_0))'\right)$$

where $\dot{w}^*(\theta) = \partial w^*(\theta)/\partial \theta'$.

4.2.2.2 Threshold estimation method

Suppose that the log asset price process $\{X_t = (X_{1,t}, \ldots, X_{m,t})'\}$ is defined by (4.2.27). In addition, we assume that the intensity parameter λ of N_t and the density $f(u)$ of U_r are parameterized by ξ (i.e., $\lambda \equiv \lambda_\xi$ and $f(u) \equiv f_\xi(u)$). Then, $M_t = \sum_{r=1}^{N_t} U_r$ is a vector-valued compound Poisson process defined in Definition 4.2.3 and has a homogeneous Poisson random measure

$$\mu_\xi(dt, du) = \sum_{s>0} 1_{\{\Delta M_s \neq 0\}} \epsilon_{(s, \Delta M_s)}(dt, du)$$

where $\Delta M_s = M_s - M_{s-}$ and ϵ_a is the Dirac measure at a. There exists the predictable compensator

$$E[\mu_\xi(dt, du)] = \nu_\xi(dz)dt = \lambda_\xi f_\xi(u) du dt.$$

In this case, we can treat the parameter vector as $\theta := (\mu', vec(\sigma)', \xi') \in \Theta$ (Θ is a compact space) and we assume that there exists the true parameter $\theta_0 := (\mu_0', vec(\sigma_0)', \xi_0')$ in Θ.
Suppose that the processes $X_t = (X_{1,t}, \ldots, X_{m,t})'$ are observed at the discrete times $0 = t_0 < t_1 < t_2 < \cdots < t_n = T$ in the interval $[0, T]$, and we assume that the sampling intervals are $t_i - t_{i-1} = h_n$ for $i = 1, \ldots, n$. We make the following assumption for the sampling scheme.

Assumption 4.2.9. $h_n \to 0$, $\quad T = t_n = nh_n \to \infty$ *and* $nh_n^2 \to 0$

According to Shimizu and Yoshida (2006), this sampling scheme is a so-called rapidly increasing experimental design. The estimation of discretely observed diffusion processes under similar sampling schemes has been studied very well by several authors, such as Prakasa Rao (1983, 1988), Florens-Zmirou (1989), Yoshida (1992) and Kessler (1997). On the other hand, the estimation of a diffusion process with jumps has been studied by, for example, Sørensen (1991), Aït-Sahalia (2004) and Shimizu and Yoshida (2006).
Shimizu and Yoshida (2006) tackled that problem and proposed an approximation of the log-likelihood of $\{\boldsymbol{X}_{t_i}\}_{i=1}^n$. They proposed an asymptotic filter to judge whether or not a jump had occurred in an observational interval $(t_{i-1}, t_i]$ of the form

$$\mathcal{H}_i^n := \{\omega \in \Omega; |\Delta_i^n \boldsymbol{X}| > h_n^\rho\}$$

where $\Delta_i^n \boldsymbol{X} = \boldsymbol{X}_{t_i} - \boldsymbol{X}_{t_{i-1}}$ and ρ is satisfied with $0 < \rho < \frac{1}{2}$. Let J_i^n be the number of jumps in the interval $(t_{i-1}, t_i]$ and for $k = 0, 1$, we denote

$$D_{i,k}^n := \mathcal{H}_i^n \cap \{J_i^n = k\}, \quad D_{i,2}^n := \mathcal{H}_i^n \cap \{J_i^n \geq 2\}$$
$$C_{i,k}^n := (\mathcal{H}_i^n)^c \cap \{J_i^n = k\}, \quad C_{i,2}^n := (\mathcal{H}_i^n)^c \cap \{J_i^n \geq 2\}.$$

Then, we can write

$$\mathcal{H}_i^n = \bigcup_{k=0}^2 D_{i,k}^n, \quad (\mathcal{H}_i^n)^c = \bigcup_{k=0}^2 C_{i,k}^n.$$

Under some regularity conditions, such as Lipschitz continuity and boundedness, Shimizu and Yoshida (2006) and Shimizu (2009) showed the following result.

Lemma 4.2.10. *For any $p \geq 1$ and some $\gamma > 0$, as $n \to \infty$*

$$\begin{aligned} &P(D_{i,0}^n|\mathcal{F}_{t_{i-1}}) = R_1(h_n^{p(1/2-\rho)}, \boldsymbol{X}_{t_{i-1}}), && P(C_{i,0}^n|\mathcal{F}_{t_{i-1}}) = e^{-\lambda_0 h_n} - P(D_{i,0}^n|\mathcal{F}_{t_{i-1}}), \\ &P(D_{i,1}^n|\mathcal{F}_{t_{i-1}}) = R_2(h_n, \boldsymbol{X}_{t_{i-1}}), && P(C_{i,1}^n|\mathcal{F}_{t_{i-1}}) = R_3(h_n^{1+\gamma}, \boldsymbol{X}_{t_{i-1}}), \\ &P(D_{i,2}^n|\mathcal{F}_{t_{i-1}}) \leq \lambda_0^2 h_n^2, && P(C_{i,0}^n|\mathcal{F}_{t_{i-1}}) \leq \lambda_0^2 h_n^2 \end{aligned}$$

where $R_j : \mathbb{R} \times \mathbb{R}^m \to \mathbb{R}$ $(j = 1, 2, 3)$ are functions for which there exists a constant C such that

$$R_j(u_n, \boldsymbol{s}) \leq u_n C(1 + |\boldsymbol{s}|)^C$$

for a real-valued sequence $\{u_n\}_{n\in\mathbb{N}}$ and all $\boldsymbol{s} \in \mathbb{R}^m$.

This result implies that for $n \to \infty$

$$1_{\mathcal{H}_i^n} \approx 1_{\{J_i^n=1\}}, \quad 1_{(\mathcal{H}_i^n)^c} \approx 1_{\{J_i^n=0\}}.$$

Based on this asymptotic filter $\mathcal{H}_i^n$, they introduced the estimator $\hat{\boldsymbol{\theta}}_{TH} = (\hat{\boldsymbol{\mu}}'_{TH}, vec(\hat{\boldsymbol{\sigma}}_{TH})', \hat{\boldsymbol{\xi}}'_{TH})'$ of $\boldsymbol{\theta}$ by

$$\ell_n(\hat{\boldsymbol{\theta}}_{TH}) = \sup_{\boldsymbol{\theta}\in\Theta} \ell_n(\boldsymbol{\theta})$$

where $\ell_n(\boldsymbol{\theta})$ is a contrast function defined by

$$\begin{aligned} \ell_n(\boldsymbol{\theta}) &= \ell_n^{(C)}(\boldsymbol{\theta}) + \ell_n^{(J)}(\boldsymbol{\theta}) \\ \ell_n^{(C)}(\boldsymbol{\theta}) &= -\frac{1}{2h_n}\sum_{i=1}^n (\Delta_i^n \boldsymbol{X} - h_n\boldsymbol{\mu})'\Sigma^{-1}(\Delta_i^n \boldsymbol{X} - h_n\boldsymbol{\mu})1_{(\mathcal{H}_i^n)^c} - \frac{1}{2}\sum_{i=1}^n \log\det(\Sigma)1_{(\mathcal{H}_i^n)^c} \\ \ell_n^{(J)}(\boldsymbol{\theta}) &= \sum_{i=1}^n \log f_{\boldsymbol{\xi}}(\Delta_i^n \boldsymbol{X})1_{\mathcal{H}_i^n} - nh_n\int_{\mathbb{R}^m} f_{\boldsymbol{\xi}}(\boldsymbol{u})d\boldsymbol{u} \end{aligned}$$

where $\Sigma = \sigma\sigma'$.

This estimation method is called the threshold estimation method, which originated in the paper by Mancini (2001) and the idea was also independently introduced by Shimizu (2002), and published later by Shimizu and Yoshida (2006). It is one of the useful techniques in inference for jump processes from discrete observations. In this method, statistics are constructed via the asymptotic filter ($\mathcal{H}_i^n$). We judge a jump has occurred in the interval $(t_{i-1}, t_i]$, if $\Delta_i^n \boldsymbol{X} \in \mathcal{H}_i^n$; otherwise we deal with no jump in this interval. In other words, there exists a threshold h_n^ρ and the judgement of the existence of the jump in this interval is determined by whether $\Delta_i^n \boldsymbol{X}$ has exceeded the threshold or not. That is why this method is called the threshold estimation method, and, based on the two groups (i.e., jump observations and no jump observations), an approximation of the log-likelihood is proposed. Then, Shimizu and Yoshida (2006) showed the following.

Lemma 4.2.11. *(Theorem 2.1. of Shimizu and Yoshida (2006) and Theorem 3.6 of Shimizu (2009)) Under Assumptions A.1–A.11 of Shimizu and Yoshida (2006) and Assumption 4.2.9, it follows that*

$$\begin{pmatrix} \sqrt{nh_n}(\hat{\mu}_{TH} - \mu_0) \\ \sqrt{n}(vec(\hat{\sigma}_{TH}) - vec(\sigma_0)) \\ \sqrt{nh_n}(\hat{\xi}_{TH} - \xi_0) \end{pmatrix} \xrightarrow{\mathcal{L}} N\left(\mathbf{0}, K(\theta_0)^{-1}\right) \tag{4.2.37}$$

as $n \to \infty$, where

$$K(\theta) = \begin{pmatrix} K_1(\theta) & 0 & 0 \\ 0 & K_2(\theta) & 0 \\ 0 & 0 & K_3(\theta) \end{pmatrix}$$

and

$$\begin{aligned} K_1(\theta) &= \Sigma \\ K_2(\theta) &= \left(\frac{1}{2} tr\left\{\left(\partial_{(vec(\sigma))_i}\Sigma\right)\Sigma^{-1}\left(\partial_{(vec(\sigma))_j}\Sigma\right)\Sigma^{-1}\right\}\right)_{i,j=1,\dots,m^2} \\ K_3(\theta) &= \int_{\mathbb{R}^m} \partial_{\xi} f_{\xi}(u)\left(\partial_{\xi} f_{\xi}(u)\right)' f_{\xi}(u) du. \end{aligned}$$

Moreover, Shimizu and Yoshida (2006) argue that the asymptotic efficiency for $\hat{\theta}_{TH}$ is obtained since K_1 and K_3 correspond to the asymptotic variance of the estimator for the continuously observed ergodic diffusion processes with jumps.

Example 4.2.12. *(CRRA power utility) Consider the case of Example 4.2.6. Let $\theta = (\mu', vec(\sigma)', \xi')'$. Then, the optimal portfolio can be written by*

$$w^*(\theta) = \arg\min_{w \in \mathbb{R}^m} g_\theta(w)$$

where

$$g_\theta(w) = -w'\left\{\mu + \frac{1}{2}(\Sigma_{11}, \dots, \Sigma_{mm})'\right\} + \frac{\gamma}{2} w'\Sigma w + \lambda \psi_\theta(w)$$

and

$$\begin{aligned} \Sigma &= \sigma\sigma' \\ \psi_\theta(w) &= -\frac{1}{1-\gamma}\int_{z>-1}\left\{(1 + w'z)^{1-\gamma} - 1\right\} f_\xi(\log(1+z_1), \dots, \log(1+z_m)) \prod_{i=1}^m \frac{1}{1+z_i} dz. \end{aligned}$$

Then, from (4.2.37) and the δ-method, we have

$$\sqrt{nh_n}(\boldsymbol{w}^*(\hat{\boldsymbol{\theta}}_{TH}) - \boldsymbol{w}^*(\boldsymbol{\theta}_0)) \xrightarrow{d} N\left(\mathbf{0}, \dot{\boldsymbol{w}}^*(\boldsymbol{\theta}_0)\tilde{K}(\boldsymbol{\theta}_0)^{-1}\left(\dot{\boldsymbol{w}}^*(\boldsymbol{\theta}_0)\right)'\right)$$

where

$$\tilde{K}(\boldsymbol{\theta})^{-1} = \begin{pmatrix} K_1(\boldsymbol{\theta})^{-1} & 0 & 0 \\ 0 & K_2(\boldsymbol{\theta})^{-1} & 0 \\ 0 & 0 & K_3(\boldsymbol{\theta})^{-1} \end{pmatrix}.$$

4.2.3 Bibliographic Notes

Merton (1969) introduced portfolio theory for continuous time models and his results are summarized in Merton (1992). In addition, Aït-Sahalia et al. (2009) expanded his results to a diffusion model with jump. Chapter 4 in Aït-Sahalia and Hansen (2009) and Heyde (1997) provide an introduction and some applications for GMM. Shimizu and Yoshida (2006) and Shimizu (2009) (in Japanese) introduced the threshold method and showed its asymptotic property.

4.3 Universal Portfolio

The mean-variance approach has two problems pointed out by finance practitioners, private investors and researchers. The first problem is related to distributional assumptions concerning the behaviour of asset prices, and the second problem is related to the selection of the optimal portfolio depending on some objective function and/or utility function defined according to the investor's goal. To overcome these problems, a different approach based on information theory to portfolio selection was proposed by Cover (1991). The main characteristic is that no distributional assumptions on the sequence of price relatives are required. To emphasize this independence from statistical assumptions, such portfolios are called "universal portfolios." Cover (1991) showed that universal portfolios achieve the same asymptotic growth rate as the best constant rebalanced portfolio in hindsight which achieves the optimal growth rate.

Moreover, Cover and Ordentlich (1996) considered adding side information such as explanatory variables and discrete ones, and obtained precise bounds on the ratio of the wealth given by the universal portfolio to the best wealth achievable by a constant rebalanced portfolio given hindsight. Gaivoronski and Stella (2000) introduced ideas of stochastic optimization into the definition of universal portfolios. Blum and Kalai (1999) addressed the original lack of consideration of transaction costs in Cover's formulation.

4.3.1 μ-Weighted Universal Portfolios

Suppose that a stock market can be represented by $\boldsymbol{x}_1, \boldsymbol{x}_2, \ldots \in \mathbb{R}_+^m$ where $m \in \mathbb{N}$ is the number of stocks in the market, $\boldsymbol{x}_i = (x_{i1}, x_{i2}, \ldots, x_{im})'$ for each $i \in \mathbb{N}$ is the vector of price relatives for all stocks on day i and $x_{ij} \geq 0$ is the price relative for stock $j(= 1, 2, \ldots, m)$ on day i. Note that the price relative x_{ij} is the ratio of the price at the end of the day i to the price at the beginning of the day i. For instance, $x_{ij} = 1.03$ means that the jth stock went up 3 percent on day i.

In this section, a portfolio $\boldsymbol{b} = (b_1, b_2, \ldots, b_m)'$, $b_i \geq 0$, $\sum_{i=1}^m b_i = 1$ is the proportion of the current wealth invested in each of the m stocks. We denote the set of $(m-1)$-dimensional portfolios by

$$\mathcal{B} = \left\{\boldsymbol{b} \in \mathbb{R}^m : \sum_{j=1}^m b_j = 1, b_j \geq 0\right\}.$$

According to Cover and Thomas (2006), this set is called the *probability simplex* in $\mathbb{R}^m$. For any $\boldsymbol{b} \in \mathcal{B}$, the wealth relative on day i (i.e., the ratio of the wealth at the end of the day i to the wealth at the beginning of the day i) is given by

$$S_i = \boldsymbol{b}'\boldsymbol{x}_i = \sum_{j=1}^{m} b_j x_{ij}.$$

Consider an arbitrary (nonrandom) sequence of stock vectors $\boldsymbol{x}_1, \boldsymbol{x}_2, \ldots \boldsymbol{x}_n \in \mathbb{R}_+^m$ for $n \in \mathbb{N}$ (i.e., the investment period is n days). According to Cover and Thomas (2006), a realistic target is the growth achieved by the best constant rebalanced portfolio strategy in hindsight (i.e., the best constant rebalanced portfolio on the known sequence of stock market vectors). For a constant rebalanced portfolio $\boldsymbol{b} \in \mathcal{B}$, the wealth is described as follows:

Definition 4.3.1. *At the end of n days, a "constant rebalanced portfolio"* $\boldsymbol{b} = (b_1, \ldots, b_m)'$ *achieves wealth:*

$$S_n(\boldsymbol{b}, \boldsymbol{x}^n) = \prod_{i=1}^{n} \boldsymbol{b}'\boldsymbol{x}_i = \prod_{i=1}^{n}\sum_{j=1}^{m} b_j x_{ij}$$

where $\boldsymbol{x}^n$ *is the sequence of n vectors (i.e.,* $\boldsymbol{x}^n = \{\boldsymbol{x}_1, \ldots, \boldsymbol{x}_n\}$*).*

Within the class of constant rebalanced portfolios $\mathcal{B}$, there exists the best constant rebalanced portfolio (in hindsight), which is defined as follows:

Definition 4.3.2. *At the end of n days, the "best constant rebalanced portfolio"* $\boldsymbol{b}^* = (b_1^*, \ldots, b_m^*)'$ *achieves wealth:*

$$S_n^*(\boldsymbol{x}^n) = \max_{\boldsymbol{b}\in\mathcal{B}} S_n(\boldsymbol{b}, \boldsymbol{x}^n) = \max_{\boldsymbol{b}\in\mathcal{B}} \prod_{i=1}^{n} \boldsymbol{b}'\boldsymbol{x}_i.$$

In this section, we will find an optimal universal strategy to achieve asymptotically the same wealth without advance knowledge of the distribution of the stock market vector. This optimal universal strategy should belong to a class of causal (nonanticipating) portfolio strategies that depend only on the past values of the stock market sequence. For a causal (nonanticipating) portfolio strategy, the wealth is described as follows:

Definition 4.3.3. *At the end of n days, a "causal (nonanticipating) portfolio" strategy* $\{\hat{\boldsymbol{b}}_i = (\hat{b}_{i1}, \ldots, \hat{b}_{im})'; i = 1, \ldots, n\}$ *achieves wealth:*

$$\hat{S}_n(\boldsymbol{x}^n) = \prod_{i=1}^{n} \hat{\boldsymbol{b}}_i'\boldsymbol{x}_i.$$

Note that the best constant rebalanced portfolio $\boldsymbol{b}^*$ is defined by the stock market sequence $\boldsymbol{x}^n = \{\boldsymbol{x}_1, \ldots, \boldsymbol{x}_n\}$, i.e. $\boldsymbol{b}^* \equiv \boldsymbol{b}^*(\boldsymbol{x}^n)$. Similarly, each causal (nonanticipating) portfolio $\hat{\boldsymbol{b}}_i$ is defined by the past stock market sequence $\boldsymbol{x}^{i-1} = \{\boldsymbol{x}_1, \ldots, \boldsymbol{x}_{i-1}\}$, i.e. $\hat{\boldsymbol{b}}_i \equiv \hat{\boldsymbol{b}}^*(\boldsymbol{x}^{i-1})$.
It would be desirable if a causal (nonanticipating) portfolio strategy $\{\hat{\boldsymbol{b}}_i\}$ yielded wealth in some sense close to the wealth obtained by means of the best constant rebalanced portfolio $\boldsymbol{b}^*$. One such strategy was proposed by Cover (1991) under the name of the universal portfolio and the definition is as follows:

Definition 4.3.4. *The universal portfolio strategy is the performance-weighted strategy specified by*

$$\hat{\boldsymbol{b}}_1 = \left(\frac{1}{m}, \frac{1}{m}, \ldots, \frac{1}{m}\right)', \quad \hat{\boldsymbol{b}}_{k+1} = \frac{\int_{\boldsymbol{b}\in\mathcal{B}} \boldsymbol{b} S_k(\boldsymbol{b}, \boldsymbol{x}^k) d\boldsymbol{b}}{\int_{\boldsymbol{b}\in\mathcal{B}} S_k(\boldsymbol{b}, \boldsymbol{x}^k) d\boldsymbol{b}}, \quad k = 1, 2, \ldots, n \tag{4.3.1}$$

where $S_k(\boldsymbol{b}, \boldsymbol{x}^k) = \prod_{i=1}^{k} \boldsymbol{b}'\boldsymbol{x}_i$ *and* $S_0(\boldsymbol{b}, \boldsymbol{x}^0) = 1$.

The following result provides the optimality of the best constant rebalanced portfolio which implies that $S_n^*(\boldsymbol{x}^n)$ (the wealth of the best constant rebalanced portfolio) exceeds $\hat{S}_n(\boldsymbol{x}^n)$ (the wealth of the universal portfolio strategy) for every sequence of outcomes from the stock market.

Theorem 4.3.5. *Let $\boldsymbol{x}_1, \boldsymbol{x}_2, \ldots, \boldsymbol{x}_n$ be a sequence of nonrandom stock vectors in $\mathbb{R}_+^m$. Let $S_n^*(\boldsymbol{x}^n)$ be the wealth achieved by the best constant rebalanced portfolio defined by Definition 4.3.2, and let $\hat{S}_n(\boldsymbol{x}^n)$ be the wealth achieved by the universal portfolio strategy defined by Definitions 4.3.3 and 4.3.4. Then*

$$\frac{S_n^*(\boldsymbol{x}^n)}{\hat{S}_n(\boldsymbol{x}^n)} \geq 1.$$

Proof: From (4.3.1), for $k = 1, 2, \ldots, n$, it follows that

$$\hat{\boldsymbol{b}}_k' \boldsymbol{x}_k = \frac{\int_{\boldsymbol{b}\in\mathcal{B}} \boldsymbol{b}'\boldsymbol{x}_k S_{k-1}(\boldsymbol{b}, \boldsymbol{x}^{k-1}) d\boldsymbol{b}}{\int_{\boldsymbol{b}\in\mathcal{B}} S_{k-1}(\boldsymbol{b}, \boldsymbol{x}^{k-1}) d\boldsymbol{b}} = \frac{\int_{\boldsymbol{b}\in\mathcal{B}} \prod_{i=1}^{k} \boldsymbol{b}'\boldsymbol{x}_i d\boldsymbol{b}}{\int_{\boldsymbol{b}\in\mathcal{B}} \prod_{i=1}^{k-1} \boldsymbol{b}'\boldsymbol{x}_i d\boldsymbol{b}} \tag{4.3.2}$$

hence, we have

$$\begin{aligned} \hat{S}_n(\boldsymbol{x}^n) &= \prod_{k=1}^{n} \hat{\boldsymbol{b}}_k' \boldsymbol{x}_k = \prod_{k=1}^{n} \frac{\int_{\boldsymbol{b}\in\mathcal{B}} \prod_{i=1}^{k} \boldsymbol{b}'\boldsymbol{x}_i d\boldsymbol{b}}{\int_{\boldsymbol{b}\in\mathcal{B}} \prod_{i=1}^{k-1} \boldsymbol{b}'\boldsymbol{x}_i d\boldsymbol{b}} = \frac{\int_{\boldsymbol{b}\in\mathcal{B}} \prod_{i=1}^{n} \boldsymbol{b}'\boldsymbol{x}_i d\boldsymbol{b}}{\int_{\boldsymbol{b}\in\mathcal{B}} d\boldsymbol{b}} \qquad (4.3.3) \\ &\leq \frac{\max_{\boldsymbol{b}\in\mathcal{B}} \prod_{i=1}^{n} \boldsymbol{b}'\boldsymbol{x}_i \int_{\boldsymbol{b}\in\mathcal{B}} d\boldsymbol{b}}{\int_{\boldsymbol{b}\in\mathcal{B}} d\boldsymbol{b}} = S_n^*(\boldsymbol{x}^n). \quad \square \end{aligned}$$

Cover (1991) and Cover and Ordentlich (1996) showed an upper bound for the ratio of the wealth by the universal portfolio strategy and the wealth by the best constant rebalanced portfolio.

Theorem 4.3.6. *(Theorem 1 of Cover and Ordentlich (1996)) Let $\boldsymbol{x}^n$ be a stock market sequence $\{\boldsymbol{x}_1, \boldsymbol{x}_2, \ldots, \boldsymbol{x}_n\}$, $\boldsymbol{x}_i \in \mathbb{R}_+^m$ of length n with m assets. Let $S_n^*(\boldsymbol{x}^n)$ be the wealth achieved by the best constant rebalanced portfolio defined by Definition 4.3.2, and let $\hat{S}_n(\boldsymbol{x}^n)$ be the wealth achieved by the universal portfolio defined by Definitions 4.3.3 and 4.3.4. Then*

$$\frac{S_n^*(\boldsymbol{x}^n)}{\hat{S}_n(\boldsymbol{x}^n)} \leq \binom{n+m-1}{m-1} \leq (n+1)^{m-1}.$$

Proof of Theorem 4.3.6. See Appendix.□

Remark 4.3.7. *Theorems 4.3.5 and 4.3.6 immediately imply that*

$$\frac{1}{n}\log 1 \leq \frac{1}{n}\log\frac{S_n^*(\boldsymbol{x}^n)}{\hat{S}_n(\boldsymbol{x}^n)} \leq \frac{m-1}{n}\log(n+1).$$

Therefore, it follows that

$$\frac{1}{n}\log \hat{S}_n(\boldsymbol{x}^n) - \frac{1}{n}\log S_n^*(\boldsymbol{x}^n) \to 0$$

as $n \to \infty$. Thus, the universal portfolio achieves the same asymptotic growth rate of wealth as the best constant rebalanced portfolio.

Cover and Ordentlich (1996) generalized the definition of the universal portfolio defined in Definition 4.3.4 as follows:

Definition 4.3.8. *The μ-weighted universal portfolio strategy is specified by*

$$\hat{\boldsymbol{b}}_{k+1} = \frac{\int_{\boldsymbol{b}\in\mathcal{B}} \boldsymbol{b} S_k(\boldsymbol{b}, \boldsymbol{x}^k) d\mu(\boldsymbol{b})}{\int_{\boldsymbol{b}\in\mathcal{B}} S_k(\boldsymbol{b}, \boldsymbol{x}^k) d\mu(\boldsymbol{b})}, \quad k = 0, 1, 2, \ldots, n \tag{4.3.4}$$

where $\int_{\boldsymbol{b}\in\mathcal{B}} d\mu(\boldsymbol{b}) = 1$, $S_k(\boldsymbol{b}, \boldsymbol{x}^k) = \prod_{i=1}^{k} \boldsymbol{b}'\boldsymbol{x}_i$ and $S_0(\boldsymbol{b}, \boldsymbol{x}^0) = 1$.

Note that if μ is symmetric, then $\hat{b}_1 = \left(\frac{1}{m}, \frac{1}{m}, \ldots, \frac{1}{m}\right)'$. The overall portfolio algorithm is specified by the choice of μ, while the portfolio action at each time depends on μ and the sequence x^k observed so far. It is easy to see that if you choose μ as $\mu(b) = (m-1)!b$, it corresponds to the uniform weighted universal portfolio defined in (4.3.1).

Remark 4.3.9. *Cover and Ordentlich (1996) considered the case when the measure μ is the Dirichlet $(1/2, 1/2, \ldots, 1/2)$ distribution, which has a density with respect to the Lebesgue (uniform) measure on the simplex $\mathcal{B}$ given by*

$$d\mu(b) = \frac{\Gamma\left(\frac{m}{2}\right)}{\left\{\Gamma\left(\frac{1}{2}\right)\right\}^m} \prod_{j=1}^{m} b_j^{-1/2} db.$$

Then, similarly to Theorem 4.3.5, it follows that

$$\frac{S_n^*(x^n)}{\hat{S}_n(x^n)} \geq 1$$

and similarly to Theorem 4.3.6, it is seen that

$$\frac{S_n^*(x^n)}{\hat{S}_n(x^n)} \leq 2(n+1)^{(m-1)/2}$$

because

$$\begin{aligned} \frac{S_n^*(x^n)}{\hat{S}_n(x^n)} &\leq \max_{j^n \in \{1,\ldots,m\}^n} \frac{\prod_{r=1}^m \left(\frac{n_r(j^n)}{n}\right)^{n_r(j^n)}}{\int_{b \in \mathcal{B}} \prod_{i=1}^n b_{j_i} d\mu(b)} \\ &\leq \max_{j^n \in \{1,\ldots,m\}^n} \frac{\prod_{r=1}^m \left(\frac{n_r(j^n)}{n}\right)^{n_r(j^n)}}{\frac{\Gamma\left(\frac{m}{2}\right)}{\Gamma\left(\sum_{r=1}^m \left(n_r(j^n)+\frac{1}{2}\right)\right)} \prod_{r=1}^m \frac{\Gamma\left(n_r(j^n)+\frac{1}{2}\right)}{\Gamma\left(\frac{1}{2}\right)}} \\ &= \max_{j^n \in \{1,\ldots,m\}^n} \frac{\left\{\Gamma\left(\frac{1}{2}\right)\right\}^m}{\Gamma\left(\frac{m}{2}\right)} \frac{\Gamma\left(n+\frac{m}{2}\right)}{n^n} \prod_{r=1}^m \frac{n_r(j^n)^{n_r(j^n)}}{\Gamma\left(n_r(j^n)+\frac{1}{2}\right)} \\ &\leq \frac{\Gamma\left(\frac{1}{2}\right)\Gamma\left(n+\frac{m}{2}\right)}{\Gamma\left(\frac{m}{2}\right)\Gamma\left(\frac{n}{2}\right)} \\ &\leq 2(n+1)^{(m-1)/2} \end{aligned} \tag{4.3.5}$$

where $n_r(j^n)$ is the number of occurrences of r in $j^n \in \{1, \ldots, m\}^n$. For details of the above proof, see Theorem 2 and Lemmas 3–5 of Cover and Ordentlich (1996). Thus the μ-weighted universal portfolio with Dirichlet $(1/2, 1/2, \ldots, 1/2)$ distribution also achieves the same asymptotic growth rate of wealth as the best constant rebalanced portfolio, i.e.,

$$\frac{1}{n} \log \hat{S}_n(x^n) - \frac{1}{n} \log S_n^*(x^n) \to 0$$

as $n \to \infty$.

4.3.2 Universal Portfolios with Side Information

So far, we have considered the portfolio selection problem in the case where no additional information is available concerning the stock market. However, it is common that investors adjust (i.e., rebalance) their portfolios by use of various sources of information concerning the stock market, which called *side information*. Following Cover and Ordentlich (1996), we model this side information as a finite-valued variable y made available at the start of each investment period. Thus the

formal domain of our market model is a sequence of pairs $\{(\boldsymbol{x}_i, y_i)\}$ where $\boldsymbol{x}_i$ is the vector of price relatives for all stocks on day i and $y_i \in \mathcal{Y} = \{1, 2, \ldots, k\}$ denotes the state of the side information on day i. One example for the side information is to set $y_i = j$ if stock j has outperformed other stocks in the previous r trading days. Another one is to use y_i to reflect whether the moving average of the last r trading days is greater or less than the average of the price relatives on the previous trading day.
The constant rebalanced portfolio defined in Definition 4.3.1 is extended to the state constant rebalanced portfolio $\boldsymbol{b}(y_i)$, which depends only on the side information $y_i \in \mathcal{Y} = \{1, 2, \ldots, k\}$, and the wealth is described as follows:

Definition 4.3.10. *At the end of n days, a "state constant rebalanced portfolio"* $\boldsymbol{b}(1), \boldsymbol{b}(2), \ldots, \boldsymbol{b}(k) \in \mathcal{B}$ *achieves wealth:*

$$S_n(\boldsymbol{b}(\cdot), \boldsymbol{x}^n | y^n) = \prod_{i=1}^{n} \boldsymbol{b}(y_i)' \boldsymbol{x}_i = \prod_{i=1}^{n} \sum_{j=1}^{m} b_j(y_i) x_{ij}$$

where $\boldsymbol{x}^n$ *is the sequence of n vectors (i.e.,* $\boldsymbol{x}^n = \{\boldsymbol{x}_1, \ldots, \boldsymbol{x}_n\}$*),* y^n *is the sequence of n state variables (i.e.,* $y^n = \{y_1, \ldots, y_n\}$*) and* $\boldsymbol{b}(\cdot)$ *is the collection of state constant rebalanced portfolios (i.e.,* $\boldsymbol{b}(\cdot) = \{\boldsymbol{b}(1), \ldots, \boldsymbol{b}(k)\} \in \mathcal{B}^k$*).*

Similarly to Definition 4.3.2, we define the best state constant rebalanced portfolio (in hindsight) within the class of the state constant rebalanced portfolio as follows:

Definition 4.3.11. *At the end of n days, the "best state constant rebalanced portfolio"* $\boldsymbol{b}^*(\cdot) = \{\boldsymbol{b}^*(1), \boldsymbol{b}^*(2), \ldots, \boldsymbol{b}^*(k)\}$ *achieves wealth:*

$$S_n^*(\boldsymbol{x}^n | y^n) = \max_{\boldsymbol{b}(\cdot) \in \mathcal{B}^k} S_n(\boldsymbol{b}(\cdot), \boldsymbol{x}^n | y^n) = \max_{\boldsymbol{b}(\cdot) \in \mathcal{B}^k} \prod_{i=1}^{n} \boldsymbol{b}'(y_i) \boldsymbol{x}_i.$$

Note that a state constant rebalanced portfolio for k states and m stocks has $k(m-1)$ degrees of freedom, and $m-1$ degrees of freedom for each of the k portfolios which must be specified. The requirement that the entries sum to one (i.e., $\sum_{j=1}^{m} b_j(y) = 1$) gives each portfolio vector $m-1$ degrees of freedom. Here m is the number of stocks and k is the number of states of side information. Cover and Ordentlich (1996) introduced μ-weighted universal portfolios with side information for $|\mathcal{Y}| = k > 1$, similarly to Definition 4.3.8, as follows:

Definition 4.3.12. *The* μ*-weighted universal portfolio with side information is specified by*

$$\hat{\boldsymbol{b}}_i(y) = \frac{\int_{\boldsymbol{b} \in \mathcal{B}} \boldsymbol{b} S_{i-1}(\boldsymbol{b}, \boldsymbol{x}^{i-1} | y) d\mu(\boldsymbol{b})}{\int_{\boldsymbol{b} \in \mathcal{B}} S_{i-1}(\boldsymbol{b}, \boldsymbol{x}^{i-1} | y) d\mu(\boldsymbol{b})}, \quad i = 1, 2, \ldots, n, \ y = 1, \ldots, k \in \mathcal{Y} \tag{4.3.6}$$

where $\int_{\boldsymbol{b} \in \mathcal{B}} d\mu(\boldsymbol{b}) = 1$ *and* $S_i(\boldsymbol{b}, \boldsymbol{x}^i | y)$ *is the wealth obtained by the constant rebalanced portfolio* $\boldsymbol{b}$ *along the subsequence* $\{j \le i : y_j = y\}$*, and is given by*

$$S_i(\boldsymbol{b}, \boldsymbol{x}^i | y) = \prod_{j \in \{j \le i, y_j = y\}} \boldsymbol{b}' \boldsymbol{x}_j$$

with $S_0(\boldsymbol{b}, \boldsymbol{x}^0 | y) = 1$.

Similarly to (4.3.2), from (4.3.6), it follows that

$$\begin{aligned}
\hat{\boldsymbol{b}}_i(y)' \boldsymbol{x}_i &= \frac{\int_{\boldsymbol{b} \in \mathcal{B}} \boldsymbol{b}' \boldsymbol{x}_i S_{i-1}(\boldsymbol{b}, \boldsymbol{x}^{i-1} | y) d\boldsymbol{b}}{\int_{\boldsymbol{b} \in \mathcal{B}} S_{i-1}(\boldsymbol{b}, \boldsymbol{x}^{i-1} | y) d\boldsymbol{b}} \\
&= \frac{\int_{\boldsymbol{b} \in \mathcal{B}} \boldsymbol{b}' \boldsymbol{x}_i \prod_{j \in \{j \le i-1, y_j = y\}} \boldsymbol{b}' \boldsymbol{x}_j d\boldsymbol{b}}{\int_{\boldsymbol{b} \in \mathcal{B}} \prod_{j \in \{j \le i-1, y_j = y\}} \boldsymbol{b}' \boldsymbol{x}_j d\boldsymbol{b}}
\end{aligned}$$

$$= \frac{\int_{b\in\mathcal{B}} \prod_{j\in\{j\le i, y_j=y\}} \boldsymbol{b}'\boldsymbol{x}_j d\boldsymbol{b}}{\int_{b\in\mathcal{B}} \prod_{j\in\{j\le i-1, y_j=y\}} \boldsymbol{b}'\boldsymbol{x}_j d\boldsymbol{b}} \tag{4.3.7}$$

for $i = 1, 2, \ldots, n$, hence, at the end of n days, the μ-weighted universal portfolio with side information $\{\hat{\boldsymbol{b}}_i(y)\}_{i=1,\ldots,n, y=1,\ldots,k}$ achieves wealth

$$\begin{aligned}
\hat{S}_n(\boldsymbol{x}^n|y^n) &= \prod_{i=1}^{n} \hat{\boldsymbol{b}}_i(y_i)'\boldsymbol{x}_i \\
&= \prod_{y=1}^{k} \prod_{i\in\{i\le n, y_i=y\}} \hat{\boldsymbol{b}}_i(y)'\boldsymbol{x}_i \\
&= \prod_{y=1}^{k} \prod_{i\in\{i\le n, y_i=y\}} \frac{\int_{b\in\mathcal{B}} \prod_{j\in\{j\le i, y_j=y\}} \boldsymbol{b}'\boldsymbol{x}_j d\mu(\boldsymbol{b})}{\int_{b\in\mathcal{B}} \prod_{j\in\{j\le i-1, y_j=y\}} \boldsymbol{b}'\boldsymbol{x}_j d\mu(\boldsymbol{b})} \\
&= \prod_{y=1}^{k} \prod_{i=1}^{n} \frac{\int_{b\in\mathcal{B}} \prod_{j\in\{j\le i, y_j=y\}} \boldsymbol{b}'\boldsymbol{x}_j d\mu(\boldsymbol{b})}{\int_{b\in\mathcal{B}} \prod_{j\in\{j\le i-1, y_j=y\}} \boldsymbol{b}'\boldsymbol{x}_j d\mu(\boldsymbol{b})} \\
&= \prod_{y=1}^{k} \frac{\int_{b\in\mathcal{B}} \prod_{j\in\{j\le n, y_j=y\}} \boldsymbol{b}'\boldsymbol{x}_j d\mu(\boldsymbol{b})}{\int_{b\in\mathcal{B}} d\mu(\boldsymbol{b})} \\
&= \prod_{y=1}^{k} \int_{b\in\mathcal{B}} S_n(\boldsymbol{b}, \boldsymbol{x}^n|y) d\mu(\boldsymbol{b}).
\end{aligned} \tag{4.3.8}$$

By using this result, we have the following.

Theorem 4.3.13. *(Theorem 3 of Cover and Ordentlich (1996)) Let $\boldsymbol{x}^n$ be a stock market sequence $\{\boldsymbol{x}_1, \boldsymbol{x}_2, \ldots, \boldsymbol{x}_n\}$, $\boldsymbol{x}_i \in \mathbb{R}_+^m$ of length n with m assets, and y^n be a state sequence $\{y_1, y_2, \ldots, y_n\}$, $y_i \in \mathcal{Y} = \{1, 2, \ldots, k\}$. Let $S_n^*(\boldsymbol{x}^n|y^n)$ be the wealth achieved by the best state constant rebalanced portfolio defined by Definition 4.3.11, and let $\hat{S}_n(\boldsymbol{x}^n|y^n)$ be the wealth achieved by the μ-weighted universal portfolio with side information defined by (4.3.8).*

(i) The uniform weighted universal portfolio with side information (i.e., $d\mu(\boldsymbol{b}) \equiv (m-1)!d\boldsymbol{b}$) satisfies

$$1 \le \frac{S_n^*(\boldsymbol{x}^n|y^n)}{\hat{S}_n(\boldsymbol{x}^n|y^n)} \le \prod_{r=1}^{k} (n_r(y^n)+1)^{m-1} \le (n+1)^{k(m-1)}$$

for all $n \in \mathbb{N}$, $\boldsymbol{x}^n \in (\mathbb{R}_+^m)^n$, $y^n \in \{1, 2, \ldots, k\}^n$.

(ii) The Dirichlet $(1/2, 1/2, \ldots, 1/2)$ weighted universal portfolio with side information (i.e., $d\mu(\boldsymbol{b}) = \frac{\Gamma\left(\frac{m}{2}\right)}{\left\{\Gamma\left(\frac{1}{2}\right)\right\}^m} \prod_{j=1}^{m} b_j^{-1/2} d\boldsymbol{b}$) satisfies

$$1 \le \frac{S_n^*(\boldsymbol{x}^n|y^n)}{\hat{S}_n(\boldsymbol{x}^n|y^n)} \le 2^k \prod_{r=1}^{k} (n_r(y^n)+1)^{(m-1)/2} \le 2^k (n+1)^{k(m-1)/2}$$

for all $n \in \mathbb{N}$, $\boldsymbol{x}^n \in (\mathbb{R}_+^m)^n$, $y^n \in \{1, 2, \ldots, k\}^n$.

Here the quantity $n_r(y^n)$ denotes the number of times $y_j = r$ in the sequence $y^n = \{y_1, \ldots, y_n\}$.

Remark 4.3.14. *Theorem 4.3.13 immediately implies that*

(i) For the uniform weighted universal portfolio with side information

$$\frac{1}{n}\log 1 \le \frac{1}{n}\log \frac{S_n^*(\boldsymbol{x}^n|y^n)}{\hat{S}_n(\boldsymbol{x}^n|y^n)} \le \frac{k(m-1)}{n}\log(n+1).$$

(ii) For the Dirichlet (1/2, 1/2, ..., 1/2) weighted universal portfolio with side information

$$\frac{1}{n}\log 1 \leq \frac{1}{n}\log\frac{S_n^*(\boldsymbol{x}^n|y^n)}{\hat{S}_n(\boldsymbol{x}^n|y^n)} \leq \frac{1}{n}\left\{k\log 2 + \frac{k(m-1)}{2}\log(n+1)\right\}.$$

Therefore, for both portfolio strategies, it follows that

$$\frac{1}{n}\log\hat{S}_n(\boldsymbol{x}^n|y^n) - \frac{1}{n}\log S_n^*(\boldsymbol{x}^n|y^n) \to 0$$

under $k, m < \infty$, as $n \to \infty$. Thus, the universal portfolios achieve the same asymptotic growth rate of wealth as the best state constant rebalanced portfolio.

4.3.3 Successive Constant Rebalanced Portfolios

Gaivoronski and Stella (2000) applied ideas of the stochastic programming problem into the definition of universal portfolios. Successive constant rebalanced portfolios (SCRP), introduced by them, are derived using a different set of ideas than Cover's universal portfolio, and originated in non-stationary and stochastic optimization. Similar to Cover's universal portfolio, SCRP do not depend on statistical assumptions about distribution of price relatives and exhibit similar asymptotic behaviour, that is, to achieve the same asymptotic growth rate of wealth as the best constant rebalanced portfolio (BCRP). Moreover, there is an advantage of SCRP compared to Cover's universal portfolio with respect to the computational cost, because Cover's universal portfolio relies on multidimensional integration, which is notoriously difficult to perform, except for few stocks; on the other hand, SCRP are computable for fairly large numbers of stocks.
Denote

$$F_n(\boldsymbol{b}) := \frac{1}{n}\sum_{i=1}^{n} f_i(\boldsymbol{b}), \quad f_i(\boldsymbol{b}) = \log(\boldsymbol{b}'\boldsymbol{x}_i) \tag{4.3.9}$$

as functions with respect to $\boldsymbol{b} \in \mathcal{B}$ under a given a sequence of price relatives $\boldsymbol{x}^n = \{\boldsymbol{x}_1, \ldots, \boldsymbol{x}_n\}$. Note that we do not need to assume the stationarity of the market; we only need the boundedness of the sequence of the price relatives $\boldsymbol{x}^n$. Then, the best constant rebalanced portfolio $\boldsymbol{b}^*$ defined in Definition 4.3.2 obtains as the solution of the following optimization problem:

$$\max_{\boldsymbol{b}\in\mathcal{B}} F_n(\boldsymbol{b}) = \max_{\boldsymbol{b}\in\mathcal{B}} \frac{1}{n}\sum_{i=1}^{n} f_k(\boldsymbol{b}) = \frac{1}{n}\log\left(\max_{\boldsymbol{b}\in\mathcal{B}}\prod_{i=1}^{n}\boldsymbol{b}'\boldsymbol{x}_k\right).$$

By using this function $F_n(\boldsymbol{b})$, the successive constant rebalanced portfolio is defined as follows:

Definition 4.3.15. *The successive constant rebalanced portfolio $\tilde{\boldsymbol{b}}^n = \{\tilde{\boldsymbol{b}}_1, \ldots, \tilde{\boldsymbol{b}}_n\}$ is defined through the following procedure:*

1. At the beginning of the first trading period take

$$\tilde{\boldsymbol{b}}_1 = \left(\frac{1}{m}, \frac{1}{m}, \ldots, \frac{1}{m}\right)'.$$

2. At the beginning of trading period $i = 2, \ldots, n$, the sequence of the price relatives $\boldsymbol{x}^{i-1} = \{\boldsymbol{x}_1, \ldots, \boldsymbol{x}_{i-1}\}$ is available. Compute $\tilde{\boldsymbol{b}}_i$ as the solution of the optimization problem

$$\max_{\boldsymbol{b}\in\mathcal{B}} F_{i-1}(\boldsymbol{b})$$

where $F_i(\boldsymbol{b})$ is defined in (4.3.9).

Let $\tilde{S}_n(\boldsymbol{x}^n)$ be the wealth by the successive constant rebalanced portfolio at the end of n days, namely,

$$\tilde{S}_n(\boldsymbol{x}^n) = \prod_{i=1}^{n} \tilde{\boldsymbol{b}}_i' \boldsymbol{x}_i.$$

Then, Gaivoronski and Stella (2000) showed the following result.

Theorem 4.3.16. *(Theorem 2 of Gaivoronski and Stella (2000)) Suppose that the sequence of the price relatives $\boldsymbol{x}^n = \{\boldsymbol{x}_1, \ldots, \boldsymbol{x}_n\}$ satisfy the following conditions:*

1. Asymptotic independence:

$$\liminf_{n \in \mathbb{N}} \lambda_{\min} \left(\frac{1}{n} \sum_{i=1}^{n} \frac{\boldsymbol{x}_i \boldsymbol{x}_i'}{\|\boldsymbol{x}_i\|^2} \right) \geq \delta > 0 \tag{4.3.10}$$

where $\lambda_{\min}(A)$ is the smallest eigenvalue of matrix A.

2. Uniform boundedness: There exist $0 < x^-, x^+ < \infty$ such that

$$0 < x^- \leq x_{ij} \leq x^+ < \infty \tag{4.3.11}$$

for $\forall i = 1, \ldots, n$ and $\forall j = 1, \ldots, m$.

Then

1. The asymptotic rate of growth of wealth $\tilde{S}_n(\boldsymbol{x}^n)$ obtained by successive constant rebalanced portfolio $\tilde{\boldsymbol{b}}^n$ coincides with asymptotic growth of wealth $S_n^*(\boldsymbol{x}^n)$ obtained by the best constant rebalanced portfolio $\boldsymbol{x}^*$ up to the first order of the exponent, i.e.,

$$\frac{1}{n} \log S_n^*(\boldsymbol{x}^n) - \frac{1}{n} \log \tilde{S}_n(\boldsymbol{x}^n) \to 0$$

as $n \to \infty$.

2. The following inequality is satisfied:

$$\frac{\tilde{S}_n(\boldsymbol{x}^n)}{S_n^*(\boldsymbol{x}^n)} \geq C(n-1)^{-2K^2/\delta}$$

where $K = \sup_{n \in \mathbb{N}, \boldsymbol{b} \in \mathcal{B}} \|\partial f_n(\boldsymbol{b})/\partial \boldsymbol{b}\|$, while C and δ are constants.

Consider the computation of the SCRP. While the data size n is not large enough, each new data point may bring about substantial change in the SCRP. In this case, some smoothing is necessary to avoid such a substantial change. To overcome this problem, Gaivoronski and Stella (2000) introduced another portfolio, which is referred as the weighted successive constant rebalanced portfolio (WSCRP). The WSCRP is defined by making a linear combination between the previous portfolio and the new SCRP.

Definition 4.3.17. *The weighted successive constant rebalanced portfolio $\tilde{\tilde{\boldsymbol{b}}}^n = \{\tilde{\tilde{\boldsymbol{b}}}_1, \ldots, \tilde{\tilde{\boldsymbol{b}}}_n\}$ is defined through the following procedure:*

1. At the beginning of the first trading period take

$$\tilde{\tilde{\boldsymbol{b}}}_1 = \left(\frac{1}{m}, \frac{1}{m}, \ldots, \frac{1}{m} \right)'.$$

2. At the beginning of trading period $i = 2, \ldots, n$, the sequence of the price relatives $\boldsymbol{x}^{i-1} = \{\boldsymbol{x}_1, \ldots, \boldsymbol{x}_{i-1}\}$ is available. Compute $\tilde{\boldsymbol{b}}_i$ as the solution of the optimization problem

$$\max_{\boldsymbol{b} \in \mathcal{B}} F_{i-1}(\boldsymbol{b})$$

where $F_i(\boldsymbol{b})$ is defined in (4.3.9).

3. *Take the current portfolio at stage i as a linear combination between previous portfolio* $\tilde{\tilde{\boldsymbol{b}}}_{i-1}$ *and* $\tilde{\boldsymbol{b}}_i$*:*

$$\tilde{\tilde{\boldsymbol{b}}}_i = \gamma\tilde{\tilde{\boldsymbol{b}}}_{i-1} + (1-\gamma)\tilde{\boldsymbol{b}}_i$$

where $\gamma \in (0,1)$ *is the weighting parameter.*

Let $\tilde{\tilde{S}}_n(\boldsymbol{x}^n)$ be the wealth by the WSCRP at the end of n days, namely,

$$\tilde{\tilde{S}}_n(\boldsymbol{x}^n) = \prod_{i=1}^{n} \tilde{\tilde{\boldsymbol{b}}}_i' \boldsymbol{x}_i.$$

The following theorem describes the asymptotic properties of the WSCRP.

Theorem 4.3.18. *(Theorem 3 of Gaivoronski and Stella (2000)) Suppose that the sequence of the price relatives* $\boldsymbol{x}^n = \{\boldsymbol{x}_1, \ldots, \boldsymbol{x}_n\}$ *satisfies the asymptotic independence (4.3.10) and the uniform boundedness (4.3.11). Then*

1. *The asymptotic rate of growth of wealth* $\tilde{\tilde{S}}_n(\boldsymbol{x}^n)$ *obtained by weighted successive constant rebalanced portfolio* $\tilde{\tilde{\boldsymbol{b}}}^n$ *coincides with the asymptotic growth of wealth* $S_n^*(\boldsymbol{x}^n)$ *obtained by the best constant rebalanced portfolio* $\boldsymbol{x}^*$ *up to the first order of the exponent, i.e.,*

$$\frac{1}{n}\log S_n^*(\boldsymbol{x}^n) - \frac{1}{n}\log \tilde{\tilde{S}}_n(\boldsymbol{x}^n) \to 0$$

as $n \to \infty$.

2. *The following inequality is satisfied:*

$$\frac{\tilde{\tilde{S}}_n(\boldsymbol{x}^n)}{S_n^*(\boldsymbol{x}^n)} \geq C(n-1)^{-5K^2/(\delta(1-\gamma))}$$

where $K = \sup_{n\in\mathbb{N}, \boldsymbol{b}\in\mathcal{B}} \|\partial f_n(\boldsymbol{b})/\partial \boldsymbol{b}\|$*, while C and* δ *are constants.*

Furthermore, Dempster et al. (2009) extend the SCRP to the case with side information. The mixture successive constant rebalanced portfolio (MSCRP) is defined by a combination of the idea of the universal portfolio with side information defined in Definition 4.3.12 and the idea of SCRP defined in Definition 4.3.15.

4.3.4 *Universal Portfolios with Transaction Costs*

Blum and Kalai (1999) extend Cover's model by adding a fixed percentage commission cost $0 \leq c \leq 1$ to each transaction. In what follows, we consider the wealth of a constant rebalanced portfolio at the end of n days including commission costs (say $S_n^c(\boldsymbol{b}, \boldsymbol{x})$). For simplicity, we will assume that the commission is charged only for purchases and not for sales (according to Blum and Kalai (1999), this can be assumed without loss of generality).

Let $\boldsymbol{p}_i = (p_{i1}, \ldots, p_{im})'$ be the vector of price for all stocks at the end of day i and at the beginning of day $i+1$.[13] Let $\boldsymbol{h}_i = (h_{i1}, \ldots, h_{im})'$ and $\boldsymbol{h}_i^- = (h_{i1}^-, \ldots, h_{im}^-)'$ be the vector of the number of shares for all stocks after and before rebalancing on day i. In the case of "constant rebalanced portfolio" strategy, the portfolio weight is rebalanced as follows:

$$b_j = \frac{p_{1j}h_{1j}}{S_1} = \cdots = \frac{p_{nj}h_{nj}}{S_n}$$

where $S_i = \sum_{l=1}^{m} p_{il}h_{il}$ for $i = 1, \ldots, n$ and $j = 1, \ldots, m$. Writing $\boldsymbol{b}_i^- = (b_{i1}^-, \ldots, b_{im}^-)'$ as the portfolio weight before rebalancing on day i, then it follows that

$$b_{ij}^- = \frac{p_{ij}h_{i-1,j}}{S_i^-}$$

[13] Note that $x_{ij} = p_{ij}/p_{i-1,j}$ for $i = 1, 2, \ldots, n$, $j = 1, \ldots, m$ with $p_{0j} = 1$.

where $S_i^- = \sum_{l=1}^m p_{il}h_{i-1,l}$. By using the above notations, the transaction cost can be written as

$$\sum_{j:h_{ij}>h_{i-1,j}} c(h_{ij}-h_{i-1,j})p_{ij} = \sum_{j:S_ib_j>S_i^-b_{ij}^-} c(S_ib_j - S_i^-b_{ij}^-).$$

Here the reason we need to consider the sum with respect to $h_{ij} > h_{i-1,j}$ is by the assumption to be charged only for purchases and not for sales (i.e., if $h_{ij} > h_{i-1,j}$, asset j is purchased on day i). In addition, it follows

$$h_{ij} > h_{i-1,j} \Leftrightarrow p_{ij}h_{ij} > p_{ij}h_{i-1,j} \Leftrightarrow S_ib_j > S_i^-b_{ij}^-.$$

Denoting $\alpha_i = S_i/S_i^-$, the transaction cost is described as

$$cS_i^- \sum_{j:\alpha_ib_j>b_{ij}^-} (\alpha_ib_j - b_{ij}^-)$$

which implies that

$$\alpha_i = \frac{S_i}{S_i^-} = \frac{S_i^- - (\text{the transaction cost})}{S_i^-} = 1 - c \sum_{j:\alpha_ib_j>b_{ij}^-} (\alpha_ib_j - b_{ij}^-). \tag{4.3.12}$$

Based on this α_i, for a constant rebalanced portfolio $\boldsymbol{b}$ ($\in \mathcal{B}$) with transaction costs, the wealth is described as follows:

Definition 4.3.19. *At the end of n days, a constant rebalanced portfolio $\boldsymbol{b} = (b_1, \ldots, b_m)'$ with transaction costs achieves wealth:*

$$S_n^c(\boldsymbol{b}, \boldsymbol{x}^n) = \prod_{i=1}^n \alpha_i \boldsymbol{b}'\boldsymbol{x}_i = \prod_{i=1}^n \alpha_i \sum_{j=1}^m b_j x_{ij}$$

where α_i is satisfied with Equation (4.3.12) for each i $= 1, 2, \ldots, n$.

In the same manner of Definition 4.3.2, we define the best constant rebalanced portfolio (in hindsight) with transaction costs as follows:

Definition 4.3.20. *At the end of n days, the best constant rebalanced portfolio with transaction costs $\boldsymbol{b}^{c*} = (b_1^{c*}, \ldots, b_m^{c*})'$ achieves wealth:*

$$S_n^{c*}(\boldsymbol{x}^n) = \max_{\boldsymbol{b}\in\mathcal{B}} S_n^c(\boldsymbol{b}, \boldsymbol{x}^n) = \max_{\boldsymbol{b}\in\mathcal{B}} \prod_{i=1}^n \alpha_i \boldsymbol{b}'\boldsymbol{x}_i.$$

Blum and Kalai (1999) proposed the universal portfolio with transaction costs by a slight modification of Definition 4.3.4:

Definition 4.3.21. *The universal portfolio with transaction costs on day i is specified by*

$$\hat{\boldsymbol{b}}_1^c = \left(\frac{1}{m}, \frac{1}{m}, \ldots, \frac{1}{m}\right)', \quad \hat{\boldsymbol{b}}_i^c = \frac{\int_{\boldsymbol{b}\in\mathcal{B}} \boldsymbol{b} S_{i-1}^c(\boldsymbol{b}, \boldsymbol{x}^{i-1})d\boldsymbol{b}}{\int_{\boldsymbol{b}\in\mathcal{B}} S_{i-1}^c(\boldsymbol{b}, \boldsymbol{x}^{i-1})d\boldsymbol{b}}, \quad i = 2, 3, \ldots, n$$

where $S_{i-1}^c(\boldsymbol{b}, \boldsymbol{x}^{i-1}) = \prod_{l=1}^{i-1} \alpha_l \boldsymbol{b}'\boldsymbol{x}_l$ and α_l is satisfied with Equation (4.3.12) for each l.

Moreover, for a causal (nonanticipating) portfolio strategy with transaction costs , the wealth is described as follows:

Definition 4.3.22. *At the end of n days, a causal (nonanticipating) portfolio strategy $\{\hat{\boldsymbol{b}}_i = (\hat{b}_{i1}, \ldots, \hat{b}_{im})'; i = 1, \ldots, n\}$ with transaction costs achieves wealth:*

$$\hat{S}_n^c(\boldsymbol{x}^n) = \prod_{i=1}^n \hat{\alpha}_i \hat{\boldsymbol{b}}_i' \boldsymbol{x}_i$$

where $\hat{\alpha}_i$ is satisfied with

$$\hat{\alpha}_i = 1 - c \sum_{j:\hat{\alpha}_i\hat{b}_{ij} > \hat{b}_{ij}^-} (\hat{\alpha}_i\hat{b}_{ij} - \hat{b}_{ij}^-)$$

$$\hat{b}_{ij}^- = \frac{p_{ij}\hat{h}_{i-1,j}}{\sum_{r=1}^m p_{ir}\hat{h}_{i-1,r}} = \frac{p_{ij}\frac{\hat{b}_{i-1,j}\hat{S}_{i-1}^c(\boldsymbol{x}^{i-1})}{p_{i-1,j}}}{\sum_{r=1}^m p_{ir}\frac{\hat{b}_{i-1,r}\hat{S}_{i-1}^c(\boldsymbol{x}^{i-1})}{p_{i-1,r}}} = \frac{x_{ij}\hat{b}_{i-1,j}}{\sum_{r=1}^m x_{ir}\hat{b}_{i-1,r}}.$$

Similarly to the case without transaction costs, Blum and Kalai (1999) showed an upper bound for the ratio $(S_n^{c*}(\boldsymbol{x}^n)/\hat{S}_n^c(\boldsymbol{x}^n))$ only by using the following natural properties of optimal rebalancing:

1. The costs paid changing from distribution $\boldsymbol{b}_1$ to $\boldsymbol{b}_3$ are no more than the costs paid changing from $\boldsymbol{b}_1$ to $\boldsymbol{b}_2$ and then from $\boldsymbol{b}_2$ to $\boldsymbol{b}_3$.
2. The cost, per dollar, of changing from a distribution $\boldsymbol{b}$ to a distribution $(1-\alpha)\boldsymbol{b} + \alpha\boldsymbol{b}_0$ (where $\boldsymbol{b}_0 \in \mathcal{B}$ and $0 < \alpha < 1$) is no more than αc, simply because we are moving at most an α fraction of our money.
3. An investment strategy I which invests an initial fraction α of its money according to investment strategy I_1 and an initial $1-\alpha$ of its money according to I_2, will achieve at least α times the wealth of I_1 plus $1-\alpha$ times the wealth of I_2. (In fact, I may do even better by occasionally saving in commission cost if, for instance, strategy I_1 says to sell stock A and strategy I_2 says to buy it.)

Theorem 4.3.23. *(Theorem 2 of Blum and Kalai (1999)) Let $\boldsymbol{x}^n$ be a stock market sequence $\{\boldsymbol{x}_1, \boldsymbol{x}_2, \ldots, \boldsymbol{x}_n\}$, $\boldsymbol{x}_i \in \mathbb{R}_+^m$ of length n with m assets. Let $S_n^{c*}(\boldsymbol{x}^n)$ be the wealth achieved by the best constant rebalanced portfolio with transaction costs defined by Definition 4.3.20, and let $\hat{S}_n^c(\boldsymbol{x}^n)$ be the wealth achieved by the universal portfolio with transaction costs defined by Definitions 4.3.21 and 4.3.22. Then*

$$\frac{S_n^{c*}(\boldsymbol{x}^n)}{\hat{S}_n^c(\boldsymbol{x}^n)} \leq \binom{(1+c)n + m - 1}{m-1} \leq ((1+c)n+1)^{m-1}.$$

Remark 4.3.24. *In the same way as Theorem 4.3.5, we have*

$$\frac{S_n^{c*}(\boldsymbol{x}^n)}{\hat{S}_n^c(\boldsymbol{x}^n)} \geq 1.$$

From this and Theorem 4.3.23, it immediately implies that

$$\frac{1}{n}\log 1 \leq \frac{1}{n}\log\frac{S_n^{c*}(\boldsymbol{x}^n)}{\hat{S}_n^c(\boldsymbol{x}^n)} \leq \frac{m-1}{n}\log((1+c)n+1).$$

Therefore, it follows that

$$\frac{1}{n}\log\hat{S}_n^c(\boldsymbol{x}^n) - \frac{1}{n}\log S_n^{c*}(\boldsymbol{x}^n) \to 0$$

as $n \to \infty$. Thus, the universal portfolio can even achieve the same asymptotic growth rate as the best constant rebalanced portfolio in the presence of commission.

4.3.5 Bibliographic Notes

Cover (1991) introduced universal portfolios, which Cover and Ordentlich (1996) extended to the case with side information. Including the introduction of information theory, these results are summarized in Cover and Thomas (2006). Regarding some properties of the Dirichlet distribution, we follow Ng et al. (2011). Gaivoronski and Stella (2000) introduced successive constant rebalanced portfolios and Dempster et al. (2009) extended their result to the case with side information. The effect of the transaction cost is addressed by Blum and Kalai (1999).

4.3.6 Appendix

Proof of Theorem 4.3.6 This proof originally comes from Cover and Thomas (2006) or Cover and Ordentlich (1996). We fill in the details or partially modify them.
Rewrite $S_n^*(\boldsymbol{x}^n)$ as

$$\begin{aligned} S_n^*(\boldsymbol{x}^n) &= \prod_{i=1}^{n} \boldsymbol{b}^{*\prime}\boldsymbol{x}_i \\ &= \prod_{i=1}^{n}\sum_{j=1}^{m} b_j^* x_{ij} \\ &= (b_1^* x_{11} + \cdots + b_m^* x_{1m}) \cdots (b_1^* x_{n1} + \cdots + b_m^* x_{nm}) \\ &= \sum_{j_1,\ldots,j_n=1}^{m} \prod_{i=1}^{n} b_{j_i}^* x_{ij_i} \\ &= \sum_{j^n=\{j_1,\ldots,j_n\}\in\{1,\ldots,m\}^n} \prod_{i=1}^{n} b_{j_i}^* x_{ij_i}. \end{aligned} \tag{4.3.13}$$

From (4.3.3), similarly to the above, we can rewrite

$$\hat{S}_n(\boldsymbol{x}^n) = \frac{\int_{\boldsymbol{b}\in\mathcal{B}} \prod_{i=1}^{n} \boldsymbol{b}'\boldsymbol{x}_i d\boldsymbol{b}}{\int_{\boldsymbol{b}\in\mathcal{B}} d\boldsymbol{b}} = \frac{\sum_{j^n\in\{1,\ldots,m\}^n} \int_{\boldsymbol{b}\in\mathcal{B}} \prod_{i=1}^{n} b_{j_i} x_{ij_i} d\boldsymbol{b}}{\int_{\boldsymbol{b}\in\mathcal{B}} d\boldsymbol{b}}. \tag{4.3.14}$$

From (4.3.13), (4.3.14) and Lemma 1 of Cover and Ordentlich (1996) (or Lemma 16.7.1 of Cover and Thomas (2006)), we have

$$\frac{S_n^*(\boldsymbol{x}^n)}{\hat{S}_n(\boldsymbol{x}^n)} \le \max_{j^n\in\{1,\ldots,m\}^n} \frac{\prod_{i=1}^{n} b_{j_i}^* \int_{\boldsymbol{b}\in\mathcal{B}} d\boldsymbol{b}}{\int_{\boldsymbol{b}\in\mathcal{B}} \prod_{i=1}^{n} b_{j_i} d\boldsymbol{b}} \tag{4.3.15}$$

since the product of the x_{ij_j}'s factors out of the numerator and denominator.
Next, we consider the upper bound of $\prod_{i=1}^{n} b_{j_i}^*$ for each j^n. For a fixed $j^n = \{j_1, \ldots, j_n\}$, let $n_r(j^n)(r = 1, \ldots, m)$ denote the number of occurrences of r in j^n. Then, for any $\boldsymbol{b} \in \mathcal{B}$, it is seen that $\prod_{i=1}^{n} b_{j_i} = \prod_{r=1}^{m} b_r^{n_r(j^n)}$. In order to find the maximizer of $\prod_{r=1}^{m} b_r^{n_r(j^n)}$, we consider the Lagrange multiplier

$$L(\boldsymbol{b}) = \prod_{r=1}^{m} b_r^{n_r(j^n)} + \lambda\left(\sum_{j=1}^{m} b_j - 1\right).$$

Differentiating this with respect to b_i yields

$$\frac{\partial L}{\partial b_i} = n_i(j^n) b_i^{n_i(j^n)-1} \prod_{r\ne i} b_r^{n_r(j^n)} + \lambda, \quad i = 1, 2, \ldots, m.$$

Setting the partial derivative equal to 0 for a maximum, we have

$$\arg\max_{\boldsymbol{b}\in\mathcal{B}} L(\boldsymbol{b}) = \left(\frac{n_1(j^n)}{n}, \ldots, \frac{n_m(j^n)}{n}\right),$$

which implies

$$\prod_{i=1}^{n} b_{j_i}^* \le \prod_{r=1}^{m} \left(\frac{n_r(j^n)}{n}\right)^{n_r(j^n)}. \tag{4.3.16}$$

Next, we consider the calculation of

$$\int_{\mathbf{b}\in\mathcal{B}} d\mathbf{b} \quad and \quad \int_{\mathbf{b}\in\mathcal{B}} \prod_{i=1}^{n} b_{j_i} d\mathbf{b} = \int_{\mathbf{b}\in\mathcal{B}} \prod_{r=1}^{m} b_r^{n_r(j^n)} d\mathbf{b}.$$

Define

$$C = \left\{ \mathbf{c} = (c_1, \ldots, c_{m-1}) \in \mathbb{R}^{m-1} : \sum_{j=1}^{m-1} c_j \leq 1, c_j \geq 0 \right\}.$$

Since a uniform distribution over $\mathcal{B}$ induces a uniform distribution over C, we have

$$\int_{\mathbf{b}\in\mathcal{B}} d\mathbf{b} = \int_{\mathbf{c}\in C} d\mathbf{c} = \int_{\mathbf{c}\in C} \prod_{r=1}^{m-1} c_r^{1-1} \left(1 - \sum_{r=1}^{m-1} c_r\right)^{1-1} \prod_{r=1}^{m-1} dc_r = \frac{\prod_{r=1}^{m} \Gamma(1)}{\Gamma\left(\sum_{r=1}^{m} 1\right)} = \frac{1}{(m-1)!} \tag{4.3.17}$$

and

$$\begin{aligned}
\int_{\mathbf{b}\in\mathcal{B}} \prod_{r=1}^{m} b_r^{n_r(j^n)} d\mathbf{b} &= \int_{\mathbf{c}\in C} \prod_{r=1}^{m} c_r^{n_r(j^n)} \left(1 - \sum_{r=1}^{m-1} c_r\right)^{n_m(j^n)} d\mathbf{c} \\
&= \int_{\mathbf{c}\in C} \prod_{r=1}^{m-1} c_r^{n_r(j^n)+1-1} \left(1 - \sum_{r=1}^{m-1} c_r\right)^{n_m(j^n)+1-1} \prod_{r=1}^{m-1} dc_r \\
&= \frac{\prod_{r=1}^{m} \Gamma\left(n_r(j^n) + 1\right)}{\Gamma\left(\sum_{r=1}^{m} (n_r(j^n) + 1)\right)} \\
&= \frac{\prod_{r=1}^{m} \Gamma\left(n_r(j^n) + 1\right)}{\Gamma(n+m)} \\
&= \frac{\prod_{r=1}^{m} n_r(j^n)!}{(n+m-1)!}
\end{aligned}$$

from the basic property of the Dirichlet distribution (see, e.g., Ng et al. (2011)). Moreover, by Theorem 11.1.3 of Cover and Thomas (2006), it follows that

$$\frac{n!}{\prod_{r=1}^{m} n_r(j^n)!} = \binom{n}{n_1(j^n), \ldots, n_m(j^n)} \leq 2^{nH(P)}$$

where $H(P)$ is the entropy of a random variable $X = \begin{cases} a_1 & with\ probability\ \frac{n_1(j^n)}{n} \\ \vdots & \vdots \\ a_m & with\ probability\ \frac{n_m(j^n)}{n} \end{cases}$, which implies that

$$H(P) = -\sum_{r=1}^{m} \frac{n_r(j^n)}{n} \log_2 \frac{n_r(j^n)}{n} = -\frac{1}{n} \log_2 \prod_{r=1}^{m} \left(\frac{n_r(j^n)}{n}\right)^{n_r(j^n)}$$

and

$$2^{nH(P)} = 2^{\log_2 \prod_{r=1}^{m} \left(\frac{n_r(j^n)}{n}\right)^{-n_r(j^n)}} = \prod_{r=1}^{m} \left(\frac{n_r(j^n)}{n}\right)^{-n_r(j^n)}.$$

Thus, we have

$$\int_{\mathbf{b}\in\mathcal{B}} \prod_{i=1}^{n} b_{j_i} d\mathbf{b} = \frac{n!}{(n+m-1)!} \frac{\prod_{r=1}^{m} n_r(j^n)!}{n!} \geq \frac{n!}{(n+m-1)!} \prod_{r=1}^{m} \left(\frac{n_r(j^n)}{n}\right)^{n_r(j^n)}. \tag{4.3.18}$$

From (4.3.15), (4.3.16), (4.3.17) and (4.3.18), it follows that

$$\begin{aligned}
\frac{S_n^*(\boldsymbol{x}^n)}{\hat{S}_n(\boldsymbol{x}^n)} &\le \max_{j^n\in\{1,\dots,m\}^n} \frac{\prod_{r=1}^m \left(\frac{n_r(j^n)}{n}\right)^{n_r(j^n)} \int_{\boldsymbol{b}\in\mathcal{B}} d\boldsymbol{b}}{\int_{\boldsymbol{b}\in\mathcal{B}} \prod_{i=1}^n b_{j_i} d\boldsymbol{b}} \\
&= \max_{j^n\in\{1,\dots,m\}^n} \frac{\prod_{r=1}^m \left(\frac{n_r(j^n)}{n}\right)^{n_r(j^n)}}{(m-1)! \int_{\boldsymbol{b}\in\mathcal{B}} \prod_{i=1}^n b_{j_i} d\boldsymbol{b}} \\
&\le \max_{j^n\in\{1,\dots,m\}^n} \frac{\prod_{r=1}^m \left(\frac{n_r(j^n)}{n}\right)^{n_r(j^n)}}{\frac{(m-1)!n!}{(n+m-1)!} \prod_{r=1}^m \left(\frac{n_r(j^n)}{n}\right)^{n_r(j^n)}} \\
&= \frac{(n+m-1)!}{(m-1)!n!} \\
&= \binom{n+m-1}{m-1} \\
&= \left(\frac{n+m-1}{m-1}\right)\left(\frac{n+m-2}{m-2}\right)\cdots\left(\frac{n+1}{1}\right) \\
&= \left(\frac{n}{m-1}+1\right)\left(\frac{n}{m-2}+1\right)\cdots\left(\frac{n}{1}+1\right) \\
&\le (n+1)(n+1)\cdots(n+1) \\
&= (n+1)^{m-1}. \quad \square
\end{aligned}$$

Problems

4.1 Verify (4.1.6).

4.2 Verify (4.2.1).

4.3 Consider why $-C_t h$ is approximated by $(H_{0,t} - H_{0,t-h})S_{0,t} + (\boldsymbol{H}_t - \boldsymbol{H}_{t-h})'\boldsymbol{S}_{t-}$ for a small period $[t-h, t)$ in the absence of any income.

4.4 Show (4.2.7) by using Ito's lemma where $d\mathcal{W}_t$ is defined by (4.2.4).

4.5 Verify (4.2.15) and (4.2.16). (Hint: Show $E(d\boldsymbol{W}_t) = \boldsymbol{0}$ and $E(d\boldsymbol{W}_t d\boldsymbol{W}_t') = I_m dt$.)

4.6 Verify (4.2.28), (4.2.34) and (4.2.35).

4.7 Verify (4.3.5) (see Theorem 2 and Lemmas 3–5 of Cover and Ordentlich (1996)).

4.8 Show Theorem 4.3.13 (see Theorem 3 of Cover and Ordentlich (1996)).

4.9 Show Theorem 4.3.23 (see Theorem 2 of Blum and Kalai (1999)).

4.10 Obtain the monthly stock prices of Toyota, Honda and Komatsu from December 1983 to January 2013 (350 observations). The simple prices are given in the file `stock-price-data.txt`.

(i) Let $\boldsymbol{x}_i = (x_{i1}, x_{i2}, x_{i3})$ be the vector of the price relative for all stocks where $i = 1, \dots, 349$, $j = 1$ (*Toyota*), 2 (*Honda*), 3 (*Komatsu*) and

$$x_{ij} = (\text{stock price of month } j + 1)/(\text{stock price of month } j).$$

Then, calculate the wealths of the constant rebalance portfolios $S_n(\boldsymbol{b}, \boldsymbol{x}^n)$ for $\boldsymbol{b} = (1, 0, 0)'$, $\boldsymbol{b} = (0, 1, 0)'$, $\boldsymbol{b} = (0, 0, 1)'$, $\boldsymbol{b} = (1/3, 1/3, 1/3)'$ and $n = 1, 2, \dots, 349$.

(ii) Calcuate the wealths of the best constant rebalanced portfolio $S_n^*(\boldsymbol{x}^n)$ for $n = 1, 2, \dots, 349$.

(iii) Calcuate the wealths of the (approximated) uniform weighted universal portfolio $S_n(\hat{\boldsymbol{b}}_n, \boldsymbol{x}^n)$ for $n = 1, 2, \dots, 349$ where an approximated form of the UP is defined by

$$\hat{\boldsymbol{b}}_{k+1} = \frac{\frac{1}{N}\sum_{b_1=0}^{N}\sum_{b_2=0}^{N-b_1}\left(\frac{b_1}{N}, \frac{b_2}{N}, 1 - \frac{b_1}{N} - \frac{b_2}{N}\right)' S_k\left(\left(\frac{b_1}{N}, \frac{b_2}{N}, 1 - \frac{b_1}{N} - \frac{b_2}{N}\right)', \boldsymbol{x}^k\right)}{\frac{1}{N}\sum_{b_1=0}^{N}\sum_{b_2=0}^{N-b_1} S_k\left(\left(\frac{b_1}{N}, \frac{b_2}{N}, 1 - \frac{b_1}{N} - \frac{b_2}{N}\right)', \boldsymbol{x}^k\right)}$$

for $k \geq 1$ and $N = 100$.

Chapter 5

Portfolio Estimation Based on Rank Statistics

5.1 Introduction to Rank-Based Statistics

5.1.1 History of Ranks

5.1.1.1 Wilcoxon's signed rank and rank sum tests

Even at the introductory level, students who learn statistics should hear about Wilcoxon's signed rank and rank sum tests. Wilcoxon (1945) described these tests for the first time to deal with the following two kinds of problems.

(a) We may have a number of paired replications for each of the two treatments, which are unpaired.

(b) We may have a number of paired comparisons leading to a series of differences, some of which may be positive and some negative.

The unpaired experiments (a) are formulated by the following two-sample location problem. Let X_i $(i = 1, \dots, N)$ be i.i.d. with unspecified (nonvanishing) density f. Under the null hypothesis $\mathcal{H}_0$, we observe $Z_i = X_i$ $(i = 1, \dots, N)$, whereas, under the alternative $\mathcal{H}_1$, $Z_i = X_i$ $(i = 1, \dots, m)$ and $Z_i = X_i - \theta$ $(i = m + 1, \dots, N)$ for some $\theta \neq 0$. As the example for the unpaired experiments, Wilcoxon (1945) introduced the following Table 5.1, which gives the results of fly spray tests on two preparations in terms of percentage mortality, where $m = 8$ and $N = 16$.

Table 5.1: Fly spray tests on two preparations

	Sample A Percent kill	Rank	Sample B Percent kill	Rank
	68	12.5	60	4
	68	12.5	67	10
	59	3	61	5
	72	15	62	6
	64	8	67	10
	67	10	63	7
	70	14	56	1
	74	16	58	2
Total	542	91	494	45

On the other hand, the paired example is described by the following symmetry test. Suppose that the independent observations $(X_1, Y_1), \dots, (X_N, Y_N)$, are given. If the observations are exchangeable, that is, the pairs (X_i, Y_i) and (Y_i, X_i) are equal in distribution, then $X_i - Y_i$ is symmetrically distributed about zero. As the example for the paired experiment, Wilcoxon (1945) also gave the data obtained in a seed treatment experiment on wheat, which is shown in Table 5.2. The data are taken from a randomized block experiment with eight replications of treatments A and B.

Table 5.2: Seed treatment experiment on wheat

Block	A	B	A–B	Rank
1	209	151	58	8
2	200	168	32	7
3	177	147	30	6
4	169	164	5	1
5	159	166	−7	−3
6	169	163	6	2
7	187	176	11	5
8	198	188	10	4

Now, we introduce the rank and signed rank statistics to test the above two types of experiments. Let $X_1, \dots, X_N$ be a set of real-valued observations which are coordinates of a random vector $\boldsymbol{X} \in \mathbb{R}^N$. Consider the space $\mathcal{Y} \subseteq \mathbb{R}^N$ defined by

$$\mathcal{Y} = \left\{ \boldsymbol{y} \mid \boldsymbol{y} \in \mathbb{R}^N,\ y_1 \le y_2 \le \cdots \le y_N \right\}.$$

The order statistic of $\boldsymbol{X}$ is the random vector $\boldsymbol{X}_{(\,)} = (X_{(1)}, \dots, X_{(N)})' \in \mathcal{Y}$ consisting of the ordered values of the coordinate of $\boldsymbol{X}$. More precisely, ith order statistic $X_{(i)}$ is ith value of the observations positioned in increasing order. The rank R_i of X_i among $X_1, \dots, X_N$ is its position number in the order statistic. That is, if $X_1, \dots, X_N$ are all different (so that with probability one if $\boldsymbol{X}$ has density), then R_i is defined by the equation

$$X_i = X_{(R_i)}.$$

Otherwise, the rank R_i is defined as the average of all indices j such that $X_i = X_{(j)}$, which is sometimes called midrank.

We write the vector of ranks as $\boldsymbol{R_N} = (R_1, \dots, R_N)'$. A rank statistic $S\,(\boldsymbol{R_N})$ is any function of the ranks $\boldsymbol{R_N}$. A linear rank statistic is a rank statistic of the special form $\sum_{i=1}^N a\,(i, R_i)$, where $(a(i, j))_{i,j=1,\dots,N}$ is a given $N \times N$ matrix. Moreover, a simple linear rank statistic belongs to the subclass of linear rank statistics, which take the form

$$\sum_{i=1}^N c_i a_{R_i},$$

where $\boldsymbol{c} = (c_1, \dots, c_N)'$ and $\boldsymbol{a} = (a_1, \dots, a_N)'$ are given vectors in $\mathbb{R}^N$ and called the coefficients and scores, respectively.

The most popular simple linear rank statistic is the Wilcoxon statistic

$$W = \sum_{i=m+1}^N R_i$$

which has coefficients $\boldsymbol{c} = (0, \dots, 0, 1, \dots, 1)'$ and scores $\boldsymbol{a} = (1, \dots, N)'$. Unlike the Student test, the Wilcoxon test does not require any assumption on the density f. Wilcoxon (1945) mainly considered his test a robust, "quick and easy" solution for the location shift problem to be used when everything else fails. For determining the significance of differences in unpaired experiments (a), he evaluated the probability of

$$\left\{ \min\left(\sum_{i=1}^{N/2} R_i,\ \sum_{i=N/2+1}^{N} R_i \right) \le x \right\}, \quad x = \frac{N(N+2)}{8}, \dots, \frac{N(N+1)}{4}, \tag{5.1.1}$$

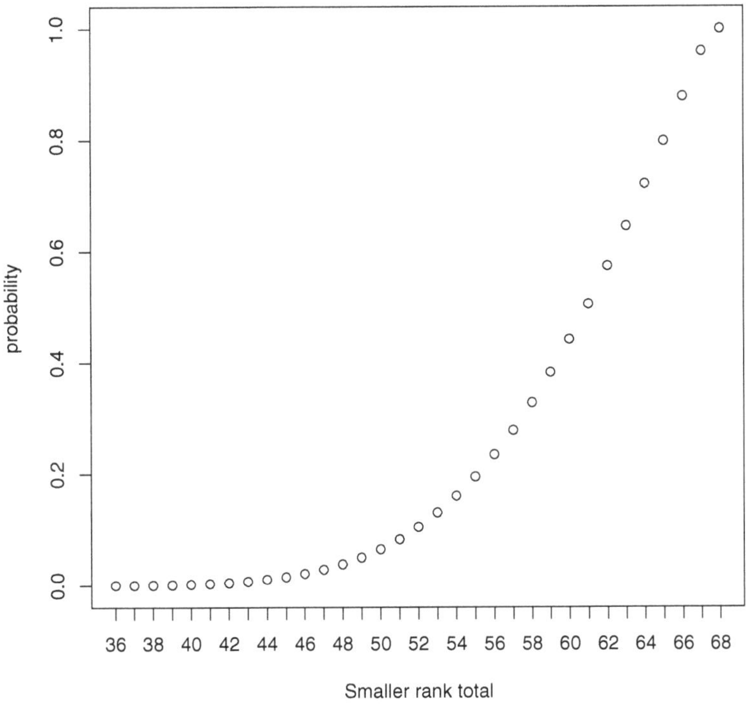

Figure 5.1: The probability (5.1.1) of the smaller rank total

which is shown in Figure 5.1 with $N = 16$. The smaller rank total of the fly spray test in Table 5.1 is 45 and the probability that this total is equal to or less than 45 is only 1.476%. To compute the probability of the smaller rank total in R, type

```
nn<-8 # value of m
mm<-nn*2 # value of N
ss<-nn*(nn+1)/2 # minimum of smaller rank total
tt<-mm*(mm+1)/2 # total of all rank
ll<-tt-ss*2
uu<-dwilcox(0:ll,nn,nn)
lll<-length(uu)
mmm<-(lll+1)/2
probability<-cumsum(c(uu[1:(mmm-1)]+uu[lll:(mmm+1)],uu[mmm]))
jjj<-(1:mmm)+ss-1
plot(probability,xaxt='n',xlab='Smaller rank total')
cc<-1:mmm
axis(1,jjj,at=cc)
PSRT<-function(x)
{
probability[x-ss+1]
}
PSRT(45)
```

The sign of a number x is defined by the function

$$\operatorname{sign}(x) = \begin{cases} 1, & \text{if x > 0,} \\ 0, & \text{if x = 0,} \\ -1, & \text{if x < 0.} \end{cases}$$

The absolute rank R_i^+ of an observation X_i in a sample $X_1, \ldots, X_N$ is defined as the rank of $|X_i|$, among the absolute values $|X_1|, \ldots, |X_N|$ of the sample. Denote the vector of absolute values, absolute ranks and signs by $|\boldsymbol{X}| = (|X_1|, \cdots, |X_N|)'$, $\boldsymbol{R}_N^+ = \left(R_1^+, \ldots, R_N^+\right)'$ and $\operatorname{sign}(|\boldsymbol{X}|) = (\operatorname{sign}(X_1), \cdots, \operatorname{sign}(X_N))'$, respectively. The ordinary ranks $\boldsymbol{R}_N$ can always be derived from the combined set $\left(\boldsymbol{R}_N^+, \operatorname{sign}(\boldsymbol{X})\right)$ of absolute ranks and signs; therefore the vector $\left(\boldsymbol{R}_N^+, \operatorname{sign}(\boldsymbol{X})\right)$ has more information of observations $\boldsymbol{X}$ than the ordinary ranks $\boldsymbol{R}_N$. Obviously, if there is no tied pair of observation (so that with probability one if $\boldsymbol{X}$ has density), then

$$X_i = |\boldsymbol{X}|_{(R_i^+)} \operatorname{sign}(X_i), \quad i = 1, \ldots, N.$$

A signed rank statistic is any function $S\left(\boldsymbol{R}_N^+, \operatorname{sign}(\boldsymbol{X})\right)$ of absolute ranks and signs. A linear signed rank statistic is a signed rank statistic of the form $\sum_{i=1}^N a\left(i, R_i^+, \operatorname{sign}(X_i)\right)$. Furthermore, a simple linear signed rank statistic has the form

$$\sum_{i=1}^{N} c_i a_{R_i^+} \operatorname{sign}(X_i).$$

The main attractive feature of signed rank statistics is their simplicity and distribution freeness over the set of all symmetric distributions.

Again, the most popular simple linear signed rank statistic is the Wilcoxon signed rank statistic

$$W = \sum_{i=1}^{N} R_i^+ \operatorname{sign}(X_i)$$

which has coefficients $\boldsymbol{c} = (1, \ldots, 1)'$ and scores $\boldsymbol{a} = (1, \ldots, N)'$. For determining the significance of differences in paired experiments (b), Wilcoxon (1945) evaluated the probability of

$$\left\{\min\left(\sum_{i=1}^{N} R_i^+ \max\{\operatorname{sign}(X_i), 0\}, \sum_{i=1}^{N} R_i^+ \min\{\operatorname{sign}(X_i), 0\}\right) \le x\right\}, \quad x = 0, \ldots, \frac{N(N+1)}{4}, \quad (5.1.2)$$

which is shown in Figure 5.2 with $N = 8$. The smaller rank total of + or − in the seed treatment experiment (Table 5.2) is 3 and the probability that this total is equal to or less than 3 is 3.906%. To compute the probability of the smaller rank total of + or − in R, type

```
nn<-8 # value of N
tt<-nn*(nn+1)/2 # total of all rank
uu<-dsignrank(0:tt,nn)
lll<-length(uu)
mmm<-(lll+1)/2
probability<-cumsum(c(uu[1:(mmm-1)]+uu[lll:(mmm+1)],uu[mmm]))
plot(probability,xaxt='n',xlab='Smaller rank total of + or -.')
jjj<-(0:(mmm-1))
cc<-1:mmm
axis(1,jjj,at=cc)
PSRN<-function(x)
{
```

```
probability[x+1]
}
PSRN(3)
```

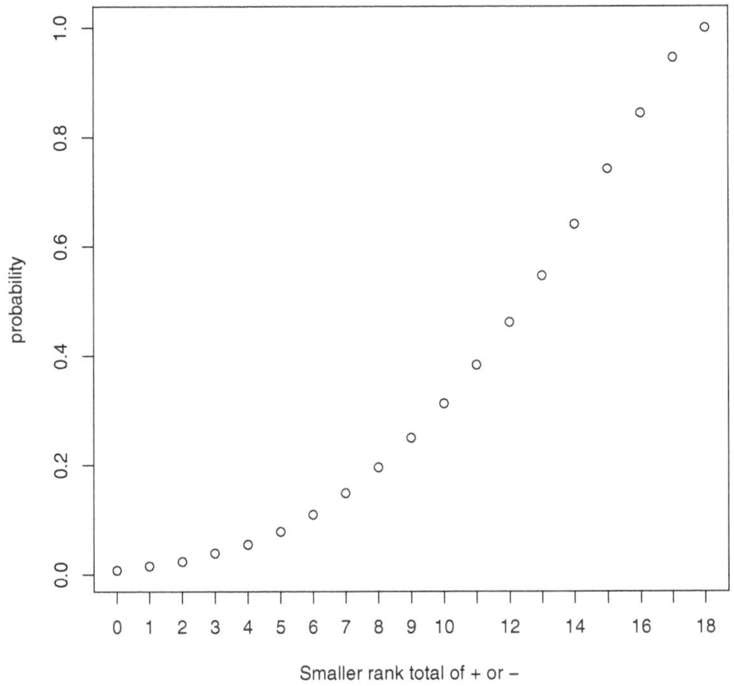

Figure 5.2: The probability (5.1.2) of the smaller rank total of + or –

5.1.1.2 Hodges–Lehmann and Chernoff–Savage

Eleven years after Wilcoxon's seminal paper (Wilcoxon (1945)), a totally unexpected result was published by Hodges and Lehmann (1956). They derived the minimum Pitman efficiency of the Wilcoxon tests. This result is based on lecture notes of Pitman (1948), and Andrews (1954), who considered the asymptotic relative efficiencies of the H test, which will defined in (5.1.3) later, proposed by Kruskal and Wallis (1952). The H test is a direct generalization of the two-sided Wilcoxon test ($\theta \neq \theta_0 = 0$) and is similar to a classical F test, with ranks replacing the original observations.

The K-sample problem is expressed formally as follows. Let $\{X_{ij}\}$ for $i = 1, \ldots, K$ and $j = 1, \ldots, n_i$ be a set of independent random variables. The probability distribution functions of X_{ij}, $j = 1, \ldots, n_i$ are common and denoted by G_i, so that $G_i(x)$ is the probabilities $P_i\left(X_{ij} \leq x\right)$, $j = 1, \ldots, n_i$. The set of admissible hypotheses designates that each G_i belongs to some class $\mathcal{G}$ of distribution functions. To avoid the problems of ties, we assume throughout that the class is the class of continuous distribution functions. The hypothesis to be tested, say H_0, specifies that for an element G of $\mathcal{G}$, we have $G_1(x) = G_2(x) = \cdots = G_K(x) = G(x)$ for all real x. Alternative to H_0 is the hypotheses that each G_i belongs to $\mathcal{G}$ but H_0 does not hold. If we consider the translation-type alternatives, the class of admissible hypotheses will be $G_i(x) = G(x + \theta_i)$, $i = 1, \ldots, K$ for some arbitrary choice of $G \in \mathcal{G}$ and real numbers $\theta_1, \ldots, \theta_K$. All $\theta_i = 0$ yield hypothesis H_0, while if $\theta_{i_1} \neq \theta_{i_2}$ for some pair (i_1, i_2) then an alternative to H_0 is given.

The H test is based on the statistic

$$H = \frac{12}{N(N+1)} \sum_{i=1}^{K} n_i \left(\overline{R}_i - \frac{N+1}{2} \right)^2, \tag{5.1.3}$$

where $\overline{R}_i$ is the average rank of the members of the ith sample obtained after ranking all of the $N = \sum_{i=1}^{K} n_i$ observations. The sequence of the local alternative hypothesis A_n specifies, for each $i = 1, \ldots, K$, that $G_i(x) = G\left(x + \theta_i/\sqrt{n}\right)$, with $G \in \mathcal{G}$, and for some pair (i_1, i_2) that $\theta_{i_1} \neq \theta_{i_2}$. The limiting probability distributions will then be found as $n \to \infty$. Therefore, for each index n, we assume that $n_i = s_i n$ with s_i a positive integer. Since the limiting distribution of H under the local alternative A_n depends on the quantity

$$\lim_{n\to\infty} \int_{\mathbb{R}} \sqrt{n}\left\{G\left(x + t/\sqrt{n}\right) - G(x)\right\} dG(x),$$

we assume the following.

Assumption 5.1.1. *The distribution function G has the density $g(x) = G'(x)$ except on a set of G-measurable zero, and for any real number t, there exists a function g which bounds the difference quotient $\left|\frac{G(x+t)-G(x)}{t}\right| \leq l(x)$ for which $\int_{\mathbb{R}} l(x)dG(x) < \infty$.*

A direct application of the Lebesgue bounded convergence theorem and the definition of the derivative prove that under Assumption 5.1.1, we have

$$\lim_{n\to\infty} \int_{\mathbb{R}} \sqrt{n}\left\{G\left(x + t/\sqrt{n}\right) - G(x)\right\} dG(x) = t\int_{\mathbb{R}} G'(x)dG(x) = t\int_{\mathbb{R}} g^2(x)dx. \tag{5.1.4}$$

Now, we have the following lemma.

Lemma 5.1.2. *Suppose that Assumption 5.1.1 is satisfied. Then, under the local hypothesis A_n, the limiting distribution of the test statistic H is $\chi^2_{K-1}\left(\lambda^H\right)$, where*

$$\lambda^H = 12\left(\int_{\mathbb{R}} g^2(x)dx\right)^2 \sum_{i=1}^{K} s_i\left(\theta_i - \overline{\theta}\right)^2, \quad \overline{\theta} = \frac{\sum_{i=1}^{K} s_i\theta_i}{\sum_{i=1}^{K} s_i}$$

and $\chi^2_r(\lambda)$ denotes the possibly noncentral chi square distribution with degrees of freedom r and noncentral parameter λ.

The concept of asymptotic relative efficiency of one consistent test with respect to another is due to Pitman (1948). We consider the comparison of the H test with respect to the ordinary analysis of variance F test, where the classical F statistic in this instance is defined by

$$F = \frac{\frac{1}{K-1}\sum_{i=1}^{K} n_i\left(\overline{X_{i\cdot}} - \overline{X}\right)^2}{\frac{1}{\sum_{i=1}^{K}(n_i-1)}\sum_{i=1}^{K}\sum_{j=1}^{n_i}\left(X_{ij} - \overline{X_{i\cdot}}\right)^2}.$$

If Assumption 5.1.1 is satisfied and $\sigma_G^2 = \int_{\mathbb{R}} x^2 dG(x) - \left(\int_{\mathbb{R}} x dG(x)\right)^2$ exists, then it is well known that the limiting distribution of F under the local alternative A_n is $\chi^2_{K-1}\left(\lambda^F\right)$, where

$$\lambda^F = \sum_{i=1}^{K} s_i\left(\frac{\theta_i - \overline{\theta}}{\sigma_G}\right)^2.$$

Suppose that two consistent tests T and T' require N and N' observations, respectively, to attain the power β at level of significance α for testing the hypothesis H_0 against hypothesis A_n. The difference in the sample sizes N and N' results from the fact that we demand that the tests yield a common power for a given alternative A_n. The asymptotic relative efficiency of T' with respect to T is defined to be

$$\varepsilon_{T',T}\left(\alpha, \beta, H_0, \{A_n\}\right) = \lim_{N\to\infty}\frac{N}{N'} = \lim_{n\to\infty}\frac{\sum_{i=1}^{K} n_i}{\sum_{i=1}^{K} n_i'} = \lim_{n\to\infty}\frac{\sum_{i=1}^{K} s_i n}{\sum_{i=1}^{K} s_i n'} = \lim_{n\to\infty}\frac{n}{n'}. \tag{5.1.5}$$

Note that under the null hypothesis H_0, we have $\theta_1 = \theta_2 = \cdots = \theta_K$, and consequently $\lambda^H = \lambda^F = 0$. That is, under H_0, both of the limiting distributions of H and F are the ordinary chi square distribution χ^2_{K-1}. The alternative hypothesis A_n states that $G_i(x) = G\left(x + \frac{\theta_i}{\sqrt{n}}\right)$ and so is characterized by the numbers $\frac{\theta_i}{\sqrt{n}}$. Let n' and n index the sample sizes for the H test and F test, respectively. To have the same alternatives for each test we must have $\frac{\theta_i}{\sqrt{n}} = \frac{\theta_i'}{\sqrt{n'}}$. If the level of significance is fixed at α and the limiting power fixed at β, then we must have $\lambda^H = \lambda^F$ to achieve the same limiting power for the two tests. The substitution $\theta_i' = \theta_i\sqrt{\frac{n'}{n}}\left(\bar{\theta}' = \bar{\theta}\sqrt{\frac{n'}{n}}\right)$ in λ^H along with the requirement $\lambda^H = \lambda^F$ gives

$$12\left(\int_{\mathbb{R}} g^2(x)dx\right)^2 \frac{n'}{n}\sum_{i=1}^{K} s_i\left(\theta_i - \bar{\theta}\right)^2 = \frac{1}{\sigma_G^2}\sum_{i=1}^{K} s_i\left(\theta_i - \bar{\theta}\right)^2,$$

and consequently the following lemma.

Lemma 5.1.3. *If Assumption 5.1.1 is satisfied and σ_F^2 exists, then the asymptotic relative efficiency of the H test with respect to the F test for testing the hypothesis H_0 against A_n is*

$$\varepsilon_{H,F}\left(\alpha, \beta, H_0, \{A_n\}\right) = 12\sigma_G^2\left(\int_{\mathbb{R}} g^2(x)dx\right)^2.$$

Note that $\varepsilon_{H,F}$ does not depend on the number K of groups, and the case with $K = 2$ corresponds to the comparison of the Wilcoxon test based on the rank sum of X_{2j}'s among the set of N-ordered observations with respect to Student's t-test. Therefore, the relative asymptotic efficiency of the Wilcoxon test relative to the t-test is also given by

$$\varepsilon_{W,t} = 12\sigma_G^2\left(\int_{\mathbb{R}} g^2(x)dx\right)^2,$$

which coincides with Pitman's computation. Some particular values for $\varepsilon_{W,t}$ are shown in Table 5.3.

Furthermore, the lower limit of the relative asymptotic efficiency $\varepsilon_{W,t}$ is given by Hodges and Lehmann (1956). Suppose that $g(x)$ has mean 0 and variance 1, and let $h(x) = \sigma^{-1}g\left(\frac{x-\mu}{\sigma}\right)$; then $h(x)$ has mean μ and variance σ^2. Since $\sigma^2\left(\int_{\mathbb{R}} h^2(x)dx\right)^2 = \left(\int_{\mathbb{R}} g^2(x)dx\right)^2$, the value of $\varepsilon_{W,t}$ is invariant under a change of location or scale, and we may take $\mu = 0$ and $\sigma^2 = 1$. Then, the problem of minimizing $\varepsilon_{W,t}$ reduces to that of minimizing

$$\left(\int_{\mathbb{R}} g^2(x)dx\right)^2$$

subject to the conditions

$$\int_{\mathbb{R}} xg(x)dx = 0, \quad \int_{\mathbb{R}} g(x)dx = \int_{\mathbb{R}} x^2 g(x)dx = 1 \text{ and } g(x) \geq 0 \text{ for all } x.$$

According to the method of undetermined multipliers, it is sufficient to minimize

$$\int_{\mathbb{R}} \left\{g^2(x) - \left(a + bx^2\right)g(x)\right\}dx.$$

For nonnegative $g(x)$, this is achieved by setting

$$g(x) = \begin{cases} -\frac{a+bx^2}{2} & a + bx^2 \leq 0 \\ 0 & \text{otherwise} \end{cases}.$$

From the conditions, we have the constants $a = -\frac{3}{2\sqrt{5}}$, $b = \frac{3}{10\sqrt{5}}$, and so

$$g(x) = -\frac{3}{20\sqrt{5}}\left(x^2 - 5\right), \quad \varepsilon_{W,t} = \frac{108}{125}.$$

Table 5.3: Some particular values for $\varepsilon_{W,t}$

Distribution	$g(x)$	σ_G^2	$\varepsilon_{W,t}$
Normal	$\frac{1}{\sqrt{2\pi\sigma^2}}\exp\left\{-\frac{(x-\mu)^2}{\sigma^2}\right\}$	σ^2	$\frac{3}{\pi} \approx 0.955$
Uniform	$\frac{1}{b-a}$ if $x \in [a, b]$	$\frac{(b-a)^2}{12}$	1
Gamma	$\frac{x^{k-1}e^{-x/\theta}}{\Gamma(k)\theta^k}$ if $x \geq 0$	$k\theta^2$	$\frac{12k\{(2k-2)!\}^2}{2^{4k-2}\{(k-1)!\}^4}$
Hodges–Lehmann	$-\frac{3}{20\sqrt{5}}\left(x^2 - 5\right)$ if $x^2 \leq 5$	1	$\frac{108}{125} = 0.864$

On the other hand, the asymptotic relative efficiency of the following c_1-test relative to the t-test has been considered in Chernoff and Savage (1958). Let $X_1, \ldots, X_m$ and $Y_1, \ldots, Y_n$ be ordered observations from the absolutely continuous cumulative distribution functions $F(x)$ and $G(x)$. And let $N = m + n$ and $\lambda_N = \frac{m}{N}$, and assume that for all N the inequalities $0 < \lambda_0 \leq \lambda_N \leq 1 - \lambda_0 < 1$ hold for some fixed $\lambda_0 \leq \frac{1}{2}$. The combined population cumulative distribution function is $H(x) = \lambda_N F(x) + (1 - \lambda_N)G(x)$.

Let $F_m(x)$ and $G_n(x)$ be the sample cumulative distribution functions of the X's and Y's, respectively, namely, $F_m(x) = (\text{the number of } X_i \leq x)/m$ and $G_n(x) = (\text{the number of } Y_i \leq x)/n$. Define the combined sample cumulative distribution function $H_N(x) = \lambda_N F_m(x) + (1 - \lambda_N)G_n(x)$. If the ith smallest in the combined sample is an X, let $z_{N_i} = 1$, and otherwise, let $z_{N_i} = 0$. Then, many nonparametric test statistics are of the form

$$mT_N = \sum_{i=1}^{N} a_{N_i} z_{N_i}, \tag{5.1.6}$$

where the a_{N_i} are given numbers. As a special case, we can consider $a_{N_i} = a\left(\frac{i}{N}\right)$ and the Wilcoxon test is given with $a_{N_i} = \frac{i}{N}$. However, the representation

$$T_N = \int_{\mathbb{R}} J_N \{H_N(x)\} dF_m(x) \tag{5.1.7}$$

is employed in Chernoff and Savage (1958). Equations (5.1.6) and (5.1.7) are equivalent when $a_{N_i} = J_N\left(\frac{i}{N}\right)$. While J_N need be defined only at $\frac{1}{N}, \frac{2}{N}, \ldots, \frac{N}{N}$, it is convenient to extend its domain of definition to (0, 1] by some convention, such as letting J_N be constant on $\left(\frac{i}{N}, \frac{i+1}{N}\right]$. And let I_N be the interval in which $0 < H_N(x) < 1$.

In the following, we will examine translation local alternatives only. That is, we assume that there is a distribution function $\Psi(x)$ which does not depend on N and has a density $\psi(x)$, and $F(x) = \Psi(x - \theta_N)$ and $G(x) = \Psi(x - \varphi_N)$. And we test the hypothesis that $\Delta_N = \theta_N - \varphi_N = 0$ versus the local alternative of the form $\Delta_N = cN^{-1/2}$. We also assume that $0 < \lim_{N\to\infty} \lambda_N = \lambda < 1$. Furthermore, we assume the following conditions.

Assumption 5.1.4. **(1)** $J(H) = \lim_{N\to\infty} J_N(H)$ *exists for* $0 < H < 1$ *and is not constant.*

(2) $\int_{I_N} \{J_N(H_N) - J(H_N)\} dF_m(x) = o_P\left(N^{-1/2}\right)$.

(3) $J_N(1) = o\left(\sqrt{N}\right)$.

(4) $\left|J^{(i)}(H)\right| = \left|\frac{d^iJ}{dH^i}\right| \leq K\{H(1-H)\}^{-i-1/2+\delta}$ *for* $i = 0, 1, 2$, *and for some* $\delta > 0$.

Now, we have the following asymptotic normality of the c_1-test which is due to Chernoff and Savage (1958).

Lemma 5.1.5. *If the conditions of Assumption 5.1.4 are satisfied, then the equation*

$$\lim_{N\to\infty} P\left(\frac{T_N - \mu_N(\Delta_N)}{\sigma_N(\Delta_N)} \leq x\right) = \Phi(x)$$

holds uniformly with respect to λ_N, θ_N, *and* φ_N *for* $\Delta_N = \theta_N - \varphi_N$ *in some neighbourhood of* 0, *where* $\Phi(x)$ *is the distribution function of the standard normal and*

$$\mu_N(\Delta_N) = \int_{\mathbb{R}} J\{H(x)\}dF(x), \tag{5.1.8}$$

$$\begin{aligned}\lim_{N\to\infty} \frac{\lambda_N N\sigma_N^2(\Delta_N)}{1-\lambda_N} &= 2\int\int_{0<x<y<1} x(1-x)J^{(1)}(x)J^{(1)}(y)dxdy \\ &= \int_0^1 J^2(x)dx - \left\{\int_0^1 J(x)dx\right\}^2. \end{aligned} \tag{5.1.9}$$

Note that if J is the inverse of some cumulative distribution function L, the right-hand side of (5.1.9) is

$$\int_0^1 \left\{L^{-1}(x)\right\}^2 dx - \left\{\int_0^1 L^{-1}(x)dx\right\}^2 = \int_{\mathbb{R}} x^2 dL(x) - \left\{\int_{\mathbb{R}} x dL(x)\right\}^2 = \sigma_L^2 \tag{5.1.10}$$

(Problem 5.2).

Although the definition (5.1.5) of asymptotic relative efficiency is easily interpreted, it is often useful to turn to equivalent expressions in order to evaluate Pitman efficiency for a given pair of test procedures in a particular setting. Assume that $T_N^{(i)}$, $i = 1, 2$ are test statistics for testing $\Delta_N = 0$ against a local alternative $\Delta_N = cN^{-1/2}$ which satisfy the following assumption.

Assumption 5.1.6. *There are functions* $\mu_N^{(i)}(\Delta_N)$ *and* $\sigma_N^{(i)}(\Delta_N)$ *such that for* (Δ_N *in some neighbourhood of* 0,

(1)

$$\lim_{N\to\infty} P_{\Delta_N}\left(\frac{T_N^{(i)} - \mu_N^{(i)}(\Delta_N)}{\sigma_N^{(i)}(\Delta_N)} \leq x\right) = \Phi(x),$$

(2) $\lim_{N\to\infty} \frac{\sigma_N^{(i)}(\Delta_N)}{\sigma_N^{(i)}(0)} = 1$,

(3) $E_{T^{(i)}} = \lim_{N\to\infty}\left(\frac{\mu_N^{(i)}(\Delta_N) - \mu_N^{(i)}(0)}{\Delta_N N^{1/2}\sigma_N^{(i)}(0)}\right)^2$ *exists and is independent of* c.

The quantity $E_{T^{(i)}}$ is called the efficacy of the procedure based on the sequence of statistics $T^{(i)}$. Let $0 < \alpha < 1$ be fixed and let $d_N^{(i)}$, $i = 1, 2$ be sequences of critical values for the respective test such that

$$\alpha = \lim_{N\to\infty} P_0\left(T_N^{(i)} \geq d_N^{(i)}\right) = P_0\left(\frac{T_N^{(i)} - \mu_N^{(i)}(0)}{\sigma_N^{(i)}(0)} \geq \frac{d_N^{(i)} - \mu_N^{(i)}(0)}{\sigma_N^{(i)}(0)}\right).$$

This requires, along with Assumption 5.1.6 (1), that

$$\lim_{N\to\infty} \frac{d_N^{(i)} - \mu_N^{(i)}(0)}{\sigma_N^{(i)}(0)} = z_\alpha,$$

where $\Phi(z_\alpha) = 1 - \alpha$. On the other hand, the respective power functions for $T_N^{(i)}$, $i = 1, 2$ are given by

$$\beta_{T_N^{(i)}}(\Delta_N) = \lim_{N\to\infty} P_{\Delta_N}\left(T_N^{(i)} \le d_N^{(i)}\right) = P_{\Delta_N}\left(\frac{T_N^{(i)} - \mu_N^{(i)}(\Delta_N)}{\sigma_N^{(i)}(\Delta_N)} \le \frac{d_N^{(i)} - \mu_N^{(i)}(\Delta_N)}{\sigma_N^{(i)}(\Delta_N)}\right).$$

In order for the tests to have the same limiting power function, i.e.,

$$\lim_{N\to\infty} \beta_{T_N^{(1)}}(\Delta_N) = \lim_{N\to\infty} \beta_{T_N^{(2)}}(\Delta_N),$$

we must have

$$\lim_{N\to\infty} \frac{d_N^{(1)} - \mu_N^{(1)}(\Delta_N)}{\sigma_N^{(1)}(\Delta_N)} = \lim_{N\to\infty} \frac{d_N^{(2)} - \mu_N^{(2)}(\Delta_N)}{\sigma_N^{(2)}(\Delta_N)}.$$

Now, note that

$$\begin{aligned}\lim_{N\to\infty} \frac{d_N^{(i)} - \mu_N^{(i)}(\Delta_N)}{\sigma_N^{(i)}(\Delta_N)} &= \lim_{N\to\infty} \frac{d_N^{(i)} - \mu_N^{(i)}(0) + \mu_N^{(i)}(0) - \mu_N^{(i)}(\Delta_N)}{\sigma_N^{(i)}(0)} \cdot \frac{\sigma_N^{(i)}(0)}{\sigma_N^{(i)}(\Delta_N)} \\ &= z_\alpha + \lim_{N\to\infty} \frac{\mu_N^{(i)}(0) - \mu_N^{(i)}(\Delta_N)}{\sigma_N^{(i)}(0)},\end{aligned}$$

hence

$$\begin{aligned}1 &= \lim_{N\to\infty} \frac{\mu_N^{(1)}(\Delta_N) - \mu_N^{(1)}(0)}{\sqrt{N^{(1)}}\Delta_N \sigma_N^{(1)}(0)} \cdot \frac{\sqrt{N^{(2)}}\Delta_N \sigma_N^{(2)}(0)}{\mu_N^{(2)}(\Delta_N) - \mu_N^{(2)}(0)} \cdot \sqrt{\frac{N^{(1)}}{N^{(2)}}} \\ &= \lim_{N\to\infty} \sqrt{\frac{E_{T^{(1)}}}{E_{T^{(2)}}}} \cdot \sqrt{\frac{N^{(1)}}{N^{(2)}}}.\end{aligned}$$

Now, we have the following lemma, which is due to Noether (1955).

Lemma 5.1.7. *The asymptotic relative efficiency (5.1.5) of $T^{(1)}$ with respect to $T^{(2)}$ satisfies the following relationship.*

$$\varepsilon_{T^{(1)},T^{(2)}}(\alpha, \beta, 0, \{\Delta_N\}) = \lim_{N\to\infty} \frac{N^{(2)}}{N^{(1)}} = \lim_{N\to\infty} \frac{E_{T^{(1)}}}{E_{T^{(2)}}}.$$

From this Lemma (5.1.8) and Equation (5.68) we see that, for c_1-test T_N,

$$\begin{aligned}\mu_{T_N}(\Delta_N) - \mu_{T_N}(0) &= \int_{\mathbb{R}} [J\{H(x - \theta_N)\} - J\{\Psi(x)\}]\, d\Psi(x) \\ &= (1 - \lambda_N)\Delta_N \int_{\mathbb{R}} J^{(1)}\{\Psi(x)\}\psi^2(x)dx + o(\Delta_N),\end{aligned}$$

since

$$H(x + \theta_N) - \Psi(x) = (1 - \lambda_N)\Psi'(x)\Delta_N + o(\Delta_N).$$

Therefore, if J is the inverse of some cumulative distribution function L which has a density $l(x)$, then the c_1-test satisfies the conditions of Assumption 5.1.6, with

$$E_{T_N} = \frac{\lambda(1-\lambda)}{\sigma_L^2}\left[\int_{\mathbb{R}} J^{(1)}\{\Psi(x)\}\psi^2(x)dx\right]^2$$

$$= \frac{\lambda(1-\lambda)}{\sigma_L^2}\left\{\int_{\mathbb{R}} \psi\left[\Psi^{-1}\{L(x)\}\right]dx\right\}^2.$$

On the other hand, if Ψ has finite second moment σ_Ψ^2, for the t-test

$$t_N = \frac{\overline{X}-\overline{Y}}{S\sqrt{\frac{m+n}{mn}}},$$

we define

$$\mu_{t_N}(\Delta_N) = \frac{\sqrt{N}\Delta_N\sqrt{\lambda(1-\lambda)}}{\sigma_\psi}, \quad \sigma_{t_N}(\Delta_N) \equiv 1$$

$$V_N = \frac{\overline{X}-\overline{Y}-\Delta_N}{S\sqrt{\frac{m+n}{mn}}},$$

then we have

$$\frac{t_N-\mu_{t_N}(\Delta_N)}{\sigma_{t_N}(\Delta_N)} - V_N = c\left(\frac{\sqrt{\lambda_N(1-\lambda_N)}}{S} - \frac{\sqrt{\lambda(1-\lambda)}}{\sigma_\Psi}\right),$$

Therefore, like V_N, $\frac{t_N-\mu_{t_N}(\Delta_N)}{\sigma_{t_N}(\Delta_N)}$ has a limiting standard normal distribution, hence Asssumption 5.1.6 is satisfied with

$$E_{t_N} = \frac{\lambda(1-\lambda)}{\sigma_\psi^2}.$$

Consequently, the asymptotic relative efficiency of the c_1-test relative to the t-test is given by

$$\varepsilon_{c_1,t} = \frac{\sigma_\Psi^2}{\sigma_L^2}\left\{\int_{\mathbb{R}} \psi\left[\Psi^{-1}\{L(x)\}\right]dx\right\}^2.$$

Suppose that both $\widetilde{\Psi}$ and $\widetilde{L}$ have mean 0 and variance 1 and let $\psi(x) = \sigma_\Psi^{-1}\widetilde{\psi}\left(\frac{x-\mu_\Psi}{\sigma_\Psi}\right)$ and $l(x) = \sigma_L^{-1}\widetilde{l}\left(\frac{x-\mu_L}{\sigma_L}\right)$. Then we see that

$$\left\{\int_{\mathbb{R}} \widetilde{\psi}\left[\widetilde{\Psi}^{-1}\{\widetilde{L}(x)\}\right]dx\right\}^2 = \frac{\sigma_\Psi^2}{\sigma_L^2}\left\{\int_{\mathbb{R}} \psi\left[\Psi^{-1}\{L(x)\}\right]dx\right\}^2,$$

so the value of $\varepsilon_{c_1,t}$ is invariant under a change of location or scale with respect to both Ψ and L.

If we take $L(x) = \Phi(x)$, the distribution function of the standard normal, the problem becomes the minimization of

$$\int_{\mathbb{R}} J^{(1)}\{\Psi(x)\}\psi^2(x)dx = \int_{\mathbb{R}} \psi\left[\Psi^{-1}\{\Phi(x)\}\right]dx$$

subject to the restrictions

$$0 = \int_{\mathbb{R}} x\psi(x)\,dx, \tag{5.1.11}$$

$$1 = \int_{\mathbb{R}} x^2\psi(x)\,dx, \tag{5.1.12}$$

where $\varphi(x)$ is the density of the standard normal distribution. Chernoff and Savage (1958) proved the following lemma using variational methods.

Lemma 5.1.8. *If $\Psi(x)$ is a cumulative distribution function which has a density and finite second-order moment, then $\varepsilon_{c_1,t} \geq 1$, and $\varepsilon_{c_1,t} = 1$ only if $\Psi(x)$ is normal.*

A simpler proof of this result has been provided by Gastwirth and Wolff (1968). Since $J^{(1)}\{\Psi(x)\} = \left\{\varphi\left(\Phi^{-1}(\Psi(x))\right)\right\}^{-1}$, if we put $Y = \frac{\varphi(\Phi^{-1}(\Psi(X)))}{\Psi(X)}$, and applying the Jensen's inequality yields

$$\int_{\mathbb{R}} J^{(1)}\{\Psi(x)\}\psi^2(x)dx = E_\psi\left(Y^{-1}\right) \geq \left\{E_\psi(Y)\right\}^{-1}$$

with equality iff $Y \equiv c_1$, a.e. Furthermore, after integrating by parts, applying the Cauchy–Schwarz inequality leads to

$$\begin{aligned} E_\psi(Y) &= \int_{\mathbb{R}} \varphi\left(\Phi^{-1}(\Psi(x))\right)dx = \int_{\mathbb{R}} x\Phi^{-1}(\Psi(x))\psi(x)dx \\ &\leq \left[\int_{\mathbb{R}} x^2\psi(x)dx \int_{\mathbb{R}} \left\{\Phi^{-1}(\Psi(x))\right\}^2 \psi(x)dx\right]^{1/2} = 1 \end{aligned} \tag{5.1.13}$$

with equality iff $\Phi^{-1}(\Psi(x)) = c_2 x$, a.e. Taking account of the restrictions (5.1.11) and (5.1.12) completes the proof.

5.1.2 *Maximal Invariants*

5.1.2.1 *Invariance of sample space, parameter space and tests*

It is natural that we impose symmetric restrictions on statistical procedures, since many statistical problems exhibit symmetries. Mathematically, symmetry is described by invariance under a suitable group of transformations (see, e.g., Lehmann and Romano (2005).) Let X be a random variable which is distributed according to a probability distribution P_θ, $\theta \in \Omega$, and let g be a transformation of the sample space $\mathcal{X}$. We assume that all transformations which are considered in connection with invariance are one-to-one transformations of $\mathcal{X}$ onto itself. The random variable denoted by gX takes on the value gx when $X = x$. Suppose that when the distribution of X is P_θ, $\theta \in \Omega$, the distribution of $Y = gX$ is $P_{\theta'}$, where θ' is also in Ω. Then, the element θ' of Ω will be denoted by $\bar{g}\theta$, so that

$$P_\theta\{gX \in A\} = P_\theta^X\{gX \in A\} = P_{\bar{g}\theta}^Y\{Y \in A\} = P_{\bar{g}\theta}\{X \in A\}.$$

This relation can be written as $P_\theta\left(g^{-1}A\right) = P_{\bar{g}\theta}(A)$ and hence $P_\theta(B) = P_{\bar{g}\theta}(gB)$.

We say that the parameter set Ω remains invariant under g (or is preserved by g) if

(i) $\bar{g}\theta \in \Omega$ for all $\theta \in \Omega$,

(ii) for any $\theta' \in \Omega$, there exists $\theta \in \Omega$ such that $\bar{g}\theta = \theta'$,

and these two conditions can be expressed by the equation

$$\bar{g}\Omega = \Omega.$$

Note that if the distributions P_θ corresponding to different values of θ are distinct, then the transformation $\bar{g}$ of Ω onto itself defined in this way is a one-to-one transformation.

We say that the problem of testing $H : \theta \in \Omega_H$ against $K : \theta \in \Omega_K$ remains invariant under a transformation g if $\bar{g}$ preserves both Ω_H and Ω_K, so that the equations

$$\bar{g}\Omega_H = \Omega_H \quad \text{and} \quad \bar{g}\Omega_K = \Omega_K$$

hold.

Any class of transformations leaving the problem invariant can be naturally extended to a group G. Furthermore, the class of induced transformations $\bar{g}$ also form a group $\bar{G}$ (see, e.g., Lemma

6.1.1 of Lehmann and Romano (2005).) In the presence of symmetries in both sample space and parameter space, which are represented by the groups G and $\overline{G}$, it is natural to restrict attention to tests ϕ, which are also symmetric, that is, which satisfy

$$\phi(gx) = \phi(x), \quad \text{for all} \quad x \in X \quad \text{and} \quad g \in G,$$

and we say that such tests are invariant under G. In such a case, the principle of invariance restricts our attentions to invariance tests and to obtain the best of these tests. For this purpose, it is convenience to characterize the totality of invariant tests.

Two points x_1, x_2 are considered equivalent under G, and denoted by

$$x_1 \sim x_2 \quad (\text{mod } G),$$

if there exists a transformation $g \in G$ for which $x_2 = gx_1$. A function M is said to be maximal invariant if it is invariant and if

$$M(x_1) = M(x_2) \quad \text{implies} \quad x_2 = gx_1 \quad \text{for some} \quad g \in G.$$

That is, the maximal invariant function M is constant on the orbits except that each orbit takes on a different value.

Example 5.1.9 (Example 6.2.2 of Lehmann and Romano (2005)). **(i)** *Let* $\boldsymbol{x} = (x_1, \ldots, x_n)'$. *And let G be the set of n! permutations of the elements of* x_i. *Since a permutation of the* x_i *only changes the order of* x_i, *the order statistics* $x_{(1)} \leq \cdots \leq x_{(n)}$ *is invariant. Furthermore, since two points which have the same order statistics can be obtained from each other by a permutation of the elements, the order statistics are maximal invariant.*

(ii) *Suppose that we restrict our attention to the points which have n different elements. Let G be the totality of transformations* $\overline{x}_i = f(x_i)$, *where f is continuous and strictly increasing. If we consider* $x_i \ldots, x_n$ *is n values on the real line, since such transformations preserve their order, the ranks* $(r_1, \ldots, r_n)'$ *of* $(x_1, \ldots, x_n)'$ *are invariant. On the other hand, we consider the case that* $x_1, \ldots, x_n$ *and* $\overline{x}_1, \ldots, \overline{x}_n$ *have the same order* $x_{i_1} < \cdots, < x_{i_n}$ *and* $\overline{x}_{i_1} < \cdots, < \overline{x}_{i_n}$. *Let* $\overline{x}_i = f(x_i)$, *for all* $i = 1, \ldots, n$, *and let* $f(x)$ *be the linear interpolation between* $f(x_{i_k})$ *and* $f(x_{i_{k+1}})$ *for* $x_{i_k} \leq x \leq x_{i_{k+1}}$. *Furthermore, let*

$$f(x) = \begin{cases} x + (\overline{x}_{i_1} - x_{i_1}) & x \leq x_{i_1}, \\ x + (\overline{x}_{i_n} - x_{i_n}) & x \geq x_{i_n}. \end{cases}$$

Then, since such f is continuous and strictly increasing, the ranks are maximal invariant.

5.1.2.2 *Most powerful invariant test*

If there exist symmetries, we may restrict our attention to invariance tests. Then, we are interested in the most powerful invariance test. We have the following result.

Theorem 5.1.10 (Theorem 6.3.1 of Lehmann and Romano (2005)). *We consider the problem of testing* Ω_0 *against* Ω_1 *remains invariant under finite group* $G = \{g_1, \ldots, g_N\}$. *Assume that* $\overline{G}$ *is transitive over* Ω_0 *and* Ω_1. *Then, there exists a uniformly most powerful (UMP) invariance test of* Ω_0 *against* Ω_1, *which rejects* Ω_0, *when*

$$\frac{\frac{1}{N}\sum_{i=1}^{N} p_{\overline{g}_i \theta_1}(x)}{\frac{1}{N}\sum_{i=1}^{N} p_{\overline{g}_i \theta_0}(x)}$$

is sufficiently large, where θ_0 *and* θ_1 *are any elements of* Ω_0 *and* Ω_1, *respectively.*

Sufficient statistics make a problem simple by reducing the sample space. On the other hand, invariance makes a problem simple by reducing the data to a maximal invariant statistic M, whose distribution may depend only on a function of the parameter. Therefore, invariance may also shrink the parameter space. This fact is described in the following.

Theorem 5.1.11 (Theorem 6.3.2 of Lehmann and Romano (2005)). *Let $M(x)$ be invariant under G, and let $v(\theta)$ be maximal invariant under the induced group $\overline{G}$. Then, the distribution of $M(X)$ depends only on $v(\theta)$.*

The two-sample problem of testing the equality of two distributions appears in the comparison of a treatment with a control. The null hypothesis of no treatment effect is tested against the alternative hypothesis, which is a beneficial effect. The observations consist of two samples $X_1, \ldots, X_m$ and $Y_1, \ldots, Y_n$ from two distributions with continuous cumulative distribution functions F and J. Then, we test the null hypothesis

$$H_1 \; : \; J = F.$$

Here, we consider testing H_1 against one-sided alternatives where the Y's are stochastically larger than the X's, i.e.,

$$K_1 \; : \; J(z) \leq F(z) \quad \text{for all } z, \quad \text{and} \quad J \neq F.$$

The two-sample problem of testing H_1 against K_1 remains invariant under the group G:

$$\widetilde{x_i} = \rho(x_i), \quad \widetilde{y_j} = \rho\left(y_j\right), \quad i = 1, \ldots, m, \; j = 1, \ldots, n,$$

where ρ is a continuous and strictly increasing function on the real line. These transformations preserve the continuity of a distribution. Furthermore, they preserve either the property of two variables being identically distributed or the property of one being stochastically larger than the other. Therefore, this test is invariant under the group G. As we saw in Example 5.1.9, a maximal invariant under G is the set of ranks

$$(\boldsymbol{R}' \; ; \; \boldsymbol{S}') = (R'_1, \ldots, R'_m \; ; \; S'_1, \ldots, S'_n)$$

of $X_1, \ldots, X_m \; ; \; Y_1, \ldots, Y_n$ in the combined sample. The distribution of $\left(R'_1, \ldots, R'_m \; ; \; S'_1, \ldots, S'_n\right)$ is symmetric in the first m variables for X's, and in the last n variables for Y's, for all distributions F and J. Therefore, sufficient statistics for $(\boldsymbol{R}' \; ; \; \boldsymbol{S}')$ are the ordered X-ranks and the ordered Y-ranks, i.e.,

$$R_1 < \cdots < R_m \quad \text{and} \quad S_1 < \cdots < S_n.$$

Since each of them determines the other, one of these sets alone is enough. Thus, any invariance test is a rank test, namely, depends only on the ranks of the observations, e.g., on $(S_1, \ldots, S_n)$.

5.1.3 *Efficiency of Rank-Based Statistics*

5.1.3.1 *Least favourable density and most powerful test*

Along with Sidak et al. (1999), we introduce the concept of the least favourable density and the most powerful test. First, we consider a simple null hypothesis p and simple alternative q. We denote the set of critical functions whose sizes are not exceeding α by $M(\alpha, p)$, namely,

$$\int \Psi(x) dP(x) \leq \alpha \quad \text{if and only if} \quad \Psi \in M(\alpha, p),$$

where $dP(x) = p(x)dx$. Furthermore, we call

$$\beta(\alpha, p, q) = \sup_{\Psi \in M(\alpha, p)} \int \Psi(x) dQ(x)$$

the envelope power function for p against q at level α, where, again, $dQ(x) = q(x)dx$.

In the case that the hypothesis is composite, that is, $H = \{p\}$, we denote by $M(\alpha, H)$ the set of critical functions such that

$$\sup_{p \in H} \int \Psi(x) dP(x) \leq \alpha,$$

and call

$$\beta(\alpha, H, q) = \sup_{\Psi \in M(\alpha, H)} \int \Psi(x) dQ(x)$$

the envelope power function for H against q at level α.

Finally, in the case that the alternative is also composite, that is, $K = \{q\}$, we call

$$\beta(\alpha, H, K) = \sup_{\Psi \in M(\alpha, H)} \inf_{q \in K} \int \Psi(x) dQ(x)$$

the envelope power function for H against K at level α.

If we consider testing p against q, a test Ψ^0 is said to be most powerful at level α if $\Psi^0 \in M(\alpha, p)$ and if

$$\beta(\alpha, p, q) = \int \Psi^0(x) dQ(x). \tag{5.1.14}$$

Furthermore, if we consider testing H against q, a test Ψ^0 is said to be most powerful at level α if $\Psi^0 \in M(\alpha, H)$ and if

$$\beta(\alpha, H, q) = \int \Psi^0(x) dQ(x). \tag{5.1.15}$$

Finally, if we consider testing H against K, a test $\Psi^0 \in M(\alpha, H)$ is called uniformly most powerful (UMP) at level α if (5.1.15) holds for all $q \in K$, while a test $\Psi^0 \in M(\alpha, H)$ is called maxmin most powerful at level α if

$$\beta(\alpha, H, K) = \inf_{q \in K} \int \Psi^0(x) dQ(x).$$

If there is a $p^0 \in H$ such that the most powerful test Ψ^0 for p^0 against q belongs to $M(\alpha, H)$, then

$$\beta\left(\alpha, p^0, q\right) = \int \Psi^0(x) dQ(x) = \sup_{\Psi \in M(\alpha, H)} \int \Psi(x) dQ(x) \leq \sup_{\Psi \in M(\alpha, p)} \int \Psi(x) dQ(x) = \beta(\alpha, p, q)$$

for every $p \in H$. Therefore, such a p^0 is called a least favourable density in H for testing H against q at level α. Similarly, if there are $p^0 \in H$ and $q^0 \in K$ such that the most powerful test Ψ^0 for p^0 against q^0 belongs to $M(\alpha, H)$ and

$$\int \Psi^0(x) dQ(x) \geq \int \Psi^0(x) dQ^0(x), \quad \text{for all } q \in K,$$

then

$$\beta\left(\alpha, p^0, q^0\right) = \int \Psi^0(x) dQ^0(x) \leq \int \Psi^0(x) dQ(x) = \sup_{\Psi \in M(\alpha, H)} \int \Psi(x) dQ(x)$$

$$\leq \sup_{\Psi \in M(\alpha, p)} \int \Psi(x) dQ(x) = \beta(\alpha, p, q)$$

for every $p \in H$ and $q \in K$. Hence, we call $\left(p^0, q^0\right)$ a least favourable pair of densities for H against K at level α.

The following basic lemma shows that the most powerful test for testing a simple hypothesis against a simple alternative may be found quite easily (see, e.g, Lemma (2.3.4.1) of Sidak et al. (1999)).

Lemma 5.1.12 (Neyman–Pearson lemma). *In testing p against q at level α,*

$$\Psi^0(x) = \begin{cases} 1, & \text{if } q(x) > kp(x), \\ 0, & \text{if } q(x) < kp(x), \end{cases}$$

is the most powerful test, where k, and the value of $\Psi^0(x)$ for x such that $q(x) = kp(x)$ is defined so that

$$\int \Psi^0(x)\, dP(x) = \alpha.$$

Example 5.1.13 (Example 3.8.1 of Lehmann and Romano (2005)). *In the case that an item should be satisfactory, some characteristic X of the item may be expected to exceed a given constant u. Let*

$$p = P(X \le u)$$

be the probability of an item being defective. We would like to test the hypothesis $H : p \ge p_0$. Furthermore, let $X_1, \ldots, X_n$ be characteristics of n sample items, that is, i.i.d. copies of X. The distributions of $X_1, \ldots, X_n$ can be determined by p and the conditional distributions

$$P_-(x) = P(X \le x \mid X \le u) \quad \text{and} \quad P_+(x) = P(X \le x \mid X > u).$$

We assume that P_- and P_+ have densities p_- and p_+. Then, the joint density of $X_1, \ldots, X_n$ is given by

$$p^m(1-p)^{n-m} p_-(x_{i_1}) \cdots p_-(x_{i_m})\, p_+\left(x_{j_1}\right) \cdots p_+\left(x_{j_{n-m}}\right),$$

where the sample points $x_1, \ldots, x_n$ satisfy

$$x_{i_1}, \ldots, x_{i_m} \le u \le x_{j_1}, \ldots, x_{j_{n-m}}.$$

Now, we consider a fixed alternative to H, say $\{p_1, P_-, P_+\}$ with $p_1 < p_0$. On the other hand, we consider $\{p_0, P_-, P_+\}$ as a candidate of the least favourable distribution. Then, the test Ψ^0 in Theorem 5.1.12 becomes

$$\Psi^0(x) = \begin{cases} 1 & \left(\frac{p_1}{p_0}\right)^m \left(\frac{1-p_1}{1-p_0}\right)^{n-m} > C, \\ 0 & \left(\frac{p_1}{p_0}\right)^m \left(\frac{1-p_1}{1-p_0}\right)^{n-m} < C, \end{cases}$$

which is equivalent to

$$\Psi^0(x) = \begin{cases} 1 & m > C', \\ 0 & m < C'. \end{cases}$$

Therefore, the test rejects when $M < C$, and when $M = C$ with probability γ, where

$$P\{M < C\} + \gamma P\{M = C\} = \alpha \quad \text{for} \quad p = p_0.$$

The distribution of M is binomial distribution $b(n, p)$, so that it does not depend on P_+ and P_-. Then, the power function of the test depends only p. Furthermore, the power function is a decreasing function of p, which takes on its maximum under H for $p = p_0$. As a consequence, $p = p_0$ is the least favourable and Ψ^0 is the most powerful. Furthermore, Ψ^0 is UMP since the test is independent of the particular choice of the alternative.

If we express the test statistic M in terms of the variables $Z_i = X_i - u$, then the test statistic M is the number of variables Z_i which are less than or equal to 0. This test is the so-called sign test.

Let H_0 be the system of densities of the form

$$p(x) = \prod_{i=1}^{N} f(x_i)$$

where $f(x)$ is an arbitrary one-dimensional density. Now, we consider testing H_0 against some simple alternative q. Assume that for some $p^0 \in H_0$, the likelihood ratio $\frac{q(\boldsymbol{x})}{p^0(\boldsymbol{x})}$ is a function of the ranks $\boldsymbol{r} = (r_1, \ldots, r_N)'$ only, that is,

$$\frac{q(\boldsymbol{x})}{p^0(\boldsymbol{x})} = h(\boldsymbol{r}).$$

Then, the test

$$\Psi^0(\boldsymbol{x}) = \begin{cases} 1, & \text{if } h(\boldsymbol{r}) > k, \\ 0, & \text{if } h(\boldsymbol{r}) < k, \end{cases}$$

is most powerful at level $\alpha = \int \Psi^0(\boldsymbol{x})\, dP^0(\boldsymbol{x})$ for p_0 against q. Moreover, the distribution of $\Psi^0(\boldsymbol{x})$ is invariant over H_0, since $\Psi^0(\boldsymbol{X}) = \Psi^0(\boldsymbol{R}_N)$ and the distribution of $\boldsymbol{R}_N = (R_1, \ldots, R_N)'$ does not depend on the choice of $p \in H_0$. Therefore, we see that

$$\int \Psi^0(\boldsymbol{x})\, dP(\boldsymbol{x}) = \alpha \quad \text{for all} \quad p \in H_0,$$

and consequently p^0 is least favourable for H_0 against q and Ψ^0 is most powerful for H_0 against q. If we put $K = \{q\}$, where q are all the densities such that

$$q(\boldsymbol{x}) = p(\boldsymbol{x})\, h(\boldsymbol{r}), \quad p \in H_0,$$

then $\int \Psi^0(\boldsymbol{x})\, dQ(\boldsymbol{x})$ does not depend on the choice of $q \in K$, hence Ψ^0 is most powerful for H_0 against K, too.

More generally, if $K = \{q_d,\ d \in W\}$ and

$$\frac{q_d(\boldsymbol{x})}{p^0(\boldsymbol{x})} = h_d(\boldsymbol{r}), \quad d \in W, \tag{5.1.16}$$

for some density p^0, then the maximin most powerful test for H_0 against K may be found within the family of rank tests. However, the relation (5.1.16) does not hold for any important practical problem.

5.1.3.2 *Asymptotically most powerful rank test*

Observing $X_i = \alpha + \beta c_i + \sigma Y_i$, Hájek (1962) considered testing the hypothesis $\beta = 0$ against the alternative $\beta > 0$. Let $\left\{(X_{m,1}, \ldots, X_{m,N_m})',\ m \geq 1\right\}$ be a sequence of random vectors, where X's are independent and

$$P(X_{m,i} \leq x \mid \alpha, \beta, \sigma) = F\left(\frac{x - \alpha - \beta c_{m,i}}{\sigma}\right), \quad 1 \leq i \leq N_m,\ m \geq 1,$$

where F is a known distribution function of the residuals $Y_{m,i}$, which has the density $f(x) = F'(x)$, $c_{m,i}$ are known constants, $\alpha \in \mathbb{R}$ and $\sigma > 0$ are nuisance parameters, and β is the parameter of interest. Suppose that the square root of the probability density $(f(x))^{1/2}$ possesses a quadratically integrable derivative, which means that

$$0 < I(f) = \int_{\mathbb{R}} \left\{\frac{f'(x)}{f(x)}\right\}^2 f(x)\, dx < \infty \tag{5.1.17}$$

(as usual, $I(f)$ is the Fisher information related to the location parameter family $\{f_\theta(x) = f(x-\theta) \mid \theta \in \mathbb{R}\}$), since

$$\left(\{f(x)\}^{1/2}\right)' = \frac{1}{2}\left\{\frac{f'(x)}{\{f(x)\}^{1/2}}\right\}.$$

Hájek (1962) defines a class of rank tests which are asymptotically most powerful for given f. We assume the following condition for the constants $c_{m,i}$.

Assumption 5.1.14. (i) Noether condition

$$\lim_{m\to\infty}\left\{\frac{\max_{1\le i\le N_m}(c_{m,i}-\bar{c}_m)^2}{\sum_{i=1}^{N_m}(c_{m,i}-\bar{c}_m)^2}\right\} = 0, \tag{5.1.18}$$

where $\bar{c}_m = N_m^{-1}\sum_{i=1}^{N_m} c_{m,i}$,

(ii) Boundedness condition

$$\sup_m \sum_{i=1}^{N_m}(c_{m,i}-\bar{c}_m)^2 < \infty. \tag{5.1.19}$$

Define

$$d_m^2 = \sum_{i=1}^{N_m}(c_{m,i}-\bar{c}_m)^2 I(f) \quad \text{and} \quad d^2 = \lim_{m\to\infty} d_m^2. \tag{5.1.20}$$

And let $F^{-1}(u) = \inf\{x \mid F(x) \ge u\}$, $0 < u < 1$, and put

$$\psi(u) = -\frac{f'\left(F^{-1}(u)\right)}{f\left(F^{-1}(u)\right)}. \tag{5.1.21}$$

From (5.1.17), it follows that

$$\int_0^1 \psi(u)\,du = \int_{\mathbb{R}} f'(x)\,dx = 0 \tag{5.1.22}$$

and

$$\int_0^1 \psi^2(u)\,du = I(f) < \infty. \tag{5.1.23}$$

If an f corresponds to ψ, $f \to \psi$, then $f_1(x) = \frac{1}{\sigma} f\left(\frac{x-\alpha}{\sigma}\right)$ corresponds to $\psi_1(u) = \frac{1}{\sigma}\psi(u)$,

$$\frac{1}{\sigma} f\left(\frac{x-\alpha}{\sigma}\right) \to \frac{1}{\sigma}\psi(u),$$

i.e., the map is invariant under translation and a change of scale only introduces a multiplicative constant, which satisfies $\sigma^2 I(f_1) = I(f)$.

Now, let $R_{m,i}^{N_m}$ be the rank of $X_{m,i}$, that is, $X_{m,i} = X_{\left(R_{m,i}^{N_m}\right)}$, and let $\psi_m(u)$ be a function defined on $(0,1)$ which is constant over intervals $\left(\frac{i}{N_m}, \frac{i+1}{N_m}\right)$, $0 \le i < N_m$, namely,

$$\psi_m(u) = \psi_m\left(\frac{i}{N_m+1}\right) := \psi_{m,i}, \quad \frac{i-1}{N_m} < u < \frac{i}{N_m}. \tag{5.1.24}$$

Consider the rank statistics S_m of the form

$$S_m = \sum_{i=1}^{N_m} (c_{m,i} - \bar{c}_m) \psi_m \left(\frac{R_{m,i}^{N_m}}{N_m + 1} \right), \quad m \geq 1. \tag{5.1.25}$$

Since the distribution of S_m does not depend on the ordering of $\{\psi_{m,i}\}$, without loss of generality, we may suppose that $\psi_m(u)$ is a nondecreasing function in u. Furthermore, assume that

$$\lim_{m \to \infty} \int_0^1 \{\psi_m(u) - \psi(u)\}^2 \, du = 0 \tag{5.1.26}$$

and $\psi_m(u)$ is nonconstant and quadratically integrable.

Here, we consider a particular alternative $\{\alpha = \alpha_0, \ \beta = \beta_0, \ \sigma = \sigma_0\}$ and associate with it the particular null hypothesis $\{\alpha = \alpha_0 + \beta_0 \bar{c}_m, \ \beta = 0, \ \sigma = \sigma_0\}$, where the distributions under the null and alternative hypotheses are denoted by P_m and Q_m, respectively, assume that

$$P_m (X_{m,i} \leq x_i, \ 1 \leq i \leq N_m) = \prod_{i=1}^{N_m} P_{m,i} (X_{m,i} \leq x_i),$$

$$Q_m (X_{m,i} \leq x_i, \ 1 \leq i \leq N_m) = \prod_{i=1}^{N_m} Q_{m,i} (X_{m,i} \leq x_i)$$

and put

$$r_{m,i} = \begin{cases} \frac{dQ_{m,i}}{dP_{m,i}} & \text{on sets } B_{m,i}, \\ 0 & \text{otherwise,} \end{cases}$$

where $B_{m,i}$ satisfies $P_{m,i}(B_{m,i}) = 1$ and $Q_{m,i}$ is absolutely continuous with respect to $P_{m,i}$ on $B_{m,i}$. If $p_{m,i}$ and $q_{m,i}$ are densities corresponding to distributions $P_{m,i}$ and $Q_{m,i}$, respectively, then

$$r_{m,i}(x) = \begin{cases} \frac{q_{m,i}(x)}{p_{m,i}(x)} & \text{if } p_{m,i}(x) > 0, \\ 0 & \text{otherwise.} \end{cases}$$

Put

$$W_m = 2 \sum_{i=1}^{N_m} \left\{ (r_{m,i})^{\frac{1}{2}} - 1 \right\}, \tag{5.1.27}$$

$$L_m = \sum_{i=1}^{N_m} \log r_{m,i}$$

and suppose that for every $\varepsilon > 0$

$$\lim_{m \to \infty} \max_{1 \leq i \leq N_m} P_{m,i} \left(\left| r_{m,i} - 1 \right| > \varepsilon \right) = 0. \tag{5.1.28}$$

We say that $\mathcal{L}(Y_m \mid P_m)$ is $AN\left(a_m, b_m^2\right)$ if the distribution of $b_m^{-1}(Y_m - a_m)$ under P_m tends to the standard normal distribution and similar notation will be used for $\mathcal{L}(Y_m \mid Q_m)$. The following lemma is due to LeCam.

Lemma 5.1.15 (LeCam is third lemma). *Assume that (5.1.28) holds true, and that $\mathcal{L}(W_m \mid P_m)$ is $AN\left(-\frac{1}{4}\sigma^2, \sigma^2\right)$. Then*

(i) *the probability distributions Q_m are contiguous to the probability measures P_m,*

(ii) $W_m - L_m \overset{P_m}{\to} \frac{1}{4}\sigma^2$ *and* $\mathcal{L}(L_m \mid P_m)$ *is* $AN\left(-\frac{1}{2}\sigma^2, \sigma^2\right)$,

(iii) *if* $\mathcal{L}(Y_m \mid P_m)$ *is* $AN\left(a, b^2\right)$, *and* $\mathcal{L}(Y_m, L_m \mid P_m)$ *tends to a bivariate normal distribution with correlation coefficient* ρ, *then*

$$\mathcal{L}(Y_m \mid Q_m) \text{ is } AN\left(a + \rho b\sigma, b^2\right).$$

Setting

$$s(x) = \{f(x)\}^{\frac{1}{2}},$$

we have

$$s'(x) = \frac{1}{2}\left[\frac{f'(x)}{\{f(x)\}^{\frac{1}{2}}}\right]$$

so that if (5.1.17) is satisfied, $s(x)$ has a quadratically integrable derivative. Concerning functions possessing a quadratically integrable derivative, Hájek (1962) formulated the following lemma.

Lemma 5.1.16. (Hájek (1962), Lemma 4.3) *Let* $s(x)$, $x \in \mathbb{R}$ *be an absolutely continuous function possessing a quadratically integrable derivative* $s'(x)$. *Then*

$$\lim_{h\to\infty} \int \left\{\frac{s(x-h) - s(x)}{h} + s'(x)\right\}^2 dx = 0.$$

We can express (5.1.27) as

$$W_m = 2\sum_{i=1}^{N_m} \left\{\frac{s(Y_{m,i} - \gamma c_{m,i} + \gamma \bar{c}_m)}{s(Y_{m,i})} - 1\right\}, \tag{5.1.29}$$

where

$$Y_{m,i} = \frac{X_{m,i} - \alpha_0 - \beta_0 \bar{c}_m}{\sigma_0}$$

and

$$\gamma = \frac{\beta_0}{\sigma_0}. \tag{5.1.30}$$

Because of Lemma 5.1.16, we can see that

$$\begin{aligned} 2E_{P_m}\left\{\frac{s(Y_{m,i} - h)}{s(Y_{m,i})} - 1\right\} &= -h^2 \int_{\mathbb{R}} \left\{\frac{s(x-h) - s(x)}{h}\right\}^2 dx \\ &\sim -h^2 \int_{\mathbb{R}} \{s'(x)\}^2 dx = -h^2 \int_{\mathbb{R}} \left\{\frac{f'(x)}{f(x)}\right\}^2 f(x)\, dx = -\frac{h^2}{4} I(f) \end{aligned}$$

where the sign $a \sim b$ means that $\lim_{h\to 0} \frac{b}{a} = 1$. Letting $h = \gamma(c_{m,i} - \bar{c}_m)$, from (5.1.29), we have

$$E_{P_m}(W_m) \sim -\frac{\gamma^2 I(f)}{4} \sum_{i=1}^{N_m} (c_{m,i} - \bar{c}_m)^2 = -\frac{\gamma^2 d_m^2}{4}, \tag{5.1.31}$$

where d_m^2 and γ are given by (5.1.20) and (5.1.30), respectively.

We now approximate W_m by

$$T_m = -\sum_{i=1}^{N_m} (c_{m,i} - \bar{c}_m) \frac{f'(Y_{m,i})}{f(Y_{m,i})}.$$

Since $\{Y_{m,i}\}$'s are independent, we have

$$\begin{aligned}\mathrm{Var}_{P_m}(T_m) &= \sum_{i=1}^{N_m}(c_{m,i}-\bar{c}_m)^2\,\mathrm{Var}_{P_m}\left\{\frac{f'(Y_{m,i})}{f(Y_{m,i})}\right\}\\ &= \sum_{i=1}^{N_m}(c_{m,i}-\bar{c}_m)^2\,E_{P_m}\left\{\frac{f'(Y_{m,i})}{f(Y_{m,i})}\right\} = d_m^2.\end{aligned}$$

Setting $h_{m,i} = \gamma(c_{m,i}-\bar{c}_m)$, we see that

$$E_{P_m}\{W_m - E_{P_m}(W_m) - \gamma T_m\}^2 = 4\sum_{i=1}^{N_m}\mathrm{Var}_{P_m}\left\{\frac{s(Y_{m,i}-h_{m,i})}{s(Y_{m,i})} - 1 + \frac{h_{m,i}}{2}\frac{f'(Y_{m,i})}{f(Y_{m,i})}\right\}$$

$$\le 4\sum_{i=1}^{N_m}E_{P_m}\left\{\frac{s(Y_{m,i}-h_{m,i})}{s(Y_{m,i})} - 1 + \frac{h_{m,i}}{2}\frac{f'(Y_{m,i})}{f(Y_{m,i})}\right\}^2$$

$$\le 4\gamma^2\sum_{i=1}^{N_m}(c_{m,i}-\bar{c}_m)^2\int\left\{\frac{s(x-h_{m,i})-s(x)}{h_{m,i}} + s'(x)\right\}^2 dx.$$

From (5.1.18), (5.1.19) and Lemma 5.1.16, it follows that

$$\lim_{m\to\infty}E_{P_m}\{W_m - E_{P_m}(W_m) - \gamma T_m\}^2 = 0. \tag{5.1.32}$$

Introduce the statistic

$$T_m^* = \sum_{i=1}^{N_m}(c_{m,i}-\bar{c}_m)\,\psi_m\{F(Y_{m,i})\},$$

where ψ_m satisfies the condition (5.1.26). Then, from Theorem 3.2 of Hájek (1961), it can be seen that

$$\lim_{m\to\infty}E_{P_m}\{S_m - T_m^*\}^2 = 0,$$

where S_m is given by (5.1.25). Therefore, it follow that

$$S_m - T_m^* \xrightarrow{P_m} 0. \tag{5.1.33}$$

Furthermore, we see that

$$\begin{aligned}\lim_{m\to\infty}E_{P_m}\{T_m - T_m^*\}^2 &= \lim_{m\to\infty}\mathrm{Var}_{P_m}\{T_m - T_m^*\}^2\\ &= \lim_{m\to\infty}\sum_{i=1}^{N_m}(c_{m,i}-\bar{c}_m)^2\,\mathrm{Var}_{P_m}\left[\frac{f'(Y_{m,i})}{f(Y_{m,i})} + \psi_m\{F(Y_{m,i})\}\right]\\ &\le \lim_{m\to\infty}\sum_{i=1}^{N_m}(c_{m,i}-\bar{c}_m)^2\,E_{P_m}\left[\frac{f'(Y_{m,i})}{f(Y_{m,i})} + \psi_m\{F(Y_{m,i})\}\right]^2\\ &= \lim_{m\to\infty}\sum_{i=1}^{N_m}(c_{m,i}-\bar{c}_m)^2\int_0^1\{-\psi(u)+\psi_m(u)\}^2\,du = 0.\end{aligned} \tag{5.1.34}$$

Now, (5.1.31), (5.1.32), (5.1.33) and (5.1.34) imply

$$W_m + \frac{\gamma^2 d_m^2}{4} - \gamma T_m \xrightarrow{P_m} 0 \tag{5.1.35}$$

and

$$T_m - T_m^* \xrightarrow{P_m} 0, \quad T_m - S_m \xrightarrow{P_m} 0. \tag{5.1.36}$$

It can be shown that S_m given by (5.1.25) satisfies Lindeberg's condition (see Lemma 4.1 of Hájek (1961)), so that

$$\mathcal{L}(S_m \mid P_m) \text{ is } AN\left(0, d_m^2\right). \tag{5.1.37}$$

Let ε_m and $Q_m(\alpha, \beta, \sigma)$ be the size and power of the test based on critical region

$$S_m > K_\varepsilon d, \quad m \geq 1, \tag{5.1.38}$$

where d is given by (5.1.20) and K_ε satisfies

$$\Phi(K_\varepsilon) = 1 - \varepsilon.$$

Definition 5.1.17. *If* $\lim_{m\to\infty} \varepsilon_m = \varepsilon$ *and*

$$\liminf_{m\to\infty} \left| Q_m(\alpha, \beta, \sigma) - Q_m^*(\alpha, \beta, \sigma) \right| \geq 0, \quad \alpha \in \mathbb{R},\ \beta > 0,\ \sigma > 0,$$

where $Q_m^*(\alpha, \beta, \sigma)$ *is the power of any other test of limiting size* ε*, then we say that the test based on (5.1.38) is asymptotically uniformly most powerful (uniformness is meant with respect to* α*,* β *and* σ*).*

From (5.1.20), (5.1.35), (5.1.36) and (5.1.37), it follows that

$$\mathcal{L}(W_m \mid P_m) \text{ is } AN\left(-\frac{\gamma^2 d^2}{4}, \gamma^2 d^2\right). \tag{5.1.39}$$

Making use of Lemma 5.1.15, we conclude that the distributions Q_m are contiguous to the distributions P_m, and from (5.1.20), (5.1.35), (5.1.36) and (5.1.39), we see that

$$L_m + \frac{\gamma^2 d^2}{2} - \gamma S_m \xrightarrow{P_m} 0. \tag{5.1.40}$$

Now the contiguity implies that

$$L_m + \frac{\gamma^2 d^2}{2} - \gamma S_m \xrightarrow{Q_m} 0.$$

Consequently, the best critical region for distinguishing the particular alternative Q_m from the particular null hypothesis P_m, which is known to be based on the statistics L_m (see, Neyman–Pearson lemma (Lemma 5.1.12)), is asymptotically approximated by the test (5.1.38). Since the statistic S_m is distribution free, the test has the correct limiting size for all possible particular hypotheses, and therefore, it is asymptotically uniformly most powerful.

Using (5.1.40), we see that $\mathcal{L}(S_m, L_m \mid P_m)$ tends to a degenerated bivariate normal distribution $N(\boldsymbol{\mu}, \boldsymbol{\Sigma})$, where $\boldsymbol{\mu} = \left(0, -\frac{\gamma^2 d^2}{2}\right)'$ and $\boldsymbol{\Sigma} = \begin{pmatrix} d^2 & \gamma d^2 \\ \gamma d^2 & \gamma^2 d^2 \end{pmatrix}$ with the correlation coefficient $\rho = 1$. Consequently, from (5.1.37), (5.1.39) and Lemma 5.1.15, it follows that

$$\mathcal{L}(S_m \mid Q_m) \text{ is } AN\left(\gamma d^2, d^2\right).$$

Now, we can state the following result.

Theorem 5.1.18 (Hájek (1962), Theorem 1.1). *Suppose that (5.1.17) and Assumption 5.1.14 hold, and let* $\psi(u)$ *given by (5.1.21) satisfy (5.1.26). Then the critical region (5.1.38) provides an asymptotically uniformly most powerful test of limiting size* ε*, and the asymptotic power equals*

$$1 - \Phi\left(K_\varepsilon - \left(\frac{\beta}{\sigma}\right) d\right). \tag{5.1.41}$$

As the candidates of step function $\psi_m(u)$ we can consider, e.g.

(i) Step functions ψ_m form an N_m-dimensional subspace in the space of quadratically integrable functions on $(0,1)$. The projection of ψ on this space is

$$\psi_m^0\left(\frac{i}{N_m+1}\right) = N_m \int_{(i-1)/N_m}^{i/N_m} \psi(u)\,du, \quad 1 \le i \le N_m.$$

(ii) Let $X_{m,(1)} < \cdots < X_{m,(N_m)}$ be the ordered sample referring to F, and $\alpha = \beta = 0$ and $\sigma = 1$, so that $Z_{m,(1)} < \cdots < Z_{m,(N_m)}$, where $Z_{m,(i)} = F(X_{m,(i)})$ is an ordered sample from the uniform distribution over $(0,1)$. Put

$$\psi_m^+\left(\frac{i}{N_m+1}\right) = E\left\{-\frac{f'(X_{m,(i)})}{f(X_{m,(i)})}\right\} = E\left\{\psi(Z_{m,(i)})\right\}, \quad 1 \le i \le N_m.$$

(iii) Another step function of particular interest is defined by

$$\psi_m^*\left(\frac{i}{N_m+1}\right) = \psi\left(\frac{i}{N_m+1}\right), \quad 1 \le i \le N_m.$$

We extend the definitions of ψ_m^0, ψ_m^+ and ψ_m^* to the whole interval $(0,1)$ according to (5.1.24). Then, it can be shown that under (5.1.17), (5.1.18) and (5.1.19), ψ_m^0 and ψ_m^+ satisfy the condition (5.1.26) (see Hájek (1961)). Moreover, suppose that (5.1.17), (5.1.18) and (5.1.19) hold, and that $\psi(u)$ is either monotone or continuous and bounded. Then, ψ_m^* also satisfies the condition (5.1.26) (see Hájek (1961)).

Under the conditions above, all the critical regions $S_m^0 > K_\varepsilon d_m$, $S_m^+ > K_\varepsilon d_m$ and $S_m^* > K_\varepsilon d_m$ provide an asymptotically most powerful test of limiting size ε and of asymptotic power (5.1.41), where

$$S_m^0 = \sum_{i=1}^{N_m} (c_{m,i} - \bar{c}_m)\,\psi_m^0\left(\frac{R_{m,i}^{N_m}}{N_m+1}\right), \quad m \ge 1,$$

$$S_m^+ = \sum_{i=1}^{N_m} (c_{m,i} - \bar{c}_m)\,\psi_m^+\left(\frac{R_{m,i}^{N_m}}{N_m+1}\right), \quad m \ge 1,$$

$$S_m^* = \sum_{i=1}^{N_m} (c_{m,i} - \bar{c}_m)\,\psi_m^*\left(\frac{R_{m,i}^{N_m}}{N_m+1}\right), \quad m \ge 1.$$

If the true density, say $g(x)$, differs from the supposed one $f(x)$, then the tests based on S_m will be no longer asymptotically most powerful. Denote

$$\varphi(u) = -\frac{g'\left(G^{-1}(u)\right)}{g\left(G^{-1}(u)\right)},$$

where $G(x) = \int_{-\infty}^{x} g(y)\,dy$. Assume that

$$\int_0^1 \varphi^2(u)\,du = \int_{\mathbb{R}} \left\{\frac{g'(x)}{g(x)}\right\}^2 g(x)\,dx = I(g) < \infty, \tag{5.1.42}$$

and introduce the correlation coefficient

$$\rho = \frac{\int_0^1 \psi(u)\varphi(u)\,du}{\left(\int_0^1 \psi^2(u)\,du \int_0^1 \varphi^2(u)\,du\right)^{1/2}} = \frac{\int_0^1 \psi(u)\varphi(u)\,du}{\sqrt{I(f)I(g)}}. \tag{5.1.43}$$

Let P_m correspond to G and $\{\alpha = \alpha_0 + \beta_0 \bar{c}_m,\ \beta = 0,\ \sigma = \sigma_0\}$, and let Q_m correspond to G and $\{\alpha = \alpha_0,\ \beta = \beta_0,\ \sigma = \sigma_0\}$. Introduce statistics

$$L_m = \sum_{i=1}^{N_m} \log \left\{ \frac{g\left(Y_{m,i} - \gamma c_{m,i} + \gamma \bar{c}_m\right)}{g\left(Y_{m,i}\right)} \right\},$$

$$\widetilde{T}_m = \sum_{i=1}^{N_m} \left(c_{m,i} - \bar{c}_m\right) \varphi \left\{G\left(Y_{m,i}\right)\right\}$$

and

$$T_m = \sum_{i=1}^{N_m} \left(c_{m,i} - \bar{c}_m\right) \psi \left\{G\left(Y_{m,i}\right)\right\}.$$

Then, the argument similar to the above leads to

$$L_m + \frac{\gamma^2 \tilde{d}^2}{2} - \gamma \widetilde{T}_m \overset{P_m}{\to} 0,$$

$$\mathcal{L}\left(\widetilde{T}_m \mid P_m\right) \text{ is } AN\left(0, \tilde{d}_m^2\right) \quad \text{and} \quad \mathcal{L}\left(T_m \mid P_m\right) \text{ is } AN\left(0, d_m^2\right),$$

where d_m^2 is given by (5.1.20),

$$\tilde{d}_m^2 = \sum_{i=1}^{N_m} \left(c_{m,i} - \bar{c}_m\right)^2 I(g) \quad \text{and} \quad \tilde{d}^2 = \lim_{m \to \infty} \tilde{d}_m^2.$$

Furthermore, since under P_m,

$$\mathrm{Cov}\left(\widetilde{T}_m, T_m\right) = \sum_{i=1}^{N_m} \left(c_{m,i} - \bar{c}_m\right)^2 \int_0^1 \psi(u) \varphi(u)\, du,$$

we can see that $\mathcal{L}\left(T_m, L_m \mid P_m\right)$ tends to a degenerated bivariate normal distribution $N(\boldsymbol{\mu}, \boldsymbol{\Sigma})$, where $\boldsymbol{\mu} = \left(0, -\frac{\gamma^2 \tilde{d}^2}{2}\right)'$ and $\boldsymbol{\Sigma} = \begin{pmatrix} d^2 & \rho\gamma\tilde{d}d \\ \rho\gamma\tilde{d}d & \gamma^2\tilde{d}^2 \end{pmatrix}$ with the correlation coefficient ρ given by (5.1.43), and consequently, it follows that

$$\mathcal{L}\left(T_m \mid Q_m\right) \text{ is } AN\left(\rho\gamma\tilde{d}d, d^2\right).$$

Now, we have the following result.

Theorem 5.1.19 (Hájek (1962), Theorem 6.1). *Let the same conditions as Theorem 5.1.18 and (5.1.42) be satisfied. Then the asymptotic power of the test $S_m > K_\varepsilon d$ equals*

$$1 - \Phi\left(K_\varepsilon - \rho\left(\frac{\beta}{\sigma}\right)\tilde{d}\right),$$

where ρ is given by (5.1.43).

In the two-sample problem where $N_m/\widetilde{N}_m \to \lambda$, $0 < \lambda < 1$, the efficiency may be interpreted as the ratio of sample sizes needed to attain the same power. However, in the general regression problem, the sample size N_m should be replaced by the sum $\sum_{i=1}^{N_m} \left(c_{m,i} - \bar{c}_m\right)^2$.

5.1.4 U-Statistics for Stationary Processes

U-statistics play an important role in nonparametric and semiparametric rank-based inference. Yoshihara (1976) investigated the asymptotic properties of U-statistics for strictly stationary, absolutely regular processes.

Let $\{\boldsymbol{X}_t, \ t \in \mathbb{Z}\}$ be a d-dimensional strictly stationary process defined on a probability space $(\Omega, \mathcal{A}, P)$. For $a \leq b$, let $\mathcal{F}_a^b$ denote the σ-algebra generated by $\{\boldsymbol{X}_a, \dots, \boldsymbol{X}_b\}$. The sequence is said to be absolutely regular if

$$\beta(N) = E\left\{\sup_{A \in \mathcal{F}_N^\infty} \left|P\left\{A \mid \mathcal{F}_{-\infty}^0\right\} - P\{A\}\right|\right\} \downarrow 0 \tag{5.1.44}$$

as $N \to \infty$. It can be shown that

$$\beta(N) = \frac{1}{2} V(P_{0,N}, P_{1,N}) = \frac{1}{2}\mathrm{Var}(P_{0,N} - P_{1,N}),$$

where $P_{0,N}$ is the measure induced by the process $\{\boldsymbol{X}_t\}$ on the σ-algebra $\mathcal{F}_{-\infty}^0 \cup \mathcal{F}_N^\infty$, and $P_{1,N}$ the measure defined for $A \in \mathcal{F}_N^\infty$, $B \in \mathcal{F}_{-\infty}^0$ by the equality

$$P_{1,N}(A \cap B) = P_{0,N}(A) P_{0,N}(B).$$

On the other hand, $\{\boldsymbol{X}_t\}$ is said to satisfy the ϕ-mixing condition if

$$\phi(N) = \sup_{B \in \mathcal{F}_{-\infty}^0, A \in \mathcal{F}_N^\infty} \frac{1}{P_{0,N}(B)} \left|P_{0,N}(A \cap B) - P_{1,N}(A \cap B)\right| \downarrow 0$$

and the strong mixing condition if

$$\alpha(N) = \sup_{B \in \mathcal{F}_{-\infty}^0, A \in \mathcal{F}_N^\infty} \left|P_{0,N}(A \cap B) - P_{1,N}(A \cap B)\right| \downarrow 0.$$

Since $\alpha(N) \leq \beta(N) \leq \phi(N)$, if $\{\boldsymbol{X}_t\}$ is ϕ-mixing, then it is absolutely regular and if $\{\boldsymbol{X}_t\}$ is absolutely regular, then it is strong mixing. Ibragimov and Solev (1969) (see also Ibragimov and Rozanov (1978), Section IV.4) introduced the complete description of stationary Gaussian processes satisfying (5.1.44).

Theorem 5.1.20 (Ibragimov and Solev (1969)). *A stationary Gaussian process $\{X_t, \ t \in \mathbb{Z}\}$ is absolutely regular if and only if this process has a spectral density $f(\lambda)$, representable as*

$$f(\lambda) = |P(\lambda)|^2 a(\lambda)$$

where $P(\lambda)$ is the polynomial with roots on the circle $|z| = 1$, and the coefficients a_j of the Fourier series $\sum_{j \in \mathbb{Z}} a_j e^{i\lambda j}$ of the function $\log a(\lambda)$ *satisfy*

$$\sum_{j \in \mathbb{Z}} |j| \left|a_j\right| < \infty.$$

We denote the distribution function of $\boldsymbol{X}_t$ by $F(\boldsymbol{x})$, $\boldsymbol{x} \in \mathbb{R}^d$ and consider a functional

$$\theta(F) = \int \cdots \int_{\mathbb{R}^{dp}} g\left(\boldsymbol{x}_1, \dots, \boldsymbol{x}_p\right) dF(\boldsymbol{x}_1) \cdots dF\left(\boldsymbol{x}_p\right),$$

where $|\theta(F)| < \infty$ and $g\left(\boldsymbol{x}_1, \dots, \boldsymbol{x}_p\right)$ is symmetric in its $m(\geq 1)$ arguments. As an estimator of $\theta(F)$, we define U-statistics

$$U_N = \frac{1}{{}_N C_p} \sum_{1 \leq t_1 < \cdots < t_p \leq N}^{N} g(\boldsymbol{X}_{t_1}, \dots, \boldsymbol{X}_{t_p}), \quad N \geq p.$$

Alternatively, we also consider a von Mises's differentiable statistical functional $\theta(F_N)$ defined by

$$\theta(F_N) = \int \cdots \int_{\mathbb{R}^{dp}} g\left(\boldsymbol{x}_1, \ldots, \boldsymbol{x}_p\right) dF_N(\boldsymbol{x}_1) \cdots dF_N\left(\boldsymbol{x}_p\right)$$
$$= \frac{1}{N^p} \sum_{t_1=1}^{N} \cdots \sum_{t_p=1}^{N} g\left(\boldsymbol{X}_{t_1}, \ldots, \boldsymbol{X}_{t_p}\right).$$

In what follows we assume that $\{\boldsymbol{X}_t\}$ is a d-dimensional strictly stationary process which satisfies absolutely regularity.

For every m $(0 \leq m \leq p)$, let

$$g_m(\boldsymbol{x}_1, \ldots, \boldsymbol{x}_m) = \int \cdots \int_{\boldsymbol{R}^{d(p-m)}} g\left(\boldsymbol{x}_1, \ldots, \boldsymbol{x}_p\right) dF(\boldsymbol{x}_{m+1}) \cdots dF\left(\boldsymbol{x}_p\right)$$

Then we have $g_0 = \theta(F)$ and $g_p = g$. Furthermore, let

$$\sigma^2 = \sigma^2(F) = \left[E\left\{g_1^2(\boldsymbol{X}_1)\right\} - \theta^2(F)\right] + 2\sum_{k=1}^{\infty}\left[E\{g_1(\boldsymbol{X}_1)g_1(\boldsymbol{X}_{k+1})\} - \theta^2(F)\right].$$

Now, we introduce the following assumption.

Assumption 5.1.21. *We assume that for some $r > 2$,*

(i)

$$\mu_r = \int \cdots \int_{\mathbb{R}^{dp}} \left|g\left(\boldsymbol{x}_1, \ldots, \boldsymbol{x}_p\right)\right|^r dF(\boldsymbol{x}_1) \cdots dF\left(\boldsymbol{x}_p\right) \leq M_0 < \infty,$$

(ii) *and for all integers $t_1 < \cdots < t_p$,*

$$\nu_r = E\left|g\left(\boldsymbol{X}_{t_1}, \ldots, \boldsymbol{X}_{t_p}\right)\right|^r \leq M_0 < \infty.$$

Under this assumption, Yoshihara (1976) proved two weak convergence results as below, namely, the central limit theorem (CLT) and the functional central limit theorem (FCLT).

Theorem 5.1.22 (Yoshihara (1976), Theorem 1). *Suppose that there is a positive number δ such that Assumption 5.1.21 holds for $r = 2 + \delta$ and*

$$\beta(N) = O\left(N^{-\frac{(2+\delta')}{\delta'}}\right)$$

for some δ' $(0 < \delta' < \delta)$. If $\sigma^2 > 0$, then we have

$$\lim_{N\to\infty} P\left\{\frac{\sqrt{N}\{U_N - \theta(F)\}}{p\sigma} \leq z\right\} = \Phi(z) \tag{5.1.45}$$

for all $z \in \mathbb{R}$. Furthermore, we have

$$\sqrt{N}\left|\theta(F_N) - U_N\right| \xrightarrow{\mathcal{P}} 0,$$

hence, (5.1.45) holds for U_N being replaced by $\theta(F_N)$.

Secondly, let C be the space of all continuous real-valued functions on $[0, 1]$, with the uniform topology, i.e., for $g, h \in C$

$$\rho(g, h) = \sup_{0\leq u\leq 1} |g(u) - h(u)|,$$

and let $W = \{W(u), \ 0 \le u \le 1\}$ be a standard Brownian motion. Suppose $\sigma^2 > 0$. For every $N \ge p$, define the random elements $X_N = \{X_N(u), \ 0 \le u \le 1\}$ and $X_N^* = \left\{X_N^*(u), \ 0 \le u \le 1\right\}$ in C by

$$X_N(u) = \begin{cases} 0 & \text{for } 0 \le u \le \frac{p-1}{N}, \\ \frac{k\{U_k - \theta(F)\}}{\sqrt{N} p \sigma} & \text{for } u = \frac{k}{N}, \ p \le k \le N, \\ \text{linearly interpolated} & \text{for } u \in \left[\frac{k}{N}, \frac{k+1}{N}\right], \ p-1 \le k \le N-1 \end{cases}$$

and

$$X_N^*(u) = \begin{cases} 0 & \text{for } u = 0, \\ \frac{k\{\theta(F_k) - \theta(F)\}}{\sqrt{N} p \sigma} & \text{for } u = \frac{k}{N}, \ 1 \le k \le N, \\ \text{linearly interpolated} & \text{for } u \in \left[\frac{k}{N}, \frac{k+1}{N}\right], \ 0 \le k \le N-1. \end{cases}$$

Theorem 5.1.23 (Yoshihara (1976), Theorem 2). *Suppose that there is a positive number δ such that Assumption 5.1.21 holds for $r = 4 + \delta$ and*

$$\beta(N) = O\left(N^{-\frac{3(4+\delta')}{(2+\delta')}}\right)$$

for some δ' $(0 < \delta' < \delta)$. Then, both X_N and X_N^* converge weakly to W. Furthermore, we have $\rho\left(X_N, X_N^*\right) \xrightarrow{\mathcal{P}} 0$.

5.2 Semiparametrically Efficient Estimation in Time Series

5.2.1 Introduction to Rank-Based Theory in Time Series

5.2.1.1 Testing for randomness against ARMA alternatives

Nonparametric methods have been developed for the analysis of univariate and multivariate observations without the distributional assumption (e.g., Gaussian assumption). A nonparametric procedure is even desired in the area of time series analysis. Hallin et al. (1985a) considered the systematic time series oriented study of testing for randomness against ARMA alternatives. They proposed statistics of the form

$$S_N = (N-p)^{-1} \sum_{t=p+1}^{N} a_N\left(R_t^{(N)}, R_{t-1}^{(N)}, \ldots, R_{t-p}^{(N)}\right), \tag{5.2.1}$$

where $a_N(\cdot)$ is some given score function and $R_t^{(N)}$ is the rank of the observation at time t in observed series $\boldsymbol{X}^{(N)} = (X_1, \ldots, X_N)$. These statistics are so-called linear serial rank statistics. As a special case, for example, we can consider:

run statistic (with respect to median)

$$a_N(i_1, i_2) = \begin{cases} 1 & \text{if } (2i_1 - N - 1)(2i_2 - N - 1) < 0, \\ 0 & \text{if } (2i_1 - N - 1)(2i_2 - N - 1) \ge 0, \end{cases}$$

turning point statistic

$$a_N(i_1, i_2, i_3) = \begin{cases} 1 & \text{if } i_1 > i_2 < i_3, \\ 1 & \text{if } i_1 < i_2 > i_3, \\ 0 & \text{elsewhere,} \end{cases}$$

Spearman's rank correlation coefficient (up to additive and multiplicative constants)

$$a_N\left(i_1,\ldots,i_{p+1}\right) = \frac{i_1 i_{p+1}}{(N+1)^2}.$$

Put $U_t = F(X_t)$ and assume that there exists a function $J = J\left(v_{p+1}, v_p, \ldots, v_1\right)$, defined over $[0,1]^{p+1}$, satisfying

$$0 < \int_{[0,1]^{p+1}} J\left(v_{p+1},\ldots,v_1\right) dv_{p+1}\cdots dv_1 < \infty$$

and

$$\lim_{N\to\infty} E\left[\left\{J\left(U_{p+1},\ldots,U_1\right) - a_N\left(R^{(N)}_{p+1},\ldots,R^{(N)}_1\right)\right\}^2 \mid H_0^{(N)}\right] = 0, \tag{5.2.2}$$

where $E\left(\cdot \mid H_0^{(N)}\right)$ indicates the expectation under the null hypothesis, which is defined below. This assumption is satisfied most of the time when a_N is of the form

$$a_N\left(i_1, i_2,\ldots,i_{p+1}\right) = J\left(\frac{i_1}{N+1}, \frac{i_2}{N+1}, \ldots, \frac{i_{p+1}}{N+1}\right).$$

Such a function J is called a score generation functon associated with the serial rank statistic S_N.

Let $\{\varepsilon_t,\ t\in\mathbb{Z}\}$ be i.i.d. random variables with $E(\varepsilon_t) = 0$, $E\left(\varepsilon_t^2\right) = \sigma^2$, and assume that it has a density $f(x)$ which has a derivative $f'(x)$ a.e. Similar to (5.1.21), we put

$$\phi\left(F^{-1}(u)\right) = -\frac{f'\left(F^{-1}(u)\right)}{f\left(F^{-1}(u)\right)},$$

that is, $\phi(x) = \psi(F(x))$ a.e. This function can also be written as $\phi(x) = -\frac{f'(x)}{f(x)}$ a.e.

Now, we assume the following condtion on the innovation process $\{\varepsilon_t,\ t\in\mathbb{Z}\}$.

Assumption 5.2.1. **(i)** *ε_t has finite moments up to the third order.*

(ii) *The derivative $f'(x)$ satisfies $\int_{\mathbb{R}} |f'(x)|\,dx < \infty$.*

(iii) *Fisher information $I(f)$ from (5.1.17) satisfies $0 < I(f) < \infty$.*

(iv) *$\phi(x)$ has a derivative $\phi'(x)$ a.e. which satisfies a Lipschitz condition $|\phi'(x) - \phi'(y)| < A|x-y|$ a.e.*

Under these condition, we have

$$\int_{\mathbb{R}} \phi(x) f(x)\,dx = 0$$

and

$$\int_{\mathbb{R}} \phi^2(x) f(x)\,dx = I(f)$$

(see also (5.1.22) and (5.1.23)). It is also easy to check that

$$\int_{\mathbb{R}} x\phi(x) f(x)\,dx = 1$$

and

$$\int_{\mathbb{R}} \phi'(x) f(x)\,dx = I(f).$$

If we put $f_\sigma(x) = \sigma^{-1} f(x/\sigma)$, then $\sigma^2 I(f_\sigma) = I(f)$; therefore, $\sigma^2 I(f)$ is independent of the scale transformation.

Let $a_1, \ldots, a_{p_1}, b_1, \ldots, b_{p_2}$ be an arbitrary $(p_1 + p_2)$-tuple of real numbers, and consider the sequence

$$X_t^{(N)} - \frac{1}{\sqrt{N}} \sum_{i=1}^{p_1} a_i X_{t-i}^{(N)} = \varepsilon_t + \frac{1}{\sqrt{N}} \sum_{i=1}^{p_2} b_i \varepsilon_{t-i}, \quad t \in \mathbb{Z},\ N \geq 1 \tag{5.2.3}$$

of stochastic difference equations. For n sufficiently large, all the roots of the characteristic equation

$$z^{p_1} - \frac{1}{\sqrt{N}} \sum_{i=1}^{p_1} a_i z^{p_1 - i} = 0, \quad z \in \mathbb{C}$$

lie inside the unit circle, and (5.2.3) generates a sequence of stationary processes $\left\{X_t := X_t^{(N)},\ t \in \mathbb{Z}\right\}$. Denote by $\boldsymbol{x}^{(N)} = \left(x_1^{(N)}, \ldots, x_N^{(N)}\right)'$ an observed realization of $\boldsymbol{X}^{(N)} = \left(X_1^{(N)}, \ldots, X_N^{(N)}\right)'$.

Denote by $H_0^{(N)}$ the sequence of simple (null) hypotheses which satisfies $a_1 = \cdots = a_{p_1} = b_1 = \cdots = b_{p_2} = 0$. Then, the processes $\left\{X_t^{(N)}\right\}$ all coincide with the white noise process $\{\varepsilon_t\}$ which has the likelihood function

$$l_N^0\left(\boldsymbol{x}^{(N)}\right) = \prod_{t=1}^{N} f\left(x_t^{(N)}\right).$$

On the other hand, denote by $H_1^{(N)}$ the corresponding sequence of alternative hypotheses which satisfies that both a_{p_1} and b_{p_2} are different from zero. Then, the processes $\left\{X_t^{(N)}\right\}$ are ARMA(p_1, p_2) processes whose likelihood function is denoted by $l_N^1\left(\boldsymbol{x}^{(N)}\right)$ (the exact expression of this likelihood function can be found in Hallin et al. (1985a), (3.3)).

In what follows, for simplicity, we write $\boldsymbol{x}$, x_t, $\boldsymbol{X}$ and X_t for $\boldsymbol{x}^{(N)}$, $x_t^{(N)}$, $\boldsymbol{X}^{(N)}$ and $X_t^{(N)}$. Now, consider the likelihood ratio

$$L_N(\boldsymbol{x}) = \begin{cases} l_N^1(\boldsymbol{x}) / l_N^0(\boldsymbol{x}) & \text{if } l_N^0(\boldsymbol{x}) > 0 \\ 1 & \text{if } l_N^1(\boldsymbol{x}) = l_N^0(\boldsymbol{x}) = 0 \\ \infty & \text{if } l_N^1(\boldsymbol{x}) > l_N^0(\boldsymbol{x}) = 0, \end{cases}$$

then, we have the following lemma.

Lemma 5.2.2 (Hallin et al. (1985a), Proposition 3.1). *Under $H_0^{(N)}$, $\log L_N(\boldsymbol{X}) = \mathcal{L}_N^0(\boldsymbol{X}) - d^2/2 + o_P(1)$, where*

$$\mathcal{L}_N^0(\boldsymbol{X}) = \frac{1}{\sqrt{N}} \sum_{t=p+1}^{N} \phi(X_t) \sum_{i=1}^{p} d_i X_{t-i},$$

$$d_i = \begin{cases} a_i + b_i & 1 \leq i \leq \min\{p_1, p_2\}, \\ a_i & p_2 < i \leq p_1 \text{ if } p_2 < p_1, \\ b_i & p_1 < i \leq p_2 \text{ if } p_1 < p_2, \end{cases}$$

$$p = \max\{p_1, p_2\} \quad \text{and} \quad d^2 = \sum_{i=1}^{p} d_i^2 \sigma^2 I(f).$$

Furthermore, $\mathcal{L}_N^0(\boldsymbol{X})$ is asymptotically normal, with mean zero and variance d^2.

From LeCam's first lemma (see, e.g., Sidak et al. (1999), Corollary 7.1.1), we can immediately see that this lemma implies that $H_1^{(N)}$ is contiguous to $H_0^{(N)}$.

Next, we consider the joint asymptotic normality of

$$\begin{pmatrix} \sqrt{N}\left\{S_N(\boldsymbol{X}) - m_N\right\} \\ \log L_N(\boldsymbol{X}) \end{pmatrix}$$

under $H_0^{(N)}$, where

$$m_N = E\left(S_N(\boldsymbol{X}) \mid H_0^{(N)}\right) = \frac{1}{N(N-1)\cdots(N-p)} \sum \cdots \sum\nolimits_{1 \le i_1 \ne \cdots \ne i_{p+1} \le N} a_N\left(i_1, \ldots, i_{p+1}\right). \quad (5.2.4)$$

Define

$$\mathcal{S}_N(\boldsymbol{X}) = \frac{1}{\sqrt{N-p}} \sum_{t=p+1}^{N} J\left(F(X_t), F(X_{t-1}), \ldots, F\left(X_{t-p}\right)\right)$$

and

$$\mathcal{M}_N(\boldsymbol{X}) = \frac{\sqrt{N-p}}{N(N-1)\cdots(N-p)} \sum \cdots \sum\nolimits_{1 \le t_1 \ne \cdots \ne t_{p+1} \le N} J\left(F(X_{t_1}), \ldots, F\left(X_{t_{p+1}}\right)\right),$$

and let

$$\Delta_N(\boldsymbol{X}) = \sqrt{N-p}\left\{S_N(\boldsymbol{X}) - m_N\right\} - \left\{\mathcal{S}_N(\boldsymbol{X}) - \mathcal{M}_N(\boldsymbol{X})\right\}.$$

Then, it can be shown that $\lim_{N\to\infty} E\left\{\Delta_N^2(\boldsymbol{X})\right\} = 0$ (see Hallin et al. (1985a), Section 4.1). Therefore, $\sqrt{N-p}\left\{S_N(\boldsymbol{X}) - m_N\right\}$ and $\left\{\mathcal{S}_N(\boldsymbol{X}) - \mathcal{M}_N(\boldsymbol{X})\right\}$ are asymptotically equivalent (under $H_0^{(N)}$).

Now, we consider U-statistics which are asymptotically equivalent to $N^{-1/2}\left\{\mathcal{S}_N(\boldsymbol{X}) - \mathcal{M}_N(\boldsymbol{X})\right\}$ and $N^{-1/2}\mathcal{L}_N^0(\boldsymbol{X})$. Define the $(p+1)$-dimensional random variables

$$\mathbf{Y}_t = \left(Y_{t,1}, \ldots, Y_{t,p+1}\right)' = \left(U_t, \ldots, U_{t-p}\right)', \quad p+1 \le t \le N.$$

The $\mathbf{Y}_t$'s are identically distributed (uniformly over $[0,1]^{p+1}$, under $H_0^{(N)}$), but, of course, they are not independent. However, they constitute an absolutely regular process since they are p-dependent (see Yoshihara (1976) and 5.1.4).

Define

$$G(\boldsymbol{y}) = G\left(y_1, \ldots, y_{p+1}\right) = \phi\left(F^{-1}(y_1)\right) \sum_{i=1}^{p} d_i F^{-1}(y_{i+1})$$

and

$$\begin{aligned} \Phi^{\mathcal{L}}\left(\mathbf{Y}_{t_1}, \ldots, \mathbf{Y}_{t_{p+1}}\right) &= \sum_{j=1}^{p+1} G\left(\mathbf{Y}_{t_j}\right) / (p+1) \\ &= \sum_{j=1}^{p+1} G\left(Y_{t_j,1}, \ldots, Y_{t_j,p+1}\right) / (p+1), \end{aligned}$$

where $\Phi^{\mathcal{L}}$ is a kernel of an U-statistics which is asymptotically equivalent to $N^{-1/2}\mathcal{L}_N^0$. Similarly, we define the kernels

$$\Phi^{S}\left(\mathbf{Y}_{t_1}, \ldots, \mathbf{Y}_{t_{p+1}}\right) = \sum_{j=1}^{p+1} J\left(\mathbf{Y}_{t_j}\right) / (p+1)$$

$$= \sum_{j=1}^{p+1} J\left(Y_{t_j,1}, \ldots, Y_{t_j,p+1}\right) / (p+1)$$

for $N^{-1/2}\mathcal{S}_N$ and

$$\Phi^{\mathcal{M}}\left(\boldsymbol{Y}_{t_1}, \ldots, \boldsymbol{Y}_{t_{p+1}}\right) = \frac{1}{(p+1)!} \sum_j J\left(Y_{j_1,1}, \ldots, Y_{j_{p+1},1}\right)$$

for $N^{-1/2}\mathcal{M}_N$, where the summation $\sum_j$ extends over all possible $(p+1)!$ permutations $\left(j_1 \ldots, j_{p+1}\right)$ of $\left(t_1 \ldots, t_{p+1}\right)$. The corresponding U-statistics satisfy

$$\frac{1}{{}_{N-p}C_{p+1}} \sum \cdots \sum\nolimits_{p+1 \le t_1 < \cdots < t_{p+1} \le N} \Phi^{\mathcal{L}}\left(\boldsymbol{Y}_{t_1}, \ldots, \boldsymbol{Y}_{t_{p+1}}\right) = N^{-1/2}\mathcal{L}_N^0 + o_P\left(N^{-1/2}\right),$$

$$\frac{1}{{}_{N-p}C_{p+1}} \sum \cdots \sum\nolimits_{p+1 \le t_1 < \cdots < t_{p+1} \le N} \Phi^{\mathcal{S}}\left(\boldsymbol{Y}_{t_1}, \ldots, \boldsymbol{Y}_{t_{p+1}}\right) = N^{-1/2}\mathcal{S}_N + o_P\left(N^{-1/2}\right),$$

$$\begin{aligned} &\frac{1}{{}_{N-p}C_{p+1}} \sum \cdots \sum\nolimits_{p+1 \le t_1 < \cdots < t_{p+1} \le N} \Phi^{\mathcal{M}}\left(\boldsymbol{Y}_{t_1}, \ldots, \boldsymbol{Y}_{t_{p+1}}\right) \\ &= \frac{1}{(N-p)\cdots(N-2p)} \sum \cdots \sum\nolimits_{p+1 \le t_1 \ne \cdots \ne t_{p+1} \le N} J\left(U_{t_1}, \ldots, U_{t_{p+1}}\right) \\ &= N^{-1/2}\mathcal{M}_N + o_P\left(N^{-1/2}\right). \end{aligned}$$

Therefore, we can see that, for arbitrary coefficents α and β,

$$\mathcal{U}_N^{\alpha\beta} = N^{-1/2}\left[\alpha\left\{\mathcal{S}_N(\boldsymbol{X}) - \mathcal{M}_N(\boldsymbol{X})\right\} + \beta\left\{\mathcal{L}_N^0(\boldsymbol{X})\right\}\right]$$

is (up to $o_P\left(N^{-1/2}\right)$ terms) a sequence of U-statistics with kernel

$$\Phi_{\alpha\beta} = \alpha\left(\Phi^{\mathcal{S}} - \Phi^{\mathcal{M}}\right) + \beta\Phi^{\mathcal{L}}.$$

From Theorem 1 of Yoshihara (1976) (see Section 4.2 of Hallin et al. (1985a) and Section 5.1.4), it follows that $\sqrt{N}\left\{\mathcal{U}_N^{\alpha\beta} - E\left(\mathcal{U}_N^{\alpha\beta}\right)\right\}$ is asymptotically normal with mean zero and variance

$$\sigma_{\alpha\beta}^2 = \alpha^2 V^2 + 2\alpha\beta\sum_{i=1}^{p} d_i C_i + \beta^2 d^2,$$

where

$$\begin{aligned} V^2 = &\int_{[0,1]^{p+1}} J^{*2}\left(v_{p+1}, \ldots, v_1\right) dv_1 \cdots dv_{p+1} \\ &+ 2\sum_{j=1}^{p} \int_{[0,1]^{p+1+j}} J^*\left(v_{p+1}, \ldots, v_1\right) J^*\left(v_{p+1+j}, \ldots, v_{1+j}\right) dv_1 \cdots dv_{p+1+j} \end{aligned}$$

and

$$C_i = \int_{[0,1]^{p+1}} J^*\left(v_{p+1}, \ldots, v_1\right) \sum_{j=0}^{p-i} \phi\left(F^{-1}\left(v_{p+1-j}\right)\right) F^{-1}\left(v_{p+1-j-i}\right) dv_1 \cdots dv_{p+1}, \qquad (5.2.5)$$

with

$$J^*\left(u_{p+1}, \ldots, u_1\right) = J\left(u_{p+1}, \ldots, u_1\right)$$

$$-\sum_{k=1}^{p+1}\int_{[0,1]^p} J\left(v_p,\ldots,v_k,u_1,v_{k-1},\ldots,v_1\right)dv_1\cdots dv_p + p\int_{[0,1]^{p+1}} J\left(v_{p+1},\ldots,v_1\right)dv_1\cdots dv_{p+1}. \tag{5.2.6}$$

Obviously, $E\left\{J^*\left(U_{p+1},\ldots,U_1\right)\right\}=0$. Now, from the Cramér–Wold device, we see that, under $H_0^{(N)}$,

$$\begin{pmatrix}\sqrt{N}\left\{S_N(\boldsymbol{X})-m_N\right\}\\ \log L_N(\boldsymbol{X})\end{pmatrix}$$

is asymptotically normal with mean $\left(0,-d^2/2\right)'$ and covariance matrix

$$\begin{pmatrix} V^2 & \sum_{i=1}^p d_iC_i \\ \sum_{i=1}^p d_iC_i & d^2\end{pmatrix}.$$

Finally, appling LeCam's third lemma (see, e.g., Sidak et al. (1999), Corollary 7.1.3), we can immediately derive the following results.

Theorem 5.2.3. *Under $H_1^{(N)}$, $\sqrt{N}\{S_N(\boldsymbol{X})-m_N\}$ is asymptotically normal with mean $\sum_{i=1}^p d_iC_i$ and variance V^2.*

Note that the score-generating function

$$\left\{\sum_{j=0}^{p-i}\phi\left(F^{-1}\left(v_{p+1-j}\right)\right)F^{-1}\left(v_{p+1-j-i}\right);\ i=1,\ldots,p\right\} \tag{5.2.7}$$

constitutes a p-tuple of L^2-functions with norms

$$\left(\int_{[0,1]^{p+1}}\left\{\sum_{j=0}^{p-i}\phi\left(F^{-1}\left(v_{p+1-j}\right)\right)F^{-1}\left(v_{p+1-j-i}\right)\right\}^2 dv_{p+1}\cdots dv_1\right)^{1/2}$$
$$=\sqrt{(p+1-i)\,\sigma^2 I(f)},\ i=1,\ldots,p.$$

From (5.2.5), it follows that

$$J_0^*\left(u_{p+1},\ldots,u_1\right)=\left\{\sigma^2 I(f)\right\}^{-1}\sum_{i=1}^{p}\frac{C_i}{(p+1-i)}\sum_{j=0}^{p-i}\phi\left(F^{-1}\left(v_{p+1-j}\right)\right)F^{-1}\left(v_{p+1-j-i}\right) \tag{5.2.8}$$

is the projection of $J^*\left(u_{p+1},\ldots,u_1\right)$ from (5.2.6) onto the linear L^2-space spanned by the functions from (5.2.7). Therefore, any square integrable score-generating function J^* of the form (5.2.6) can be decomposed into $J^*=J_0^*+J_\perp^*$, where $J_\perp^*$ satisfies

$$\int_{[0,1]^{p+1}} J_\perp^*\left(u_{p+1},\ldots,u_1\right)\sum_{j=0}^{p-i}\phi\left(F^{-1}\left(v_{p+1-j}\right)\right)F^{-1}\left(v_{p+1-j-i}\right)dv_{p+1}\cdots dv_1=0$$

for $i=1,\ldots,p$. This leads to the same values of the C_i's if we use J^* or J_0^* in (5.2.5). Denote by S_N, S_N^0 and $S_N^\perp$ linear serial rank statistics associated with J^*, J_0^* and $J_\perp^*$; denote by m_N, m_N^0 and $m_N^\perp$ their means; denote by V^2, V_0^2 and $V_\perp^2$ the asymptotic variances of $\sqrt{N}(S_N-m_N)$, $\sqrt{N}\left(S_N^0-m_N^0\right)$ and $\sqrt{N}\left(S_N^\perp-m_N^\perp\right)$, respectively. It is easy to check that

$$\sqrt{N}(S_N-m_N)=\sqrt{N}\left(S_N^0-m_N^0\right)+\sqrt{N}\left(S_N^\perp-m_N^\perp\right)+o_P(1)$$

with $\lim_{N\to\infty} NE\left\{\left(S_N^0 - m_N^0\right)\left(S_N^{\perp} - m_N^{\perp}\right)\right\}$ (under $H_0^{(N)}$) Hence we have $V^2 = V_0^2 + V_{\perp}^2$.

Now, we can evaluate the asymptotic relative efficiency of two linear serial rank statistics. Let $S_N^{(1)}$ and $S_N^{(2)}$ have asymptotic normal distributions under $H_1^{(N)}$ with means $\sum_{i=1}^p d_i C_i^{(1)}$ and $\sum_{i=1}^p d_i C_i^{(2)}$, and with variances $V^{(1)2}$ and $V^{(2)2}$. Then, the ARE of $S_N^{(1)}$ with respect to $S_N^{(2)}$ is given by

$$e\left(S_N^{(1)}, S_N^{(2)}\right) = \left\{\frac{V^{(2)} \sum_{i=1}^p d_i C_i^{(1)}}{V^{(1)} \sum_{i=1}^p d_i C_i^{(2)}}\right\}^2$$

(see also (5.1.5)). We shall denote by $H_{\underline{d}}^{(N)}$ a sequence of alternatives characterized by the coefficients $\underline{d} = \left(d_1, \ldots, d_p\right)'$. Now, we have the following result.

Lemma 5.2.4 (Hallin et al. (1985a), Lemma 5.1). *Let S_N and S_N^0 be linear rank statistics with score-generating functions $J^*\left(u_{p+1}, \ldots, u_1\right)$ (from (5.2.6)) and $J_0^*\left(u_{p+1}, \ldots, u_1\right)$ (from (5.2.8)), respectively. Then, we have*

$$e\left(S_N^0, S_N\right) = \frac{V^2\left(\sum_{i=1}^p d_i C_i\right)^2}{V_0^2\left(\sum_{i=1}^p d_i C_i\right)^2} = \frac{V_0^2 + V_{\perp}^2}{V_0^2} \geq 1$$

for any alternative $H_{\underline{d}}^{(N)}$.

From this lemma we can restrict ourselves to the score functions $J_0^*\left(u_{p+1}, \ldots, u_1\right)$ of the form (5.2.8). Straightforward calculation leads to

$$V_0^2 = \left\{\sigma^2 I(f)\right\}^{-1} \sum_{i=1}^p C_i^2.$$

Thus, the values of $\left(C_1, \ldots, C_p\right)'$ that maximize

$$\frac{\left(\sum_{i=1}^p d_i C_i\right)^2}{V_0^2} = \frac{\sigma^2 I(f)\left(\sum_{i=1}^p d_i C_i\right)^2}{\sum_{i=1}^p C_i^2}$$

are parallel to $\left(d_1, \ldots, d_p\right)'$.

A test statistics $\overline{S}_N$ such that $e\left(\overline{S}_N, S_N\right) \geq 1$ for any serial rank statistic S_N is asymptotically the most efficient statistic (in Pitman's sense) within the class of linear serial rank statistics for testing randomness $H_0^{(N)}$ against ARMA alternative $H_{\underline{d}}^{(N)}$. Therefore, we can establish the following result.

Theorem 5.2.5 (Hallin et al. (1985a), Proposition 5.1). *An asymptotically most efficient linear serial rank test for $H_0^{(N)}$ against $H_{\underline{d}}^{(N)}$ is any statistic $S_N^{\underline{d}}$ with score-generating function (up to additive and multiplicative constants) given by*

$$J_{\underline{d}}^*\left(u_{p+1}, \ldots, u_1\right) = \sum_{i=1}^p \frac{d_i}{(p+1-i)} \sum_{j=0}^{p-i} \phi\left(F^{-1}\left(v_{p+1-j}\right)\right) F^{-1}\left(v_{p+1-j-i}\right).$$

It should also be noted that under $H_{\widetilde{d}}^{(N)}$ $\left(\widetilde{d} = \left(\widetilde{d}_1, \ldots, \widetilde{d}_1\right)' \in \mathbb{R}^p\right)$, $\sqrt{N}\left(S_N^{\underline{d}} - m_N^{\underline{d}}\right)$ is asymptotically normal with mean $\sigma^2 I(f) \sum_{i=1}^p \widetilde{d}_i d_i$ and variance $\sigma^2 I(f) \sum_{i=1}^p d_i^2$.

As alternative approach, the dependence coefficient that measures all types of dependence between random vectors $\boldsymbol{X}$ and $\boldsymbol{Y}$ in arbitrary dimension has recently proposed. Distance correlation and distance covariance (Székely et al. (2007)), and Brownian covariance (Székely and Rizzo (2009)) provide a new approach to the problem of measuring dependence and testing the joint independence of random vectors in arbitrary dimension.

5.2.1.2 Testing an ARMA model against other ARMA alternatives

Testing for randomness is also important in the econometric field to check hypotheses, such as market efficiency, life cycle permanent income, etc. However, testing a given ARMA model against other ARMA models is more important in the various identification and validation steps in a time series model building procedure. Regarding to the testing one ARMA model against another ARMA model, the most popular one is based on the Akaike information criteria (AIC). If we would like to compare two models, the one with lower AIC is generally better. Alternatively, Hallin and Puri (1988), Hallin and Puri (1992) considered the following rank-based methods.

For simplicity, we assume that all densities f, g are absolutely continuous with distribution functions F, G, so that the derivatives $f'(x) = df(x)/dx$ exist for almost all x. Furthermore, assume that all of them are of the form

$$f(x) = f_\sigma(x) = \frac{1}{\sigma} f_1\left(\frac{x}{\sigma}\right) > 0, \quad x \in \mathbb{R}$$

with

$$\int_{\mathbb{R}} x f_1(x)\, dx = 0, \qquad \int_{\mathbb{R}} x^2 f_1(x)\, dx = 1.$$

Defining $\phi_f(x) = -\frac{f'(x)}{f(x)}$, we also assume that the Fisher information

$$\sigma^2 I(f) = \int_{\mathbb{R}} \phi_f^2(x) f(x)\, dx = \int_{\mathbb{R}} \phi_{f_1}^2(x) f_1(x)\, dx = I(f_1)$$

is finite. Since f is always strictly positive, F is strictly increasing, and the inverse F^{-1} is well defined. Denote by $H^{(N)}(\boldsymbol{A}, \boldsymbol{B}; f)$ the hypothesis under which the observed series $\boldsymbol{X}^{(N)} = \left(X_1^{(N)}, \ldots, X_N^{(N)}\right)'$ is generated by the ARMA(p_1, q_1) model

$$A(L) X_t = B(L) \varepsilon_t, \quad t \in \mathbb{Z}, \tag{5.2.9}$$

where $\{\varepsilon_t\}$ is a white noise process with density function f and

$$A(L) = 1 - A_1 L - \cdots - A_{p_1} L^{p_1}, \quad B(L) = 1 + B_1 L + \cdots + B_{q_1} L^{q_1}.$$

Assume that under the null hypothesis f remains unspecified, which is denoted by $H^{(N)}(\boldsymbol{A}, \boldsymbol{B}; \cdot) = \cup_f H^{(N)}(\boldsymbol{A}, \boldsymbol{B}; f)$. Consider the filtered series $\boldsymbol{Z}^{(N)} = \left(Z_1^{(N)}, \ldots, Z_N^{(N)}\right)'$ associated with the observed series $\boldsymbol{X}^{(N)}$, where $\left\{Z_t = B(L)^{-1} A(L) X_t\right\}$ is the filtered process. Clearly, $\boldsymbol{X}^{(N)}$ is an ARMA(p_1, q_1) model from (5.2.9) if and only if $\boldsymbol{Z}^{(N)}$ is a white noise series.

Here, the alternative hypotheses of interest are those under which $\boldsymbol{X}^{(N)}$ is an observed series from some ARMA model

$$\alpha(L) X_t = \beta(L) \varepsilon_t, \quad t \in \mathbb{Z}$$

distinct from (5.2.9). In order to investigate locally optimal procedures, Hallin and Puri (1988, 1992) considered the sequences of alternatives that are contiguous to the null hypotheses. Let

$$\alpha^{(N)}(L) X_t = \beta^{(N)}(L) \varepsilon_t, \quad t \in \mathbb{Z} \tag{5.2.10}$$
$$\alpha^{(N)}(L) = 1 - \alpha_1^{(N)} L - \cdots - \alpha_{p_2}^{(N)} L^{p_2}, \quad \beta^{(N)}(L) = 1 + \beta_1^{(N)} L + \cdots + \beta_{q_2}^{(N)} L^{q_2}$$

be a sequence of ARMA(p_2, q_2) model, with $p_2 \geq p_1$, $q_2 \geq p_2$ and

$$\alpha_i^{(N)} = \begin{cases} A_i + N^{-1/2} \gamma_i & 1 \leq i \leq p_1, \\ N^{-1/2} \gamma_i & p_1 + 1 \leq i \leq p_2, \end{cases}$$

$$\beta_i^{(N)} = \begin{cases} B_i + N^{-1/2}\delta_i & 1 \le i \le q_1, \\ N^{-1/2}\delta_i & q_1 + 1 \le i \le q_2. \end{cases}$$

Denote $\boldsymbol{A} = \left(A_1, \ldots, A_{p_1}, 0, \ldots, 0\right)' \in \mathbb{R}^{p_2}$, $\boldsymbol{B} = \left(B_1, \ldots, B_{q_1}, 0, \ldots, 0\right)' \in \mathbb{R}^{q_2}$, $\boldsymbol{\theta} = (\boldsymbol{A}', \boldsymbol{B}')' \in \mathbb{R}^{p_2+q_2}$. And let $\boldsymbol{\Theta}$ be the open subset for which (5.2.9) constitutes a valid, causal and invertible ARMA model of order p_1 and q_1, i.e., satisfies $A_{p_1} \neq 0$, $B_{q_1} \neq 0$, no common roots, causality and invertibility. Furthermore, let $\boldsymbol{\gamma} = \left(\gamma_1, \ldots, \gamma_{p_2}\right)' \in \mathbb{R}^{p_2}$, $\boldsymbol{\delta} = \left(\delta_1, \ldots, \delta_{q_2}\right)' \in \mathbb{R}^{q_2}$ and $\boldsymbol{\tau} = (\boldsymbol{\gamma}', \boldsymbol{\delta}')' \in \mathbb{R}^{p_2+q_2}$. Denote by $H^{(N)}\left(\boldsymbol{\theta} + N^{-1/2}\boldsymbol{\tau}; f\right)$ a sequence of hypotheses under which $\boldsymbol{X}^{(N)}$ is generated by some ARMA(p, q) model with $p_1 \le p \le p_2$, $q_1 \le p \le q_2$, coefficients of the form $\boldsymbol{A} + N^{-1/2}\boldsymbol{\gamma}$, $\boldsymbol{B} + N^{-1/2}\boldsymbol{\delta}$ and innovation density f, which is approaching in some sense to $H^{(N)}(\boldsymbol{\theta}; f)$. If $\boldsymbol{X}^{(N)}$ is an observed series from (5.2.10), then the filtered series $\boldsymbol{Z}^{(N)}$ is generated by the model

$$\alpha^{(N)}(L)\{A(L)\}^{-1} Z_t = \beta^{(N)}(L)\{B(L)\}^{-1} \varepsilon_t, \quad t \in \mathbb{Z}.$$

Using the relation

$$\alpha^{(N)}(L) = A(L) - N^{-1/2}\gamma(L), \quad \beta^{(N)}(L) = B(L) + N^{-1/2}\delta(L)$$

with $\gamma(L) = \sum_{i=1}^{p_2} \gamma_i L^i$ and $\delta(L) = \sum_{i=1}^{q_2} \delta_i L^i$, we can rewrite

$$Z_t - N^{-1/2}a(L) Z_t = \varepsilon_t + N^{-1/2}b(L) N^{-1/2}\varepsilon_t, \tag{5.2.11}$$

where

$$a(L) = \sum_{i=1}^{\infty} a_i L^i = \gamma(L)\{A(L)\}^{-1},$$

$$b(L) = \sum_{i=1}^{\infty} b_i L^i = \delta(L)\{B(L)\}^{-1}.$$

Let $\widetilde{a}_i$ and $\widetilde{b}_i$ be the Green's functions associated with $A(L)$ and $B(L)$, namely,

$$\sum_{i=0}^{\infty} \widetilde{a}_i L^i = \{A(L)\}^{-1} \text{ and } \sum_{i=0}^{\infty} \widetilde{b}_i L^i = \{B(L)\}^{-1}.$$

Then we have the relations for the coefficients

$$a_i = \sum_{j=1}^{\min(p_2, i)} \gamma_j \widetilde{a}_{i-j} \text{ and } b_i = \sum_{j=1}^{\min(q_2, i)} \delta_j \widetilde{b}_{i-j}. \tag{5.2.12}$$

Consider the white noise hypothesis with specified density f, $H^{(N)}(\boldsymbol{0}; f)$, and the generalized alternative ARMA hypothesis $H^{(N)}\left(N^{-1/2}\mathbf{c}; f\right)$ with $\boldsymbol{a} = (a_1, a_2, \ldots)'$, $\boldsymbol{b} = (b_1, b_2, \ldots)'$ and $\mathbf{c} = (\boldsymbol{a}', \boldsymbol{b}')'$, under which $\boldsymbol{Z}^{(N)}$ is given by the realization of the generalized ARMA process (5.2.11). Now, we can reduce testing $H^{(N)}(\boldsymbol{\theta}; f)$ against $H^{(N)}\left(\boldsymbol{\theta} + N^{-1/2}\boldsymbol{\tau}; f\right)$ to the problem of testing $H^{(N)}(\boldsymbol{0}; f)$ against $H^{(N)}\left(N^{-1/2}\mathbf{c}; f\right)$.

Denote by $L_{\boldsymbol{\theta};f}^{(N)}$ and $L_{\boldsymbol{\theta}+N^{-1/2}\boldsymbol{\tau};f}^{(N)}$ the likelihood functions of $\boldsymbol{X}^{(N)}$ under $H^{(N)}(\boldsymbol{\theta}; f)$ and $H^{(N)}\left(\boldsymbol{\theta} + N^{-1/2}\boldsymbol{\tau}; f\right)$, respectively. Furthermore, let $L_{\boldsymbol{0};f}^{(N)}$ and $L_{N^{-1/2}\mathbf{c};f}^{(N)}$ be the likelihood functions of $\boldsymbol{Z}^{(N)}$ under $H^{(N)}(\boldsymbol{0}; f)$ and $H^{(N)}\left(N^{-1/2}\mathbf{c}; f\right)$, respectively. Then, the log likelihood ratios $\Lambda_{\boldsymbol{\theta},\boldsymbol{\tau};f}^{(N)}$ and $\Lambda_{\boldsymbol{0},\mathbf{c};f}^{(N)}$ satisfy the relation

$$\Lambda_{\boldsymbol{\theta},\boldsymbol{\tau};f}^{(N)} := \log \frac{L_{\boldsymbol{\theta}+N^{-1/2}\boldsymbol{\tau};f}^{(N)}\left(\boldsymbol{X}^{(N)}\right)}{L_{\boldsymbol{\theta};f}^{(N)}\left(\boldsymbol{X}^{(N)}\right)}$$

$$=\Lambda^{(N)}_{\mathbf{0},\mathbf{c};f} := \log \frac{L^{(N)}_{N^{-1/2}\mathbf{c};f}\left(\boldsymbol{Z}^{(N)}\right)}{L^{(N)}_{\mathbf{0};f}\left(\boldsymbol{Z}^{(N)}\right)}.$$

Then, the result of Lemma 5.2.2 can be extended as follows.

Lemma 5.2.6. (Proposition 2.2 of Hallin and Puri (1988)) *Under $H^{(N)}(\mathbf{0};f)$, the log-likelihood ratio $\Lambda^{(N)}_{\mathbf{0},\mathbf{c};f}$ is asymptotically normal, with mean $-\frac{1}{2}\sum_{i=1}^{\infty}(a_i+b_i)^2\sigma^2 I(f)$ and variance $\sum_{i=1}^{\infty}(a_i+b_i)^2\sigma^2 I(f)$. Therefore, the sequence of hypotheses $H^{(N)}\left(N^{-1/2}\mathbf{c};f\right)$ and $H^{(N)}(\mathbf{0};f)$ is contiguous.*

Now, we consider, again, the linear serial rank statistics S_N of order p defined in (5.2.1), whose expectation under $H^{(N)}(\mathbf{0};f)$ is given by m_N in (5.2.4). We assume that there exists a function $J = J\left(v_{p+1}, v_p, \ldots, v_1\right)$, defined over $[0,1]^{p+1}$, satisfying

$$0 < \int_{[0,1]^{p+1}} \left|J\left(v_{p+1},\ldots,v_1\right)\right|^{2+\delta} dv_{p+1}\cdots dv_1 < \infty$$

for some $\delta > 0$, and the relationship of the approximation (5.2.2) under $H^{(N)}(\mathbf{0};f)$. Furthermore, we can assume without loss of generality that

$$\int_{[0,1]^p} J\left(v_{p+1},\ldots,v_1\right)\prod_{j\neq i} dv_j = 0, \quad i = 1,\ldots,p+1.$$

Then, we have the following generalization of the above result.

Proposition 5.2.7. (Proposition 2.3 of Hallin and Puri (1988)) *The linear serial rank statistics S_N of order p satisfies that $\sqrt{N}(S_N - m_N)$ is asymptotically normal, with mean 0 and variance V^2 under $H^{(N)}(\mathbf{0};f)$, while it is asymptotically normal with mean $\sum_{i=1}^{p}(a_i+b_i)C_i$ and variance V^2 under $H^{(N)}\left(N^{-1/2}\mathbf{c};f\right)$, where*

$$V^2 = \int_{[0,1]^{2p+1}} \left\{J^2\left(v_{p+1},\ldots,v_1\right) + 2\sum_{j=1}^{p} J\left(v_{p+1},\ldots,v_1\right)J\left(v_{p+1+j},\ldots,v_{1+j}\right)\right\} dv_1\cdots dv_{2p+1}$$

and

$$C_i = \int_{[0,1]^{p+1}} J\left(v_{p+1},\ldots,v_1\right)\sum_{j=0}^{p-i}\phi\left(F^{-1}\left(v_{1+j}\right)\right)F^{-1}\left(v_{1+j+i}\right)dv_1\cdots dv_{p+1}.$$

Note that this result implies that the asymptotic distribution of S_N under $H^{(N)}\left(N^{-1/2}\mathbf{c};f\right)$ is exactly the same as that under the pth- order truncation of $H^{(N)}\left(N^{-1/2}\mathbf{c};f\right)$ which is obtained by letting $a_{p+1} = a_{p+2} = \cdots = b_{p+1=b_{p+2}=\cdots=0}$ in (5.2.11). That is, a finite-order serial rank statistic cannot contain all the information for testing $H^{(N)}(\mathbf{0};f)$ against $H^{(N)}\left(N^{-1/2}\mathbf{c};f\right)$ in general. Therefore, the asymptotically most powerful test for $H^{(N)}(\mathbf{0};f)$ against $H^{(N)}\left(N^{-1/2}\mathbf{c};f\right)$ does not belong to the class of linear serial rank statistics. However, we can still construct the asymptotically most powerful test based on ranks.

For $\boldsymbol{a},\boldsymbol{b} \in l_2$, we denote their inner product by $\langle\boldsymbol{a},\boldsymbol{b}\rangle = \sum_{i=1}^{\infty} a_i b_i$ and the corresponding norm by $\|\boldsymbol{a}\| = \langle\boldsymbol{a},\boldsymbol{a}\rangle^{1/2}$. We define the rank autocorrelation coefficient associated with density f (f-rank autocorrelation) of order i as

$$r^{(N)}_{i;f} = \frac{1}{s^{(N)}}\left\{\frac{1}{N-i}\sum_{t=i+1}^{N}\phi\left(F^{-1}\left(\frac{R^{(N)}_t}{N+1}\right)\right)F^{-1}\left(\frac{R^{(N)}_{t-i}}{N+1}\right) - m^{(N)}\right\},$$

where

$$m^{(N)} = \frac{1}{N(N-1)} \sum \sum_{1 \le i_1 \ne i_2 \le N} \phi\left(F^{-1}\left(\frac{i_1}{N+1}\right)\right) F^{-1}\left(\frac{i_2}{N+1}\right)$$

and

$$\begin{aligned}\left\{s^{(N)}\right\}^2 &= \frac{1}{N(N-1)} \sum \sum_{1 \le i_1 \ne i_2 \le N} \left\{\phi\left(F^{-1}\left(\frac{i_1}{N+1}\right)\right) F^{-1}\left(\frac{i_2}{N+1}\right)\right\}^2 \\ &+ \frac{2(N-2i)}{(N-i)N(N-1)(N-2)} \sum \sum \sum_{1 \le i_1 \ne i_2 \ne i_3 \le N} \phi\left(F^{-1}\left(\frac{i_1}{N+1}\right)\right) \\ &\times \phi\left(F^{-1}\left(\frac{i_2}{N+1}\right)\right) F^{-1}\left(\frac{i_2}{N+1}\right) F^{-1}\left(\frac{i_3}{N+1}\right) \\ &+ \frac{N^2 - N(2i+3) + i^2 + 5i}{N(N-1)(N-2)(N-3)(N-i)} \sum \sum \sum \sum_{1 \le i_1 \ne i_2 \ne i_3 \ne i_4 \le N} \phi\left(F^{-1}\left(\frac{i_1}{N+1}\right)\right) \\ &\times \phi\left(F^{-1}\left(\frac{i_2}{N+1}\right)\right) F^{-1}\left(\frac{i_3}{N+1}\right) F^{-1}\left(\frac{i_4}{N+1}\right) \\ &- (N-i)\left\{m^{(N)}\right\}^2.\end{aligned}$$

Then, we have the following result.

Proposition 5.2.8. (Proposition 2.4 of Hallin and Puri (1988)) *We consider the test statistics*

$$S^{(N)*} = \frac{1}{\sqrt{N\left\{\sum_{i=1}^{N-1}(a_i+b_i)^2\right\}}} \sum_{i=1}^{N-1} \sqrt{N-i}\,(a_i+b_i)\, r_{i;f}^{(N)}.$$

Then, we have

(i) $\sqrt{N} S^{(N)*}$ *is asymptotically normal, with mean* 0 *and variance* 1 *under* $H^{(N)}(\mathbf{0}; f)$*, and with mean* $\frac{\langle (\boldsymbol{a}+\boldsymbol{b}), (\widetilde{\boldsymbol{a}}+\widetilde{\boldsymbol{b}}) \rangle \sqrt{\sigma^2 I(f)}}{\|\boldsymbol{a}+\boldsymbol{b}\|}$ *and variance* 1 *under* $H^{(N)}\left(N^{-1/2}\widetilde{\mathbf{c}}; f\right)$ *(with* $\widetilde{\boldsymbol{a}}, \widetilde{\boldsymbol{b}} \in l_2$, $\widetilde{\mathbf{c}} = \left(\widetilde{\boldsymbol{a}}', \widetilde{\boldsymbol{b}}'\right)'$*).*

(ii) $S^{(N)*}$ *provides an asymptotically most powerful test for* $H^{(N)}(\mathbf{0}; f)$ *against the general linear alternative* $H^{(N)}\left(N^{-1/2}\mathbf{c}; f\right)$ *(among all tests of given level* α*), that is,*

$$\varphi_\alpha^{(N)*}(\boldsymbol{\alpha}, \boldsymbol{\beta}, f) = 1$$

$$\text{if} \quad \sum_{i=1}^{N-1} \sqrt{N-i}\,(a_i+b_i)\, r_{i;f}^{(N)} > \sqrt{\sum_{i=1}^{N-1}(a_i+b_i)^2}\, k_{1-\alpha},$$

where $k_{1-\alpha}$ *is the* $(1-\alpha)$*-quantile of the standard normal distribution.*

5.2.2 Tangent Space

Now, we briefly review the calculation of tangent space (see Bickel et al. (1998), Chapter 3 for details).

Let μ be a fixed σ-finite measure on $(\boldsymbol{X}, \mathcal{B})$, $\boldsymbol{M}$ be all probability measures on $(\boldsymbol{X}, \mathcal{B})$ and $\boldsymbol{M}_\mu$ be that dominated by μ, that is,

$$\boldsymbol{M}_\mu = \{P \in \boldsymbol{M} \mid P << \mu\}.$$

Furthermore, let $X_1, \ldots, X_n$ be i.i.d. with common distribution $P \in \boldsymbol{P}$ where $\boldsymbol{P}$ is parametric, so that

$$\boldsymbol{P} = \left\{P_{\boldsymbol{\theta}} \mid \boldsymbol{\theta} \in \boldsymbol{\Theta} \subset \mathbb{R}^k\right\}.$$

Then, we can write

$$p(\boldsymbol{\theta}) = p(\cdot, \boldsymbol{\theta}) = \frac{dP_{\boldsymbol{\theta}}}{d\mu}(\cdot), \quad l(\boldsymbol{\theta}) = \log p(\boldsymbol{\theta}),$$

which are the density and log-likelihood of $P_{\boldsymbol{\theta}}$, respectively. It is convenient to consider $\boldsymbol{P}$ as a subset of $L_2(\mu)$ in terms of embedding $P \to s$, where

$$p \equiv \frac{dP}{d\mu}, \quad s \equiv \sqrt{p}, \quad p(\boldsymbol{\theta}) \to s(\boldsymbol{\theta}).$$

Then, we have the following definitions.

Definition 5.2.9. *We say that $\boldsymbol{\theta}_0$ is a regular point of the parametrization $\boldsymbol{\theta} \to P_{\boldsymbol{\theta}}$ if $\boldsymbol{\theta}_0$ is an interior point of $\boldsymbol{\Theta}$, and*

(i) *the map $\boldsymbol{\theta} \to s(\boldsymbol{\theta})$ from $\boldsymbol{\Theta}$ to $L_2(\mu)$ is Fréchet differentiable at $\boldsymbol{\theta}_0$, that is, there exists a vector $\dot{\boldsymbol{s}}(\boldsymbol{\theta}_0) = (\dot{s}_1(\boldsymbol{\theta}_0), \ldots, \dot{s}_k(\boldsymbol{\theta}_0))'$ of elements of $L_2(\mu)$ such that*

$$\left\| s(\boldsymbol{\theta}_0 + \boldsymbol{h}) - s(\boldsymbol{\theta}_0) - \dot{\boldsymbol{s}}(\boldsymbol{\theta}_0)' \boldsymbol{h} \right\| = o(|\boldsymbol{h}|) \quad \text{as } \boldsymbol{h} \to 0,$$

(ii) *the $k \times k$ matrix $\int \dot{\boldsymbol{s}}(\boldsymbol{\theta}_0)\, \dot{\boldsymbol{s}}(\boldsymbol{\theta}_0)'\, d\mu$ is nonsingular,*

where $|\cdot|$ is the Euclidean norm and $\|\cdot\|$ is the Hilbert norm in $L_2(\mu)$.

Definition 5.2.10. *A parametrization $\boldsymbol{\theta} \to P_{\boldsymbol{\theta}}$ is regular if*

(i) *every point of $\boldsymbol{\Theta}$ is regular,*

(ii) *the map $\boldsymbol{\theta} \to \dot{s}_i(\boldsymbol{\theta}_0)$ is continuous from $\boldsymbol{\Theta}$ to $L_2(\mu)$ for $i = 1, \ldots, k$.*

Define the Fisher information matrix of $\boldsymbol{\theta}$ by

$$\boldsymbol{I}(\boldsymbol{\theta}) = 4 \int \dot{\boldsymbol{s}}(\boldsymbol{\theta})\, \dot{\boldsymbol{s}}(\boldsymbol{\theta})'\, d\mu$$

and the score function $\dot{\boldsymbol{l}}$ of an observation by

$$\dot{\boldsymbol{l}}(\boldsymbol{\theta}) = 2 \frac{\dot{\boldsymbol{s}}(\boldsymbol{\theta})}{s(\boldsymbol{\theta})} \chi\{s(\boldsymbol{\theta}) > 0\} = \frac{\dot{\boldsymbol{p}}(\boldsymbol{\theta})}{p(\boldsymbol{\theta})} \chi\{p(\boldsymbol{\theta}) > 0\}, \tag{5.2.13}$$

where $\dot{\boldsymbol{p}}(\boldsymbol{\theta}) = 2s(\boldsymbol{\theta})\, \dot{\boldsymbol{s}}(\boldsymbol{\theta})$ and χ denotes the indicator function. If $\boldsymbol{\theta}$ is a regular point, then $\left|\dot{\boldsymbol{l}}(\boldsymbol{\theta})\right| \in L_2(P_{\boldsymbol{\theta}})$, and therefore we have another representation of the Fisher information matrix for $\boldsymbol{\theta}$ as

$$\boldsymbol{I}(\boldsymbol{\theta}) = 4 \int \dot{\boldsymbol{l}}(\boldsymbol{\theta})\, \dot{\boldsymbol{l}}(\boldsymbol{\theta})'\, dP_{\boldsymbol{\theta}}.$$

Furthermore, we can consider another local embedding of $\boldsymbol{P}$ into $L_2(P_{\boldsymbol{\theta}_0})$ by

$$P_{\boldsymbol{\theta}} \leftrightarrow r(\boldsymbol{\theta}) \equiv 2\left(\frac{s(\boldsymbol{\theta})}{s(\boldsymbol{\theta}_0)} - 1\right) \chi\{s(\boldsymbol{\theta}_0) > 0\}.$$

Suppose $\boldsymbol{\nu}$ is a Euclidean parameter defined on a regular parametric model $\boldsymbol{P} = \{P_{\boldsymbol{\theta}} \mid \boldsymbol{\theta} \in \boldsymbol{\Theta}\}$. We can identify $\boldsymbol{\nu}$ with the parametric function $q : \boldsymbol{\Theta} \to \mathbb{R}^m$ defined by

$$q(\boldsymbol{\theta}) = \nu(P_{\boldsymbol{\theta}}).$$

Fix $P = P_{\boldsymbol{\theta}}$ and suppose q has a total differential matrix $\dot{q}_{m \times k}$ at $\boldsymbol{\theta}$. Then, we can define the information bound and the efficient influence function for $\boldsymbol{\nu}$ in terms of

$$\boldsymbol{I}^{-1}(P \mid \boldsymbol{\nu}, \boldsymbol{P}) = \dot{q}(\boldsymbol{\theta})\, \boldsymbol{I}(\boldsymbol{\theta})\, \dot{q}(\boldsymbol{\theta})' \quad \text{and} \quad \widetilde{\boldsymbol{l}}(\cdot, P \mid \boldsymbol{\nu}, \boldsymbol{P}) = \dot{q}(\boldsymbol{\theta})\, \boldsymbol{I}(\boldsymbol{\theta})\, \dot{\boldsymbol{l}}(\boldsymbol{\theta}),$$

respectively.

Now, we can define the tangent space in Hilbert space, which plays an important role in semi-parametric inference. Let $\boldsymbol{H}$ be a Hilbert space with norm $\|\cdot\|$ and inner product $< \cdot, \cdot >$, and $\boldsymbol{V}$ be a subset of $\boldsymbol{H}$. We use the notation

$$\boldsymbol{a}_n = \boldsymbol{b}_n + o(\varepsilon_n)$$

for $\boldsymbol{a}_n, \boldsymbol{b}_n \in \boldsymbol{H}$, which implies

$$\|\boldsymbol{a}_n - \boldsymbol{b}_n\| = o(\varepsilon_n).$$

We call $\boldsymbol{V}$ a k-dimensional surface in $\boldsymbol{H}$ if it can be represented as the image of the open unit sphere in $\mathbb{R}^k$ under a continuously Fréchet differentiable map which is of rank k. Namely, we can write

$$\boldsymbol{V} = \{v(\boldsymbol{\eta}) \mid |\boldsymbol{\eta}| < 1\}$$

with

(i) $v(\boldsymbol{\eta} + \boldsymbol{\Delta}) = v(\boldsymbol{\eta}) + \boldsymbol{\Delta}'\dot{v}(\boldsymbol{\eta}) + o(|\boldsymbol{\Delta}|)$,

(ii) $\dot{\boldsymbol{v}} = (\dot{v}_1, \ldots, \dot{v}_k)' \in \boldsymbol{H}^k$,

(iii) $\dim[\dot{\boldsymbol{v}}] = k$,

where $[\boldsymbol{W}]$ denotes the closed linear span of $\boldsymbol{W}$.

Any surface may have many representations (parametrizations). However, if $\boldsymbol{\eta} \to v(\boldsymbol{\eta})$ and $\boldsymbol{\gamma} \to g(\boldsymbol{\gamma})$ are two representations of $\boldsymbol{V}$ with $g(\boldsymbol{\gamma}_0) = v(\boldsymbol{\eta}_0)$ and $v^{-1}\{v(\boldsymbol{\eta}_0)\} = \boldsymbol{\eta}_0$, then the application of the chain rule and inverse function theorem leads to $\dot{\boldsymbol{v}}(\boldsymbol{\eta}_0) = \boldsymbol{M}\dot{\boldsymbol{g}}(\boldsymbol{\gamma}_0)$, where $\boldsymbol{M}$ is a $k \times k$ nonsingular matrix. Therefore, $[\dot{\boldsymbol{v}}(\boldsymbol{\eta}_0)] = [\dot{\boldsymbol{g}}(\boldsymbol{\gamma}_0)]$ is independent of the parametrization. We call it the tangent space of $\boldsymbol{V}$ at $v_0 = v(\boldsymbol{\eta}_0)$, and write it $\dot{\boldsymbol{V}}$ (or $\dot{\boldsymbol{V}}(v_0)$).

Now, we first study $\boldsymbol{P}$ as a subset $\boldsymbol{S}$ of Hilbert space $L_2(\mu)$ via the correspondence $P \leftrightarrow s$. Next, we will consider $\boldsymbol{P}$ as a subset of $L_2(P_0)$ in terms of the correspondence $P \leftrightarrow r$. If we take $\boldsymbol{H} = L_2(\mu)$ and $\boldsymbol{V} = \boldsymbol{S}$, we have the following. We call $\boldsymbol{P}$ a regular parametric model if it has a regular parametrization in the sense of Definition 5.2.10.

Proposition 5.2.11. (Proposition 3.2.1 of Bickel et al. (1998))

(i) $\boldsymbol{P} = \left\{P_{\boldsymbol{\theta}} \mid \boldsymbol{\theta} \in \boldsymbol{\Theta} \subset \mathbb{R}^k\right\}$ *is a regular parametric model and* $\boldsymbol{\Theta}$ *is a surface in* $\mathbb{R}^k$ *if and only if* $\boldsymbol{S}$ *is a k-dimensional surface in* $L_2(\mu)$.

(ii) *Then, we can represent the projection operator onto* $\dot{\boldsymbol{S}}$ *by*

$$\boldsymbol{\Pi}\left(h \mid \dot{\boldsymbol{S}}\right) = 4\langle h, \dot{\boldsymbol{s}}\rangle' \boldsymbol{I}^{-1}\dot{\boldsymbol{s}},$$

where $\dot{\boldsymbol{s}} = (\dot{s}_1, \ldots, \dot{s}_k)'$ *and* $\langle h, \dot{\boldsymbol{s}}\rangle = (\langle h, \dot{s}_1\rangle, \ldots, \langle h, \dot{s}_k\rangle)'$.

Next, we consider the image of $\boldsymbol{S}$ under the mapping

$$s \to r = 2\left(\frac{s}{s_0} - 1\right)\chi\{s_0 > 0\}$$

which maps s_0 into zero. This image is a subset of $L_2(P_0)$. However, we call this set $\boldsymbol{P}$ following a familiar abuse of notation, since we do not lose anything by identifying this image with $\boldsymbol{P}$. By (5.2.13), we have

$$\dot{\boldsymbol{P}} = \left[\frac{2\dot{s}}{s_0}\right] = \left[\dot{l}_1, \ldots, \dot{l}_k\right]$$

and

$$\boldsymbol{\Pi}_0\left(h \mid \dot{\boldsymbol{P}}\right) = \left\langle h, \dot{\boldsymbol{l}}\right\rangle_0' \boldsymbol{I}^{-1}\dot{\boldsymbol{l}},$$

where $\dot{I} = \left(\dot{I}_1, \ldots, \dot{I}_k\right)'$ and

$$\left\langle h, \dot{I}\right\rangle_0 = \left(\left\langle h, \dot{I}_1\right\rangle_0, \ldots, \left\langle h, \dot{I}_k\right\rangle_0\right)' = \left(\int h\dot{I}_1 dP_0, \ldots, \int h\dot{I}_k dP_0\right)'.$$

Note that for any $\dot{S}$ and corresponding $\dot{P}$, these projection operators are related by the identities

$$\Pi_0\left(h \mid \dot{P}\right) = s_0^{-1}\Pi\left(hs_0 \mid \dot{S}\right) \quad \text{for} \quad h \in L_2(P_0)$$

$$\Pi\left(t \mid \dot{S}\right) = s_0\Pi_0\left(\frac{t}{s_0} \mid \dot{P}\right) \quad \text{for} \quad t \in L_2(\mu).$$

For the general nonparametric case, we have the following definition.

Definition 5.2.12. *If $v_0 \in V \subset H$, then let the tangent set at v_0 be the union of all the (1-dimensional) tangent spaces of surfaces (curves) $C \subset V$ passing through v_0, and denote it by $\dot{V}^0$. Furthermore, we call the closed linear span $\left[\dot{V}^0\right]$ of the tangent set $\dot{V}^0$ the tangent space of V, and denote it by $\dot{V}$.*

Suppose that P is a regular parametric, semiparametric or nonparametric model, and $\nu : P \to \mathbb{R}^m$ is a Euclidean parameter. Let ν also denote the corresponding map from S to $\mathbb{R}^m$ given by $\nu(s(P)) = \nu(P)$. Now, we give the following definition.

Definition 5.2.13. *For $m = 1$, we say that ν is pathwise differentiable on S at s_0 if there exists a bounded linear functional $\dot{\nu}(s_0) : \dot{S} \to \mathbb{R}$ such that*

$$\nu(s(\eta)) = \nu(s_0) + \eta\dot{\nu}(s_0)(t) + o(|\eta|) \tag{5.2.14}$$

for any curve $s(\cdot)$ in S which passes through $s_0 = s(0)$ and $\dot{s}(0) = t$.

Defining the bounded linear function $\dot{\nu}(P_0) : \dot{P} \to \mathbb{R}$ by

$$\dot{\nu}(P_0)(h) = \dot{\nu}(s_0)\left(\frac{1}{2}hs_0\right),$$

(5.2.14) holds if and only if

$$\nu\left(P_\eta\right) = \nu(P_0) + \eta\dot{\nu}(P_0)(h) + o(|\eta|), \tag{5.2.15}$$

where $\left\{P_\eta\right\}$ is the curve in P corresponding to $\{s(\eta)\}$ in S and $h = 2t/s_0$. We call (5.2.15) pathwise differentiability of ν on P at P_0. To simplify, we write $\dot{\nu}(h)$ for $\dot{\nu}(P_0)(h)$ and $\dot{\nu}(t)$ for $\dot{\nu}(s_0)(t)$, ignoring the dependence of $\dot{\nu}(P_0)$ on P_0. The functional $\dot{\nu}$ is uniquely determined by (5.2.15) on $\dot{P}^0$ and hence on $\dot{P}$. Naturally, we often describe the parameter ν as the restriction to P of a parameter ν_e defined on a larger model $M \supset P$. That is, if ν_e is pathwise differentiable on M with derivative $\dot{\nu}_e : \dot{M} \to \mathbb{R}$, then we have

$$\dot{\nu}_e = \dot{\nu} \quad \text{on} \quad \dot{P}.$$

Now, we have the following results (see Section 3.3 of Bickel et al. (1998) for the proofs).

Theorem 5.2.14. (Theorem 3.3.1 of Bickel et al. (1998)) *For $m = 1$, let ν be pathwise differentiable on P at P_0. For any regular parametric submodel Q, where $\mathcal{I}^{-1}(P_0 \mid \nu, Q)$ is defined, the efficient influence function $\widetilde{I}(\cdot, P_0 \mid \nu, Q)$ satisfies*

$$\widetilde{I}(\cdot, P_0 \mid \nu, Q) = \Pi_0\left(\dot{\nu}(P_0) \mid \dot{Q}\right) \tag{5.2.16}$$

Therefore,

$$\mathcal{I}^{-1}(P_0 \mid \nu, Q) = \left\|\Pi_0\left(\dot{\nu}(P_0) \mid \dot{Q}\right)\right\|_0^2 \leq \|\dot{\nu}(P_0)\|_0^2 \tag{5.2.17}$$

with equality if and only if

$$\dot{\nu}(P_0) \in \dot{Q}. \tag{5.2.18}$$

Next, we consider the case that $\boldsymbol{\nu} = (\nu_1, \ldots, \nu_m)'$ is m-dimensional. If each ν_j is pathwise differentiable with derivative $\dot{\nu}_j = \dot{\nu}_j(P_0)$, then we call $\boldsymbol{\nu}$ pathwise differentiable on $\boldsymbol{P}$ at P_0 with derivative $\dot{\boldsymbol{\nu}} = (\dot{\nu}_1, \ldots, \dot{\nu}_m)'$. Now, we extend the above results to $m \geq 1$. Let

$$\widetilde{I}_j = \dot{\nu}_j(P_0), \; j = 1, \ldots, m \quad \text{and} \quad \widetilde{\boldsymbol{I}} = \left(\widetilde{I}_1, \ldots, \widetilde{I}_m\right)'. \tag{5.2.19}$$

Corollary 5.2.15. (Corollary 3.3.1 of Bickel et al. (1998)) *Suppose that*
(i) *$\boldsymbol{\nu}$ is pathwise differentiable on $\boldsymbol{P}$ at P_0,*
(ii) *$\boldsymbol{Q}$ is any regular parametric submodel where $\boldsymbol{\mathcal{I}}^{-1}(P_0 \mid \boldsymbol{\nu}, \boldsymbol{Q})$ is defined.*
Then, the efficient influence function $\widetilde{\boldsymbol{I}}(\cdot, P_0 \mid \boldsymbol{\nu}, \boldsymbol{Q})$ for $\boldsymbol{\nu}$ in $\boldsymbol{Q}$ satisfies (5.2.16), and therefore,

$$\begin{aligned}\boldsymbol{\mathcal{I}}^{-1}(P_0 \mid \nu, \boldsymbol{Q}) &= \langle \Pi_0\left(\dot{\boldsymbol{\nu}}(P_0) \mid \dot{\boldsymbol{Q}}\right), \Pi_0\left(\dot{\boldsymbol{\nu}}'(P_0) \mid \dot{\boldsymbol{Q}}\right)\rangle_0 \\ &\leq \langle \widetilde{\boldsymbol{I}}, \widetilde{\boldsymbol{I}}'\rangle_0,\end{aligned}$$

where the order $\boldsymbol{A} \leq \boldsymbol{B}$ implies that $\boldsymbol{B} - \boldsymbol{A}$ is nonnegative definite. The equality holds if and only if

$$\widetilde{I}_j \in \dot{\boldsymbol{Q}}, \; j = 1, \ldots, m,$$

and in this case, the efficient influence function $\widetilde{\boldsymbol{I}}(\cdot, P_0 \mid \boldsymbol{\nu}, \boldsymbol{Q})$ equals $\widetilde{\boldsymbol{I}}$.

In addition, if
(iii) $\left[\widetilde{I}_j \;:\; j = 1, \ldots, m\right] \subset \dot{\boldsymbol{P}}^0$,
then for each $\boldsymbol{a} \in \mathbb{R}^m$ there exists a one-dimensional regular parametric submodel $\boldsymbol{Q}$ such that

$$\boldsymbol{a}'\boldsymbol{\mathcal{I}}^{-1}(P_0 \mid \boldsymbol{\nu}, \boldsymbol{Q})\,\boldsymbol{a} = \boldsymbol{a}'E\left(\widetilde{\boldsymbol{I}}\widetilde{\boldsymbol{I}}'\right)\boldsymbol{a}.$$

If $\boldsymbol{Q}$ is a regular parametric submodel of $\boldsymbol{P}$ satisfying (5.2.18), that is, it gives the equality in (5.2.17), we call $\boldsymbol{Q}$ least favourable. For $m = 1$, from Theorem 5.2.14, under the condition $\dot{\nu}(P_0) \in \dot{\boldsymbol{P}}$, there exists a least favourable $\boldsymbol{Q}$ which may be chosen to be one dimensional. We extend this idea for the general m-dimensional case in which no least favourable submodel $\boldsymbol{Q}$ exists; instead, a least favourable sequence $\boldsymbol{Q}_j$ does.

Suppose that $\boldsymbol{\nu}$ is pathwise differentiable as in Theorem 5.2.14 and Corollary 5.2.15, and
(iii)′

$$\left[\dot{\nu}_j(P_0) \;:\; j = 1, \ldots, m\right] \subset \overline{\dot{\boldsymbol{P}}^0}, \tag{5.2.20}$$

where $\overline{W}$ denotes the closure of W.

Under these conditions, we have the following definition.

Definition 5.2.16. **(i)** *We call $\widetilde{\boldsymbol{I}} = \widetilde{\boldsymbol{I}}(\cdot, P_0 \mid \boldsymbol{\nu}, \boldsymbol{P})$ defined by (5.2.19) the efficient influence function for $\boldsymbol{\nu}$, that is,*

$$\widetilde{\boldsymbol{I}} = \widetilde{\boldsymbol{I}}(\cdot, P_0 \mid \boldsymbol{\nu}, \boldsymbol{P}) = \dot{\boldsymbol{\nu}}(P_0).$$

(ii) *The information matrix $\boldsymbol{\mathcal{I}}(P_0 \mid \boldsymbol{\nu}, \boldsymbol{P})$ for $\boldsymbol{\nu}$ in $\boldsymbol{P}$ is defined as the inverse of*

$$\boldsymbol{\mathcal{I}}^{-1}(P_0 \mid \boldsymbol{\nu}, \boldsymbol{P}) = \langle \widetilde{\boldsymbol{I}}, \widetilde{\boldsymbol{I}}'\rangle_0 = E_0\left(\widetilde{\boldsymbol{I}}\widetilde{\boldsymbol{I}}'\right).$$

The parameters are often defined implicity through the parametrization in semiparametric models. We consider the information bounds and efficient influence functions for such parameters. This approach is equivalent to the approach for parameters which are explicitly defined as functions of $\boldsymbol{P}$. Each of them has its own advantages depending on the example considered. Here, it is convenient to present the semiparametric model in terms of $\boldsymbol{P}$ rather than $\boldsymbol{S}$. Let

$$\boldsymbol{P} = \left\{P_{(\boldsymbol{\nu}, \boldsymbol{g})} \mid \boldsymbol{\nu} \in \boldsymbol{N}, \boldsymbol{g} \in \boldsymbol{G}\right\},$$

where $\boldsymbol{N}$ is an open subset of $\mathbb{R}^m$ and $\boldsymbol{G}$ is general. Fix $P_0 = P_{(\boldsymbol{\nu}_0, \boldsymbol{g}_0)}$ and let $\dot{\boldsymbol{l}}_1$ denote the vectors of partial derivatives of $\log p(\cdot, \boldsymbol{\nu}, \boldsymbol{g})$ with respect to $\boldsymbol{\nu}$ at P_0, which are just the score functions for the parametric model

$$\boldsymbol{P}_1 = \left\{P_{(\boldsymbol{\nu}, \boldsymbol{g}_0)} \mid \boldsymbol{\nu} \in \boldsymbol{N}\right\}.$$

Also, let

$$\boldsymbol{P}_2 = \left\{P_{(\boldsymbol{\nu}_0, \boldsymbol{g})} \mid \boldsymbol{g} \in \boldsymbol{G}\right\}$$

Then we have the following results (see Section 3.4 of Bickel et al. (1998) for the proofs).

Theorem 5.2.17. (Theorem 3.4.1 of Bickel et al. (1998)) *Suppose that*

(i) $\boldsymbol{P}_1$ *is regular,*

(ii) ν *is a* 1*-dimensional parameter,*

and let

$$\dot{l}_1^* = \dot{l}_1 - \Pi_0\left(\dot{l}_1 \mid \dot{\boldsymbol{P}}_2\right).$$

Then, we have the following.

(A) *If* $\boldsymbol{Q} = \left\{P_{(\nu, g_\gamma)} \mid \nu \in N, \gamma \in \boldsymbol{\Gamma}\right\}$ *is a regular parametric submodel of* $\boldsymbol{P}$ *with* $P_0 \in \boldsymbol{Q}$, *then*

$$\mathcal{I}(P_0 \mid \nu, \boldsymbol{Q}) \geq \left\|\dot{l}_1^*\right\|_0^2 = E_0\left(\dot{l}_1^{*2}\right)$$

with equality if and only if $\dot{l}_1^* \in \dot{\boldsymbol{Q}}$.

(B) *If* $\dot{\boldsymbol{P}} = \dot{\boldsymbol{P}}_1 + \dot{\boldsymbol{P}}_2$, $\dot{l}_1^* \neq 0$, *and* ν *is pathwise differentiable on* $\boldsymbol{P}$ *at* P_0, *then the efficient influence function for* ν *is given by*

$$\widetilde{l}_1 = \left\|\dot{l}_1^*\right\|_0^{-2} \dot{l}_1^* \equiv \mathcal{I}^{-1}(P_0 \mid \nu, \boldsymbol{P})\, \dot{l}_1^*.$$

From these results, we have the following definition.

Definition 5.2.18. *The efficient score function for* ν *in* $\boldsymbol{P}$ *is given by* $\dot{l}_1^*$ *and written as* $\dot{l}^*(\cdot, P_0 \mid \nu, \boldsymbol{P})$.

Now, we consider the case that $\boldsymbol{\nu}$ is m-dimensional with $m > 1$. The efficient score function vector is given by

$$\dot{\boldsymbol{l}}_1^* = \left(\dot{l}_{11}^*, \ldots, \dot{l}_{1m}^*\right)' = \dot{\boldsymbol{l}}_1 - \Pi_0\left(\dot{\boldsymbol{l}}_1 \mid \dot{\boldsymbol{P}}_2\right).$$

Then, we have the following results.

Corollary 5.2.19 (Corollary 3.4.1 of Bickel et al. (1998)). *Suppose* $\boldsymbol{P}_1$ *is regular.*

(A) *If* $\boldsymbol{Q}$ *is a regular parametric submodel as in Theorem 5.2.17 where* $\boldsymbol{\mathcal{I}}^{-1}(P_0 \mid \boldsymbol{\nu}, \boldsymbol{Q})$ *is defined, then we have*

$$\boldsymbol{\mathcal{I}}(P_0 \mid \boldsymbol{\nu}, \boldsymbol{Q}) \geq E_0\left(\dot{\boldsymbol{l}}_1^* \dot{\boldsymbol{l}}_1^{*\prime}\right)$$

with equality if and only if $\left[\dot{\boldsymbol{l}}_1^*\right]$ *is contained in* $\dot{\boldsymbol{Q}}$.

(B) *If* $\dot{\boldsymbol{P}} = \dot{\boldsymbol{P}}_1 + \dot{\boldsymbol{P}}_2$, $\boldsymbol{\nu} : \boldsymbol{P} \to \mathbb{R}^m$ *is pathwise differentiable on* $\boldsymbol{P}$ *at* P_0 *and (5.2.20) holds, then the efficient influence function for* $\boldsymbol{\nu}$ *is given by*

$$\widetilde{\boldsymbol{l}}_1 = \boldsymbol{\mathcal{I}}^{-1}(P_0 \mid \boldsymbol{\nu}, \boldsymbol{P})\, \dot{\boldsymbol{l}}_1^*$$

and

$$\boldsymbol{\mathcal{I}}(P_0 \mid \boldsymbol{\nu}, \boldsymbol{P}) = E_0\left(\dot{\boldsymbol{l}}_1^* \dot{\boldsymbol{l}}_1^{*\prime}\right).$$

5.2.3 Introduction to Semiparametric Asymptotic Optimal Theory

The assumption that the innovation densities underlying these models are known seems quite unrealistic. If these densities remain unspecified, the model becomes a semiparametric one, and rank-based inference methods naturally come into the picture. Rank-based inference methods under very general conditions are known to achieve semiparametric efficiency bounds.

Here, we briefly review semiparametrically efficient estimation theory. Consider the sequence of "semiparametric models"

$$\mathcal{E}^{(N)} := \left\{\mathcal{X}^{(N)}, \mathcal{A}^{(N)}, \mathcal{P}^{(N)} = \left\{\mathrm{P}^{(N)}_{\eta,g} : \eta \in \mathbf{\Theta}, g \in \mathcal{G}\right\}\right\} \quad N \in \mathbb{N},$$

where η denotes a finite-dimensional parameter with $\mathbf{\Theta}$ an open subset of $\mathbb{R}^k$, g an infinite-dimensional nuisance and $\mathcal{G}$ an arbitrary set. This sequence of models is studied in the neighbourhood of arbitrary parameter values $(\eta_0, g_0) \in \mathbf{\Theta} \times \mathcal{G}$. Denote expectations under $\mathrm{P}^{(N)}_{\eta,g}$ by $E^{(N)}_{\eta,g}$. Associated with $\mathcal{E}^{(N)}$, consider the fixed-g sequences of "parametric submodels"

$$\mathcal{E}^{(N)}_g := \left\{\mathcal{X}^{(N)}, \mathcal{A}^{(N)}, \mathcal{P}^{(N)}_g = \left\{\mathrm{P}^{(N)}_{\eta,g} : \eta \in \mathbf{\Theta}\right\}\right\} \quad N \in \mathbb{N}.$$

Throughout this section we assume that the sequence of parametric models $\left\{\mathcal{E}^{(N)}_{g_0}\right\}$ is locally asymptotic normal (LAN) at η_0, with central sequence $\mathbf{\Delta}^{(N)}(\eta_0, g_0)$ and Fisher information $\boldsymbol{I}(\eta_0, g_0)$. More precisely, we have the following assumption.

Assumption 5.2.20 (Hallin and Werker (2003), Assumption A). *For any bounded sequence $\{\boldsymbol{\tau}_N\}$ in $\mathbb{R}^k$, we have*

$$\frac{d\mathrm{P}^{(N)}_{\eta_0 + \frac{\tau_N}{\sqrt{N}}, g_0}}{d\mathrm{P}^{(N)}_{\eta_0, g_0}} = \exp\left\{\boldsymbol{\tau}'_N \mathbf{\Delta}^{(N)}(\eta_0, g_0) + \frac{1}{2}\boldsymbol{\tau}'_N \boldsymbol{I}(\eta_0, g_0)\boldsymbol{\tau}_N + r_N\right\},$$

where, under $\mathrm{P}^{(N)}_{\eta_0,g_0}$, $\mathbf{\Delta}^{(N)}(\eta_0, g_0) \xrightarrow{\mathcal{L}} N(\mathbf{0}, \boldsymbol{I}(\eta_0, g_0))$ and $r_N \xrightarrow{\mathrm{P}} 0$, as $N \to \infty$.

More generally, we consider the sequences of "parametric experiments"

$$\mathcal{E}^{(N)}_q(g_0) := \left\{\mathcal{X}^{(N)}, \mathcal{A}^{(N)}, \mathcal{P}^{(N)}_q = \left\{\mathrm{P}^{(N)}_{\eta, q(\boldsymbol{\lambda})} : \eta \in \mathbf{\Theta}, \boldsymbol{\lambda} \in (-1,1)^k\right\}\right\} \quad N \in \mathbb{N}, \tag{5.2.21}$$

with maps of the form $q : (-1,1)^k \to \mathcal{G}$. Let $\mathcal{Q}$ denote the set of all maps q such that

(i) $q(\mathbf{0}) = g_0$, and

(ii) the sequence $\mathcal{E}^{(N)}_q(g_0)$ is locally asymptotic normal (LAN) with respect to $(\eta', \boldsymbol{\lambda}')'$ at $\left(\eta'_0, \mathbf{0}'\right)'$.

The central sequence and Fisher information matrix associated with the LAN (at $\left(\eta'_0, \mathbf{0}'\right)'$) property (ii) naturally decompose into η- and $\boldsymbol{\lambda}$-parts, namely, into

$$\left(\mathbf{\Delta}^{(N)}(\eta_0, g_0)', \boldsymbol{H}^{(N)}_q(\eta_0, g_0)'\right)' \tag{5.2.22}$$

and

$$\boldsymbol{I}_q(\eta_0, g_0) := \begin{pmatrix} \boldsymbol{I}(\eta_0, g_0) & \boldsymbol{C}_q(\eta_0, g_0)' \\ \boldsymbol{C}_q(\eta_0, g_0) & \boldsymbol{I}_{\boldsymbol{H}_q}(\eta_0, g_0) \end{pmatrix}.$$

Then, the "influence function" for parameter η is defined as the η-component of

$$\boldsymbol{I}_q(\eta_0, g_0)^{-1}\left(\mathbf{\Delta}^{(N)}(\eta_0, g_0)', \boldsymbol{H}^{(N)}_q(\eta_0, g_0)'\right)'.$$

After elementary algebra, that influence function for η takes the form

$$\left\{\boldsymbol{I}(\eta_0, g_0) - \boldsymbol{C}_q(\eta_0, g_0)'\boldsymbol{I}^{-1}_{\boldsymbol{H}_q}(\eta_0, g_0)\boldsymbol{C}_q(\eta_0, g_0)\right\}^{-1}\mathbf{\Delta}^{(N)}_q(\eta_0, g_0), \tag{5.2.23}$$

with

$$\Delta_q^{(N)}(\eta_0, g_0) := \Delta^{(N)}(\eta_0, g_0) - C_q(\eta_0, g_0)' I_{H_q}^{-1}(\eta_0, g_0) H_q^{(N)}(\eta_0, g_0).$$

"Asymptotically efficient inference" on η (in the parametric submodel $\mathcal{E}_q^{(N)}(g_0)$, at $(\eta_0', \mathbf{0}')'$) should be based on this influence function. Note that $\Delta_q^{(N)}(\eta_0, g_0)$ corresponds to the residual of the regression, with respect to the λ-part $H_q^{(N)}$, of the η-part $\Delta^{(N)}$ of the central sequence (5.2.22), in the covariance $I_q(\eta_0, g_0)$.

The "residual Fisher information" (the asymptotic covariance matrix of $\Delta_q^{(N)}(\eta_0, g_0)$) is given by

$$I(\eta_0, g_0) - C_q(\eta_0, g_0)' I_{H_q}^{-1}(\eta_0, g_0) C_q(\eta_0, g_0);$$

hence the "loss of η-information" (due to the nonspecification of λ in $\mathcal{E}_q^{(N)}(g_0)$) is

$$C_q(\eta_0, g_0)' I_{H_q}^{-1}(\eta_0, g_0) C_q(\eta_0, g_0).$$

Does there exist a $q_{lf} \in \mathcal{Q}$ such that the corresponding $\Delta_{q_{lf}}^{(N)}$ is asymptotically orthogonal to any $H_q(\eta_0, g_0)$, $q \in \mathcal{Q}$? If such a q_{lf} exists, it can be considered "least favourable," since the corresponding loss of information is maximal. We then say that $H_{q_{lf}}$ is a "least favourable direction," and $\Delta_{q_{lf}}^{(N)}(\eta_0, g_0)$ a "semiparametrically efficient (at (η_0, g_0)) central sequence" for the full experiment $\mathcal{E}^{(N)}$.

Consider fixed η subexperiments, denoted by

$$\mathcal{E}_{\eta_0}^{(N)} := \left\{\mathcal{X}^{(N)}, \mathcal{A}^{(N)}, \mathcal{P}_{\eta_0}^{(N)} = \left\{\mathrm{P}_{\eta_0,g}^{(N)} : g \in \mathcal{G}\right\}\right\} \; N \in \mathbb{N},$$

corresponding to the sequence of nonparametric experiments of a fixed value $\eta_0 \in \Theta$. Here we focus on the situation where these experiments $\left\{\mathcal{E}_{\eta_0}^{(N)}, N \in \mathbb{N}\right\}$ for some distribution freeness or invariance structure. More precisely, assume that there exists, for all η_0, a sequence of σ-fields $\left\{B^{(N)}(\eta_0), N \in \mathbb{N}\right\}$ with $B^{(N)}(\eta_0) \subset \mathcal{A}^{(N)}$, such that the restriction of $\mathrm{P}_{\eta_0,g}^{(N)}$ to $B^{(N)}(\eta_0)$ does not depend on $g \in \mathcal{G}$. We denote this restriction by $\mathrm{P}_{\eta_0,g|B^{(N)}(\eta_0)}^{(N)}$. This distribution freeness of $B^{(N)}(\eta_0)$ generally follows from some invariance property, under which $B^{(N)}(\eta_0)$ is generated by the orbits of some group acting on $\left\{\mathcal{X}^{(N)}, \mathcal{A}^{(N)}\right\}$. Hallin and Werker (2003) have shown how this invariance property leads to the fundamental and quite appealing consequence that semiparametric efficiency bounds in this context can be achieved by distribution-free, hence rank-based, procedures.

Proposition 5.2.21 (Hallin and Werker (2003), Proposition 2.1). *Fix $q \in \mathcal{Q}$, and consider the model $\mathcal{E}_q^{(N)}(g_0)$ in (5.2.21). Let $B^{(N)}(\eta_0) \subset \mathcal{A}^{(N)}$ be a sequence of σ-fields such that $\mathrm{P}_{\eta_0,q(\lambda)|B^{(N)}(\eta_0)}^{(N)}$ does not depend on λ. If the λ-part $H_q^{(N)}(\eta_0, g_0)$ of the central sequence is uniformly integrable under $\mathrm{P}_{(\eta_0,g_0)}^{(N)}$, we have*

$$E_{\eta_0,g_0}^{(N)}\left\{H_q^{(N)}(\eta_0, g_0) \mid B^{(N)}(\eta_0)\right\} = o_{L_1}(1).$$

Note that this proposition states that, if $\mathrm{P}_{\eta_0,g|B^{(N)}(\eta_0)}^{(N)}$ does not depend on $g \in \mathcal{G}$, then under $\mathrm{P}_{\eta_0,g_0}^{(N)}$, the conditional (upon $B^{(N)}(\eta_0)$) expectation of the λ-part $H_q^{(N)}(\eta_0, g_0)$ of the central sequence associated with experiment (5.2.21) converges to zero in L_1 norm, as $N \to \infty$. Moreover, a uniformly integrable version of $H_q^{(n)}(\eta_0, g_0)$ always exists.

Lemma 5.2.22 (Hallin and Werker (2003), Lemma 2.2). *Denote by $\Delta_\eta^{(N)}$ an arbitrary central sequence in some locally asymptotically normal sequence of experiments $\left\{\mathcal{E}^{(N)}\right\}$, with parameter η and probability distributions $\mathrm{P}_\eta^{(N)}$. Then, there exists a sequence $\overline{\Delta}_\eta^{(N)} = \left(\overline{\Delta}_{\eta,1}^{(N)}, \ldots, \overline{\Delta}_{\eta,k}^{(N)}\right)'$, $N \in \mathbb{N}$ such that*

(i) *for any $p \in (-1, \infty)$, the sequences $\left|\overline{\Delta}^{(N)}_{\eta,i}\right|^p$, $n \in \mathbb{N}$, $i = 1, \ldots, k$, are uniformly integrable;*

(ii) $\boldsymbol{\Delta}^{(N)}_{\eta} - \overline{\boldsymbol{\Delta}}^{(N)}_{\eta} = o_{\mathrm{P}}(1)$ *under* $\mathrm{P}^{(N)}_{\eta}$ *as* $N \to \infty$; *therefore, for any* $p \in (-1, \infty)$, $\overline{\boldsymbol{\Delta}}^{(N)}_{\eta}$ *constitutes a uniformly pth-order integrable version of the central sequence.*

From Proposition 5.2.21, we see that for all $q \in \mathcal{Q}$, the $\boldsymbol{\lambda}$-part $\boldsymbol{H}^{(N)}_q(\eta_0, g_0)$ of the central sequence is asymptotically orthogonal to the entire space of $\boldsymbol{B}^{(N)}(\eta_0)$-measurable variables. Therefore,

$$\boldsymbol{H}^{(N)}_{lf}(\eta_0, g_0) := \boldsymbol{\Delta}^{(N)}(\eta_0, g_0) - E^{(N)}_{\eta_0, g_0}\left\{\boldsymbol{\Delta}^{(N)}(\eta_0, g_0) \mid \boldsymbol{B}^{(N)}(\eta_0)\right\}$$

is an obvious candidate as the least favourable direction, and, automatically,

$$\boldsymbol{\Delta}^{(N)}_{lf}(\eta_0, g_0) := E^{(N)}_{\eta_0, g_0}\left\{\boldsymbol{\Delta}^{(N)}(\eta_0, g_0) \mid \boldsymbol{B}^{(N)}(\eta_0)\right\} \tag{5.2.24}$$

is a candidate as a semiparametrically efficient (at η_0, g_0) central sequence. Note that the efficient score is obtained by taking the residual of the projection of the parametric score on the tangent space generated by g_0. The following condition is thus sufficient for $\boldsymbol{H}^{(N)}_{lf}(\eta_0, g_0)$ to be least favourable.

Condition 5.2.23 (LF1). $\boldsymbol{H}^{(N)}_{lf}(\eta_0, g_0)$ *can be obtained as the $\boldsymbol{\lambda}$-part of the central sequence* $\left(\boldsymbol{\Delta}^{(N)}(\eta_0, g_0)', \boldsymbol{H}^{(N)}_{lf}(\eta_0, g_0)'\right)'$ *of some parametric submodel* $\mathcal{E}^{(N)}_{lf}(g_0)$.

More precisely, we have the following result.

Proposition 5.2.24 (Hallin and Werker (2003), Proposition 2.4). *Assume that condition (LF1) holds. Then $\boldsymbol{H}^{(N)}_{lf}(\eta_0, g_0)$ is a least favourable direction, and $\boldsymbol{\Delta}^{(N)}_{lf}(\eta_0, g_0)$ a semiparametrically efficient (at g_0) central sequence.*

Furthermore, assuming that condition (LF1) holds, the information matrix in $\mathcal{E}^{(N)}_{lf}(g_0)$ has the form

$$\tilde{\boldsymbol{I}}(\eta_0, g_0) = \begin{pmatrix} \boldsymbol{I}(\eta_0, g_0) & \boldsymbol{I}(\eta_0, g_0) - \boldsymbol{I}(\eta_0, g_0)^* \\ \boldsymbol{I}(\eta_0, g_0) - \boldsymbol{I}(\eta_0, g_0)^* & \boldsymbol{I}(\eta_0, g_0) - \boldsymbol{I}(\eta_0, g_0)^* \end{pmatrix}, \tag{5.2.25}$$

where $\boldsymbol{I}^*(\eta, g)$ is the variance of the limiting distribution of $E^{(N)}_{\eta, g}\left\{\boldsymbol{\Delta}^{(N)}(\eta, g) \mid \boldsymbol{B}^{(N)}(\eta)\right\}$. Therefore, up to $o_{\mathrm{P}}(1)$ terms, the efficient influence function is the $\boldsymbol{\lambda}$-part of

$$\begin{pmatrix} \boldsymbol{I}(\eta, g) & \boldsymbol{I}(\eta, g) - \boldsymbol{I}(\eta, g)^* \\ \boldsymbol{I}(\eta, g) - \boldsymbol{I}(\eta, g)^* & \boldsymbol{I}(\eta, g) - \boldsymbol{I}(\eta, g)^* \end{pmatrix}^{-1} \begin{pmatrix} \boldsymbol{\Delta}^{(N)}(\eta, g) \\ \boldsymbol{H}^{(N)}_{lf}(\eta, g) \end{pmatrix}.$$

Still under (LF1), this admits the asymptotic representation

$$\boldsymbol{I}(\eta, g)^{*-1} E^{(N)}_{\eta, g}\left\{\boldsymbol{\Delta}^{(N)}(\eta, g) \mid \boldsymbol{B}^{(N)}(\eta)\right\} + o_{\mathrm{P}}(1),$$

as $N \to \infty$, under $P^{(N)}_{\eta, g}$, for any $(\eta, g) \in \Theta \times \mathcal{G}$.

Proposition 5.2.24 deals with a fixed g value, hence with semiparametric efficiency at given g. If efficiency is to be attained over some class $\mathcal{F}$ of values of g, the following additional requirement is needed.

Condition 5.2.25 (LF2). *For all $g \in \mathcal{F} \subset \mathcal{G}$, there exists a version of the influence function (5.2.23) for η in $\mathcal{E}^{(N)}_{lf}(g)$ that does not depend on g.*

The conditions under which condition (LF2) can be satisfied in the general framework cannot be easily found. We will discuss them for the specific case of rank-based inference in the next section.

Now we give a precise statement of a set of assumptions that are sufficient for condition (LF1) to hold. As before, fix $\eta_0 \in \Theta$ and $g_0 \in \mathcal{G}$, and consider the following set of assumptions.

Assumption 5.2.26 (Hallin and Werker (2003), Assumption B). **(i)** *Let $\boldsymbol{H}^{(N)}_{lf}(\eta_0, g_0)$ be such that $\left(\boldsymbol{\Delta}^{(N)}(\eta_0, g_0)', \boldsymbol{H}^{(N)}_{lf}(\eta_0, g_0)'\right)'$ is asymptotically normal (automatically the limiting distribution is $N\left(\boldsymbol{0}, \tilde{\boldsymbol{I}}(\eta_0, g_0)\right)$), under $\mathrm{P}^{(N)}_{\eta_0, g_0}$, as $N \to \infty$.*

(ii) *There exists a function* $q_{lf} : (-1,1)^r \to \mathcal{G}$ *such that, for any sequence* $\eta_N = \eta_0 + O\left(N^{-1/2}\right)$, *the sequence of experiments*

$$\left\{\mathcal{X}^{(N)}, \mathcal{A}^{(N)}, \left\{\mathrm{P}^{(N)}_{\eta_N, q_{lf}(\boldsymbol{\lambda})} : \boldsymbol{\lambda} \in (-1,1)^r\right\}\right\}$$

is LAN at $\boldsymbol{\lambda} = \mathbf{0}$ *with central sequence* $\boldsymbol{H}^{(N)}_{lf}(\eta_N, g_0)$ *and Fisher information matrix* $\boldsymbol{I}(\eta_0, g_0) - \boldsymbol{I}(\eta_0, g_0)^*$.

(iii) *The sequence* $\boldsymbol{H}^{(N)}_{lf}(\eta_0, g_0)$ *satisfies a local asymptotic linearity property in the sense that, for any bounded sequence* $\boldsymbol{\tau}_N$ *in* $\mathbb{R}^r$, *we have*

$$\boldsymbol{H}^{(N)}_{lf}(\eta_0, g_0) - \boldsymbol{H}^{(N)}_{lf}\left(\eta_0 + \frac{\boldsymbol{\tau}_N}{\sqrt{N}}, g_0\right) = \left[\boldsymbol{I}(\eta_0, g_0) - \boldsymbol{I}(\eta_0, g_0)^*\right]\boldsymbol{\tau}_N + o_{\mathrm{P}_{\eta_0, g_0}}(1)$$

under $\mathrm{P}^{(N)}_{\eta_0, g_0}$ *as* $N \to \infty$.

Let $\boldsymbol{\tau}_N$ and $\boldsymbol{\rho}_N$ be bounded sequences in $\mathbb{R}^k$. The contiguity of $\mathrm{P}^{(N)}_{\eta_0 + \frac{\tau_N}{\sqrt{N}}, g_0}$ and $\mathrm{P}^{(N)}_{\eta_0, g_0}$ follows from Assumption 5.2.20 via LeCam's first lemma. Using Assumption 5.2.26 (ii) and (iii), observe that

$$\begin{aligned}
\log \frac{d\mathrm{P}^{(N)}_{\eta_0 + \frac{\tau_N}{\sqrt{N}}, q\left(\frac{\rho_N}{\sqrt{N}}\right)}}{d\mathrm{P}^{(N)}_{\eta_0, g_0}} &= \log \frac{d\mathrm{P}^{(N)}_{\eta_0 + \frac{\tau_N}{\sqrt{N}}, q\left(\frac{\rho_N}{\sqrt{N}}\right)}}{d\mathrm{P}^{(N)}_{\eta_0 + \frac{\tau_N}{\sqrt{N}}, g_0}} + \log \frac{d\mathrm{P}^{(N)}_{\eta_0 + \frac{\tau_N}{\sqrt{N}}, g_0}}{d\mathrm{P}^{(N)}_{\eta_0, g_0}} \\
&= \boldsymbol{\rho}_N' \boldsymbol{H}^{(N)}_{lf}\left(\eta_0 + \frac{\boldsymbol{\tau}_N}{\sqrt{N}}, g_0\right) - \frac{1}{2}\boldsymbol{\rho}_N' \left[\boldsymbol{I}(\eta_0, g_0) - \boldsymbol{I}(\eta_0, g_0)^*\right]\boldsymbol{\rho}_N \\
&\quad + \boldsymbol{\tau}_N' \boldsymbol{\Delta}^{(N)}(\eta_0, g_0) - \frac{1}{2}\boldsymbol{\tau}_N' \boldsymbol{I}(\eta_0, g_0)\boldsymbol{\tau}_N + o_{\mathrm{P}}(1) \\
&= \boldsymbol{\rho}_N' \boldsymbol{H}^{(N)}_{lf}(\eta_0, g_0) + \boldsymbol{\tau}_N' \boldsymbol{\Delta}^{(N)}(\eta_0, g_0) - \frac{1}{2}\boldsymbol{\rho}_N' \left[\boldsymbol{I}(\eta_0, g_0) - \boldsymbol{I}(\eta_0, g_0)^*\right]\boldsymbol{\tau}_N \\
&\quad - \frac{1}{2}\boldsymbol{\rho}_N' \left[\boldsymbol{I}(\eta_0, g_0) - \boldsymbol{I}(\eta_0, g_0)^*\right]\boldsymbol{\rho}_N - \frac{1}{2}\boldsymbol{\tau}_N' \boldsymbol{I}(\eta_0, g_0)\boldsymbol{\tau}_N + o_{\mathrm{P}}(1).
\end{aligned}$$

Assumption 5.2.26 (i) asserts the asymptotic normality of the central sequence. Therefore, we have the following result.

Proposition 5.2.27 (Hallin and Werker (2003), Proposition 2.5). *Suppose that Assumptions 5.2.20 and 5.2.26 are satisfied. Then the model*

$$\left\{\mathcal{X}^{(N)}, \mathcal{A}^{(N)}, \left\{\mathrm{P}^{(N)}_{\eta, q_{lf}(\boldsymbol{\lambda})} : \eta \in \boldsymbol{\Theta},\ \boldsymbol{\lambda} \in (-1,1)^r\right\}\right\}$$

is locally asymptotically normal at $(\eta_0, \mathbf{0})$ *with central sequence* $\left(\boldsymbol{\Delta}^{(N)}(\eta_0, g_0)', \boldsymbol{H}^{(N)}_{lf}(\eta_0, g_0)'\right)'$ *and the Fisher information matrix (5.2.25), and condition (LF1) is satisfied.*

An important property of some semiparametric models is adaptiveness. If inference problems are, locally and asymptotically, no more difficult in the semiparametric model than in the underlying parametric model, then adaptiveness holds. In the above notation, adaptiveness occurs at (η_0, g_0), if $\boldsymbol{I}(\eta_0, g_0)^* = \boldsymbol{I}(\eta_0, g_0)$. Hence, if, under $\mathrm{P}^{(N)}_{\eta_0, g_0}$,

$$\boldsymbol{\Delta}^{(N)}(\eta_0, g_0) = E^{(N)}_{\eta_0, g_0}\left\{\boldsymbol{\Delta}^{(N)}(\eta_0, g_0) \mid \boldsymbol{B}^{(N)}(\eta_0)\right\} + o_{\mathrm{P}}(1), \tag{5.2.26}$$

or if there exists any $\boldsymbol{B}^{(N)}(\eta_0)$-measurable version of the central sequence $\boldsymbol{\Delta}^{(N)}(\eta_0, g_0)$, then adaptiveness holds. These statements are made rigorous in the following proposition.

Proposition 5.2.28 (Hallin and Werker (2003), Proposition 2.6). *Fix* $q \in \mathcal{Q}$, *that is, let* $\left\{\mathcal{X}^{(N)}, \mathcal{A}^{(N)}, \mathcal{P}^{(N)}_q = \left\{\mathrm{P}^{(N)}_{\eta, q(\boldsymbol{\lambda})} : \eta \in \boldsymbol{\Theta}, \boldsymbol{\lambda} \in (-1,1)^k\right\}\right\}$ *denote an arbitrary submodel of the* *semiparametric model that satisfies the LAN propety at* $\left(\eta_0', \mathbf{0}'\right)'$ *with central sequence* $\left(\mathbf{\Delta}^{(N)}(\eta_0, g_0)', \mathbf{H}_q^{(N)}(\eta_0, g_0)'\right)'$. *If a* $\mathbf{B}^{(N)}(\eta_0)$*-measurable version* $\mathbf{\Delta}_B^{(N)}(\eta_0, g_0)$ *of the central sequence* $\mathbf{\Delta}^{(N)}(\eta_0, g_0)$ *exists, then, under* $\mathrm{P}_{\eta_0, g_0}^{(N)}$*, we have, for any* $q \in \mathcal{Q}$,

$$\begin{pmatrix} \mathbf{\Delta}^{(N)}(\eta_0, g_0) \\ \mathbf{H}_q^{(N)}(\eta_0, g_0) \end{pmatrix} \xrightarrow{\mathcal{L}} N\left(\mathbf{0}, \begin{pmatrix} \mathbf{I}(\eta_0, g_0) & \mathbf{0} \\ \mathbf{0} & \mathbf{I}_{\mathbf{H}_q}(\eta_0, g_0) \end{pmatrix}\right)$$

as $N \to \infty$. *Therefore, the model* $\mathcal{E}^{(N)}$ *is adaptive at* (η_0, g_0).

Note that under condition (LF1), adaptiveness holds at (η_0, g_0) if and only if (5.2.26) holds, that is, if and only if a $\mathbf{B}^{(N)}(\eta_0)$-measurable version of the central sequence $\mathbf{\Delta}^{(N)}(\eta_0, g_0)$ exists.

5.2.4 Semiparametrically Efficient Estimation in Time Series, and Multivariate Cases

In this section, we deal with the special cases of the general results of the previous section, namely, we consider semiparametrically efficient estimation in univariate time series, and the multivariate case.

5.2.4.1 Rank-based optimal influence functions (univariate case)

First, we consider semiparametric models for univariate time series. In this case, the randomness is described by the unspecified probability density f of a sequence of i.i.d. innovation random variables $\varepsilon_1^{(N)}, \ldots, \varepsilon_N^{(N)}$. Therefore, the sequence of semiparametric models is given by

$$\mathcal{E}^{(N)} := \left\{\mathcal{X}^{(N)}, \mathcal{A}^{(N)}, \mathcal{P}^{(N)} = \left\{\mathrm{P}_{\boldsymbol{\theta}, f}^{(N)} : \boldsymbol{\theta} \in \mathbf{\Theta}, f \in \mathcal{F}\right\}\right\} \quad N \in \mathbb{N} \tag{5.2.27}$$

with $\mathbf{\Theta}$ an open subset of $\mathbb{R}^k$ and $\mathcal{F}$ is a set of densities. We assume that the parametric model is locally asymptotic normal (LAN) if we fix the density $f \in \mathcal{F}$. Furthermore, we assume that $\mathcal{F}$ is a subset of the class $\mathcal{F}_0$ of all nonvanishing densities over a real line, so that a rank-based optimal inference function will be yielded. (If we restrict ourselves to the class of non-vanishing densities which are symmetric with respect to the origin instead, a signed rank-based inference function will be yielded, and the class of nonvanishing densities with median zero will yield a sign and rank-based inference function). As a consequence we suppose the following assumption.

Assumption 5.2.29 (Hallin and Werker (2003), Assumption I). **(i)** *For all* $f \in \mathcal{F}$ *the parametric submodel* $\mathcal{E}_f^{(N)} := \left\{\mathcal{X}^{(N)}, \mathcal{A}^{(N)}, \mathcal{P}_f^{(N)} = \left\{\mathrm{P}_{\boldsymbol{\theta}, f}^{(N)} : \boldsymbol{\theta} \in \mathbf{\Theta}\right\}\right\}$ *is locally asymptotically normal (LAN) in* $\boldsymbol{\theta}$ *at all* $\boldsymbol{\theta} \in \mathbf{\Theta}$*, with central sequence* $\Delta^{(N)}(\boldsymbol{\theta}, f)$ *of the form*

$$\Delta^{(N)}(\boldsymbol{\theta}, f) := \frac{1}{\sqrt{N-p}} \sum_{t=p+1}^{N} \mathbf{J}_{\boldsymbol{\theta}, f}\left(\varepsilon_t^{(N)}(\boldsymbol{\theta}), \ldots, \varepsilon_{t-p}^{(N)}(\boldsymbol{\theta})\right), \tag{5.2.28}$$

where the function $\mathbf{J}_{\boldsymbol{\theta}, f}$ *is allowed to depend on* $\boldsymbol{\theta}$ *and* f*, and the residuals* $\varepsilon_t^{(N)}(\boldsymbol{\theta})$, $t = 1, \ldots, N$ *are an invertible function of the observations such that* $\varepsilon_1^{(N)}(\boldsymbol{\theta}), \ldots, \varepsilon_N^{(N)}(\boldsymbol{\theta})$ *under* $\mathrm{P}_{\boldsymbol{\theta}, f}^{(N)}$ *are i.i.d. with density* f. *The Fisher information matrix corresponding to this LAN condition is denoted as* $\mathbf{I}(\boldsymbol{\theta}, f)$.

(ii) *The class* $\mathcal{F}$ *is a subset of the class* $\mathcal{F}_0$ *of all densities* f *over the real line such that* $f(x) > 0$, $x \in \mathbb{R}$.

Under Assumption 5.2.29(ii), for fixed $\boldsymbol{\theta}$ and N, the nonparametric model

$$\mathcal{E}_{\boldsymbol{\theta}}^{(N)} := \left\{\mathcal{X}^{(N)}, \mathcal{A}^{(N)}, \mathcal{P}_{\boldsymbol{\theta}}^{(N)} = \left\{\mathrm{P}_{\boldsymbol{\theta}, f}^{(N)} : f \in \mathcal{F}\right\}\right\}$$

is generated by the group of continuous order-preserving transformations acting on $\varepsilon_1^{(N)}(\boldsymbol{\theta}), \ldots, \varepsilon_N^{(N)}(\boldsymbol{\theta})$, namely, transformations $\mathcal{G}_h^{(N)}$ defined by

$$\mathcal{G}_h^{(N)}\left\{\varepsilon_1^{(N)}(\boldsymbol{\theta}), \ldots, \varepsilon_N^{(N)}(\boldsymbol{\theta})\right\} := \left[h\left\{\varepsilon_1^{(N)}(\boldsymbol{\theta})\right\}, \ldots, h\left\{\varepsilon_N^{(N)}(\boldsymbol{\theta})\right\}\right],$$

where h belongs to the set $\mathcal{H}$ of all functions $h : \mathbb{R} \to \mathbb{R}$ that are continuous, strictly increasing, and satisfy $\lim_{x\to\pm\infty} h(x) = \pm\infty$. The corresponding maximal invariant σ-algebra is the σ-algebra $\sigma\left\{R_1^{(N)}(\boldsymbol{\theta}), \ldots, R_N^{(N)}(\boldsymbol{\theta})\right\}$ generated by the ranks $R_1^{(N)}(\boldsymbol{\theta}), \ldots, R_N^{(N)}(\boldsymbol{\theta})$ of the residuals $\varepsilon_1^{(N)}(\boldsymbol{\theta}), \ldots, \varepsilon_N^{(N)}(\boldsymbol{\theta})$. This σ-algebra is distribution free under $\mathcal{E}_{\boldsymbol{\theta}}^{(N)}$ and plays the role of $\mathcal{B}^{(N)}(\boldsymbol{\theta})$.

Here, we need the conditions on the function $\boldsymbol{J}_{\boldsymbol{\theta},f}$ ensuring that either condition (LF1) alone or conditions (LF1) and (LF2) together hold.

Let $\varepsilon_0, \varepsilon_1, \ldots, \varepsilon_p$ be an arbitrary $(p+1)$-tuple of i.i.d. random variables with density f and independent of the residuals $\varepsilon_t^{(N)}(\boldsymbol{\theta})$. Expectations with respect to $\varepsilon_0, \varepsilon_1, \ldots, \varepsilon_p$ are denoted by $E_f(\cdot)$, which does not depend on N and $\boldsymbol{\theta}$. Now, we introduce the following set of assumptions on the function $\boldsymbol{J}_{\boldsymbol{\theta},f}$.

Assumption 5.2.30 (Hallin and Werker (2003), Assumption J). **(i)** *The function $\boldsymbol{J}_{\boldsymbol{\theta},f}$ is such that* $0 < E_f\left\{\left\|\boldsymbol{J}_{\boldsymbol{\theta},f}\left(\varepsilon_0, \varepsilon_1, \ldots, \varepsilon_p\right)\right\|^2\right\} < \infty$ *and*

$$E_f\left\{\boldsymbol{J}_{\boldsymbol{\theta},f}\left(\varepsilon_0, \varepsilon_1, \ldots, \varepsilon_p\right) \mid \varepsilon_1, \ldots, \varepsilon_p\right\} = \mathbf{0}, \quad \left(\varepsilon_1, \ldots, \varepsilon_p\right)\text{-a.e.} \tag{5.2.29}$$

(ii) *The function $\boldsymbol{J}_{\boldsymbol{\theta},f}$ is componentwise monotone increasing with respect to all its arguments or a linear combination of such functions.*

(iii) *For all sequences $\boldsymbol{\theta}_N = \boldsymbol{\theta}_0 + O\left(1/\sqrt{N}\right)$, we have, under $\mathrm{P}_{\boldsymbol{\theta}_0,f}^{(N)}$, as $N \to \infty$,*

$$\begin{aligned}&\frac{1}{\sqrt{N-p}} \sum_{t=p+1}^{N} \left\{\boldsymbol{J}_{\boldsymbol{\theta},f}\left(\varepsilon_t^{(N)}(\boldsymbol{\theta}_N), \ldots, \varepsilon_{t-p}^{(N)}(\boldsymbol{\theta}_N)\right) - \boldsymbol{J}_{\boldsymbol{\theta},f}\left(\varepsilon_t^{(N)}(\boldsymbol{\theta}_0), \ldots, \varepsilon_{t-p}^{(N)}(\boldsymbol{\theta}_0)\right)\right\}\\ &= -\boldsymbol{I}_f(\boldsymbol{\theta}_0)\sqrt{N}(\boldsymbol{\theta}_N - \boldsymbol{\theta}_0) + o_P(1),\end{aligned} \tag{5.2.30}$$

where the matrix-valued function $\boldsymbol{\theta} \mapsto \boldsymbol{I}_f(\boldsymbol{\theta}) := E_f\left\{\boldsymbol{J}_{\boldsymbol{\theta},f}\left(\varepsilon_0, \varepsilon_1, \ldots, \varepsilon_p\right)\boldsymbol{J}_{\boldsymbol{\theta},f}\left(\varepsilon_0, \varepsilon_1, \ldots, \varepsilon_p\right)'\right\}$ is continuous in $\boldsymbol{\theta}$ for all f. Furthermore, let

$$\begin{aligned}&\boldsymbol{J}_{\boldsymbol{\theta},f}^*\left(\varepsilon_t^{(N)}(\boldsymbol{\theta}), \ldots, \varepsilon_{t-p}^{(N)}(\boldsymbol{\theta})\right)\\ &:= \boldsymbol{J}_{\boldsymbol{\theta},f}\left(\varepsilon_t^{(N)}(\boldsymbol{\theta}), \ldots, \varepsilon_{t-p}^{(N)}(\boldsymbol{\theta})\right) - E_f\left\{\boldsymbol{J}_{\boldsymbol{\theta},f}\left(\varepsilon_t^{(N)}(\boldsymbol{\theta}), \varepsilon_1, \ldots, \varepsilon_p\right) \mid \varepsilon_t^{(N)}(\boldsymbol{\theta})\right\}.\end{aligned} \tag{5.2.31}$$

Then, we have, under $\mathrm{P}_{\boldsymbol{\theta}_0,f}^{(N)}$,

$$\begin{aligned}&\frac{1}{\sqrt{N-p}} \sum_{t=p+1}^{N} \left\{\boldsymbol{J}_{\boldsymbol{\theta},f}^*\left(\varepsilon_t^{(N)}(\boldsymbol{\theta}_N), \ldots, \varepsilon_{t-p}^{(N)}(\boldsymbol{\theta}_N)\right) - \boldsymbol{J}_{\boldsymbol{\theta},f}^*\left(\varepsilon_t^{(N)}(\boldsymbol{\theta}_0), \ldots, \varepsilon_{t-p}^{(N)}(\boldsymbol{\theta}_0)\right)\right\}\\ &= -\boldsymbol{I}_f^*(\boldsymbol{\theta}_0)\sqrt{N}(\boldsymbol{\theta}_N - \boldsymbol{\theta}_0) + o_P(1),\end{aligned}$$

where $\boldsymbol{\theta} \mapsto \boldsymbol{I}_f^*(\boldsymbol{\theta}) := E_f\left\{\boldsymbol{J}_{\boldsymbol{\theta},f}^*\left(\varepsilon_0, \varepsilon_1, \ldots, \varepsilon_p\right)\boldsymbol{J}_{\boldsymbol{\theta},f}^*\left(\varepsilon_0, \varepsilon_1, \ldots, \varepsilon_p\right)'\right\}$ is also continuous in $\boldsymbol{\theta}$ for all f.

Equation (5.2.29) in Assumption 5.2.30(i) with the martingale central limit theorem leads to the asymptotic normality of the parametric central sequence $\Delta^{(N)}(\boldsymbol{\theta}, f)$, i.e.,

$$\frac{1}{\sqrt{N-p}} \sum_{t=p+1}^{N} \boldsymbol{J}_{\boldsymbol{\theta},f}\left(\varepsilon_t^{(N)}(\boldsymbol{\theta}), \ldots, \varepsilon_{t-p}^{(N)}(\boldsymbol{\theta})\right) \xrightarrow{\mathcal{L}} N\left(\mathbf{0}, \boldsymbol{I}_f(\boldsymbol{\theta})\right).$$

The existence of rank-based versions of the score functions is also obtained from Assumption 5.2.30(ii), namely, the existence of functions $\boldsymbol{a}_f^{(N)} : \{1, \ldots, N\}^{p+1} \to \mathbb{R}^k$ such that

$$\lim_{N\to\infty} E_f\left\{\left\|\boldsymbol{J}_{\boldsymbol{\theta},f}\left(\varepsilon_1^{(N)}(\boldsymbol{\theta}), \ldots, \varepsilon_{p+1}^{(N)}(\boldsymbol{\theta})\right) - \boldsymbol{a}_f^{(N)}\left(R_1^{(N)}, \ldots, R_{p+1}^{(N)}\right)\right\|^2\right\} = 0, \tag{5.2.32}$$

as $N \to \infty$, where $R_1^{(N)}, \ldots, R_N^{(N)}$ are the ranks of the residuals $\varepsilon_1^{(N)}(\boldsymbol{\theta}), \ldots, \varepsilon_N^{(N)}(\boldsymbol{\theta})$. The classical choices for $\boldsymbol{a}_f^{(N)}$ are exact scores

$$\boldsymbol{a}_f^{(N)}\left(R_t^{(N)}, \ldots, R_{t-p}^{(N)}\right) := E_f\left\{\boldsymbol{J}_{\boldsymbol{\theta},f}\left(\varepsilon_t^{(N)}(\boldsymbol{\theta}), \ldots, \varepsilon_{t-p}^{(N)}(\boldsymbol{\theta})\right) \mid R_1^{(N)}, \ldots, R_N^{(N)}\right\}$$

and approximate scores

$$\boldsymbol{a}_f^{(N)}\left(R_t^{(N)}, \ldots, R_{t-p}^{(N)}\right) := \boldsymbol{J}_{\boldsymbol{\theta},f}\left(F^{-1}\left(\frac{R_t^{(N)}}{N+1}\right), \ldots, F^{-1}\left(\frac{R_{t-p}^{(N)}}{N+1}\right)\right),$$

where F is the distribution function corresponding to f. The smoothness conditions for $\boldsymbol{J}_{\boldsymbol{\theta},f}$ and $\boldsymbol{J}^*_{\boldsymbol{\theta},f}$ follow from Assumption 5.2.30(iii) which are needed for Assumption 5.2.26(iii). Recall that the LAN condition with (5.2.30) is equivalent to the ULAN condition which is the LAN condition uniformly over $1/\sqrt{N}$-neighbourhoods of $\boldsymbol{\theta}$.

In what follows, we write $\varepsilon_t^{(N)}$ for $\varepsilon_t^{(N)}(\boldsymbol{\theta}_0)$ when it will not cause confusion. Recall that $\varepsilon_1^{(N)}, \ldots, \varepsilon_N^{(N)}$ are i.i.d. with density f under $\mathrm{P}^{(N)}_{\boldsymbol{\theta}_0,f}$. Now, we have the following results (see Section 4.1 of Hallin et al. (1985a)).

Proposition 5.2.31 (Hallin and Werker (2003), Proposition 3.1). *Suppose that Assumptions 5.2.30(i), (ii) are satisfied, and let $\boldsymbol{a}_f^{(N)}$ be any function satisfying (5.2.32). Writing*

$$\boldsymbol{S}_f^{(N)}(\boldsymbol{\theta}_0) := \frac{1}{N-p}\sum_{t=p+1}^{N} \boldsymbol{a}_f^{(N)}\left(R_t^{(N)}, \ldots, R_{t-p}^{(N)}\right),$$
$$\boldsymbol{m}_f^{(N)} := \{N(N-1)\cdots(N-p)\}^{-1}\sum\cdots\sum_{1\le i_0\ne\cdots\ne i_p\le N} \boldsymbol{a}_f^{(N)}\left(i_0, \ldots, i_p\right),$$

define

$$\widetilde{\Delta}^{(N)}(\boldsymbol{\theta}_0, f) := \sqrt{N-p}\left\{\boldsymbol{S}_f^{(N)}(\boldsymbol{\theta}_0) - \boldsymbol{m}_f^{(N)}\right\}. \tag{5.2.33}$$

Then,

$$\widetilde{\Delta}^{(N)}(\boldsymbol{\theta}_0, f) = \frac{1}{\sqrt{N-p}}\sum_{t=p+1}^{N} \boldsymbol{J}_{\boldsymbol{\theta},f}\left(\varepsilon_t^{(N)}, \ldots, \varepsilon_{t-p}^{(N)}\right)$$
$$- \frac{\sqrt{N-p}}{N(N-1)\cdots(N-p)}\sum\cdots\sum_{1\le i_0\ne\cdots\ne i_p\le N} \boldsymbol{J}_{\boldsymbol{\theta},f}\left(\varepsilon_{i_0}^{(N)}, \ldots, \varepsilon_{i_p}^{(N)}\right) + o_{L_2}(1),$$

under $\mathrm{P}^{(N)}_{\boldsymbol{\theta}_0,f}$, as $N \to \infty$.

The U-statistics with kernel $\boldsymbol{J}_{\boldsymbol{\theta},f}$ can be rewritten in a simpler form.

Corollary 5.2.32. *Uner the same conditions as in Proposition 5.2.31, we also have*

$$\widetilde{\Delta}^{(N)}(\boldsymbol{\theta}_0, f) = \frac{1}{\sqrt{N-p}}\sum_{t=p+1}^{N} \boldsymbol{J}^*_{\boldsymbol{\theta},f}\left(\varepsilon_t^{(N)}, \ldots, \varepsilon_{t-p}^{(N)}\right) + o_P(1)$$

with $\boldsymbol{J}^*_{\boldsymbol{\theta},f}$ defined in (5.2.31).

Then, we can construct a candidate of the least favourable direction in the present case which is generated by the sequence

$$\boldsymbol{H}_{lf}^{(N)}(\boldsymbol{\theta}, f) = \Delta^{(N)}(\boldsymbol{\theta}, f) - E^{(N)}_{\boldsymbol{\theta},f}\left\{\Delta^{(N)}(\boldsymbol{\theta}, f) \mid \mathcal{B}^{(N)}(\boldsymbol{\theta})\right\}$$

$$= \frac{1}{\sqrt{N-p}} \sum_{t=p+1}^{N} \boldsymbol{J}_{\boldsymbol{\theta},f}\left(\varepsilon_t^{(N)}(\boldsymbol{\theta}), \ldots, \varepsilon_{t-p}^{(N)}(\boldsymbol{\theta})\right)$$

$$- E_{\boldsymbol{\theta},f}^{(N)}\left\{\frac{1}{\sqrt{N-p}} \sum_{t=p+1}^{N} \boldsymbol{J}_{\boldsymbol{\theta},f}\left(\varepsilon_t^{(N)}(\boldsymbol{\theta}), \ldots, \varepsilon_{t-p}^{(N)}(\boldsymbol{\theta})\right) \mid R_1^{(N)}(\boldsymbol{\theta}), \ldots, R_N^{(N)}(\boldsymbol{\theta})\right\}$$

$$= \frac{1}{\sqrt{N-p}} \sum_{t=p+1}^{N} \left[\boldsymbol{J}_{\boldsymbol{\theta},f}\left(\varepsilon_t^{(N)}(\boldsymbol{\theta}), \ldots, \varepsilon_{t-p}^{(N)}(\boldsymbol{\theta})\right)\right.$$

$$\left. - E_{\boldsymbol{\theta},f}^{(N)}\left\{\boldsymbol{J}_{\boldsymbol{\theta},f}\left(\varepsilon_t^{(N)}(\boldsymbol{\theta}), \ldots, \varepsilon_{t-p}^{(N)}(\boldsymbol{\theta})\right) \mid R_1^{(N)}(\boldsymbol{\theta}), \ldots, R_N^{(N)}(\boldsymbol{\theta})\right\}\right].$$

Now, we have the following result.

Proposition 5.2.33 (Hallin and Werker (2003), Proposition 3.3). *Consider the semiparametric model (5.2.27). Assume that Assumptions 5.2.29 and 5.2.30 are satisfied. Then, there exists a mapping* $q : (-1,1)^k \to \mathcal{F}$ *such that the parametric model*

$$\mathcal{E}_q^{(N)} := \left\{\mathcal{X}^{(N)}, \mathcal{A}^{(N)}, \mathcal{P}^{(N)} = \left\{\mathrm{P}_{\boldsymbol{\theta},q(\boldsymbol{\lambda})}^{(N)} : \boldsymbol{\theta} \in \Theta, \boldsymbol{\lambda} \in (-1,1)^k\right\}\right\} \quad N \in \mathbb{N}$$

is locally asymptotic normal at $(\boldsymbol{\theta}_0, \mathbf{0})$ *with central sequence*

$$\begin{pmatrix} \Delta^{(N)}(\boldsymbol{\theta}_0, f) \\ \boldsymbol{H}_{lf}^{(N)}(\boldsymbol{\theta}_0, f) \end{pmatrix} \xrightarrow{\mathcal{L}} N\left(\mathbf{0}, \begin{pmatrix} \boldsymbol{I}_f(\boldsymbol{\theta}_0) & \boldsymbol{I}_f(\boldsymbol{\theta}_0) - \boldsymbol{I}_f^*(\boldsymbol{\theta}_0) \\ \boldsymbol{I}_f(\boldsymbol{\theta}_0) - \boldsymbol{I}_f^*(\boldsymbol{\theta}_0) & \boldsymbol{I}_f(\boldsymbol{\theta}_0) - \boldsymbol{I}_f^*(\boldsymbol{\theta}_0) \end{pmatrix}\right),$$

where $\boldsymbol{I}_f^*(\boldsymbol{\theta}_0)$ *is the variance of the limiting distribution of* $E_{\boldsymbol{\theta},f}^{(N)}\left\{\Delta^{(N)}(\boldsymbol{\theta}, f) \mid \mathcal{B}^{(N)}(\boldsymbol{\theta})\right\}$. *Hence,* $\boldsymbol{H}_{lf}^{(N)}(\boldsymbol{\theta}, f)$ *satisfies condition (LF1) and the corresponding inference function for* $\boldsymbol{\theta}$ *is* $\boldsymbol{I}_f^*(\boldsymbol{\theta}_0)^{-1}\widetilde{\Delta}^{(N)}(\boldsymbol{\theta}, f)$. *It follows that semiparametrically efficient inference at* $\boldsymbol{\theta}$ *and* f *can be based on the efficient central sequence*

$$\Delta^{(N)^*}(\boldsymbol{\theta}, f) := \frac{1}{\sqrt{N-p}} \sum_{t=p+1}^{N} \boldsymbol{J}_{\boldsymbol{\theta},f}^*\left(\varepsilon_t^{(N)}, \ldots, \varepsilon_{t-p}^{(N)}\right),$$

of which $\widetilde{\Delta}^{(N)}(\boldsymbol{\theta}, f)$ *defined in (5.2.33) provides a rank-based version.*

This result is obtained without explicitly using the tangent space arguments in Section 5.2.2 and without computing the efficient score function. The efficient score is obtained by taking the residual of the projection of the parametric score on the tangent space generated by f. However, in this case f is completely unrestricted, so that the tangent space consists of all square integrable functions of $\varepsilon_t^{(N)}$ which have mean 0 and vanish when f vanishes. Therefore, the projection of $\boldsymbol{J}_{\boldsymbol{\theta},f}\left(\varepsilon_t^{(N)}(\boldsymbol{\theta}), \ldots, \varepsilon_{t-p}^{(N)}(\boldsymbol{\theta})\right)$ onto this space is given by $E_f\left\{\boldsymbol{J}_{\boldsymbol{\theta},f}\left(\varepsilon_t^{(N)}(\boldsymbol{\theta}), \varepsilon_1, \ldots, \varepsilon_p\right) \mid \varepsilon_t^{(N)}(\boldsymbol{\theta})\right\}$, which is the least favourable submodel. As a consequence, $\boldsymbol{J}_{\boldsymbol{\theta},f}^*$ indeed becomes the efficient score function.

Since it depends on unknown density f, we have to substitute some adequate estimator $\widehat{f}_N$ for f in the influence function $\boldsymbol{I}_f^*(\boldsymbol{\theta}_0)^{-1}\widetilde{\Delta}^{(N)}(\boldsymbol{\theta}, f)$ in order to satisfy condition (LF2). Define C as the class of all densities f in $\mathcal{F}$ which satisfy the following assumption.

Assumption 5.2.34 (Assumption in Section 3.3 of Hallin and Werker (2003)). **(i)** $\boldsymbol{J}_{\boldsymbol{\theta},f}$ *satisfies Assumption 5.2.30.*

(ii) *There exists an estimator* $\widehat{f}_N$ *which is measurable with respect to the order statistics of the residuals* $\varepsilon_1^{(N)}(\boldsymbol{\theta}), \ldots, \varepsilon_N^{(N)}(\boldsymbol{\theta})$, *such that*

$$E_f\left\{\left\|\boldsymbol{a}_{\widehat{f}_N}^{(N)}\left(R_1^{(N)}, \ldots, R_{p+1}^{(N)}\right) - \boldsymbol{a}_f^{(N)}\left(R_1^{(N)}, \ldots, R_{p+1}^{(N)}\right)\right\|^2 \mid \widehat{f}_N\right\} = o_P(1)$$

as $N \to \infty$, *where the rank scores* $\boldsymbol{a}_f^{(N)}$ *are defined in (5.2.32).*

Therefore, the class C contains all densities $f \in \mathcal{F}$ such that Assumption 5.2.30 holds. Furthermore, the rank scores $\boldsymbol{a}_f^{(N)}$ can be estimated consistently. A possible estimator is given in (1) of Sidak et al. (1999), Section 8.5.2, as follows. Let $\boldsymbol{X} = (X_1, \ldots, X_N)'$ be a random sample from the density f and the score $\varphi(u, f) = -\frac{f'(F^{-1}(u))}{f(F^{-1}(u))}$, $0 < u < 1$, and let $X_{(1)}, \ldots, X_{(N)}$ be its order statistics. Put $m_N = \left[N^{3/4}\epsilon_N^{-2}\right]$, $n_N = \left[N^{1/4}\epsilon_N^{3}\right]$, where $\{\epsilon_N\}$ is some sequence of positive numbers such that

$$\epsilon_N \to 0, \quad N^{1/4}\epsilon_N^3 \to \infty.$$

Furthermore, put

$$h_{N_j} = \left[\frac{jN}{n_N + 1}\right], \quad 1 \le j \le n_N,$$

where $[x]$ denotes the largest integer not exceeding x. Let

$$\widetilde{\varphi}_N(u, \boldsymbol{X}) = \begin{cases} 2\frac{m_N n_N}{N+1}\left\{\left[Z_{\left(h_{N_j}+m_N\right)} - Z_{\left(h_{N_j}-m_N\right)}\right]^{-1} - \left[Z_{\left(h_{N_{j+1}}+m_N\right)} - Z_{\left(h_{N_{j+1}}-m_N\right)}\right]^{-1}\right\}, \\ \left(\frac{h_{N_j}}{N} \le u < \frac{h_{N_{j+1}}}{N}, \quad 1 \le j \le n_N\right), \\ 0, \qquad (otherwise). \end{cases}$$

Then, $\widetilde{\varphi}_N(u, \boldsymbol{X})$ is a consistent estimate of $\varphi(u, f)$ in the sense that

$$\lim_{N\to\infty} P\left\{\int_0^1 |\widetilde{\varphi}_N(u, \boldsymbol{X}) - \varphi(u, f)|^2\, du > \epsilon\right\} = 0$$

for every $\epsilon > 0$ and the density f satisfying $I(f) < \infty$.

We denote the covariance matrix of $\widetilde{\Delta}^{(N)}\left(\boldsymbol{\theta}, \widehat{f}_N\right)$ conditional on $\widehat{f}_N$ by $\widehat{\boldsymbol{I}}^{(N)*}_{\widehat{f}_N}(\boldsymbol{\theta}) := \text{Var}\left(\widetilde{\Delta}^{(N)}\left(\boldsymbol{\theta}, \widehat{f}_N\right) \mid \widehat{f}_N\right)$. Since $\widetilde{\Delta}^{(N)}\left(\boldsymbol{\theta}, \widehat{f}_N\right)$ is conditionally distribution free, $\widehat{\boldsymbol{I}}^{(N)*}_{\widehat{f}_N}(\boldsymbol{\theta})$ does not depend on f and can be computed from the observations. Moreover, $\widehat{\boldsymbol{I}}^{(N)*}_{\widehat{f}_N}(\boldsymbol{\theta})$ becomes the consistent estimator of $\boldsymbol{I}_f^*(\boldsymbol{\theta})$. The following result shows that conditions (LF1) and (LF2) are satisfied for all $f \in C$.

Proposition 5.2.35 (Hallin and Werker (2003), Proposition 3.4). *For all $f \in C$ and $\boldsymbol{\theta} \in \boldsymbol{\Theta}$, under $\mathrm{P}^{(N)}_{\boldsymbol{\theta},f}$,*

$$\widehat{\boldsymbol{I}}^{(N)*}_{\widehat{f}_N}(\boldsymbol{\theta})^{-1}\,\widetilde{\Delta}^{(N)}\left(\boldsymbol{\theta}, \widehat{f}_N\right) = \boldsymbol{I}_f^*(\boldsymbol{\theta})\,\widetilde{\Delta}^{(N)}(\boldsymbol{\theta}, f) + o_p(1).$$

Therefore, conditions (LF1) and (LF2) are satisfied and $\Delta^{(N)^*}\left(\boldsymbol{\theta}, \widehat{f}_N\right)$ (as well as $\widetilde{\Delta}^{(N)}\left(\boldsymbol{\theta}, \widehat{f}_N\right)$) are versions of the efficient central sequence for $\mathcal{E}^{(N)}$ at f and $\boldsymbol{\theta}$.

As a simple example, we consider the following MA(1) process.

Example 5.2.36 (Hallin and Werker (2003), Example 3.1; MA(1) process). *Let $Y_1^{(N)}, \cdots, Y_N^{(N)}$ be a finite realization of the MA(1) process*

$$Y_t = \varepsilon_t + \theta\varepsilon_{t-1}, \quad t \in \mathbb{N},$$

where $\theta \in (-1, 1)$ and $\{\varepsilon_t, t \le 1\}$ are i.i.d. random variables with common density f, and assume that the starting value ε_0 is observed. Furthermore, we have to assume that f is absolutely continuous with finite variance σ_f^2 and finite Fisher information for location $I(f) :=$

$\int_{\mathbb{R}} \{f'(x)/f(x)\}^2 f(x)\,dx < \infty$. Under these assumptions, this MA(1) model is locally asymptotically normal (LAN) with univariate central sequence

$$\Delta^{(N)}(\theta, f) = \frac{1}{\sqrt{N-1}} \sum_{t=2}^{N} \frac{-f'}{f} \left\{\varepsilon_t^{(N)}(\theta)\right\} \varepsilon_{t-1}^{(N)}(\theta),$$

where the residuals $\left\{\varepsilon_t^{(N)}(\theta), 1 \le t \le N\right\}$ are recursively computed from $\varepsilon_t^{(N)}(\theta) := Y_t^{(N)} - \theta\varepsilon_{t-1}^{(N)}(\theta)$ with initial value $\varepsilon_0^{(N)}(\theta) = \varepsilon_0$. We see that this central sequence is of the form (5.2.28).

For this model, the function $J_{\theta,f}$ is given by

$$J_{\theta,f}(\varepsilon_0, \varepsilon_1) := \frac{-f'}{f}(\varepsilon_0)\,\varepsilon_1,$$

which obviously satisfies Assumption 5.2.30(i). Let $\mu_f := \int_{\mathbb{R}} x f(x)\,dx$. Then, we immediately obtain the efficient score function

$$J^*_{\theta,f}(\varepsilon_0, \varepsilon_1) = \frac{-f'}{f}(\varepsilon_0)\left(\varepsilon_1 - \mu_f\right).$$

Observing that

$$\varepsilon_t^{(N)}(\theta_N) = Y_t^{(N)} - \theta_N\varepsilon_{t-1}^{(N)}(\theta_N) = \varepsilon_t^{(N)}(\theta_0) + \theta_0\varepsilon_{t-1}^{(N)}(\theta_0) - \theta_N\varepsilon_{t-1}^{(N)}(\theta_N),$$

Assumption 5.2.30(iii) follows from standard arguments.

In the case where $\mu_f = 0$, the efficient score function $J^*_{\theta,f}$ and the parametric score function $J_{\theta,f}$ coincide, and the model is adaptive, which is a well-known result.

5.2.4.2 *Rank-based optimal estimation for elliptical residuals*

We call the distribution of a k-dimensional random variable $\boldsymbol{X}$ spherical if for some $\boldsymbol{\theta} \in \mathbb{R}^k$, the distribution of $\boldsymbol{X} - \boldsymbol{\theta}$ is invariant under orthogonal transformations. For multivariate normal variable cases, the sphericity is equivalent to the covariance matrix $\boldsymbol{\Sigma}$ of $\boldsymbol{X}$ being proportional to the identity matrix $\boldsymbol{I}_k$. For elliptical random variables, in which the finite second-order moments need not exist, the sphericity is equivalent to the shape matrix $\boldsymbol{V}$ being equal to the unit matrix $\boldsymbol{I}_k$. Let $\boldsymbol{X}^{(n)} = \left(\boldsymbol{X}_1^{(n)\prime}, \ldots, \boldsymbol{X}_n^{(n)\prime}\right)'$, $n \in \mathbb{N}$ be a triangle array of k-dimensional observations, where $\boldsymbol{X}_1^{(n)\prime}, \ldots, \boldsymbol{X}_n^{(n)\prime}$ are i.i.d. random variables with ellipical density

$$f_{(\boldsymbol{\theta},\sigma^2,\boldsymbol{V};f_1)}(\boldsymbol{x}) := c_{k,f_1}\frac{1}{\sigma^k |\boldsymbol{V}|^{1/2}} f_1\left[\frac{1}{\sigma}\left\{(\boldsymbol{x}-\boldsymbol{\theta})'\boldsymbol{V}^{-1}(\boldsymbol{x}-\boldsymbol{\theta})\right\}^{1/2}\right], \quad \boldsymbol{x} \in \mathbb{R}^k, \tag{5.2.34}$$

where $\boldsymbol{\theta} \in \mathbb{R}^k$ is a location parameter, $\sigma^2 \in \mathbb{R}^+ := (0, \infty)$ is a scale parameter, $\boldsymbol{V} := \left(V_{ij}\right)$ is a shape parameter which is a symmetric positive definite real $k \times k$ matrix with $V_{11} = 1$, $f_1 : \mathbb{R}_0^+ := (0, \infty) \to \mathbb{R}^+$ is the infinite-dimensional parameter which is an a.e. strictly positive function, and the constant c_{k,f_1} is a normalization factor depending on the dimension k and f_1. The function f_1 is called a radial density, since f_1 does not integrate to one, and therefore is not a probability density. We denote the modulus of the centred and sphericized observations $\boldsymbol{Z}_i^{(n)} = \boldsymbol{Z}_i^{(n)}(\boldsymbol{\theta}, \boldsymbol{V}) := \boldsymbol{V}^{-1/2}\left(\boldsymbol{X}^{(n)} - \boldsymbol{\theta}\right)$, $i = 1, \ldots, n$ by $d_i^{(n)} = d_i^{(n)}(\boldsymbol{\theta}, \boldsymbol{V}) := \left\|\boldsymbol{Z}_i^{(n)}(\boldsymbol{\theta}, \boldsymbol{V})\right\|$. If $\left\{\boldsymbol{X}^{(n)}\right\}$ have the density (5.2.34), then moduli $\left\{d_i^{(n)}\right\}$ are i.i.d. with density function

$$r \mapsto \frac{1}{\sigma}\bar{f}_{1,k}(r/\sigma) := \frac{1}{\sigma\mu_{k-1:f_1}}\left(\frac{r}{\sigma}\right)^{k-1} f_1(r/\sigma)\chi_{\{r>0\}}$$

and distribution function

$$r \mapsto \overline{F}_{1,k}(r/\sigma) := \int_0^{r/\sigma} \overline{f}_{1,k}(s)\,ds, \tag{5.2.35}$$

provided

$$\mu_{k-1:f_1} := \int_0^\infty r^{k-1} f_1(r)\,dr < \infty.$$

Let $\text{vech}\boldsymbol{M} = \left(M_{11}, \left(\overset{\circ}{\text{vech}}\,\boldsymbol{M}\right)'\right)'$ be the $\frac{k(k+1)}{2}$-dimensional vector of the upper triangle elements of a $k \times k$ symmetric matrix $\boldsymbol{M} = (M_{ij})$, and $\boldsymbol{\eta} = \left(\boldsymbol{\theta}', \sigma^2, \left(\overset{\circ}{\text{vech}}\,\boldsymbol{V}\right)'\right)'$. Therefore, the parameter space becomes $\boldsymbol{\Theta} := \mathbb{R}^k \times \mathbb{R}^+ \times \mathcal{V}_k$, where $\mathcal{V}_k$ is the set of values of $\overset{\circ}{\text{vech}}\,\boldsymbol{V}$ in $\mathbb{R}^{\frac{k(k-1)}{2}}$. We denote the distribution $\boldsymbol{X}^{(n)}$ under given values of $\boldsymbol{\eta} = \left(\boldsymbol{\theta}', \sigma^2, \left(\overset{\circ}{\text{vech}}\,\boldsymbol{V}\right)'\right)'$ and f_1 by $P^{(n)}_{\boldsymbol{\eta}:f_1}$ or $P^{(n)}_{\boldsymbol{\theta},\sigma^2,\boldsymbol{V}:f_1}$.

We denote the rank of $d_i^{(n)} = d_i^{(n)}(\boldsymbol{\theta}, \boldsymbol{V})$ among $d_1^{(n)}, \ldots, d_n^{(n)}$ by $R_i^{(n)} = R_i^{(n)}(\boldsymbol{\theta}, \boldsymbol{V})$. Under $P^{(n)}_{\boldsymbol{\eta}:f_1}$, the vector $\left(R_1^{(n)}, \ldots, R_n^{(n)}\right)$ is uniformly distributed over the $n!$ permutations of $(1, \ldots, n)$. Furthermore, let $\boldsymbol{U}_i^{(n)} = \boldsymbol{U}_i^{(n)}(\boldsymbol{\theta}, \boldsymbol{V}) := \boldsymbol{Z}_i^{(n)}/d_i^{(n)}$. Then under $P^{(n)}_{\boldsymbol{\eta}:f_1}$, the vectors $\boldsymbol{U}_i^{(n)}$ are i.i.d., uniformly distributed over the unit sphere and independent of the ranks $R_i^{(n)}$. The vectors $\boldsymbol{U}_i^{(n)}$ are considered as multivariate signs of the centred observations $\boldsymbol{X}_i - \boldsymbol{\theta}$ since they are invariant under transformations of $\boldsymbol{X}_i - \boldsymbol{\theta}$ which preserve half lines through the origin (Problem 5.6).

Hallin and Paindaveine (2006) considered testing the hypothesis that the shape matrix $\boldsymbol{V}$ is equal to some given value $\boldsymbol{V}_0$. As a special case, we can consider the case that $\boldsymbol{V}_0 = \boldsymbol{I}_k$, which yields the testing of sphericity. Therefore, the shape matrix $\boldsymbol{V}$ is the parameter of interest, while $\boldsymbol{\theta}$, σ^2 and f_1 are nuisance parameters. We would like to employ the test statistics whose null distributions remain invariant under variations of $\boldsymbol{\theta}$, σ^2 and f_1. When $\boldsymbol{\theta}$ is known, this is achieved by test statistics based on the signs $\boldsymbol{U}_i^{(n)}$ and ranks $R_i^{(n)}$ computed from $\boldsymbol{Z}_i^{(n)}(\boldsymbol{\theta}, \boldsymbol{V}_0)$, $i = 1, \ldots, n$. These tests are invariant under monotone radial transformations, including scale transformations, rotations and reflections of the observations with respect to $\boldsymbol{\theta}$.

The statistical experiments involve the nonparametric family

$$\begin{aligned}\mathcal{P}^{(n)} &:= \cup_{f_1 \in \mathcal{F}_A} \mathcal{P}^{(n)}_{f_1} := \cup_{f_1 \in \mathcal{F}_A} \cup_{\sigma>0} \mathcal{P}^{(n)}_{\sigma^2:f_1} \\ &:= \cup_{f_1 \in \mathcal{F}_A} \cup_{\sigma>0} \left\{\mathcal{P}^{(n)}_{\boldsymbol{\theta},\sigma^2,\boldsymbol{V}:f_1} \mid \boldsymbol{\theta} \in \mathbb{R}^k, \boldsymbol{V} \in \mathcal{V}_k\right\},\end{aligned}$$

where f_1 ranges over the set $\mathcal{F}_A$ of standardized densities satisfying the assumptions below (Assumptions (A1) and (A2) of Hallin and Paindaveine (2006)). The semiparametric structure is induced by the partition of $\mathcal{P}^{(n)}$ into a collection of parametric subexperiments $\mathcal{P}^{(n)}_{\sigma^2:f_1}$.

We need uniform local asymptotic normality (ULAN) with respect to $\boldsymbol{\eta} = \left(\boldsymbol{\theta}', \sigma^2, \left(\overset{\circ}{\text{vech}}\,\boldsymbol{V}\right)'\right)'$ of the family $\mathcal{P}^{(n)}_{f_1}$ (see Hallin and Paindaveine (2006)). For this purpose, denote the space of measurable functions $h : \Omega \to \mathbb{R}$ satisfying $\int_\Omega \left\{h(x)^2\right\} d\lambda(x) < \infty$ by $L_2(\Omega, \lambda)$. Especially, we consider the space $L_2(\mathbb{R}^+, \mu_l)$ of measurable functions $h : \mathbb{R}^+ \to \mathbb{R}$ satisfying $\int_0^\infty \left\{h(x)^2\right\} x^l dx < \infty$ and the space $L_2(\mathbb{R}, \nu_l)$ of measurable functions $h : \mathbb{R} \to \mathbb{R}$ satisfying $\int_{-\infty}^\infty \left\{h(x)^2\right\} e^{lx} dx < \infty$. Note that $g \in L_2(\Omega, \lambda)$ admits weak derivative T iff $\int_\Omega g(x)\phi'(x)\,dx = -\int_\Omega T(x)\phi(x)\,dx$ for all infinitely differentiable compactly supported functions ϕ on Ω. Define the mappings $x \mapsto f_1^{1/2}(x)$ and $x \mapsto f_{1:exp}^{1/2}(x)$, and let $\psi_{f_1}(r) := -2\frac{\left(f_1^{1/2}\right)'(r)}{f_1^{1/2}(r)}$ and $\varphi_{f_1}(r) := -2\frac{\left(f_{1:exp}^{1/2}\right)'(\log r)}{f_{1:exp}^{1/2}(\log r)}$, where $\left(f_1^{1/2}\right)'$ and

$\left(f_{1:exp}^{1/2}\right)'$ stand for the weak derivatives of $f_1^{1/2}$ in $L_2(\mathbb{R}^+, \mu_{k-1})$ and $f_{1:exp}^{1/2}$ in $L_2(\mathbb{R}, \nu_k)$, respectively. The following assumptions (Assumptions (A1) and (A2) of Hallin and Paindaveine (2006)) imply the finiteness of radial Fisher information for location

$$\mathcal{I}_k(f_1) := E_{P^{(n)}_{\eta:f_1}}\left[\varphi_{f_1}^2\left(d_i^{(n)}(\boldsymbol{\theta}, \boldsymbol{V})/\sigma\right)\right] = \int_0^1 \varphi_{f_1}^2\left(\overline{F}_{1,k}^{-1}(u)\right)du$$

and radial Fisher information for shape (and scale)

$$\mathcal{J}_k(f_1) := E_{P^{(n)}_{\eta:f_1}}\left[\psi_{f_1}^2\left(d_i^{(n)}(\boldsymbol{\theta}, \boldsymbol{V})/\sigma\right)\left\{d_i^{(n)}(\boldsymbol{\theta}, \boldsymbol{V})/\sigma\right\}^2\right] = \int_0^1 K_{f_1}^2(u)\,du,$$

where $K_{f_1}(u) := \psi_{f_1}\left(\overline{F}_{1,k}^{-1}(u)\right)\overline{F}_{1,k}^{-1}(u)$, respectively.

Assumption 5.2.37 ((A1)). *The mapping $x \mapsto f_1^{1/2}(x)$ is in the Sobolev space $W_{1,2}(\mathbb{R}^+, \mu_{k-1})$ of order 1 on $L_2(\mathbb{R}^+, \mu_{k-1})$.*

Assumption 5.2.38 ((A2)). *The mapping $x \mapsto f_{1:exp}^{1/2}(x)$ is in the Sobolev space $W_{1,2}(\mathbb{R}, \nu_k)$ of order 1 on $L_2(\mathbb{R}, \nu_k)$.*

Furthermore, we use the following notation. The lth vector in the canonical basis of $\mathbb{R}^k$ is denoted by $\boldsymbol{e}_l$ and the $k \times k$ identical matrix is denoted by $\boldsymbol{I}_k$. And let

$$\boldsymbol{K}_k := \sum_{i,j=1}^{k} \left(\boldsymbol{e}_i\boldsymbol{e}_j'\right) \otimes \left(\boldsymbol{e}_j\boldsymbol{e}_i'\right),$$

which is the so-called the commutation matrix,

$$\boldsymbol{J}_k := \sum_{i,j=1}^{k} \left(\boldsymbol{e}_i\boldsymbol{e}_j'\right) \otimes \left(\boldsymbol{e}_i\boldsymbol{e}_j'\right) = (\mathrm{vec}\boldsymbol{I}_k)(\mathrm{vec}\boldsymbol{I}_k)',$$

$\boldsymbol{M}_k$ be the $\left(\frac{k(k-1)}{2}\right) \times k^2$ matrix such that $\boldsymbol{M}_k'\left(\mathrm{v\mathring{e}ch}\,\boldsymbol{V}\right) = \mathrm{vec}\boldsymbol{V}$ for any symmetric $k \times k$ matrix $\boldsymbol{V} = \left(v_{ij}\right)$ with $v_{11} = 0$ and $\boldsymbol{V}^{\otimes 2}$ be the Kroneker product $\boldsymbol{V} \otimes \boldsymbol{V}$. Then, we have the following ULAN result.

Theorem 5.2.39 (Proposition 2.1 of Hallin and Paindaveine (2006)). *Under Assumptions (A1) and (A2), the family $\mathcal{P}_{f_1}^{(n)} = \left\{P_{\eta:f_1}^{(n)} \mid \boldsymbol{\eta} \in \boldsymbol{\Theta}\right\}$ is ULAN with central sequence*

$$\boldsymbol{\Delta}_{f_1}^{(n)}(\boldsymbol{\eta}) := \begin{pmatrix} \boldsymbol{\Delta}_{f_1:1}^{(n)}(\boldsymbol{\eta}) \\ \Delta_{f_1:2}^{(n)}(\boldsymbol{\eta}) \\ \boldsymbol{\Delta}_{f_1:3}^{(n)}(\boldsymbol{\eta}) \end{pmatrix}$$

$$:= \begin{pmatrix} n^{-1/2}\frac{1}{\sigma}\sum_{i=1}^{n}\varphi_{f_1}\left(\frac{d_i^{(n)}}{\sigma}\right)\boldsymbol{V}^{-1/2}\boldsymbol{U}_i^{(n)} \\ \frac{1}{2}n^{-1/2}\begin{pmatrix} \sigma^{-2}(\mathrm{vec}\boldsymbol{I}_k)' \\ \boldsymbol{M}_k\left(\boldsymbol{V}^{\otimes 2}\right)^{-1/2} \end{pmatrix}\sum_{i=1}^{n}\mathrm{vec}\left(\psi_{f_1}\left(\frac{d_i^{(n)}}{\sigma}\right)\frac{d_i^{(n)}}{\sigma}\boldsymbol{U}_i^{(n)}\boldsymbol{U}_i^{(n)\prime} - \boldsymbol{I}_k\right) \end{pmatrix}$$

and full-rank information matrix

$$\boldsymbol{\Gamma}_{f_1}(\boldsymbol{\eta}) := \begin{pmatrix} \boldsymbol{\Gamma}_{f_1:11}(\boldsymbol{\eta}) & \boldsymbol{0} & \boldsymbol{0} \\ \boldsymbol{0} & \Gamma_{f_1:22}(\boldsymbol{\eta}) & \boldsymbol{\Gamma}_{f_1:32}'(\boldsymbol{\eta}) \\ \boldsymbol{0} & \boldsymbol{\Gamma}_{f_1:32}(\boldsymbol{\eta}) & \boldsymbol{\Gamma}_{f_1:33}(\boldsymbol{\eta}) \end{pmatrix},$$

where

$$\boldsymbol{\Gamma}_{f_1:11}(\boldsymbol{\eta}) := \frac{1}{k\sigma^2}\mathcal{I}_k(f_1)\boldsymbol{V}^{-1},$$

$$\Gamma_{f_1:22}(\boldsymbol{\eta}) := \frac{1}{4\sigma^2}\left(\mathcal{J}_k(f_1) - k^2\right),$$

$$\boldsymbol{\Gamma}_{f_1:32}(\boldsymbol{\eta}) := \frac{1}{4k\sigma^2}\left(\mathcal{J}_k(f_1) - k^2\right)\boldsymbol{M}_k \text{vec}\boldsymbol{V}^{-1}$$

and

$$\boldsymbol{\Gamma}_{f_1:33}(\boldsymbol{\eta}) := \boldsymbol{M}_k\left(\boldsymbol{V}^{\otimes 2}\right)^{-1/2}\left\{\frac{\mathcal{J}_k(f_1)}{k(k+2)}\left(\boldsymbol{I}_{k^2} + \boldsymbol{K}_k + \boldsymbol{J}_k\right) - \boldsymbol{J}_k\right\}\left(\boldsymbol{V}^{\otimes 2}\right)^{-1/2}\boldsymbol{M}_k'.$$

More precisely, for any $\boldsymbol{\eta}^{(n)} = \left(\boldsymbol{\theta}^{(n)\prime}, \sigma^{2(n)}, \left(\overset{\circ}{\text{vech}}\ \boldsymbol{V}^{(n)}\right)'\right)' = \boldsymbol{\eta} + O\left(n^{-1/2}\right)$ *and any bounded sequence* $\boldsymbol{\tau}^{(n)} := \left(\boldsymbol{t}^{(n)\prime}, s^{(n)}, \left(\overset{\circ}{\text{vech}}\ \boldsymbol{v}^{(n)}\right)'\right)' = \left(\boldsymbol{\tau}_1^{(n)\prime}, \tau_2^{(n)}, \boldsymbol{\tau}_3^{(n)\prime}\right)' = in\ \mathbb{R}^{k+1+\frac{k(k-1)}{2}}$, *we have*

$$\Lambda^{(n)}_{\boldsymbol{\eta}^{(n)}+n^{-1/2}\boldsymbol{\tau}^{(n)}/\boldsymbol{\eta}^{(n)}:f_1} := \log\frac{dP^{(n)}_{\boldsymbol{\eta}^{(n)}+n^{-1/2}\boldsymbol{\tau}^{(n)}:f_1}}{dP^{(n)}_{\boldsymbol{\eta}^{(n)}:f_1}}$$
$$= \left(\boldsymbol{\tau}^{(n)}\right)'\boldsymbol{\Delta}^{(n)}_{f_1}\left(\boldsymbol{\eta}^{(n)}\right) - \frac{1}{2}\left(\boldsymbol{\tau}^{(n)}\right)'\boldsymbol{\Gamma}_{f_1}(\boldsymbol{\eta})\boldsymbol{\tau}^{(n)} + o_P(1)$$

and

$$\boldsymbol{\Delta}^{(n)}_{f_1}\left(\boldsymbol{\eta}^{(n)}\right) \xrightarrow{\mathcal{L}} N\left(\boldsymbol{0}, \boldsymbol{\Gamma}_{f_1}(\boldsymbol{\eta})\right)$$

under $P^{(n)}_{\boldsymbol{\eta}^{(n)}:f_1}$ *as* $n \to \infty$.

Test statistics based on efficient central sequences are only valid under correctly specified radial densities. However, a correct specification f_1 of the actual radial density g_1 is unrealistic. Therefore, we have to consider the problem from a semiparametric point of view, where g_1 is a nuisance parameter. Within the family of distributions $\cup_{\sigma^2}\cup_{\boldsymbol{V}}\cup_{g_1}\left\{P^{(n)}_{\boldsymbol{\theta},\sigma^2,\boldsymbol{V}:g_1}\right\}$, where $\boldsymbol{\theta}$ is fixed, we consider the null hypothesis $\mathcal{H}_0(\boldsymbol{\theta}, \boldsymbol{V}_0)$ under which we have $\boldsymbol{V} = \boldsymbol{V}_0$. Namely, $\boldsymbol{\theta}$ and $\boldsymbol{V} = \boldsymbol{V}_0$ are fixed, and σ^2 and the radial density g_1 remain unspecified. For the scalar nuisance σ^2, we have the efficient central sequence by means of a simple projection. For the infinite-dimensional nuisance g_1, we can take the projection of the central sequences along adequate tangent space in principle. Here, however, we consider the appropriate group invariance structures which allow for the same result by conditioning central sequences with respect to maximal invariants such as ranks or signs in line with Hallin and Werker (2003).

The null hypothesis $\mathcal{H}_0(\boldsymbol{\theta}, \boldsymbol{V}_0)$ is invariant under the following two groups of transformations acting on the observations $\boldsymbol{X}_1, \ldots, \boldsymbol{X}_n$.

(i) The group $\mathcal{G}^{\text{orth}(n)}, \circ := \mathcal{G}^{\text{orth}(n)}_{\boldsymbol{\theta},\boldsymbol{V}_0}, \circ$ of $\boldsymbol{V}_0$-orthogonal transformations centered at $\boldsymbol{\theta}$ consisting of all transformations of the form

$$\boldsymbol{X}^{(n)} \mapsto \mathcal{G}_{\boldsymbol{O}}(\boldsymbol{X}_1, \ldots, \boldsymbol{X}_n)$$
$$= \mathcal{G}_{\boldsymbol{O}}\left(\boldsymbol{\theta} + d_1^{(n)}(\boldsymbol{\theta}, \boldsymbol{V}_0)\boldsymbol{V}_0^{1/2}\boldsymbol{U}_1^{(n)}(\boldsymbol{\theta}, \boldsymbol{V}_0), \ldots, \boldsymbol{\theta} + d_n^{(n)}(\boldsymbol{\theta}, \boldsymbol{V}_0)\boldsymbol{V}_0^{1/2}\boldsymbol{U}_n^{(n)}(\boldsymbol{\theta}, \boldsymbol{V}_0)\right)$$
$$:= \left(\boldsymbol{\theta} + d_1^{(n)}(\boldsymbol{\theta}, \boldsymbol{V}_0)\boldsymbol{V}_0^{1/2}\boldsymbol{O}\boldsymbol{U}_1^{(n)}(\boldsymbol{\theta}, \boldsymbol{V}_0), \ldots, \boldsymbol{\theta} + d_n^{(n)}(\boldsymbol{\theta}, \boldsymbol{V}_0)\boldsymbol{V}_0^{1/2}\boldsymbol{O}\boldsymbol{U}_n^{(n)}(\boldsymbol{\theta}, \boldsymbol{V}_0)\right),$$

where $\boldsymbol{O}$ is an arbitrary $k \times k$ orthogonal matrix. This group contains rotations around $\boldsymbol{\theta}$ in the metric associated with $\boldsymbol{V}_0$. Furthermore, it also contains the reflection with respect to $\boldsymbol{\theta}$, namely, the mapping $(\boldsymbol{X}_1, \ldots, \boldsymbol{X}_n) \mapsto (\boldsymbol{\theta} - (\boldsymbol{X}_1 - \boldsymbol{\theta}), \ldots, \boldsymbol{\theta} - (\boldsymbol{X}_n - \boldsymbol{\theta}))$.

(ii) The group $\mathcal{G}^{(n)}, \circ := \mathcal{G}^{(n)}_{\boldsymbol{\theta},\boldsymbol{V}_0}, \circ$ of continuous monotone radial transformations consisting of all transformations of the form

$$\boldsymbol{X}^{(n)} \mapsto \mathcal{G}_h(\boldsymbol{X}_1, \ldots, \boldsymbol{X}_n)$$

$$= \mathcal{G}_h\left(\theta + d_1^{(n)}(\theta, V_0)\, V_0^{1/2} U_1^{(n)}(\theta, V_0), \ldots, \theta + d_n^{(n)}(\theta, V_0)\, V_0^{1/2} U_n^{(n)}(\theta, V_0)\right)$$
$$:= \left(\theta + h\left(d_1^{(n)}(\theta, V_0)\right) V_0^{1/2} U_1^{(n)}(\theta, V_0), \ldots, \theta + h\left(d_n^{(n)}(\theta, V_0)\right) V_0^{1/2} U_n^{(n)}(\theta, V_0)\right),$$

where $h : \mathbb{R}_0^+ \to \mathbb{R}_0^+$ is continuous, monotone increasing with $h(0) = 0$ and $\lim_{r\to\infty} h(r) = \infty$. This group contains the subgroup of scale transformations $(X_1, \ldots, X_n) \mapsto (\theta + \alpha(X_1 - \theta), \ldots, \theta + \alpha(X_n - \theta))$ with $\alpha > 0$.

The group $\mathcal{G}^{(n)}, \circ$ of continuous monotone radial transformations is a generating group for the family of distributions $\cup_{\sigma^2} \cup_{f_1} \left\{P^{(n)}_{\theta,\sigma^2,V_0:f_1}\right\}$, namely, a generating group for the null hypothesis $\mathcal{H}_0(\theta, V_0)$ which is under consideration. Therefore, from the invariant principle, we can consider the test statistics which are measurable with respect to the corresponding maximal invariant. In this case, it is the vector $\left(R_1^{(n)}(\theta, V_0), \ldots, R_n^{(n)}(\theta, V_0), U_1^{(n)}(\theta, V_0), \ldots, U_n^{(n)}(\theta, V_0)\right)$, where $R_i^{(n)}(\theta, V_0)$ denotes the rank of $d_i^{(n)}(\theta, V_0)$ among $d_1^{(n)}(\theta, V_0), \ldots, d_n^{(n)}(\theta, V_0)$. Then, we have the signed rank test statistics which are invariant under $\mathcal{G}^{(n)}, \circ$ so that they are distribution free under $\mathcal{H}_0(\theta, V_0)$.

Combining invariance and optimality arguments by considering a signed rank-based version of the f_1-efficient central sequences for shape, we construct the test statistics for the null hypothesis $\mathcal{H}_0(\theta, V_0)$. Note that the central sequences are always defined up to $o_P(1)$ under $P^{(n)}_{\eta:f_1}$ as $n \to \infty$. We consider the signed rank version $\underline{\Delta}^{(n)}_{f_1}(\eta)$ of the shape efficient central sequence

$$\Delta^{*(n)}_{f_1}(\eta) = \Delta^{(n)}_{f_1:3}(\eta) - \Gamma_{f_1:32}(\eta)\Gamma^{-1}_{f_1:22}(\eta)\Delta^{(n)}_{f_1:2}(\eta)$$
$$= \frac{1}{2} n^{-1/2} M_k \left(V^{\otimes 2}\right)^{-1/2} J_k^{\perp} \sum_{i=1}^{n} \psi_{f_1}\left(\frac{d_i^{(n)}}{\sigma}\right) \frac{d_i^{(n)}}{\sigma} \mathrm{vec}\left(U_i^{(n)} U_i^{(n)\prime}\right),$$

where $J_k^{\perp} := I_{k^2} - \frac{1}{k} J_k$, in our nonparametric test. For this purpose, we use the f_1-score version, based on the scores $K = K_{f_1}$, of statistic

$$\underline{\Delta}^{(n)}_{K}(\eta) := \frac{1}{2} n^{-1/2} M_k \left(V^{\otimes 2}\right)^{-1/2} J_k^{\perp} \sum_{i=1}^{n} K\left(\frac{R_i^{(n)}}{n+1}\right) \mathrm{vec}\left(U_i^{(n)} U_i^{(n)\prime}\right)$$
$$= \frac{1}{2} n^{-1/2} M_k \left(V^{\otimes 2}\right)^{-1/2} \sum_{i=1}^{n} K\left(\frac{R_i^{(n)}}{n+1}\right) \mathrm{vec}\left(U_i^{(n)} U_i^{(n)\prime} - \frac{1}{k} I_k\right)$$
$$= \frac{1}{2} n^{-1/2} M_k \left(V^{\otimes 2}\right)^{-1/2} \left\{\sum_{i=1}^{n} K\left(\frac{R_i^{(n)}}{n+1}\right) \mathrm{vec}\left(U_i^{(n)} U_i^{(n)\prime}\right) - \frac{m_K^{(n)}}{k} \mathrm{vec} I_k\right\},$$

where $R_i^{(n)} = R_i^{(n)}(\theta, V)$ denotes the rank of $d_i^{(n)} = d_i^{(n)}(\theta, V)$ among $d_1^{(n)}, \ldots, d_n^{(n)}$, $U_i^{(n)} = U_i^{(n)}(\theta, V)$ and $m_K^{(n)} = \sum_{i=1}^{n} K\left(\frac{i}{n+1}\right)$. The following asymptotic representation result shows that $\underline{\Delta}^{(n)}_{f_1}(\eta)$ is indeed another version of the efficient central sequence $\Delta^{*(n)}_{f_1}(\eta)$.

Lemma 5.2.40 (Lemma 4.1 of Hallin and Paindaveine (2006)). *Assume that the score function $K : (0,1) \to \mathbb{R}$ is continuous and square integrable. And assume that it can be expressed as the difference of two monotone increasing functions. Then, defining*

$$\Delta^{*(n)}_{K:g_1}(\eta) := \frac{1}{2} n^{-1/2} M_k \left(V^{\otimes 2}\right)^{-1/2} J_k^{\perp} \sum_{i=1}^{n} K\left(\overline{G}_{1k}\left(\frac{d_i^{(n)}}{\sigma}\right)\right) \mathrm{vec}\left(U_i^{(n)} U_i^{(n)\prime}\right),$$

we have $\underline{\Delta}^{(n)}_{K}(\eta) = \Delta^{*(n)}_{K:g_1}(\eta) + o_{L_2}(1)$ as $n \to \infty$, under $P^{(n)}_{\eta:g_1}$, where $\overline{G}_{1k}$ is defined in the same manner as $\overline{F}_{1k}$ in (5.2.4.2).

Now, we can propose the class of test statistics. Let $K : (0,1) \to \mathbb{R}$ be some score function as in Lemma 5.2.40, and write $E\{K(U)\}$ and $E\{K^2(U)\}$ for $\int_0^1 K(u)\,du$ and $\int_0^1 K^2(u)\,du$, respectively. Then, the K-score version of the statistics for testing $\mathcal{H} : \boldsymbol{V} = \boldsymbol{V}_0$ is given by

$$\underline{Q}_K = \underline{Q}_K^{(n)} := \frac{k(k+2)}{2nE\{K^2(U)\}} \sum_{i,j=1}^{n} K\left(\frac{R_i^{(n)}}{n+1}\right) K\left(\frac{R_j^{(n)}}{n+1}\right) \left\{\left(\boldsymbol{U}_i^{(n)\prime} \boldsymbol{U}_j^{(n)}\right)^2 - \frac{1}{k}\right\},$$

where $R_i^{(n)} = R_i^{(n)}(\boldsymbol{\theta}, \boldsymbol{V}_0)$ and $\boldsymbol{U}_i^{(n)} = \boldsymbol{U}_i^{(n)}(\boldsymbol{\theta}, \boldsymbol{V}_0)$. These tests can be rewritten as

$$\begin{aligned}
\underline{Q}_K &= \frac{nk(k+2)}{2E\{K^2(U)\}} \left(\mathrm{tr}\boldsymbol{S}_K^2 - \frac{1}{k}\mathrm{tr}^2 \boldsymbol{S}_K\right) \\
&= \frac{nk(k+2)E^2\{K(U)\}}{2E\{K^2(U)\}} \left\| \frac{\boldsymbol{S}_K}{\mathrm{tr}\boldsymbol{S}_K} - \frac{1}{k}\boldsymbol{I}_k \right\|^2 + o_P(1) \\
&= \frac{k(k+2)}{2E\{K^2(U)\}} \left\| n^{-1/2} \sum_{i=1}^{n} K\left(\frac{R_i^{(n)}}{i+1}\right) \left(\boldsymbol{U}_i^{(n)} \boldsymbol{U}_i^{(n)\prime} - \frac{1}{k}\boldsymbol{I}_k\right) \right\|^2 + o_P(1),
\end{aligned}$$

as $n \to \infty$ under any distribution, where

$$\boldsymbol{S}_K = \boldsymbol{S}_K^{(n)} := \frac{1}{n} \sum_{i=1}^{n} K\left(\frac{R_i^{(n)}}{i+1}\right) \boldsymbol{U}_i^{(n)} \boldsymbol{U}_i^{(n)\prime}.$$

On the other hand, the local optimality under radial density f_1 is achieved by the test based on the test statistics taking the form

$$\underline{Q}_{f_1} = \underline{Q}_{K_{f_1}}^{(n)} := \frac{k(k+2)}{2n\mathcal{I}_k(f_1)} \sum_{i,j=1}^{n} K_{f_1}\left(\frac{R_i^{(n)}}{n+1}\right) K_{f_1}\left(\frac{R_j^{(n)}}{n+1}\right) \left\{\left(\boldsymbol{U}_i^{(n)\prime} \boldsymbol{U}_j^{(n)}\right)^2 - \frac{1}{k}\right\},$$

which can be rewritten as

$$\begin{aligned}
\underline{Q}_{f_1} &= \frac{nk(k+2)}{2\mathcal{I}_k(f_1)} \left(\mathrm{tr}\boldsymbol{S}_{f_1}^2 - \frac{1}{k}\mathrm{tr}^2 \boldsymbol{S}_{f_1}\right) \\
&= \frac{nk(k+2)}{2\mathcal{I}_k(f_1)} \left\| \frac{\boldsymbol{S}_{f_1}}{\mathrm{tr}\boldsymbol{S}_{f_1}} - \frac{1}{k}\boldsymbol{I}_k \right\|^2 + o_P(1),
\end{aligned}$$

as $n \to \infty$ under any distribution, where

$$\boldsymbol{S}_{f_1} = \boldsymbol{S}_{f_1}^{(n)} := \frac{1}{n} \sum_{i=1}^{n} K_{f_1}\left(\frac{R_i^{(n)}}{i+1}\right) \boldsymbol{U}_i^{(n)} \boldsymbol{U}_i^{(n)\prime}.$$

To describe the asymptotic behaviour of $\underline{Q}_{f_1}$ and $\underline{Q}_K$, we define the quantities

$$\mathcal{I}_k(f_1, g_1) := \int_0^1 K_{f_1}(u) K_{g_1}(u)\,du,$$

which is a measure of cross information, and

$$\mathcal{I}_k(K; g_1) := \int_0^1 K(u) K_{g_1}(u)\,du.$$

We define the rank-based test $\underline{\phi}_{f_1}^{(n)}$ and $\underline{\phi}_K^{(n)}$, which consists in rejecting $\mathcal{H} : \boldsymbol{V} = \boldsymbol{V}_0$ when $\underline{Q}_{f_1}$ and $\underline{Q}_K$ exceed the α-upper quantile of a chi-square distribution with $\frac{k(k-1)}{2}$ degrees of freedom, respectively. Now, we have the following result.

Theorem 5.2.41 (Proposition 4.1 of Hallin and Paindaveine (2006)). *Let K be a continuous, square integrable score function defined on $(0,1)$ that can be expressed as the difference of two monotone increasing functions. Furthermore, let f_1, satisfying Assumptions (A1) and (A2), be such that K_{f_1} is continuous and can be expressed as the difference of two monotone increasing functions. Then, we have the following.*

(i) *Test statistics $\underline{Q}_{f_1}$ and $\underline{Q}_K$ are asymptotically chisquare with $\frac{k(k-1)}{2}$ degrees of freedom under $\cup_{\sigma^2}\cup_{g_1}\left\{P^{(n)}_{\theta,\sigma^2,V_0:g_1}\right\}$. On the other hand, under $\cup_{\sigma^2}\left\{P^{(n)}_{\theta,\sigma^2,V_0+n^{-1/2}v:g_1}\right\}$, $\underline{Q}_{f_1}$ and $\underline{Q}_K$ are asymptotically noncentral chisquare with $\frac{k(k-1)}{2}$ degrees of freedom with noncentrality parameters*

$$\frac{\mathcal{I}_k^2(f_1,g_1)}{2k(k+2)\mathcal{I}_k(f_1)}\left[\mathrm{tr}\left\{\left(V_0^{-1}v\right)^2\right\}-\frac{1}{k}\left\{\mathrm{tr}\left(V_0^{-1}v\right)\right\}^2\right]$$

and

$$\frac{\mathcal{I}_k^2(K,g_1)}{2k(k+2)E\{K^2(U)\}}\left[\mathrm{tr}\left\{\left(V_0^{-1}v\right)^2\right\}-\frac{1}{k}\left\{\mathrm{tr}\left(V_0^{-1}v\right)\right\}^2\right],$$

respectively.

(ii) *The sequences of tests $\underline{\phi}^{(n)}_{f_1}$ and $\underline{\phi}^{(n)}_K$ have asymptotic level α under $\cup_{\sigma^2}\cup_{g_1}\left\{P^{(n)}_{\theta,\sigma^2,V_0:g_1}\right\}$.*

(ii) *Furthermore, the sequence of tests $\underline{\phi}^{(n)}_{f_1}$ is locally asymptotically maximin efficient for $\cup_{\sigma^2}\cup_{g_1}\left\{P^{(n)}_{\theta,\sigma^2,V_0:g_1}\right\}$ against alternatives of the form $\cup_{\sigma^2}\cup_{V\neq V_0}\left\{P^{(n)}_{\theta,\sigma^2,V:f_1}\right\}$.*

5.3 Asymptotic Theory of Rank Order Statistics for ARCH Residual Empirical Processes

This section develops the asymptotic theory of a class of rank order statistics $\{T_N\}$ for a two-sample problem pertaining to empirical processes based on the squared residuals from two classes of ARCH models. The results can be applied to testing the innovation distributions of two financial returns generated by different mechanisms (e.g., different countries and/or different industries, etc.).

An important result is that, unlike the residuals of ARMA models, the asymptotics of $\{T_N\}$ depend on those of ARCH volatility estimators. Such asymptotics provide a useful guide to the reliability of confidence intervals, asymptotics relative efficiency and ARCH affection.

As we explained in the previous chapters, the ARCH model is one of the most fundamental econometric models to describe financial returns. For an ARCH(p) model, Horváth et al. (2001) derived the asymptotic distribution of the empirical process based on the squared residuals. Then they showed that, unlike the residuals of ARMA models, those residuals do not behave like asymptotically independent random variables, and the asymptotic distribution involves a term depending on estimators of the volatility parameters of the model.

In i.i.d. settings, a two-sample problem is one of the important statistical problems. For this problem, a class of rank order statistics plays a principal role since it provides locally most powerful rank tests. The study of the asymptotic properties is a fundamental and essential part of nonparametric statistics. Many authors have contributed to the development, and numerous theorems have been formulated to show the asymptotic normality of a properly normalized rank order statistic in many testing problems. The classical limit theorem is the celebrated Chernoff and Savage (1958) theorem, which is widely used to study the asymptotic power and power efficiency of a class of two-sample tests. Further refinements of the theorem are given by Hájek and Šidák (1967) and Puri and Sen (1971). Specifically, Puri and Sen (1971) formulated the Chernoff–Savage theorem under less stringent conditions on the score-generating functions.

This section discusses the asymptotic theory of the two-sample rank order statistics $\{T_N\}$ for ARCH residual empirical processes based on the techniques of Puri and Sen (1971) and Horváth et al. (2001). Since the asymptotics of the residual empirical processes are different from those for the usual ARMA processes, the limiting distribution of $\{T_N\}$ is greatly different from that of the ARMA case (of course the i.i.d. case) .

Suppose that a class of ARCH(p) models is characterized by the equations

$$X_t = \begin{cases} \sigma_t(\theta_X)\epsilon_t \,, \sigma_t^2(\theta_X) = \theta_X^0 + \sum_{i=1}^{p_X} \theta_X^i X_{t-i}^2, & for\ t = 1, \cdots, m, \\ 0, & for\ t = -p_X + 1, \cdots, 0, \end{cases} \tag{5.3.1}$$

where $\{\epsilon_t\}$ is a sequence of i.i.d. (0, 1) random variables with fourth-order cumulant $k_4^X, \theta_X = (\theta_X^0, \theta_X^1, \cdots, \theta_X^{p_X})' \in \Theta_X \subset \vec{R}^{p_x+1}$ is an unknown parameter vector satisfying $\theta_X^0 > 0, \theta_X^i \geq 0, i = 1, \cdots, p_X - 1, \theta^{p_X} > 0$, and ϵ_t is independent of $X_s, s < t$. Denote by $F(x)$ the distribution function of ϵ_t^2 and we assume that $f(x) = F'(x)$ exists and is conditioned on $(0, \infty)$.

Suppose that another class of ARCH(p) models, independent of $\{X_t\}$, is characterized similarly by the equations

$$Y_t = \begin{cases} \sigma_t(\theta_Y)\xi_t \,, \sigma_t^2(\theta_Y) = \theta_Y^0 + \sum_{i=1}^{p_Y} \theta_Y^i Y_{t-i}^2, & for\ t = 1, \cdots, m, \\ 0, & for\ t = -p_Y + 1, \cdots, 0, \end{cases} \tag{5.3.2}$$

where $\{\xi_t\}$ is a sequence of i.i.d. (0, 1) random variables with fourth-order cummulant $k_4^Y, \theta_Y = (\theta_Y^0, \theta_Y^1, \cdots, \theta_Y^{p_Y})' \in \Theta_Y \subset \vec{R}^{p_Y+1}, \theta_Y^0 > 0, \theta_Y^i \geq 0, i = 1, \cdots, p_Y - 1, \theta_Y^{p_Y} > 0$, are unknown parameters, and ξ_T is independent of Y_s, $s < t$. The distribution function of ξ_t^2 is denoted by $G(x)$ and we assume that $g(x) = G'(x)$ exists and is continuous on $(0, \infty)$. For (5.3.1) and (5.3.2), we assume that $\theta_X^1 + \cdots + \theta_X^{p_X} < 1$ and $\theta_Y^1 + \cdots + \theta_Y^{p_Y} < 1$ for stationarity (see Milhøj (1985)).

Now we are interested in the two-sample problem of testing

$$H_0 : F(x) = G(x)\ for\ all\ x\ against\ H_A : F(x) \neq G(x)\ for\ some\ x. \tag{5.3.3}$$

First, consider the estimation of θ_X and θ_Y. Letting

$$Z_{X,t} = X_t^2, Z_{Y,t} = Y_t^2, \zeta_{X,t} = (\epsilon_t^2 - 1)\sigma_t^2(\theta_X), \zeta_{Y,t} = (\xi_t^2 - 1)\sigma_t^2(\theta_Y),$$
$$W_{X,t} = (1, Z_{X,t}, \cdots, Z_{X,t-P_X+1})'\ and\ W_{Y,t} = (1, Z_{Y,t}, \cdots, Z_{Y,t-P_Y+1})',$$

we have the following autoregressive representations :

$$Z_{X,t} = \theta_X' W_{X,t-1} + \zeta_{X,t} \,, Z_{Y,t} = \theta_Y' W_{Y,t-1} + \zeta_{Y,t}. \tag{5.3.4}$$

Here, note that $E[\zeta_{X,t}|\mathcal{F}_{t-1}^X] = E[\zeta_{Y,t}|\mathcal{F}_{t-1}^Y] = 0$, where $\mathcal{F}_t^X = \sigma\{X_t, X_{t-1}, \cdots\}$ and $\mathcal{F}_t^Y = \sigma\{Y_t, Y_{t-1}, \cdots\}$. Suppose that $Z_{X,1}, \cdots, Z_{X,m}$ and $Z_{Y,1}, \cdots, Z_{Y,n}$ are observed stretches from $\{Z_{X,t}\}$ and $\{Z_{Y,t}\}$, respectively. Let

$$L_m(\theta_X) = \sum_{t=1}^{m} (Z_{X,t} - \theta_X' W_{X,t-1})^2\ and\ L_n(\theta_Y) = \sum_{t=1}^{n} (Z_{Y,t} - \theta_Y' W_{Y,t-1})^2 \tag{5.3.5}$$

be the least squares penalty functions. Then the conditional least-squares estimators of θ_X and θ_Y are given by

$$\hat{\theta}_{X,m} = \arg \min_{\theta_X \in \Theta_X} L_m(\theta_X), \quad \hat{\theta}_{Y,n} = \arg \min_{\theta_Y \in \Theta_Y} L_n(\theta_Y),$$

respectively. In what follows we assume that $\hat{\theta}_{X,m}$ and $\hat{\theta}_{Y,n}$ are asymptotically consistent and normal with rate $m^{-1/2}$ and $n^{-1/2}$, respectively, i.e.,

$$\sqrt{m}||\hat{\theta}_{X,m} - \theta_X|| = 0_p(1), \quad \sqrt{n}||\hat{\theta}_{Y,n} - \theta_Y|| = 0_p(1), \tag{5.3.6}$$

where $||.||$ denotes the Euclidean norm. For the validity of (5.3.6), Tjøstheim (1986, pp. 254–256) gave a set of sufficient conditions.
The corresponding empirical squared residuals are

$$\hat{\epsilon}_t^2 = X_t^2/\hat{\sigma}_t^2(\hat{\theta}_X,\ t = 1, \cdots, m \text{ and } \hat{\xi}_t^2 = Y_t^2/\sigma_t^2(\hat{\theta}_Y)\ , t = 1, \cdots, n \tag{5.3.7}$$

where $\hat{\sigma}_\tau^2(\hat{\theta}_X) = \hat{\theta}_X^0 + \sum_{i=1}^{p_x} \hat{\theta}_X^i X_{t-i}^2$ and $\hat{\sigma}_t^2(\hat{\theta}_Y) = \hat{\theta}_Y^0 + \sum_{i=1}^{p_y} \hat{\theta}_Y^i Y_{t-i}^2$.
In line with Puri and Sen (1971) we describe the setup. Let $N = m + n$ and $\lambda_N = m/N$. For (5.3.7) the m and n are assumed to be such that $0 < \lambda_0 \le \lambda_N \le 1 - \lambda_0 < 1$ hold for some fixed $\lambda_0 \le \frac{1}{2}$. Then the combined distribution is defined by

$$H_N(x) = \lambda_N F(x) + (1 - \lambda_N)G(x).$$

If $\hat{F}_m(x)$ and $\hat{G}_n(x)$ denote the empirical distribution functions of $\{\hat{\epsilon}_t^2\}$ and $\{\hat{\xi}_t^2\}$, the corresponding empirical distribution to $H_N(x)$ is

$$\hat{H}_N(x) = \lambda_N \hat{F}_m(x) + (1 - \lambda_N)\hat{G}_n(x). \tag{5.3.8}$$

Write $\hat{B}_m(x) = \sqrt{m}(\hat{F}_m(x) - F(x))$ and $\hat{B}_n(x) = \sqrt{n}(\hat{G}_n(x) - G(x))$. Then they are expressed as

$$\hat{B}_m(x) = \frac{1}{\sqrt{m}} \sum_{t=1}^{m} [\chi(\hat{\epsilon}_t^2 \le x) - F(x)], \tag{5.3.9}$$

and

$$\hat{B}_n(x) = \frac{1}{\sqrt{n}} \sum_{t=1}^{m} [\chi(\hat{\xi}_t^2 \le x) - G(x)], \tag{5.3.10}$$

where $\chi(\cdot)$ is the indicator function of $(\cdot)$. Horváth et al. (2001) showed that (5.3.9) has the representation

$$\hat{B}_m(x) = \epsilon_m(x) + A_X x f(x) + \eta_m(x), \tag{5.3.11}$$

where

$$\epsilon_m(x) = \frac{1}{\sqrt{m}} \sum_{t=1}^{m} [\chi(\epsilon_t^2 \le x) - F(x)], A_X = \sum_{i=o}^{p_X} \sqrt{m}(\hat{\theta}_{X,m}^i \theta_X^i)\tau_{i,X}$$

and

$$sup_x|\eta_m(x)| = o_p(1) \quad with$$
$$\tau_{0,X} = E[1/\sigma_t^2(\theta_X)] \ and \ \tau_{i,X} = E[\sigma_{t-i}^2(\theta_X)\epsilon_{t-i}^2/\sigma^2(\theta_X)], \quad 1 \le i \le p_X.$$

Similarly

$$\hat{B}_n(x) = \epsilon_n(x) + A_Y x g(x) + \eta_n(x), \tag{5.3.12}$$

where

$$\epsilon_n(x) = \frac{1}{\sqrt{n}} \sum_{t=1}^{n} [\chi(\xi_t^2 \le x) - G(x)]\ , A_Y = \sum_{i=0}^{p_Y} \sqrt{n}(\hat{\theta}_{Y,n}^i - \theta_Y^i)\tau_{i,Y} \tag{5.3.13}$$

and

$$sup_x|\eta_n(x)| = o_p(1) \ with$$

$$\tau_{0,Y} = E[1/\sigma_t^2(\theta_Y)] \text{ and } \tau_{i,Y} = E[\sigma_{t-i}^2(\theta_Y)\xi_{t-i}^2/\sigma_t^2(\theta_Y)], \quad 1 \le i \le p_Y.$$

Let $F_m(x) = m^{-1}\sum_{t=1}^m \chi(\epsilon_t^2 \le x)$, $G_n(x) = n^{-1}\sum_{t=1}^n \chi(\xi_t^2 \le x)$ and $\widetilde{H}_N(x) = \lambda_N F_m(x) + (1-\lambda_N)G_n(x)$. From (5.3.10) and (5.3.11) it follows that

$$\widetilde{\mathcal{H}}_N(x) = \widetilde{\mathcal{H}}_N(x) + m^{-1/2}\lambda_N A_X x f(x) + n^{-1/2}(1-\lambda_N)A_Y x g(x) + \eta_N^*(x), \tag{5.3.14}$$

where $\eta_N^*(x) = m^{-1/2}\lambda_N\eta_m(x) + n^{-1/2}(1-\lambda_N)\eta_n(x)$. The expression (5.3.14) is fundamental and will be used later.

Define $S_{N,i} = 1$, if the ith smallest one in the combined residuals $\hat{\epsilon}_1^2, \cdots, \hat{\epsilon}_m^2, \hat{\xi}_1^2, \cdots, \hat{\xi}_n^2$ is from $\hat{\epsilon}_1^2, \cdots, \hat{\epsilon}_m^2$, and otherwise define $S_{N,i} = 0$. For the testing problem (5.3.3), consider a class of rank order statistics of the form

$$T_N = \frac{1}{m}\sum_{i=1}^N \varphi_{N,i} S_{N,i},$$

where the $\varphi_{N,i}$ are given constants called weights or scores. This definition is conventionally used. However, we use its equivalent form given by

$$T_N = \int J\left[\frac{N}{N+1}\widetilde{\mathcal{H}}_N(x)\right] d\hat{F}_m(x), \tag{5.3.15}$$

where $\varphi_{N,i} = J(j/(N+1))$, and $J(u)$, $0 < u < 1$, is a continuous function. The class of T_N is sufficiently rich since it includes the following typical examples:

(i) Wilcoxon's two-sample test with $J(u) = u$, $0 < u < 1$,
(ii) Van der Waerden's two-sample test with $J(u) = \Phi^{-1}(u), 0 < u < 1$ where $\Phi(x) = (2\pi)^{-1/2}\int_{-\infty}^x e^{-t^2/2}dt$,
(iii) Mood's two-sample test with $J(u) = (u - \frac{1}{2})^2$, $0 < u < 1$,
(iv) Klotz's normal scores test with $J(u) = (\Phi^{-1}(u))^2$, $0 < u < 1$.

Examples (i) and (ii) are the tests for location, and (iii) and (iv) are tests for scale when $\{\hat{\epsilon}_t^2\}$ and $\{\hat{\xi}_t^2\}$ are assumed to have the some median. The purpose of this section is to elucidate the asymptotics of (5.3.15). For this we need the following regularity conditions.

Assumption 5.3.1. *(i) $J(u)$ is not constant and has a continuous derivative $J'(u)$ on* $(0, 1)$.
(ii) There exists a constant $K > 0$ such that

$$|J| \le K[u(1-u)]^{-(1/2)+\delta} \text{ and } |J'| \le K[u(1-u)]^{-(3/2)+\delta}$$

for some $\delta > 0$.
(iii) $xf(x)$, $xg(x)$, $xf'(x)$ and $xg'(x)$ are uniformly bounded continuous, and integrable functions on $(0, \infty)$.
(iv) There exists a constant $c > 0$ such that $F(x) \ge c\{xf(x)\}$ and $G(x) \ge c\{xg(x)\}$ for all $x > 0$.

Returning to the model of $\{X_t\}$ and $\{Y_t\}$, we impose a further condition. Write

$$A_{X,t} = \begin{pmatrix} \theta_X^1\epsilon_t^2 & \cdots & \theta_X^{p_X-1}\epsilon_t^2 & \theta_X^{p_X}\epsilon_t^2 \\ 1 & \cdots & 0 & 0 \\ \ddots & \vdots & \ddots & \ddots \\ 0 & \cdots & 1 & 0 \end{pmatrix}, \quad \text{and}$$

$$A_{Y,t} = \begin{pmatrix} \theta_Y^1\xi_t^2 & \cdots & \theta_Y^{p_Y-1}\xi_t^2 & \theta_Y^{p_Y}\xi_t^2 \\ 1 & \cdots & 0 & 0 \\ \ddots & \vdots & \ddots & \ddots \\ 1 & \cdots & 1 & 0 \end{pmatrix}.$$

We introduce the tensor product $A_{X,t}^{\otimes 3} = A_{X,t} \otimes A_{X,t} \otimes A_{X,t}$ (e.g., Hannan (1970, pp. 516–518)), and define $\sum_{X,3} = E[A_{X,t}^{\otimes 3}]$ and $\sum_{Y,3} = E[A_{Y,t}^{\otimes 3}]$.

Assumption 5.3.2. *$E|\epsilon_t|^8 < \infty$ and $\| \sum_{X,3} \| < 1$, $E|\xi_t|^8 < \infty$ and $\| \sum_{Y,3} \| < 1$, where $\| \cdot \|$ is the spectral matrix norm.*

From Chen and An (1998), we see that Assumption 5.3.2 implies $E(Z_{X,t}^4) < \infty$ and $E(Z_{Y,t}^4) < \infty$. In the case when $p_X = 1$ and $\{\epsilon_t\}$ is Gaussian, if $\| \sum_{X,3} \| < 1$, then $\theta_X^1 < 15^{-1/3} \approx 0.4$.
Before we state the main result, note that the matrices

$$\mathcal{U}_X = E[W_{X,t-1} W'_{X,t-1}], \quad \mathcal{U}_Y = E[W_{Y,t-1} W'_{Y,t-1}],$$

$$\mathcal{R}_X = 2E[\sigma_t^4(\theta_X) W_{X,t-1} W'_{X,t-1}] \textit{ and } \mathcal{R}_Y = 2E[\sigma_t^4(\theta_Y) W_{Y,t-1} W'_{Y,t-1}] \tag{5.3.16}$$

are positive definite (see Exercise 5.1).
Recalling (5.3.4), we observe that

$$\frac{\partial}{\partial \theta_X^0} \mathcal{L}_m(\theta_X) = -2 \sum_{t=1}^m (\epsilon_t^2 - 1)\sigma_t^2(\theta_X) = -2 \sum_{t=1}^m \phi_X(\epsilon_t^2) \vartheta_{t,X}^0 \ \ (say),$$

$$\frac{\partial}{\partial \theta_X^i} \mathcal{L}_m(\theta_X) = -2 \sum_{t=1}^m (\epsilon_t^2 - 1)\sigma_t^2(\theta_X) Z_{X,t-i} = -2 \sum_{t=1}^m \phi_X(\epsilon_t^2) \vartheta_{t,X}^i, \ \ 1 \le i \le p_X \ \ (say),$$

$$\frac{\partial}{\partial \theta_Y^0} \mathcal{L}_n(\theta_Y) = -2 \sum_{t=1}^m (\xi_t^2 - 1)\sigma_t^2(\theta_Y) = -2 \sum_{t=1}^n \phi_Y(\epsilon_t^2) \vartheta_{t,Y}^0 \ \ (say),$$

$$\frac{\partial}{\partial \theta_Y^i} \mathcal{L}_n(\theta_Y) = -2 \sum_{t=1}^n (\xi_t^2 - 1)\sigma_t^2(\theta_Y) Z_{Y,t-i} = -2 \sum_{t=1}^m \phi_Y(\xi_t^2) \vartheta_{t,Y}^i, \ \ 1 \le i \le p_Y \ \ (say)$$

where $\phi_\cdot(u) = u - 1$. Write $\vartheta_{t,X} = (\vartheta_{t,X}^0, \cdots, \vartheta_{t,X}^{p_X})'$ and $\vartheta_{t,Y} = (\vartheta_{t,Y}^0, \cdots, \vartheta_{t,p_Y,Y}^{p_Y})'$. Then, using standard arguments, it is seen that the ith element of $\hat{\theta}_{X,m}$ and $\hat{\theta}_{Y,n}$ admits the stochastic expansions

$$\hat{\theta}_{X,m}^i - \theta_X^i = \frac{1}{m} \sum_{t=1}^m U_{t,X}^i \phi_X(\epsilon_t^2) + o_p(m^{-1/2}), \ \ 0 \le i \le p_X \ \ and$$

$$\hat{\theta}_{Y,n}^i - \theta_Y^i = \frac{1}{n} \sum_{t=1}^n U_{t,Y}^i \phi_Y(\epsilon_t^2) + o_p(n^{-1/2}), \ \ 0 \le i \le p_Y \tag{5.3.17}$$

where $U_{t,X}^i$ and $U_{t,Y}^i$ are the ith elements of $\mathcal{U}_X^{-1} \vartheta_{t,X}$ and $\mathcal{U}_Y^{-1} \vartheta_{t,Y}$, respectively. Write $\alpha_X^i = E(U_{t,X}^i)$, $0 \le i \le p_X$, $\alpha_Y^i = E(U_{t,Y}^i)$, $0 \le i \le p_Y$, and $\tau_X = (\tau_{0,X}, \tau_{1,X}, \cdots, \tau_{p_X,X})'$ (recall (5.3.11) and (5.3.12)). The following theorem is due to Chandra and Taniguchi (2003). Here we just give the main line of the proof because the detailed one is very technical.

Theorem 5.3.3. *Suppose that Assumptions 5.3.1 and 5.3.2 hold and that $\hat{\theta}_{X,m}$ and $\hat{\theta}_{Y,n}$ are the conditional least-squares estimators of θ_X and θ_Y satisfying (5.3.6). Then*

$$N^{1/2}(T_N - \mu_N)/\sigma_N \xrightarrow{d} N(0, 1) \ \ as \ \ N \to \infty,$$

where

$$\mu_N = \int J[H_N(x)] dF(x) \ \ and \ \ \sigma_N^2 = \sigma_{1N}^2 + \sigma_{2N}^2 + \sigma_{3N}^2 + \gamma_N \neq 0 \ \ with$$

$$\sigma_{1N}^2 = 2(1 - \lambda_N)\{ \int \int_{x<y} A_N(x, y) dF(x) dF(y)$$

$$+\frac{1-\lambda_N}{\lambda_N}\int\int_{x<y} B_N(x,y)dG(x)dG(y)\},$$

$$\sigma_{2N}^2 = \omega'_{X,N}\mathcal{V}_X^{-1}\mathcal{B}_X\mathcal{U}_X^{-1}\omega_{X,N}\ ,\sigma_{3N}^2 = \omega'_{Y,N}\mathcal{V}_Y^{-1}\mathcal{R}_Y\mathcal{U}_Y^{-1}\omega_{Y,N}$$

and

$$\gamma_N = 2(1-\lambda_N)\left\{\frac{1-\lambda_N}{\lambda_N}\sum_{i=0}^{p_X}\tau_{i,X}\iint h_X^i(x)\delta_{f,N}(x,y)dG(x)dG(y)\right.$$
$$\left.+\sum_{i=0}^{p_Y}\tau_{i,Y}\iint h_Y^i(x)\delta_{g,N}(x,z)dF(x)dF(z)\right\}, \tag{5.3.18}$$

where

$$A_N(x,y) = G(x)(1-G(y))J'[H_N(x)]J'[H_N(y)],$$
$$B_N(x,y) = F(x)(1-F(y))J'[H_N(x)]J'[H_N(y)],$$
$$\omega_{X,N} = -(\lambda_N)^{-1/2}(1-\lambda_N)\int xf(x)J'[H_N(x)]dG(x)\times\tau_X,$$
$$\omega_{Y,N} = (1-\lambda)^{1/2}\int zg(z)J'[H_N(z)]dF(z)\times\tau_Y,$$
$$\delta_{f,N}(x,y) = yf(y)J'[H_N(x)]J'[H_N(y)],$$
$$\delta_{g,N}(x,z) = zg(z)J'[H_N(x)]J'[H_N(z)],$$
$$h_X^i(x) = \alpha_X^i\int_0^x\phi_X(u)f(u)du, \quad 0\le i\le p_X,$$
$$h_Y^i(x) = \alpha_Y^i\int_0^x\phi_Y(u)g(u)du, \quad 0\le i\le p_Y.$$

Sketch of Proof. First, we rewrite the integrand of T_N as

$$J\left[\frac{N}{N+1}\hat{\mathcal{H}}_N\right] = J[H_N] + (\hat{\mathcal{H}}_N - H_N)J'[H_N] - \frac{\hat{\mathcal{H}}_N}{N+1}J'[H_N]$$
$$+J\left[\frac{N}{N+1}\hat{\mathcal{H}}_N\right] - J[H_N] - \left(\frac{N}{N+1}\hat{\mathcal{H}}_N - H_N\right)J'[H_N],$$

and $d\hat{F}_m = d(\hat{F}_m - F + F)$. Then it is seen that T_N can be expressed as

$$T_N = \mu_N + \mathcal{B}_{1N} + \mathcal{B}_{2N} + \sum_{i=1}^{3} C_{iN}, \tag{5.3.19}$$

where

$$\mathcal{B}_{1N} = \int J[H_N]d(\hat{F}_m - F)(x),$$
$$\mathcal{B}_{2N} = \int(\hat{\mathcal{H}}_N - H_N)J'[H_N]dF(x),$$
$$C_{1N} = -\frac{1}{N+1}\int\hat{\mathcal{H}}_N J'[H_N]d\hat{F}_m(x),$$
$$C_{2N} = \int(\hat{\mathcal{H}}_N - H_N)J'[H_N]d(\hat{F}_m - F)(x),$$
$$C_{3N} = \int J\left[\frac{N}{N+1}\hat{\mathcal{H}}_N\right] - J[H_N] - \left(\frac{N}{N+1}\hat{\mathcal{H}}_N - H_N\right)d\hat{F}_m(x).$$

The theorem is established if we show

$$(i)\ \sqrt{N}(\mathcal{B}_{1N}+\mathcal{B}_{2N})/\sigma_N \xrightarrow{d} N(0,1)\ \ as\ \ N\to\infty, \tag{5.3.20}$$
$$(ii)\ C_{iN}=o_p(N^{-1/2}),\ \ i=1,2,3.$$

Although the proof of (ii) is technically very complicated, it is understood intuitively. Hence we omit it (see Chandra and Taniguchi (2003) for details). In what follows we prove (i). From (5.3.11) we observe that

$$\mathcal{B}_{1N}=\int J[H_N]d(F_m-F)(x)+m^{-1/2}A_x\int J[H_N]d[xf(x)]$$
$$+\ lower\ order\ terms. \tag{5.3.21}$$

Then, integrating $\mathcal{B}_{2N}$ by parts, and adding it to (5.3.21), we obtain

$$\begin{aligned}&\sqrt{N}(\mathcal{B}_{1N}+\mathcal{B}_{2N})\\ &=\sqrt{N}(1-\lambda_N)\left\{\int s(x)d(F_m-F)-\int s^*(x)d(G_n-G)\right.\\ &\left.-m^{-1/2}A_X\int xf(x)J'[H_N]dG(x)+n^{-1/2}A_Y\int zg(z)J'[H_N]dF(z)\right\}\\ &+\ lower\ order\ terms\\ &=a_N+b_N+c_N+d_N+lower\ order\ terms,\ \ (say),\end{aligned} \tag{5.3.22}$$

where

$$s(x)=\int_{x_0}^{x}J'[H_N(y)]dG(y),\ \ s^*(x)=\int_{x_0}^{x}J'[H_N(y)]dF(y)$$

and $\lambda_N s^*(x)+(1-\lambda_N)s(x)=J[H_N(x)]-J[H_N(x_0)]$ *with* x_0 determined arbitrarily, say, by $H_N(x_0)=1/2$. Since a_N and b_N are mutually independent, using the result by Puri and Sen (1971, pp. 97–99), we obtain

$$\sigma_{1N}^2=Var(a_N+b_N). \tag{5.3.23}$$

From Tjøstheim (1986) it follows that

$$Var[\sqrt{m}(\hat{\theta}_{X,m}-\theta_X)]=\mathcal{U}_X^{-1}\mathcal{R}_X\mathcal{U}_X^{-1}$$

and

$$Var[\sqrt{n}(\hat{\theta}_{Y,n}-\theta_Y)]=\mathcal{U}_Y^{-1}\mathcal{R}_Y\mathcal{U}_Y^{-1}. \tag{5.3.24}$$

Recalling (5.3.11), (5.3.12) and (5.3.22) we get

$$\sigma_{2N}^2=Var(c_N)\ \ and\ \ \sigma_{3N}^2=Var(d_N). \tag{5.3.25}$$

Next, we compute the covariance terms. Since $\{X_t\}$ and $\{Y_t\}$ are independent, we have only to evaluate

$$L_{1N}=2E[a_Nc_N]\ \ and\ \ L_{2N}=2E[b_Nd_N].$$

From (5.3.22) we obtain

$$L_{1N}=2Nm^{-1}(1-\lambda_N)^2\iint E\left\{(\sqrt{m}(F_m-F)(x))A_X\right\}$$

$$\times \delta_{f,N}(x,y)dG(x)dG(y), \tag{5.3.26}$$

for which; it is necessary to find E{·}. Using the result by Horváth et al. (2001), it follows from (5.3.11) and (5.3.17) that

$$E\{(\sqrt{m}(F_m - F)(x))A_X\} = \sum_{i=0}^{p_X} \tau_{i,X} h^i_X(x).$$

Thus,

$$L_{1N} = 2\frac{(1-\lambda_N)^2}{\lambda_N}\sum_{i=0}^{p_X} \tau_{i,X} \iint h^i_X(x)\delta_{f,N}(x,y)dG(x)dG(y) \tag{5.3.27}$$

and analogously

$$L_{2N} = 2(1-\lambda_N)\sum_{i=0}^{p_Y} \tau_{i,Y} \iint h^i_Y(x)\delta_{g,N}(x,z)dF(x)dF(z). \tag{5.3.28}$$

Adding (5.3.27) and (5.3.28) yields γ_N.

For (5.3.20), using the central limit theorems given by Horváth et al. (2001) and Tjøstheim (1986), we may conclude that

$$\sqrt{N}(\mathcal{B}_{1N} + \mathcal{B}_{2N})/\sigma_N \xrightarrow{d} N(0,1) \ as \ N \to \infty, \tag{5.3.29}$$

leading to (i). □

The asymptotic distribution of $\{T_N\}$ given in the theorem provides useful information for the reliability of confidence intervals, asymptotic relative efficiency (ARE) and ARCH affection.

The test statistic T_N was constructed from the empirical residuals $\{\hat{\epsilon}_t^2\}$ and $\{\hat{\xi}_t^2\}$. Likewise, if we construct it replacing $\{\hat{\epsilon}_t^2\}$ and $\{\hat{\xi}_t^2\}$ by $\{\epsilon_t^2\}$ and $\{\xi_t^2\}$, respectively, then it becomes the usual rank order statistic given in Puri and Sen (1971), which is denoted by T_N^{PS}. Here we consider the problem of constructing approximate confidence intervals based on T_N^{PS} and T_N in the i.i.d. and in our ARCH residual settings, respectively. Some important feature of T_N will be elucidated.

For this, let us consider the ARCH(1) model:

$$X_t = \begin{cases} \sigma_t(\theta_X)\epsilon_t, \quad \sigma_t^2(\theta_X) = \theta_X^0 + \theta_X^1 X_{t-1}^2, & for\ t = 1, \cdots, m, \\ 0, & for\ t \le 0, \end{cases}$$

where $\theta_X^0 > 0$, $0 \le \theta_X^1 < 1$, $\{\epsilon_t\}$ is a sequence of i.i.d. (0,1) random variables with fourth-order cumulant κ_4^X, and ϵ_t is independent of X_s, $s < t$.

Another ARCH(1) model, independent of $\{X_t\}$, is given by

$$Y_t = \begin{cases} \sigma_t(\theta_Y)\xi_t, \quad \sigma_t^2(\theta_Y) = \theta_Y^0 + \theta_Y^1 Y_{t-1}^2, & for\ t = 1, \cdots, n, \\ 0, & for\ t \le 0, \end{cases}$$

where $\theta_Y^0 > 0$, $0 \le \theta_Y^1 < 1$, $\{\xi_t\}$ is a sequence of i.i.d. (0,1) random variables with fourth-order cumulant κ_4^Y, and ξ_t is independent of Y_s, $s < t$.

Recall that F(x) and G(x) are the distribution functions of ϵ_t^2 and ξ_t^2, respectively. Consider the scale problem in the case of $G(x) = F(\theta x)$, $\theta \in [1,\infty)$. Henceforth it is assumed that F is arbitrary and has finite variance $\sigma^2(F)$. The two-sample testing problem for scale is described as

$$H_0 \ : \ \theta = 1 \quad vs \quad H_A \ : \ \theta > 1. \tag{5.3.30}$$

For this, we use Wilcoxon's test with $J(u) = u$, $0 < u < 1$. Assume that $m = n = N/2$. Then, from Theorem 5.3.3 it follows that the asymptotic mean and variance under H_0 are, respectively, given by

$$\mu(\theta) = \int F(x)dF(x), \quad and \quad \sigma^2(1) = \sigma_1^2(1) + \sigma_2^2(1) + \sigma_3^2(1) + \gamma(1),$$

where

$$\sigma_1^2(1) = \int_0^1 u^2 du - \left(\int_0^1 u du\right)^2 = \frac{1}{12}, \quad \sigma_2^2(1) = \frac{C_X}{2}\left(\int x f^3(x)dx\right)^2,$$
$$\sigma_3^2(1) = \frac{C_Y}{2}\left(\int z f^3(z)dz\right)^2 \quad and$$
$$\gamma(1) = k_1 \iint p(x)yf^2(x)f^3(y)dxdy + k_2 \iint p(x)zf^2(x)f^3(z)dxdz,$$

with

$$C_x = \tau_X' \mathcal{U}_X^{-1} \mathcal{R}_X \mathcal{U}_X^{-1} \tau_x, \quad C_Y = \tau_Y' \mathcal{U}_Y^{-1} \mathcal{R}_Y \mathcal{U}_Y^{-1} \tau_Y,$$
$$k_1 = (\tau_{0,X} + \tau_{1,X})(\alpha_X^0 + \alpha_X^1), \quad k_2 = (\tau_{0,Y} + \tau_{1,Y})(\alpha_Y^0 + \alpha_Y^1) \quad and$$
$$p(x) = \int_0^x (u-1)f(u)du,$$

where $\tau_X = (\tau_{0,X}, \tau_{1,X})'$ and $\tau_Y = (\tau_{0,Y}, \tau_{1,Y})'$. Writing $\mu = 1/2$, $\sigma^2 = \sigma^2(1)$ and $\sigma_1^2 = 1/12$, we have, under H_0,

$$\sqrt{N}(T_N^{PS} - \mu)/\sigma_1 \xrightarrow{d} N(0,1) \quad and \quad \sqrt{N}(T_N - \mu)/\sigma \xrightarrow{d} N(0,1). \tag{5.3.31}$$

For $\alpha \in (0,1)$, write $\lambda_\alpha = \Phi^{-1}(1-\alpha)$, where $\Phi(\cdot)$ is the distribution function of $N(0,1)$. The $(1-\alpha)$ asymptotic coverage probabilities are expressed as

$$P\left\{\mu - N^{-1/2}\sigma_1\lambda_{\alpha/2} \le T_N^{PS} \le \mu + N^{-1/2}\sigma_1\lambda_{\alpha/2}\right\} \approx 1-\alpha,$$
$$P\left\{\mu - N^{-1/2}\sigma\lambda_{\alpha/2} \le T_N \le \mu + N^{-1/2}\sigma\lambda_{\alpha/2}\right\} \approx 1-\alpha.$$

Hence the approximate theoretical $(1-\alpha)$ confidence intervals are

$$\left(1/2 - N^{-1/2}\frac{1}{2\sqrt{3}}\lambda_{\alpha/2}\ ,\ 1/2 + N^{-1/2}\frac{1}{2\sqrt{3}}\lambda_{\alpha/2}\right) \quad and \tag{5.3.32}$$

$$\left(1/2 - N^{-1/2}\sigma\lambda_{\alpha/2}\ ,\ 1/2 + N^{-1/2}\sigma\lambda_{\alpha/2}\right), \tag{5.3.33}$$

respectively.

Chandra and Taniguchi (2003) evaluated the interval (5.3.32) numerically, for three types of the distribution function F^* of ϵ_t:

$$(i)\ F^* = F_N^* \ (normal); \quad F_N^*(y) = \int_{-\infty}^{y} (2\pi)^{-1/2} exp\left(-\frac{t^2}{2}\right) dt.$$

$$(ii)\ F^* = F_{DE}^* \ (double\ exponential); \quad F_{DE}^*(y) = \frac{1}{4}\int_{-\infty}^{y} e^{-|t|} dt.$$

$$(iii)\ F^* = F_L^* \ (logistic); \quad F_L^*(y) = \frac{3}{\pi^2}(1 + e^{-y}).$$

If ϵ_t has the distribution F^*, then ϵ_t^2 has the distribution

$$F(x) \equiv P(\epsilon_t^2 \le x) = \begin{cases} 2F^*(\sqrt{x}) - 1, & x \ge 0, \\ 0, & x < 0. \end{cases} \tag{5.3.34}$$

For (i)–(iii), the corresponding distribution F is, respectively,

$$\begin{aligned} F_N(x) &= 2F_N^*(\sqrt{x}) - 1, \\ F_{DE}(x) &= \frac{1}{2}(1 - e^{-\sqrt{x}}), \\ F_L(x) &= 3\{1 - 2/(1 + e^{\sqrt{x}})\}/\pi^2, \end{aligned} \tag{5.3.35}$$

(see Exercise 5.2).

For $\alpha = 0.05$ and for $F = F_N$, F_{DE} and F_L, the confidence intervals (5.3.31) are

$$\begin{aligned} &(0.4747,\ 0.5253), \quad N = 1000 \\ &(0.4821,\ 0.5179), \quad N = 2000. \end{aligned} \tag{5.3.36}$$

In the case of ARCH, we have to evaluate σ in (5.3.32). For this, set $\theta^0 \equiv \theta_X^0 = \theta_Y^0 = 30$ and $\theta^1 \equiv \theta_X^1 = \theta_Y^1 = 0.1, 0.5$. For $m = n = 500, 1000$ (i.e., $N = 1000, 2000$), we generated realizations of X_t and Y_t with 100 replications. Chandra and Taniguchi (2003) evaluated the sample version of σ, and gave

Table 5.4: 95% Confidence interval based on T_N in (5.3.32)

Dist.	$m = n = 500, \theta^0 = 30$		$m = n = 1000, \theta^0 = 30$	
	$\theta^1 = 0.1$	$\theta^1 = 0.5$	$\theta^1 = 0.1$	$\theta^1 = 0.5$
F_N	(0.3642, 0.6344)	(0.3216, 0.6672)	(0.3746, 0.6240)	(0.3632, 0.6354)
F_{DE}	(0.4411, 0.5590)	(0.4163, 0.5838)	(0.4580, 0.5420)	(0.4401, 0.5599)
F_L	(0.4745, 0.5254)	(0.4735, 0.5264)	(0.4818, 0.5181)	(0.4740, 0.5259)

The difference between (5.3.36) and (5.3.37) is due to the effect of the volatility estimators $\hat{\theta}_{X,m}$ and $\hat{\theta}_{Y,m}$. Table 5.4 implies that the Wilcoxon test is preferable if the underlying distribution is logistic.

Next we discuss the asymptotic relative efficiency (ARE), due to Pitman, of T_N for distributions F_N, F_{DE} and F_L. Suppose that T_N is a test statistic based on N observations for testing $H_0 : \theta = \theta_0$, vs $H_A : \theta > \theta_0$ with critical region $T_N \ge \lambda_{N,\alpha}$. Further, we assume that
(i) $\lim_{N\to\infty} P_{\theta_0}(T_N \ge \lambda_{\mu,\alpha}) = \alpha$, where $0 < \alpha < 1$ is a given level,
(ii) there exist functions $\mu_N(\theta)$ and $\sigma_N(\theta)$ such that

$$\sqrt{N}\big(T_N - \mu_N(\theta)\big)/\sigma_N(\theta) \xrightarrow{d} N(0, 1) \tag{5.3.37}$$

uniformly in $\theta \in [\theta_0, \theta_0 + \epsilon]$, $\epsilon > 0$,
(iii) $\mu_N'(\theta_0) > 0$,
(iv) for a sequence $\{\theta_N = \theta_0 + N^{-1/2}\delta,\ \delta > 0\}$ such that $\theta_N \to \theta_0$ as $N \to \infty$,

$$\begin{aligned} &\lim_{N\to\infty} [\mu_N'(\theta_N)/\mu_N'(\theta_0)] = 1, \quad \lim_{N\to\infty} [\sigma_N(\theta_N)/\sigma_N(\theta_0)] = 1, \\ &\lim_{N\to\infty} [\mu_N'(\theta_0)/\sigma_N(\theta_0)] = c > 0. \end{aligned} \tag{5.3.38}$$

Then, the asymptotic power is given by $1 - \Phi(\lambda_\alpha - \delta_c)$, where $\lambda_\alpha = \Phi^{-1}(1-\alpha)$. The quantity c defined by (5.3.38) is called the efficacy of T_N. Let $T_N^{(1)}$ and $T_N^{(2)}$ be sequences of statistics with efficacies c_1 and c_2, respectively, and define $e(T_N^{(2)}, T_N^{(1)}) \equiv c_2^2/c_1^2$. We call $e(T_N^{(2)}, T_N^{(1)})$ the asymptotic relative efficiency (ARE) of $T_N^{(2)}$ relative to $T_N^{(1)}$. Especially, if we use the Wilcoxon test $T_N = T_N(F)$ based on an innovation distribution F, $e(F_2, F_1)$ means the ARE of $T_N(F_2)$ relative to $T_N(F_1)$.

In our setting, Chandra and Taniguchi (2003) evaluated $e(\cdot,\cdot)$ and $\delta_F = \sigma_1/\sigma$ (recall (5.3.31)) numerically, in the case of $m = n = 500$ and $\theta^0 = 300$.

Table 5.5: ARE

ARE	$\theta^1 = 0.1$	$\theta^1 = 0.5$
$e(F_N, F_L)$	0.4408	0.1972
$e(F_{DE}, F_L)$	0.9698	0.4820

Table 5.6: ARCH affection

	$\theta^1 = 0.1$	$\theta^1 = 0.5$
δ_{F_N}	0.5685	0.3795
$\delta_{F_{DE}}$	0.8292	0.7921
δ_{F_L}	0.9684	0.9432

Table 5.5 shows that the Wilcoxon test is good in the case of F_L. Also δ_F shows an ARCH affection. For the Wilcoxon test, the ARCH affection becomes large if $F = F_N$ and if θ^1 becomes large. But for $F = F_L$, the ARCH affection becomes small. (See Table 5.6.)

5.4 Independent Component Analysis

5.4.1 *Introduction to Independent Component Analysis*

5.4.1.1 *The foregoing model for financial time series*

In many financial applications, e.g., asset pricing, portfolio selection, hedging and risk management, a modeling of the time varying conditional covariance matrix for asset returns is widely invoked. However, multivariate volatility models have two main difficulties. Namely, they have the high-dimensional problem and the restriction of positive definite at every point.

The K-factor GARCH model can reduce the number of parameters and parameter constraints (see, e.g., Lin (1992)). Let $\boldsymbol{\varepsilon}_t$ be a $d \times 1$ vector of random variables, $\mathcal{F}_t$ be the σ-field generated by $\{\boldsymbol{\varepsilon}_s\}_{s=-\infty}^t$, and $\boldsymbol{H}_t$ be a covariance matrix of $\boldsymbol{\varepsilon}_t$ that is measurable with respect to $\mathcal{F}_{t-1}$. Then, the conditional distribution of a multivariate GARCH model can be expressed as

$$\boldsymbol{\varepsilon}_t \mid \mathcal{F}_{t-1} \sim (\boldsymbol{0}, \boldsymbol{H}_t), \tag{5.4.1}$$

namely, it is some arbitrary multivariate distribution which has the mean vector $\boldsymbol{0}$ and the covariance matrix $\boldsymbol{H}_t$. There are many possible ways to parametrize $\boldsymbol{H}_t$ as a function of past values. Bollerslev et al. (1988) gave one simple formulation where each element of the covariance matrix depends solely on its own past values. That is, let $h_{ij,t}$ be the (i, j)th element of $\boldsymbol{H}_t$; then $h_{ij,t}$ can be represented by

$$h_{ij,t} = c_{ij} + a_{ij}\varepsilon_{i,t-1}\varepsilon_{j,t-1} + b_{ij}h_{ij,t-1} \quad \text{for all } i, j.$$

This model is the so-called the "diagonal multivariate GARCH" since it is obtained as a diagonal representation in the *vec* model (see, e.g., Engle and Kroner (1995)). We may require that $\boldsymbol{H}_t$ be positive definite. However, it can be hard to check even in the diagonal representation. Alternatively, Engle and Kroner (1995) proposed a multivariate GARCH (p, q, K) model defined by

$$\boldsymbol{H}_t = \boldsymbol{\Omega} + \sum_{k=1}^{K}\sum_{j=1}^{q} \boldsymbol{A}'_{k,j}\boldsymbol{\varepsilon}_{t-j}\boldsymbol{\varepsilon}'_{t-j}\boldsymbol{A}_{k,j} + \sum_{k=1}^{K}\sum_{j=1}^{p} \boldsymbol{B}'_{k,j}\boldsymbol{H}_{t-j}\boldsymbol{B}_{k,j}, \tag{5.4.2}$$

where $\boldsymbol{A}_{k,j}$ and $\boldsymbol{B}_{k,j}$ are $d \times d$ parameter matrices and $\boldsymbol{\Omega}$ is a $d \times d$ symmetric parameter matrix. They showed that this will be positive definite under very weak conditions, and this representation is sufficiently general since it includes all positive definite diagonal representations. However, the number of parameters easily becomes huge, as the dimension of variables d increases. Engle (1987) suggested a K-factor structure to the covariance matrix and Lin (1992) adapted it to the following K-factor GARCH model.

Definition 5.4.1. *A multivariate GARCH (p, q, K) model in (5.4.1) and (5.4.2) is said to be a K-factor model if it can be represented as*

$$\boldsymbol{A}_{kj} = \alpha_{kj}\boldsymbol{f}_k\boldsymbol{g}'_k, \quad \text{and} \quad \boldsymbol{B}_{kj} = \beta_{kj}\boldsymbol{f}_k\boldsymbol{g}'_k$$

with $d \times 1$ vectors $\boldsymbol{f}_k$, $\boldsymbol{g}_k$, $(k = 1, \ldots, K)$ which satisfy

$$\boldsymbol{f}'_k\boldsymbol{g}_l = \begin{cases} 0 & (k \neq l) \\ 1 & (k = l). \end{cases}$$

Therefore, a K-factor GARCH (p, q) model can be rewritten as

$$\boldsymbol{H}_t = \boldsymbol{\Omega} + \sum_{k=1}^{K} \boldsymbol{g}_k\boldsymbol{g}'_k \left(\sum_{j=1}^{q} \alpha_{kj}^2 \boldsymbol{f}'_k\boldsymbol{\varepsilon}_{t-j}\boldsymbol{\varepsilon}'_{t-j}\boldsymbol{f}_k + \sum_{k=1}^{K}\sum_{j=1}^{p} \beta_{kj}^2 \boldsymbol{f}'_k\boldsymbol{H}_{t-j}\boldsymbol{f}_k \right).$$

Hence, the time-varying part of the covariance matrix in the K-factor GARCH (p, q) model is reduced to the rank K. The following properties of the K-factor GARCH (p, q) model were listed in Lin (1992).

Property 5.4.2. *Let the kth factor be $\eta_{k,t} = \boldsymbol{f}'_k\boldsymbol{\varepsilon}_t$. Then, we have the following.*

(i) *The kth factor $\eta_{k,t}$ follows the GARCH (p, q) process*

$$\eta_{k,t} \mid \mathcal{F}_{t-1} \sim (0, h_{k,t}),$$

$$h_{k,t} = \omega_k + \sum_{j=1}^{q} \alpha_{k,j}^2\eta_{k,t-j}^2 + \sum_{j=1}^{p} \beta_{k,j}^2 h_{k,t-j},$$

where $h_{k,t} = \boldsymbol{f}'_k\boldsymbol{H}_t\boldsymbol{f}_k$ and $\omega_k = \boldsymbol{f}'_k\boldsymbol{\Omega}\boldsymbol{f}_k$. Furthermore, the conditional covariance of $\boldsymbol{\varepsilon}_t$ can be rewritten as

$$\boldsymbol{H}_t = \boldsymbol{\Omega}^* + \sum_{k=1}^{K} \boldsymbol{g}_k\boldsymbol{g}'_k h_{k,t},$$

where $\boldsymbol{\Omega}^* = \boldsymbol{\Omega} - \sum_{k=1}^{K} \boldsymbol{g}_k\boldsymbol{g}'_k\omega_k$, and therefore it depends only on the K conditional variance of common factors.

(ii) *The conditional covariance of any two factors satisfies*

$$E\left(\eta_{k,t}\eta_{j,t} \mid \mathcal{F}_{t-1}\right) = \boldsymbol{f}'_k\boldsymbol{\Omega}\boldsymbol{f}_j \quad \text{for } k \neq j,\ k, j = 1, \ldots, K,$$

so that it is time invariant.

(iii) *The linear combinations of ε_t still follow the K-factor GARCH (p, q) process with the same common factors of ε_t. Namely, let $\boldsymbol{P}$ be a $d \times m$ matrix with full column rank m, then,*

$$\boldsymbol{P}'\varepsilon_t \mid \mathcal{F}_{t-1} \sim \left(\mathbf{0}, \widetilde{\boldsymbol{H}}_t\right),$$

$$\widetilde{\boldsymbol{H}}_t = \widetilde{\boldsymbol{\Omega}} + \sum_{k=1}^{K} \widetilde{\boldsymbol{g}}_k \widetilde{\boldsymbol{g}}_k' h_{k,t},$$

where $\widetilde{\boldsymbol{\Omega}} = \boldsymbol{P}'\boldsymbol{\Omega}^*\boldsymbol{P}$ and $\widetilde{\boldsymbol{g}}_k = \boldsymbol{P}'\boldsymbol{g}_k$.

(iv) *The covariance stationary condition of the K-factor GARCH (p, q) process is given by*

$$\sum_{j=1}^{q} \alpha_{kj}^2 + \sum_{j=1}^{p} \beta_{kj}^2 < 1 \quad \text{for } k = 1, \dots, K.$$

Moreover, the stationary covariance matrix of ε_t is obtained as

$$E\left(\varepsilon_t \varepsilon_t'\right) = \boldsymbol{\Omega}^* + \sum_{k=1}^{K} \boldsymbol{g}_k \boldsymbol{g}_k' \sigma_k^2,$$

where $\sigma_k^2 = \omega_k / \left(1 - \sum_{j=1}^{q} \alpha_{kj}^2 - \sum_{j=1}^{p} \beta_{kj}^2\right)$.

Lin (1992) proposed four alternative estimators for the K-factor GARCH (p, q) model, and gave the Monte Carlo comparison of their finite sample properties.

Alternatively, the observations $\boldsymbol{y}_t$ $(t = 1, \dots, T)$ of the multivariate full-factor GARCH model are given by the following equations (see Vrontos et al. (2003)):

$$\boldsymbol{y}_t = \boldsymbol{\mu} + \varepsilon_t,$$
$$\varepsilon_t = \boldsymbol{W}\boldsymbol{x}_t,$$
$$\boldsymbol{x}_t \mid \mathcal{F}_{t-1} \sim N_N\left(\mathbf{0}, \boldsymbol{\Sigma}_t\right),$$

where $\boldsymbol{\mu}$ is the $N \times 1$ constant mean vector, ε_t is the $N \times 1$ innovation vector, $\boldsymbol{W}$ is the $N \times N$ parameter matrix and $\boldsymbol{x}_t$ is the $N \times 1$ factor vector. Furthermore, the kth element $x_{k,t}$ $(k = 1, \dots, N)$ of the factors $\boldsymbol{x}_t$ are GARCH (1, 1) processes; therefore $\boldsymbol{\Sigma} = \operatorname{diag}\left(\sigma_{1,t}^2, \dots, \sigma_{N,t}^2\right)$ is the $N \times N$ diagonal covariance matrix which satisfies

$$\sigma_{k,t}^2 = \omega_k + \alpha_k x_{k,t-1}^2 + \beta_k \sigma_{k,t-1}^2, \qquad k = 1, \dots, N, \ t = 1, \dots, T,$$

where $\sigma_{k,t}^2$ is the conditional variance of the kth factor $x_{k,t}$ at time t and coefficients satisfy $\omega_k > 0$, $\alpha_k \geq 0$, $\beta_k > 0$, $k = 1, \dots, N$. Since the factor vector $\boldsymbol{x}_t$ follows a conditional normal distribution, the innovation vector ε_t satisfies $\varepsilon_t \mid \mathcal{F}_{t-1} \sim N_N\left(\mathbf{0}, \boldsymbol{H}_t\right)$ with

$$\boldsymbol{H}_t = \boldsymbol{W}\boldsymbol{\Sigma}_t\boldsymbol{W}' = \boldsymbol{W}\boldsymbol{\Sigma}_t^{1/2}\boldsymbol{\Sigma}_t^{1/2}\boldsymbol{W}' = \left(\boldsymbol{W}\boldsymbol{\Sigma}_t^{1/2}\right)\left(\boldsymbol{W}\boldsymbol{\Sigma}_t^{1/2}\right)' := \boldsymbol{L}_t\boldsymbol{L}_t'.$$

In this case we can naturally assume that $\boldsymbol{W}$ is triangular and has unit diagonal elements, namely, $w_{ij} = 0$ for $j > i$ and $w_{ii} = 1$, $(i = 1, \dots, N)$. Under these assumptions, the conditional covariance matrix $\boldsymbol{H}_t$ has the (i, j)th element

$$h_{ij,t} = \sum_{k=1}^{\min\{i,j\}} w_{ik} w_{jk} \sigma_{k,t}^2,$$

and is always positive definite if the factor conditional variance $\sigma_{k,t}^2$ $(k = 1, \dots, N)$ is well defined. In this model the conditional correlations are given by

$$\rho_{ij,t} = \frac{\sum_{k=1}^{\min\{i,j\}} w_{ik} w_{jk} \sigma_{k,t}^2}{\left(\sum_{k=1}^{i} w_{ik}^2 \sigma_{k,t}^2\right)^{1/2} \left(\sum_{k=1}^{j} w_{jk}^2 \sigma_{j,t}^2\right)^{1/2}}.$$

Therefore, it can be considered as a generalization of the constant correlation coefficient model. For the sake of convenience, we additionally assume that $x_{k,0}$ and $\sigma^2_{k,0}$ are known, and $\alpha_k = \alpha$, $\beta_k = \beta$ $(k = 1, \ldots, N)$. Then, the unconditional covariances are given by $\boldsymbol{U} = E(\boldsymbol{H}_t)$ with the (i, j)th element

$$u_{ij} = \frac{\sum_{k=1}^{\min\{i,j\}} w_{ik} w_{jk} \omega_k}{1 - \alpha - \beta}.$$

The likelihood of the multivariate full-factor GARCH model for observations $\boldsymbol{y} = (\boldsymbol{y}_1, \ldots, \boldsymbol{y}_T)$ can be written as

$$l(\boldsymbol{y} \mid \boldsymbol{\theta}) := (2\pi)^{-\frac{TN}{2}} \prod_{t=1}^{T} |\boldsymbol{H}_t|^{-1/2} \exp\left[-\frac{1}{2} \sum_{t=1}^{T} (\boldsymbol{y}_t - \boldsymbol{\mu})' \boldsymbol{H}_t^{-1} (\boldsymbol{y}_t - \boldsymbol{\mu})\right],$$

where the number of parameters is $2N + 2 + \frac{N(N-1)}{2}$ and the parameter vector is

$$\boldsymbol{\theta} = (\mu_1, \ldots, \mu_N, \omega_1, \ldots, \omega_N, \alpha, \beta, w_{21}, w_{31}, w_{32}, \ldots, w_{N1}, \ldots, w_{NN-1})'.$$

Note that the factors $x_{k,t}$ are not parameters to be estimated but are given by $\boldsymbol{x}_t = \boldsymbol{W}^{-1}\boldsymbol{\varepsilon}_t$. Vrontos et al. (2003) used the classical and Bayesian techniques for the estimation of the model parameters. They showed that maximum likelihood estimation and the construction of well-mixing MCMC algorithms are easily performed.

Another type of the multivariate GARCH model in financial time series is described by the orthogonal GARCH type. If we assume that $\boldsymbol{x}_t$ is normally distributed and has the conditional covariance matrix $\boldsymbol{V}_t$ which is measurable with respect to $\mathcal{F}_{t-1}$, then the multivariate GARCH model is given by

$$\boldsymbol{x}_t \mid \mathcal{F}_{t-1} \sim N(\boldsymbol{0}, \boldsymbol{V}_t),$$

where we also assume that $\boldsymbol{x}_t$ is second order stationary and so that $\boldsymbol{V} = E(\boldsymbol{V}_t)$ exists. The problem in multivariate GARCH modeling is how to parametrize $\boldsymbol{V}_t$ as a function of $\mathcal{F}_{t-1}$ where the model is sufficiently general while feasible estimation methods can be applied. For this purpose van der Weide (2002) suggested the generalized orthogonal GARCH (GO-GARH) model. The key assumption of GO-GARH modeling is given by the following.

Assumption 5.4.3. *The observed process $\boldsymbol{x}_t$ is given by a linear combination of uncorrelated components $\boldsymbol{y}_t$, namely,*

$$\boldsymbol{x}_t = \boldsymbol{Z}\boldsymbol{y}_t,$$

where the linear link map $\boldsymbol{Z}$ is constant over time and invertible.

We can assume the unobserved components satisfy $E(\boldsymbol{y}_t\boldsymbol{y}_t') = \boldsymbol{I}_N$ without loss of generality. Therefore, the observed process satisfies

$$\boldsymbol{V} = E(\boldsymbol{x}_t\boldsymbol{x}_t') = \boldsymbol{Z}\boldsymbol{Z}'. \tag{5.4.3}$$

A simple example is the GO-GARCH (1, 1) model

$$\begin{aligned} \boldsymbol{x}_t &= \boldsymbol{Z}\boldsymbol{y}_t \\ \boldsymbol{y}_t \mid \mathcal{F}_{t-1} &\sim N(\boldsymbol{0}, \boldsymbol{H}_t), \end{aligned}$$

where each component is a GARCH (1, 1) process, so that

$$\boldsymbol{H}_t = \mathrm{diag}(h_{1,t}, \ldots, h_{N,t})$$

$$h_{k,t} = (1 - \alpha_k - \beta_k) + \alpha_k y_{k,t-1}^2 + \beta_k h_{k,t-1}, \quad k = 1, \ldots, N.$$

Additionally, if we assume $\boldsymbol{H}_0 = \boldsymbol{I}_N$, then the unobserved conponents have unit variances, i.e., $E(\boldsymbol{y}_t \boldsymbol{y}_t') = \boldsymbol{I}_N$. Therefore, the unconditional covariance $\boldsymbol{V}$ of the observed process $\boldsymbol{x}_t$ is given by (5.4.3). On the other hand, the conditional covariances of $\boldsymbol{x}_t$ satisfy

$$\boldsymbol{V}_t = \boldsymbol{Z}\boldsymbol{H}_t\boldsymbol{Z}'.$$

Let $\boldsymbol{P}$ and $\boldsymbol{\Lambda} = \text{diag}(\lambda_1, \ldots, \lambda_N)$ denote matrices of the orthonormal eigenvectors and the eigenvalues of the unconditional covariance matrix $\boldsymbol{V} = \boldsymbol{Z}\boldsymbol{Z}'$, respectively. Then, the following properties are useful for the estimation procedure.

Property 5.4.4 (Lemma 2 of van der Weide (2002)). *Let $\boldsymbol{Z}$ be the map that links the uncorrelated components $\boldsymbol{y}_t$ with the observed process $\boldsymbol{x}_t$. Then, there exists an orthogonal matrix $\boldsymbol{U}$ which satisfies*

$$\boldsymbol{P}\boldsymbol{\Lambda}^{1/2}\boldsymbol{U} = \boldsymbol{Z}.$$

For the estimation of $\boldsymbol{U}$, we can restrict the determinant of $\boldsymbol{U}$ to be 1 without loss of generality. Then, we also have the following property.

Property 5.4.5 (Lemma 3 of van der Weide (2002)). *Every N-dimensional orthogonal matrix $\boldsymbol{U}$ with* $\det\{\boldsymbol{U}\} = 1$ *can be represented as a product of $\frac{N(N-1)}{2}$ rotation matrices, namely,*

$$\boldsymbol{U} = \prod_{i<j} \boldsymbol{R}_{ij}(\theta_{ij}) \qquad -\pi < \theta_{ij} \le \pi,$$

where $\boldsymbol{R}_{ij}(\theta_{ij})$ is a rotation matrix in the plane spanned by the unit vectors $\mathbf{e}_i$ and $\mathbf{e}_j$ over an angle θ_{ij}.

The matrices $\boldsymbol{P}$ and $\boldsymbol{\Lambda}$ will be estimated from the sample version of the covariance matrix $\boldsymbol{V}$. On the other hand, conditional information is required for estimation of $\boldsymbol{U}$. The parameters which will be estimated from the conditional information include the vector $\boldsymbol{\theta}$ of rotation matrices and the parameters $(\boldsymbol{\alpha}, \boldsymbol{\beta})$ for the N univariate GARCH (1, 1) processes. The log-likelihood for the GO-GARCH model is given by

$$\begin{aligned}
L_{\boldsymbol{\theta},\boldsymbol{\alpha},\boldsymbol{\beta}} &= -\frac{1}{2}\sum_{t=1}^{T}\left\{N\log(2\pi) + \log|\boldsymbol{V}_t| + \boldsymbol{x}_t'\boldsymbol{V}_t^{-1}\boldsymbol{x}_t\right\} \\
&= -\frac{1}{2}\sum_{t=1}^{T}\left\{N\log(2\pi) + \log\left|\boldsymbol{Z}_{\boldsymbol{\theta}}\boldsymbol{H}_t\boldsymbol{Z}_{\boldsymbol{\theta}}'\right| + \boldsymbol{y}_t'\boldsymbol{Z}_{\boldsymbol{\theta}}'\left(\boldsymbol{Z}_{\boldsymbol{\theta}}\boldsymbol{H}_t\boldsymbol{Z}_{\boldsymbol{\theta}}'\right)^{-1}\boldsymbol{Z}_{\boldsymbol{\theta}}\boldsymbol{y}_t\right\} \\
&= -\frac{1}{2}\sum_{t=1}^{T}\left\{N\log(2\pi) + \log\left|\boldsymbol{Z}_{\boldsymbol{\theta}}\boldsymbol{Z}_{\boldsymbol{\theta}}'\right| + \log|\boldsymbol{H}_t| + \boldsymbol{y}_t'\boldsymbol{H}_t^{-1}\boldsymbol{y}_t\right\},
\end{aligned}$$

where $\boldsymbol{Z}_{\boldsymbol{\theta}}\boldsymbol{Z}_{\boldsymbol{\theta}}' = \boldsymbol{P}\boldsymbol{\Lambda}\boldsymbol{P}'$ is independent of $\boldsymbol{\theta}$.

The orthogonal GARCH model uses principle components which are uncorrelated unconditionally. On the other hand, Fan et al. (2008) proposed an alternative approach which is based on the conditionally uncorrelated components (CUCs). Let $\boldsymbol{Z}_t = (Z_{1,t}, \ldots, Z_{N,t})'$ be N CUCs which satisfy conditions

$$\begin{aligned}
&E(Z_{k,t} \mid \mathcal{F}_{t-1}) = 0, \\
&\text{Var}(Z_{k,t}) = 1, \\
&E(Z_{k,t}Z_{l,t} \mid \mathcal{F}_{t-1}) = 0 \quad \text{for all } k \ne l.
\end{aligned} \tag{5.4.4}$$

Put $\sigma_{k,t}^2 = \text{Var}(Z_{k,t} \mid \mathcal{F}_{t-1})$. Then we have

$$\text{Var}(\boldsymbol{Z}_t \mid \mathcal{F}_{t-1}) = \text{diag}\left(\sigma_{1,t}^2, \ldots, \sigma_{N,t}^2\right).$$

We assume that each component of ε_t is a linear combination of N CUCs $\boldsymbol{Z}_t$, namely,

$$\varepsilon_t = \boldsymbol{A}\boldsymbol{Z}_t, \tag{5.4.5}$$

where $\boldsymbol{A}$ is a constant matrix. For simplicity we suppose that $\text{Var}(\varepsilon_t) = \boldsymbol{I}_N$. In practice, this can be achieved by ε_t by $\boldsymbol{S}^{-1/2}\varepsilon_t$, where $\boldsymbol{S}$ is a sample covariance matrix of ε_t. Therefore, $\boldsymbol{A}$ is a $N \times N$ orthogonal matrix with $\frac{N(N-1)}{2}$ free elements so that $\boldsymbol{Z}_t = \boldsymbol{A}'\varepsilon_t$, since

$$\boldsymbol{I}_N = \text{Var}(\varepsilon_t) = \boldsymbol{A}\text{Var}(\boldsymbol{Z}_t)\boldsymbol{A}' = \boldsymbol{A}\boldsymbol{A}'.$$

This assumption is not essential and introduced to reduce the free parameters in $\boldsymbol{A}$ from N^2 to $\frac{N(N-1)}{2}$. Let $\zeta_{1,t} = \boldsymbol{b}_1'\varepsilon_t$ and $\zeta_{2,t} = \boldsymbol{b}_2'\varepsilon_t$ be any two portfolios and $\boldsymbol{b}_j'\boldsymbol{A} = \left(b_{1,j}, \ldots, b_{N,j}\right)$, $(j = 1, 2)$. Then we have

$$\text{Var}(\zeta_{1,t} \mid \mathcal{F}_{t-1}) = \sum_{k=1}^{N} b_{k,1}^2\sigma_{k,t}^2,$$

$$\text{Cov}(\zeta_{1,t}, \zeta_{2,t} \mid \mathcal{F}_{t-1}) = \sum_{k=1}^{N} b_{k,1}b_{k,2}\sigma_{k,t}^2,$$

so that the volatilities for any portfolio can be deduced to N univariate volatility models. Let $\boldsymbol{A} = (\boldsymbol{a}_1, \ldots, \boldsymbol{a}_N)$. Since $Z_{k,t} = \boldsymbol{a}_k'\varepsilon_t$ and $\boldsymbol{a}_1, \ldots, \boldsymbol{a}_N$ are N orthogonal vectors, our goal is the estimation for the orthogonal matrix $\boldsymbol{A}$. The condition (5.4.4) is equivalent to

$$\sum_{B \in \mathcal{B}_t} \left|E\{Z_{k,t}Z_{l,t}I(B)\}\right| = 0 \tag{5.4.6}$$

for any π-class (closed under finite intersections) $\mathcal{B}_t \subset \mathcal{F}_{t-1}$ such that the σ-algebra which is generated by $\mathcal{B}_t$ is equal to $\mathcal{F}_{t-1}$. In practice, we use some simple $\mathcal{B}_t$, namely, $\mathcal{B}$ which is the collection of all the balls that are centred at the origin in $\mathbb{R}^N$. Let u_0 be a prescribed integer and

$$\begin{aligned}\Psi(\boldsymbol{A}) &= \sum_{1 \le k < l \le N} \sum_{B \in \mathcal{B}} \omega(B) \sum_{u=1}^{u_0} \left|E\left\{\boldsymbol{a}_k'\varepsilon_t\varepsilon_t'\boldsymbol{a}_l I(\varepsilon_{t-u} \in B)\right\}\right| \\ &= \sum_{1 \le k < l \le N} \sum_{B \in \mathcal{B}} \omega(B) \sum_{u=1}^{u_0} \left|\text{Corr}\left\{\boldsymbol{a}_k'\varepsilon_t, \boldsymbol{a}_l'\varepsilon_t \mid \varepsilon_{t-u} \in B\right\}\right| P(\varepsilon_{t-u} \in B),\end{aligned}$$

where $\omega(\cdot)$ is a weight function which satisfies $\sum_{B \in \mathcal{B}} \omega(B) < \infty$. We can understand $\Psi(\boldsymbol{A})$ as a collective conditional correlation measure among the N directions $\boldsymbol{a}_1, \ldots, \boldsymbol{a}_N$. Since the order of $\boldsymbol{a}_1, \ldots, \boldsymbol{a}_N$ is arbitrary and $\boldsymbol{a}_k$ may be replaced by $-\boldsymbol{a}_l$, we measure the distance between two orthogonal matrices $\boldsymbol{A}$ and $\boldsymbol{B}$ by

$$D(\boldsymbol{A}, \boldsymbol{B}) = 1 - \frac{1}{N} \sum_{k=1}^{N} \max_{1 \le l \le N} \left|\boldsymbol{a}_k'\boldsymbol{b}_l\right|.$$

For any two orthogonal matrices $\boldsymbol{A}$, $\boldsymbol{B}$, $D(\boldsymbol{A}, \boldsymbol{B}) \ge 0$ and if the columns of $\boldsymbol{A}$ are obtained from a permutation of the columns of $\boldsymbol{B}$ or their reflections, then $D(\boldsymbol{A}, \boldsymbol{B}) = 0$. The sample version of $\Psi(\boldsymbol{A})$ is given by

$$\Psi_T(\boldsymbol{A}) = \sum_{1 \le k < l \le N} \sum_{B \in \mathcal{B}} \omega(B) \sum_{u=1}^{u_0} \frac{1}{T-u} \left|\boldsymbol{a}_k'\left\{\sum_{t=u+1}^{T} \varepsilon_t\varepsilon_t' I(\varepsilon_{t-u} \in B)\right\}\boldsymbol{a}_l\right|. \tag{5.4.7}$$

Let $\boldsymbol{A}_0$ be a unique, under the D-distance, minimizer of $\Psi(\boldsymbol{A})$. We can say the estimator $\widehat{\boldsymbol{A}}$ of $\boldsymbol{A}_0$ is consistent if the D-distance between $\widehat{\boldsymbol{A}}$ and $\boldsymbol{A}_0$ converges to 0 in probability, since $\Psi_T(\boldsymbol{A})$ does not give any difference between orthogonal matrices $\boldsymbol{A}$ and $\boldsymbol{B}$ as long as $D(\boldsymbol{A}, \boldsymbol{B}) = 0$. Fan et al. (2008) proposed estimation methods using $\Psi_T(\boldsymbol{A})$ and iterative algorithms and showed that the resulting estimator is consistent.

Once the CUCs are identified, we can fit each $\sigma^2_{k,t}$ with an appropriate univariate volatility such as the following extended GARCH (1, 1) model.

$$Z_{k,t} = \sigma_{k,t}\eta_{k,t}$$
$$\sigma^2_{k,t} = \omega_k + \sum_{l=1}^{N} \alpha_{kl} Z^2_{l,t-1} + \beta_k \sigma^2_{k,t-1},$$

where $\{\eta_{k,t}\}$ is a sequence of i.i.d. random variables with mean 0 and variance 1, and $\eta_{k,t}$ is independent of $\mathcal{F}_{t-1}$. The coefficients satisfy $\omega_k > 0$, $\alpha_{kl} \geq 0$, $\beta_k \geq 0$, and $\omega_k = 1 - \sum_{l=1}^{N} \alpha_{kl} - \beta_k$ to eusure that $\mathrm{Var}(Z_{k,t}) = 1$. The extended GARCH (1, 1) model contains extra $N - 1$ terms $\sum_{l \neq k} \alpha_{kl} Z^2_{l,t-1}$ compared with the standard GARCH (1, 1) model. They can capture the dependence between the kth CUC and the other CUCs, which is called the causal components if $\alpha_{kl} > 0$. On the other hand, the conditional uncorrelated condition (5.4.4) still holds. If $0 \leq \beta_k < 1$, the extended GARCH (1, 1) model can be rewritten as

$$\begin{aligned}\sigma^2_{k,t} = \mathrm{Var}(Z_{k,t} \mid \mathcal{F}_{t-1}) &= \frac{\omega_k}{1 - \beta_k} + \sum_{l=1}^{N} \alpha_{kl} \sum_{s=1}^{\infty} \beta_k^{s-1} Z^2_{l,t-1} \\ &= 1 - \frac{\sum_{l=1}^{N} \alpha_{kl}}{1 - \beta_k} + \sum_{l=1}^{N} \alpha_{kl} \sum_{s=1}^{\infty} \beta_k^{s-1} Z^2_{l,t-1}. \end{aligned} \tag{5.4.8}$$

The quasi-Gaussian log-likelihood for $\boldsymbol{\theta}_k = (\alpha_{k1}, \ldots, \alpha_{kN}, \beta_k)'$ is given by

$$l_k(\boldsymbol{\theta}_k) = -\sum_{t=\nu+1}^{T} \left[\log\left\{\sigma^2_{k,t}(\boldsymbol{\theta}_k)\right\} + \frac{Z^2_{k,t}}{\sigma^2_{k,t}(\boldsymbol{\theta}_k)} \right]$$

for a prescribed integer ν, where $\sigma^2_{k,t}(\boldsymbol{\theta}_k)$ is given by the truncated version of Equation (5.4.8). For the selection of causal components, a sequential method can be used. We start with the standard GARCH (1, 1) model, namely, with the constraints $\alpha_{kk} > 0$ and $\alpha_{kl} = 0$ for $k \neq l$. Next, we compute the QMLE $\widehat{\boldsymbol{\theta}}^{(1)}_k$ with the constraints $\alpha_{kk} > 0$, $\alpha_{kl} > 0$, $\alpha_{km} = 0$ for $m \notin \{k, l\}$. Then, we choose $l_1 \neq k$ to maximize $l_k\left(\widehat{\boldsymbol{\theta}}^{(1)}_k\right)$ and denote it by $l_k(1)$. If the model already contains $d - 1$ terms $Z_{l_1,t-1}, \ldots, Z_{l_{d-1},t-1}$, then we choose an additional term $Z_{l_d,t-1}$ among $l_d \notin \{k, l_1, \ldots, l_{d-1}\}$ which maximizes the quasi-Gaussian log-likelihood $l_k(d) = l_k\left(\widehat{\boldsymbol{\theta}}^{(d)}_k\right)$ in the same manner. Finally, putting

$$\mathrm{BIC}_k(d) = -l_k(d) + (d + 2)\log(T - \nu),$$

we choose d_k which minimizes $\mathrm{BIC}_k(d)$ over $0 \leq d \leq N - 1$.

A natural question is whether the CUCs $Z_{1,t}, \ldots, Z_{1,t}$ exist or not. To test the existence of the CUCs, we consider the null hypothesis

$$H_0 : \boldsymbol{\varepsilon}_t = \boldsymbol{A}\boldsymbol{Z}_t \quad \text{and } \boldsymbol{\varepsilon}_t = \mathrm{diag}(\sigma_{1,t}, \ldots, \sigma_{N,t})\,\boldsymbol{\eta}_t,$$

where $\boldsymbol{A}'\boldsymbol{A} = \boldsymbol{I}_N$, $\boldsymbol{\eta}_t = (\eta_{1,t}, \ldots \eta_{N,t})'$, and $\{\eta_{1,t}, \ldots \eta_{N,t}\}$ are mutually independent and each of them is a sequence of i.i.d. random variables. This assumption is too strong for the conditionally uncorrelated condition (5.4.4). The assumption of independence needs a bootstrap method. Let $\widehat{\boldsymbol{A}} = (\widehat{\boldsymbol{a}}_1, \ldots, \widehat{\boldsymbol{a}}_N)'$ be the estimator which minimizes $\Psi_T(\boldsymbol{A})$ in (5.4.7), $Z_{k,t} = \widehat{\boldsymbol{a}}'_k \boldsymbol{\varepsilon}_t$ and $\widehat{\boldsymbol{\theta}}_k$ be an estimator for $\boldsymbol{\theta}_k$. The standardized residuals $\widehat{\eta}_{k,t}$, $t = \nu + 1, \ldots, T$ are obtained by standardizing the raw residuals, that is,

$$\widehat{\eta}_{k,t} = \frac{Z_{k,t}}{\sigma_{k,t}\left(\widehat{\boldsymbol{\theta}}_k\right)}, \qquad t = \nu + 1, \ldots, T.$$

Then, the following bootstrap sampling scheme can be executed.

(i) Draw $\eta^*_{k,t}$, $t = -\infty, \ldots, T$ by random sampling with replacement from the standardized residuals $\{\widehat{\eta}_{k,\nu+1}, \ldots, \widehat{\eta}_{k,T}\}$ for $k = 1, \ldots, N$.

(ii) Draw $Z^*_{k,t} = \sigma^*_{k,t}\eta^*_{k,t}$, $t = -\infty, \ldots, T$ for $k = 1, \ldots, N$, where

$$\sigma^{*\,2}_{k,t} = 1 - \widehat{\beta}_k - \sum_{l=1}^{N} \widehat{\alpha}_{kl} + \sum_{l=1}^{N} \widehat{\alpha}_{kl} Z^{*\,2}_{l,t-1} + \widehat{\beta}_k \sigma^{*\,2}_{k,t-1}.$$

(iii) Let $\varepsilon^*_t = \widehat{\boldsymbol{A}}\left(Z^*_{1,t}, \ldots, Z^*_{N,t}\right)'$ for $t = 1, \ldots, T$.

Note that the bootstrap sample $\{\varepsilon^*_t\}$ is drawn from the model in which $\boldsymbol{a}'_k\varepsilon^*_t$ is its genuine CUC. Furthermore, if $Z_{k,t}$ and $Z_{l,t}$ are not conditionally uncorrelated, the left hand side of Equation (5.4.6) is equal to a positive constant (instead of 0), so that larger values of $\Psi_T\left(\widehat{\boldsymbol{A}}\right)$ imply that the CUCs do not exist. We define $\Psi^*_T(\boldsymbol{A})$ from replacing ε_t by ε^*_t in Equation (5.4.7), and compute the bootstrap estimator $\boldsymbol{A}^* = \left(\boldsymbol{a}^*_1, \ldots, \boldsymbol{a}^*_N\right)$ in the same manner as $\widehat{\boldsymbol{A}}$ with $\Psi_T(\boldsymbol{A})$ replaced by $\Psi^*_T(\boldsymbol{A})$. Then, we reject H_0 if $\Psi_T\left(\widehat{\boldsymbol{A}}\right)$ is greater than the $[n\alpha]$th largest value of $\Psi^*_T(\boldsymbol{A}^*)$ in a replication of the above bootstrap resampling for n times, where $\alpha \in (0, 1)$ is the size of the test. Similarly, a bootstrap approximation for a $1 - \alpha$ confidence set of the transformation matrix $\boldsymbol{A}$ can be obtained from

$$\left\{\boldsymbol{A} \mid D\left(\boldsymbol{A}, \widehat{\boldsymbol{A}}\right) \leq c_\alpha,\ \boldsymbol{A}'\boldsymbol{A} = \boldsymbol{I}_N\right\}$$

where c_α is the $[n\alpha]$th largest value of $D\left(\boldsymbol{A}^*, \widehat{\boldsymbol{A}}\right)$. A bootstrap confidence interval for each element of $\boldsymbol{\theta}_k$ can be constructed from the same policy.

5.4.1.2 ICA modeling for financial time series

The sequence of innovations from financial asset returns can be regarded as an additive mixture of several independent latent economic variables which act on the financial market. Independent component analysis (ICA) provides a solution to extract these latent sources from the observed data under few assumptions on these sources. Let $\boldsymbol{s}_t$ be mutually independent latent sources. For simplicity, we assume a full rank linear mixing transformation $\boldsymbol{M}_0$ of the latent sources $\boldsymbol{s}_t$ exists and gives an empirically sufficient representation of observations $\boldsymbol{e}_t$, namely, $\boldsymbol{e}_t = \boldsymbol{M}_0\boldsymbol{s}_t$. Similarly, the separating matrix $\boldsymbol{W}_0$ is defined by $\boldsymbol{s}_t = \boldsymbol{W}_0\boldsymbol{e}_t$. The traditional independent component estimation is achieved by maximization of non-Gaussianity. This concept is motivated by the central limit theorem, which confirms that the sums of independent non-Gaussian random variables are closer to Gaussian than the original random variables. For given observations $\boldsymbol{e}_t$, we seek the components of $\boldsymbol{W}\boldsymbol{e}_t$ which are mutually independent (non-Gaussian) for some separating matrix $\boldsymbol{W}$. This estimation of $\boldsymbol{W}$ can be carried out by minimizing the mutual information of $\boldsymbol{W}\boldsymbol{e}_t$. For instance, the Kullback–Leibler divergence measures the difference between the joint distribution of its components and the product of the marginal distributions of its components.

The approach to multivariate volatility modeling based on the concept of dynamic orthogonal components (DOCs) was proposed by Matteson and Tsay (2011). This approach is an extension of PCA in which orthogonality in cross dependence over time is also sought. However, it is not as stringent as ICA, since it focuses on orthogonality only up to the fourth order. Let $\boldsymbol{y}_t = (y_{1,t}, \ldots, y_{N,t})'$ denote a vector of returns, or log returns, of N assets at time t. The return vector can be partitioned as

$$\boldsymbol{y}_t = \boldsymbol{\mu}_t + \boldsymbol{e}_t, \tag{5.4.9}$$

in which $\mu_t = E(y_t \mid \mathcal{F}_{t-1})$ is the conditional mean of y_t given $\mathcal{F}_{t-1}$, and e_t is the mean zero innovation vector. The conditional covariance matrix of the returns given $\mathcal{F}_{t-1}$, $\Sigma_t = \mathrm{Cov}(y_t \mid \mathcal{F}_{t-1}) = \mathrm{Cov}(e_t \mid \mathcal{F}_{t-1})$ is often referred to as a volatility matrix in the finance literature.

For a stationary vector time series x_t, assume that there exist DOCs s_t such that $x_t = M s_t$ where M is a nonsingular mixing matrix. Furthermore, let U denote an uncorrelating matrix so that $z_t = U x_t$ is an uncorrelated observation. It is straightward to find an invertible, linear transform U which provides the uncorrelated random variables. Namely, let $\Sigma = \mathrm{Cov}(x_t)$ denote the positive definite unconditional covariance matrix of x_t and define $z_t = U x_t$. Then $\mathrm{Cov}(z_t)$ equals the $N \times N$ identity matrix I_N. Let Γ and Λ denote the orthogonal matrix of eigenvectors and the diagonal matrix of corresponding eigenvalues of Σ, respectively. Then, the invertible matrix U can be chosen as the inverse symmetric square root of the unconditional covariance matrix defined as $U = \Sigma^{-1/2} = \Gamma\Lambda^{-1/2}\Gamma'$, or simply in connection with principle component analysis as $U = \Lambda^{-1/2}\Gamma'$. The relationship between s_t and z_t is given by

$$s_t = M^{-1} x_t = M^{-1} U^{-1} z_t \equiv W z_t, \tag{5.4.10}$$

where $W = M^{-1}U^{-1}$.

We begin with the specification of DOCs in mean for time series with multivariate serial correlation. To simplify the problem, we assume that we observe $x_t = y_t - E(y_t)$ instead of y_t, and Σ_t as constant. Let M_1 denote the mixing matrix associated with DOCs in mean, namely, $x_t = M_1 s_t$. Then, combining (5.4.9) and (5.4.10), we have

$$s_t = M_1^{-1} x_t = M_1^{-1} \mu_t + M_1^{-1} e_t = \widetilde{\mu}_t + \widetilde{e}_t. \tag{5.4.11}$$

Here, the components of s_t are dynamically orthogonal in mean, that is $\mathrm{Cov}(s_t, s_{t-l})$ is diagonal for all lags $l = 0, 1, 2, \ldots$. The remaining marginal serial correlation can be modeled by using the univariate ARMA model. In the corresponding vector ARMA model, we can obtain the version of $\widetilde{e}_t$ in Equation (5.4.11), in which $\mathrm{Cov}(\widetilde{e}_t)$ is diagonal. Therefore, we can obtain the version of $\widetilde{\mu}_t$ in which the coefficient matrices specified by $\widetilde{\mu}_t$ are also diagonal.

Next, we focus on volatility modeling. To simplify the discussion, we assume that μ_t is known, namely, we assume that $x_t = e_t = y_t - E(y_t \mid \mathcal{F}_{t-1})$ is observed. Let M_2 denote the mixing matrix associated with DOCs in volatility, namely, $x_t = M_2 s_t$. Then, combining (5.4.9) and (5.4.10), we have

$$\Sigma_t = \mathrm{Cov}(x_t \mid \mathcal{F}_{t-1}) = M_2 \mathrm{Cov}(s_t \mid \mathcal{F}_{t-1}) M_2'.$$

Here, the components of s_t are dynamically orthogonal in volatility, that is, $\mathrm{Cov}\left(s_t^2, s_{t-l}^2\right)$ is diagonal, and $\mathrm{Cov}\left(s_{i,t}s_{j,t}, s_{i,t-l}s_{j,t-l}\right) = 0$ for $i \neq j$, for all lags $l = 0, 1, 2, \ldots$. For linear generalized autoregressive conditional heteroscedastic time series, we can obtain the version of the conditional covariance process $\mathrm{Cov}(s_t \mid \mathcal{F}_{t-1})$ in which $\mathrm{Cov}(s_t \mid \mathcal{F}_{t-1})$ is diagonal at all time points. The remaining marginal conditional heteroscedasticity can be modeled by univariate models such as the GARCH model. Then, we obtain the version of $\mathrm{Cov}(s_t \mid \mathcal{F}_{t-1})$ in which all coefficient matrices for the corresponding matrix models of $\mathrm{Cov}(s_t \mid \mathcal{F}_{t-1})$ are diagonal.

For any DOC specification we require the x_t satisfies the standard regularity condition.

Assumption 5.4.6. *The process x_t is stationary and ergodic with $E\|x_t\|^2 < \infty$.*

Without loss of generality, we can assume that s_t is standardized, that is, $E(s_{i,t}) = 0$ and $\mathrm{Var}(s_{i,t}) = 1$ for $i = 1, \ldots, N$. The fundamental motivation is the fact that, empirically, the dynamics of x_t can often be well approximated by an invertible linear combination of DOCs, namely, $x_t = M s_t$. For theoretical and practical considerations, it is convenient to work with uncorrelated random variables, so that it is useful to start with uncorrelated process z_t instead of x_t in Equation (5.4.10).

We test for significant lagged cross correlations in s_t for DOCs in mean and significant lagged cross correlations in s_t^2 for DOCs in volatility. Let $\boldsymbol{h}$ denote either the identity transformation $\boldsymbol{h}(\boldsymbol{x}) = \boldsymbol{x}$ for DOCs in mean or the square transformation $\boldsymbol{h}(\boldsymbol{x}) = \boldsymbol{x}^2$ for DOCs in volatility, and let $h_{i,t-l} = h_i(\boldsymbol{s}_{t-l})$ and $\rho_{i,j}^h(l) = \text{Corr}\left(h_{i,t}, h_{j,t-l}\right)$. Then, the joint lag m null and alternative hypotheses to test for the existence of DOCs are given by

$$H_0 : \rho_{i,j}^h(l) = 0 \quad \text{for all } i \neq j, l = 0, \ldots, m,$$
$$H_A : \rho_{i,j}^h(l) = 0 \quad \text{for some } i \neq j, l = 0, \ldots, m.$$

Matteson and Tsay (2011) suggested using Ljung–Box-type test statistic

$$\underline{Q}_N^0(m) = T \sum_{i<j} \rho_{i,j}^h(0)^2 + T(T+1) \sum_{k=1}^{m} \sum_{i \neq j} \frac{\rho_{i,j}^h(k)^2}{T-k},$$

where T is the sample size of the observations. Under H_0, $\underline{Q}_N^0$ is asymptotically distributed as a chi square distribution with $\frac{N(N-1)}{2} + mN(N-1)$ degrees of freedom. The null hypothesis is rejected for a large value of $\underline{Q}_N^0$. If H_0 is rejected, we should consider an alternative modeling procedure.

Given the uncorrelated process $\boldsymbol{z}_t$, the separating matrix $\boldsymbol{W}$ is orthogonal since, from Equation (5.4.10), we have

$$\boldsymbol{I}_N = \text{Var}(\boldsymbol{s}_t) = \boldsymbol{W} \text{Var}(\boldsymbol{z}_t) \boldsymbol{W}' = \boldsymbol{W}\boldsymbol{W}'.$$

Therefore, $\boldsymbol{W}$ has $p = \frac{N(N-1)}{2}$ free elements, and the orthogonality of $\boldsymbol{W}$ implies that it represents a rotation in the N-dimensional space and can be parametrized by the vector $\boldsymbol{\theta}$ of the rotation angles with length p.

Let $O(N)$ denote the group of all $N \times N$ orthogonal matrices, $SO(N)$ denote the subgroup (rotation group) whose determinants equal to 1, and $\boldsymbol{\xi}_1, \ldots, \boldsymbol{\xi}_N$ denote the canonical basis of $\mathbb{R}^N$ for $N \geq 2$. Furthermore, let $\boldsymbol{Q}_{i,j}(\psi)$ denote a rotation of all vectors lying in the $\left(\boldsymbol{\xi}_i, \boldsymbol{\xi}_j\right)$ plane of $\mathbb{R}^N$ by angle ψ which is oriented such that the rotation from $\boldsymbol{\xi}_i$ to $\boldsymbol{\xi}_j$ becomes positive, namely, $\boldsymbol{Q}_{i,j}(\psi)$ is given by replacing the elements of identity matrix $\boldsymbol{I}_N$ in which the (i, i)th and (j, j)th elements are replaced by $\cos\psi$, the (i, j)th element is replaced by $-\sin\psi$ and the (j, i)th element is replaced by $\sin\psi$.

If $\boldsymbol{\theta}$ is a vector of rotation angles with length $p = \frac{N(N-1)}{2}$, indexed by $i, j : 1 \leq i < j \leq N$, then any rotation $\boldsymbol{W} \in SO(N)$ can be written in the form

$$\boldsymbol{W}_{\boldsymbol{\theta}} = \boldsymbol{Q}^{(N)} \cdots \boldsymbol{Q}^{(2)}$$

in which $\boldsymbol{Q}^{(k)} = \boldsymbol{Q}_{1,k}\left(\theta_1^k\right) \cdots \boldsymbol{Q}_{k-1,k}\left(\theta_{k-1}^k\right)$. Moreover, we assume that there exists a unique inverse mapping of $\boldsymbol{W} \in SO(N)$ into $\boldsymbol{\theta} \in [-\pi, \pi]^p$ where the mapping is continuous.

Given an uncorrelated mean 0 vector time series $\boldsymbol{z}_t$, $t = 1, \ldots, T$, define $\boldsymbol{s}_t(\boldsymbol{\theta}) = \boldsymbol{W}_{\boldsymbol{\theta}} \boldsymbol{z}_t$ and let $\widehat{E}\{\cdot\}$ denote the sample expectation operator. We begin by estimating the parameter vector of rotation angles $\boldsymbol{\theta}$. Let $\boldsymbol{h}$ be a suitably chosen vector function $\boldsymbol{h} : \mathbb{R}^N \to \mathbb{R}^N$ with $h_i(\boldsymbol{s}) = h_i(s_i)$, $i = 1, \ldots, N$. Our objective in estimating $\boldsymbol{\theta}$ is to make the lagged sample cross covariance function of the transformed process

$$\widehat{\Gamma}^{\boldsymbol{h}(\boldsymbol{s}(\boldsymbol{\theta}))} = \widehat{E}\{\boldsymbol{h}(\boldsymbol{s}_t(\boldsymbol{\theta}))\, \boldsymbol{h}(\boldsymbol{s}_{t-l}(\boldsymbol{\theta}))'\} - \widehat{E}\{\boldsymbol{h}(\boldsymbol{s}_t(\boldsymbol{\theta}))\}\, \widehat{E}\{\boldsymbol{h}(\boldsymbol{s}_{t-l}(\boldsymbol{\theta}))\}'$$

as close to diagonal as possible for some finite set of lags $l \in \overline{\mathbb{N}}_0 \subset \mathbb{N}_0$, in which lag 0 is included.

To estimate DOCs in mean, let $h_i(\boldsymbol{s}) = s_i$, namely, the identity function of each element. On the other hand, to estimate DOCs in volatility, let $h_i(\boldsymbol{s}) = s_i^2$, or if the observations $\boldsymbol{z}_t$ exhibit heavy tails, let

$$h_i(\boldsymbol{s}) = \text{Huber}_c(\boldsymbol{s}) = \begin{cases} s_i^2 & \text{if } |s_i| \leq c, \\ 2|s_i|c - c^2 & \text{if } |s_i| > c \end{cases}$$

for some $0 < c < \infty$, which is a continuously differentiable version of Huber's function.

Now, we define the estimator in line with the generalized method of moments (GMM) literature. Let $\boldsymbol{f}(\boldsymbol{z}_t, \boldsymbol{\theta})$ denote a vectorized array of orthogonal constraints whose elements are given by

$$f_{i,j}^{l}(\boldsymbol{z}_t, \boldsymbol{\theta}) = h_i(\boldsymbol{s}_t(\boldsymbol{\theta}))\, h_j(\boldsymbol{s}_{t-l}(\boldsymbol{\theta})) - \widehat{E}\{h_i(\boldsymbol{s}_t(\boldsymbol{\theta}))\}\, \widehat{E}\left\{h_j(\boldsymbol{s}_{t-l}(\boldsymbol{\theta}))\right\}.$$

Note that the constraint $\boldsymbol{f}$ is indexed by $i < j$ for $l = 0$ and $i \neq j$ for $l \neq 0$, since the cross covariance function is symmetric at lag 0. Therefore, the length of $\boldsymbol{f}$ is $q = p\left(2\left|\overline{\mathbb{N}}_0\right| - 1\right)$, where $\left|\overline{\mathbb{N}}_0\right|$ denotes the number of elements of $\overline{\mathbb{N}}_0$. We apply the weights

$$\phi_l = \begin{cases} \dfrac{1 - l/\left|\overline{\mathbb{N}}_0\right|}{p\sum_l\left(1 - l/\left|\overline{\mathbb{N}}_0\right|\right)} & \text{for } l = 0, \\ \dfrac{1 - l/\left|\overline{\mathbb{N}}_0\right|}{2p\sum_l\left(1 - l/\left|\overline{\mathbb{N}}_0\right|\right)} & \text{for } l \neq 0, \end{cases}$$

which are arranged into a diagonal weight matrix as

$$\boldsymbol{\Phi} = \operatorname{diag}\left(\phi_{l_1}, \ldots, \phi_{l_1}, \ldots, \phi_{l_{\left|\overline{\mathbb{N}}_0\right|}}, \ldots, \phi_{l_{\left|\overline{\mathbb{N}}_0\right|}}\right).$$

Letting $\widetilde{\boldsymbol{f}}_T(\boldsymbol{\theta}) = \widehat{E}\{\boldsymbol{f}(\boldsymbol{z}_t, \boldsymbol{\theta})\}$, define the objective function as

$$\mathcal{J}_T(\boldsymbol{\theta}) = \widetilde{\boldsymbol{f}}_T(\boldsymbol{\theta})'\, \boldsymbol{\Phi}\, \widetilde{\boldsymbol{f}}_T(\boldsymbol{\theta})$$

and the estimator of $\boldsymbol{\theta}$ as $\widehat{\boldsymbol{\theta}}_T = \arg\min_{\boldsymbol{\theta}} \mathcal{J}_T(\boldsymbol{\theta})$. Then, the estimator of the separating matrix is given by $\boldsymbol{W}_{\widehat{\boldsymbol{\theta}}_T}$ and the estimator of the vector time series of DOCs is given by $\widehat{\boldsymbol{s}}_t = \boldsymbol{W}_{\widehat{\boldsymbol{\theta}}_T} \boldsymbol{z}_t$.

The magnitude, sign and order of DOCs are ambiguous when they are associated with estimating $\boldsymbol{W}$ and $\boldsymbol{s}$. The magnitude of the DOCs has already been fixed by assuming $\mathrm{Var}(\boldsymbol{s}) = \boldsymbol{I}_N$. On the other hand, letting $\boldsymbol{P}_{\pm}$ denote a signed permutation matrix, the mixing model $\boldsymbol{x} = \boldsymbol{M}\boldsymbol{s}$ is equivalent to

$$\boldsymbol{x} = \boldsymbol{M}\boldsymbol{P}_{\pm}'\boldsymbol{P}_{\pm}\boldsymbol{s} = (\boldsymbol{M}\boldsymbol{P}_{\pm}')(\boldsymbol{P}_{\pm}\boldsymbol{s})$$

in which $\boldsymbol{P}_{\pm}\boldsymbol{s}$ are new DOCs and $\boldsymbol{M}\boldsymbol{P}_{\pm}'$ is the new mixing matrix. However, the identification of the DOCs up to a signed permutation is sufficient for the purpose of extracting the independent univariate time series.

In the following, without loss of generality, we assume that $E(\boldsymbol{x}_t)$ and $\mathrm{Cov}(\boldsymbol{x}_t) = \boldsymbol{I}_N$. Let

$$\mathcal{J}(\boldsymbol{\theta}) = E\{\boldsymbol{f}(\boldsymbol{x}_t, \boldsymbol{\theta})\}'\, \boldsymbol{\Phi} E\{\boldsymbol{f}(\boldsymbol{x}_t, \boldsymbol{\theta})\} \tag{5.4.12}$$

and let $\overline{\boldsymbol{\Theta}}$ denote a sufficiently large compact subset of the space $\boldsymbol{\Theta}$ (see the online supplementary materials of Matteson and Tsay (2011) for the detailed definition). We assume the following regularity conditions.

Assumption 5.4.7.

$$\sup_{\theta \in \overline{\boldsymbol{\Theta}}} E\left\|\boldsymbol{h}(\boldsymbol{W}_{\boldsymbol{\theta}}\boldsymbol{x}_t)^2\right\|^2 < \infty$$

and

$$\sup_{\theta \in \overline{\boldsymbol{\Theta}}} E\left\|\frac{\partial \boldsymbol{h}(\boldsymbol{W}_{\boldsymbol{\theta}}\boldsymbol{x}_t)}{\partial \theta}\boldsymbol{x}_t\right\|^2 < \infty.$$

Now, we have the following results.

Theorem 5.4.8 (Theorem 1 of Matteson and Tsay (2011)). *Suppose that $\overline{\mathbb{N}}_0 \subset \mathbb{N}_0$ is fixed and finite, $\boldsymbol{\Phi}$ is constant and positive definite, and $\boldsymbol{h}$ is measurable and continuously differentiable. Furthermore, suppose that there exists a unique minimizer $\boldsymbol{\theta}^0 \in \overline{\boldsymbol{\Theta}}$ of Equation (5.4.12) and $\boldsymbol{W}_{\boldsymbol{\theta}^0}$ satisfies the conditions for a unique continuous inverse to exist. If Assumption 5.4.6 holds for $\boldsymbol{x}_t$ and $\boldsymbol{h}$ is chosen for which Assumption 5.4.7 is satisfied, then we have*

$$\widehat{\theta}_T \overset{a.s.}{\to} \boldsymbol{\theta}^0 \quad as \quad T \to \infty.$$

Note that if DOCs exist, then there exists $\boldsymbol{\theta}^0 \in \boldsymbol{\Theta}$ such that $E\left\{\boldsymbol{f}\left(\boldsymbol{x}_t, \boldsymbol{\theta}^0\right)\right\} = \mathbf{0}$.

Also note that $\widetilde{\boldsymbol{f}}_T(\boldsymbol{\theta})$ is continuously differentiable with respect to $\boldsymbol{\theta}$ on $\boldsymbol{\Theta}$ by assumption. Denote $q \times p$ gradient matrix of $\widetilde{\boldsymbol{f}}_T(\boldsymbol{\theta})$ by $\overline{\boldsymbol{F}}_T(\boldsymbol{\theta}) = \frac{\partial \widetilde{\boldsymbol{f}}_T(\boldsymbol{\theta})}{\partial \boldsymbol{\theta}}$ and let $\overline{\boldsymbol{V}}_T = T\text{Var}\left\{\widetilde{\boldsymbol{f}}_T\left(\boldsymbol{\theta}^0\right)\right\}$ be a nonrandom and positive definite $q \times q$ matrix. Since both $\overline{\boldsymbol{V}}_T$ and $\boldsymbol{\theta}^0$ are unknown, we need the weak consistent estimator to use the following theorem for statistical inference.

Theorem 5.4.9 (Theorem 2 of Matteson and Tsay (2011)). *Suppose that $\boldsymbol{h}(\boldsymbol{W}_{\boldsymbol{\theta}}\boldsymbol{x}_t)$ is strong mixing with geometric rate, $E\left\{\frac{\partial}{\partial \boldsymbol{\theta}}\boldsymbol{f}\left(\boldsymbol{x}_t, \boldsymbol{\theta}^0\right)\right\}$ has linearly independent columns and there exists an estimator $\widehat{V}_T$ of $\overline{\boldsymbol{V}}_T$ such that $\widehat{V}_T \overset{\mathcal{P}}{\to} \overline{\boldsymbol{V}}_T$ as $T \to \infty$. Then, under the conditions of Theorem 5.4.8, we have*

$$\left\{\overline{\boldsymbol{F}}_T\left(\widehat{\theta}_T\right)' \boldsymbol{\Phi}\widehat{V}_T\boldsymbol{\Phi}\overline{\boldsymbol{F}}_T\left(\widehat{\theta}_T\right)\right\}^{-1/2}\left\{\overline{\boldsymbol{F}}_T\left(\widehat{\theta}_T\right)' \boldsymbol{\Phi}\overline{\boldsymbol{F}}_T\left(\widehat{\theta}_T\right)\right\}\sqrt{T}\left(\widehat{\theta}_T - \boldsymbol{\theta}^0\right) \overset{\mathcal{L}}{\to} N\left(\mathbf{0}, \boldsymbol{I}_p\right)$$

as $T \to \infty$.

Note that $\boldsymbol{x}_t$ is strong mixing with geometric rate if there exists constant $K > 0$ and $\alpha \in (0, 1)$ such that

$$\sup_{A \in \sigma\{\boldsymbol{x}_t | t \le 0\}, B \in \sigma\{\boldsymbol{x}_t | t > k\}} |P(A \cap B) - P(A)P(B)| = \phi_k \le K\alpha^k$$

in which ϕ_k is referred to as the mixing rate function.

5.4.1.3 *ICA modeling in frequency domain for time series*

Regarding analysis of multiple time series, only a few ICA methods have been applied satisfactorily. Pham and Garat (1997) suggested ICA based on a quasi-maximum likelihood approach in the setting of independent non-Gaussian sources and correlated sources. Denote the observation record by $\boldsymbol{X}(1), \ldots, \boldsymbol{X}(T)$ where each $\boldsymbol{X}(t)$ is a random vector of K components $X_1(t), \ldots, X_K(t)$ corresponding to K observation channel. Assume that each observation $X_i(t)$ is a linear combination of K independent sources, namely, $\boldsymbol{X}(t) = \boldsymbol{AS}(t)$, where $\boldsymbol{A}$ is some unknown matrix. Furthermore, $\{\boldsymbol{S}(t), t = 1, \ldots, T\}$ is a stationary sequence of random vectors with K components $S_1(t), \ldots, S_K(t)$ and the sequences $\{S_k, (1), \ldots, S_k(T)\}$, $k = 1, \ldots, K$ are mutually independent. The source separation problem is to reconstruct the sources $\{S_k, (1), \ldots, S_k(T)\}$ from the observations $\{\boldsymbol{X}(1), \ldots, \boldsymbol{X}(T)\}$ based only on the assumption of independence between the sources. To write down the likelihood and derive a separation procedure, first, Pham and Garat (1997) assume that the sources are white in the sense that the $S_i(t)$ at different times t are independent and identically distributed. Of cause this assumption is unrealistic, but it will be used only as a working assumption. Furthermore, one needs the density function of the sources, which is unknown in a blind context. However, we assume for the moment that these densities are known up to a scale factor to avoid scaling ambiguity. Namely, the source $S_k(t)$ has density $f_k(\cdot/\sigma_k)/\sigma_k$, where $f_k(\cdot)$ is known and σ_k is unknown. Then, $\boldsymbol{S}(t)$ has density $\prod_{k=1}^{K} f_k\{s_k(t)/\sigma_k\}/\sigma_k$ and $\boldsymbol{X}(t)$ is related to $\boldsymbol{S}(t)$ through $\boldsymbol{X}(t) = \boldsymbol{AS}(t)$. Here, we assume that $\boldsymbol{A}$ is invertible. Now, we can write down the log-likelihood of the observation data as

$$L_T = T\left[\sum_{k=1}^{K} \widehat{E}\left[\log\left\{\frac{1}{\sigma_k} f_k\left(\frac{\boldsymbol{e}_k' \boldsymbol{A}^{-1}\boldsymbol{X}}{\sigma_k}\right)\right\}\right] - \log|\det \boldsymbol{A}|\right], \tag{5.4.13}$$

where $\widehat{E}$ is the time average operator $\widehat{E}\{g(\boldsymbol{X})\} = [g\{\boldsymbol{X}(1)\} + \cdots + g\{\boldsymbol{X}(T)\}]/T$ and e_k is the kth column of the identity matrix of order K. Denoting $d\boldsymbol{A}^{-1} = -\boldsymbol{A}^{-1}d\boldsymbol{A}\boldsymbol{A}^{-1}$, $d\log|\det \boldsymbol{A}| = \mathrm{tr}\left\{\boldsymbol{A}^{-1}d\boldsymbol{A}\right\}$ and putting $\widetilde{\psi}_k(x) = -\frac{d\log\{f_k(x)\}}{dx}$, we see that

$$T^{-1}dL_T = \sum_{k=1}^{K}\widehat{E}\left\{\widetilde{\psi}_k\left(\frac{e_k'\boldsymbol{A}^{-1}\boldsymbol{X}}{\sigma_k}\right)\frac{e_k'\boldsymbol{A}^{-1}d\boldsymbol{A}\boldsymbol{A}^{-1}\boldsymbol{X}}{\sigma_k}\right\} - \mathrm{tr}\left\{\boldsymbol{A}^{-1}d\boldsymbol{A}\right\}$$
$$+\sum_{k=1}^{K}\widehat{E}\left\{\widetilde{\psi}_k\left(\frac{e_k'\boldsymbol{A}^{-1}\boldsymbol{X}}{\sigma_k}\right)\frac{e_k'\boldsymbol{A}^{-1}\boldsymbol{X}}{\sigma_k^2} - \frac{1}{\sigma_k}\right\}d\sigma_k.$$

Letting ∂_{ij} be the (i, j)th elements of matrix $\boldsymbol{A}^{-1}d\boldsymbol{A}$, the first two terms of the above right hand side can be rewritten by

$$\sum_{i=1}^{K}\sum_{j=1}^{K}\widehat{E}\left\{\widetilde{\psi}_i\left(\frac{e_i'\boldsymbol{A}^{-1}\boldsymbol{X}}{\sigma_i}\right)\frac{e_j'\boldsymbol{A}^{-1}\boldsymbol{X}}{\sigma_i}\right\}\partial_{ij} - \sum_{i=1}^{K}\partial_{ii}.$$

Since dL_T must vanish for all $d\boldsymbol{A}$ (or equivalently ∂_{ij}) and $d\sigma_i$ at the maximum likelihood estimators $\widehat{\boldsymbol{A}}$ and $\widehat{\sigma}_i$ of $\boldsymbol{A}$ and σ_i, we have the estimating equation

$$\widehat{E}\left\{\widetilde{\psi}_i\left(\frac{e_i'\widehat{\boldsymbol{A}}^{-1}\boldsymbol{X}}{\widehat{\sigma}_i}\right)e_j'\widehat{\boldsymbol{A}}^{-1}\boldsymbol{X}\right\} = 0, \quad i \neq j = 1, \ldots, K,$$
$$\widehat{\sigma}_i = \widehat{E}\left\{\widetilde{\psi}_i\left(\frac{e_i'\widehat{\boldsymbol{A}}^{-1}\boldsymbol{X}}{\widehat{\sigma}_i}\right)e_i'\widehat{\boldsymbol{A}}^{-1}\boldsymbol{X}\right\}, \quad i = 1, \ldots, K.$$

Note that post-multiplying by a diagonal matrix and dividing the σ_i by its corresponding diagonal elements does not change the likelihood; therefore, the parameter set $\{\boldsymbol{A}, \sigma_1, \ldots, \sigma_K\}$ is redundant. The first $K(K-1)$ equations above in fact determine $\widehat{\boldsymbol{A}}$ up to scaling and once some scaling convention is adopted to make it unique, the last K equations merely serve to estimate the σ_i. Since we are not interested in values of σ_i, we just drop these equations. On the other hand, since the densities of the sources are unknown in the blind separation context, we have to take some a priori function ψ_i, instead of $\widetilde{\psi}_i$, which also includes the factor $\overline{\sigma}_i$. Thus, the separation procedure will be proposed to solve the system of estimation equations

$$\widehat{E}\left\{\psi_i\left(e_i'\widehat{\boldsymbol{A}}^{-1}\boldsymbol{X}\right)e_j'\widehat{\boldsymbol{A}}^{-1}\boldsymbol{X}\right\} = 0, \quad i \neq j = 1, \ldots, K \tag{5.4.14}$$

with respect to $\widehat{\boldsymbol{A}}$. Note that by choosing the ψ_i a priori, the above procedure is no longer maximum likelihood. Nevertheless, the procedure will be justified. For instance, the sources satisfy being centred and possessing stationarity and ergodicity, the left hand sides of (5.4.14) converge to the expectations of $\psi_i(S_i(t)/\widetilde{\sigma}_i)S_j(t)$, $i \neq j$, where $\widetilde{\sigma}_i$ are some scaling factors, and these expectations vanish since $S_i(t)$ has zero mean and $S_i(t)$ are mutually independent.

Next, Pham and Garat (1997) considered the derivation of the separation procedure with the maximum likelihood approach for temporally correlated sources. A well-known technique to decorrelate a stationary signal is given by performing a discrete Fourier transform, which transforms $\boldsymbol{X}(1), \ldots, \boldsymbol{X}(T)$ into

$$\boldsymbol{d}_X\left(\frac{k}{T}\right) = \frac{1}{\sqrt{T}}\sum_{t=1}^{T}\boldsymbol{X}(t)e^{-i2\pi tk/T}, \quad k = 0, \ldots, T-1.$$

It is well known that for $k \neq l$, $E\left\{\boldsymbol{d}_X\left(\frac{k}{T}\right)\overline{\boldsymbol{d}_X\left(\frac{l}{T}\right)}\right\} \to 0$ as $T \to \infty$, and $E\left\{\boldsymbol{d}_X\left(\frac{k}{T}\right)\overline{\boldsymbol{d}_X\left(\frac{k}{T}\right)}\right\}$ converge to the spectral density matrices of the process $\boldsymbol{f}_X(\lambda)$ at $\frac{k}{T}$. It is also known that these random vectors are asymptotically Gaussian under mild conditions (see, e.g., Brillinger (2001c)). Therefore, we can regard $\left\{\boldsymbol{d}_X\left(\frac{k}{T}\right),\ 0 \le k \le \frac{T}{2}\right\}$ as independent Gaussian random vectors as is the usual practice in time series literature. The joint probability density function of $\boldsymbol{d}_X(0)$, $\boldsymbol{d}_X\left(\frac{1}{T}\right), \ldots$ for T even is given by

$$\frac{1}{\sqrt{\det\left\{4\pi^2 \boldsymbol{f}_X(0)\,\boldsymbol{f}_X\left(\frac{1}{2}\right)\right\}}}\exp\left\{-\frac{\boldsymbol{d}_X(0)'\,\boldsymbol{f}_X(0)^{-1}\,\boldsymbol{d}_X(0)+\boldsymbol{d}_X\left(\frac{1}{2}\right)'\,\boldsymbol{f}_X\left(\frac{1}{2}\right)^{-1}\boldsymbol{d}_X\left(\frac{1}{2}\right)}{2}\right\}$$

$$\cdot\prod_{k=1}^{\frac{T}{2}-1}\frac{1}{\det\left\{\pi \boldsymbol{f}_X\left(\frac{k}{T}\right)\right\}}\exp\left\{-\overline{\boldsymbol{d}_X\left(\frac{k}{T}\right)}'\,\boldsymbol{f}_X\left(\frac{k}{T}\right)^{-1}\boldsymbol{d}_X\left(\frac{k}{T}\right)\right\},$$

whereas the terms corresponding to $\boldsymbol{d}_X\left(\frac{1}{2}\right)$ simply disappear when T is odd. Since $\boldsymbol{f}_X(\lambda) = \boldsymbol{A}\mathrm{diag}\{g_1(\lambda),\ldots,g_K(\lambda)\}\,\boldsymbol{A}'$, where the g_i are the spectral densities of the sources, taking the logarithm reduces to

$$L_T = \frac{1}{2}\sum_{i=1}^{K}\sum_{k=0}^{T-1}\left\{\frac{\left|\mathrm{e}_i'\boldsymbol{A}^{-1}\boldsymbol{d}_X\left(\frac{k}{T}\right)\right|^2}{g_i\left(\frac{k}{T}\right)}+\log g_i\left(\frac{k}{T}\right)\right\} - T\log\left|\det \boldsymbol{A}\right| + constant.$$

For the moment, we make the working assumption that the spectral densities of the sources are up to constant, so that $g_i(\lambda) = \eta_i h_i(\lambda)$, where h_i are known functions and η_i are unknown parameters. The ML approach maximizes L_T with respect to $\eta_1,\ldots,\eta_K$ and $\boldsymbol{A}$, hence we have

$$dL_T = \sum_{i=1}^{K}\sum_{k=0}^{T-1}\frac{\mathrm{e}_i'\boldsymbol{A}^{-1}d\boldsymbol{A}\boldsymbol{A}^{-1}\boldsymbol{d}_X\left(\frac{k}{T}\right)\mathrm{e}_i'\boldsymbol{A}^{-1}\overline{\boldsymbol{d}_X\left(\frac{k}{T}\right)}}{\eta_i h_i\left(\frac{k}{T}\right)} - T\mathrm{tr}\left\{\boldsymbol{A}^{-1}d\boldsymbol{A}\right\}$$

$$+\frac{1}{2}\sum_{i=1}^{K}\left\{\sum_{k=0}^{T-1}\frac{\left|\mathrm{e}_i'\boldsymbol{A}^{-1}\boldsymbol{d}_X\left(\frac{k}{T}\right)\right|^2}{\eta_i h_i\left(\frac{k}{T}\right)} - T\right\}\frac{d\eta_i}{\eta_i}.$$

Again, introducing the element ∂_{ij} of the matrix $\boldsymbol{A}^{-1}d\boldsymbol{A}$, we can rewrite the first two terms as

$$\sum_{i=1}^{K}\sum_{j=1}^{K}\sum_{k=0}^{T-1}\frac{\mathrm{e}_j'\boldsymbol{A}^{-1}\boldsymbol{d}_X\left(\frac{k}{T}\right)\mathrm{e}_i'\boldsymbol{A}^{-1}\overline{\boldsymbol{d}_X\left(\frac{k}{T}\right)}}{\eta_i h_i\left(\frac{k}{T}\right)}\partial_{ij} - T\sum_{i=1}^{K}\partial_{ii}.$$

The differential of L_T must vanish for all infinitesimal increments $d\boldsymbol{A}, d\eta_1,\ldots,d\eta_K$, at the ML estimators $\widehat{\boldsymbol{A}}, \widehat{\eta}_1,\ldots,\widehat{\eta}_K$. Therefore the estimating functions are given by

$$\sum_{k=0}^{T-1}\frac{\mathrm{e}_j'\widehat{\boldsymbol{A}}^{-1}\boldsymbol{d}_X\left(\frac{k}{T}\right)\mathrm{e}_i'\widehat{\boldsymbol{A}}^{-1}\overline{\boldsymbol{d}_X\left(\frac{k}{T}\right)}}{h_i\left(\frac{k}{T}\right)} = 0, \quad i \neq j = 1,\ldots,K,$$

$$\widehat{\eta}_i = \frac{1}{T}\sum_{k=0}^{T-1}\frac{\left|\mathrm{e}_i'\widehat{\boldsymbol{A}}^{-1}\boldsymbol{d}_X\left(\frac{k}{T}\right)\right|^2}{h_i\left(\frac{k}{T}\right)}, \quad i = 1,\ldots,K.$$

The second set of equations determines the scale factors and are not of interest. The estimated mixing matrix is led from the first set of equations (up to the scale factor). However, we approximate sums by integrals, so that

$$\int_0^1\frac{\mathrm{e}_j'\widehat{\boldsymbol{A}}^{-1}\boldsymbol{d}_X(\lambda)\,\mathrm{e}_i'\widehat{\boldsymbol{A}}^{-1}\overline{\boldsymbol{d}_X(\lambda)}}{h_i(\lambda)}d\lambda = 0, \quad i \neq j = 1,\ldots,K.$$

Let $\phi_{i,t}$, $t \in \mathbb{Z}$ denote the impulse response of the filter ϕ_i whose frequency response is $h_i(u)^{-1}$, namely,

$$h_i(\lambda)^{-1} = \sum_{t \in \mathbb{Z}} \phi_{i,t} e^{-i2\pi t \lambda}.$$

Then $\sqrt{T} \boldsymbol{d}_X(\lambda) h_i(\lambda)^{-1}$ becomes the Fourier series whose coefficients are the convolution of $\boldsymbol{X}(t)$ with ϕ_i, where $\boldsymbol{X}(t) = 0$ for $t < 1$ or $t > T$ by convention.

Therefore, using the Parseval equality, the above equations are rewritten as

$$\widehat{E}\left\{e_j' \widehat{\boldsymbol{A}}^{-1} \boldsymbol{X} \left(\phi_i * e_i' \widehat{\boldsymbol{A}}^{-1} \boldsymbol{X}\right)\right\} = 0, \quad i \neq j = 1, \ldots, K, \tag{5.4.15}$$

where $\widehat{E}$ is the time average operator as in (5.4.13) and $*$ denotes the convolution operator. The equivalent form of these equations is given by

$$\frac{1}{T} \sum_{k=1-T}^{T-1} \phi_{i,k} \sum_{t=\max\{1,1-k\}}^{\min\{T,T-k\}} e_j' \widehat{\boldsymbol{A}}^{-1} \boldsymbol{X}(t) e_i' \widehat{\boldsymbol{A}}^{-1} \boldsymbol{X}(t-k).$$

Using the sample autocovariance matrices of the observed process

$$\boldsymbol{R}_T(k) \frac{1}{T} \sum_{t=\max\{1,1-k\}}^{\min\{T,T-k\}} \boldsymbol{X}(t) \boldsymbol{X}(t-k), \quad (\boldsymbol{R}_T(k) = 0, \text{ for } |k| \geq T),$$

the estimating equation for $\boldsymbol{A}$ is given by

$$\sum_{k=1-T}^{T-1} \phi_{i,k} e_j' \widehat{\boldsymbol{A}}^{-1} \boldsymbol{R}_T(k) \widehat{\boldsymbol{A}}^{-1\prime} e_i = 0, \quad i \neq j = 1, \ldots, K.$$

If we permute and scale both $\widehat{\boldsymbol{A}}$ and $\boldsymbol{A}$ by the same convention, we can expect that $\boldsymbol{\delta} = \boldsymbol{I} - \widehat{\boldsymbol{A}}^{-1} \boldsymbol{A}$ is small. Therefore, $e_i' \widehat{\boldsymbol{A}}^{-1} \boldsymbol{X}(t) = S_i(t) - \sum_{j=1}^{K} \delta_{ij} S_j(t)$, where $S_i(t) = \boldsymbol{A}^{-1} \boldsymbol{X}(t)$ and δ_{ij} denotes the (i, j)th elements of $\boldsymbol{\delta}$. Hence, we can write (5.4.15) approximately as

$$\widehat{E}\left\{S_j(\phi_i * S_i)\right\} - \sum_{k=1}^{K} \left[\widehat{E}\{S_k(\phi_i * S_i)\} \delta_{jk} + \widehat{E}\left\{S_j(\phi_i * S_k)\right\} \delta_{ik}\right] = 0, \quad i \neq j = 1, \ldots, K.$$

If we assume ergodicity, $\widehat{E}\left\{S_j(\phi_i * S_k)\right\}$ converges to $E\left\{S_j(t)(\phi_{i,t} * S_k(t))\right\}$ as $T \to \infty$, which vanishes for $j \neq k$ and equals $\int_{-1/2}^{1/2} f_j^*(\lambda) h_i^{-1}(\lambda) d\lambda$ for $j = k$, where $f_j^*(\lambda)$ is the true spectral density of the process $S_j(t)$. Therefore, the system of equations (5.4.15) can be further written approximately as

$$\widehat{E}\left\{S_j(\phi_i * S_i)\right\} \approx \int_{-1/2}^{1/2} f_i^*(\lambda) h_i^{-1}(\lambda) d\lambda \delta_{ji} + \int_{-1/2}^{1/2} f_j^*(\lambda) h_i^{-1}(\lambda) d\lambda \delta_{ij}, \quad i \neq j = 1, \ldots, K.$$

Let $\boldsymbol{\Delta}$ and $\boldsymbol{\Psi}$ be the $K(K-1)$-vectors $(\delta_{12}, \delta_{21}, \delta_{13}, \delta_{31}, \ldots)'$ and $(\Psi_{12}, \Psi_{21}, \Psi_{13}, \Psi_{31}, \ldots)'$ with $\Psi_{ij} = S_j(\phi_i * S_i)$. Then the above system of equations is gathered up to $\boldsymbol{\Delta} \approx \boldsymbol{H}^{-1} \widehat{E}(\boldsymbol{\Psi})$, where $\boldsymbol{H}$ is the block diagonal matrix with blocks

$$\boldsymbol{H}_{(ij)} = \begin{pmatrix} \int_{-1/2}^{1/2} f_j^*(\lambda) h_i^{-1}(\lambda) d\lambda & \int_{-1/2}^{1/2} f_i^*(\lambda) h_i^{-1}(\lambda) d\lambda \\ \int_{-1/2}^{1/2} f_j^*(\lambda) h_j^{-1}(\lambda) d\lambda & \int_{-1/2}^{1/2} f_i^*(\lambda) h_j^{-1}(\lambda) d\lambda \end{pmatrix}$$

and the pairs (i, j), $i \neq j = 1, \ldots, K$. Since $\widehat{E}(\Psi)$ is the average of the stationary process $\Psi(t)$, with the mild condition, the central limit theorem leads to $\sqrt{T}\widehat{E}(\Psi) \overset{\mathcal{D}}{\to} N(\mathbf{0}, \boldsymbol{G})$, where

$$\boldsymbol{G} = \lim_{T\to\infty} \frac{1}{T} E\left\{\sum_{t=1}^{T}\sum_{s=1}^{T} \Psi(t)\Psi(s)'\right\}$$

is the block diagonal matrix with blocks

$$\boldsymbol{G}_{(ij)} = \sum_{k=-\infty}^{\infty} \begin{pmatrix} \gamma_j(k)\widetilde{\gamma}_i(k) & \overline{\gamma}_j(k)\overline{\gamma}_i(-k) \\ \overline{\gamma}_i(k)\overline{\gamma}_j(-k) & \gamma_i(k)\widetilde{\gamma}_j(k) \end{pmatrix}$$

and the pairs (i, j), $i \neq j = 1, \ldots, K$,

$$\gamma_i(k) = E\{S_i(t)S_i(t+k)\}, \quad \widetilde{\gamma}_i(k) = E\{(\phi_{i,t} * S_i(t))(\phi_{i,t+k} * S_i(t+k))\}$$

are autocovariance functions and

$$\overline{\gamma}_i = E\{S_i(t)(\phi_{i,t+k} * S_i(t+k))\}$$

is the cross-covariance function of the process $S_i(t)$ and $(\phi_{i,t} * S_i(t))$. Note that the spectral density of the process $(\phi_{i,t} * S_i(t))$ is $f_i^*(\lambda)h_i^{-2}(\lambda)$, and the cross-spectral density between $S_i(t)$ and $(\phi_{i,t} * S_i(t))$ is $f_i^*(\lambda)h_i^{-1}(\lambda)$. Therefore, Parseval's identity leads to

$$\boldsymbol{G}_{(ij)} = \begin{pmatrix} \int_{-1/2}^{1/2} f_j^*(\lambda)f_i^*(\lambda)h_i^{-2}(\lambda)d\lambda & \int_{-1/2}^{1/2} f_j^*(\lambda)f_i^*(\lambda)h_j^{-1}(\lambda)h_i^{-1}(\lambda)d\lambda \\ \int_{-1/2}^{1/2} f_i^*(\lambda)f_j^*(\lambda)h_i^{-1}(\lambda)h_j^{-1}(\lambda)d\lambda & \int_{-1/2}^{1/2} f_i^*(\lambda)f_j^*(\lambda)h_j^{-2}(\lambda)d\lambda \end{pmatrix}$$

with the pairs (i, j), $i \neq j = 1, \ldots, K$. Since $\boldsymbol{H}$ and $\boldsymbol{G}$ are block diagonal, the pairs $\left(\delta_{ij}, \delta_{ji}\right)$ are asymptotically independent for different $(i_1, j_1) \neq (i_2, j_2)$ and have covariance matrices $\boldsymbol{H}_{(ij)}^{-1}\boldsymbol{G}_{(ij)}\boldsymbol{H}_{(ij)}^{-1\prime}/T$. Furthermore, defining

$$\boldsymbol{J}_{(ij)} = \begin{pmatrix} \int_{-1/2}^{1/2} f_j^*(\lambda)f_i^{*-1}(\lambda)d\lambda & 1 \\ 1 & \int_{-1/2}^{1/2} f_i^*(\lambda)f_j^{*-1}(\lambda)d\lambda \end{pmatrix},$$

we have

$$\begin{pmatrix} \boldsymbol{G}_{(ij)} & \boldsymbol{H}_{(ij)} \\ \boldsymbol{H}'_{(ij)} & \boldsymbol{J}_{(ij)} \end{pmatrix} = \int_{-1/2}^{1/2} \begin{pmatrix} h_i^{-1}(\lambda) \\ h_j^{-1}(\lambda) \\ f_i^{*-1}(\lambda) \\ f_j^{*-1}(\lambda) \end{pmatrix} \begin{pmatrix} h_i^{-1}(\lambda) & h_j^{-1}(\lambda) & f_i^{*-1}(\lambda) & f_j^{*-1}(\lambda) \end{pmatrix} f_i^*(\lambda)f_j^*(\lambda)d\lambda,$$

where the right hand side of this equation is nonnegative definite, so that $\boldsymbol{H}_{(ij)}^{-1}\boldsymbol{G}_{(ij)}\boldsymbol{H}_{(ij)}^{-1\prime} \geq \boldsymbol{J}_{(ij)}$. The equality holds in the case that each of the $h_i(\lambda)$ is proportional to the corresponding $f_i^*(\lambda)$. In such a case, the asymptotic covariance matrix of the estimator becomes minimum and is given by $\boldsymbol{J}/T$, where $\boldsymbol{J}$ is the block diagonal matrix whose blocks are $\boldsymbol{J}_{(ij)}$ with the pairs (i, j), $i \neq j = 1, \ldots, K$.

To derive an iterative algorithm for the solution of (5.4.15), we expand (5.4.15) around an initial estimator $\widetilde{\boldsymbol{A}}$ of $\boldsymbol{A}$ and construct the next step estimator $\widetilde{\boldsymbol{A}}\left(\boldsymbol{I} - \widetilde{\delta}\right)^{-1}$ as the solution of

$$\widehat{E}\left\{\widetilde{S}_j\left(\phi_i * \widetilde{S}_i\right)\right\} = \sum_{k=1}^{K}\left[\widehat{E}\left\{\widetilde{S}_j\left(\phi_i * \widetilde{S}_k\right)\right\}\widetilde{\delta}_{ik} + \widehat{E}\left\{\widetilde{S}_k\left(\phi_i * \widetilde{S}_i\right)\right\}\widetilde{\delta}_{jk}\right],$$

where $\widetilde{S}_i(t) = e_i'\widetilde{A}^{-1}X(t)$ and $\widetilde{\delta}_{ij}$ denotes the (i, j)th elements of $\widetilde{\delta}$. We can expect $\widetilde{S}_i(t)$ is close to $S_i(t)$ and approximate $\widehat{E}\{\widetilde{S}_j(\phi_i * \widetilde{S}_k)\}$ by $E\{S_j(t)(\phi_{i,t} * S_k(t))\}$, which vanishes for $j \neq k$. Therefore, $\widetilde{\delta}_{ij}$ can be computed from the system of equations

$$\widehat{E}\{\widetilde{S}_j(\phi_i * \widetilde{S}_i)\} = \widehat{E}\{\widetilde{S}_j(\phi_i * \widetilde{S}_j)\}\widetilde{\delta}_{ij} + \widehat{E}\{\widetilde{S}_i(\phi_i * \widetilde{S}_i)\}\widetilde{\delta}_{ji}, \quad i \neq j = 1, \dots, K,$$

which yields the new estimator of A, and we can iterate the procedure until convergence.

Shumway and Der (1985) consider the problem of estimating simultaneously two convolved vector time series with additive noise. For instance, the observed time series recorded by seismometers contains convolutions in the problem of monitoring a nuclear test. We discuss the problem of estimating an $n \times 1$ stochastic vector $x(t) = (x_1(t), \dots, x_n(t))'$ in terms of the linear model

$$y(t) = a(t) \otimes x(t) + v(t)$$

for $t = 0 \pm 1, \pm 2, \dots$, where $a(t)$ is an $N \times n$ matrix of fixed functions which satisfy the condition

$$\sum_{t=-\infty}^{\infty} |t| \left|a_{ij}(t)\right| < \infty \tag{5.4.16}$$

for $i = 1, \dots, N$, $j = 1, \dots, n$, and the notation $\otimes$ denotes the convolution. Furthermore, the $n \times 1$ signal process $x(t)$ and the $N \times 1$ noise process $v(t)$ are zero mean strong mixing stationary processes and uncorrelated with each other. The spectral density matrices of $y(t)$, $x(t)$ and $v(t)$ are absolutely continous positive definite and denoted by $f_y(\lambda)$, $f_x(\lambda)$ and $f_v(\lambda)$, respectively. The problem of estimating the vector process $x(t)$ with minimum mean squared error can be achieved by a linear estimator of the form

$$\widehat{x}(t) = \sum_{u=-\infty}^{\infty} h(u)\, y(t-u), \tag{5.4.17}$$

which satisfies $\langle x(t) - \widehat{x}(t), y(t)\rangle = 0$, so that $\langle x(t), y(t)\rangle = \langle \widehat{x}(t), y(t)\rangle$ from orthogonality. Therefore, the Fourier transform of $h(t)$ is given by the $n \times N$ matrix

$$\begin{aligned} H(\lambda) &= f_x(\lambda) A^*(\lambda) f_y^{-1}(\lambda) \\ &= \left(f_x^{-1}(\lambda) + A^*(\lambda) f_v^{-1}(\lambda) A(\lambda)\right)^{-1} A^*(\lambda) f_v^{-1}(\lambda), \end{aligned} \tag{5.4.18}$$

where A^* denotes the complex conjugate transpose of the matrix A, and the matrix $A(\lambda)$ is defined as the Fourier transform

$$A(\lambda) = \sum_{t=-\infty}^{\infty} a(t) \exp(-i\lambda t), \quad -\pi \leq \lambda \leq \pi, \tag{5.4.19}$$

of the $N \times n$ matrix valued function $a(t)$. The mean squared error of the estimator is

$$\begin{aligned} &\langle x(t) - \widehat{x}(t), x(t) - \widehat{x}(t)\rangle = E\left[\{x(t) - \widehat{x}(t)\}\{x(t) - \widehat{x}(t)\}'\right] \\ &= \int_{-\pi}^{\pi} \left\{H(\lambda) f_y(\lambda) H^*(\lambda) - 2 f_x(\lambda) A^*(\lambda) H^*(\lambda) + f_x(\lambda)\right\} \frac{d\lambda}{2\pi} \\ &= \int_{-\pi}^{\pi} f_x(\lambda)\{1 - A^*(\lambda) H^*(\lambda)\} \frac{d\lambda}{2\pi} \\ &= \int_{-\pi}^{\pi} \left(f_x^{-1}(\lambda) + A^*(\lambda) f_v^{-1}(\lambda) A(\lambda)\right)^{-1} \frac{d\lambda}{2\pi}. \end{aligned} \tag{5.4.20}$$

Note that the filter response (5.4.18) becomes the best linear unbiased estimator when $f_x^{-1}(\lambda) = 0$. Replacing $\boldsymbol{h}(u)$ by

$$\boldsymbol{h}_M(u) = \frac{1}{M}\sum_{m=0}^{M-1} \boldsymbol{H}(\lambda_m)\exp(i\lambda_m u)$$

for $\lambda_m = \frac{2\pi m}{M}$, the approximate version of (5.4.17) can be given.

To construct the approximation of likelihood, we define the discrete Fourier transform (DFT) of the vector series $\boldsymbol{y}(t)$, $\boldsymbol{x}(t)$ and $\boldsymbol{v}(t)$ as

$$\boldsymbol{Y}(k) = T^{-\frac{1}{2}}\sum_{t=0}^{T-1} \boldsymbol{y}(t)\exp(-i\lambda_k t), \quad \boldsymbol{X}(k) = T^{-\frac{1}{2}}\sum_{t=0}^{T-1} \boldsymbol{x}(t)\exp(-i\lambda_k t)$$

$$\text{and } \boldsymbol{V}(k) = T^{-\frac{1}{2}}\sum_{t=0}^{T-1} \boldsymbol{v}(t)\exp(-i\lambda_k t)$$

for $\lambda_k = \frac{2\pi k}{T}$, $k = 0, 1, \ldots, T-1$, respectively. Then, we can see that

$$\boldsymbol{Y}(k) = \boldsymbol{A}(\lambda_k)\boldsymbol{X}(k) + \boldsymbol{V}(k) + o_P(1), \tag{5.4.21}$$

where $\boldsymbol{A}(\lambda)$ is the usual both side infinite Fourier transform defined in (5.4.19). If signal process $\boldsymbol{x}(t)$ and noise process $\boldsymbol{v}(t)$ are strictly stationary and mixing, the vectors $\boldsymbol{X}(k)$ and $\boldsymbol{V}(k)$ are approximately independent and complex normal variables whose covariance matrices are given by $f_x(\lambda_k)$ and $f_v(\lambda_k)$ for $k = 1, \ldots, \frac{T}{2} - 1$, and real multivariate normal for $k = 0, \frac{T}{2}$, when T is even. As before, the estimator of $\boldsymbol{X}(k)$ can be constructed as

$$\widehat{\boldsymbol{X}}(k) = \boldsymbol{H}(\lambda_k)\boldsymbol{Y}(k), \tag{5.4.22}$$

where $\boldsymbol{H}(\lambda)$ is defined in (5.4.18). Since the approximation in (5.4.21) is obtained by DFT, we can suggest inverting to obtain the approximation to (5.4.17), namely,

$$\widehat{\boldsymbol{x}}(t) \approx T^{-\frac{1}{2}}\sum_{k=0}^{T-1} \widehat{\boldsymbol{X}}(k)\exp(i\lambda_k t).$$

Now, we consider the application of this general theory to the particular model of interest

$$y_{ij}(t) = w_{ij}(t) \otimes r_i(t) \otimes x_j(t) + v_{ij}(t)$$

with $w_{ij}(t)$ a known function, which is rewritten in the frequency domain as

$$Y_{ij}(k) = W_{ij}(\lambda_k)R_i(\lambda_k)X_j(k) + V_{ij}(K) + o_P(1), \tag{5.4.23}$$

where the argument λ_k implies the both side infinite Fourier transform and k implies the DFT. To simplify, we assume that the series $x_j(t)$ are uncorrelated and stationary with spectral density $f_{x_j}(\lambda)$ and that the noise series are uncorrelated and stationary with identical spectral density $f_v(\lambda)$. Let $\boldsymbol{y}_j(t) = \left(y_{1j}(t), \ldots, y_{mj}(t)\right)'$, $\boldsymbol{h}_j(t) = \left(h_{1j}(t), \ldots, h_{mj}(t)\right)$ and

$$\widehat{x}_j(t) = \sum_{u=-\infty}^{\infty} \boldsymbol{h}_j(u)\boldsymbol{y}_j(t-u),$$

where $\boldsymbol{h}_j(t)$ is the Fourier transform of

$$\boldsymbol{H}_j(\lambda) = \left(f_{x_j}^{-1}(\lambda) + \boldsymbol{A}_j^*(\lambda)f_v^{-1}(\lambda)\boldsymbol{A}_j(\lambda)\right)^{-1}\boldsymbol{A}_j^*(\lambda)f_v^{-1}(\lambda)$$

$$= \left(f_{x_j}^{-1}(\lambda) f_v(\lambda) + \boldsymbol{A}_j^*(\lambda) \boldsymbol{A}_j(\lambda)\right)^{-1} \boldsymbol{A}_j^*(\lambda) \tag{5.4.24}$$

with $\boldsymbol{A}_j(\lambda) = \left(W_{1j}(\lambda) R_1(\lambda), \ldots, W_{mj}(\lambda) R_m(\lambda)\right)'$. Therefore, in this case, the estimator in (5.4.22) becomes

$$\widehat{X}_j(k) = D_j^{-1}(\lambda_k) \sum_{i=1}^{m} \overline{W_{ij}(\lambda_k) R_i(\lambda_k)} Y_{ij}(k), \tag{5.4.25}$$

where

$$D_j(\lambda_k) = \sum_{i=1}^{m} \left|W_{ij}(\lambda_k)\right|^2 |R_i(\lambda_k)|^2 + \theta_j(\lambda_k)$$

and

$$\theta_j(\lambda_k) = f_{x_j}^{-1}(\lambda_k) f_v(\lambda_k) \tag{5.4.26}$$

is the inverse of the signal to noise ratio at frequency λ_k. Furthermore, the mean squared error (5.4.20) in this case becomes

$$\begin{aligned} E\left[\{\boldsymbol{x}(t) - \widehat{\boldsymbol{x}}(t)\}\{\boldsymbol{x}(t) - \widehat{\boldsymbol{x}}(t)\}'\right] &= \int_{-\pi}^{\pi} \left(f_x^{-1}(\lambda) + \boldsymbol{A}^*(\lambda) f_v^{-1}(\lambda) \boldsymbol{A}(\lambda)\right)^{-1} \frac{d\lambda}{2\pi} \\ &= \int_{-\pi}^{\pi} D_j^{-1}(\lambda) f_v(\lambda) \frac{d\lambda}{2\pi} := \int_{-\pi}^{\pi} \sigma_j^2(\lambda) \frac{d\lambda}{2\pi} \end{aligned} \tag{5.4.27}$$

with

$$\sigma_j^2(\lambda) = D_j^{-1}(\lambda) f_v(\lambda). \tag{5.4.28}$$

Up to now, we assumed the fixed functions $r_i(t)$, $i = 1, \ldots, m$ and the spectral density functions $f_v(\lambda)$ and $f_{x_j}(\lambda)$ are known in advance. However, we must indeed estimate these parameters, e.g., by maximizing the likelihood function. We can consider the frequency domain approximation of likelihood function. Namely, we maximize the likelihood function corresponding to the approximated model (5.4.23) with respect to the parameters $R_i(\lambda_k)$, $f_v(\lambda_k)$ and $f_{x_j}(\lambda_k)$ for $i = 1, \ldots, m$, $j = 1, \ldots, n$, $k = 0, 1, \ldots, \frac{T}{2}$, which all together is denoted by $\boldsymbol{\Phi}$. From (5.4.23), the frequency domain observations $Y_{ij}(k)$ are approximately independent complex Gaussian random variables which have zero means and variances

$$\eta_{ij}^2(\lambda_k) = \left|W_{ij}(\lambda_k)\right|^2 |R_i(\lambda_k)|^2 f_{x_j}(\lambda_k) + f_v(\lambda_k)$$

under the condition (5.4.16) for $a_{ij}(t) = w_{ij}(t) \otimes r_i(t)$. Therefore, the approximation of the log-likelihood is given by

$$L(\boldsymbol{Y}; \boldsymbol{\Phi}) \approx -\sum_{ijk} \delta_k \log\left\{\eta_{ij}^2(\lambda_k)\right\} - \sum_{ijk} \delta_k \frac{\left|Y_{ij}(k)\right|^2}{\eta_{ij}^2(\lambda_k)}, \tag{5.4.29}$$

where

$$\delta_k = \begin{cases} \frac{1}{2} & \left(k = 0, \frac{T}{2}\right), \\ 1 & \text{(otherwise)} \end{cases}$$

and (5.4.29) is termed the incomplete-data likelihood in Dempster et al. (1977). On the other hand, if the joint likelihood of the unobserved component $X_j(k)$ and $V_{ij}(K)$ in (5.4.23) is available, it should be of the form

$$L(\boldsymbol{X}, \boldsymbol{V}; \boldsymbol{\Phi}) = -\sum_{jk} \delta_k \log\left\{f_{x_j}(\lambda_k)\right\} - \sum_{jk} \delta_k \frac{|X_j(k)|^2}{f_{x_j}(\lambda_k)}$$

$$-\sum_{ijk} \delta_k \log\{f_v(\lambda_k)\} - \sum_{ijk} \delta_k \frac{|Y_{ij}(k) - W_{ij}(\lambda_k) R_i(\lambda_k) X_j(k)|^2}{f_v(\lambda_k)}, \qquad (5.4.30)$$

which is termed the complete-data likelihood in Dempster et al. (1977), and will be maximized as a function of the unobserved parameters $f_{x_j}(\lambda_k)$, $f_v(\lambda_k)$ and $R_i(\lambda_k)$.

Dempster et al. (1977) proposed and EM algorithm which involves two steps, namely, the expectation step (E-step) and the maximization step (M-step). Assume that there exist two sample spaces $\mathcal{Y}$ and $\mathcal{X}$ and the observed data $\boldsymbol{y}$ are a realization from $\mathcal{Y}$ and termed incomplete-data. The corresponding $\boldsymbol{x}$ in $\mathcal{X}$ is not observed directly, but only indirectly through $\boldsymbol{y}$. That is, there is a mapping $\boldsymbol{x} \to \boldsymbol{y}(\boldsymbol{x})$ from $\mathcal{X}$ to $\mathcal{Y}$, and $\boldsymbol{x}$ is known only to lie in $\mathcal{X}(\boldsymbol{y})$, which is the subset of $\mathcal{X}$ determined by the equation $\boldsymbol{y} = \boldsymbol{y}(\boldsymbol{x})$. We refer to $\boldsymbol{x}$ as the complete-data even though in certain examples $\boldsymbol{x}$ includes the parameters.

We assume that a family of the sampling densities $f(\boldsymbol{x} \mid \boldsymbol{\Phi})$ depends on $r \times 1$ parameter vector $\boldsymbol{\Phi} \in \Omega$ and derive its corresponding family of sampling densities $g(\boldsymbol{y} \mid \boldsymbol{\Phi})$. The relationship between the complete-data specification $f(\boldsymbol{x} \mid \boldsymbol{\Phi})$ and the incomplete-data specification $g(\boldsymbol{x} \mid \boldsymbol{\Phi})$ is given by

$$g(\boldsymbol{y} \mid \boldsymbol{\Phi}) = \int_{\mathcal{X}(\boldsymbol{y})} f(\boldsymbol{x} \mid \boldsymbol{\Phi})\, d\boldsymbol{x}.$$

Our purpose is to find the value $\boldsymbol{\Phi}^*$ of $\boldsymbol{\Phi}$ that maximizes

$$L(\boldsymbol{\Phi}) = \log g(\boldsymbol{y} \mid \boldsymbol{\Phi}). \qquad (5.4.31)$$

The conditional density of $\boldsymbol{x}$ given $\boldsymbol{y}$ and $\boldsymbol{\Phi}$ is denoted by

$$k(\boldsymbol{x} \mid \boldsymbol{y}, \boldsymbol{\Phi}) = \frac{f(\boldsymbol{x} \mid \boldsymbol{\Phi})}{g(\boldsymbol{y} \mid \boldsymbol{\Phi})}.$$

Therefore, (5.4.1.3) can be written as

$$L(\boldsymbol{\Phi}) = \log f(\boldsymbol{x} \mid \boldsymbol{\Phi}) - \log k(\boldsymbol{x} \mid \boldsymbol{y}, \boldsymbol{\Phi}).$$

We now define a new function

$$Q(\boldsymbol{\Phi}' \mid \boldsymbol{\Phi}) = E\{\log f(\boldsymbol{x} \mid \boldsymbol{\Phi}') \mid \boldsymbol{y}, \boldsymbol{\Phi}\}, \qquad (5.4.32)$$

which is assumed to exist for all pairs $(\boldsymbol{\Phi}', \boldsymbol{\Phi})$. Furthermore, we assume $f(\boldsymbol{x} \mid \boldsymbol{\Phi}) > 0$ almost everywhere in $\mathcal{X}$ for all $\boldsymbol{\Phi} \in \Omega$.

Now we introduce the EM iteration $\boldsymbol{\Phi}^{(p)} \to \boldsymbol{\Phi}^{(p+1)}$ as follows:

E-step Compute $Q\left(\boldsymbol{\Phi} \mid \boldsymbol{\Phi}^{(p)}\right)$.

M-step Choose $\boldsymbol{\Phi}^{(p+1)}$ which maximizes $Q\left(\boldsymbol{\Phi} \mid \boldsymbol{\Phi}^{(p)}\right)$ among $\boldsymbol{\Phi} \in \Omega$.

We would like to choose $\boldsymbol{\Phi}^*$ to maximize $\log f(\boldsymbol{x} \mid \boldsymbol{\Phi})$. However, since we do not know $\log f(\boldsymbol{x} \mid \boldsymbol{\Phi})$, we maximize instead its current expectation given the data $\boldsymbol{y}$ and the current fit $\boldsymbol{\Phi}^{(p)}$.

In addition, it is convenient to write

$$H(\boldsymbol{\Phi}' \mid \boldsymbol{\Phi}) = E\{\log k(\boldsymbol{x} \mid \boldsymbol{y}, \boldsymbol{\Phi}') \mid \boldsymbol{y}, \boldsymbol{\Phi}\}.$$

Therefore, (5.4.32) becomes

$$Q(\boldsymbol{\Phi}' \mid \boldsymbol{\Phi}) = L(\boldsymbol{\Phi}') + H(\boldsymbol{\Phi}' \mid \boldsymbol{\Phi}).$$

Then, we have the following lemma.

Lemma 5.4.10 (Lemma 1 of Dempster et al. (1977)). *For any pair* $(\mathbf{\Phi}', \mathbf{\Phi})$ *in* $\Omega \times \Omega$, *we have*

$$H(\mathbf{\Phi}' \mid \mathbf{\Phi}) \leq H(\mathbf{\Phi} \mid \mathbf{\Phi}),$$

where the equality holds if and only if $k(\boldsymbol{x} \mid \boldsymbol{y}, \mathbf{\Phi}') = k(\boldsymbol{x} \mid \boldsymbol{y}, \mathbf{\Phi})$ *almost everywhere.*

In general, the iterative algorithm should be applicable to any starting point. Therefore, we define a mapping $\mathbf{\Phi} \to \boldsymbol{M}(\mathbf{\Phi})$ from Ω to Ω such that each step $\mathbf{\Phi}^{(p)} \to \mathbf{\Phi}^{(p+1)}$ is given by

$$\mathbf{\Phi}^{(p+1)} = \boldsymbol{M}\left(\mathbf{\Phi}^{(p)}\right).$$

Definition 5.4.11. *An iterative algorithm with mapping* $\boldsymbol{M}(\mathbf{\Phi})$ *is called a generalized EM (GEM) algorithm if*

$$Q(\boldsymbol{M}(\mathbf{\Phi}) \mid \mathbf{\Phi}) \geq Q(\mathbf{\Phi} \mid \mathbf{\Phi})$$

for every $\mathbf{\Phi} \in \Omega$.

It should be noted that the definition of the EM algorithm requires

$$Q(\boldsymbol{M}(\mathbf{\Phi}) \mid \mathbf{\Phi}) \geq Q(\mathbf{\Phi}' \mid \mathbf{\Phi})$$

for every pair $(\mathbf{\Phi}', \mathbf{\Phi})$ in $\Omega \times \Omega$, that is, $\mathbf{\Phi}' = \boldsymbol{M}(\mathbf{\Phi})$ maximizes $Q(\mathbf{\Phi}' \mid \mathbf{\Phi})$. Now, we have the following results.

Theorem 5.4.12 (Theorem 1 of Dempster et al. (1977)). *For every GEM algorithm*

$$L(\boldsymbol{M}(\mathbf{\Phi})) \geq L(\mathbf{\Phi}) \text{ for all } \mathbf{\Phi} \in \Omega$$

where the equality holds if and only if

$$Q(\boldsymbol{M}(\mathbf{\Phi}) \mid \mathbf{\Phi}) = Q(\mathbf{\Phi} \mid \mathbf{\Phi})$$

and

$$k(\boldsymbol{x} \mid \boldsymbol{y}, \boldsymbol{M}(\mathbf{\Phi})) = k(\boldsymbol{x} \mid \boldsymbol{y}, \mathbf{\Phi})$$

almost everywhere.

Corollary 5.4.13 (Corollary 1 of Dempster et al. (1977)). *Assume that there exists some* $\mathbf{\Phi}^* \in \Omega$ *such that* $L(\mathbf{\Phi}^*) \geq L(\mathbf{\Phi})$ *for all* $\mathbf{\Phi} \in \Omega$. *Then, for every GEM algorithm, we have*

(i) $L(\boldsymbol{M}(\mathbf{\Phi}^*)) = L(\mathbf{\Phi}^*)$

(ii) $Q(\boldsymbol{M}(\mathbf{\Phi}^*) \mid \mathbf{\Phi}^*) = Q(\mathbf{\Phi}^* \mid \mathbf{\Phi}^*)$

(iii) $k(\boldsymbol{x} \mid \boldsymbol{y}, \boldsymbol{M}(\mathbf{\Phi}^*)) = k(\boldsymbol{x} \mid \boldsymbol{y}, \mathbf{\Phi}^*)$ *almost everywhere.*

Corollary 5.4.14 (Corollary 2 of Dempster et al. (1977)). *Assume that there exists some* $\mathbf{\Phi}^* \in \Omega$ *such that* $L(\mathbf{\Phi}^*) > L(\mathbf{\Phi})$ *for all* $\mathbf{\Phi} \in \Omega$, $\mathbf{\Phi} \neq \mathbf{\Phi}^*$. *Then, for every GEM algorithm, we have*

$$\boldsymbol{M}(\mathbf{\Phi}^*) = \mathbf{\Phi}^*.$$

Theorem 5.4.15 (Theorem 2 of Dempster et al. (1977)). *Let* $\mathbf{\Phi}^{(p)}$, $p = 0, 1, 2, \ldots$ *be a sequence of a GEM algorithm such that*

(i) *the sequence* $L\left(\mathbf{\Phi}^{(p)}\right)$ *is bounded,*

(ii) $Q\left(\mathbf{\Phi}^{(p+1)} \mid \mathbf{\Phi}^{(p)}\right) - Q\left(\mathbf{\Phi}^{(p)} \mid \mathbf{\Phi}^{(p)}\right) \geq \lambda \left\|\mathbf{\Phi}^{(p+1)} - \mathbf{\Phi}^{(p)}\right\|^2$ *for some scalar* $\lambda > 0$ *and all* p.

Then, the sequence $\mathbf{\Phi}^{(p)}$ *converges to some* $\mathbf{\Phi}^*$ *in the closure of* Ω.

Shumway and Der (1985) applied the EM algorithm in the form of

$$Q\left(\mathbf{\Phi} \mid \mathbf{\Phi}^{(p)}\right) = E_p\left\{L(\boldsymbol{X}, \boldsymbol{V}; \mathbf{\Phi}) \mid \boldsymbol{Y}\right\} \tag{5.4.33}$$

using (5.4.30), where E_p denotes the expectation with respect to the density determined by the parameter $\mathbf{\Phi}^{(p)}$. From the orthogonal arguments in (5.4.25) and (5.4.27), we have

$$E\left(X_j(k) \mid \boldsymbol{Y}\right) = \widehat{X}_j(k)$$

and

$$Var\left(X_j(k) \mid \boldsymbol{Y}\right) = \sigma_j^2(\lambda_k).$$

Therefore, we can rewrite (5.4.33) as

$$Q\left(\mathbf{\Phi} \mid \mathbf{\Phi}^{(p)}\right) = -\sum_{jk} \delta_k \log\left\{f_{x_j}(\lambda_k)\right\} - \sum_{jk} \delta_k \frac{\left|\widehat{X}_j(k)\right|^2 + \sigma_j^2(\lambda_k)}{f_{x_j}(\lambda_k)}$$

$$-\sum_{ijk} \delta_k \log\left\{f_v(\lambda_k)\right\} - \sum_{ijk} \delta_k \frac{\left|Y_{ij}(k) - W_{ij}(\lambda_k) R_i(\lambda_k) \widehat{X}_j(k)\right|^2 + \left|W_{ij}(\lambda_k) R_i(\lambda_k)\right|^2 \sigma_j^2(\lambda_k)}{f_v(\lambda_k)}$$

and so the maximization step leads to

$$\widehat{R}_i(\lambda_k) = C_i(k)^{-1} \sum_{j=1}^{n} \overline{W_{ij}(\lambda_k) \widehat{X}_j(k)} Y_{ij}(k), \tag{5.4.34}$$

where

$$C_i(k) = \sum_{j=1}^{n} \left\{\left|\widehat{X}_j(k)\right|^2 + \sigma_j^2(\lambda_k)\right\} \left|W_{ij}(\lambda_k)\right|^2 \tag{5.4.35}$$

for the $i = 1, \ldots, m$ receiver functions. On the other hand, the signal and noise spectral densities are estimated by

$$\widehat{f_{x_j}}(\lambda_k) = \left|\widehat{X}_j(k)\right|^2 + \sigma_j^2(\lambda_k) \tag{5.4.36}$$

for $j = 1, \ldots, n$ and

$$\widehat{f_v}(\lambda_k) = \frac{1}{mn}\left\{\sum_{ij}\left|Y_{ij}(k) - W_{ij}(\lambda_k) \widehat{R}_i(\lambda_k) \widehat{X}_j(k)\right|^2 + \left|W_{ij}(\lambda_k) \widehat{R}_i(\lambda_k)\right|^2 \sigma_j^2(\lambda_k)\right\}. \tag{5.4.37}$$

Shumway and Der (1985) suggested the EM algorithm for estimating the unknown parameters and getting the deconvolved series $\widehat{x}_j(t)$, $j = 1, \ldots, n$.

(i) Specify the initial guess of the receiver functions $R_i(t)$, which can be taken as simple delta function for the first iteration.

(ii) Initialize $f_v(\lambda)$ and the noise-to-signal ratios $\theta_j(\lambda)$, $j = 1, \ldots, n$ in Equation (5.4.26).

(iii) Calculate $\widehat{X}_j(k)$ and $\sigma_j^2(\lambda)$ using Equations (5.4.25)–(5.4.28).

(iv) Calculate $\widehat{R}_i(\lambda)$ using Equations (5.4.34) and (5.4.35).

(v) Compute $\widehat{f_{x_j}}(\lambda)$ and $\widehat{f_v}(\lambda)$ using (5.4.36) and (5.4.37).

(vi) Obtain the time versions of $\widehat{X}_j(k)$ and $\widehat{R}_i(\lambda)$ by DFT. Then, return to Step 3 for the next iteration.

5.5 Rank-Based Optimal Portfolio Estimation

In this subsection, we propose two different multivariate rank-based portfolio estimations. Let $\boldsymbol{y}_t = (y_{1,t}, \ldots, y_{N,t})'$ be the vector of returns (or log returns) of an N-tuple of assets at time point t. Assume that $\boldsymbol{y}_t$ is of the form

$$\boldsymbol{y}_t = \boldsymbol{\mu}_t + \boldsymbol{e}_t,$$

where $\boldsymbol{\mu}_t = E\{\boldsymbol{y}_t \mid \mathcal{F}_{t-1}\}$ is the conditional expectation of $\boldsymbol{y}_t$ given the sigma-field $\mathcal{F}_{t-1}$ generated by the past returns $\boldsymbol{y}_{t-1}, \boldsymbol{y}_{t-2}, \ldots$, and $\{\boldsymbol{e}_t\}$ is the mean zero innovation process. For simplicity, assume that $\boldsymbol{\mu}_t$ is known or, equivalently, that we directly observe $\boldsymbol{x}_t = \boldsymbol{e}_t = \boldsymbol{y}_t - E\{\boldsymbol{y}_t \mid \mathcal{F}_{t-1}\}$. In practice, of course, $\boldsymbol{\mu}_t$ is unknown and has to be estimated in some way, for instance by the DOCs in mean discussed in Section 5.4.1.2. However, we concentrate on volatility issues.

5.5.1 *Portfolio Estimation Based on Ranks for Independent Components*

For a stationary vector time series $\boldsymbol{x}_t$, assume that there exist DOCs in volatility $\boldsymbol{s}_t$ such that $\boldsymbol{x}_t = \boldsymbol{M}\boldsymbol{s}_t$ where $\boldsymbol{M}$ is a nonsingular mixing matrix. Furthermore, let $\boldsymbol{L}$ denote an uncorrelating matrix so that $\boldsymbol{z}_t = \boldsymbol{L}\boldsymbol{x}_t$ is an uncorrelated observation. Namely, let $\boldsymbol{\Sigma} = \mathrm{Cov}(\boldsymbol{x}_t)$ denote the positive definite unconditional covariance matrix of $\boldsymbol{x}_t$, and let $\boldsymbol{\Gamma}$ and $\boldsymbol{\Lambda}$ denote the orthogonal matrix of eigenvectors and the diagonal matrix of corresponding eigenvalues of $\boldsymbol{\Sigma}$, respectively. Then, the invertible matrix $\boldsymbol{L}$ can be chosen as the inverse symmetric square root of the unconditional covariance matrix defined as $\boldsymbol{L} = \boldsymbol{\Sigma}^{-1/2} = \boldsymbol{\Gamma}\boldsymbol{\Lambda}^{-1/2}\boldsymbol{\Gamma}'$, or simply in connection with principle component analysis as $\boldsymbol{L} = \boldsymbol{\Lambda}^{-1/2}\boldsymbol{\Gamma}'$. The relationship between $\boldsymbol{s}_t$ and $\boldsymbol{z}_t$ is given by

$$\boldsymbol{s}_t = \boldsymbol{M}^{-1}\boldsymbol{x}_t = \boldsymbol{M}^{-1}\boldsymbol{L}^{-1}\boldsymbol{z}_t \equiv \boldsymbol{W}\boldsymbol{z}_t,$$

where $\boldsymbol{W} = \boldsymbol{M}^{-1}\boldsymbol{L}^{-1}$.

The volatility modeling is given by

$$\begin{aligned}\boldsymbol{\Sigma}_t &= \mathrm{Cov}(\boldsymbol{x}_t \mid \mathcal{F}_{t-1}) = \boldsymbol{M}\mathrm{Cov}(\boldsymbol{s}_t \mid \mathcal{F}_{t-1})\boldsymbol{M}' \\ &= \boldsymbol{L}^{-1}\boldsymbol{W}^{-1}\mathrm{Cov}(\boldsymbol{s}_t \mid \mathcal{F}_{t-1})\boldsymbol{W}^{-1\prime}\boldsymbol{L}^{-1\prime} = \boldsymbol{L}^{-1}\boldsymbol{W}^{-1}\boldsymbol{A}_t\boldsymbol{W}^{-1\prime}\boldsymbol{L}^{-1\prime},\end{aligned}$$

where $\boldsymbol{s}_t$ is assumed to be conditionally uncorrelated given $\mathcal{F}_{t-1}$ for every time point t, and therefore, $\boldsymbol{A}_t = \mathrm{Cov}(\boldsymbol{s}_t \mid \mathcal{F}_{t-1}) = \mathrm{diag}\left(\sigma^2_{1,t}, \ldots, \sigma^2_{N,t}\right)$ with $A_{ii,t} = \sigma^2_{i,t} = \mathrm{Cov}(s_{i,t} \mid \mathcal{F}_{t-1})$. Otherwise, $\boldsymbol{s}_t$ is assumed to be dynamically orthogonal in volatility as in Subsection 5.4.1.2.

We can obtain the invertible consistent estimator $\widehat{\boldsymbol{L}}$ of the uncorrelating transformation $\boldsymbol{L}$ from orthogonal eigenvectors and eigenvalues of the estimated sample covariance matrix $\widehat{\boldsymbol{\Sigma}}$. Furthermore, we can apply the estimator $\boldsymbol{W}_{\widehat{\boldsymbol{\theta}}_T}$ of the separating matrix given by DOCs in volatility as in Section 5.4.1.2, and the estimator of the vector time series of DOCs $\widehat{\boldsymbol{s}}_t = \boldsymbol{W}_{\widehat{\boldsymbol{\theta}}_T}\boldsymbol{z}_t$.

For the conditionally uncorrelated process $\boldsymbol{A}_t = \mathrm{Cov}(\boldsymbol{s}_t \mid \mathcal{F}_{t-1}) = \mathrm{diag}\left(\sigma^2_{1,t}, \ldots, \sigma^2_{N,t}\right)$, univariate models such as GARCH can be used for modeling. Hirukawa et al. (2010) considered that each conditionally uncorrelated signal $\{s_{i,t} \mid t \in \mathbb{Z}\}$, $i = 1, \ldots, N$ is a univariate LARCH (∞) process

$$s_{i,t} = \sigma_{i,t}\varepsilon_{it} \tag{5.5.1}$$

$$\sigma_{i,t} = b_{i,0} + \sum_{j=1}^{\infty} b_{i,j}s_{i,t-j}, \quad t \in \mathbb{Z}.$$

In order to characterize semiparametric efficiency, we need the uniform local asymptotic normality (ULAN) result for LARCH models. We consider a class of semiparametric LARCH models of the form

$$s_t = \sigma_t(\eta)\,\varepsilon_t$$

$$\sigma_t(\eta) = b_0(\eta) + \sum_{j=1}^{\infty} b_j(\eta)\, s_{t-j}, \quad t \in \mathbb{Z}$$

where the ε_t's are i.i.d., with mean 0, variance 1, and unspecified density g, $b_0(\eta) \neq 0$, and $\eta \in \mathcal{H} \subset \mathbb{R}^r$. Denote by $\mathrm{P}^{(T)}_{\eta,g}$ the probability distribution of the observation under parameter value η and density g. Let η_0 be some fixed value of η, i.e., the true parameter value, and consider sequences $\tilde{\eta}_T$ such that, for some bounded sequence $\boldsymbol{h}_T$, $\sqrt{T}(\tilde{\eta}_T - \eta_0) - \boldsymbol{h}_T \to \mathbf{0}$. Then, the log-likelihood ratio $\Lambda_T(\tilde{\eta}_T, \eta_0)$ for $\mathrm{P}^{(T)}_{\tilde{\eta}_T,g}$ with respect to $\mathrm{P}^{(T)}_{\eta_0,g}$ is given by

$$\Lambda_T(\tilde{\eta}_T, \eta_0) = \sum_{t=1}^{T} \left[l_{1+\boldsymbol{W}'_{Tt}(\tilde{\eta}_T-\eta_0)}(\varepsilon_t(\eta_0)) - l_1(\varepsilon_t(\eta_0)) \right],$$

where $l_\sigma(x) := \log\left(\sigma^{-1} g\left(\frac{x}{\sigma}\right)\right)$, $\varepsilon_t(\eta) := \frac{s_t}{\sigma_t(\eta)}$, $\sigma_t(\eta) := b_0(\eta) + \sum_{j=1}^{\infty} b_j(\eta)\, s_{t-j}$, and

$$\begin{aligned}
&\boldsymbol{W}_{Tt} := (W_{Tt1}, \ldots, W_{Ttr}), \text{ with} \\
&W_{Ttk} := W_t(\eta_0, \tilde{\eta}_T)_k = \sigma_t(\eta_0)^{-1} (\tilde{\eta}_{T,k} - \eta_{0,k})^{-1} \left(\sigma_t\left(\tilde{\eta}^k_T\right) - \sigma_t\left(\tilde{\eta}^{k-1}_T\right)\right), \text{ and} \\
&\tilde{\eta}^k_T := (\tilde{\eta}_{T,1}, \ldots, \tilde{\eta}_{T,k}, \eta_{0,k+1}, \ldots, \eta_{0,r})', \ k = 0, \ldots, r.
\end{aligned}$$

Then, we can easily see that

$$\boldsymbol{W}'_{Tt}(\tilde{\eta}_T - \eta_0) = \sigma_t(\eta_0)^{-1} (\sigma_t(\tilde{\eta}_T) - \sigma_t(\eta_0)).$$

Define

$$\boldsymbol{W}_t := \boldsymbol{W}_t(\eta_0) := \sigma_t(\eta_0)^{-1} \operatorname{grad}_\eta \{\sigma_t(\eta)\}\big|_{\eta=\eta_0}$$

and assume that $x \mapsto g(x)$ admits a derivative $x \mapsto Dg(x)$ and the score for scale $\psi(x) = -\left(1 + x\frac{Dg(x)}{g(x)}\right)$ satisfies

$$0 < I_s(g) := E\left\{\psi^2(\varepsilon_t)\right\} = \int \left(1 + x\frac{Dg(x)}{g(x)}\right)^2 g(x)\, dx < \infty,$$

where $I_s(g)$ is the Fisher information of scale.

Furthermore, denote the score and Fisher information matrix by

$$\dot{\boldsymbol{l}}_t(\eta) = \boldsymbol{W}_t(\eta)\psi(\varepsilon_t(\eta))$$

and $\boldsymbol{I}_g(\eta) = E_{\eta,g}\left\{\dot{\boldsymbol{l}}_t(\eta)\dot{\boldsymbol{l}}'_t(\eta)\right\}$, respectively, and assume that several regularity conditions hold (see Drost et al. (1997) and Hirukawa et al. (2010) for details). Then, we have the following ULAN result for LARCH models.

Theorem 5.5.1 (ULAN). *Let the several regularity conditions as in Drost et al. (1997) hold. Then, under $\mathrm{P}^{(T)}_{\eta_0,g}$, for any bounded sequence $\boldsymbol{h}_T \in \mathbb{R}^r$, we have*

$$\Lambda_T(\tilde{\eta}_T, \eta_0) = \boldsymbol{h}'_T T^{-1/2} \sum_{t=1}^{T} \dot{\boldsymbol{l}}_t(\eta_0) - \frac{1}{2} T^{-1} \sum_{t=1}^{T} \left|\boldsymbol{h}'_T \dot{\boldsymbol{l}}_t(\eta_0)\right|^2 + R_T, \tag{5.5.2}$$

with

$$R_T \xrightarrow{P} 0 \quad \text{and} \quad \boldsymbol{\Delta}^{(T)}(\eta_0, g) := T^{-1/2} \sum_{t=1}^{T} \dot{\boldsymbol{l}}_t(\eta_0) \xrightarrow{\mathcal{D}} N\left(\mathbf{0}, \boldsymbol{I}_g(\eta_0)\right).$$

Furthermore, the quadratic approximation (5.5.2) holds uniformly over $T^{-1/2}$-neighbourhoods of η_0.

Since $\sigma_t(\eta)$ admits the Volterra series expansion

$$\sigma_t(\eta) = b_0(\eta)\left(1 + \sum_{k=1}^{\infty} \sum_{j_1,\dots,j_k=1}^{\infty} b_{j_1}(\eta)\cdots b_{j_k}(\eta)\,\varepsilon_{t-j_1}(\eta)\cdots\varepsilon_{t-j_1-\dots-j_k}(\eta)\right),$$

the central sequence

$$\begin{aligned}\boldsymbol{\Delta}^{(T)}(\eta_0, g) &= T^{-1/2}\sum_{t=1}^{T} \dot{\boldsymbol{l}}_t(\eta_0)\\ &= T^{-1/2}\sum_{t=1}^{T} \boldsymbol{W}_t(\eta_0)\psi(\varepsilon_t(\eta_0))\\ &= T^{-1/2}\sum_{t=1}^{T} \sigma_t(\eta_0)^{-1}\,\mathrm{grad}_\eta\{\sigma_t(\eta)\}\big|_{\eta=\eta_0}\psi(\varepsilon_t(\eta_0))\end{aligned}$$

depends on infinitely many lagged values of ε_t. Therefore, we replace them with an approximate central sequence of the form

$$\begin{aligned}\tilde{\boldsymbol{\Delta}}^{(T)}(\eta_0, g) &= (T-p)^{-1/2}\sum_{t=p+1}^{T} \tilde{\sigma}_t(\eta_0)^{-1}\,\mathrm{grad}_\eta\{\tilde{\sigma}_t(\eta)\}\big|_{\eta=\eta_0}\psi(\varepsilon_t(\eta_0))\\ &=: (T-p)^{-1/2}\sum_{t=p+1}^{T} \boldsymbol{J}_{\eta_0,g}\left(\varepsilon_t(\eta_0),\dots,\varepsilon_{t-p}(\eta_0)\right),\end{aligned}$$

where, with p sufficiently large,

$$\tilde{\sigma}_t(\eta) := b_0(\eta)\left(1 + \sum_{k=1}^{\infty} \sum_{1\le j_1+\dots+j_k\le p} b_{j_1}(\eta)\cdots b_{j_k}(\eta)\,\varepsilon_{t-j_1}(\eta)\cdots\varepsilon_{t-j_1-\dots-j_k}(\eta)\right).$$

Finally, we can estimate the $\sigma_{i,t}(\eta)$, $i = 1,\dots,N$ corresponding to the N univariate LARCH (∞) models (5.5.1) from the estimated DOCs $\widehat{s}_{i,t}$, $i = 1,\dots,N$, which are elements of $\widehat{\boldsymbol{s}}_t = \boldsymbol{W}_{\widehat{\boldsymbol{\theta}}_T}\boldsymbol{z}_t$, via the univariate rank-based estimation methods developed in Section 5.2.4.1. This implies, of course, N distinct rankings, i.e., one for each univariate LARCH process. Then, the estimated volatility is given by

$$\widehat{\boldsymbol{\Sigma}}_t = \widehat{\boldsymbol{L}}^{-1}\boldsymbol{W}_{\widehat{\boldsymbol{\theta}}_T}{}^{-1}\widehat{\boldsymbol{A}}_t\boldsymbol{W}_{\widehat{\boldsymbol{\theta}}_T}{}^{-1\prime}\widehat{\boldsymbol{L}}^{-1\prime},$$

where $\widehat{\boldsymbol{A}}_t = \mathrm{diag}\left\{\widehat{\sigma}^2_{1,t},\dots,\widehat{\sigma}^2_{N,t}\right\}$.

5.5.2 *Portfolio Estimation Based on Ranks for Elliptical Residuals*

Assume that the joint distribution $\mathrm{P}^{(T)}_{\eta,f}$ of the observations is characterized by a model involving a parameter $\boldsymbol{\eta}$ and some unobserved multivariate white noise process $\boldsymbol{\varepsilon}_t$ with density f. More specifically, in the present context, assume that the observation at time t is given by

$$\boldsymbol{e}_t = \boldsymbol{\Sigma}_t(\boldsymbol{\eta})\,\boldsymbol{\varepsilon}_t,$$

where $\boldsymbol{\Sigma}_t$ is some unobserved positive definite matrix-valued process characterizing the joint volatility of $\boldsymbol{e}_t$, and the $\boldsymbol{\varepsilon}_t$'s are i.i.d. with density f.

Furthermore, we assume that the density f of the noise $\{\boldsymbol{\varepsilon}_t\}$ is elliptically symmetric. That is, let $\boldsymbol{V}$ be a symmetric positive definite $N \times N$ matrix, $f_1 : \mathbb{R}_0^+ \to \mathbb{R}^+$ be such that $f_1 > 0$ a.e., and $\int_0^\infty r^{N-1} f_1(r)\,dr < \infty$, then we assume that the innovation density is of the form

$$f(\boldsymbol{x}; \boldsymbol{V}, f_1) := c_{N,f_1} (\det \boldsymbol{V})^{1/2} f_1(\|\boldsymbol{x}\|_{\boldsymbol{V}}), \quad \boldsymbol{x} \in \mathbb{R}^N \tag{5.5.3}$$

where $\|\boldsymbol{x}\|_{\boldsymbol{V}} := \left(\boldsymbol{x}'\boldsymbol{V}^{-1}\boldsymbol{x}\right)^{1/2}$ denotes the norm of $\boldsymbol{x}$ in the metric associated with $\boldsymbol{V}$, the constant $c_{N,f_1} := \left(\omega_N \mu_{N-1:f_1}\right)^{-1}$ is the normalization factor, ω_N stands for the $(N-1)$-dimensional Lebesgue measure of the unit sphere $\mathcal{S}^{N-1} \subset \mathbb{R}^N$ and $\mu_{l:f_1} := \int_0^\infty r^l f_1(r)\,dr$.

Moreover, assume that there exists a residual function $\boldsymbol{\varepsilon}^{(T)}(\boldsymbol{\eta}) := (\varepsilon_1(\boldsymbol{\eta}), \ldots, \varepsilon_T(\boldsymbol{\eta}))'$, where $\varepsilon_t(\boldsymbol{\eta})$ is measurable with respect to $\boldsymbol{\eta}$, the past observations and a p-tuple of initial values, such that, under $\mathrm{P}^{(T)}_{\boldsymbol{\eta},f}$, $\varepsilon_t^{(T)}(\boldsymbol{\eta})$ coincides with ε_t.

Define the $\boldsymbol{V}$-standardized residuals as

$$\boldsymbol{Z}_t^{(T)}(\boldsymbol{\eta}, \boldsymbol{V}) := \boldsymbol{V}^{-1/2}\varepsilon_t^{(T)}(\boldsymbol{\eta}).$$

Let $R_t = R_t^{(T)}(\boldsymbol{\eta}, \boldsymbol{V})$ denote the ranks of $d_t = d_t^{(T)}(\boldsymbol{\eta}, \boldsymbol{V}) := \left\|\boldsymbol{Z}_t^{(T)}\right\|$ among $d_1, \ldots, d_T$, and define the multivariate signs $\boldsymbol{U}_t$ as

$$\boldsymbol{U}_t = \boldsymbol{U}_t^{(T)}(\boldsymbol{\eta}, \boldsymbol{V}) := \|\boldsymbol{Z}_t\|^{-1} \boldsymbol{Z}_t.$$

Under $\mathrm{P}^{(T)}_{\boldsymbol{\eta},f}$, the signs $\boldsymbol{U}_t^{(T)}(\boldsymbol{\eta}, \boldsymbol{V})$ are i.i.d. and uniformly distributed over the unit sphere, the ranks $R_t^{(T)}(\boldsymbol{\eta}, \boldsymbol{V})$ are uniformly distributed over the $T!$ permutations of $\{1, \ldots, T\}$, and signs and ranks are mutually independent.

However, since the true value of the shape matrix $\boldsymbol{V}$ is unknown, the genuine ranks $R_t^{(T)}(\boldsymbol{\eta}_0, \boldsymbol{V})$ and signs $\boldsymbol{U}_t^{(T)}(\boldsymbol{\eta}_0, \boldsymbol{V})$ associated with some given parameter value $\boldsymbol{\eta}_0$ cannot be computed from the residuals $(\varepsilon_1(\boldsymbol{\eta}_0), \ldots, \varepsilon_T(\boldsymbol{\eta}_0))$. Therefore, we need some estimator $\widehat{\boldsymbol{V}}$ of the shape matrix $\boldsymbol{V}$ which satisfies the following assumption.

Assumption 5.5.2 (Assumptions (D1)–(D2) of Hallin and Paindaveine (2005)). *Let $\varepsilon_1, \ldots, \varepsilon_T$ be i.i.d. with density f satisfying (5.5.3). The sequence $\widehat{\boldsymbol{V}} = \widehat{\boldsymbol{V}}^{(T)} := \widehat{\boldsymbol{V}}^{(T)}(\varepsilon_1, \ldots, \varepsilon_T)$ of estimators of $\boldsymbol{V}$ is such that*

(i) $\sqrt{T}\left(\widehat{\boldsymbol{V}}^{(T)} - a\boldsymbol{V}\right) = O_P(1)$ *as $T \to \infty$ for some positive real a,*

(ii) $\widehat{\boldsymbol{V}}^{(T)}$ *is invariant under permutations and reflections, with respect to the origin in $\mathbb{R}^N$, of the ε_t's, and*

(iii) *the estimator $\widehat{\boldsymbol{V}}$ is quasi-affine-equivariant, in the sense that, for all T, all $N \times N$ full-rank matrices $\boldsymbol{M}$, $\widehat{\boldsymbol{V}}(\boldsymbol{M}) = d\boldsymbol{M}\widehat{\boldsymbol{V}}\boldsymbol{M}'$, where $\widehat{\boldsymbol{V}}(\boldsymbol{M})$ stands for the statistic $\widehat{\boldsymbol{V}}$ computed from the T-tuple $(\boldsymbol{M}\varepsilon_1, \ldots, \boldsymbol{M}\varepsilon_T)$, and d denotes some positive scalar that may depend on $\boldsymbol{M}$ and $(\varepsilon_1, \ldots, \varepsilon_T)$.*

Then, we can, instead, use the pseudo-Mahalanobis signs and ranks defined in Hallin and Paindaveine (2002) as

$$\boldsymbol{W}_t(\boldsymbol{\eta}_0) = \boldsymbol{W}_t^{(T)}(\boldsymbol{\eta}_0) := \widehat{\boldsymbol{V}}^{-1/2}\varepsilon_t(\boldsymbol{\eta}_0) \Big/ \left\|\widehat{\boldsymbol{V}}^{-1/2}\varepsilon_t(\boldsymbol{\eta}_0)\right\|,$$

where $\widehat{\boldsymbol{V}}$ is an appropriate estimator of the shape $\boldsymbol{V}$ which satisfies Assumption 5.5.2. The pseudo-Mahalanobis ranks

$$\widehat{R}_t(\boldsymbol{\eta}_0) := \widehat{R}_t^{(T)}(\boldsymbol{\eta}_0)$$

similarly are defined as the ranks of the pseudo-Mahalanobis distances

$$d_t\left(\boldsymbol{\eta}_0, \widehat{\boldsymbol{V}}\right) := \left\|\widehat{\boldsymbol{V}}^{-1/2}\varepsilon_t(\boldsymbol{\eta}_0)\right\|.$$

Finally, nonparametric signed rank test statistics can be based on lag-i rank-based generalized cross-covariance matrices of the form

$$\widetilde{\mathbf{\Gamma}}^{(T)}_{i;J}\{\eta\} := \widehat{\boldsymbol{V}}'^{-1/2}\left(\frac{1}{T-i}\sum_{t=i+1}^{T} J_1\left(\frac{\widehat{R}_t(\eta_0)}{T+1}\right) J_2\left(\frac{\hat{R}_{t-i}(\eta_0)}{T+1}\right)\boldsymbol{W}_t(\eta_0)\,\boldsymbol{W}'_{t-i}(\eta_0)\right)\widehat{\boldsymbol{V}}'^{1/2},$$

where the score functions J_1 and J_2 satisfy the following assumption.

Assumption 5.5.3 (Assumptions (C) of Hallin and Paindaveine (2005)). *The score functions* $J_l : (0,1) \to \mathbb{R}$, $l = 1,2$ *are continuous differences of two monotone increasing functions, and satisfy* $\int_0^1 \{J_l(u)\}^2\, du < \infty$, $l = 1,2$.

Let $\varphi_f := -2\frac{(f^{1/2})'}{f^{1/2}}$, where $\left(f^{1/2}\right)'$ is the weak derivative of $f^{1/2}$ in $L_2\left(\mathbb{R}_0^+, \mu_{N-1}\right)$ which stands for the space of all measurable functions $h : \mathbb{R}_0^+ \to \mathbb{R}$ satisfying $\int_0^\infty \{h(r)\}^2 r^{N-1} dr < \infty$. The score functions which yield locally and asymptotically optimal procedures are of the form $J_1 := \varphi_{\widetilde{f}_1} \circ \widetilde{F}_N^{-1}$ and $J_2 := \widetilde{F}_N^{-1}$ for some radial density $\widetilde{f}_1$, where $\widetilde{F}_N$ stands for the cdf associated with the radial pdf $\widetilde{f}_N = \left(\mu_{N-1:\widetilde{f}_1}\right)^{-1} r^{N-1}\widetilde{f}_1(r)\chi_{\{r>0\}}$, $r \in \mathbb{R}$. Then, the above assumption can be rewritten as follows.

Assumption 5.5.4 (Assumptions (C') of Hallin and Paindaveine (2005)). *The radial density* $\widetilde{f}_1$ *is such that* $\varphi_{\widetilde{f}_1}$ *is the continuous difference of two monotone increasing functions,* $\mu_{N-1:\widetilde{f}_1} < \infty$, *and* $\int_0^\infty \left\{\varphi_{\widetilde{f}_1}(r)\right\}^2 r^{N-1}\widetilde{f}_1(r)\, dr < \infty$.

Problems

5.1 Show (5.1.4) by applying the Lebesgue bounded convergence theorem.

5.2 Show (5.1.10), where σ_L^2 corresponds to the variance of the random variable X which has distribution function L.

5.3 Integrating by part, and then applying the Cauchy–Schwarz inequality, show (5.1.13).

5.4 Verify (5.1.22) and (5.1.23).

5.5 Verify the relations for the coefficients (5.2.12).

5.6 Show that the vectors $\boldsymbol{U}_i^{(n)}$ are invariant under transformations of $\boldsymbol{X}_i - \boldsymbol{\theta}$ which preserve half lines through the origin.

5.7 Show that matrices $\mathcal{U}_X$ and $\mathcal{R}_X$ in (5.3.16) are positive definite.

5.8 Show (5.3.35).

5.9 Using the Parseval equality, verify (5.4.15).

5.10 Verify (5.4.24) and (5.4.27).

Chapter 6

Portfolio Estimation Influenced by Non-Gaussian Innovations and Exogenous Variables

The traditional theory of portfolio estimation assumes that the asset returns are independent normal random variables. However, as we saw in the previous chapters, the returns are not necessarily either normal or independent. Also Gouriéroux and Jasiak (2001) studied financial time series returns, and showed how these series can be modeled as ARMA processes. Aït-Sahalia et al. (2011) discussed how stock returns sampled at a high frequency show AR dependence. Discussion on the asymmetry of the distribution of returns can be found in Adcock and Shutes (2005) and the references therein. De Luca et al. (2008) observed that the distribution of the European market index returns conditional on the sign of the one-day lagged US return is skew-normal. Hence, time series models with skew-normal innovations are important. In view of this we assume that the return processes are generated by non-Gaussian linear processes.

In Section 6.1, we suppose that the return process is generated by a linear process with skew-normal distribution characterized by the skewness parameter δ. Let $\hat{g}$ be the generalized mean-variance portfolio estimator (recall Section 3.2). Theorem 3.2.5 showed that $\hat{g}$ is asymptotically normal with zero mean and variance matrix $V(\delta)$. We evaluate the influence of δ on $V(\delta)$, and investigate the robustness of the estimator.

If the return process is non-Gaussian, it depends on higher-order cumulant spectra. Hence, it is natural to introduce a class $\mathcal{G}$ of portfolio estimators which depend on higher-order sample cumulant spectra. Then, we derive the asymptotic distribution of $\tilde{g} \in \mathcal{G}_1$, and compare it with that of $\hat{g}$. Numerical studies illuminate some interesting features of $\tilde{g}$ in comparison with $\hat{g}$.

When we construct a portfolio on some assets $\boldsymbol{X}$, we often use the information of exogenous variables $\boldsymbol{Z}$. Section 6.3 introduces a class $\mathcal{G}_2$ of portfolio estimators which depend on $\widehat{E(\boldsymbol{X})}$, $\widehat{\text{cov}}(\boldsymbol{X}, \boldsymbol{X})$, $\widehat{\text{cov}}(\boldsymbol{X}, \boldsymbol{Z})$, $\widehat{\text{cum}}(\boldsymbol{X}, \boldsymbol{X}, \boldsymbol{Z})$, where $\widehat{(\cdot)}$ means the sample version of $(\cdot)$. Then we derive the asymptotic distribution of $\breve{g} \in \mathcal{G}_2$, and elucidate the influence of exogenous variables on it.

6.1 Robust Portfolio Estimation under Skew-Normal Return Processes

De Luca et al. (2008) introduced a class of skew-GARCH models in view of empirical research on European stock markets. Hence, it is natural to introduce a class of time series models generated by skew-normal innovations. Suppose that $\{\boldsymbol{X}(t) = (X_1(t), \cdots, X_m(t))'; t \in \boldsymbol{Z}\}$ is the return process of m assets generated by

$$\boldsymbol{X}(t) = \sum_{j=0}^{\infty} A(j)\boldsymbol{U}(t-j) + \boldsymbol{\mu} \tag{6.1.1}$$

where $\{U(t)\}$ ~i.i.d. $(\mathbf{0}, K)$ and the A(j)'s are $m \times m$ matrices satisfying Assumption 3.2.1. Hence $\mu = \mathrm{E}\{X(t)\}$. Denoting $\Sigma = \mathrm{Var}\{X(t)\}$, we define $\theta = (\theta_1, \ldots, \theta_{m+r})'$ by

$$\theta = (\mu', \mathrm{vech}(\Sigma)')'$$

where $r = m(m+1)/2$. As in Section 3.2, we assume that the optimal portfolio weight on $X(t)$ is written as $g(\theta)$ where $g : \theta \to \mathbb{R}^{m-1}$ is a smooth function. From the partial realization $\{X(1), \ldots, X(n)\}$, a natural estimator of θ is

$$\hat{\theta} = \left(\hat{\mu}', \mathrm{vech}(\hat{\Sigma})'\right)'$$

where $\hat{\mu} = n^{-1} \sum_{t=1}^{n} X(t)$ and $\hat{\Sigma} = n^{-1} \sum_{t=1}^{n} \{X(t) - \hat{\mu}\}\{X(t) - \hat{\mu}\}'$. Then we use $g(\hat{\theta})$ as an estimator of $g(\theta)$, which is called a generalized mean-variance portfolio estimator. Under Assumptions 3.2.1 and 3.2.4, Theorem 3.2.5 yields

$$\sqrt{n}\left\{g(\hat{\theta}) - g(\theta)\right\} \xrightarrow{\mathcal{L}} N\left[\mathbf{0}, \left(\frac{\partial g}{\partial \theta'}\right) \Omega \left(\frac{\partial g}{\partial \theta'}\right)'\right]. \tag{6.1.2}$$

In what follows we introduce the multivariate skew-normal distribution. An m-dimensional random vector U is said to have a multivariate skew-normal distribution, denoted by $SN_m(\mu, \Gamma, \alpha)$, if its probability density function (with respect to the Lebesgue measure) is given by

$$2\phi_m(u; \mu, \Gamma) \Phi(\alpha'(u - \mu)), \quad u \in \mathbb{R}^m, \tag{6.1.3}$$

where $\phi_m(u; \mu, \Gamma)$ is the probability density of $N_m(\mu, \Gamma)$, $\Phi(\cdot)$ is the distribution function of $N(0, 1)$ and α is an m-dimensional vector. Let

$$\delta = (1 + \alpha' \Gamma \alpha)^{-1/2} \Gamma \alpha,$$

which regulates the skewness of the distribution. If $\delta = \mathbf{0}$, then it becomes the ordinal $N_m(\mu, \Gamma)$.

The moment generating function of $SN_m(\mu, \Gamma, \alpha)$ is

$$\begin{aligned} \mathrm{M}(s) &= 2 \int_{\mathbb{R}^m} \exp(s'u)\, \phi_m(u; \mu, \Gamma)\; \Phi\; (\alpha'(u - \mu))\, du \\ &= 2 \exp(s'\mu) \exp\{(s'\Gamma s)/2\}\, \Phi(\delta' s). \end{aligned} \tag{6.1.4}$$

The cumulants of $SN_m(\mu, \Gamma, \alpha)$ can be computed from the moment generating function. Let κ_j denote the jth-order cumulant matrix ($j = 1, \ldots, 4$), so that κ_1 is $m \times 1$, κ_2 is $m \times m$, κ_3 is $m^2 \times m$ and κ_4 is $m^2 \times m^2$. Concretely,

$$\kappa_1 = \frac{\partial}{\partial s} \log \mathrm{M}(s)\Big|_{s=0} = \mu + \sqrt{\frac{2}{\pi}}\delta, \tag{6.1.5}$$

$$\kappa_2 = \frac{\partial^2}{\partial s \partial s'} \log \mathrm{M}(s)\Big|_{s=0} = \Gamma - \frac{2}{\pi}\delta \otimes \delta', \tag{6.1.6}$$

$$\kappa_3 = \frac{\partial^3}{\partial s \partial s' \partial s} \log \mathrm{M}(s)\Big|_{s=0} = \frac{1}{\sqrt{2\pi}}\left(\frac{1}{\pi} - 2\right)(I_m \otimes \delta)(\delta \otimes \delta'), \tag{6.1.7}$$

$$\kappa_4 = \frac{\partial^4}{\partial s \partial s' \partial s \partial s'} \log \mathrm{M}(s)\Big|_{s=0} = \left(\frac{4}{\pi} - \frac{3}{\pi^2}\right)\delta \otimes \delta' \otimes \delta \otimes \delta'. \tag{6.1.8}$$

Next we evaluate the derivative of the asymptotic variance of $g(\hat{\boldsymbol{\theta}})$ with respect to the skew-normal perturbation $\boldsymbol{\delta}$. Suppose that the innovations $\boldsymbol{U}(t)$ in (6.1.1) are i.i.d. $SN_m(\boldsymbol{\mu}, \boldsymbol{\Gamma}, \boldsymbol{\alpha})$ with $\boldsymbol{\mu} = -\sqrt{2/\pi}\boldsymbol{\delta}$. Then $\mathrm{E}\{\boldsymbol{U}(t)\} = 0$ and $\mathrm{Var}\{\boldsymbol{U}(t)\} = \Gamma - (2/\pi)\boldsymbol{\delta}\boldsymbol{\delta}'$. Write $\boldsymbol{\delta}$ as $\boldsymbol{\delta} = \delta\boldsymbol{d}$ where $\boldsymbol{d} = (d_1, \ldots, d_m)'$ with $\|\boldsymbol{d}\| = 1$, and δ is a positive scalar. Here $\boldsymbol{d}$ represents the direction of the skewness, and δ represents the intensity. In this setting, regarding the matrix Ω as a function of δ, we write

$$\Omega = \Omega(\delta) = \begin{pmatrix} \Omega_1(\delta) & \Omega_3(\delta) \\ \Omega_3(\delta)' & \Omega_2(\delta) \end{pmatrix}.$$

Let

$$V(\delta) = \left(\frac{\partial g}{\partial \boldsymbol{\theta}'}\right)\Omega(\delta)\left(\frac{\partial g}{\partial \boldsymbol{\theta}'}\right)',$$
$$\boldsymbol{D} = \boldsymbol{d}\boldsymbol{d}',$$

and

$$\boldsymbol{A}(\lambda) = \sum_{j=0}^{\infty} A(j)e^{ij\lambda}.$$

The following two propositions are due to Taniguchi et al. (2012).

Proposition 6.1.1. *Under Assumptions 3.2.1 and 3.2.4,*

$$\tilde{\Omega}(\delta) \equiv \frac{\partial}{\partial \delta}\Omega(\delta) = \begin{pmatrix} \tilde{\Omega}_1(\delta) & \tilde{\Omega}_3(\delta) \\ \tilde{\Omega}_3(\delta)' & \tilde{\Omega}_2(\delta) \end{pmatrix},$$

where

$$\tilde{\Omega}_1(\delta) = -\frac{4\delta}{\pi}\boldsymbol{A}(0)\boldsymbol{D}\boldsymbol{A}(0)',$$
$$\tilde{\Omega}_2(\delta) = \Bigg\{-2\frac{\delta^2}{\pi^2}\sum_{p,q,u,v=1}^{m}\left(D_{pq}K_{uv} + D_{uv}K_{pq}\right)$$
$$\times \int_{-\pi}^{\pi}\left\{A_{a_1 p}(\lambda)\overline{A_{a_3 q}(\lambda)A_{a_2 u}(\lambda)}A_{a_4 v}(\lambda) + A_{a_1 p}(\lambda)\overline{A_{a_4 q}(\lambda)A_{a_2 u}(\lambda)}A_{a_3 v}(\lambda)\right\}d\lambda$$
$$+\left(\frac{1}{2\pi}\right)^2\sum_{b_1,b_2,b_3,b_4=1}^{m}\tilde{c}^U_{b_1b_2b_3b_4}\int\int_{-\pi}^{\pi}A_{a_1b_1}(\lambda_1)A_{a_2b_2}(-\lambda_1)A_{a_3b_3}(\lambda_2)$$
$$\times A_{a_4b_4}(-\lambda_2)\,d\lambda_1\,d\lambda_2;\ a_1,a_2,a_3,a_4 = 1,\ldots,m,\ a_1 \geq a_2,\ a_3 \geq a_4\Bigg\},$$
$$\tilde{\Omega}_3(\delta) = \Bigg\{\frac{1}{(2\pi)^2}\sum_{b_1,b_2,b_3=1}^{m}\tilde{c}^U_{b_1b_2b_3}\int_{-\pi}^{\pi}A_{a_1b_1}(\lambda_1+\lambda_2)A_{a_2b_2}(-\lambda_1)A_{a_3b_3}(-\lambda_2)d\lambda_1 d\lambda_2;$$
$$a_1,a_2,a_3 = 1,\ldots,m,\ a_2 \geq a_3\Bigg\}.$$

Here, $D_{p_1p_2}$, $K_{p_1p_2}$ and $A_{p_1p_2}(\lambda)$ are the (p_1, p_2)-components of $\boldsymbol{D}$, $\boldsymbol{K}$ and $\boldsymbol{A}(\lambda)$, respectively, and the quantities $\tilde{c}^U_{b_1b_2b_3}$, $\tilde{c}^U_{b_1b_2b_3b_4}$ can be obtained from the matrices

$$\tilde{\boldsymbol{\kappa}}_3 = \frac{3\delta^2}{\sqrt{2\pi}}\left(\frac{1}{\pi} - 2\right)(\boldsymbol{I}_m \otimes \boldsymbol{d}')\,\boldsymbol{D}.$$

Differentiation of $V(\delta)$ *yields*

$$\frac{\partial}{\partial\delta}V(\delta) = \left(\frac{\partial g}{\partial\boldsymbol{\theta}'}(\delta)\right)\tilde{\Omega}(\delta)\left(\frac{\partial g}{\partial\boldsymbol{\theta}'}(\delta)\right)' + \left(\frac{\partial^2 g}{\partial\delta\partial\boldsymbol{\theta}'}(\delta)\right)\Omega(\delta)\left(\frac{\partial g}{\partial\boldsymbol{\theta}'}(\delta)\right)' + \left(\frac{\partial g}{\partial\boldsymbol{\theta}'}(\delta)\right)\Omega(\delta)\left(\frac{\partial^2 g}{\partial\delta\partial\boldsymbol{\theta}'}(\delta)\right)'. \tag{6.1.9}$$

From Proposition 6.1.1 and (6.1.9) we have,

Proposition 6.1.2. *Under Assumptions 3.2.1 and 3.2.4, if*

$$\left(\frac{\partial^2 g}{\partial\delta\partial\boldsymbol{\theta}'}(\delta)\right)_{\delta=0} = \mathbf{0}, \tag{6.1.10}$$

then $\left.\frac{\partial}{\partial\delta}V(\delta)\right|_{\delta=0} = \mathbf{0}$*, i.e., the asymptotic distribution of portfolio estimator* $g(\hat{\boldsymbol{\theta}})$ *is robust for small deviations from normality.*

The classical mean-variance portfolio coefficient is given by

$$\begin{cases} \max_{\boldsymbol{\theta},\,\theta_0} \boldsymbol{\theta}'\boldsymbol{\mu} + \theta_0 R_0 - \beta\boldsymbol{\theta}'\Sigma\boldsymbol{\theta} \\ \text{subject to } \sum_{j=0}^{m}\theta_j = 1, \end{cases} \tag{6.1.11}$$

where R_0 is the risk-free interest rate and β is a positive constant representing risk aversion. Then the solution of (6.1.11) is given by

$$\begin{cases} g_{MV}(\boldsymbol{\theta}) = \frac{1}{2\beta}\Sigma^{-1}(\boldsymbol{\mu} - R_0\boldsymbol{e}), \\ \theta_0 = 1 - \boldsymbol{e}'g(\boldsymbol{\theta}), \end{cases} \tag{6.1.12}$$

where $\boldsymbol{e} = (\overbrace{1,\ldots,1}^{m})'$. It is not difficult to show (see Problem 6.3)

$$\frac{\partial g_{MV}(\boldsymbol{\theta})}{\partial\delta\partial\theta_j} = O(\delta), \quad j = 1,\ldots,r, \tag{6.1.13}$$

which, together with Proposition 6.1.2, implies that the classical mean-variance portfolio estimator $g_{ML}(\hat{\boldsymbol{\theta}})$ is robust for a small perturbation with respect to skew parameter δ.

Further, Taniguchi et al. (2012) evaluated

$$\left.\frac{\partial^2}{\partial\delta^2}V(\delta)\right|_{\delta=0}, \tag{6.1.14}$$

which measures a sort of second-order sensitivity with respect to the asymmetry parameter δ. They observed that if $\boldsymbol{A}(\lambda)^{-1}$ or $\boldsymbol{A}(\lambda)$ has near unit roots, (6.1.14) shows high sensitivity with respect to δ.

In Theorem 8.3.6, we will establish the LAN for vector-valued non-Gaussian linear processes. If we assume that the innovation distribution is SN_m with asymmetry $\boldsymbol{\delta}$, then the central sequence $\Delta_n(\boldsymbol{h};\boldsymbol{\theta})$ in (8.3.14) depends on $\boldsymbol{\delta}$. Denoting $\Delta_n(\boldsymbol{h};\boldsymbol{\theta})$ by $\Delta_n(\boldsymbol{\delta})$, we expand it as

$$\Delta_n(\boldsymbol{\delta}) = \Delta_n(\mathbf{0}) + (A) + \text{higher-order terms in } \boldsymbol{\delta}. \tag{6.1.15}$$

It is seen that $\mathrm{E}\{(A)\} = 0$, and (A) is of the form $\boldsymbol{\delta}'(\cdot)\boldsymbol{\delta}$. Because the central sequence is a key quantity which describes the optimality of various statistical procedures, Taniguchi et al. (2012) introduced an influence measure criterion (IMC) of $\boldsymbol{\delta}$ on the central sequence by IMC=Var$\{(A)\}$. They observed that, if $\boldsymbol{A}(\lambda)^{-1}$ or $\boldsymbol{A}(\lambda)$ has near unit roots, IMC becomes large.

6.2 Portfolio Estimators Depending on Higher-Order Cumulant Spectra

In the previous section we introduced the generalized mean variance portfolio $g(\boldsymbol{\theta})$ for a non-Gaussian linear process (6.1.1). Throughout this section we assume that the coefficient matrices $\{A(j)\}$ satisfy Assumption 3.2.1 and $\{U(t)\}$ has the kth-order cumulant for $k = 2, 3, \ldots$. Then the process $\{\boldsymbol{X}(t)\}$ becomes a kth-order stationary one (e.g., Brillinger (2001b, p. 34) For $2 \le l \le k$, denote the lth-order cumulant of $\{X_{a_1}(0), X_{a_2}(t_1), \ldots, X_{a_l}(t_{l-1})\}$ by $c^{(l)}_{a_1\cdots a_l}(t_1, \ldots, t_{l-1})$. The lth-order cumulant spectral density is defined as

$$f^{(l)}_{a_1\cdots a_l}(\lambda_1, \ldots, \lambda_{l-1}) = (2\pi)^{-l+1} \sum_{t_1, t_2, \ldots, t_{l-1} = -\infty}^{\infty} \exp\left\{ i \sum_{j=1}^{l-1} \lambda_j t_j \right\} c^{(l)}_{a_1\cdots a_l}(t_1, \ldots, t_{l-1}). \tag{6.2.1}$$

The lth-order cumulant spectral distribution is defined as

$$F^{(l)}_{a_1\cdots a_l}(A) = \int_A f^{(l)}_{a_1\cdots a_l}(\lambda_1, \ldots, \lambda_{l-1}) d\lambda_1 \cdots d\lambda_{l-1}, \tag{6.2.2}$$

where A is a Borel set in $S^{l-1} = [-\pi, \pi]^{l-1}$.

Let us consider the problem of constructing a portfolio for the return process (6.1.1). Suppose that $\mathcal{U}(\cdot)$ is a utility function and $\boldsymbol{\alpha} = (\alpha_1, \ldots, \alpha_m)'$ is a portfolio coefficient vector satisfying $\alpha_1 + \cdots + \alpha_m = 1$. We seek the optimal portfolio $\boldsymbol{\alpha}$ which maximizes $\mathrm{E}[\mathcal{U}(\boldsymbol{\alpha}'\boldsymbol{X}(t))]$, i.e.,

$$\boldsymbol{\alpha}_{\text{opt}} \equiv \arg\max_{\boldsymbol{\alpha}} \mathrm{E}\left[\mathcal{U}\left(\boldsymbol{\alpha}'\boldsymbol{X}(t)\right)\right]. \tag{6.2.3}$$

Expanding $\mathcal{U}(\cdot)$ in a Taylor expansion, we obtain the following approximation:

$$\begin{aligned} \mathrm{E}[\mathcal{U}(M(t))] \approx \mathcal{U}(c_1^M(t)) &+ \frac{1}{2!} D^2\mathcal{U}(c_1^M(t)) c_2^M(t) + \frac{1}{3!} D^3\mathcal{U}(c_1^M(t)) c_3^M(t) \\ &+ \cdots + \frac{1}{k!} D^k\mathcal{U}(c_1^M(t)) c_k^M(t), \end{aligned} \tag{6.2.4}$$

where $M(t) = \boldsymbol{\alpha}'\boldsymbol{X}(t)$ and $c_l^M(t)$ is the lth-order cumulant of $M(t)$. If we use the right-hand-side approximation of (6.2.4), then it is seen that $\boldsymbol{\alpha}_{\text{opt}}$ depends on the mean μ and the higher-order cumulant spectral densities $\boldsymbol{f}^{(l)} = \{f^{(l)}_{a_1\cdots a_l}; 1 \le a_j \le m\}$, $2 \le l \le k$. For $2 \le l \le k$, write

$$\boldsymbol{F}^{(l)} = \{F^{(l)}_{a_1\cdots a_l}; 1 \le a_j \le m\}, \tag{6.2.5}$$

$$\begin{aligned} \boldsymbol{\theta}_l\left(\boldsymbol{F}^{(l)}\right) &= \left\{ \int \cdots \int_{-\pi}^{\pi} h_l(\lambda_1, \ldots, \lambda_{l-1}) dF^{(l)}_{a_1\cdots a_l}(\lambda_1, \ldots, \lambda_{l-1}); 1 \le a_j \le m \right\} \\ &= \left\{ \theta_l\left(F^{(l)}_{a_1\cdots a_l}\right); 1 \le a_j \le m \right\} (say), \end{aligned} \tag{6.2.6}$$

where $h_l(\cdot)$'s are real-valued functions whose total variations are bounded. Therefore we can describe the optimal portfolio $\boldsymbol{\alpha}_{\text{opt}}$ as a function of μ and $\boldsymbol{F}^{(l)}$, i.e.,

$$g : \left(\boldsymbol{\mu}, \boldsymbol{\theta}_2\left(\boldsymbol{F}^{(2)}\right), \ldots, \boldsymbol{\theta}_k\left(\boldsymbol{F}^{(k)}\right)\right) \to \mathbb{R}^m. \tag{6.2.7}$$

To estimate $g(\cdot)$, we construct estimators of $\boldsymbol{\theta}_l(\boldsymbol{F}^{(l)})$. Suppose that an observed stretch $\{\boldsymbol{X}(1), \ldots, \boldsymbol{X}(n)\}$ from (6.1.1) is available. For this we introduce finite Fourier transforms:

$$d_a^n(\lambda) \equiv \sum_{t=0}^{n-1} X_a(t) \exp\{-i\lambda t\}. \tag{6.2.8}$$

Define $F^{n,(l)}_{a_1,a_2\cdots a_l}$ as

$$F^{n,(l)}_{a_1\cdots a_l}(\lambda_1,\ldots,\lambda_{l-1}) = \left(\frac{2\pi}{n}\right)^{l-1} \sum_{\substack{2\pi r_j/n\le\lambda_j;\\ j=1,\ldots,l-1}} \boldsymbol{I}^n_{a_1,\ldots,a_l}\left(\frac{2\pi r_1}{n},\ldots,\frac{2\pi r_l}{n}\right), \tag{6.2.9}$$

where

$$\boldsymbol{I}^n_{a_1\cdots a_l}\left(\frac{2\pi r_1}{n},\ldots,\frac{2\pi r_l}{n}\right) = (2\pi)^{-k+1}n^{-1}\eta\left(\sum_{j=1}^{l}\lambda_j\right)\prod_{j=1}^{l} d^n_{a_j}\left(\frac{2\pi r_j}{n}\right), \tag{6.2.10}$$

(lth-order periodogram). Here $\eta(\cdot)$ is the Kronecker comb:

$$\eta(\lambda) = \begin{cases} 1, & \lambda \equiv 0 \ (\mathrm{mod}\, 2\pi), \\ 0, & \lambda \not\equiv 0 \ (\mathrm{mod}\, 2\pi). \end{cases} \tag{6.2.11}$$

To estimate $\theta_l\left(F^{(l)}_{a_1\cdots a_l}\right)$ we introduce

$$\begin{aligned}\theta_l\left(F^{n,(l)}_{a_1\cdots a_l}\right) &= \int\cdots\int_{-\pi}^{\pi} h_l(\lambda_1,\ldots,\lambda_{l-1})dF^{n,(l)}_{a_1\cdots a_l}(\lambda_1,\ldots,\lambda_{l-1})\\ &= \left(\frac{2\pi}{n}\right)^{l-1}\sum{}^{*} h_l\left(\frac{2\pi r_1}{n},\ldots,\frac{2\pi r_l}{n}\right)\boldsymbol{I}^n_{a_1\cdots a_l}\left(\frac{2\pi r_1}{n},\ldots,\frac{2\pi r_l}{n}\right),\end{aligned} \tag{6.2.12}$$

where $\sum^*$ extends over $(r_1,\ldots,r_l)$ satisfying

$$-\left[\frac{n}{2}\right]+1 \le r_1 \le \left[\frac{n}{2}\right],\ldots,-\left[\frac{n}{2}\right]+1 \le r_l \le \left[\frac{n}{2}\right]$$

and $r_1+\cdots+r_l = 0$ (mod n). The mean vector $\boldsymbol{\mu} = (\mu_1,\ldots,\mu_m)'$ is estimated by $\hat{\boldsymbol{\mu}} \equiv (\hat{\mu}_1,\ldots,\hat{\mu}_m)'$ with $\hat{\mu}_j = n^{-1}d^n_j(0),\ j = 1,\ldots,m$.

To evaluate the expectation of $\theta_l\left(F^{n,(l)}_{a_1\cdots a_l}\right)$, we prepare the following lemmas.

Lemma 6.2.1. *(Theorem 2.3.2 of Brillinger (2001b, p. 21)) For $Y_{ij},\ j = 1,2,\ldots,J_i,\ i = 1,2,\ldots,I$, a two-way array of random variables and*

$$Z_i = \prod_{j=1}^{J_i} Y_{ij},$$

the joint cumulant of $\{Z_1,Z_2,\ldots,Z_I\}$ is given by

$$\sum_{\nu}\left[cum(Y_{ij};\ ij\in\nu_1)\cdots cum(Y_{ij};\ ij\in\nu_p)\right], \tag{6.2.13}$$

where the summation is over all indecomposable partitions $\boldsymbol{\nu} = \nu_1\cup\cdots\cup\nu_p$ of the two-way array

$$\{(i,j);\ j=1,\ldots,J_i,\ i=1,\ldots,I\}.$$

Lemma 6.2.2. *(Theorem 4.3.1 of Brillinger (2001b, p. 92)) Under Assumption 3.2.1,*

$$cum\{d^n_{a_1}(\lambda_1),d^n_{a_2}(\lambda_2),\ldots,d^n_{a_l}(\lambda_l)\} = (2\pi)^{l-1}\Delta_n\left\{\sum_{j=1}^{l}\lambda_j\right\}\eta\left\{\sum_{j=1}^{l}\lambda_j\right\}f^{(l)}_{a_1\cdots a_l}(\lambda_1,\ldots,\lambda_{l-1}) + O(1),$$

where

$$\lambda_j = \frac{2\pi r_j}{n},\quad r_j\in\mathbb{Z},$$

$$\Delta_n(\lambda) = \sum_{t=0}^{n-1}\exp(-i\lambda t) = \begin{cases} 0, & \lambda\not\equiv 0\ (mod\ 2\pi),\\ n, & \lambda\equiv 0\ (mod\ 2\pi),\end{cases}$$

and the error term $O(1)$ is uniform for all $\lambda_1,\ldots,\lambda_l$.

From Lemmas 6.2.1 and 6.2.2, it follows that

$$\mathrm{E}\left[\theta_l\left(F^{n,(l)}_{a_1\cdots a_l}\right)\right] = \theta_l\left(F^{(l)}_{a_1\cdots a_l}\right) + \sum_{\nu}\int_{S^l} h(\lambda_1,\ldots,\lambda_l)\left[\prod_{j=1}^{p} f_{\{\alpha_j;\alpha_j\in\nu_j\}}(\lambda_{j_i};\ j_i\in\nu_j)\right]$$
$$\times\left[\prod_{j=1}^{p}\eta\left\{\sum_{i=1}^{|\nu_j|}\lambda_{j_i}\right\}\right]\prod_{k=1}^{l} d\lambda_k + O(n^{-1}), \tag{6.2.14}$$

where $\sum_{\nu}$ is over all $\nu = (\nu_1,\nu_2,\ldots,\nu_p)$, $p > 1$ and $|\nu_j| > 1$, and $|\nu_j|$ denotes the number of elements in ν_j. Then we can see that

$$\mathrm{E}\left[\theta_l\left(F^{n,(l)}_{a_1\cdots a_l}\right)\right] = \theta_l\left(F^{(l)}_{a_1\cdots a_l}\right) + O(1) + O(n^{-1}), \tag{6.2.15}$$

which implies that $\theta_l\left(F^{n,(l)}_{a_1\cdots a_l}\right)$ is not asymptotically unbiased. Keenan (1987) introduced a modification of $\theta_l\left(F^{n,(l)}_{a_1\cdots a_l}\right)$ to be asymptotically unbiased as follows.

Let $\Omega^{(l)}$ be the collection of $(\lambda_1,\ldots,\lambda_l)$ in $[-\pi,\pi]$ such that $\sum_{j=1}^{l}\lambda_j \equiv 0 \pmod{2\pi}$, $\lambda_j = 2\pi r_j/n$, but is not contained in any further integral region of the form $\sum_{j=1}^{s}\lambda_{i_j} \equiv 0 \pmod{2\pi}$ where $\{i_j;\, j = 1,\ldots,s\}$ is a proper subset of $\{1,2,\ldots,l\}$. Let $\theta_l\left(\hat{F}^{n,(l)}_{a_1\cdots a_l}\right)$ be defined the same as $\theta_l\left(F^{n,(l)}_{a_1\cdots a_l}\right)$, except that the sum is now over $(2\pi r_1/n,\ldots,2\pi r_l/n)$, which are contained in $\Omega^{(l)}$.

The following three propositions are due to Hamada and Taniguchi (2014).

Proposition 6.2.3. *Under Assumption 3.2.1, it holds that*

$$E\left[\theta_l\left(\hat{F}^{n,(l)}_{a_1\cdots a_l}\right)\right] = \theta_l\left(F^{(l)}_{a_1\cdots a_l}\right) + O(n^{-1}), \tag{6.2.16}$$

i.e., $\theta_l\left(\hat{F}^{n,(l)}_{a_1\cdots a_l}\right)$ *is an asymptotically unbiased estimator of* $\theta_l\left(F^{(l)}_{a_1\cdots a_l}\right)$.

Write

$$\begin{aligned} \boldsymbol{\theta} &= \mathrm{vec}\left(\boldsymbol{\mu}, \boldsymbol{\theta}_2(\boldsymbol{F}^{(2)}),\ldots,\boldsymbol{\theta}_l(\boldsymbol{F}^{(l)})\right), \\ \hat{\boldsymbol{\theta}} &= \mathrm{vec}\left(\hat{\boldsymbol{\mu}}, \boldsymbol{\theta}_2(\hat{\boldsymbol{F}}^{(2)}_n),\ldots,\boldsymbol{\theta}_l(\hat{\boldsymbol{F}}^{(l)}_n)\right), \end{aligned} \tag{6.2.17}$$

where $\boldsymbol{\theta}_l(\hat{\boldsymbol{F}}^{(l)}_n) = \left\{\theta_l\left(\hat{F}^{n,(l)}_{a_1\cdots a_l}\right);\, 1 \le a_j \le m\right\}$. By use of Lemmas 6.2.1 and 6.2.2, it is not difficult to show the following.

Proposition 6.2.4. *Under Assumption 3.2.1, it holds that*

$$\sqrt{n}(\hat{\boldsymbol{\theta}} - \boldsymbol{\theta}) \xrightarrow{d} N(\boldsymbol{0},\Sigma), \quad (n\to\infty), \tag{6.2.18}$$

where the elements of Σ *corresponding to those of the variance-covariance matrix between* $\hat{\boldsymbol{\mu}}$ *and* $\hat{\boldsymbol{\mu}}$, $\hat{\boldsymbol{\mu}}$ *and* $\boldsymbol{\theta}_p(\hat{\boldsymbol{F}}^{(p)}_n)$, $\boldsymbol{\theta}_p(\hat{\boldsymbol{F}}^{(p)}_n)$ *and* $\boldsymbol{\theta}_q(\hat{\boldsymbol{F}}^{(q)}_n)$, $p,q \le l$, *are, respectively, given by*

$$n\,Cov(\hat{\mu}_a,\hat{\mu}_b) = 2\pi f^{(2)}_{ab}(0),$$
$$n\,Cov(\hat{\mu}_a,\theta_q(\hat{F}^{n,(q)}_{b_1\cdots b_q})) = \int\cdots\int_{\Omega^{(q)}} h_q(\lambda_1,\ldots,\lambda_{q-1})$$
$$\times f^{(q+1)}_{ab_1\cdots b_q}(0,\lambda_1,\ldots,\lambda_{q-1})d\lambda_1\cdots d\lambda_{q-1},$$
$$n\,Cov(\theta_p(\hat{F}^{n,(p)}_{a_1\cdots a_p}),\theta_q(\hat{F}^{n,(q)}_{b_1\cdots b_q})) = \sum_{\Delta}\int\cdots\int_{M^{p+q-r-2}} h_p(\lambda_1,\ldots,\lambda_p)b_q(\lambda_{p+1},\ldots,\lambda_{p+q})$$
$$\times\eta\left(\sum_{j=1}^{|\Delta_1|}\lambda_{1_j}\right) f_{\{a_{1_j};a_{1_j}\in\Delta_1\}}(\lambda_{1_j};\lambda_{1_j}\in\Delta_1)\cdots$$

$$\times \eta\left(\sum_{j=1}^{|\Delta_r|} \lambda_{r_j}\right) f_{\{a_{r_j};a_{r_j}\in\Delta_r\}}(\lambda_{r_j}; \lambda_{r_j} \in \Delta_r) d\lambda_1 \cdots d\lambda_{p+q},$$

where $M_{p+q-r-2} = [-\pi, \pi]^{p+q-r-2}$, *and the summation is extended over all indecomposable portions* $\Delta = (\Delta_1, \dots, \Delta_r)$ *of the two-way array* $\{(i, j_i); i = 1, 2, j_1 = 1, 2, \dots, p, j_2 = 1, 2, \dots, q\}$.

Proposition 6.2.4, together with the δ-method (e.g., Proposition 6.4.3 of Brockwell and Davis (2006)), yields the following.

Proposition 6.2.5. *Under Assumptions 3.2.1 and 3.2.4,*

$$\sqrt{n}[g(\hat{\boldsymbol{\theta}}) - g(\boldsymbol{\theta})] \xrightarrow{d} N\left(\mathbf{0}, \left(\frac{\partial g}{\partial \boldsymbol{\theta}'}\right) \Sigma \left(\frac{\partial g}{\partial \boldsymbol{\theta}'}\right)'\right), \tag{6.2.19}$$

as $n \to \infty$.

In what follows we provide a numerical example.

Example 6.2.6. *Let the return process* $\{\boldsymbol{X}(t) = (X_1(t), X_2(t))\}$ *be generated by*

$$\begin{cases} X_1(t) = \mu_1, \\ X_2(t) - \mu_2 = \rho(X_2(t-1) - \mu_2) + u(t), \end{cases} \tag{6.2.20}$$

where $\{u(t)\}$*'s are i.i.d.* $t(\nu)$ *(Student t-distribution with* ν *degrees of freedom). The portfolio is defined as*

$$M(t) = (1 - \alpha)X_1(t) + \alpha X_2(t),$$

whose cumulants are $c_1^M = (1-\alpha)\mu_1 + \alpha\mu_2$, $c_2^M = \alpha^2 c_2^{X_2} = \alpha^2\sigma_2^2$ *and* $c_3^M = \alpha^3 c_3^{X_2}$ *where* $\sigma_2^2 = Var\{X_2(t)\}$ *and* $c_3^{X_2} = \int\int_{-\pi}^{\pi} f_{222}^{(3)}(\lambda_1, \lambda_2) d\lambda_1 d\lambda_2$. *We are interested in the optimal portfolio* α_{opt} *by the following cubic utility:*

$$\alpha_{opt} = \arg\max_{\alpha} E[M(t) - \gamma\alpha^2\{(X_2(t) - \mu_2)^2 - \frac{1}{\sqrt{n}} h_3(X_2(t) - \mu_2)\}], \tag{6.2.21}$$

where $\gamma > 0$ *is a risk aversion coefficient and* $h_3(\cdot)$ *is the Hermite polynomial of order 3. We approximate the expectation in (6.2.21) by Edgeworth expansion as*

$$\begin{aligned}
(6.2.21) &\approx (1-\alpha)\mu_1 + \alpha\mu_2 - \gamma\alpha^2\sigma_2^2 + \int_{-\infty}^{\infty} \left\{-\gamma\alpha^2\sigma_2^2 h_2(y) + \gamma\alpha^2 \frac{1}{\sqrt{n}} h_3(y)\right\} \\
&\qquad\qquad \times \left\{1 + \frac{\beta}{6} h_3(y)\right\} \phi(y) dy \\
&\approx (1-\alpha)\mu_1 + \alpha\mu_2 - \gamma\alpha^2\sigma_2^2 + \gamma\alpha^2 \frac{\beta}{\sqrt{n}},
\end{aligned} \tag{6.2.22}$$

which imply that α_{opt} *for sufficiently large n is*

$$\alpha_{opt} \approx \frac{\mu_2 - \mu_1}{2\gamma\sigma_2^2} + \frac{\beta}{\sqrt{n}}\left(\frac{\mu_2 - \mu_1}{2\gamma\sigma_2^4}\right). \tag{6.2.23}$$

while we know that the classical mean-variance optimal portfolio is given by

$$\alpha_{opt}^c = \frac{\mu_2 - \mu_1}{2\gamma\sigma_2^2}, \tag{6.2.24}$$

Returning to (6.2.20), we set $\mu_1 = 0.3, \mu_2 = 0.4, \gamma = 0.5$ and $\nu = 10$. For each $\rho = -0.9(0.12)0.9$, we generated $\{\boldsymbol{X}(1), \dots, \boldsymbol{X}(100)\}$ from (6.2.20), and estimated (μ, σ_2^2, β) with $\mu = \mu_2 - \mu_1$ by

$$\hat{\mu} = \frac{1}{100} \sum_{t=1}^{100} \{X_2(t) - X_1(t)\},$$

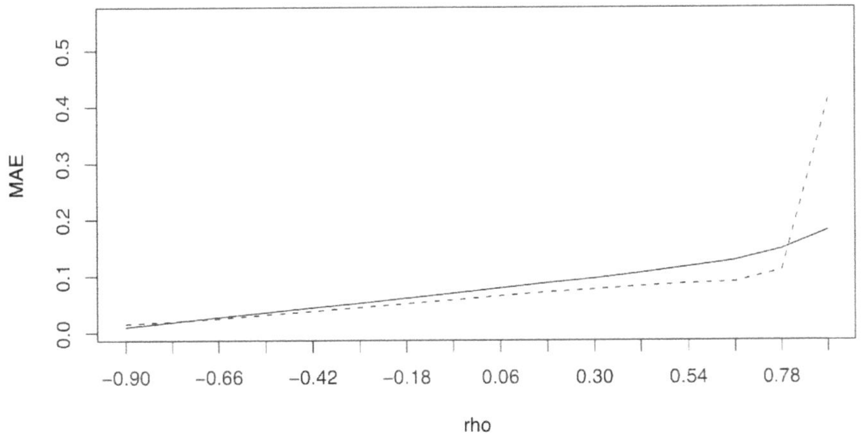

Figure 6.1: Values of MAE and MAE_c

$$\hat{\sigma}_2^2 = \frac{1}{100}\sum_{t=1}^{100}\{X_2(t) - \hat{\mu}_2\}^2, \tag{6.2.25}$$

$$\hat{\beta} = \frac{1}{100}\sum_{t=1}^{100}\{X_2(t)\}^3 - 2\hat{\mu}_2\hat{\sigma}_2^2 - \hat{\mu}_2^3,$$

where $\hat{\mu}_2 = 100^{-1}\sum_{t=1}^{100} X_2(t)$. The estimated portfolios are

$$\hat{\alpha}_{\mathrm{opt}} = \frac{\hat{\mu}}{\hat{\sigma}_2^2} + \frac{\hat{\beta}\hat{\mu}}{10\hat{\sigma}_2^4}, \tag{6.2.26}$$

$$\hat{\alpha}^c_{\mathrm{opt}} = \frac{\hat{\mu}}{\hat{\sigma}_2^2}. \tag{6.2.27}$$

Then we repeated the procedure above 10,000 times in simulation. We write $\hat{\alpha}_{\mathrm{opt},m}$ and $\hat{\alpha}^c_{\mathrm{opt},m}$, $\hat{\alpha}_{\mathrm{opt}}$ and $\hat{\alpha}^c_{\mathrm{opt}}$ in the mth replication, respectively. Their mean absolute errors (MAE) are

$$\mathrm{MAE} = \frac{1}{10,000}\sum_{m=1}^{10,000}\left|\hat{\alpha}_{\mathrm{opt},m} - \alpha_{\mathrm{opt}}\right|, \tag{6.2.28}$$

and

$$\mathrm{MAE}_c = \frac{1}{10,000}\sum_{m=1}^{10,000}\left|\hat{\alpha}^c_{\mathrm{opt},m} - \alpha^c_{\mathrm{opt}}\right|. \tag{6.2.29}$$

We plotted MAE_c and MAE solid line and dotted line with a, respectively, in Figure 6.1. The figure shows that $\hat{\alpha}_{\mathrm{opt}}$ is better than $\hat{\alpha}^c_{\mathrm{opt}}$ except for the values of ρ around the unit root ($\rho = 1$).

6.3 Portfolio Estimation under the Utility Function Depending on Exogenous Variables

In the estimation of portfolios, it is natural to assume that the utility function depends on exogenous variables. From this point of view, in this section, we develop estimation under this setting, where the optimal portfolio function depends on moments of the return process, and cumulants between the return process and exogenous variables. Introducing a sample version of it as a portfolio estimator, we derive the asymptotic distribution. Then, an influence of exogenous variables on the return process is illuminated by a few examples.

Suppose the existence of a finite number of assets indexed by i, $(i = 1, \ldots, m)$. Let $\boldsymbol{X}(t) = (X_1(t), \ldots, X_m(t))'$ denote the random returns on m assets at time t, and let $\boldsymbol{Z}(t) = (Z_1(t), \ldots, Z_q(t))'$ denote the exogenous variables influencing the utility function at time t. We write

$$\boldsymbol{Y}(t) \equiv (\boldsymbol{X}(t)', \boldsymbol{Z}(t)')' = (Y_1(t), \ldots, Y_{m+q}(t))' \quad \text{(say)}.$$

Since it is empirically observed that $\{\boldsymbol{X}(t)\}$ is non-Gaussian and dependent, we will assume that it is a non-Gaussian vector-valued linear process with third-order cumulants. Also, suppose that there exists a risk-free asset whose return is denoted by $Y_0(t)$. Let α_0 and $\boldsymbol{\alpha} = (\overbrace{\alpha_1, \ldots, \alpha_m}^{m}, \overbrace{0, \ldots, 0}^{q})'$ be the portfolio weights at time t, and the portfolio is $M(t) = \boldsymbol{Y}(t)'\boldsymbol{\alpha} + Y_0(t)\alpha_0$, whose higher-order cumulants are written as

$$\begin{aligned} c_1^M(t) &= \mathrm{cum}\{M(t)\} = c^{a_1}\alpha_{a_1} + Y_0(t)\alpha_0, \\ c_2^M(t) &= \mathrm{cum}\{M(t), M(t)\} = c^{a_2 a_3}\alpha_{a_2}\alpha_{a_3}, \\ c_3^M(t) &= \mathrm{cum}\{M(t), M(t), M(t)\} = c^{a_4 a_5 a_6}\alpha_{a_4}\alpha_{a_5}\alpha_{a_6}, \end{aligned} \tag{6.3.1}$$

where $c^{a_1} = \mathrm{E}\{Y_{a_1}(t)\}$, $c^{a_2 a_3} = \mathrm{cum}\{Y_{a_2}(t), Y_{a_3}(t)\}$ and $c^{a_4 a_5 a_6} = \mathrm{cum}\{Y_{a_4}(t), Y_{a_5}(t), Y_{a_6}(t)\}$. Here we use Einstein's summation conversion. For a utility function $\mathcal{U}(\cdot)$, the expected utility can be approximated as

$$\mathrm{E}[\mathcal{U}(M(t))] \approx \mathcal{U}(c_1^M(t)) + \frac{1}{2!}D^2\mathcal{U}(c_1^M(t))c_2^M(t) + \frac{1}{3!}D^3\mathcal{U}(c_1^M(t))c_3^M(t), \tag{6.3.2}$$

by Taylor expansion of order 3. The approximate optimal portfolio may be described as

$$\begin{cases} \max\limits_{\alpha_0, \boldsymbol{\alpha}} [\text{the right hand side of (6.3.2)}], \\ \text{subject to } \alpha_0 + \sum_{j=1}^m \alpha_j = 1. \end{cases} \tag{6.3.3}$$

If the utility function $\mathcal{U}(\cdot)$ depends on the exogenous variable $\boldsymbol{Z}$, then the solution $\boldsymbol{\alpha} = \boldsymbol{\alpha}_{\mathrm{opt}}$ of (6.3.3) depends on $\boldsymbol{Z}$ and the return of assets $\boldsymbol{X}$, i.e., $\boldsymbol{\alpha}_{\mathrm{opt}} = \boldsymbol{\alpha}_{\mathrm{opt}}(\boldsymbol{X}, \boldsymbol{Z})$. However, the form of $\boldsymbol{\alpha}_{\mathrm{opt}}(\boldsymbol{X}, \boldsymbol{Z})$ is too general to handle. Hence, we restrict ourselves to the case when $\boldsymbol{\alpha}_{\mathrm{opt}}$ is an m-dimensional smooth function:

$$g(\boldsymbol{\theta}) : g[\mathrm{E}(\boldsymbol{X}), \mathrm{cov}(\boldsymbol{X}, \boldsymbol{X}), \mathrm{cov}(\boldsymbol{X}, \boldsymbol{Z}), \mathrm{cum}(\boldsymbol{X}, \boldsymbol{X}, \boldsymbol{Z})] \to \mathbb{R}^m. \tag{6.3.4}$$

We estimate $g(\boldsymbol{\theta})$ by its sample version:

$$\hat{g} \equiv g(\hat{\boldsymbol{\theta}}) = g[\widehat{\mathrm{E}(\boldsymbol{X})}, \widehat{\mathrm{cov}}(\boldsymbol{X}, \boldsymbol{X}), \widehat{\mathrm{cov}}(\boldsymbol{X}, \boldsymbol{Z}), \widehat{\mathrm{cum}}(\boldsymbol{X}, \boldsymbol{X}, \boldsymbol{Z})], \tag{6.3.5}$$

where $\widehat{(\cdot)}$ is a sample version of $(\cdot)$.

Let $\{\boldsymbol{Y}(t) = (Y_1(t), \ldots, Y_{m+q}(t))'\}$ be an $(m+q)$-dimensional linear process defined by

$$\boldsymbol{Y}(t) = \sum_{j=0}^{\infty} G(j)\boldsymbol{e}(t-j) + \boldsymbol{\mu}, \tag{6.3.6}$$

where $\{\boldsymbol{e}(t)\}$ is an $(m+q)$-dimensional stationary process such that $\mathrm{E}\{\boldsymbol{e}(t)\} = \boldsymbol{0}$ and $\mathrm{E}\{\boldsymbol{e}(s)\boldsymbol{e}(t)'\} = \delta(s,t)K$, with K a nonsingular $(m+q) \times (m+q)$ matrix, $G(j)$'s are $(m+q) \times (m+q)$ matrices and $\boldsymbol{\mu} = (\mu_1, \ldots, \mu_{m+q})'$. Assuming that $\{\boldsymbol{e}(t)\}$ has all order cumulants, let $Q^e_{a_1 \cdots a_j}(t_1, \ldots, t_{j-1})$ be the joint jth-order cumulant of $e_{a_1}(t), e_{a_2}(t+t_1), \ldots, e_{a_j}(t+t_{j-1})$. In what follows we impose

Assumption 6.3.1. **(i)** *For each* $j = 1, 2, 3, \ldots,$

$$\sum_{t_1,\cdots,t_{j-1}=-\infty}^{\infty} \sum_{a_1,\ldots,a_j=1}^{m+q} \left| Q^e_{a_1\cdots a_j}(t_1, \ldots, t_{j-1}) \right| < \infty, \tag{6.3.7}$$

(ii)

$$\sum_{j=0}^{\infty} \|G(j)\| < \infty. \tag{6.3.8}$$

Letting $Q^Y_{a_1\cdots a_j}(t_1, \ldots, t_{j-1})$ be the joint jth-order cumulant of $Y_{a_1}(t), Y_{a_2}(t+t_1), \ldots, Y_{a_j}(t+t_{j-1})$, we define the jth-order cumulant spectral density of $\{\boldsymbol{Y}(t)\}$ by

$$f_{a_1\cdots a_j}(\lambda_1, \ldots, \lambda_{j-1}) = \left(\frac{1}{2\pi}\right)^{j-1} \sum_{t_1,\ldots,t_{j-1}=-\infty}^{\infty} \exp\{-i(\lambda_1 t_1 + \cdots + \lambda_{j-1} t_{j-1})\} \times Q^Y_{a_1\cdots a_j}(t_1, \ldots, t_{j-1}). \tag{6.3.9}$$

For an observed stretch $\{\boldsymbol{Y}(1), \ldots, \boldsymbol{Y}(n)\}$, we introduce

$$\begin{aligned}
\hat{c}^{a_1} &= \frac{1}{n}\sum_{s=1}^{n} Y_{a_1}(s),\\
\hat{c}^{a_2 a_3} &= \frac{1}{n}\sum_{s=1}^{n} (Y_{a_2}(s) - \hat{c}^{a_2})(Y_{a_3}(s) - \hat{c}^{a_3}),\\
\hat{c}^{a_4 a_5} &= \frac{1}{n}\sum_{s=1}^{n} (Y_{a_4}(s) - \hat{c}^{a_4})(Y_{a_5}(s) - \hat{c}^{a_5}),\\
\hat{c}^{a_6 a_7 a_8} &= \frac{1}{n}\sum_{s=1}^{n} (Y_{a_6}(s) - \hat{c}^{a_6})(Y_{a_7}(s) - \hat{c}^{a_7})(Y_{a_8}(s) - \hat{c}^{a_8}),
\end{aligned}$$

where $1 \le a_1, a_2, a_3, a_4, a_6, a_7 \le m$ and $m+1 \le a_5, a_8 \le m+q$. Write the quantities that appeared in (6.3.4) and (6.3.5) by

$$\begin{aligned}
\hat{\boldsymbol{\theta}} &= (\hat{c}^{a_1}, \hat{c}^{a_2 a_3}, \hat{c}^{a_4 a_5}, \hat{c}^{a_6 a_7 a_8}),\\
\boldsymbol{\theta} &= (c^{a_1}, c^{a_2 a_3}, c^{a_4 a_5}, c^{a_6 a_7 a_8}),
\end{aligned} \tag{6.3.10}$$

where $c^{a_1\cdots a_j} = Q^Y_{a_1\cdots a_j}(0, \ldots, 0)$. Then, $\dim \boldsymbol{\theta} = \dim \hat{\boldsymbol{\theta}} = a+b+c+d$, where $a = m$, $b = m(m+1)/2$, $c = mq$, $d = m(m+1)q/2$. As in Proposition 6.2.5, it is not difficult to show the following Proposition 6.3.2, whose proof, together with that of Propositions 6.3.3 and 6.3.5, is given in Hamada et al. (2012).

Proposition 6.3.2. *Under Assumptions 3.2.4 and 6.3.1,*

$$\sqrt{n}(g(\hat{\boldsymbol{\theta}}) - g(\boldsymbol{\theta})) \xrightarrow{d} N(\boldsymbol{0}, (Dg)\Omega(Dg)'), \quad (n \to \infty), \tag{6.3.11}$$

where

$$\Omega = \begin{pmatrix} \Omega_{11} & \Omega_{12} & \Omega_{13} & \Omega_{14}\\ \Omega_{21} & \Omega_{22} & \Omega_{23} & \Omega_{24}\\ \Omega_{31} & \Omega_{32} & \Omega_{33} & \Omega_{34}\\ \Omega_{41} & \Omega_{42} & \Omega_{43} & \Omega_{44} \end{pmatrix}, \tag{6.3.12}$$

and the typical element of Ω_{ij} corresponding to the covariance between $\hat{c}^{\Delta}$ and $\hat{c}^{\nabla}$ is denoted by $V\{(\Delta)(\nabla)\}$, and

$$
\begin{aligned}
V\{(a_1)(a_1')\} &= \Omega_{11} = 2\pi f_{a_1 a_1'}(0),\\
V\{(a_2a_3)(a_1')\} &= \Omega_{12} = 2\pi \int_{-\pi}^{\pi} f_{a_2a_3a_1'}(\lambda, 0)d\lambda (= \Omega_{21}),\\
V\{(a_4a_5)(a_1')\} &= \Omega_{13} = 2\pi \int_{-\pi}^{\pi} f_{a_4a_5a_1'}(\lambda, 0)d\lambda (= \Omega_{31}),\\
V\{(a_6a_7a_8)(a_1')\} &= \Omega_{14} = 2\pi \int\int_{-\pi}^{\pi} f_{a_6a_7a_8a_1'}(\lambda_1, \lambda_2, 0)d\lambda_1 d\lambda_2 (= \Omega_{41}),\\
V\{(a_2a_3)(a_2'a_3')\} &= \Omega_{22} = 2\pi \int\int_{-\pi}^{\pi} f_{a_2a_3a_2'a_3'}(\lambda_1, \lambda_2, -\lambda_2)d\lambda_1 d\lambda_2\\
&\quad + 2\pi \int\int_{-\pi}^{\pi} \left\{ f_{a_2a_2'}(\lambda) f_{a_3a_3'}(-\lambda) + f_{a_2a_3'}(\lambda) f_{a_3a_2'}(-\lambda) \right\} d\lambda,\\
V\{(a_2a_3)(a_4'a_5')\} &= \Omega_{23} = 2\pi \int\int_{-\pi}^{\pi} f_{a_2a_3a_4'a_5'}(\lambda_1, \lambda_2, -\lambda_2)d\lambda_1 d\lambda_2\\
&\quad + 2\pi \int\int_{-\pi}^{\pi} \left\{ f_{a_2a_4'}(\lambda) f_{a_3a_5'}(-\lambda) + f_{a_2a_5'}(\lambda) f_{a_3a_4'}(-\lambda) \right\} d\lambda (= \Omega_{32}),\\
V\{(a_2a_3)(a_6'a_7'a_8')\} &= \Omega_{24} = 2\pi \int\int\int_{-\pi}^{\pi} f_{a_2a_3a_6'a_7'a_8'}(\lambda_1, \lambda_2, \lambda_3, -\lambda_3)d\lambda_1 d\lambda_2 d\lambda_3 (= \Omega_{42}),\\
V\{(a_4a_5)(a_4'a_5')\} &= \Omega_{33} = 2\pi \int\int_{-\pi}^{\pi} f_{a_4a_5a_4'a_5'}(\lambda_1, \lambda_2, -\lambda_2)d\lambda_1 d\lambda_2\\
&\quad + 2\pi \int\int_{-\pi}^{\pi} \left\{ f_{a_4a_4'}(\lambda) f_{a_5a_5'}(-\lambda) + f_{a_4a_5'}(\lambda) f_{a_5a_4'}(-\lambda) \right\} d\lambda,\\
V\{(a_4a_5)(a_6'a_7'a_8')\} &= \Omega_{34} = 2\pi \int\int\int_{-\pi}^{\pi} f_{a_4a_5a_6'a_7'a_8'}(\lambda_1, \lambda_2, \lambda_3, -\lambda_3)d\lambda_1 d\lambda_2 d\lambda_3 (= \Omega_{43}),\\
V\{(a_6a_7a_8)(a_6'a_7'a_8')\} &= \Omega_{44} = 2\pi \int\int\int\int_{-\pi}^{\pi} f_{a_6a_7a_8a_6'a_7'a_8'}(\lambda_1, \lambda_2, \lambda_3, -\lambda_3 - \lambda_4)d\lambda_1 \cdots d\lambda_4\\
&\quad + 2\pi \int\int\int_{-\pi}^{\pi} \sum_{\nu_1} f_{a_{i_1}a_{i_2}a_{i_3}a_{i_4}}(\lambda_1, \lambda_2, \lambda_3)\\
&\qquad \times f_{a_{i_5}a_{i_6}}(-\lambda_2 - \lambda_3)d\lambda_1 d\lambda_2 d\lambda_3\\
&\quad + 2\pi \int\int\int_{-\pi}^{\pi} \sum_{\nu_2} f_{a_{i_1}a_{i_2}a_{i_3}}(\lambda_1, \lambda_2)\\
&\qquad \times f_{a_{i_4}a_{i_5}a_{i_6}}(\lambda_3, -\lambda_2 - \lambda_3)d\lambda_1 d\lambda_2 d\lambda_3\\
&\quad + 2\pi \int\int\int_{-\pi}^{\pi} \sum_{\nu_3} f_{a_{i_1}a_{i_2}}(\lambda_1)\\
&\qquad \times f_{a_{i_3}a_{i_4}}(\lambda_2) f_{a_{i_5}a_{i_6}}(-\lambda_1 - \lambda_2)d\lambda_1 d\lambda_2.
\end{aligned}
$$

Next we investigate an influence of the exogenous variables $\boldsymbol{Z}(t)$ on the asymptotics of the portfolio estimator $g(\hat{\boldsymbol{\theta}})$. Assume that the exogenous variables have "shot noise" in the frequency domain, i.e.,

$$
Z_{a_j}(\lambda) = \delta(\lambda_{a_j} - \lambda), \tag{6.3.13}
$$

where $\delta(\cdot)$ is the Dirac delta function with period 2π, and $\lambda_{a_j} \neq 0$, hence $Z_{a_j}(\lambda)$ has one peak at $\lambda + \lambda_{a_j} \equiv 0 \pmod{2\pi}$.

Proposition 6.3.3. *For (6.3.13), denote Ω_{ij} and $V\{(\Delta)(\nabla)\}$ in Proposition 6.3.2 by Ω'_{ij} and $V'\{(\Delta)(\nabla)\}$, respectively. That is, Ω'_{ij} and $V'\{(\Delta)(\nabla)\}$ represent the asymptotic variances when the exogenous variables are shot noise. Then,*

$$\begin{aligned}
V'\{(a_4a_5)(a'_1)\} &= \Omega'_{13} = 0 (= \Omega'_{31}),\\
V'\{(a_6a_7a_8)(a'_1)\} &= \Omega'_{14} = 2\pi f_{a_6a_7a'_1}(\lambda_{a_8}, 0)(= \Omega'_{41}),\\
V'\{(a_2a_3)(a'_4a'_5)\} &= \Omega'_{23} = 2\pi f_{a_2a_3a'_4}(\lambda_{a'_5}, 0) + 2\pi f_{a_2a'_4}(\lambda_{a_5}) f_{a_3a'_5}(-\lambda_{a'_5})\\
&\qquad + 2\pi f_{a_2a'_5}(-\lambda_{a_5}) f_{a_3a'_4}(\lambda_{a_5})(= \Omega'_{32}),\\
V'\{(a_2a_3)(a'_6a'_7a'_8)\} &= \Omega'_{24} = 2\pi \int\int_{-\pi}^{\pi} f_{a_2a_3a'_6a'_7a'_8}(-\lambda_{a'_8}, \lambda_1, \lambda_2, -\lambda_2) d\lambda_1 d\lambda_2 (= \Omega'_{42}),\\
V'\{(a_4a_5)(a'_4a'_5)\} &= \Omega'_{33} = 2\pi f_{a_4a_5a'_4a'_5}(\lambda_{a'_5}, \lambda_{a_5}, -\lambda_{a_5}) + 2\pi f_{a_4a'_4}(\lambda_{a_5}) f_{a_5a'_5}(-\lambda_{a_5})\\
&\qquad + 2\pi f_{a_4a'_5}(\lambda_{a_5}) f_{a_5a'_4}(-\lambda_{a_5}),\\
V'\{(a_4a_5)(a'_6a'_7a'_8)\} &= \Omega'_{34} = 2\pi \int_{-\pi}^{\pi} f_{a_4a_5a'_6a'_7a'_8}(-\lambda_{a'_8}, \lambda_{a_5}, \lambda, -\lambda) d\lambda (= \Omega'_{43}).
\end{aligned} \tag{6.3.14}$$

In (6.3.14), Ω'_{33} shows an influence of exogenous variables. It is easily seen that if $\{\boldsymbol{X}(t)\}$ has near unit roots, and if λ_{a_j} is near zero, then the magnitude of Ω'_{33} becomes large, hence, in such cases, the asymptotics of $g(\hat{\boldsymbol{\theta}})$ become sensitive.

In what follows we consider the case when the sequence of exogenous variables $\{\boldsymbol{Z}(t) = (Z_1(t), \ldots, Z_q(t))'\}$ is nonrandom, and satisfies Grenander's conditions (G1)–(G4) with

$$a^{(n)}_{j,k}(h) = \sum_{t=1}^{n-h} Z_j(t+h) Z_k(t), \tag{6.3.15}$$

(G1) $\lim_{n\to\infty} a^{(n)}_{j,j}(0) = \infty$, $(j = 1, 2, \ldots, q)$,

(G2) $\lim_{n\to\infty} Z_j(n+1)^2 / a^{(n)}_{j,j}(0) = 0$, $(j = 1, 2, \ldots, q)$,

(G3) $a^{(n)}_{j,k}(h) / \sqrt{a^{(n)}_{j,j}(0) a^{(n)}_{k,k}(0)} = \rho_{j,k}(h) + o(n^{-1/2})$, for $j, k = 1, \ldots, q$, $h \in \mathbb{Z}$,

(G4) the matrix $\Phi(0) = \{\rho_{j,k}(0) : j, k = 1, \ldots, q\}$ is regular.

Under Grenander's conditions, there exists a Hermitian matrix function $\boldsymbol{M}(\lambda) = \{M_{j,k}(\lambda) : j, k = 1, \ldots, q\}$ with positive semidefinite increments such that

$$\Phi(h) = \int_{-\pi}^{\pi} e^{ih\lambda} d\boldsymbol{M}(\lambda). \tag{6.3.16}$$

Here $\boldsymbol{M}(\lambda)$ is called the regression spectral measure of $\{\boldsymbol{Z}(t)\}$.

We discuss the asymptotics for sample versions of $\text{cov}(\boldsymbol{X}, \boldsymbol{Z})$ and $\text{cum}(\boldsymbol{X}, \boldsymbol{X}, \boldsymbol{Z})$ which are defined by

$$\begin{aligned}
\hat{A}_{j,k} &= \frac{\sum_{t=1}^{n} X_j(t) Z_k(t)}{\sqrt{n \sum_{t=1}^{n} Z_k^2(t)}},\\
\hat{B}_{j,m,k} &= \frac{\sum_{t=1}^{n} X_j(t) X_m(t) Z_k(t)}{\sqrt{n \sum_{t=1}^{n} Z_k^2(t)}},
\end{aligned} \tag{6.3.17}$$

respectively. For this we need the following assumption.

Assumption 6.3.4. *There exists a constant $b > 0$ such that*

$$det\{f(\lambda)\} \geq b, \quad for\ all\ \lambda \in [-\pi, \pi], \tag{6.3.18}$$

where $f(\lambda)$ is the spectral density matrix of $\{\boldsymbol{X}(t)\}$.

Proposition 6.3.5. *Suppose that Assumptions 6.3.1 and 6.3.4 and Grenander's conditions are satisfied. Then the following hold.*

(i)

$$\{\sqrt{n}(\hat{A}_{j,k} - A_{j,k})\} \xrightarrow{d} N(\mathbf{0}, \Omega_A), \tag{6.3.19}$$

where the covariance between (j,k)*th and* (j',k')*th elements in* Ω_A *is given by* $2\pi \int_{-\pi}^{\pi} f_{jj'}(\lambda) dM_{k,k'}(\lambda)$.

(ii)

$$\{\sqrt{n}(\hat{B}_{j,m,k} - B_{j,m,k})\} \xrightarrow{d} N(\mathbf{0}, \Omega_B), \tag{6.3.20}$$

where $\Omega_B = \{V(j,m,k : j',m',k')\}$ *with*

$$V(j,m,k : j',m',k') = 2\pi \int_{-\pi}^{\pi} \Bigg[\int_{-\pi}^{\pi} \{f_{jm'}(\lambda - \lambda_1) f_{mj'}(\lambda_1) + f_{jj'}(\lambda - \lambda_1) f_{mm'}(\lambda_1)\} d\lambda_1$$
$$+ \int\int_{-\pi}^{\pi} f_{jmj'm'}(\lambda_1, \lambda_2 - \lambda, \lambda_2) d\lambda_1 d\lambda_2 \Bigg] dM_{k,k'}(\lambda).$$

The results above show an influence of $\{\boldsymbol{Z}(t)\}$ on the return process $\{\boldsymbol{X}(t)\}$, which reflects that for the asymptotics of $g(\hat{\boldsymbol{\theta}})$. As a simple example, let $\{\boldsymbol{X}(t)\}$ and $\{\boldsymbol{Z}(t)\}$ be generated by

$$\begin{cases} X(t) - aX(t-1) = u(t), \\ Z(t) = \cos(bt) + \cos(0.25bt), \end{cases} \tag{6.3.21}$$

where $u(t)$'s are i.i.d. N(0, 1) variables. Figure 6.2 plots $V = V(j,m,k : j',m',k')$ in Proposition 6.3.5 with respect to a and b. The figure shows that if $b \approx 0$ and $|a| \approx 1$, then V becomes large,

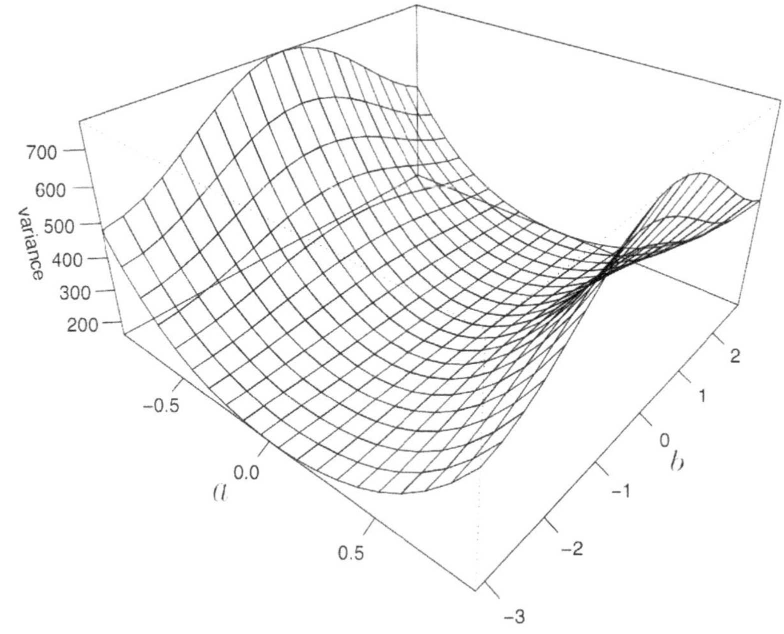

Figure 6.2: Values of V for $a = -0.9(0.1)0.9$, $b = -3.14(0.33)3.14$

which implies that if the return process contains near unit roots, and if the exogenous variables are constant, then the asymptotics of $g(\hat{\boldsymbol{\theta}})$ will be sensitive for such a situation.

6.4 Multi-Step Ahead Portfolio Estimation

This section addresses the problem of estimation for multi-step ahead portfolio predictors in the case when the asset return process is a vector-valued non-Gaussian process. First, to grasp the idea simply, we mention the multi-step prediction for scalar-valued stationary processes.

Let $\{X_t : t \in \mathbb{Z}\}$ be a stationary process with mean zero and spectral density $g(\lambda)$. Write the spectral representation as

$$X_t = \int_{-\pi}^{\pi} e^{-it\lambda} dZ(\lambda), \qquad \mathrm{E}\{|dZ(\lambda)|^2\} = g(\lambda)d\lambda,$$

and assume

$$\int_{-\pi}^{\pi} \log g(\lambda) d\lambda > -\infty.$$

Let $A_g(z) = \sum_{j=0}^{\infty} a_j^{(g)} z^j$, $z \in \mathbb{C}$, be the response function satisfying

$$g(\lambda) = \frac{1}{2\pi} |A_g(e^{i\lambda})|^2.$$

We are now interested in linear prediction of X_t based on $X_{t-1}, X_{t-2}, \ldots$, and seek the linear predictor $\hat{X}_t = \sum_{j\geq 1} b_j X_{t-j}$ which minimizes $\mathrm{E}|X_t - \hat{X}_t|^2$. The best linear predictor is given by

$$\hat{X}_t^{\text{best}} = \int_{-\pi}^{\pi} e^{-it\lambda} \frac{A_g(e^{i\lambda}) - A_g(0)}{A_g(e^{i\lambda})} dZ(\lambda) \tag{6.4.1}$$

(e.g., Chapter 8 of Grenander and Rosenblatt (1957)). Although $\hat{X}_t^{\text{best}}$ is the most plausible predictor, in practice, model selection and estimation for $g(\lambda)$ are needed. Hence the true spectral density $g(\lambda)$ is likely to be misspecified. This leads us to a misspecified predicition problem when a conjectured spectral density

$$f(\lambda) = \frac{1}{2\pi} \left| A_f(e^{i\lambda}) \right|^2$$

is fitted to $g(\lambda)$. From (6.4.1) the best linear predictor computed on the basis of the conjectured spectral density $f(\lambda)$ is

$$\hat{X}_t^f = \int_{-\pi}^{\pi} e^{-it\lambda} \frac{A_f(e^{i\lambda}) - A_f(0)}{A_f(e^{i\lambda})} dZ(\lambda). \tag{6.4.2}$$

Grenander and Rosenblatt (1957) evaluated the prediction error as

$$\mathrm{E}\left| \{X_t - \hat{X}_t^f\}^2 \right| = \frac{|A_f(0)|^2}{2\pi} \int_{-\pi}^{\pi} \frac{g(\lambda)}{f(\lambda)} d\lambda. \tag{6.4.3}$$

The setting is very wide, and is applicable to the problem of h-step ahead prediction, i.e., prediction of X_{t+h} based on a linear combination of $X_t, X_{t-1}, \ldots$. This can be grasped by fitting the misspecified spectral density

$$f_{\theta}(\lambda) = \frac{1}{2\pi} |1 - \theta_1 e^{ih\lambda} - \theta_2 e^{i(h+1)\lambda} - \cdots|^{-2}, \tag{6.4.4}$$
$$\boldsymbol{\theta} = (\theta_1, \theta_2, \ldots)',$$

to $g(\lambda)$. The best h-step ahead linear predictor of X_{t+h} is given in the form

$$\underline{\theta}_1 X_t + \underline{\theta}_2 X_{t-1} + \underline{\theta}_3 X_{t-2} + \cdots, \tag{6.4.5}$$

where $\underline{\theta} = (\underline{\theta}_1, \underline{\theta}_2, \ldots)$ is determined by the value which minimizes the misspecified prediction error (6.4.3):

$$\frac{1}{2\pi} \int_{-\pi}^{\pi} \frac{g(\lambda)}{f_{\theta}(\lambda)} d\lambda, \tag{6.4.6}$$

with respect to $\boldsymbol{\theta}$. Since $\underline{\boldsymbol{\theta}}$ is actually unknown, we estimate it by a Whittle estimator. In what follows, we describe the details in vector form. Let $\{\boldsymbol{X}(t) = (X_1(t), \ldots, X_m(t))' : t \in \mathbb{Z}\}$ be a zero mean stationary process with $m \times m$ spectral density matrix

$$\boldsymbol{g}(\lambda) = \{g_{jk}(\lambda) : j, k = 1, 2, \ldots, m\} = \frac{1}{2\pi} \boldsymbol{A}_g(e^{i\lambda}) \boldsymbol{K} \boldsymbol{A}_g(e^{i\lambda})^*,$$

where $\boldsymbol{A}_g(e^{i\lambda}) = \sum_{j=0}^{\infty} \boldsymbol{A}(j) e^{ij\lambda}$, and $\boldsymbol{K}$ is a nonsingular $m \times m$ matrix. Let the spectral representation of $\{\boldsymbol{X}(t)\}$ be

$$\boldsymbol{X}(t) = \int_{-\pi}^{\pi} e^{-it\lambda} d\boldsymbol{Z}(\lambda),$$

where $\{\boldsymbol{Z}(\lambda) = (Z_1(\lambda), \ldots, Z_m(\lambda))' : -\pi \leq \lambda \leq \pi\}$ is an orthogonal increment process satisfying

$$\mathrm{E}[dZ_j(\lambda)\overline{dZ_k(\mu)}] = \begin{cases} g_{jk}(\lambda) d\lambda & (\text{if } \lambda = \mu) \\ 0 & (\text{if } \lambda \neq \mu). \end{cases}$$

Hannan (1970) (Theorem 1 in Chapter III) gave the best linear predictor of $\boldsymbol{X}(t+1)$ based on $\boldsymbol{X}(t), \boldsymbol{X}(t-1), \ldots$ by

$$\hat{\boldsymbol{X}}(t)^{\text{best}} = \int_{-\pi}^{\pi} e^{it\lambda} \{\boldsymbol{A}_g(e^{i\lambda}) - \boldsymbol{A}_g(0)\} \boldsymbol{A}_g(e^{-i\lambda})^{-1} d\boldsymbol{Z}(\lambda). \tag{6.4.7}$$

It is well known that the prediction error is

$$\begin{aligned} &\mathrm{tr}\, \mathrm{E}\left[\{\boldsymbol{X}(t+1) - \hat{\boldsymbol{X}}(t)^{\text{best}}\} \{\boldsymbol{X}(t+1) - \hat{\boldsymbol{X}}(t)^{\text{best}}\}' \right] \\ &\qquad = \exp\left[\frac{1}{2\pi m} \int_{-\pi}^{\pi} \mathrm{tr}\{\log \det(2\pi \boldsymbol{g}(\lambda))\} d\lambda \right]. \end{aligned} \tag{6.4.8}$$

As in the scalar case, we consider the problem of misspecified prediction based on a conjectured spectral density matrix

$$\boldsymbol{f}(\lambda) = \frac{1}{2\pi} \boldsymbol{A}_{\boldsymbol{f}}(e^{i\lambda}) \boldsymbol{A}_{f}(e^{i\lambda})^*,$$

where $\boldsymbol{A}_{\boldsymbol{f}}(z) = \sum_{j=0}^{\infty} \boldsymbol{a}_j^{(\boldsymbol{f})} z^j$, $(z \in \mathbb{C})$. Then the pseudo best linear predictor computed on the basis of $\boldsymbol{f}(\lambda)$ is given by

$$\hat{\boldsymbol{X}}(t)^{p\text{-best}(\boldsymbol{f})} = \int_{-\pi}^{\pi} e^{-it\lambda} \{\boldsymbol{A}_{\boldsymbol{f}}(e^{i\lambda}) - \boldsymbol{A}_{\boldsymbol{f}}(0)\} \boldsymbol{A}_{\boldsymbol{f}}(e^{i\lambda})^{-1} d\boldsymbol{Z}(\lambda),$$

where $\boldsymbol{A}_{\boldsymbol{f}}(0) = \boldsymbol{I}_m$ ($m \times m$ identity matrix). As (6.4.3) the prediction error is

$$\mathrm{tr}\, \mathrm{E}\left[\left\{ \boldsymbol{X}(t+1) - \hat{\boldsymbol{X}}(t)^{p\text{-best}(\boldsymbol{f})} \right\} \left\{ \boldsymbol{X}(t+1) - \hat{\boldsymbol{X}}(t)^{p\text{-best}(\boldsymbol{f})} \right\}' \right] \propto \int_{-\pi}^{\pi} \mathrm{tr}\{\boldsymbol{f}^{-1}(\lambda) \boldsymbol{g}(\lambda)\} d\lambda. \tag{6.4.9}$$

We can apply this to the h-step ahead prediction. Consider the problem of prediction for $\boldsymbol{X}_{t+h}$ by linear combination

$$\hat{\boldsymbol{X}}(t+h) = \sum_{j=1}^{L} \boldsymbol{\Phi}(j) \boldsymbol{X}(t-j+1), \tag{6.4.10}$$

where $\mathbf{\Phi}(j)$'s are $m \times m$ matrices. This problem can be understood if we fit

$$f_{\theta}(\lambda) = \frac{1}{2\pi}\left\{\Gamma_{\theta}^{-1}(\lambda)\Gamma_{\theta}^{-1}(\lambda)^*\right\} \tag{6.4.11}$$

to $g(\lambda)$, where $\Gamma_{\theta}^{-1}(\lambda) = I_m - \sum_{j=1}^{L} \mathbf{\Phi}(j)e^{i(h+j-1)\lambda}$. Here

$$\theta = (\text{vec}(\mathbf{\Phi}(1))', \ldots, \text{vec}(\mathbf{\Phi}(L))')' \in \Theta \subset \mathbb{R}^r,$$

where $\gamma = \dim \theta = L \times m^2$. The best h-step ahead linear predictor is given by $\hat{X}(t)^{p\text{-best}(f_{\underline{\theta}})}$, where $\underline{\theta}$ is defined by

$$\underline{\theta} = \arg\min_{\theta} \int_{-\pi}^{\pi} \text{tr}\,\{f_{\theta}(\lambda)^{-1}g(\lambda)\}d\lambda.$$

Because $\underline{\theta}$ is unknown, we estimate it by a Whittle estimator. Suppose that an observed stretch $\{X(t-n+1), X(t-n+2), \ldots, X(t)\}$ is available. Let

$$I_n(\lambda) = \frac{1}{2\pi n}\{\sum_{j=1}^{n} X(t-n+j)e^{ij\lambda}\}\{\sum_{j=1}^{n} X(t-n+j)e^{ij\lambda}\}^*.$$

The Whittle estimator for $\underline{\theta}$ is defined by

$$\hat{\theta}_W = \arg\min_{\theta\in\Theta} \int_{-\pi}^{\pi} \text{tr}\,\{f_{\theta}(\lambda)^{-1}I_n(\lambda)\}d\lambda.$$

To state the asymptotics of $\hat{\theta}_W$, we assume

Assumption 6.4.1. **(i)** *All the components of $g(\lambda)$ are square integrable.*

(ii) *$g(\lambda) \in Lip(\alpha)$, $\alpha > 1/2$.*

(iii) *The fourth-order cumulants of $\{X(t)\}$ exist.*

(iv) *$\underline{\theta}$ uniquely, and $\underline{\theta} \in Int\ \Theta$.*

(v) *The $r \times r$ matrix*

$$\Sigma_{\underline{\theta}} = \left(\int_{-\pi}^{\pi}\left[\frac{\partial^2}{\partial\theta_i\partial\theta_j} tr\{f_{\theta}(\lambda)^{-1}g(\lambda)\}\right]_{\theta=\underline{\theta}} d\lambda;\ i,j = 1,\ldots,r\right)$$

is nonsingular.

As an estimator of $\hat{X}(t+h)^{p\text{-best}(f_{\underline{\theta}})}$, we can use

$$\hat{X}(t+h)^{p\text{-best}(f_{\hat{\theta}_W})}. \tag{6.4.12}$$

Next we are interested in the construction of a portfolio on $\{X(t)\}$. For a utility function we can define the optimal portfolio weight $\alpha_{\text{opt}} = (\alpha_1, \ldots, \alpha_m)'$ and its estimator $\hat{\alpha}_{\text{opt}}$ based on $X(t), \ldots, X(t-n+1)$ (recall Section 3.2). The asymptotics of $\hat{\theta}_W$ and $\hat{\alpha}_{\text{opt}}$ are established in Theorems 8.2.2 and 3.2.5, respectively. Then we can estimate the h-step ahead portolio value $\alpha_{\text{opt}}'X(t+h)$ by

$$\hat{\alpha}_{\text{opt}}'\hat{X}(t+h)^{p\text{-best}(f_{\hat{\theta}_W})}. \tag{6.4.13}$$

Letting PE $\equiv \alpha_{\text{opt}}'\hat{X}(t+h)^{p\text{-best}(f_{\underline{\theta}})} - \alpha_{\text{opt}}'\hat{X}(t+h)$, we can evaluate the accuracy $\hat{\alpha}_{\text{opt}}'\hat{X}(t+h)^{p\text{-best}(f_{\hat{\theta}_W})} - \alpha_{\text{opt}}'X(t+h)$ as follows.

Proposition 6.4.2. *(Hamada and Taniguchi (2014)) Under Assumption 6.4.1, it holds that*

(i) $\hat{\alpha}_{opt}'\hat{X}(t+h)^{p\text{-}best(f_{\hat{\theta}_W})} - \alpha_{opt}'X(t+h) = PE + o_p(1)$,

(ii) $E\{PE^2\} = \alpha_{opt}' \int_{-\pi}^{\pi} \Gamma_{\underline{\theta}}(\lambda)g(\lambda)\Gamma_{\underline{\theta}}^*(\lambda)d\lambda\, \alpha_{opt}$.

If one wants to evaluate a long-run behaviour of a pension portfolio, this result will be useful.

6.5 Causality Analysis

Granger (1969) introduced a useful way to describe the ralationship between two variables when one is causing the other. The concept is called Granger causality (for short G-causality), which has been populatized and developed by a variety of works, e.g., Sims (1972), Geweke (1982), Hosoya (1991), Lütkepohl (2007), Jeong et al. (2012), etc.The applications are swelling, e.g., Sato et al. (2010) in fMRI data analysis, Shojaie and Michailidis (2010) in game analysis, etc.

Let $\boldsymbol{X} = \{\boldsymbol{X}(t) = (X_1(t), \ldots, X_p(t))' : t \in \boldsymbol{Z}\}$ and $\boldsymbol{Y} = \{\boldsymbol{Y}(t) = (Y_1(t), \ldots, Y_q(t))' : t \in \boldsymbol{Z}\}$ be two time series which constitute a $(p+q)$-dimensional stationary process $\{\boldsymbol{S}(t) = (\boldsymbol{X}(t)', \boldsymbol{Y}(t)')' : t \in \boldsymbol{Z}\}$ with mean zero and spectral density matrix

$$\boldsymbol{f}_s(\lambda) = \begin{pmatrix} \boldsymbol{f}_{XX}(\lambda) & \boldsymbol{f}_{XY}(\lambda) \\ \boldsymbol{f}_{YX}(\lambda) & \boldsymbol{f}_{YY}(\lambda) \end{pmatrix}.$$

Now we are interested in whether $\boldsymbol{Y}$ causes $\boldsymbol{X}$ or not. In what follows we denote by $\sigma\{\cdots\}$ the σ-field generated by $\{\cdots\}$. Let $\widehat{\boldsymbol{X}}(t|\sigma\{\cdots\})$ be the linear projection (predictor) of $\boldsymbol{X}(t)$ on $\sigma\{\cdots\}$.

Definition 6.5.1. *If*

$$\begin{aligned} &\mathrm{tr}E[\{\boldsymbol{X}(t) - \widehat{\boldsymbol{X}}(t|\sigma\{\boldsymbol{S}(t-1), \boldsymbol{S}(t-2), \ldots\})\}\{\boldsymbol{X}(t) - \widehat{\boldsymbol{X}}(t|\sigma\{\boldsymbol{S}(t-1), \boldsymbol{S}(t-2), \ldots\})\}'] \\ &\quad < \mathrm{tr}E[\{\boldsymbol{X}(t) - \widehat{\boldsymbol{X}}(t|\sigma\{\boldsymbol{X}(t-1), \boldsymbol{X}(t-2), \ldots\})\}\{\boldsymbol{X}(t) - \widehat{\boldsymbol{X}}(t|\sigma\{\boldsymbol{X}(t-1), \boldsymbol{X}(t-2), \ldots\})\}'], \end{aligned} \tag{6.5.1}$$

we say that $\boldsymbol{Y}$ Granger causes $\boldsymbol{X}$ denoted by $\boldsymbol{Y} \Rightarrow \boldsymbol{X}$. If the equality holds in (6.5.1), we say that $\boldsymbol{Y}$ does not Granger cause $\boldsymbol{X}$ denoted by $\boldsymbol{Y} \not\Rightarrow \boldsymbol{X}$.

Next we express (6.5.1) in terms of $\boldsymbol{f}_s(\lambda)$. For this we assume that $\{\boldsymbol{S}(t)\}$ is generated by

$$\boldsymbol{S}(t) = \sum_{j=0}^{\infty} \boldsymbol{A}(j)\boldsymbol{U}(t-s), \tag{6.5.2}$$

where $\boldsymbol{A}(0) = \boldsymbol{I}_{p+q}$ and $\boldsymbol{U}(t) \equiv (\boldsymbol{\epsilon}(t)', \boldsymbol{\eta}(t)')'$s are i.i.d. $(\boldsymbol{0}, \boldsymbol{G})$ random vectors. Here $\boldsymbol{\epsilon}(t) = (\epsilon_1(t), \ldots, \epsilon_p(t))'$, $\boldsymbol{\eta}(t) = (\eta_1(t), \ldots, \eta_q(t))'$ and the covariance matrix $\boldsymbol{G}$ is of the form

$$\boldsymbol{G} = \begin{pmatrix} G_{11} & G_{12} \\ G_{21} & G_{22} \end{pmatrix}, \tag{6.5.3}$$

where $G_{11} : p \times p$, $G_{12} : p \times q$, $G_{21} : q \times p$, $G_{22} : q \times q$.

Assumption 6.5.2. *(i)* $\det\{\sum_{j=0}^{\infty} \boldsymbol{A}(j)z^j\} \neq 0$ *for all* $z \in \mathbb{C}$ *and* $|z| \leq 1$.

(ii) $\sum_{j=0}^{\infty} \|\boldsymbol{A}(j)\| < \infty$.

Now we evaluate the right hand side (RHS) of (6.5.1). We use $[\boldsymbol{B}(z)]_+$ to mean that we retain only the terms from the expansion for positive powers of z. $[\boldsymbol{B}(z)]_-$ is formed similarly, taking only nonpositive powers of z, so that $\boldsymbol{B}(z) = [\boldsymbol{B}(z)]_+ + [\boldsymbol{B}(z)]_-$. Let $\boldsymbol{h}_{11}(\lambda) = \sum_{j=1}^{\infty} \boldsymbol{H}_{11}(j)e^{-ij\lambda}$, where the $\boldsymbol{H}_{11}(j)$'s are $p \times p$ matrices. Then,

$$\begin{aligned} RHS\ of\ (6.5.1) &= (2\pi)^{-1}\mathrm{tr}\int_{-\pi}^{\pi} \{\boldsymbol{I}_p - \boldsymbol{h}_{11}(\lambda)\}\boldsymbol{A}_{11}(\lambda)G_{11}^{1/2}\left\{\{\boldsymbol{I}_p - \boldsymbol{h}_{11}(\lambda)\}\boldsymbol{A}_{11}(\lambda)G_{11}^{1/2}\right\}^* d\lambda \\ &= (2\pi)^{-1}\mathrm{tr}\int_{-\pi}^{\pi} \{[e^{i\lambda}\boldsymbol{A}_{11}(\lambda)]_+ + [e^{i\lambda}\boldsymbol{A}_{11}(\lambda)]_- - e^{i\lambda}\boldsymbol{h}_{11}(\lambda)\boldsymbol{A}_{11}(\lambda)\}G_{11}\{\quad " \quad\}^*. \end{aligned} \tag{6.5.4}$$

Because $[e^{i\lambda}\boldsymbol{A}_{11}(\lambda)]_+ \in \mathcal{H}(e^{i\lambda}, e^{2i\lambda}, \ldots)$, $[e^{i\lambda}\boldsymbol{A}_{11}(\lambda)]_- \in \mathcal{H}(1, e^{-i\lambda}, e^{-2i\lambda}, \ldots)$, and $e^{i\lambda}\boldsymbol{h}_{11}(\lambda)\boldsymbol{A}_{11}(\lambda) \in \mathcal{H}(1, e^{-i\lambda}, e^{-2i\lambda}, \ldots)$, where $\mathcal{H}(\ldots)$ is the Hilbert space generated by $(\ldots)$, it is seen that if

$$[e^{i\lambda}\boldsymbol{A}_{11}(\lambda)]_- - e^{i\lambda}\boldsymbol{h}_{11}(\lambda)\boldsymbol{A}_{11}(\lambda) = \boldsymbol{0}, \tag{6.5.5}$$

then (6.3.4) attains the minimum

$$\frac{1}{2\pi}\mathrm{tr}\int_{-\pi}^{\pi}[e^{i\lambda}\boldsymbol{A}_{11}(\lambda)]_+G_{11}[e^{i\lambda}\boldsymbol{A}_{11}(\lambda)]_+^* d\lambda = \mathrm{tr}G_{11}. \tag{6.5.6}$$

Next, the variance matrix of $\epsilon(t) - G_{12}G_{22}^{-1}\eta(t)$ is given by

$$G_{11} - G_{12}G_{22}^{-1}G_{21} \tag{6.5.7}$$

Then, $(RHS\ -\ LHS)$ of (6.5.1) is $\mathrm{tr}G_{12}G_{22}^{-1}G_{21}$. Hence, if

$$\mathrm{tr}G_{12}G_{22}^{-1}G_{21} = 0, \tag{6.5.8}$$

$\boldsymbol{Y} \not\Rightarrow \boldsymbol{X}$.

It is known that

$$\det\boldsymbol{G} = \exp\left[\frac{1}{2\pi}\int_{-\pi}^{\pi}\log\{\det 2\pi\boldsymbol{f}_s(\lambda)\}d\lambda\right] \tag{6.5.9}$$

(e.g., Hannan (1970, p.162)). We observe that

$$\begin{aligned}\det\{&\boldsymbol{I} - G_{11}^{-1/2}G_{12}G_{22}^{-1}G_{21}G_{11}^{-1/2}\}\\ &= \exp\left[\frac{1}{2\pi}\int_{-\pi}^{\pi}\log\{\det 2\pi\boldsymbol{f}_s(\lambda)\}d\lambda\right]\\ &\quad - \exp\left[\frac{1}{2\pi}\int_{-\pi}^{\pi}\log\{\det 2\pi\boldsymbol{f}_{XX}(\lambda)\det 2\pi\boldsymbol{f}_{YY}(\lambda)\}d\lambda\right].\end{aligned} \tag{6.5.10}$$

If (6.5.10) $= 0$, then $\boldsymbol{Y} \not\Rightarrow \boldsymbol{X}$. To estimate the LHS of (6.5.10), we substitute a nonparametric spectral estimator $\widehat{\boldsymbol{f}}_n(\lambda)$ to the RHS. We use

$$\widehat{\boldsymbol{f}}_n(\lambda) = \int_{-\pi}^{\pi}W_n(\lambda-\mu)\boldsymbol{I}_n(\mu)d\mu, \tag{6.5.11}$$

where $\boldsymbol{I}_n(\mu) = \frac{1}{2\pi n}\left\{\sum_{t=1}^{n}\boldsymbol{S}(t)e^{it\lambda}\right\}\left\{\sum_{t=1}^{n}\boldsymbol{S}(t)e^{it\lambda}\right\}^*$, and $W_n(\cdot)$ satisfies the following.

Assumption 6.5.3. *(i) $W(x)$ is bounded, even, nonnegative and such that*

$$\int_{-\infty}^{\infty}W(x)dx = 1.$$

(ii) For $M = O(n^\alpha)$, $(1/4 < \alpha < 1/2)$, the function $W_n(x) \equiv MW(M\lambda)$ can be expanded as

$$W_n(\lambda) = \frac{1}{2\pi}\sum_l w\left(\frac{l}{M}\right)\exp(-il\lambda),$$

where $w(x)$ is a continuous, even function with $w(0) = 1$, $|w(x)| \le 1$, and $\int_{-\infty}^{\infty}w(x)^2dx < \infty$, and satisfies

$$\lim_{x\to 0}\frac{1 - w(x)}{|x|^2} = \kappa_2 < \infty.$$

Under this assumption we can check the assumptions of Theorems 9 and 10 in Hannan (1970), whence

$$\widehat{\boldsymbol{f}}_n(\lambda) - \boldsymbol{f}(\lambda) = O_p\{\sqrt{M/n}\}, \tag{6.5.12}$$

uniformly in $\lambda \in [-\pi, \pi]$.

Let $m = p + q$, and let $K(\cdot)$ be a function defined on $\boldsymbol{f}(\lambda)$.

Assumption 6.5.4. *$K : D \Rightarrow \mathbb{R}$ is holomorphic, where D is an open subset of $\mathbb{C}^{m^2}$.*

Theorem 6.5.5. *Suppose that (HT1)–(HT6) in Section 8.1 hold. Under Assumptions 6.5.2–6.5.4, it holds that*

$$T_n \equiv \sqrt{n}\left[\int_{-\pi}^{\pi} K\{\widehat{\boldsymbol{f}}_n(\lambda)\}d\lambda - \int_{-\pi}^{\pi} K\{\boldsymbol{f}(\lambda)\}d\lambda\right] \tag{6.5.13}$$

is asymptotically normal with mean zero and variance $\nu_1(\boldsymbol{f}) + \nu_2\{Q^s\}$, where

$$\nu_1(\boldsymbol{f}) = 4\pi \int_{-\pi}^{\pi} \mathrm{tr}\left[\boldsymbol{f}(\lambda)K^{(1)}\{\boldsymbol{f}(\lambda)\}\right]^2 d\lambda, \tag{6.5.14}$$

and

$$\nu_2(Q^s) = 2\pi \sum_{r,t,u,s=1}^{m} \iint_{-\pi}^{\pi} K_{rt}^{(1)}(\lambda_1)K_{us}^{(1)}(\lambda_2)Q_{rtus}^s(-\lambda_1, \lambda_2, -\lambda_2)d\lambda_1 d\lambda_2.$$

Here $K^{(1)}\{\boldsymbol{f}(\lambda)\}$ is the first derivative of $K\{\boldsymbol{f}(\lambda)\}$ at $\boldsymbol{f}(\lambda)$ (see Magnus and Neudecker (1999)), and $K_{rt}^{(1)}(\lambda)$ is the (r,t)th conponent of $K^{(1)}(\lambda)$. Also $Q_{rtus}^s(\cdot)$ is the (r,t,u,s) conponent of the fourth-order cumulant spectral density (see Theorem 8.1.1.).

Proof. Expanding $K\{\widehat{\boldsymbol{f}}_n(\lambda)\}$ in Taylor series at $\boldsymbol{f}(\lambda)$, the proof follows from 6.5.12 and Theorem 8.1.1 (see also Taniguchi et al. (1996)). □

Recalling 6.5.10, we see that if

$$H: \int_{-\pi}^{\pi} \log\det\{\boldsymbol{I} - \boldsymbol{f}_{XX}^{-1/2}(\lambda)\boldsymbol{f}_{XY}(\lambda)\boldsymbol{f}_{YY}(\lambda)^{-1}\boldsymbol{f}_{YX}(\lambda)\boldsymbol{f}_{XX}^{-1/2}(\lambda)\}d\lambda = 0, \tag{6.5.15}$$

then, $\boldsymbol{Y} \not\Rightarrow \boldsymbol{X}$. Let

$$K_G\{\boldsymbol{f}_s(\lambda)\} = \log\det\{\boldsymbol{I} - \boldsymbol{f}_{XX}^{-1/2}(\lambda)\boldsymbol{f}_{XY}(\lambda)\boldsymbol{f}_{YY}(\lambda)^{-1}\boldsymbol{f}_{YX}(\lambda)\boldsymbol{f}_{XX}^{-1/2}(\lambda)\}d\lambda = 0.$$

In view of Theorem 6.5.5, we introduce

$$GT \equiv \sqrt{n}\left[\int_{-\pi}^{\pi} K_G\{\widehat{\boldsymbol{f}}_n(\lambda)\}d\lambda - \int_{-\pi}^{\pi} K_G\{\boldsymbol{f}_s(\lambda)\}d\lambda\right] \Big/ \sqrt{\nu_1(\boldsymbol{f}_s) + \nu_2\{Q^s\}}. \tag{6.5.16}$$

Under H, it is seen that

$$GT_0 \equiv \sqrt{n}\int_{-\pi}^{\pi} K_G\{\widehat{\boldsymbol{f}}_n(\lambda)\}d\lambda \Big/ \sqrt{\nu_1(\boldsymbol{f}_s) + \nu_2\{Q^s\}} \tag{6.5.17}$$

is asymptotically $N(0, 1)$. Because the denominator is unknown, we may propose

$$\widehat{GT} \equiv \sqrt{n}\int_{-\pi}^{\pi} K_G\{\widehat{\boldsymbol{f}}_n(\lambda)\}d\lambda \Big/ \sqrt{\nu_1(\widehat{\boldsymbol{f}}_n) + \widehat{\nu_2(Q^s)}}, \tag{6.5.18}$$

where $\widehat{v_2(Q^s)}$ is a consistent estimator of $v_2(Q^s)$, which is given by Taniguchi et al. (1996). If the process $\{S(t)\}$ is Gaussian, we observe that

$$\widehat{GT}' \equiv \sqrt{n} \int_{-\pi}^{\pi} K_G\{\widehat{f}_n(\lambda)\}d\lambda \Big/ \sqrt{v_1(\widehat{f}_n)} \tag{6.5.19}$$

is asymptotically $N(0, 1)$ under H. Hence, we propose the test of H given by the critical region

$$[|\widehat{GT}'| > t_\alpha], \tag{6.5.20}$$

where t_α is defined by

$$\int_{t_\alpha}^{\infty} (2\pi)^{-1/2} \exp\left(-\frac{x^2}{2}\right) dx = \frac{\alpha}{2}.$$

6.6 Classification by Quantile Regression

Discriminant analysis has been developed for various fields, e.g., multivariate data, regression and time series data, etc. Taniguchi and Kakizawa (2000) develop discriminant analysis for time series in view of the statistical asymptotic theory based on local asymptotic normality. Shumway (2003) introduces the use of time-varying spectra to develop a discriminant procedure for nonstationary time series. Taniguchi et al. (2008) provide much detail in discriminant analysis for financial time series. Hirukawa (2006) examines hierarchical clustering by a disparity between spectra for daily log-returns of 13 companies. The results show that the proposal method can classify the type of industry clearly.

In discriminant analysis, with observations in hand, we try to find which category a sample belongs to. Without loss of generality, we discuss the case of two categories Π_1 and Π_2. First, assuming that the two categories are mutually exclusive, we evaluate the misclassification probability $P(1|2)$ when a sample is misclassified into Π_1 with the fact that it belongs to Π_2, and vice versa for $P(2|1)$. Then we discuss the goodness by minimizing the misclassification probability:

$$P(1|2) + P(2|1). \tag{6.6.1}$$

Quantile regression is a methodology for estimating models of conditional quantile functions, and examines how covariates influence the entire response distribution. In this field, there is much theoretical literature. Koenker and Bassett (1978) propose quantile regression methods for non-random regressors. Koenker and Xiao (2006) extend the result to the case of linear quantile autoregression models. Koenker (2005) provides a comprehensive content of his research for 25 years on quantiles and great contributions to statistics and econometrics.

In this section we discuss discriminant analysis for quantile regression models. First, we introduce a classification statistic for two categories, and propose the classification rule. Second, we evaluate the misclassification probabilities $P(i|j), i \neq j$, which will be shown to converge to zero. Finally, we examine a delicate goodness of the statistic by evaluating $P(i|j)$ when the categories are contiguous.

Let $\{Y_t\}$ be generated by

$$Y_t = \beta_0 + \beta_1 Y_{t-1} + \cdots + \beta_p Y_{t-p} + u_t, \tag{6.6.2}$$

where the u_t's are i.i.d. random variables with mean 0, variance σ^2, and probability distribution function $F(\cdot)$, and $f(u) = dF(u)/du$. Here we assume that $B(z) \equiv 1 - \sum_{j=1}^{p} \beta_j z^j \neq 0$ on $D = \{z \in \mathbb{C} : |z| \leq 1\}$. Letting $\rho_\tau(u) = u(\tau - \chi(u < 0))$, where $\chi(\cdot)$ is the indicator function, we define

$$\beta(\tau) = \big(\beta_0(\tau), \beta_1(\tau), \ldots, \beta_p(\tau)\big)'$$

by

$$\beta(\tau) \equiv \arg\min_{\beta} E\{\rho_\tau(Y_t - \beta' \mathbf{Y}_{t-1})\}, \tag{6.6.3}$$

where $\mathbf{Y}_{t-1} \equiv (1, Y_{t-1}, \ldots, Y_{t-p})'$.

Suppose that $Y_1, \ldots, Y_n$ belong to one of two categories Π_1 or Π_2 as follows:

$$\begin{aligned} \Pi_1 &: \beta(\tau) = \beta_1(\tau) \\ \Pi_2 &: \beta(\tau) = \beta_2(\tau)\ (\not\equiv \beta_1(\tau)). \end{aligned} \tag{6.6.4}$$

We introduce the following as a classification statistic:

$$D \equiv \sum_{t=p+1}^{n} \{\rho_\tau (Y_t - \beta_2(\tau)' \mathbf{Y}_{t-1}) - \rho_\tau (Y_t - \beta_1(\tau)' \mathbf{Y}_{t-1})\}. \tag{6.6.5}$$

We classify $(Y_1, \ldots, Y_n)$ into Π_1 if $D > 0$, and into Π_2 if $D \le 0$.

Theorem 6.6.1. *The classification rule D is consistent, i.e.,*

$$\lim_{n\to\infty} P(1|2) = \lim_{n\to\infty} P(2|1) = 0. \tag{6.6.6}$$

Proof. We use Knight's indentity (Knight (1998)):

$$\rho_\tau(u - v) - \rho_\tau(u) = -v\psi_\tau(u) + \int_0^v \{I(u \le s) - I(u < 0)\}\, ds, \tag{6.6.7}$$

where $\psi_\tau(u) = \tau - \chi(u < 0)$. Under Π_1, write $u_{t\tau} \equiv Y_t - \beta_1(t)' \mathbf{Y}_{t-1}$, and $\boldsymbol{d}(t) \equiv \beta_2(\tau) - \beta_1(\tau)$. Then

$$\begin{aligned} D &= \sum_{t=p+1}^{n} \{\rho_\tau (u_{tt} - \boldsymbol{d}(t)' \mathbf{Y}_{t-1}) - \rho_\tau(u_{t\tau})\}, && (6.6.8) \\ &= -\sum_{t=p+1}^{n} \boldsymbol{d}(t)' \mathbf{Y}_{t-1} \phi_\tau(u_{t\tau}) && (6.6.9) \\ &\quad + \sum_{t=p+1}^{n} \int_0^{\boldsymbol{d}(t)' \mathbf{Y}_{t-1}} \{I(u_{t\tau} \le s) - I(u_{t\tau} < 0)\}\, ds. && (6.6.10) \end{aligned}$$

Since $\{\boldsymbol{d}(t)' \mathbf{Y}_{t-1} \phi_\tau(u_{t\tau})\}$ is a martingale difference sequence, it is seen that (6.6.9) $= O_p(\sqrt{n})$. Let $\mathscr{F}_t$ be the σ-field generated by $Y_t, Y_{t-1}, \ldots$. Then we observe that

$$\begin{aligned} A_t &\equiv E[\int_0^{\boldsymbol{d}(t)' \mathbf{Y}_{t-1}} \{I(u_{t\tau} \le s) - I(u_{t\tau} < 0)\}\, ds | \mathscr{F}_{t-1}] \\ &= \int_0^{\boldsymbol{d}(t)' \mathbf{Y}_{t-1}} \{F(s) - F(0)\}\, ds\ > 0\ \ a.e., \end{aligned} \tag{6.6.11}$$

and

$$E|A_t|^2 \le E|\boldsymbol{d}(t)' \mathbf{Y}_{t-1}|^2 < \infty. \tag{6.6.12}$$

Now (6.6.10) is written as

$$\sum_{t=p+1}^{n} [\int_0^{\boldsymbol{d}(t)' \mathbf{Y}_{t-1}} \{I(u_{t\tau} \le s) - I(u_{t\tau} < 0)\}\, ds - A_t] \tag{6.6.13}$$

$$+\sum_{t=p+1}^{n} A_t. \tag{6.6.14}$$

The term (6.6.13) becomes a martingale, whose order is of $O_p(\sqrt{n})$, and (6.6.14) $= O_p(\sqrt{n}) > 0$ a.e.

Under Π_1,

$$\begin{aligned} P(2|1) &= P(D \leq 0) = P(6.6.9 \leq -6.6.10) \\ &= P\left(o_p(\sqrt{n}) \leq -O_p(n)\right) \longrightarrow 0 \ as \ n \to \infty, \end{aligned} \tag{6.6.15}$$

which proves the theorem. □

Next we evaluate a delicate goodness of D. For this we suppose that Π_1 and Π_2 are contiguous, i.e.,

$$\beta_2(\tau) - \rho_1(\tau) = \frac{1}{\sqrt{n}}\delta. \tag{6.6.16}$$

Lemma 6.6.2. *(Koenker and Xiao (2006)) Under* Π_1,

$$D \xrightarrow{d} -\delta' \boldsymbol{W}(\tau) + \frac{1}{2}\delta' \boldsymbol{\Psi}(\tau)\delta, \tag{6.6.17}$$

where $\boldsymbol{W}(\tau) \sim N_p(\boldsymbol{0},\ \tau(1-\tau)\Omega_0)$, *with* $\Omega_0 = E[\boldsymbol{Y}_{t-1}\boldsymbol{Y}'_{t-1}]$, *and*

$$\boldsymbol{\Psi}(\tau) = f\left[F^{-1}(\tau)\right]\Omega_0.$$

From this we obtain

Theorem 6.6.3. *Under (6.1.16),*

$$\lim_{n\to\infty} P(2|1) = \lim_{n\to\infty} P(1|2) = F_{\delta,\tau}[-\frac{1}{2}\delta' \boldsymbol{\Psi}(\tau)\delta], \tag{6.6.18}$$

where $F_{\delta,\tau}(\cdot)$ *is the distribution function of* $N(\boldsymbol{0},\ \tau(1-\tau)\delta'\Omega_0\delta)$.

In practice we do not know $\beta_1(\tau)$ and $\beta_2(\tau)$. But if we can use the training samples $\{Y_t^{(1)};\ t = 1, \ldots, n_1\}$ and $\{Y_t^{(2)};\ t = 1, \ldots, n_2\}$ from Π_1 and Π_2, respectively, the estimators are given by

$$\widehat{\beta_i}(\tau) = \arg\min_{\beta_i(\tau)} \sum_{t=1}^{n_i} \rho_\tau(Y_t^{(i)} - \beta_i(\tau)'\boldsymbol{Y}_{t-1}^{(i)}), \quad i = 1, 2.$$

Then we introduce an estimated version $\widehat{D}$ of D by

$$\widehat{D} \equiv \sum_{t=p+1}^{n} \left\{\rho_\tau(Y_t - \widehat{\beta_2}(\tau)'Y_{t-1}) - \rho_\tau(Y_t - \widehat{\beta_1}(\tau)'Y_{t-1})\right\}. \tag{6.6.19}$$

Chen et al. (2016) applied the discriminant method above to the monthly mean maximum temperature at Melbourne for 1994–2015. The results illuminate interesting features of weather change.

6.7 Portfolio Estimation under Causal Variables

This section discusses the problem of portfolio estimation under causal variables, and its application to the Japanese pension investment portfolio, whose main stream is mainly based on Kobayashi et al. (2013).

First, we mention canonical correlation analysis (CCA). Let

$\boldsymbol{X} = \{(X_1, \ldots, X_p)'\}$ and $\boldsymbol{Y} = \{(Y_1, \ldots, Y_p)'\}$ be random vectors with

$$E\left\{\begin{pmatrix}\boldsymbol{X}\\ \boldsymbol{Y}\end{pmatrix}\right\} = 0, \quad Cov\left\{\begin{pmatrix}\boldsymbol{X}\\ \boldsymbol{Y}\end{pmatrix}, \begin{pmatrix}\boldsymbol{X}\\ \boldsymbol{Y}\end{pmatrix}\right\} = \begin{pmatrix}\Sigma_{XX} & \Sigma_{XY}\\ \Sigma_{YX} & \Sigma_{YY}\end{pmatrix}, (p+q)\times(p+q) - matrix.$$

We want to find linear combinations $\boldsymbol{\alpha}'\boldsymbol{X}$ and $\boldsymbol{\beta}'\boldsymbol{Y}$ whose correlation is the largest, say $\xi_1 = \boldsymbol{\alpha}'\boldsymbol{X}, \eta_1 = \boldsymbol{\beta}'\boldsymbol{Y}$ and $\lambda_1 = \mathrm{Corr}(\xi_1, \eta_1)$. We call λ_1 the first canonical correlation, and ξ_1, η_1 the first canonical variates. Next we consider finding second linear combinations $\xi_2 = \boldsymbol{\alpha}_2\boldsymbol{X}$ and $\eta_2 = \boldsymbol{\beta}_2\boldsymbol{Y}$ such that of all combinations uncorrelated with ξ_1 and η_1, these have maximum correlation λ_2. This procedure is continued. Then we get $\xi_j = \boldsymbol{\alpha}_j'\boldsymbol{X}$ and $\eta_j = \boldsymbol{\beta}_j'\boldsymbol{Y}$ whose correlation is λ_j. We call λ_j the jth canonical correlation, and ξ_j, η_j the jth canonical variates. It is known that $\lambda_j's(\lambda_1 \geq \lambda_2 \geq \cdots)$ are the jth eigenvalues of

$$\Sigma_{XX}^{-1/2}\Sigma_{XY}\Sigma_{YY}^{-1}\Sigma_{YX}\Sigma_{XX}^{-1/2} \tag{6.7.1}$$

$$\Sigma_{YY}^{-1/2}\Sigma_{YX}\Sigma_{XX}^{-1}\Sigma_{XY}\Sigma_{YY}^{-1/2} \tag{6.7.2}$$

The vectors $\boldsymbol{\alpha}_j$ and $\boldsymbol{\beta}_j$ are the jth eigenvectors of (6.7.1) and (6.7.2), respectively, corresponding to the jth eigenvalue λ_j (e.g., Anderson (2003), Brillinger (2001a)).

In what follows we apply the CCA above to the problem of portfolios. Let $\boldsymbol{X} = \{\boldsymbol{X}(t) = (X_1(t), \ldots, X_p(t))'\}$ be an asset return process which consists of p asset returns. We write $\boldsymbol{\mu} = E\{\boldsymbol{X}(t)\}$, and $\Sigma_{XX} = E[(\boldsymbol{X}(t) - \boldsymbol{\mu}_X)(\boldsymbol{X}(t) - \boldsymbol{\mu}_X)']$. For a portfolio coefficient vector $\boldsymbol{\alpha} = (\alpha_1, \ldots, \alpha_p)'$, the expectation and variance of portfolio $\boldsymbol{\alpha}'\boldsymbol{X}(t)$ are, respectively, given by $\mu(\boldsymbol{\alpha}) = \boldsymbol{\mu}_X'\boldsymbol{\alpha}$ and $\eta^2 = \boldsymbol{\alpha}'\Sigma_{XX}\boldsymbol{\alpha}$. The classical mean-variance portfolio is defined by the optimization problem

$$\begin{cases}\max_{\boldsymbol{\alpha}}\{\mu(\boldsymbol{\alpha}) - \beta\eta(\boldsymbol{\alpha})\}\\ \text{subject to } \boldsymbol{e}'\boldsymbol{\alpha} = 1,\end{cases} \tag{6.7.3}$$

where $\boldsymbol{e} = (1, \ldots, 1)'$ (p × 1 − vector), and β is a given positive number. The solution for $\boldsymbol{\alpha}$ is given by

$$\boldsymbol{\alpha}_{MV} = \frac{1}{2\beta}\{\Sigma_{XX}^{-1}\boldsymbol{\mu}_X - \frac{\boldsymbol{e}'\Sigma_{XX}^{-1}\boldsymbol{\mu}_X}{\boldsymbol{e}'\Sigma_{XX}^{-1}\boldsymbol{e}}\Sigma_{XX}^{-1}\boldsymbol{e}\} + \frac{\Sigma_{XX}^{-1}\boldsymbol{e}}{\boldsymbol{e}'\Sigma_{XX}^{-1}\boldsymbol{e}}, \tag{6.7.4}$$

which is specified by the asset return $\boldsymbol{X}$.

However, in the real world, if we observe some other causal variables, we may use their information to construct the optimal portfolio. Let $\boldsymbol{Y} = \{\boldsymbol{Y}(t) = (Y_1(t), \ldots, Y_q(t))'\}$ be a q-causal return process with $\boldsymbol{\mu}_Y = E\{\boldsymbol{Y}(t)\}, \Sigma_{YY} = E[(\boldsymbol{Y}(t) - \boldsymbol{\mu}_Y)(\boldsymbol{Y}(t) - \boldsymbol{\mu}_Y)']$. If we want to construct a portfolio $\boldsymbol{\alpha}'\boldsymbol{X}(t)$ including the best linear information from $\boldsymbol{Y}(t)$, we are led to find the CCA variate $\boldsymbol{\alpha}$ given by

$$\max_{\boldsymbol{\alpha}, \boldsymbol{\beta}} Corr\{\boldsymbol{\alpha}'\boldsymbol{X}(t), \boldsymbol{Y}(t)'\boldsymbol{\beta}\}. \tag{6.7.5}$$

This is exactly a CCA portfolio estimation by use of causal variables.

To develop the asymptotic theory we assume

Assumption 6.7.1. *The process $\{\boldsymbol{Z}(t) = (\boldsymbol{X}(t)', \boldsymbol{Y}(t)')'\}$ is generated by (p+q)-dimensional linear process*

$$Z(t) - \mu = \sum_{j=1}^{\infty} A(j)U(t-j), \tag{6.7.6}$$

where $A(j)$*'s satisfy* $\sum_{j=0}^{\infty} \|A(j)\|^2 < \infty$*, and* $E[U(t)] = 0$*,* $E[U(t)U(s)'] = \delta(t,s)K$*. Here* $\delta(t,s)$ *is the Kroneker delta function,* $\mu = (\mu_X', \mu_Y')'$ *and* K *is a positive definite matrix.*

Write

$$\Sigma = E[\{Z(t) - E(Z(t))\}\{Z(t) - E(Z(t))\}'] = \begin{pmatrix} \Sigma_{XX} & \Sigma_{XY} \\ \Sigma_{YX} & \Sigma_{YY} \end{pmatrix}.$$

Let λ_j's$(\lambda_1 \geq \lambda_2 \geq \cdots)$ be the jth eigenvalues of (6.7.1) and (6.7.2), and let α_j and β_j the jth eigenvectors of (6.7.1) and (6.7.2), respectively. Then we can see that the solutions of (6.7.5) are

$$\lambda = \lambda_1 \text{ and } \beta = \beta_1.$$

Suppose that the observed stretch $\{Z(1), \ldots, Z(n)\}$ from (6.7.6) is available. Then, the sample versions of Σ and $(\mu_X', \mu_Y',) = E(Z(t))$ are, respectively, given by

$$\begin{aligned} \hat{\Sigma} &= \frac{1}{n}\sum_{t=1}^{n}\{Z(t) - \bar{Z}\}\{Z(t) - \bar{Z}\}' \\ &= \begin{pmatrix} \hat{\Sigma}_{XX} & \hat{\Sigma}_{XY} \\ \hat{\Sigma}_{YX} & \hat{\Sigma}_{YY} \end{pmatrix}, \end{aligned} \tag{6.7.7}$$

$$\bar{Z} \equiv \frac{1}{n}\sum_{t=1}^{n} Z(t) = \begin{pmatrix} \hat{\mu}_X \\ \hat{\mu}_Y \end{pmatrix}. \tag{6.7.8}$$

Naturally we can introduce the following estimator for α_{MV} given by (6.7.4):

$$\hat{\alpha}_{MV} = \frac{1}{2\beta}\{\hat{\Sigma}_{XX}\hat{\mu} - \frac{e'\hat{\Sigma}_{XX}^{-1}\hat{\mu}_X}{e'\hat{\Sigma}_{XX}^{-1}e}\hat{\Sigma}_{XX}^{-1}e\} + \frac{\hat{\Sigma}_{XX}^{-1}e}{e'\hat{\Sigma}_{XX}^{-1}e}. \tag{6.7.9}$$

Also we estimate α_1 by solving

$$\hat{\Sigma}_{XX}^{-1/2}\hat{\Sigma}_{XY}\hat{\Sigma}_{YY}^{-1}\hat{\Sigma}_{YX}\hat{\Sigma}_{XX}^{-1/2}\hat{\alpha}_1 = \hat{\lambda}_1\hat{\alpha}_1. \tag{6.7.10}$$

If we use information from $Y(t)$, we may introduce the following estimator for the portfolio coefficient α:

$$\hat{\alpha} = (1-\varepsilon)\hat{\alpha}_{MV} + \varepsilon\hat{\alpha}_1, \tag{6.7.11}$$

where $\varepsilon > 0$ is chosen appropriately. Because the length of the portfolio is 1, the final form of the causal portfolio estimator is given by

$$\hat{\alpha}_\varepsilon = \frac{\hat{\alpha}}{\|\hat{\alpha}\|}. \tag{6.7.12}$$

Next we consider the asymptotic properties of $\hat{\alpha}_{MV}$ and $\hat{\alpha}_1$. Recalling (6.7.6), assume that the innovation process $\{\boldsymbol{U}(t)\}$ satisfies the conditions (HT1) – (HT6) in Section 8.1 Then, it holds that

$$\hat{\boldsymbol{\Sigma}} - \boldsymbol{\Sigma} = \int_{-\pi}^{\pi} \{\boldsymbol{I}_Z(\lambda) - \boldsymbol{f}_Z(\lambda)\}\, d\lambda, \tag{6.7.13}$$

where $\boldsymbol{I}_Z(\lambda)$ is the periodogram matrix of $\{\boldsymbol{Z}(1), \ldots, \boldsymbol{Z}(n)\}$, and $\boldsymbol{f}_Z(\lambda)$ is the spectral density matrix of $\{\boldsymbol{Z}(t)\}$. For simplicity, in what follows we assume $\boldsymbol{\mu} = 0$. Then,

$$\boldsymbol{\alpha}_{MV} = \frac{1}{\boldsymbol{e}'\Sigma_{XX}^{-1}\boldsymbol{e}}\Sigma_{XX}^{-1}\boldsymbol{e}, \quad \hat{\boldsymbol{\alpha}}_{MV} = \frac{1}{\boldsymbol{e}'\hat{\Sigma}_{XX}^{-1}\boldsymbol{e}}\hat{\Sigma}_{XX}^{-1}\boldsymbol{e}. \tag{6.7.14}$$

Letting $L(\hat{\boldsymbol{\Sigma}}) = \hat{\Sigma}_{XX}^{-1/2}\hat{\Sigma}_{XY}\hat{\Sigma}_{YY}^{-1}\hat{\Sigma}_{YX}\hat{\Sigma}_{XX}^{-1/2}$, it follows from the proof of Theorem 9.2.4 of Brillinger (2001a) that

$$\begin{aligned}\hat{\boldsymbol{\alpha}}_1 - \boldsymbol{\alpha}_1 = \sum_{k \neq 1} [\boldsymbol{\alpha}_k'\{L(\hat{\boldsymbol{\Sigma}}) - L(\boldsymbol{\Sigma})\}\boldsymbol{\alpha}_1]\boldsymbol{\alpha}_k/(\lambda_1 - \lambda_k) \\ + \text{lower-order terms.}\end{aligned} \tag{6.7.15}$$

In view of (6.7.14) and (6.7.15), we may write

$$\begin{aligned}\sqrt{n}\begin{pmatrix}\hat{\boldsymbol{\alpha}}_{MV} - \boldsymbol{\alpha}_{MV} \\ \hat{\boldsymbol{\alpha}}_1 - \boldsymbol{\alpha}_1\end{pmatrix} &= \sqrt{n}\{H(\hat{\boldsymbol{\Sigma}}) - H(\boldsymbol{\Sigma})\} + o_p(1) \\ &= \sqrt{n}\partial H(\boldsymbol{\Sigma})\{vech(\hat{\boldsymbol{\Sigma}}) - vech(\boldsymbol{\Sigma})\} + o_p(1),\end{aligned} \tag{6.7.16}$$

where $H(\cdot)$ is a smooth function, and $\partial H(\boldsymbol{\Sigma})$ is the derivative with respect to $\boldsymbol{\Sigma}$. Noting (6.7.13), (6.7.16) and Theorem 8.1.1 in Chapter 8, we can see that

$$\sqrt{n}\begin{pmatrix}\hat{\boldsymbol{\alpha}}_{MV} - \boldsymbol{\alpha}_{MV} \\ \hat{\boldsymbol{\alpha}}_1 - \boldsymbol{\alpha}_1\end{pmatrix} \xrightarrow{L} N(\boldsymbol{0}, \boldsymbol{V}), (n \to \infty), \tag{6.7.17}$$

where $\boldsymbol{V}$ is expressed as integral functionals of $\boldsymbol{f}_Z(\lambda)$ and the fourth-order cumulant spectra of $\{\boldsymbol{Z}(t)\}$. Hence,

$$\sqrt{n}(\hat{\boldsymbol{\alpha}} - \boldsymbol{\alpha}) \xrightarrow{L} N(\boldsymbol{0}, \boldsymbol{F}'\boldsymbol{V}\boldsymbol{F}), (n \to \infty), \tag{6.7.18}$$

where $\boldsymbol{F} = [(1 - \varepsilon)\boldsymbol{I}_p, \varepsilon\boldsymbol{I}_p]'$.

We next apply the results above to the problem of the portfolio for the Japanese government pension investment fund (GPIF). In 2012, GPIF invested the pension reserve fund in the following five assets: S_1 (government bond), S_2 (domestic stock), S_3 (foreign bond), S_4 (foreign stock) and S_5 (short-term fund), with portfolio coefficient $\hat{\boldsymbol{\alpha}}_{MV}^{GPIF} = (0.67, 0.11, 0.08, 0.09, 0.05)'$. We write the price processes at the time t by $S_1 = \{S_1(t)\}, S_2 = \{S_2(t)\}, S_3 = \{S_3(t)\}, S_4 = \{S_4(t)\}$ and $S_5 = \{S_5(t)\}$. Letting $X_i(t) = log S_i(t+1)/S_i(t), i = 1, \ldots, 4$, we introduce the return process on the five assets:

$$\boldsymbol{X}(t) = (X_1(t), X_2(t), X_3(t), X_4(t))'. \tag{6.7.19}$$

As exogenous variables, we take the consumer price index $S_6(t)$ and the average wege $S_7(t)$. Hence the return process is

$$\boldsymbol{Y}(t) = (logS_6(t+1)/S_6(t), logS_7(t+1)/S_7(t))'. \tag{6.7.20}$$

Suppose that the monthly observations of $\{\boldsymbol{X}(t)\}$ and $\{\boldsymbol{Y}(t)\}$ are available for $t = 1, \dots, 502$ (December 31, 1970 – October 31, 2012).

Considering the variables $\boldsymbol{Y}_1(t)$ and $\boldsymbol{Y}_2(t)$, Kobayashi et al. (2013) calculated $\hat{\alpha}(\varepsilon)$ in (6.7.12). For $\varepsilon = 0.01$,

$$\hat{\alpha}(0.01) = (0.902, 0.013, 0.084, 0.001)'. \tag{6.7.21}$$

Next we investigate the predictive property of 2012's GPIF portfolio coefficient (0.67, 0.11, 0.08, 0.09)'. For this, let

$$\boldsymbol{\alpha}_{GPIF}(\varepsilon) = (0.67 + \varepsilon, 0.11 + \varepsilon, 0.08 + \varepsilon, 0.09 + \varepsilon)'. \tag{6.7.22}$$

Introduce the following ε-perturbed portfolio:

$$\varphi_\varepsilon(t) \equiv \boldsymbol{\alpha}'_{GPIF}(\varepsilon)\boldsymbol{X}(t). \tag{6.7.23}$$

We measure the predictive goodness of 2012's GPIF portfolio by

$$\underset{\beta_1,\dots,\beta_p}{E}[\{\varphi_0(t) - \beta_1\varphi_\varepsilon(t-1) - \cdots - \beta_p\varphi_\varepsilon(t-p)\}^2]. \tag{6.7.24}$$

The estimated version of the expectation of (6.7.24) is given by

$$PE(\varepsilon, m, p) \equiv \frac{1}{n-p}\sum_{t=p+1}^{502}\{\varphi_0(t) - \hat{\beta}'_{LS}\begin{pmatrix}\varphi_\varepsilon(t-1)\\ \vdots \\ \varphi_\varepsilon(t-p)\end{pmatrix}\}^2, \tag{6.7.25}$$

where

$$\hat{\beta}_{LS} = [\sum_{t=502-m}^{502}\{\begin{pmatrix}\varphi_\varepsilon(t-1)\\ \vdots \\ \varphi_\varepsilon(t-p)\end{pmatrix}(\varphi_\varepsilon(t-1), \dots, \varphi_\varepsilon(t-p))\}]^{-1}$$

$$\times \left[\sum_{t=502-m}^{502} \begin{pmatrix} \varphi_\varepsilon(t-1) \\ \vdots \\ \varphi_\varepsilon(t-p) \end{pmatrix} \varphi_0(t) \right].$$

For $m = 50, n = 502, p = 1, 2, \ldots,$ and $|\varepsilon| \leq 0.05$, Kobayashi et al. (2013) calculated $\mathrm{PE}(\varepsilon, m, p)$ and showed that $\mathrm{PE}(\varepsilon, 50, p)$ is minimized at $\varepsilon = 0$ and $p = 3$, which implies that 2012's GPIF portfolio has an optimal predictive property for $p = 3$.

Problems

6.1. Verify (6.1.4).

6.2. Verify (6.1.5)–(6.1.8).

6.3. Concrete expression is possible for (6.1.13). Let $\Sigma = \{\Sigma_{ij}; i, j = 1, \ldots, m\}$. Then, show that

$$\frac{\partial^2 g_{MV}(\boldsymbol{\theta})}{\partial\delta\partial\boldsymbol{\mu}'} = 2\frac{\delta}{\pi\beta}\Sigma^{-1}\tilde{\boldsymbol{D}}\Sigma^{-1},$$

$$\frac{\partial^2 g_{MV}(\boldsymbol{\theta})}{\partial\delta\partial\Sigma_{ii}} = -2\frac{\delta}{\pi\beta}\Sigma^{-1}\left[\tilde{\boldsymbol{D}}\Sigma^{-1}\boldsymbol{E}_{ii} + \boldsymbol{E}_{ii}\Sigma^{-1}\tilde{\boldsymbol{D}}\right]\Sigma^{-1}(\boldsymbol{\mu} - R_0\boldsymbol{e}), \quad i = 1, \ldots, m,$$

$$\frac{\partial^2 g_{MV}(\boldsymbol{\theta})}{\partial\delta\partial\Sigma_{ij}} = -2\frac{\delta}{\pi\beta}\Sigma^{-1}\left[\tilde{\boldsymbol{D}}\Sigma^{-1}\boldsymbol{E}_{ij} + \boldsymbol{E}_{ij}\Sigma^{-1}\tilde{\boldsymbol{D}}\right]\Sigma^{-1}(\boldsymbol{\mu} - R_0\boldsymbol{e}), \quad i, j = 1, \ldots, m, \ i > j,$$

where $\boldsymbol{E}_{ii}$ is an $m \times m$ matrix whose elements are all 0 except for the (i, i)th element, which is 1, $\boldsymbol{E}_{ij}$ is an $m \times m$ matrix whose elsments are all 0 except for the (i, j)th and (j, i)th elements, which are 1, and

$$\tilde{\boldsymbol{D}} = \sum_{j=0}^{\infty} A(j)\boldsymbol{D}A(j)'. \tag{6.7.26}$$

6.4. In Proposition 6.3.2, verify the expressions for Ω_{12}, Ω_{14} and Ω_{22}.

6.5. In Proposition 6.3.3, verify the expressions (6.3.14).

6.6. Show that the best linear predictor $\hat{X}_t^{best}$ is given by (6.4.1).

6.7. Show that the prediction error of the pseudo best linear predictor is given by (6.4.9).

6.8. Prove that the RHS of (6.5.1) is equal to (6.5.4).

6.9. Prove Theorem 6.6.3.

6.10. Derive the formula (6.7.4) for the classical mean-variance portfolio α_{MV}.

6.11. Establish the result (6.7.17).

Chapter 7

Numerical Examples

This chapter summarizes numerical examples, applying theories introduced in previous chapters. We also introduce additional practical new theories relevant to portfolio analysis in this chapter. The numerical examples include the following five real data analyses: Section 7.1, based on the theory of portfolio estimators discussed in Chapter 3, time series data analysis for six Japanese companies' monthly log-returns have been investigated; Section 7.2 as an application for the multiperiod portfolio optimization problem introduced in Chapter 4, we performed portfolio analysis for the Japanese government pension investment fund (GPIF); Section 7.3, Microarray data analysis by using rank order statistics, which was introduced in Chapter 5; Section 7.4, DNA sequencing data analysis by using a mean-variance portfolio, Spectral envelope (Stoffer et al. (1993)) and mean-diversification portfolio; corticomuscular coupling analysis by the CHARN model discussed in Chapter 2 is introduced, and we apply a mean-variance portfolio, Spectrum envelope and mean-diversification portfolio to the analysis.

7.1 Real Data Analysis for Portfolio Estimation

This section provides various numerical studies to investigate the accuracy of the portfolio estimators discussed in Chapter 3. We provide real data analysis based on Toyota, Honda, Nissan, Fanuc, Denso and Komatsu's monthly log-return. We introduce a procedure to construct a confidence region of the efficient frontier by using the AR bootstrap method. In addition, assuming a locally stationary process, nonparametric portfolio estimators are provided.

7.1.1 Introduction

As we mentioned in Chapter 3, the mean-variance portfolio is identified by the mean vector (μ) and the covariance matrix (Σ) of the asset returns. Since we cannot know μ and Σ, we construct these estimators $\hat{\mu}$ and $\hat{\Sigma}$. This study aims to evaluate the effect of the major portfolio quantities (such as the efficient frontier, portfolio reconstruction) caused by the estimators $\hat{\mu}$ and $\hat{\Sigma}$. We first consider the selection of the underlying model for the asset returns. Then, based on the selected model, the confidence region for a point of an estimated efficient frontier is constructed. On the other hand, assuming a locally stationary process, we consider a sequence of estimated portfolio weights to optimize the estimated time-varying mean vector and covariance matrix.

7.1.2 Data

The data consists of Toyota, Honda, Nissan, Fanuc, Denso and Komatsu's monthly log-return from December 1983 to January 2013, inclusive, so there are 350 months of data.

Figure 7.1 shows the time series plots of the returns for each company. The plots look stationary but they may not bc independent and identically distributed (i.i.d.). Figure 7.2 shows the autocorrelation plots of the returns. There is no evidence of long memory effect but there exist some short-term autocorrelations. Figure 7.3 shows normal probability plots of the returns. There is evidence of heavy tails.

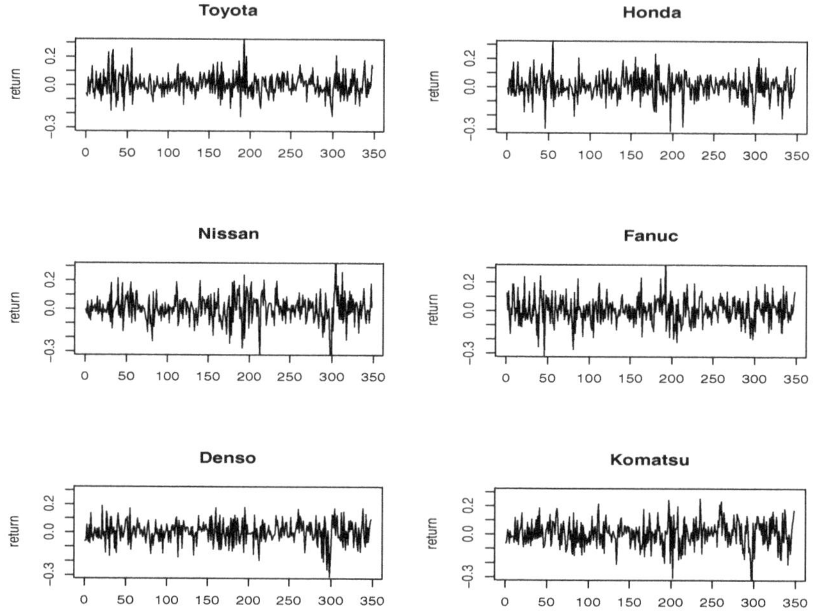

Figure 7.1: Time series plots of monthly log-returns for six companies

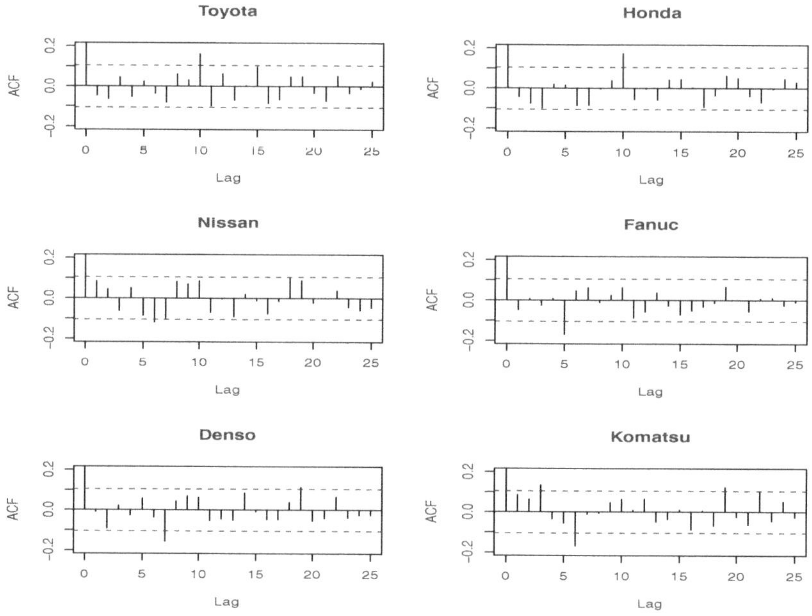

Figure 7.2: Autocorrelation plots of monthly log-returns for six companies

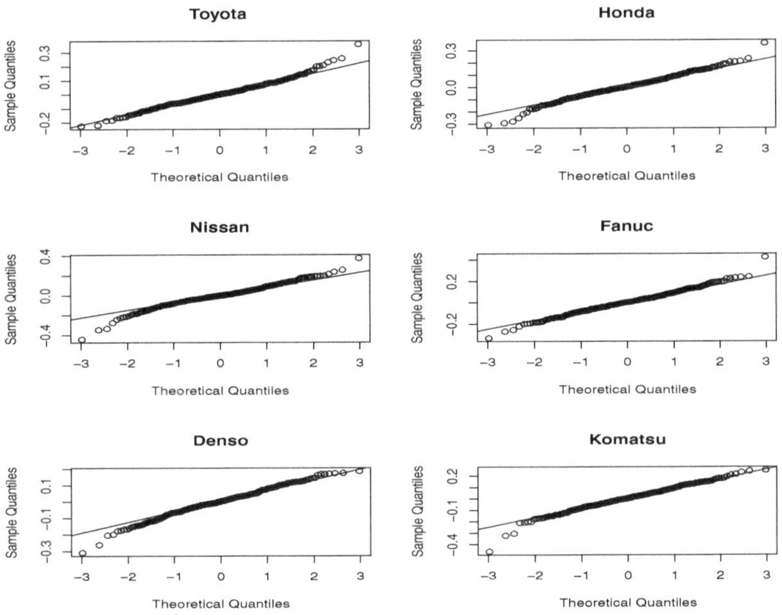

Figure 7.3: Normal probability plots of monthly log-returns for six companies

7.1.3 Method

7.1.3.1 Model selection

Here we discuss the procedure for order specification, estimation and model checking to build a vector AR model for a given time series. Suppose that $\{\boldsymbol{X}(t) = (X_1(t), \ldots, X_m(t))'; t \in \mathbb{N}\}$ follows a VAR(i) model:

$$\boldsymbol{X}(t) = \boldsymbol{\mu} + \boldsymbol{U}(t) + \sum_{j=1}^{i} \Phi^{(i)}(j)\{\boldsymbol{X}(t-j) - \boldsymbol{\mu}\}.$$

Given $\{\boldsymbol{X}(1), \ldots, \boldsymbol{X}(n)\}$, the residual can be written as

$$\hat{\boldsymbol{U}}^{(i)}(t) = \hat{\boldsymbol{Y}}(t) - \sum_{j=1}^{i} \hat{\Phi}^{(i)}(j)\hat{\boldsymbol{Y}}(t-j), \quad i+1 \le t \le n$$

where $\hat{\boldsymbol{Y}}(t) = \boldsymbol{X}(t) - \hat{\boldsymbol{\mu}}$ with $\hat{\boldsymbol{\mu}} = 1/n \sum_{t=1}^{n} \boldsymbol{X}(t)$ and $\hat{\Phi}^{(i)}(j)$ for $j = 1, \ldots, i$ is the OLS estimator of $\Phi^{(i)}(j)$. By using the residual, the following information criterion and the final prediction error (FPE) are available for the selection of order i.

$$AIC(i) = \ln\left(|\tilde{\Sigma}_i|\right) + \frac{2m^2 i}{n}, \tag{7.1.1}$$

$$BIC(i) = \ln\left(|\tilde{\Sigma}_i|\right) + \frac{m^2 i \ln(n)}{n}, \tag{7.1.2}$$

$$HQ(i) = \ln\left(|\tilde{\Sigma}_i|\right) + \frac{2m^2 i \ln(\ln(n))}{n}, \tag{7.1.3}$$

$$FPE(i) = \left(\frac{n+i^*}{n-i^*}\right)^m |\tilde{\Sigma}_i| \tag{7.1.4}$$

where

$$\tilde{\Sigma}_i = \frac{1}{n} \sum_{t=i+1}^{n} \hat{\boldsymbol{U}}^{(i)}(t)\hat{\boldsymbol{U}}^{(i)}(t)', \quad i \geq 0$$

and i^* is the total number of the parameters in each model and i assigns the lag order. The Akaike information criterion (AIC), the Schwarz–Bayesian criterion (BIC) and the Hannan and Quinn criterion (HQ) are proposed by Akaike (1973), Schwarz (1978) and Hannan and Quinn (1979), respectively.

7.1.3.2 *Confidence region*

Next, we discuss the procedure for construction of the confidence region for an estimated efficient frontier. In general, when statistics are computed from a randomly chosen sample, these statistics are random variables and we cannot evaluate the inference of the randomness exactly if we do not know the population. However, if the sample is a good representative of the population, we can simulate sampling from the population by sampling from the sampling, which is usually called "resampling." The "bootstrap" introduced by Efron (1979) is one of the most popular resampling methods and has found application to a number of statistical problems, including construction of the confidence region. However, it is difficult to apply a bootstrap to dependent data, because the idea is to simulate sampling from the population by sampling from the sample under the i.i.d. assumption. Here we apply a "(stationary) AR bootstrap." The idea is to resample the residual from the fitted model assuming the AR model.[1] In the autoregressive bootstrap (ARB) method, we assume that $X_1, \ldots, X_n$ are observed from the AR(p) model. Then, based on $X_1, \ldots, X_n$ and the least squares estimators of the AR coefficients, the "residuals" from the fitted model are constructed. Under the correct fitted model, the residuals turn out to be "approximately independent" and then we resample the residuals to define the bootstrap observations through an estimated version of the structural equation of the fitted AR(p) model. Properties of the ARB have been investigated by many authors. Especially, Bose (1988) showed that the ARB approximation is second-order correct. Although the above method assumes stationarity for the underlying AR model, nonstationary cases are also discussed. As for bootstrapping an explosive autoregressive AR process, Datta (1995) showed that the validity of the ARB critically depends on the initial values. As for bootstrapping an unstable autoregressive AR process (e.g., unit root process), Datta (1996) provides conditions on the resample size m for the "m out of n" ARB to ensure validity of the bootstrap approximation. In a similar way, the validity of the bootstrapping approximation for the stationary autoregressive and moving average (ARMA) process is discussed by Kreiss and Franke (1992) and Allen and Datta (1999), among others. For details about the validity of the ARB, see, e.g., Lahiri (2003).

Suppose that a vector stochastic process which describes Toyota, Denso, and Komatsu's returns follows VAR(2)[2] as follows:

$$\boldsymbol{X}(t) = \boldsymbol{\mu} + \boldsymbol{U}(t) + \Phi^{(2)}(1)\{\boldsymbol{X}(t-1) - \boldsymbol{\mu}\} + \Phi^{(2)}(2)\{\boldsymbol{X}(t-2) - \boldsymbol{\mu}\}.$$

Suppose that $\{\boldsymbol{X}(t)\}$ follows the VAR(2) model. Then residual can be written as

$$\hat{\boldsymbol{U}}^{(2)}(t) = \hat{\boldsymbol{Y}}(t) - \hat{\Phi}^{(2)}(1)\hat{\boldsymbol{Y}}(t-1) - \hat{\Phi}^{(2)}(2)\hat{\boldsymbol{Y}}(t-2), \quad 3 \leq t \leq n.$$

Let $G_n(\cdot)$ denote the distribution which puts mass $1/n$ at $\hat{\boldsymbol{U}}^{(2)}(t)$. Then, the centering adjusted distribution is given by $F_n(\boldsymbol{x}) = G_n(\boldsymbol{x} + \bar{\boldsymbol{U}})$ where $\boldsymbol{x} \in \mathbb{R}^3$ and $\bar{\boldsymbol{U}} = 1/(n-2)\sum_{t=3}^{n} \hat{\boldsymbol{U}}^{(2)}(t)$. Let

[1] Other resampling methods for dependent data are known as the Jackknife, the (Moving) Block bootstrap, the Model-Based bootstrap (resampling including AR bootstrap, Sieve bootstrap) and the Wild bootstrap.

[2] See Table 7.1.

$\{U^*(t); t = 1, \ldots, n\}$ be a sequence of i.i.d. observations from $F_n(\cdot)$. Given $\{U^*(t)\}$, we can generate a sequence of the resampled return $\{X^*(t) : t = 1, \ldots, n\}$ by

$$X^*(t) = \hat{\mu} + U^*(t) + \hat{\Phi}^{(2)}(1)\{X^*(t-1) - \hat{\mu}\} + \hat{\Phi}^{(2)}(1)\{X^*(t-2) - \hat{\mu}\} \tag{7.1.5}$$

where $X^*(1) = \hat{\mu} + U^*(1)$ and $X^*(2) = \hat{\mu} + U^*(2) + \hat{\Phi}^{(2)}(1)\{X^*(1) - \hat{\mu}\}$.

For each sequence of the resampled return $\{X^*(t) : t = 1, \ldots, n\}$, the sample mean vector μ^* and the sample covariance matrix Σ^* are calculated, and the optimal portfolio weight with a mean return of μ_P is estimated by

$$w^*_{\mu_P} = \frac{\mu_P\alpha^* - \beta^*}{\delta^*}(\Sigma^*)^{-1}\mu^* + \frac{\gamma^* - \mu_P\beta^*}{\delta^*}(\Sigma^*)^{-1}e \tag{7.1.6}$$

with $\alpha^* = e'(\Sigma^*)^{-1}e$, $\beta^* = \mu^{*\prime}(\Sigma^*)^{-1}e$, $\gamma^* = \mu^{*\prime}(\Sigma^*)^{-1}\mu^*$ and $\delta^* = \alpha^*\gamma^* - (\beta^*)^2$.

For N times resampling of $\{X^*(t)\}$, N kinds of the mean vector μ^* and the covariance matrix Σ^* are obtained. Based on these statistics, N kinds of the resampled efficient frontiers can be calculated by

$$(\mu_P - \mu^*_{GMV})^2 = s^*(\sigma_P^2 - \sigma^*_{GMV}{}^2) \tag{7.1.7}$$

where

$$\mu^*_{GMV} = \frac{\beta^*}{\alpha^*}, \quad \sigma^*_{GMV}{}^2 = \frac{1}{\alpha^*} \quad and \quad s^* = \frac{\delta^*}{\alpha^*}.$$

Moreover, plugging fixed σ_P in (7.1.7), N kinds of μ_P are obtained. Denoting the μ_P in descending order as $\mu_P^{(b)}(\sigma_P); b = 1, \ldots, N$ (i.e., $\mu_P^{(1)}(\sigma_P) \geq \mu_P^{(2)}(\sigma_P) \geq \cdots \geq \mu_P^{(N)}(\sigma_P)$), we can obtain the p-percentile of μ_P by $\mu_P^{([pN/100])}(\sigma_P)$.

7.1.3.3 *Locally stationary estimation*

Dahlhaus (1996c) estimated, nonparametrically, the covariance of lag k at time u of locally stationary processes with mean vector $\mu = 0$. Shiraishi and Taniguchi (2007) generalized his results to the case of nonzero mean vector $\mu = \mu(u)$. In this section, we suppose that $X(t,n) = (X_1(t,n), \ldots, X_m(t,n))'(t = 1, \ldots, n; n \in \mathbb{N})$ follow the locally stationary process

$$X(t,n) = \mu\left(\frac{t}{n}\right) + \int_{-\pi}^{\pi} \exp(i\lambda t)A^\circ(t,n,\lambda)d\xi(\lambda). \tag{7.1.8}$$

Let $\mu(u), \Sigma(u)$ denote the local mean vector and local covariance matrix at time $u = t/n \in [0,1]$. Then, the nonparametric estimators are given by

$$\hat{\mu}\left(\frac{t}{n}\right) := \frac{1}{b_n n}\sum_{s=[t-b_n n/2]+1}^{[t+b_n n/2]} K\left(\frac{t-s}{b_n n}\right)X(s,n) \tag{7.1.9}$$

$$\hat{\Sigma}\left(\frac{t}{n}\right) := \frac{1}{b_n n}\sum_{s=[t-b_n n/2]+1}^{[t+b_n n/2]} K\left(\frac{t-s}{b_n n}\right)\left\{X(s,n) - \hat{\mu}\left(\frac{t}{n}\right)\right\}\left\{X(s,n) - \hat{\mu}\left(\frac{t}{n}\right)\right\}', \tag{7.1.10}$$

respectively, where $K : \mathbb{R} \to [0, \infty)$ is a kernel function and b_n is a bandwidth. Here, for $n < t < n + [b_n n/2]$, we define $X(t,n) \equiv X(2n-t,n)$, and, for $1 > t > -[b_n n/2]$, we define $X(t,n) \equiv X(2-t,n)$.

By using $\hat{\mu}(t/n)$ and $\hat{\Sigma}(t/n)$ for $t = 1, \ldots, n$, a (local) global minimum variance (GMV) portfolio weight estimator is given by

$$\hat{w}_{GMV}\left(\frac{t}{n}\right) := \frac{\hat{\Sigma}^{-1}(t/n)e}{e'\hat{\Sigma}^{-1}(t/n)e}. \tag{7.1.11}$$

7.1.4 *Results and Discussion*

Results for Model Selection Table 7.1 shows the selected orders of the VAR model for 3 returns within 6 returns based on AIC, HQ, BIC and FPE. In addition, we plot $AIC(i), HQ(i), BIC(i)$ and

Table 7.1: Selected orders for 3 returns within T (Toyota), H (Honda), N (Nissan), F (Funac), D (Denso) and K (Komatsu) based on AIC, HQ, BIC and FPE

Asset1	T	T	T	T	T	T	T	T	T	**T**	H	H	H	H	H	H	N	N	N	F
Asset2	H	H	H	H	N	N	N	F	F	**D**	N	N	N	F	F	D	F	F	D	D
Asset3	N	F	D	K	F	D	K	D	K	**K**	F	D	K	D	K	K	D	K	K	K
AIC	2	1	2	3	1	2	1	1	1	**2**	2	2	2	1	2	2	2	1	7	2
HQ	1	1	1	1	1	1	1	1	1	**2**	1	1	1	1	1	2	1	1	2	2
BIC	1	1	1	1	1	1	1	1	1	**1**	1	1	1	1	1	1	1	1	1	1
FPE	2	1	2	3	1	2	1	1	1	**2**	2	2	2	1	2	2	2	1	6	2

$FPE(i)$ for $i = 1, \dots, 10$, for the returns of Toyota, Denso and Komatsu in Figure 7.4. You can see that in the case of $i = 2$, the criterion have the smallest value for AIC, HQ and FPE, while for BIC $BIC(1)$ is the smallest but $BIC(2)$ is close to that.

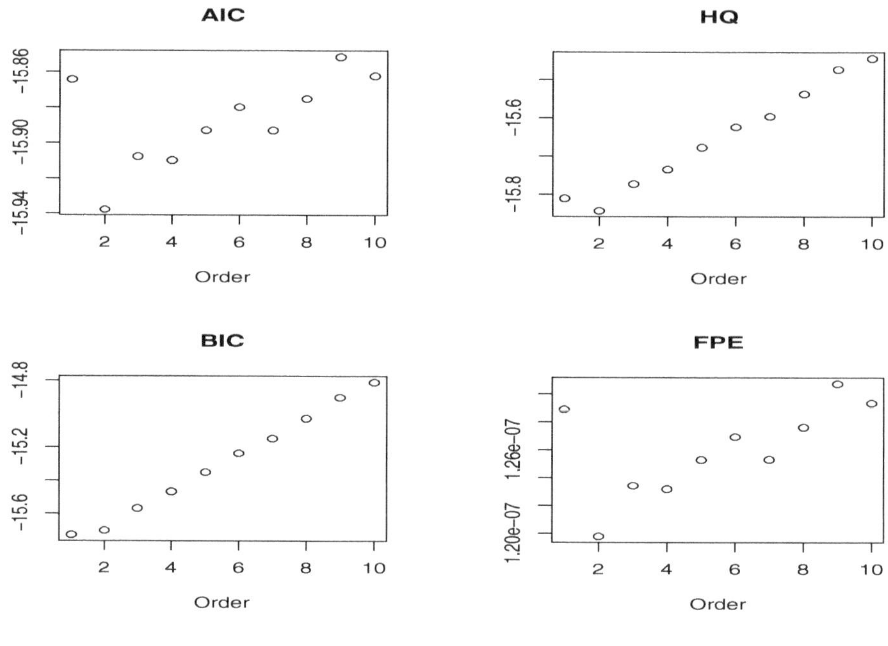

Figure 7.4: AIC, HQ, BIC and FPE for the returns of Toyota, Denso and Komatsu

In particular, the fitted model for the returns of Toyota, Denso, Komatsu is a VAR(2) model with the following coefficient matrices.

$$\hat{\Phi}^{(2)}(1) = \begin{bmatrix} -0.1122 & -0.0108 & 0.1465 \\ 0.0997 & -0.1898 & 0.1551 \\ -0.0038 & 0.1203 & 0.0468 \end{bmatrix}, \quad \hat{\Phi}^{(2)}(2) = \begin{bmatrix} -0.1265 & 0.0236 & 0.0736 \\ -0.0294 & -0.1773 & 0.2037 \\ 0.0837 & -0.0038 & 0.0258 \end{bmatrix},$$

$$\hat{\mu} = \begin{bmatrix} 0.0039 \\ 0.0020 \\ 0.0041 \end{bmatrix}.$$

Results for Confidence Region In Figure 7.5, the points $(\sigma^*(\mu_P), \mu^*(\mu_P))$ are plotted from 10,000 bootstrap resamples where

$$\mu^*(\mu_P) = \boldsymbol{w}_{\mu_P}^{*}{}'\hat{\boldsymbol{\mu}}, \quad \sigma^*(\mu_P) = \sqrt{\boldsymbol{w}_{\mu_P}^{*}{}'\hat{\Sigma}\boldsymbol{w}_{\mu_P}^{*}} \tag{7.1.12}$$

and $\mu_P = 0.008$.

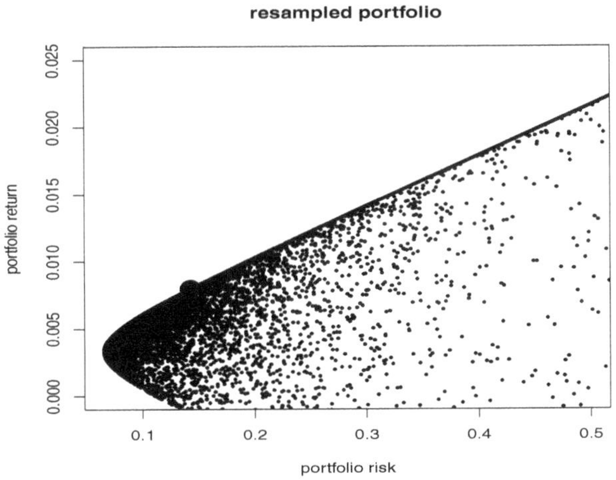

Figure 7.5: Results from 10,000 bootstrap resamples of a VAR(2) model. For each resample, the optimal portfolio with a mean return of 0.008 is estimated. The resampled points are plotted as a small dot. The large dot is the estimated point and the solid line is the (estimated) efficient frontier.

In Figure 7.6, for $N = 10{,}000$ resampling results, we plot the 90, 95 and 99 percentile confidence region (i.e., $\mu_P^{(9000)}(\sigma_P)$, $\mu_P^{(9500)}(\sigma_P)$ and $\mu_P^{(9900)}(\sigma_P)$ for each σ_P), respectively.

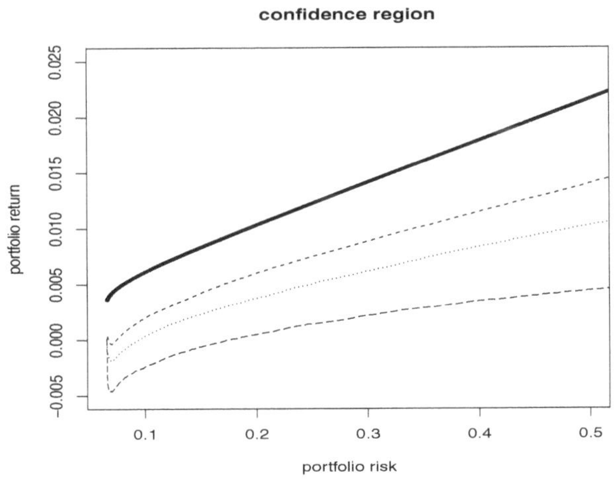

Figure 7.6: Confidence region of efficient frontier. The solid line is the estimated efficient frontier. The upper, middle and lower dotted lines are the 90, 95 and 99 percentile confidence regions, respectively.

Results for Locally Stationary Estimation Figure 7.7 shows (local) GMV portfolio weight estimators ($\hat{\boldsymbol{w}}_{GMV}(t/n)$) and traditional GMV portfolio weight

$$\hat{\boldsymbol{w}}_{GMV}^{\text{(traditional)}} := \frac{\hat{\Sigma}^{-1}e}{e'\hat{\Sigma}^{-1}e} \tag{7.1.13}$$

where $\hat{\Sigma}$ is the sample covariance matrix. The data are Toyota, Honda, Nissan, Fanuc, Denso and Komatsu's monthly log-returns. Note that we apply the kernel function as $K(x) = 6\left(\frac{1}{4} - x^2\right)$ for $-1/2 \leq x \leq 1/2$ (this function satisfies Assumption 3 of Shiraishi and Taniguchi (2007)), and the bandwidth as $b_n = 0.1, 0.3, 0.5$ (according to Shiraishi and Taniguchi (2007), the optimal bandwidth satisfies $O(n^{-1/5})$). You can see that the portfolio weights are dramatically changed as the bandwidth becomes small.

Based on these portfolio weights, the sequences of the portfolio return (i.e., $\boldsymbol{X}(t,n)'\hat{\boldsymbol{w}}_{GMV}\left(\frac{t}{n}\right)$ or $\boldsymbol{X}(t,n)'\hat{\boldsymbol{w}}_{GMV}^{\text{(traditional)}}$) are plotted in Figure 7.8. It looks like the variance of the portfolio returns is smaller as the bandwidth becomes small. Table 7.2 shows the mean and standard deviation of the portfolio returns. As the bandwidth becomes small, the mean tends to be small but the standard deviation also tends to be small. This phenomenon implies that locally stationary fitting works well.

Figure 7.7 and Table 7.2 show the merits and drawbacks for investigation of portfolio selection problems under the locally stationary model. The merit is that we can catch the actual phenomenon rather than the analysis under the stationary model. The drawback is the difficulty in determining the bandwidth. In addition, we have to take care that we do not catch any jump phenomenon under the locally stationary model.

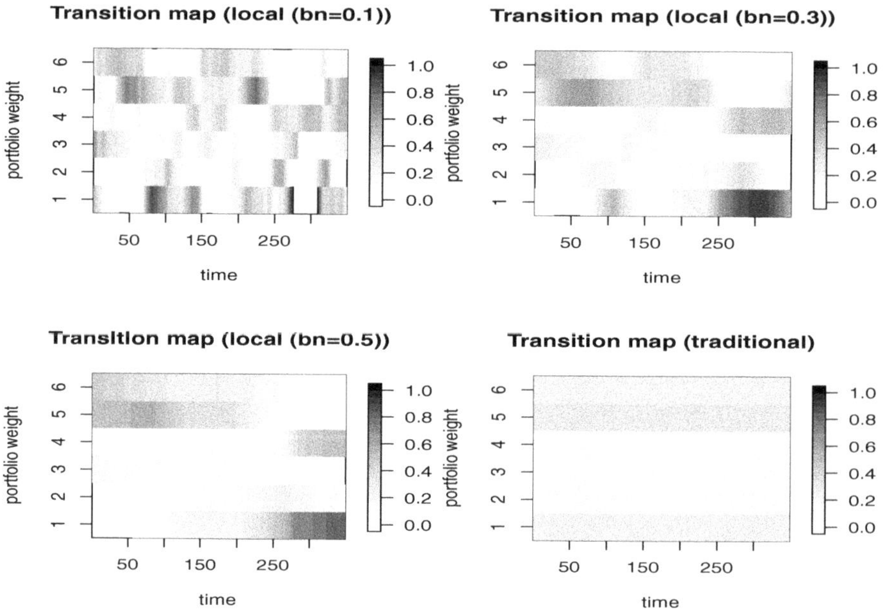

Figure 7.7: Transition map for (local) GMV portfolio weight (1:Toyota, 2:Honda, 3:Nissan, 4:Fanuc, 5:Denso and 6:Komatsu)

7.1.5 *Conclusions*

In this section, two types of portfolio analysis have been investigated. The first one is a construction of a confidence region for an estimated efficient frontier. Model selection for Toyota, Denso and Komatsu's returns is considered from a VAR(p) model and $p = 2$ is selected based on some

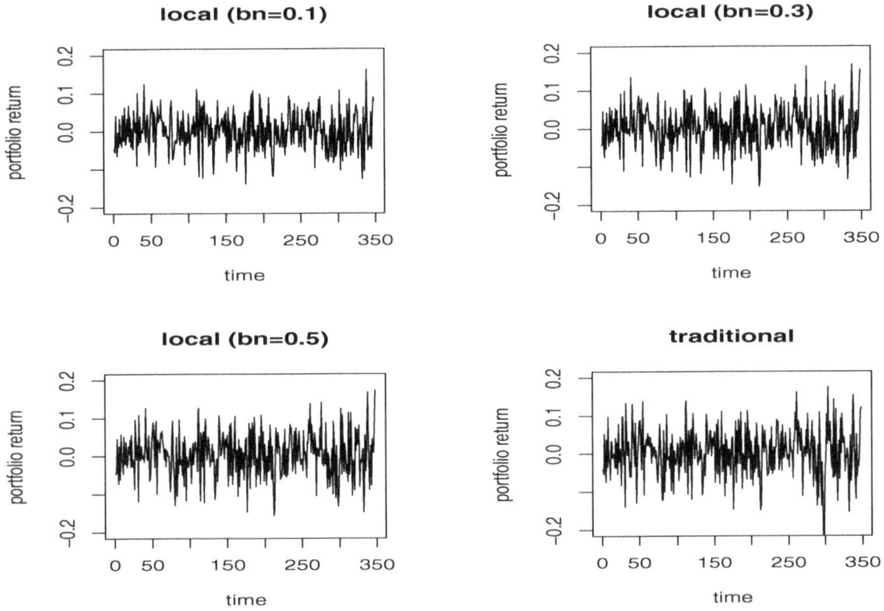

Figure 7.8: Time series plots of returns for (local) GMV portfolio

Table 7.2: Mean and standard deviation (SD) of returns for (local) GMV portfolio

Portfolio Type	b_n	Mean	SD
local	0.1	0.00114	0.04978
	0.3	0.00238	0.05417
	0.5	0.00345	0.05798
traditional	-	0.00354	0.05914

information criteria. Assuming the VAR(2) model for these asset returns, a confidence region for an estimated efficient frontier is constructed by using the autoregressive bootstrap (ARB). The result shows that the wideness of the confidence regions tends to increase as the portfolio risk is increasing. The second one is a construction of a sequence of estimated portfolio weights under an assumption of a locally stationary process. By using the nonparametric estimation method, the local mean vector and local covariance matrix for each time are estimated. As a result, we found that the sequence of the estimated portfolios is able to grasp the actual phenomenon but there exists a difficulty in how to determine the bandwidth.

7.1.6 Bibliographic Notes

Lahiri (2003) introduces various resampling methods for dependent data. Hall (1992) gives the accuracy of the bootstrap theoretically by using the Edgeworth expansion.

7.2 Application for Pension Investment

In this section, we evaluate a multiperiod effect in portfolio analysis, so that this study is an application for the multiperiod portfolio optimization problem in Chapter 4.

7.2.1 Introduction

A portfolio analysis for the Japanese government pension investment fund (GPIF) is discussed in this section. In actuarial science, "required reserve fund" for pension insurance has been investigated for a long time. The traditional approach focuses on the expected value of future obligations and interest rate. Then, the investment strategy is determined for exceeding the expected value of interest rate. Recently, solvency for the insurer is defined in terms of random values of future obligations (e.g., Olivieri and Pitacco (2000)). In this section, assuming that the reserve fund is defined in terms of the random interest rate and the expected future obligations, we consider optimal portfolios by optimizing the randomized reserve fund. Shiraishi et al. (2012) proposed three types of the optimal portfolio weights based on the maximization of the Sharpe ratio under the functional forms for the portfolio mean and variance, two of them depending on the Reserve Fund at the end-of-period target (e.g., 100 years). Regarding the structure of the return processes, we take account of the asset structure of the GPIF. On the asset structure of the GPIF, the asset classes are split into cash, domestic and foreign bonds, domestic and foreign equity. The first three classes (i.e., cash, domestic and foreign bonds) are assumed to be short memory while the other classes (i.e., domestic and foreign equity) are long memory. To obtain the optimal portfolio weights, we rely on a bootstrap. For the short memory returns, we use a wild bootstrap (WB). Early work focuses on providing first- and second-order theoretical justification for WB in the classical linear regression model (see, e.g., Wu (1986)). Gonçalves and Kilian (2004) show that WB is applicable for the linear regression model with a conditional heteroscedastic such as stationary ARCH, GARCH, and stochastic volatility effects. For the long memory returns, we apply a sieve bootstrap (SB). Bühlmann (1997) establishes consistency of the autoregressive sieve bootstrap. Assuming that the long memory process can be written as AR(∞) and MA(∞) processes, we estimate the long memory parameter by means of Whittle's approximate likelihood (Beran (1994)). Given this estimator, the residuals are computed and resampled for the construction of the bootstrap samples, from which the optimal portfolio estimated weights are obtained.

7.2.2 Data

The Government Pension Investment Fund (GPIF) of Japan was established on April 1, 2006 as an independent administrative institution with the mission of managing and investing the Reserve Fund of the employees' pension insurance and the national pension (www.gpif.go.jp for more information). It is the world's largest pension fund ($1.4 trillion in assets under management as of December 2009) and its mission is managing and investing the Reserve Funds in *safe and efficient investment with a long-term perspective*. Business management targets to be achieved by GPIF are set by the Minister of Health, Labour and Welfare based on the law on the general rules of independent administrative agencies.

Available data is monthly log-returns from January 31, 1971 to October 31, 2009 (466 observations) of the five types of assets: Domestic Bond (DB), Domestic Equity (DE), Foreign Bond (FB), Foreign Equity (FE) and cash. We suppose that cash and bonds are gathered in the short-memory panel $\boldsymbol{X}_t^S = (X_t^{(DB)}, X_t^{(FB)}, X_t^{(cash)})$ and follow (7.2.6), and equities are gathered into the long-memory panel $\boldsymbol{X}_t^L = (X_t^{(DE)}, X_t^{(FE)})$ and follow (7.2.7).

7.2.3 Method

Let $S_{i,t}$ be the price of the ith asset at time t ($i = 1, \ldots, m$), and $X_{i,t}$ be its log-return. Time runs from 0 to T. In this section, we consider that today is T_0 and T is the end-of-period target. Hence the past and present observations run for $t = 0, \ldots, T_0$, and the future until the end-of-period target for

$t = T_0 + 1, \ldots, T$. The price $S_{i,t}$ can be written as

$$S_{i,t} = S_{i,t-1}\exp(X_{i,t}) = S_{i,0}\exp\left(\sum_{s=1}^{t} X_{i,s}\right), \tag{7.2.1}$$

where $S_{i,0}$ is the initial price. Let $F_{i,t}$ denote the Reserve Fund on asset i at time t and defined by

$$F_{i,t} = F_{i,t-1}\exp(X_{i,t}) - c_{i,t},$$

where $c_{i,t}$ denotes the maintenance cost at time t. By recursion $F_{i,t}$ can be written as

$$\begin{aligned} F_{i,t} &= F_{i,t-1}\exp(X_{i,t}) - c_{i,t} \\ &= F_{i,t-2}\exp\left(\sum_{s=t-1}^{t} X_{i,s}\right) - \sum_{s=t-1}^{t} c_{i,s}\exp\left(\sum_{s'=s+1}^{t} X_{i,s'}\right) \\ &= F_{i,0}\exp\left(\sum_{s=1}^{t} X_{i,s}\right) - \sum_{s=1}^{t} c_{i,s}\exp\left(\sum_{s'=s+1}^{t} X_{i,s'}\right), \end{aligned} \tag{7.2.2}$$

where $F_{i,0} = S_{i,0}$.

We gather the Reserve Funds in the vector $\boldsymbol{F}_t = (F_{1,t}, \ldots, F_{m,t})$. Let $F_t(\boldsymbol{w}) = \boldsymbol{w}'\boldsymbol{F}_t$ be a portfolio formed by the m Reserve Funds, which depend on the vector of weights $\boldsymbol{w} = (w_1, \ldots, w_m)'$. The portfolio Reserve Fund can be expressed as a function of all past returns:

$$\begin{aligned} F_t(\boldsymbol{w}) &:= \sum_{i=1}^{m} w_i F_{i,t} \\ &= \sum_{i=1}^{m} w_i \left(F_{i,0}\exp\left(\sum_{s=1}^{t} X_{i,s}\right) - \sum_{s=1}^{t} c_{i,s}\exp\left(\sum_{s'=s+1}^{t} X_{i,s'}\right)\right). \end{aligned} \tag{7.2.3}$$

We are interested in maximizing $F_t(\boldsymbol{w})$ at the end-of-period target $F_T(\boldsymbol{w})$:

$$F_T(\boldsymbol{w}) = \sum_{i=1}^{m} \alpha_i \left(F_{i,T_0}\exp\left(\sum_{s=T_0+1}^{T} X_{i,s}\right) - \sum_{s=T_0+1}^{T} c_{i,s}\exp\left(\sum_{s'=s+1}^{T} X_{i,s'}\right)\right). \tag{7.2.4}$$

It depends on the future returns, the maintenance cost and the portfolio weights. While the first two are assumed to be constant from T_0 to T (the constant return can be seen as the average return over the $T - T_0$ periods), we focus on the optimality of the weights that we denote by $\boldsymbol{w}^{opt}$.

In the first specification, the estimation of the optimal portfolio weights is based on the maximization of the Sharpe ratio:

$$\boldsymbol{w}^{opt} = \arg\max_{\boldsymbol{w}} \frac{\mu(\boldsymbol{w})}{\sigma(\boldsymbol{w})}, \tag{7.2.5}$$

under different functional forms of the expectation $\mu(\boldsymbol{w})$ and the risk $\sigma(\boldsymbol{w})$ of the portfolio. Shiraishi et al. (2012) proposed three functional forms, two of the them depending on the Reserve Fund.

Definition 7.2.1. *The first one is the traditional form based on the returns:*

$$\mu(\boldsymbol{w}) = \boldsymbol{w}'E(\boldsymbol{X}_T) \quad \text{and} \quad \sigma(\boldsymbol{w}) = \sqrt{\boldsymbol{w}'V(\boldsymbol{X}_T)\boldsymbol{w}},$$

where $E(\boldsymbol{X}_T)$ and $V(\boldsymbol{X}_T)$ are the expectation and the covariance matrix of the returns at the end-of-period target.

Definition 7.2.2. *The second form for the portfolio expectation and risk is based on the vector of Reserve Funds:*

$$\mu(\boldsymbol{w}) = \boldsymbol{w}'E(\boldsymbol{F}_T) \quad \text{and} \quad \sigma(\boldsymbol{w}) = \sqrt{\boldsymbol{w}'V(\boldsymbol{F}_T)\boldsymbol{w}},$$

where $E(\boldsymbol{F}_T)$ and $V(\boldsymbol{F}_T)$ indicate the mean and covariance of the Reserve Funds at time T.

Definition 7.2.3. *Last, we consider the case where the portfolio risk depends on the lower partial moments of the Reserve Funds at the end-of-period target:*

$$\mu(\boldsymbol{w}) = \boldsymbol{w}'E(\boldsymbol{F}_T) \quad \text{and} \quad \sigma(\boldsymbol{w}) = E\left\{\left(\tilde{F} - F_T(\boldsymbol{w})\right)\mathbb{I}\left(F_T(\boldsymbol{w}) < \tilde{F}\right)\right\},$$

where $\tilde{F}$ is a given value.

The m returns split into p- and q-dimensional vectors $\{\boldsymbol{X}_t^S; t \in \mathbb{Z}\}$ and $\{\boldsymbol{X}_t^L; t \in \mathbb{Z}\}$, where S and L stand for short and long memory, respectively. The short memory returns correspond to cash and domestic and foreign bonds, which we generically denote by bonds. The long memory returns correspond to domestic and foreign equity, which we denote as equity. Cash and bonds follow the nonlinear model

$$\boldsymbol{X}_t^S \quad = \quad \boldsymbol{\mu}^S + \boldsymbol{H}(\boldsymbol{X}_{t-1}^S, \ldots, \boldsymbol{X}_{t-m}^S)\boldsymbol{\epsilon}_t^S, \tag{7.2.6}$$

where $\boldsymbol{\mu}^S$ is a vector of constants, $\boldsymbol{H} : \mathbb{R}^{kp} \to \mathbb{R}^p \times \mathbb{R}^p$ is a positive definite matrix-valued measurable function, and $\boldsymbol{\epsilon}_t^S = (\epsilon_{1,t}^S, \ldots, \epsilon_{p,t}^S)'$ are i.i.d. random vectors with mean $\mathbf{0}$ and covariance matrix $\boldsymbol{\Sigma}^S$. By contrast, equity returns follow a long memory nonlinear model

$$\boldsymbol{X}_t^L = \sum_{\nu=0}^{\infty} \phi_\nu \boldsymbol{\epsilon}_{t-\nu}^L, \quad \boldsymbol{\epsilon}_t^L = \sum_{\nu=0}^{\infty} \psi_\nu \boldsymbol{X}_{t-\nu}^L, \tag{7.2.7}$$

where

$$\phi_\nu = \frac{\Gamma(\nu + d)}{\Gamma(\nu + 1)\Gamma(d)} \quad \text{and} \quad \psi_\nu = \frac{\Gamma(\nu - d)}{\Gamma(\nu + 1)\Gamma(-d)}$$

with $-1/2 < d < 1/2$, and $\boldsymbol{\epsilon}_t^L = (\epsilon_{1,t}^L, \ldots, \epsilon_{p,t}^L)'$ are i.i.d. random vectors with mean $\mathbf{0}$ and covariance matrix $\boldsymbol{\Sigma}^L$. We estimate the optimal portfolio weights by means of a bootstrap. Let the super-indexes (S, b) and (L, b) denote the bootstrapped samples for the bonds and equity, respectively, and B the total number of bootstrapped samples. Below we show the bootstrap procedure for both types of assets.

Bootstrap procedure for $\boldsymbol{X}_t^{(S,b)}$

Step 1: Generate the i.i.d. sequences $\{\boldsymbol{\epsilon}_t^{(S,b)}\}$ for $t = T_0 + 1, \cdots, T$ and $b = 1, \ldots, B$ from $N(\mathbf{0}, \boldsymbol{I}_p)$.

Step 2: Let $\boldsymbol{Y}_t^S = \boldsymbol{X}_t^S - \hat{\boldsymbol{\mu}}^S$, where $\hat{\boldsymbol{\mu}}^S = \frac{1}{T_0}\sum_{s=1}^{T_0} \boldsymbol{X}_s^S$. Generate the i.i.d. sequences $\{\boldsymbol{Y}_t^{(S,b)}\}$ for $t = T_0 + 1, \cdots, T$ and $b = 1, \ldots, B$ from the empirical distribution of $\{\boldsymbol{Y}_t^S\}$.

Step 3: Compute $\{\boldsymbol{X}_t^{(S,b)}\}$ for $t = T_0 + 1, \cdots, T$ and $b = 1, \ldots, B$ as

$$\boldsymbol{X}_t^{(S,b)} = \hat{\boldsymbol{\mu}}^S + \boldsymbol{Y}_t^{(S,b)} \odot \boldsymbol{\epsilon}_t^{(S,b)},$$

where $\odot$ denotes the cellwise product.

■

Bootstrap procedure for $\boldsymbol{X}_t^{(L,b)}$

Step 1: Estimate $\hat{d}$ from the observed returns by means of Whittle's approximate likelihood:

$$\hat{d} = \arg\min_{d \in (0,1/2)} L(d, \Sigma)$$

where

$$
\begin{aligned}
L(d,\Sigma) &= \frac{2}{T_0}\sum_{j=1}^{(T_0-1)/2}\left\{\log\det f(\lambda_{j,T_0},d,\Sigma)+tr\left(f(\lambda_{j,T_0},d,\Sigma)^{-1}I(\lambda_{j,T_0})\right)\right\},\\
f(\lambda,d,\Sigma) &= \frac{|1-\exp(i\lambda)|^{-2d}}{2\pi}\Sigma,\\
I(\lambda) &= \left|\frac{1}{\sqrt{2\pi T_0}}\sum_{t=1}^{T_0}\boldsymbol{X}_t^L e^{it\lambda}\right|^2 \quad \text{and}\\
\lambda_{j,T_0} &= 2\pi j/T_0.
\end{aligned}
$$

Step 2: Compute $\{\hat{\epsilon}_t^L\}$ for $t = 1,\dots,T_0$:

$$
\hat{\epsilon}_t^L = \sum_{k=0}^{t-1}\pi_k \boldsymbol{X}_{t-k}^L \quad \text{where} \quad \pi_k = \begin{cases} \frac{\Gamma(k-\hat{d})}{\Gamma(k+1)\Gamma(-\hat{d})} & k \le 100 \\ \frac{k^{-\hat{d}-1}}{\Gamma(-\hat{d})} & k > 100. \end{cases}
$$

Step 3: Generate $\{\epsilon_t^{(L,b)}\}$ for $t = T_0+1,\dots,T$ and $b = 1,\dots,B$ from the empirical distribution of $\{\hat{\epsilon}_t^L\}$.

Step 4: Generate $\{\boldsymbol{X}_t^{(L,b)}\}$ for $t = T_0+1,\dots,T$ and $b = 1,\dots,B$ as

$$
\boldsymbol{X}_t^{(L,b)} = \sum_{k=0}^{t-T_0-1}\tau_k\epsilon_{t-k}^{(L,b)} + \sum_{k=t-T_0}^{t-1}\tau_k\hat{\epsilon}_{t-k}.
$$

■

We gather $\boldsymbol{X}_t^{(S,b)}$ and $\boldsymbol{X}_t^{*(L,b)}$ into $\boldsymbol{X}_t^{(b)} = (\boldsymbol{X}_t^{(S,b)}, \boldsymbol{X}_t^{(L,b)}) = (X_{1,t}^{*(b)},\dots,X_{p+q,t}^{*(b)})$ for $t = T_0 + 1,\dots,T$ and $b = 1,\dots,B$. The bootstrapped Reserve Funds $\boldsymbol{F}_T^{(b)} = (F_{1,T}^{(b)},\dots,F_{p+q,T}^{(b)})$

$$
F_{i,T}^{(b)} = F_{T_0}\exp\left(\sum_{s=T_0+1}^{T}X_{i,s}^{(b)}\right) - \sum_{s=T_0+1}^{T}c_{i,s}\exp\left(\sum_{s'=s+1}^{T}X_{i,s'}^{(b)}\right).
$$

And the bootstrapped Reserve Fund portfolio is

$$
F_T^{(b)}(\boldsymbol{w}) = \boldsymbol{w}'\boldsymbol{F}_T^{(b)} = \sum_{i=1}^{p+q}w_iF_{i,T}^{(b)}.
$$

Finally, the estimated portfolio weights that give the optimal portfolio are

$$
\hat{\boldsymbol{w}}^{opt} = \arg\max_{\boldsymbol{w}}\frac{\mu^{(b)}(\boldsymbol{w})}{\sigma^{(b)}(\boldsymbol{w})},
$$

where $\mu^{(b)}(\boldsymbol{w})$ and $\sigma^{(b)}(\boldsymbol{w})$ may take any of the three forms introduced earlier, but be evaluated in the bootstrapped returns or Reserve Funds.

Based on the observed data $\{(\boldsymbol{X}_1^S,\boldsymbol{X}_1^L),\dots,(\boldsymbol{X}_{466}^S,\boldsymbol{X}_{466}^L)\}$ (where $\boldsymbol{X}_t^S = (X_t^{(DB)},X_t^{(FB)},X_t^{(cash)})$ and $\boldsymbol{X}_t^L = (X_t^{(DE)},X_t^{(FE)})$), we estimate the optimal portfolio weights, $\hat{\boldsymbol{w}}^{opt1},\hat{\boldsymbol{w}}^{opt2}$ and $\hat{\boldsymbol{w}}^{opt3}$, corresponding to the three forms (Definition 7.2.1–Definition 7.2.3) for the expectation (i.e., $\mu^{(b)}(\boldsymbol{w}) \equiv \boldsymbol{w}'E^{(b)}(\boldsymbol{X}_T^{(b)})$ or $\mu^{(b)}(\boldsymbol{w}) \equiv \boldsymbol{w}'E^{(b)}(\boldsymbol{F}_T^{(b)})$) and risk (i.e., $\sigma^{(b)}(\boldsymbol{w}) \equiv \sqrt{\boldsymbol{w}'V^{(b)}(\boldsymbol{X}_T^{(b)})\boldsymbol{w}}$ or $\sigma^{(b)}(\boldsymbol{w}) \equiv \sqrt{\boldsymbol{w}'V^{(b)}(\boldsymbol{F}_T^{(b)})\boldsymbol{w}}$ or $\sigma^{(b)}(\boldsymbol{w}) \equiv E^{(b)}\left\{\left(\tilde{F} - F_T^{(b)}(\boldsymbol{w})\right)\mathbb{I}\left(F_T^{(b)}(\boldsymbol{w}) < \tilde{F}\right)\right\}$) of the Sharpe ratio, where

$$
E^{(b)}(\boldsymbol{T}^{(b)}) := \frac{1}{B}\sum_{b=1}^{B}\boldsymbol{T}^{(b)} \quad \text{and} \quad V^{(b)}(\boldsymbol{T}^{(b)}) := \frac{1}{B}\sum_{b=1}^{B}\left\{\boldsymbol{T}^{(b)} - E^{(b)}(\boldsymbol{T}^{(b)})\right\}\left\{\boldsymbol{T}^{(b)} - E^{(b)}(\boldsymbol{T}^{(b)})\right\}'
$$

for any bootstrapped sequence $\{T^{(b)}\}$. In addition, we compute the trajectory of the optimal Reserve Fund for $t = T_0+1, \dots, T$. Because of liquidity reasons, the portfolio weight for cash is kept constant at 5%. The target period is fixed at 100 years, and the maintenance cost is based on the 2004 Pension Reform.

7.2.4 *Results and Discussion*

Table 7.3 shows the estimated optimal portfolio weights for the three different choices of the portfolio expectation and risk. The weight of domestic bonds is very high and clearly dominates over the other assets. Domestic bonds are low risk and medium return, which is in contrast with equity, which shows a higher return but also higher risk, and with foreign bonds, which show a low return and risk. Therefore, in a sense, domestic bonds are a compromise between the characteristics of the four equities and bonds. Figure 7.9 shows the trajectory of the future Reserve Fund for different

Table 7.3: Estimated optimal portfolio weights

	DB	DE	FB	FE	cash
Returns ($\hat{\boldsymbol{w}}^{opt1}$)	0.95	0.00	0.00	0.00	0.05
Reserve Fund ($\hat{\boldsymbol{w}}^{opt2}$)	0.75	0.00	0.20	0.00	0.05
Low Partial ($\hat{\boldsymbol{w}}^{opt3}$)	0.85	0.10	0.00	0.00	0.05

values of the yearly return (assumed to be constant from $T_0 + 1$ to T) ranging from 2.7% to 3.7%. Since the investment term is extremely long, 100 years, the Reserve Fund is quite sensitive to the choice of the yearly return. In the 2004 Pension Reform, authorities assumed a yearly return of the portfolio of 3.2%, which corresponds to the middle thick line of the figure.

This simulation result shows that the portfolio weight of domestic bonds is quite high. The reason is that the investment term is extremely long (100 years). Because the investment risk for the Reserve Fund is exponentially amplified year by year, the portfolio selection problem for the Reserve Fund is quite sensitive to the year-based portfolio risk. Note that the above result is obtained only under the assumption without reconstruction of the portfolio weights. If we can relax this assumption, the result would be significantly changed. This assumption depends on the liquidity of the portfolio assets.

7.2.5 *Conclusions*

In this section, optimal portfolio weights to invest for five asset categories such as Domestic Bond (DB), Domestic Equity (DE), Foreign Bond (FB), Foreign Equity (FE) and cash are investigated. In this study, selected optimal portfolio weights are fixed at the end-of-period target (e.g., 100 years), and the objective function for the optimization is defined based on the randomized (future) Reserve Fund. The future expected return and the future risk are estimated by using some bootstrap methods. The result shows that the weight of DB is very high because DB is very low risk and risk is amplified highly in this study.

7.3 Microarray Analysis Using Rank Order Statistics for ARCH Residual

Statistical two-group comparisons are widely used to identify the significant differentially expressed (DE) signatures against a therapy response for microarray data analysis. To DE analysis, we applied rank order statistics based on an autoregressive conditional heteroskedasticity (ARCH) residual empirical process introduced in Chapter 5. This approach was considered for publicly available datasets and compared with two-group comparison by original data and an autoregressive (AR) residual (Solvang and Taniguchi (2017a)) . The significant DE genes by the ARCH and AR residuals were

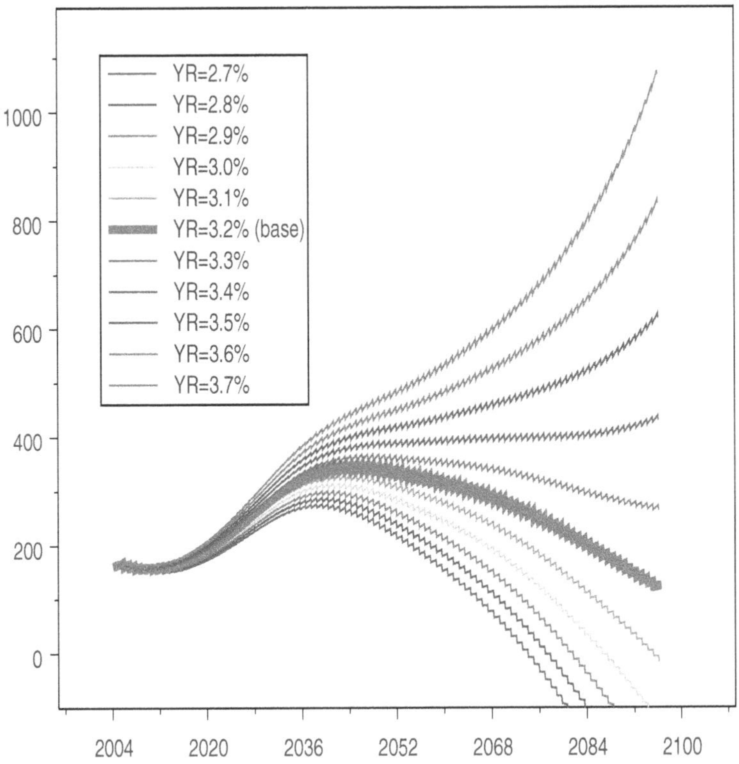

Figure 7.9: Trajectory of the Future Reserve Fund for different values of the yearly return ranging from 2.7% to 3.7%.

reduced by about 20–30% of the genes by the original data. Almost 100% of the genes by ARCH are covered by the genes by the original data, unlike the genes by AR residuals. GO enrichment and pathway analyses indicate the consistent biological characteristics between genes by ARCH residuals and original data. ARCH residuals array data might contribute to refining the number of significant DE genes to detect the biological feature as well as ordinal microarray data.

7.3.1 *Introduction*

Microarray technology provides a high-throughput way to simultaneously investigate gene expression information on a whole genome level. In the field of cancer research, the genome-wide expression profiling of tumours has become an important tool to identify gene sets and signatures that can be used as clinical endpoints, such as survival and therapy response (Zhao et al. (2011)). When we are contrasting expressions between different groups or conditions (i.e., the response is polytumous), such important genes are described as differentially expressed (DE) (Yang et al. (2005)). To identify important genes, a statistical scheme is required that measures and captures evidence for a DE per gene. If the response consists of binary data, the DE is measured using a two-group comparison for which such statistical methods as t-statistics, the statistical analysis of microarrays (SAM) (Tusher et al. (2001)), fold change, and B statistics have been proposed (Campain and Yang (2010)). The p-value of the statistics is calculated to assess the significance of the DE genes. The p-value per gene is ranked in ascending order; however, selecting significant genes must be considered by multiple testing corrections, e.g., false discovery rate (FDR) (Benjamini and Hochberg (1995)), to avoid type I errors. Even if significant DE genes are identified by the FDR procedure, the gene list may still include too many to apply a statistical test for a substantial number of probes through whole genomic locations. Such a long list of significant DE genes complicates capturing gene signatures that should provide the availability of robust clinical and pathological prognostic and predictive factors to guide patient decision-making and the selection of treatment options.

As one approach for this challenge, based on the residuals from the autoregressive conditional heteroskedasticity (ARCH) models, the proposed rank order statistic for two-sample problems pertaining to empirical processes refines the significant DE gene list. The ARCH process was proposed by Engle (1982), and the model was developed in much research to investigate a daily return series from finance domains. The series indicate time-inhomogeneous fluctuations and sudden changes of variance called volatility in finance. Financial analysts have attempted more suitable time series modeling for estimating this volatility. Chandra and Taniguchi (2003) proposed rank order statistics and the theory provided an idea for applying residuals from two classes of ARCH models to test the innovation distributions of two financial returns generated by such varied mechanisms as different countries and/or industries. Empirical residuals called "innovation" generally perturb systems behind data. Theories of innovation approaches to time series analysis have historically been closely related to the idea of predicting dynamic phenomena from time series observations. Wiener's theory is a well-known example that deems prediction error to be a source of information for improving the predictions of future phenomena. In a sense, innovation is a more positive label than prediction error (Ozaki and Iino (2001)). As we see in innovation distribution for ARCH processes, it resembles the sequential expression level based on the whole genomic location. For applying time indices of ARCH model to the genomic location, time series mining has been practically used in DNA sequence data analysis (Stoffer et al. (2000)) and microarray data analysis (Koren et al. (2007)). To investigate the data's properties, we believe that innovation analysis is more effective than analysis just based on the original data. While the original idea in Chandra and Taniguchi (2003) was based on squared residuals from an ARCH model, not-squared empirical residuals are also theoretically applicable, as introduced in Lee and Taniguchi (2005). In this study, we apply this idea to test DEs between two sample groups in microarray datasets that we assume to be generated by different biological conditions.

To investigate whether ARCH residuals can consistently refine a list of significant DE genes, we apply publicly available datasets called Affy947 (van Vliet et al. (2008)) for breast cancer research to compare significant gene signatures. As a statistical test for two-group comparisons, the estrogen receptor (ER) is applied in clinical outcomes to identify prognostic gene expression signatures. Estrogen is an important regulator of the development, the growth, and the differentiation of normal mammary glands. It is well documented that endogenous estrogen plays a major role in the development and progression of breast cancer. ER expression in breast tumours is frequently used to group breast cancer patients in clinical settings, both as a prognostic indicator and to predict the likelihood of response to treatment with antiestrogen (Rezaul et al. (2010)). If the cancer is ER+, hormone therapy using medication slows or stops the growth of breast cancer cells. If the cancer is ER−, then hormonal therapy is unlikely to succeed. Based on these two categorical factors for ER status, we applied our proposed statistical test to the expression levels for each genomic location. After identifying significant DE genes, biological enrichment analyses use the gene list and seek biological processes and interconnected pathways. These analyses support the consistency for refined gene lists obtained by ARCH residuals.

7.3.2 Data

Due to the extensive usage of microarray technology, in recent years publicly available datasets have exploded (Campain and Yang (2010)), including the Gene Expression Omnibus (GEO, http://www.ncbi.nlm.nih.gov/geo/) (Edgar et al. (2002)) and ArrayExpress (http://www.evi.ac.uk/microarray-as/ae/). In this study for breast cancer research, we used five different expression datasets, collectively called the Affy947 expression dataset (van Vliet et al. (2008)). These datasets, which all measure the Human Genome HG U133A Affymetrix arrays, are normalized using the same protocol and are assessable from GEO with the following identifiers:

GSE6532 for the Loi et al. dataset (Loi et al. (2007), Loi), GSE3494 for the Miller et al. dataset (Miller et al. (2005), Mil), GSE7390 for the Desmedt et al. dataset (Desmedt et al. (2007)), Des), and GSE5327 for the Minn et al. dataset (Minn et al. (2005), Min). The Chin et al. dataset (Chin et al. (2006), Chin) is available from ArrayExpress. This pooled dataset was preprocessed and normalized, as described in Zhao et al. (2014). Microarray quality control assessment was carried out using the R AffyPLM package from the Bioconductor web site (http://www.bioconductor.org, Bolstad et al. (2005)). The Relative Log Expression (RLE) and Normalized Unscaled Standard Errors (NUSE) tests were applied. Chip pseudo-images were produced to assess artifacts on the arrays that did not pass the preceding quality control tests. The selected arrays were normalized by a three-step procedure using the RMA expression measure algorithm (http://www.bioconductor.org, Irizarry et al. (2003)): RMA background correction convolution, the median centering of each gene separately across arrays for each dataset, and the quantile normalization of all arrays. Gene mean centering effectively removes many dataset specific biases, allowing for effective integration of multiple datasets (Sims et al. (2008)). The total number of probes is 22,268 for these microarray data.

Against all probes that covered the whole genome, we use probes that correspond to the intrinsic signatures that were obtained by classifying breast tumours into five molecular subtypes (Perou et al. (2000)). We extracted 777 probes from the whole 22K probes for the microarray data-sets using the intrinsic annotation included in the R codes in Zhao et al. (2014). As the response contrasting expression between two groups, we used a hormone receptor called ER, which indicates whether a hormone drug works as well for treatment as a progesterone receptor and is critical to determine prognosis and predictive factors. ERs are used for classifying breast tumours into ER-positive (ER+) and ER-negative (ER−) diseases. The two upper figures in Figure 7.10 present the mean of the microarray data by averaging all of the previously obtained samples (Desmedt et al. (2007)). The left and right plots correspond to a sample indicating ER+ and ER−. The data for ER− show more fluctuation than for ER+. The two lower figures illustrate histograms of the averaged data for ER+ and ER− and present sharper peakedness and heavier tails than the shape of an ordinary Gaussian distribution.

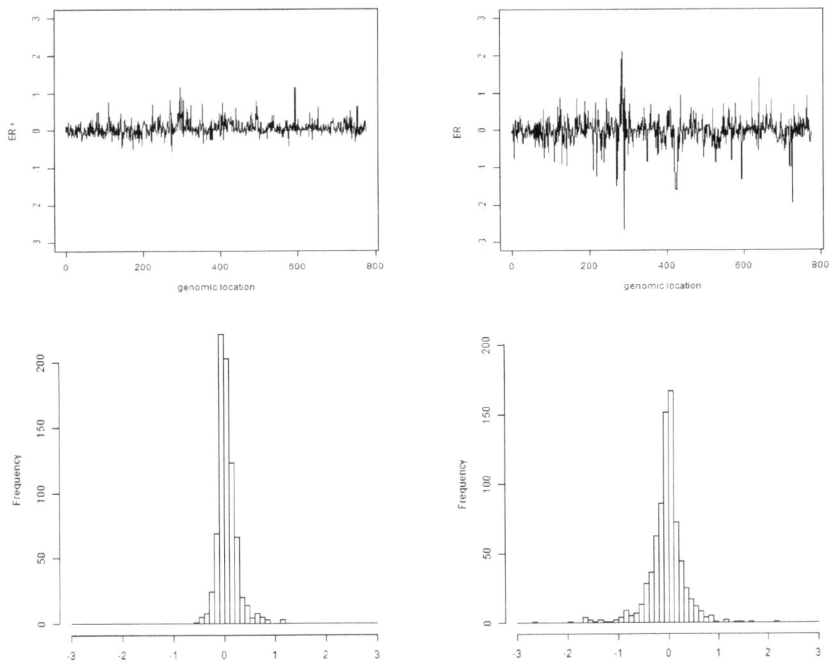

Figure 7.10: Upper: Mean of expression levels for ER+ (left) and ER− (right) across all Des samples. Lower: Histograms of expression levels for ER+ (left) and ER− (right)

7.3.3 Method

Denote the sample and the genomic location by i and j in microarray data x_{ij}. The samples for the microarray data are divided by two biologically different groups; one group is for breast cancer tumours driven by ER+ and another group is for breast cancer tumours driven by ER−. We apply two-group comparison testing to identify significantly different expression levels between two groups of ER+ and ER− samples for each gene (genomic location). As the statistical test, we propose the rank order statistics for the ARCH residual empirical process introduced in Section 7.3.3.1. For comparisons with the ARCH model's performance, we consider applying the two-group comparison testing to original array data and applying the test to the residuals obtained by the ordinal AR (autoregressive) model. The details about both methods are summarized in Section 7.3.3.2. For the obtained significant DE gene lists, biologists or medical scientists require further analysis for their biological interpretation to investigate the biological process or biological network. In this article, we apply GO (gene ontology) analysis, shown in Section 7.3.3.3, and pathway analysis, shown in Section 7.3.3.4, which methods are generally used to investigate specific genes or relationships among gene groups. Our proposed algorithm is summarized in Figure 7.11.

7.3.3.1 The rank order statistic for the ARCH residual empirical process

Suppose that a classes of ARCH(p) models is generated by following Equations (5.3.1) and another class of ARCH(p) models, independent of $\{X_t\}$, is generated similarly by Equations (5.3.2). We are interested in the two-sample problem of testing given by (5.3.3). In this section, $F(x)$ and $G(x)$ correspond to the distribution for the expression data of samples driven by ER+ and ER−, individually.

For this testing problem, we consider a class of rank order statistics, such as Wilcoxon's two-sample test. The form is derived from the empirical residuals $\hat{\varepsilon}_t^2 = X_t^2/\sigma_t^2(\hat{\theta}_X), t = 1, \cdots, n$ and $\hat{\xi}_t^2 = Y_t^2/\sigma_t^2(\hat{\theta}_Y), t = 1, \cdots, n$. Because Lee and Taniguchi (2005) developed the asymptotic theory for not squared empirical residuals, we may apply the results to $\hat{\varepsilon}_t$ and $\hat{\xi}_t$.

7.3.3.2 Two-group comparison for microarray data

To obtain the empirical residuals, the ARCH model is applied to a vector $\{x_{i1}, x_{i2}, \cdots, x_{iL}\}$ for the ith sample, where L is the total number of genomic locations in the microarray data. Assuming that the ER+ and ER− samples correspond to distributions $F(x)$ and $G(x)$ as above, orders p_X and p_Y of the ARCH model are identified by model selection using the Akaike Information Criterion (AIC), where the model with the minimum AIC is defined as the best fit model (Akaike (1974)) (see 1 in Figure 7.11). According to those responses, the empirical residuals are grouped as $\varepsilon_{ij}^{+}$and ξ_{ij}^{-}. The Wilcoxon statistic is applied as the order statistic to those two groups for each genomic location j, and the p-value is calculated (see 2 in Figure 7.11). The p-value vector is adjusted for multiple testing corrections using FDR (Benjamini and Hochberg (1995)) (see 3 in Figure 7.11).

For comparisons with the ARCH model's performance, we applied the same procedure to the original data and the residuals by the Autoregressive (AR) model. The AR model represents the current value using the weighted average of the past values as $x_{ij} = \sum_{k=1}^{K} \beta_i x_{ij-k} + w_{ij}$, where β_i, k, and w_{ij} are the AR coefficient, the AR order, and the error terms. The AR model is widely applied in time series analysis and signal processing of economics, engineering, and science. In this study, we apply it to the expression data for two ER+ and ER− groups. The AR order for the best fit model is identified by the AIC. Empirical residuals w_{ij}^{+} for ER+ and w_{ij}^{-} for ER− are subtracted from the data by predictions. These procedures are summarized as follows: 1) take the original microarray and ER data from one study; 2) apply the ARCH and AR models to the original data for each sample and identify the best fit model among the model candidates with 1–10 time lags; 3) subtract the residuals by the best fit model from the original data; 4) apply the Wilcoxon statistic to the original data and to the ARCH and AR residuals; 5) list the p-values and identify the significant FDR (5%) corrected genes. 6) apply gene ontology analysis and pathway analysis (see the details in Sections

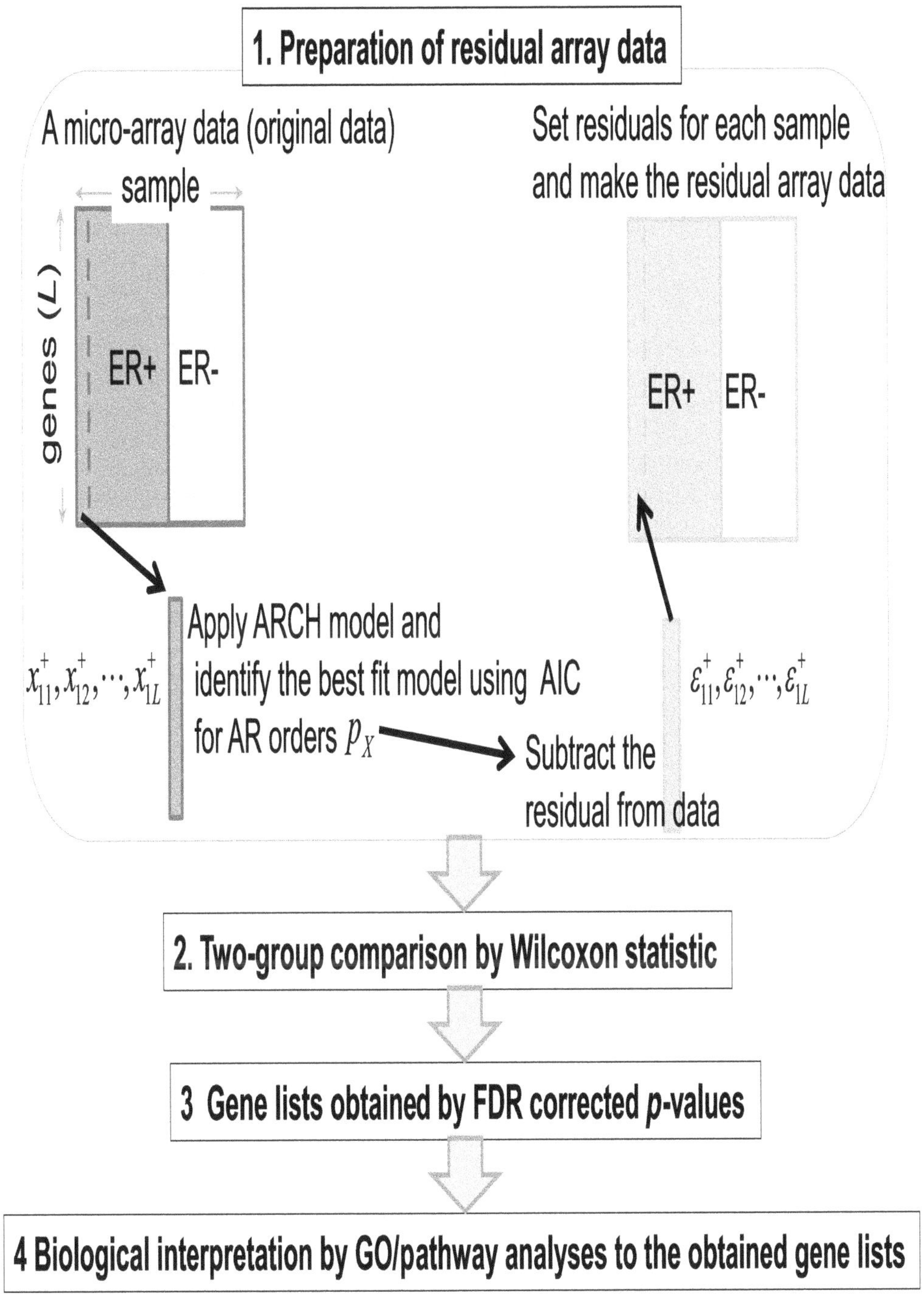

Figure 7.11: Procedures for generating residual array data and analysis by rank order statistics

7.3.3.3 and 7.3.3.4) for biological interpretation to the obtained gene list (see 4 in Figure 7.11). The computational programs were done by the *garchFit* function for ARCH fitting, by the *ar.ols* function for AR fitting, the *wilcox.test* as a rank-sum test, and *fdr.R* for the FDR adjustment in the R package.

7.3.3.3 GO analysis

To investigate the gene product attributes from the gene list, gene ontology (GO) analysis was performed to find specific gene sets that are statistically associated among several biological categories. GO is designed as a functional annotation database to capture the known relationships between biological terms and all the genes that are instances of those terms. It is widely used by many functional enrichment tools and is highly regarded both for its comprehensiveness and its unified approach for annotating genes in different species to the same basic set of underlying functions (Glass and Girvan (2012)). Many tools have been developed to explore, filter, and search the GO database. In our study, GOrilla (Eden et al. (2009)) was used as a GO analysis tool. GOrilla is an efficient web-based interactive user interface that is based on a statistical framework called minimum hypergenometric (mHG) for enrichment analysis in ranked gene lists, which are naturally represented as functional genomic information. For each GO term, the method independently identifies the threshold at which the most significant enrichment is obtained. The significant mHG scores are accurately and tightly corrected for threshold multiple testing without time-consuming simulations (Eden et al. (2009)). The tool identifies enriched GO terms in ranked gene lists for background gene sets which are obtained by the whole genomic location of microarray data. GO consists of three hierarchically structured vocabularies (ontologies) that describe gene products in terms of their associated biological processes, cellular components, and molecular functions. The building blocks of GO are called terms, and the relationship among them can be described by a directed acyclic graph (DAG), which is a hierarchy where each gene product may be annotated to one or more terms in each ontology (Glass and Girvan (2012)). GOrilla requires a list of gene symbols as input data. The obtained significant Etrez gene lists by FDR correction are converted into gene symbols using a web-based database called SOURCE (Diehn et al. (2003)), which was developed by the Genetics Department of Stanford University.

7.3.3.4 Pathway analysis

For GO analysis, the identified genes are mapped to the well-defined biological pathways. Pathway analysis determines which pathways are overrepresented among genes that present significant variations. The difference from GO analysis is that pathway analysis includes interactions among a given set of genes. Several tools for pathway analysis have been published. In this study, we used a web-based analysis tool called REACTOME, which is a manually curated open-source open-data resource of human pathways and reactions Joshi–Tops et al. (2005). REACTOME is a recent fast and sophisticated tool that has grown to include annotations for 7088 of the 20,774 protein-coding genes in the current Ensembl human genome assembly, 15,107 literature references, and 1,421 small molecules organized into 6,744 reactions collected in 1,481 pathways Joshi–Tops et al. (2005).

7.3.4 Simulation Study

To investigate the performance of our proposed algorithm, we performed a simulation study. We first prepared the clinical indicator like ER+ and ER−. The artificial indicator includes "1" for 50 samples and "0" for 50 samples. Next, we considered two types of artificial 1,000-array and 100 samples: one array data (A) generating by normal distributions was set. The mean and variance values of the distribution were set as 1.0 to generate overall array data at once. In addition, the array data for the 201-400 array and the 601-600 array were replaced with data generating a different normal distribution with 1.8 mean and 10 variance; another array data (B) was generated by an ARCH model. The model was applied to real array (DES data, see details in Chapter 4) and the parameters (mu: the intercept, omega: the constant coefficient of the variance equation, alpha: the coefficients of the variance equation, skew: the skewness of the data, shape: the shape parameter of the conditional distribution setting as 3) for the model were estimated for ER+ and ER−, respectively. We used these parameters and random numbers to generate the simulation data. For the computational programs, we conducted normrnd of MATLAB® command to generate random variables by normal distributions for array data A, and conducted garchSim of the R package fGARCH for array data B. We iterated 100 times to generate the two array datasets. To 100 datasets for A and B, we applied two-group comparison for the original simulation data and the ARCH residuals of them and identified 5% FDR significant parts.

7.3.5 Results and Discussion

7.3.5.1 Simulation data

For the simulation data and ARCH residuals, we summarized the average of the number of the identified 5% FDR significant parts and the number of overlapping parts in Table 7.4. In the case of the simulation data generated by normal distribution, the significant number for original series and ARCH residuals in array sets A and B did not differ. The parts identified in A were mostly the same as ones in B. On the other hand, in the simulation data generated by the ARCH model, the ARCH residuals identified more significant parts from the data than the original series. The number of significant parts for ARCH residuals was about 30% less than the number of significant parts for the original series. The overlapping number was less than in case A; however, over 50% parts were covered.

Table 7.4: Summary of the average for the identified significant parts

	1. Original series	2. ARCH residuals	Overlapped ♯ of 2 with 1	Ratio for the overlapped ♯
Array set A	30.2	29.8	29.3	98.2%
Array set B	512.2	338.9	212.1	62.6%

7.3.5.2 Affy947 expression dataset

Based on the method explained in Section 3.2, the best fit AR and GARCH models were selected by AIC for each sample. The estimated orders of all the best fit models of all the studies are summarized in Supplementary Table 1 https://www.dropbox.com/sh/sotz8jufje73eg6/AACKHD-tXqB02h_rxlvVlXqsa?dl=0. Figure 7.10 summarizes the ratio of the sample numbers for each selected order against the total number of samples. These figures suggest that the most often selected orders were one while ER+ samples tended to take more complicated models than for the ER− samples.

Using residuals obtained by the best fit ARCH and AR models and the original data, we applied the Wilcoxon statistic to compare DEs between two groups divided by ER+ and ER−. The significant genomic locations were assessed by an FDR. The locations were mapped on Entrez gene IDs according to the Affy probes presented in the original microarray data and converted into gene symbols by SOURCE. The identified genes in the original data and the ARCH residual analyses are listed in Supplementary Table 2 https://www.dropbox.com/sh/sotz8jufje73eg6/AACKHD-tXqB02h_rxlvVlXqsa?dl=0. Based on these gene lists, we investigated the overlapped significant genes for the original data with significant genes for the ARCH and AR residuals and summarized the results in Table 7.5. About 200–280 significant DE genes in the studies of Des, Mil, Min, and Chin were identified with FDR correction. For Loi, the significant genes were fewer than the other studies in all cases. Except for Loi, the number of significant genes for the ARCH residuals was reduced by about 20–35% less than the number of genes for the original data. The estimated genes

Table 7.5: Summary for FDR 5% adjusted Entrez genes of five datasets

Model	Data	Des	Mil	Min	Loi	Chin
-	Original #EntrezID (unique)	245 (186)	238 (176)	274 (195)	53 (47)	277 (201)
ARCH	Residual #EntrezID (unique)	193 (152)	175 (139)	207 (154)	46 (41)	177 (133)
	Overlapped with original [%]	100	100	100	98	100
AR	Residual #EntrezID (unique)	194 (152)	161 (131)	183 (141)	37 (34)	178 (139)
	Overlapped with original [%]	95	99	87	92	98

Note: Values in parentheses indicate number of unique genes to avoid duplicate and multiple genes from the obtained gene list. Percentages for overlapped with original indicate ratios for overlapped significant genes for ARCH or AR residuals with significant genes in original data.

Table 7.6: Identified differentially expressed genes (FDR 5%) and cytobands for ER status in original data and empirical ARCH residuals

Studies	Original		ARCH residuals	
	cytoband	genes	cytoband	genes
	1p13.3	VAV3, GSTM3, CHI3L2	1p13.3	VAV3, GSTM3
	1p32.3	ECHDC2		
	1p34.1	CTPS1		
	1p35	IFI6	1p35	IFI6
	1p35.3	ATPIF1	1p35.3	ATPIF1
	1p35.3-p33	MEAF6		
	1q21	S100A11, S100A1	1q21	S100A1, S100A11
	1q21.1	PEA15	1q21.1	PEA15
	1q21.3	CRABP2	1q21.3	CRABP2
	1q23.2	COPA	1q23.2	COPA
	1q24-q25	CACYBP	1q24-q25	CACYBP
...				

Note: Complete table is shown in Table 3 in Solvang and Taniguchi (2017a).

for all the datasets (except for Loi's case) overlapped 100% with the estimated genes for the original data. The number of significant genes for the AR residuals in the cases of Mil, Min, and Loi was less than the number of genes by ARCH and resulted in about a 20–35% reduction of significant genes for the original data except for Loi. However, these genes for the AR residuals did not completely 100% overlap with the genes for the original data, unlike the case of the ARCH residuals, suggesting that empirical ARCH residuals might be more effective to correctly specify important genes from a list of long genes than AR residuals.

To investigate the overlapping genes by the ARCH residuals with genes by the original data, the corresponding cytoband and gene symbols are summarized in Table 7.6. The total numbers of common genes by the original and ARCH residuals in four studies were 132 and 99. If we take into account Loi's case, the total numbers of common genes across all studies for the original and ARCH residuals are 12 and 9. The genes obtained by the ARCH residuals were completely covered by the genes obtained by the original data. The results by the ARCH residuals covered several important genes for breast cancer, such as TP53 in the chromosome 1q region, ERBB2 in the chromosome 17q region, and ESR1 in the chromosome 6q region, even if the number of identified Entrez genes was less than the number of identified genes from the original data.

Next, we performed GO enrichment analysis using significant DE gene lists for the original data and ARCH residual analyses in all studies. To correctly find the enriched GO terms for the associated genes, a background list was prepared of all the probes included in the original microarray data. The Entrez genes in the background list were converted into 13,177 gene symbols without any duplication by SOURCE. As the input gene lists to GOrilla, the numbers of summarized unique genes are shown in the parentheses of Table 7.5. All the associated GO terms for the original and ARCH residuals in all the studies are summarized in Supplementary Table 3 https://www.dropbox.com/sh/sotz8jufje73eg6/AACKHD-tXqB02h_rxlvVlXqsa?dl=0. Since the estimated gene symbols in Loi's case were less than half of the amount taken in other studies, few associated GO terms were identified in the biological process and cellular component and no GO terms in the molecular function. Also, significant DE genes for the ARCH residuals contributed to finding additional associated GO terms that did not appear in the GO terms for the original data, e.g., mammary gland epithelial cell proliferation for Des, a single-organism metabolic process for Des, an organonitrogen compound metabolic process for Mil and Min, and a single-organism developmental process for Min and Chin, all of which are related to meaningful biological associations like cellular differentiation, proliferation, and metabolic pathways in cancer cells (Glass and Girvan (2012)).

Table 7.7 summarizes the common GO terms of the biological processes for Des, Mil, Min, and Chin and presents 13 terms for the original data. The terms for the ARCH residuals mostly overlapped with them except for Mil's case. As shown in Supplementary Table 3 https://www.dropbox.com/sh/sotz8jufje73eg6/AACKHD-tXqB02h_rxlvVlXqsa?dl=0, two terms in the molecular function and eight in the cellular components were commonly identified by the original data. The GO terms for the ARCH residuals covered them, and more terms were shown in the molecular function. Furthermore, to investigate the consistency of the refined significant gene lists, we applied pathway analysis to the significant DE genes for the original and ARCH residuals listed in Table 7.6. All the identified pathways with Entities FDR (<1.0) and associated genes are summarized in Supplementary Table 4 https://www.dropbox.com/sh/sotz8jufje73eg6/AACKHD-tXqB02h_rxlvVlXqsa?dl=0. In the pathway components shown in Supplementary Table 3 https://www.dropbox.com/sh/sotz8jufje73eg6/AACKHD-tXqB02h_rxlvVlXqsa?dl=0, ERBB2 signaling, EGFR, cell-cycle, immune system, metabolic pathway, AKT signaling and Wnt pathway are well-known important breast cancer signaling pathways (Teschendorff et al. (2007)). We took them to be representative of important pathways and counted the number of identified pathways related to these components in the case of the original and ARCH residuals. The number and associated gene symbols are summarized in Table 7.8. The representative pathways were mostly covered by the significant DE genes for the ARCH residuals. This result supports that the refined gene lists obtained by the ARCH residuals generally captured the differentiating breast tumours based on ER status and did not overlook any important biological information by the limited DE gene lists for the ARCH residuals.

7.3.6 Conclusions

We applied a rank order statistic for an ARCH residual empirical process to refine significant DE genes by two-group comparison in microarray analysis. Our approach considered publicly available gene expression datasets and the clinical output for ER in addition to the simulation study. We compared the analysis performances by the ARCH residuals with the AR residuals and the ordinal original microarray data. While the genes for the AR residuals did not cover 100% of the genes for

Table 7.7: Common associated biological processes among Des, Mil, Min, and Chin for original and ARCH residuals

Associated GO terms	Des		Mil		Min		Chin	
	Orig	Arch	Orig	Arch	Orig	Arch	Orig	Arch
epithelial cell proliferation	+	+	+	+	+	+	+	+
response to estrogen	+	+	+	+	+	+	+	+
epidermis development	+	+	+		+	+	+	
regulation of phosphatidylonositol 3-kinase activity	+	+	+	+	+		+	+
erythropoietin-mediated signaling pathway	+	+	+		+		+	+
regulation of lipid kinase activity	+	+	+	+	+	+	+	+
phosphatidylinostitol 3-kinase signaling	+	+	+	+	+	+	+	+
positive regulation of phosphatidylinositol 3-kinase activity	+	+	+	+	+	+	+	+
phenylpropanoid catabolic process	+	+	+	+	+	+	+	+
mast cell differentiation	+	+	+	+	+	+	+	+
extracellular vesicle	+	+	+	+		+	+	+
extracellular vesicular exosome	+	+	+	+		+	+	+
extracellular region part	+	+	+	+		+	+	+

Table 7.8: Identified important breast cancer signaling pathways and associated gene symbols obtained from original data and ARCH residuals

Pathways	Original		ARCH residuals	
	number	Gene symbol	number	Gene symbol
ERBB2 signaling	7	ERBB2, KIT, NRG1	6	ERBB2
EGFR pathways	11	FLT1, KIT, PIK3R1, VAV3	10	ERBB2, FLT1, PIK3R1
Cell cycle	5	CDH1, MCM3	3	MCM3, NUMA1
Immune system	5	CDH1, KIT, PIK3R1, STAT6	5	ERBB2, IFI6, STAT6
Metabolic disorder	1	SAT1	1	FZD1
PI3K/AKT signaling	5	KIT, PIK3R1	5	ERBB2, PIK3R1
Wnt pathway	5	FZD1	5	FZD1

the original data analysis, the genes by the ARCH residuals were mostly 100% overlapping with the original data, and the gene lists were reduced about 30% from the gene lists obtained by the original data analysis. We confirmed a similar property for the 30% reduction in the simulation study. In GO enrichment and pathway analyses, the result by the ARCH residuals was mostly covered with associated biological terms obtained by the original data analysis and presented additional important GO terms in biological processes. These results suggest that data processing using ARCH residuals array data could contribute to refining significant DE genes that follow the required gene signatures and provide prognostic accuracy and guide clinical decisions.

7.4 Portfolio Estimation for Spectral Density of Categorical Time Series Data

Stoffer et al. (1993) proposed a way of scaling categorical time series data and presented the frequency domain framework called Spectrum Envelope (SpecEnv). SpecEnv can be understood as the largest proportion of the total power, with the maximum achieved by the scaling parameters. We consider this a similar concept to the portfolio approach to measuring risk diversification proposed by Meucci (2010). In this article, we explore the relationship between SpecEnv and the diversification index and, moreover, provide a portfolio estimation of spectral density. Furthermore, we apply this portfolio estimation to the study of DNA sequences and confirm its consistency with the results shown in Stoffer et al. (2000), Solvang and Taniguchi (2017b).

7.4.1 *Introduction*

Stoffer et al. (1993) presented a method of scaling categorical time series data and a framework of spectral analysis called Spectral Envelope (SpecEnv). The SpecEnv approach envelops the standardized spectrum of any scaled process. For a categorical time series, the spectral envelope discovers the efficient periodic components in it. The method was applied to DNA sequence data for a specific gene and used to explore the emphasized periodic feature, the strong-weak bonding alphabet of the sequence, and the coding/noncoding areas (Stoffer et al. (2000)). The SpecEnv and the optimum scales correspond to the largest eigenvalue and the eigenvector of the matrix for the standardized spectral density of the scaled categorical data. In other words, since the scaling characterizes the frequency-based categorical contribution to maximize the diversification of the data, it comes to serve, for example, as principal component analysis of multivariate time series. We find similar attempts to this approach in the principal portfolio proposed by Rudin and Morgan (2006) and the diversification analysis proposed by Meucci (2010). They introduced a diversification index that represents the effective number of bets in a portfolio based on maximizing the entropy, calculated by the contribution from each asset to the total portfolio. The principles of maximum entropy and minimum cross-entropy have been extensively applied in finance (Zhou et al. (2013)). For a robust portfolio problem, Xu et al. (2014) pointed out that entropy serves as a measure of portfolio diversification and suggested that the maximum entropy method provides better performance than a classical mean-variance model. In this section, we explore the relationship between SpecEnv and the diversification index and provide a procedure to estimate a portfolio for the spectral density of categorical time series data. We also apply the proposed procedure to the DNA sequence data used for the analysis in Stoffer et al. (2000).

7.4.2 *Method*

7.4.2.1 *Spectral Envelope*

First, we briefly summarize the fundamental theory of SpecEnv according to Stoffer et al. (1993). A stationary time series $\{X_t, t = 0, \pm 1, \pm 2, \cdots\}$ is decomposed into orthogonal components at frequency λ measured in cycles per unit of time, where $-\pi \leq \lambda \leq \pi$. When we observe a numerical time series, X_t, $t = 1, \cdots, n$, the periodogram is defined as $I_n(\lambda) = \left|(2\pi n)^{-1/2} \sum_{t=1}^{n} X_t e^{it\lambda}\right|^2$. Since the periodogram is not consistent with the spectral density, the periodogram is smoothed as $f(\lambda_j) = \sum_{q=-m}^{m} h_q I_n(\lambda_{j+q})$ where $j = 1, 2, \ldots, [n/2]$ ($[n/2]$ is the greatest integer less than or equal to $n/2$), the weights are chosen for $h_q = h_{-q} \geq 0$ and $\sum_{q=-m}^{m} h_q = 1$. We use the case where $h_q = 1/(2m+1)$ for $q = 0, \pm 1, \cdots, \pm m$, known as the Daniell kernel smoother (moving average).

Now, let X_t, $t = 0, \pm 1, \pm 2, \ldots$, be a categorical-valued time series with finite state-space $c = \{c_1, c_2, \cdots, c_k\}$. Assume that X_t is stationary and $p_j = pr\left\{X_t = c_j\right\} > 0$ for $\lambda \in (-\pi, \pi]$. For $\beta = (\beta_1, \beta_2, \cdots, \beta_k)^t \in \mathbb{R}^k$, $X_t(\beta)$, corresponds to the scaling that assigns the category c_j the numerical value β_j, which is called the scaling parameter and used to detect a sort of weight for each category. The spectral envelope aims to find the optimal scaling β maximizing the power at each frequency λ

for $\lambda \in (-\pi, \pi]$, relative to the total power $\sigma^2(\beta) = \text{var}\{X_t(\beta)\}$. That is, we choose $\beta(\lambda)$, at each of interest λ, so that

$$s(\lambda) = \sup_{\beta} \left\{ \frac{f(\lambda;\beta)}{\sigma^2(\beta)} \right\}. \tag{7.4.1}$$

over all β not proportional to 1_k, the $k \times 1$ vector of ones. The variable $s(\lambda)$ is called the Spectral Envelope (SpecEnv) for a frequency λ. When $X_t = c_j$, we then define a k-dimensional time series $\boldsymbol{Y}_t$ by $\boldsymbol{Y}_t = \boldsymbol{e}_j$, where $\boldsymbol{e}_j$ indicates the $k \times 1$ vector with 1 in the jth row and zeros elsewhere. The time series $X_t(\beta)$ can be obtained from $\boldsymbol{Y}_t$ by the relationship $X_t(\beta) = \beta^t \boldsymbol{Y}_t$. Assume that the vector process $\boldsymbol{Y}_t$ has a continuous spectral density matrix $f_Y(\lambda)$ for each λ. The relationship $X_t(\beta) = \beta^t \boldsymbol{Y}_t$ implies that $f_X(\lambda;\beta) = \beta^t f_Y^{\text{Re}}(\lambda)\beta$, where $f_Y^{\text{Re}}(\lambda)$ denotes the real part of $f_Y(\lambda)$. Denoting the variance-covariance matrix of $\boldsymbol{Y}_t$ by $\boldsymbol{V}$, formula (7.4.1) can be expressed as

$$s(\lambda) = \sup_{\beta} \left\{ \frac{\beta^t f_Y^{\text{Re}}(\lambda)\beta}{\beta^t \boldsymbol{V} \beta} \right\}$$

In practice, $\boldsymbol{Y}_t$ is formed by the given categorical time series data, the fast Fourier transform of $\boldsymbol{Y}_t$ is calculated, the periodogram should be smoothed, the variance-covariance matrix $\boldsymbol{S}$ of $\boldsymbol{Y}_t$ is calculated, and, finally, the largest eigenvalue and the corresponding eigenvector of the matrix $2n^{-1}\boldsymbol{S}^{-1/2}\boldsymbol{f}^{re}\boldsymbol{S}^{-1/2}$ are obtained. The sample spectral envelope corresponds to the eigenvalue. Furthermore, the optimal sample scaling is the multiplication of $\boldsymbol{S}^{-1}$ and the eigenvector.

7.4.2.2 *Diversification analysis*

Meucci (2010) provided a diversification index as the measurement for risk diversification management. He first represented a generic portfolio as the weights of each asset, the N-dimensional return of assets $\boldsymbol{R}$, and the return on the portfolio $\boldsymbol{R}_w = \boldsymbol{w}^t \boldsymbol{R}$. For the covariance matrix Σ with the N-dimensional return of assets, principal component analysis (PCA) was applied using $\Sigma = \boldsymbol{E}\boldsymbol{\Lambda}\boldsymbol{E}^t$, where $\boldsymbol{\Lambda} = \text{diag}(v_1, \cdots, v_N)$ denotes the eigenvalue and $\boldsymbol{E} = (\boldsymbol{e}_1, \cdots, \boldsymbol{e}_N)$ denotes the eigenvector. The eigenvectors define a set of N uncorrelated portfolios, and the returns of the principal portfolios are represented by $\tilde{\boldsymbol{R}} = \boldsymbol{E}^{-1}\boldsymbol{R}$. Then, the generic portfolio can be considered a combination of the original assets with weights $\tilde{\boldsymbol{w}} = \boldsymbol{E}^{-1}\boldsymbol{w}$. Since the principal portfolios are uncorrelated, the total portfolio variance is obtained by $\text{Var}(\boldsymbol{R}_w) = \sum_{i=1}^N w_i^2 v_i$. Furthermore, the risk contribution from each principle portfolio to total portfolio variance is $p_i = \tilde{w}_i^2 v_i / \text{Var}(\boldsymbol{R}_w)$. Using the contribution p_i, he introduced an entropy as the portfolio diversification by

$$N_{\text{ENT}} \equiv \exp\left(-\sum_{i=1}^{N} p_i \ln p_i \right).$$

If all of the risk is completely due to a single principal portfolio, that is, $p_i = 1$ and $p_j = 0$ $(i \neq j)$, we obtain $N_{\text{ENT}} = 1$. On the other hand, if the risk is homogeneously spread among N (that is, $p_i = p_j = 1/N$), we obtain $N_{\text{ENT}} = N$. To optimize diversification, Meucci presented the mean-diversification efficient frontier

$$\boldsymbol{w}_\varphi \equiv \arg\max_{\boldsymbol{w} \in C} \left\{ \varphi \boldsymbol{w}^t \boldsymbol{\mu} + (1 - \varphi) N_{\text{ENT}}(\boldsymbol{w}) \right\} \tag{7.4.2}$$

where $\boldsymbol{\mu}$ denotes the estimated expected returns, C is a set of investment constraints, and $\varphi \in [0, 1]$. For small values of φ, diversification is the main concern, whereas as $\varphi \to 1$, the focus shifts on the expected returns (Meucci (2010)).

7.4.2.3 *An extension of SpecEnv to the mean-diversification efficient frontier*

For the mean-diversification efficient frontier in Meucci (2010), if we take $\varphi = 0$, and if $N_{\text{ENT}}(\cdot)$ is the spectral density of the portfolio, then it loads to the spectral envelope. If $\varphi \neq 0$ and the entropy

$N_{\text{ENT}}(\cdot)$ is the variance of the portfolio, Equation (7.4.2) leads to the mean-variance portfolio to the periodogram at λ as

$$\boldsymbol{w}_\varphi \equiv \arg\max_{\boldsymbol{w}\in C} \left\{\varphi \boldsymbol{w}^t \boldsymbol{\mu}(\lambda) + (1-\varphi)\boldsymbol{w}^t \text{Var}(\tilde{\boldsymbol{f}}_Y^{\text{Re}}(\lambda))\boldsymbol{w}\right\} \tag{7.4.3}$$

where $\boldsymbol{\mu}(\lambda)$ denotes the mean vector of the periodgram smoothed by Daniell kernel around λ and $\text{Var}(\tilde{\boldsymbol{f}}_Y^{\text{Re}}(\lambda))$ denotes the variance of the periodogram. Although Meucci (2010) estimated the optimum portfolio for diversification by the principal portfolios, we simplify using the following mean-variance analysis (Taniguchi et al. (2008)):

$$\begin{cases} \max_{\boldsymbol{w}} \{\boldsymbol{\mu}^t \boldsymbol{w} - \beta \boldsymbol{w}^t \boldsymbol{V} \boldsymbol{w}\} \\ \text{subject to } \boldsymbol{e}^t \boldsymbol{w} = 1 \end{cases} \tag{7.4.4}$$

where $\boldsymbol{e} = (1, 1, \cdots, 1)^t$ ($m \times 1$ vector) and β is a given number. If we substitute (7.4.3) to (7.4.4), the solution for $\boldsymbol{w}$ is given by

$$\boldsymbol{w}(\lambda) = \frac{\varphi}{1-\varphi}\left\{\text{Var}(\tilde{\boldsymbol{f}}_Y^{\text{Re}}(\lambda))^{-1}\boldsymbol{\mu}(\lambda) - \frac{\boldsymbol{e}^t\text{Var}(\tilde{\boldsymbol{f}}_Y^{\text{Re}}(\lambda))^{-1}\boldsymbol{\mu}(\lambda)}{\boldsymbol{e}^t\text{Var}(\tilde{\boldsymbol{f}}_Y^{\text{Re}}(\lambda))^{-1}\boldsymbol{e}}\text{Var}(\tilde{\boldsymbol{f}}_Y^{\text{Re}}(\lambda))^{-1}\boldsymbol{e}\right\} + \frac{\text{Var}(\tilde{\boldsymbol{f}}_Y^{\text{Re}}(\lambda))^{-1}\boldsymbol{e}}{\boldsymbol{e}^t\text{Var}(\tilde{\boldsymbol{f}}_Y^{\text{Re}}(\lambda))^{-1}\boldsymbol{e}} \tag{7.4.5}$$

For a small value φ, we obtain the portfolio return and variance by $\boldsymbol{\mu}(\tilde{\boldsymbol{f}}_Y^{\text{Re}})^t\boldsymbol{w}(\lambda)$ and $\boldsymbol{w}(\lambda)^t\text{Var}(\tilde{\boldsymbol{f}}_Y^{\text{Re}}(\lambda))^t\boldsymbol{w}(\lambda)$.

In applying SpecEnv to categorical data, one category was fixed as a base and the number of variables in the scaling parameter vector $\boldsymbol{\beta}$ should be the total number of categories -1. $\boldsymbol{Y}_t$ should be formed as the matrix for the number of data $\times$ the total number of the categories -1. On the other hand, in applying a portfolio study, we should not take a "base" for the factorization and the number of weights estimated by Equation (7.4.5) should be equivalent to the number of categories.

7.4.3 Data

7.4.3.1 Simulation data

To confirm the performance of our proposed extension of SpecEnv to the mean-diversification efficiency frontier, we conducted a simulation study. We considered sequence data including four numbers (1, 2, 3 and 4) . We first generated 1,000 random four-character sequences. Next, the number "3" was placed instead of the number located at every ten and every five positions in the entire sequence. For the procedure, we used the following R code:

```
tmp <- sample(c(1,2,3,4),1000,replace=TRUE)
l31 <- seq(from = 1, to = 1000, by = 5 ); tmp[l31] <- 3
l32 <- seq(from = 1, to = 1000, by = 10 ); tmp[l32] <- 3.
```

7.4.3.2 DNA sequence data

As a case of real data analysis in this section, we used the DNA sequence for Epstein–Barr virus (EBV), which was demonstrated in Stoffer et al. (2000) as an example of categorical time series data. A DNA sequence is represented by four letters, termed base pairs (bp), from the finite set of alphabetic characters related to two purines, adenine (A) and guanine (G), and two pyrimidines, cytosine (C) and thymine (T). The order of the alphabet (nucleotide) contains the genetic information specific to the organism. Translating the information stored in the protein-coding sequences (CDS) of the DNA is an important process. CDS disperse throughout the sequence and are separated by regions of noncoding (Stoffer (2005)). Stoffer et al. (2000) clarified the optimal periodic length of the DNA sequence for specific genes and identified the most favourable alphabets, such as the strong-weak (S-W) alphabet related to regional fluctuations or codon context, for use in individual amino acid shifting. The entire EMV DNA sequence they applied consists of approximately 172,000 bp.

We use a part of the sequence in the region of 1736 – 5689 bp related to a gene labeled BNRF1 shown in Stoffer et al. (2000). They analyzed the entire 3954-bp sequence and its four 1000-bp windows of the gene.

7.4.4 *Results and Discussion*

7.4.4.1 *Simulation study*

The generated sequence is summarized in Box 1. The total number of generated data for each category was 209 for 1, 184 for 2, 414 for 3 and 193 for 4.

Box 1. Generated artificial sequence data

3 3 3 2 3 3 3 3 4 3 3 2 3 2 3 1 1 3 2 3 4 4 2 1 3 2 2 2 4 3 4 3 2 2 3 4 4 2 4 3 2 3 3 2 3 3 2 2 2 3
1 1 3 1 3 3 3 4 2 3 1 3 3 2 3 4 4 2 3 3 3 1 4 3 3 1 2 4 4 3 4 1 3 4 3 2 3 4 4 3 3 1 3 3 3 2 1 1 2 3 3
4 3 3 3 4 4 1 4 3 4 1 1 3 3 2 1 1 1 3 3 4 1 1 3 4 3 1 1 3 1 1 1 1 3 1 2 1 2 3 3 1 1 3 3 4 4 4 2 3 4 2
1 4 3 1 4 1 2 3 4 4 2 1 3 1 3 1 2 3 4 3 4 4 3 1 2 4 2 3 2 1 2 2 3 1 3 3 1 3 3 1 3 1 3 4 1 3 3 3 3 4 2
1 3 3 4 2 3 3 1 2 3 1 3 4 2 1 1 3 1 3 3 4 3 3 1 3 3 3 3 1 4 3 3 1 1 2 2 3 4 3 4 2 3 2 2 4 2 3 4 3 2 4
3 2 3 4 1 3 3 3 3 2 3 2 2 2 1 3 2 4 2 1 3 2 1 3 4 3 3 4 4 3 3 4 4 1 1 3 3 1 4 4 3 3 1 1 3 3 2 1 2 3 3
3 3 2 3 3 3 4 3 2 3 3 1 1 4 3 3 3 2 3 3 3 1 2 4 3 1 1 4 4 3 4 1 3 3 3 2 1 2 3 3 1 1 1 4 3 3 1 1 2 3 2
1 2 4 3 1 1 3 1 3 3 2 2 2 3 3 3 4 1 3 1 4 2 4 3 4 3 2 3 3 3 3 3 3 3 1 4 3 3 3 1 4 2 3 3 2 1 3 1 3 2 3
2 2 3 2 2 4 4 3 2 2 1 1 3 4 1 4 3 3 3 1 1 1 3 3 3 2 1 3 4 4 4 3 3 2 2 1 4 3 1 3 2 2 3 1 3 4 4 3 2 2 2
3 3 3 4 2 2 3 2 4 2 4 3 2 4 2 2 3 1 1 3 2 3 4 4 1 2 3 1 1 3 2 3 2 4 3 2 3 1 1 3 2 3 3 3 1 1 3 1 2 1 2
3 4 1 4 4 3 1 2 1 1 3 2 1 3 3 3 3 4 3 1 3 2 1 2 3 3 3 4 2 1 3 3 4 2 2 3 2 1 4 4 3 1 1 4 3 3 4 2 4 3 3
1 3 3 2 3 2 4 1 1 3 3 2 3 4 3 3 3 3 1 3 2 1 1 2 3 4 1 2 3 3 2 3 4 2 3 1 3 2 4 3 1 1 4 1 3 4 4 3 3 3 4
4 4 4 3 4 1 3 3 3 3 3 1 2 3 4 1 1 4 3 3 3 1 3 3 4 3 2 3 3 1 4 1 3 3 1 2 4 2 3 4 3 4 2 3 3 1 4 3 3 4 1
4 1 3 3 1 4 4 3 1 1 4 3 3 1 1 1 3 3 1 4 4 2 3 3 2 4 2 3 2 2 3 4 3 3 2 3 4 3 4 2 2 1 3 2 4 4 1 3 2 1 4
1 3 3 4 1 2 3 4 2 1 1 3 4 3 1 2 3 2 2 3 2 3 3 4 2 3 3 4 3 1 3 3 3 1 1 2 3 1 3 3 3 3 1 4 2 2 3 2 4 4 3
3 2 2 1 1 3 1 1 4 1 3 2 2 4 3 3 4 1 3 4 3 1 3 2 1 3 4 4 4 3 3 3 2 3 3 3 3 3 1 4 3 1 1 4 3 3 4 2 2 3 3
4 4 2 4 3 3 3 3 2 3 2 4 1 1 3 2 2 1 1 3 4 4 4 4 3 2 4 2 2 3 1 3 4 2 3 2 2 3 3 3 4 2 1 3 3 2 3 4 1 3 1
3 3 2 3 1 4 3 1 3 4 4 4 4 3 1 4 2 3 3 1 3 2 3 3 4 3 3 1 3 1 2 1 3 3 3 1 4 1 3 2 3 4 2 3 4 2 4 4 3 2 3
3 4 3 3 1 1 1 3 1 3 3 1 3 1 3 4 1 3 3 3 4 3 3 4 3 4 1 3 3 2 4 3 3 4 3 4 1 3 1 1 4 2 3 1 3 2 2 3 4 2 3
3 3 4 1 2 1 3 1 2 1 3 3 4 3 1 4 3 1 3 4 3 3 2 3 4 4 3 1 1 4 1

For the data, we first applied the SpecEnv approach. We summarize the calculated periodogram and SpecEnv in Figures 7.12. The obtained SpecEnv contained two clear peaks at the 0.2 and 0.4 frequencies (202nd and 404th positions). In this calculation, the factor for the fourth category was treated as a base. The obtained scaling parameters were 0.018 for 1, 0.087 for 2, and 0.996 for 3 at the 0.2 frequency, and 0.013 for 1, −0.002 for 2, and 1.00 for 3 at the 0.4 frequency. Next, we applied the portfolio study to the data. The portfolio weights were obtained for the two clear peaks at the 0.2 and 0.4 frequency. The portfolio returns $\boldsymbol{\mu}(\tilde{\boldsymbol{f}}_Y^{\mathrm{Re}})^t\boldsymbol{w}(\lambda)$ and variances by the obtained weights were plotted for the small value $\varphi = 0, \cdots 0.1$ in Figures 7.13 (middle and lower). The configurations of those plots were basically not different according to the degree φ for dividing mean and diversification. The portfolio weights for categories 1 and 2 at the first peaks are close, and the weights for category 3 are far from the weights for categories 1 and 2 at the second peaks. The tendency of the obtained weights is consistent with the obtained scaling parameters of SpecEnv. The portfolio variances indicate two large peaks at the 0.2 and 0.4 frequencies, with the second peak twice as high as the first. The frequencies of the maximum two peaks are consistent with the frequencies of the two largest SpecEnv. The configurations for the obtained portfolio return are also consistent for $\varphi = 0, \cdots 0.1$.

As we mentioned in Section 7.4.2.3, the $\boldsymbol{\mu}(\lambda)$ was the mean vector of the periodogram smoothed by Daniell kernel around λ. If we simply used the periodogram without smoothing by Daniel kernel, the portfolio weights would not become consistent for the increases of φ, as shown in Figure 7.14

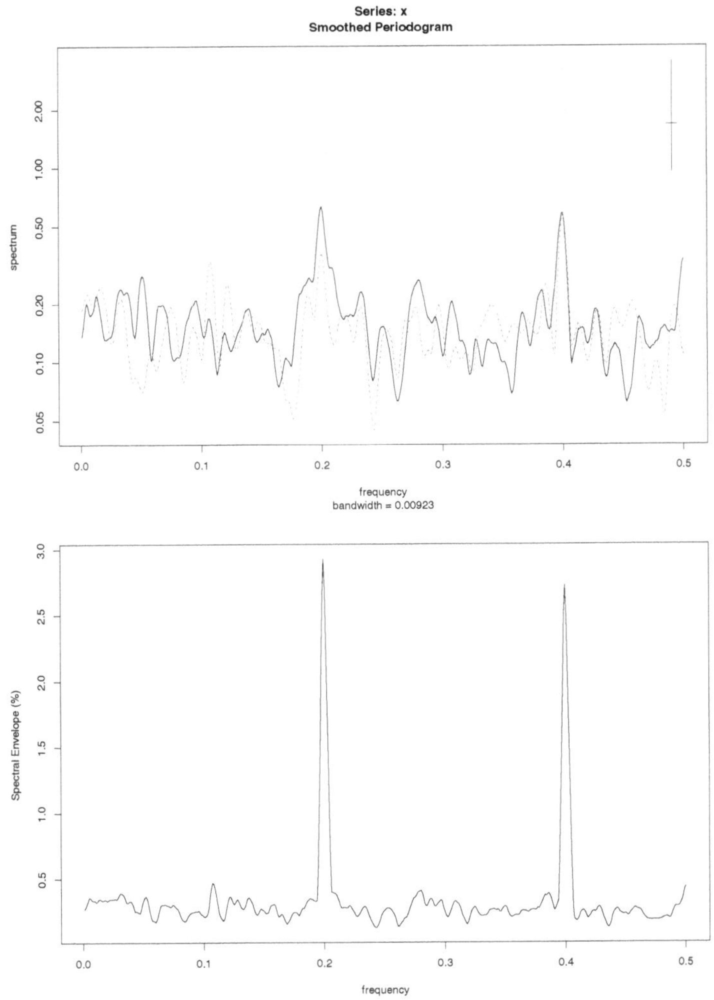

Figure 7.12: Plots for the periodogram and calculated SpecEnv for the simulation data.(Upper) Periodogram for the simulation data. Solid line for 1, dotted line for 2 and broken line for 3. (Lower) SpecEnv for the simulation data. Two peaks appeared at 0.2 and 0.4 frequencies.

(Upper). The estimates for portfolio variance and returns summarized in Figure 7.14 (Middle) and (Lower) also came to fluctuate for $\varphi > 0.075$. The results would suggest that the portfolio estimation depends on the smoothing for the periodogram. To confirm this, we should compare the portfolio estimates for the periodogram while applying another smoothing function as a further consideration.

7.4.4.2 DNA sequence for the BNRF1 genes

According to Stoffer's article, we first performed SpecEnv on the entire 3954 bp for the BNRF1 gene and on four 1000-bp sections, using (formula) with $m = 7$. The estimated SpecEnv are illustrated in Figure 7.15 for the entire sequence and in the four panels of Figures 7.16, 7.17, 7.18 and 7.19 for four sections. Just as they found, we confirmed the strong peak at frequency $1/3$ for the entire coding sequence in Figure 7.15, and this same peak appears in the first three sections in Figures 7.16, 7.17, and 7.18, while there is no peak in Figure 7.19. This raises the conjecture that the fourth section was actually protein noncoding. The optimal scalings for the first three sections were (b) A = 0.05, C = −0.68, G = −0.72, T = 0; (c) A = −0.06, C = −0.69, G = −0.71, T = 0; (d) A = −0.17, C

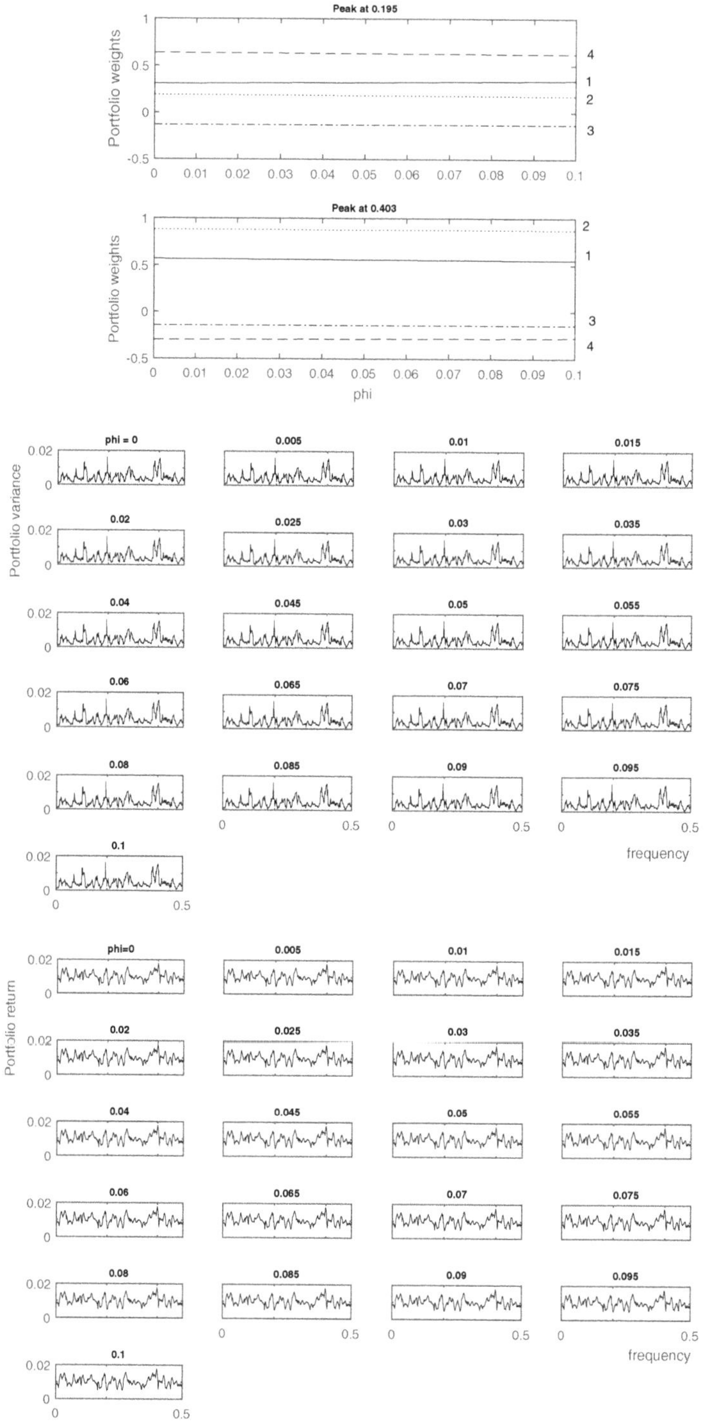

Figure 7.13: Portfolio estimates for the simulation data. (Upper) Estimated portfolio weights of simulation data for $\varphi = 0, \cdots, 0.1$. (Middle) Portfolio variances for $\varphi = 0, \cdots, 0.1$. (Lower) Portfolio returns for $\varphi = 0, \cdots, 0.1$.

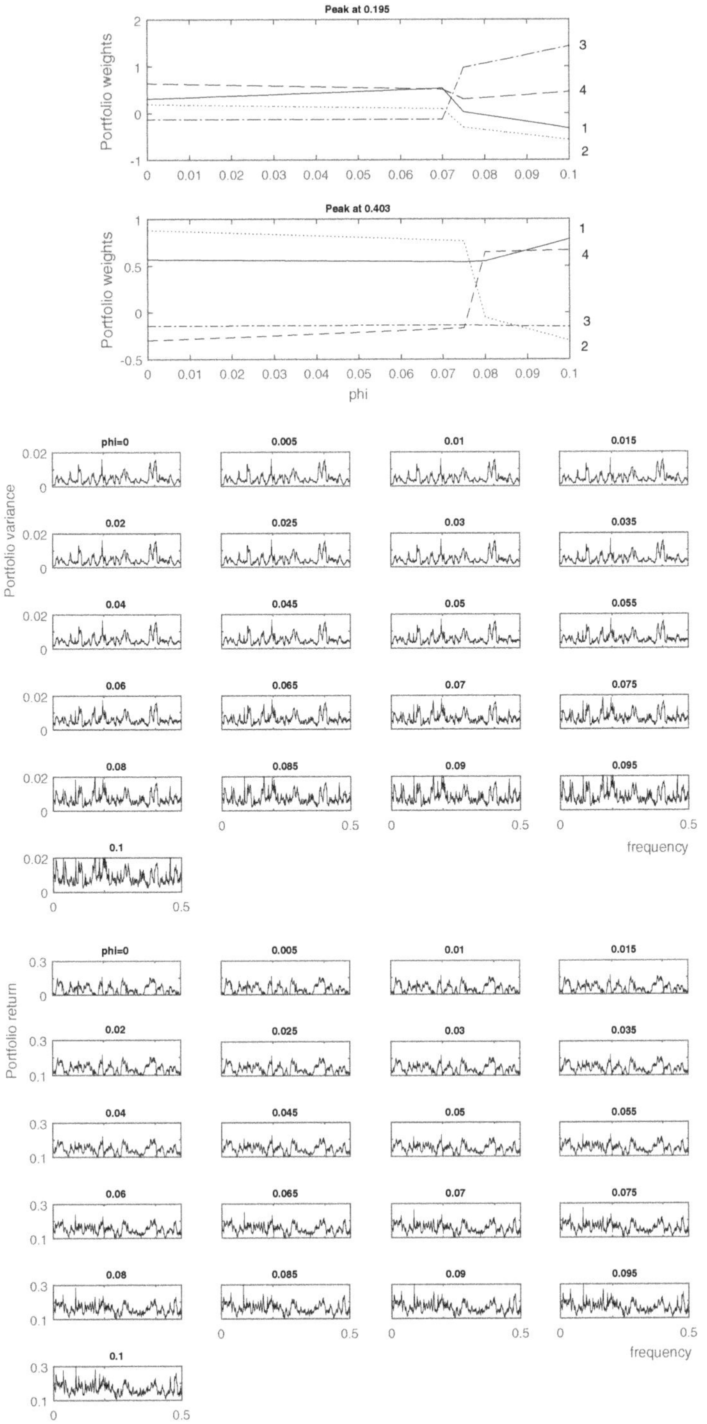

Figure 7.14: Results of the portfolio estimates using $\mu(\lambda)$ by periodogram without a Daniell smoother. (Upper) Estimated portfolio weights for $\varphi = 0, \cdots, 0.1$. (Middle) Portfolio variances for $\varphi = 0, \cdots, 0.1$. (Lower) Portfolio returns for $\varphi = 0, \cdots, 0.1$.

= −0.58, G = −0.80, T = 0. The first two sections show the consistent strong-weak bonding alphabet for the combination of C and G toward the combination of A and T (strong for C&G and weak for A&T), and the third shows consistent minor bonding, as mentioned in Stoffer et al. (2000).

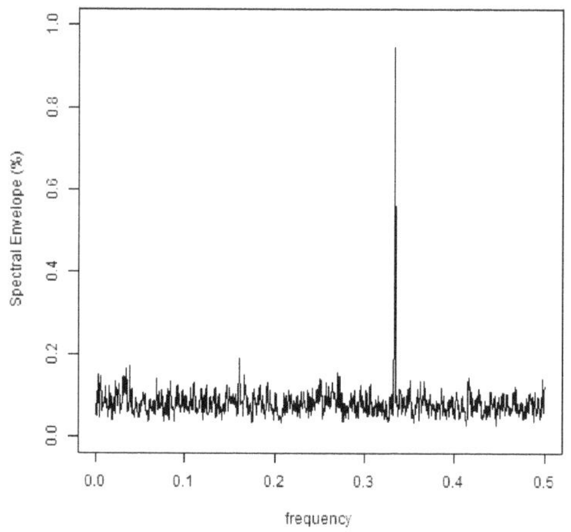

Figure 7.15: Estimated SpecEnv for BNRF1 gene of EBV virus (a. overall set of 3945 sequences1)

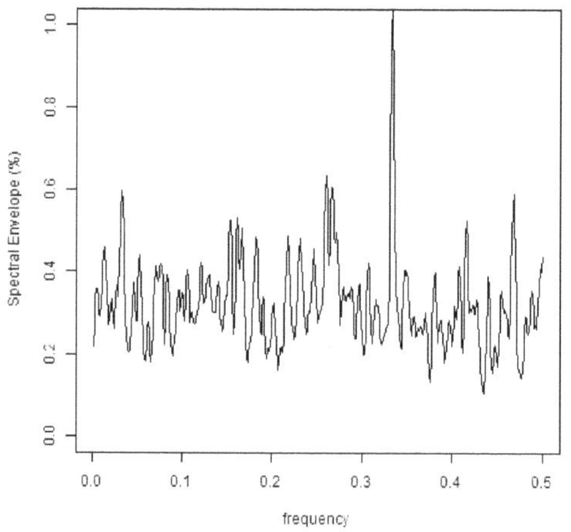

Figure 7.16: Estimated SpecEnv for BNRF1 gene of EBV virus (b. first 1000 sections)

Next, we applied the portfolio estimation to the fourth section DNA sequence data. The results are summarized in Figures 7.20 – 7.23. For $\varphi = 0, \dots, 0.1$, the configurations of the obtained variance indicated close similarity with the obtained SpecEnv for each φ in the corresponding section. The maximum peaks in the first, second and third sections stood at the $1/3$ frequency as well as SpecEnv. Furthermore, we compared the portfolio weights of the four categories based on the maximum peaks at the $1/3$ frequency. The results are summarized in Figures 7.24 – 7.27. We found that the weights for C and G were close in the first and second sections in Figures 7.24 and 7.25. This suggests a strong combination of C and G in the sequence sections. These results are consistent with those of Stoffer et al. (2000). On the other hand, the weights in the third section, as shown in Figure 7.26, did not indicate a strong relationship between C and G. For the third section, Stoffer

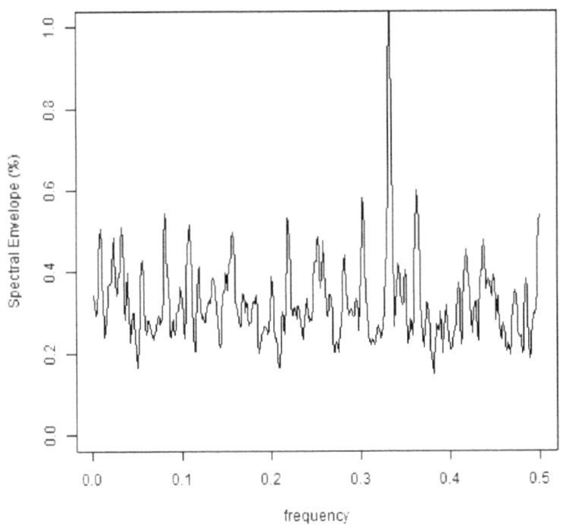

Figure 7.17: Estimated SpecEnv for BNRF1 gene of EBV virus (c. second 1000 sections)

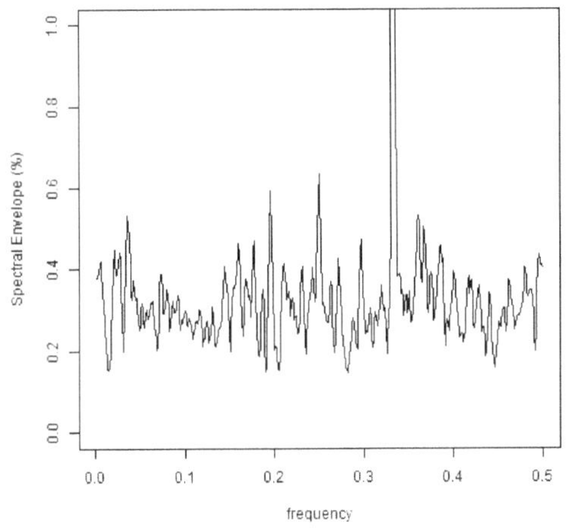

Figure 7.18: Estimated SpecEnv for BNRF1 gene of EBV virus (d. third 1000 sections)

et al. (2000) showed some minor departure from the strong-weak bonding alphabet. In the fourth section, shown in Figure 7.27, the weights for A-C and G-T were close at $\varphi = 0$. However, when φ increases, the disparity between the weights for A and C also increases. For the fourth section, Stoffer et al. (2000) also showed that no signal was present in the section and concluded that the fourth section of BNRF1 of Epstein–Barr was actually noncoding.

In the case of SpecEnv for the second section, the obtained scaling parameters indicated strong C-G bonding. The portfolio weights in Figure 7.25 also indicated the same characteristics. The portfolio variance for the second section showed consistency for $\varphi = 0, \cdots, 0.1$. To confirm the possible consistency for larger φ, we calculated the variations for $0.1 \leq \varphi \leq 0.9$, shown in Figure 7.28. The peaks at the $1/3$ frequency were not different until $\varphi = 0.6$; however, as shown in Figure 7.29 the portfolio weights actually changed from $\varphi = 0.3$. For $\varphi > 0.5$, the divergence between the weights became apparent. The portfolio variance in the same range for φ showed several peaks.

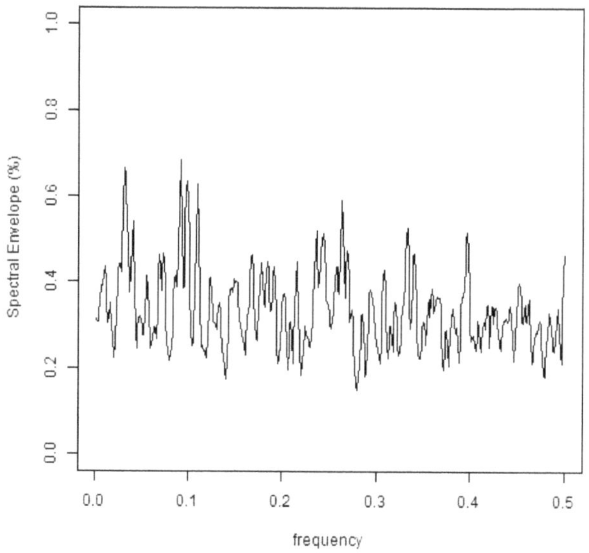

Figure 7.19: Estimated SpecEnv for BNRF1 gene of EBV virus (e. fourth 1000 sections)

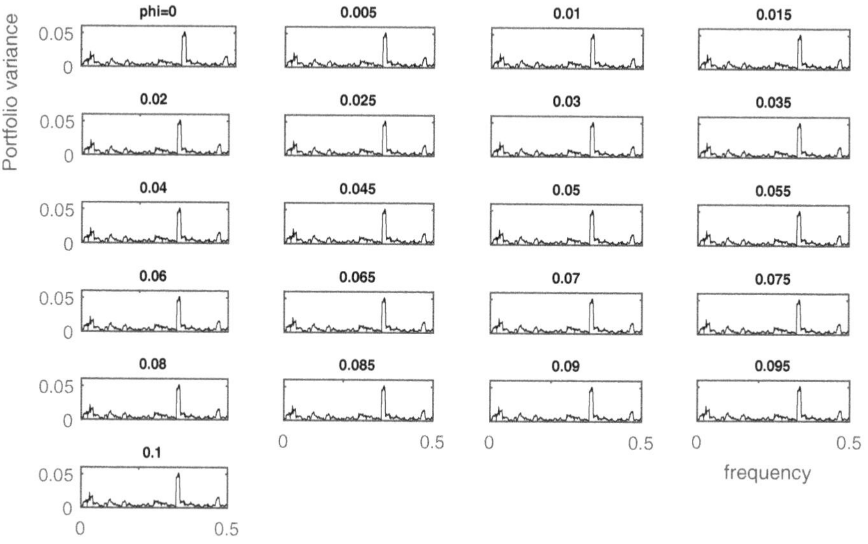

Figure 7.20: Estimated portfolio variance plots for first 1000 sections

7.4.5 Conclusions

We considered an extension to the current mean-diversification efficient frontier that is based on using SpecEnv. A simulation study was performed to explore the relationship between SpecEnv and the proposed portfolio approach. Furthermore, we applied the proposed approach to real DNA

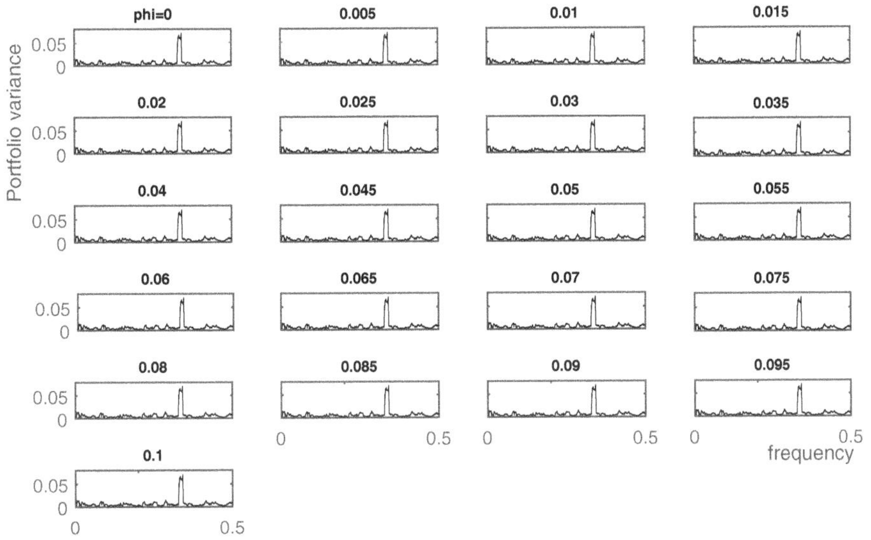

Figure 7.21: Estimated portfolio variance plots for the second 1000 sections

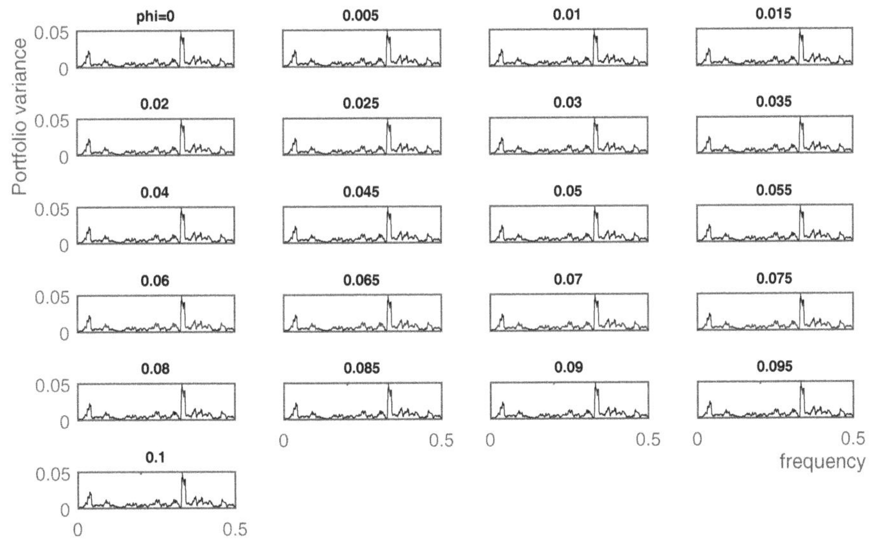

Figure 7.22: Estimated portfolio variance plots for the third 1000 sections

sequence data of the BNRF gene for Epstein–Barr virus (EBV). We found a consistent tendency in the results between SpecEnv and portfolio estimation. Our findings for the proposed approach support the idea that the scaling parameters for SpecEnv correspond to the portfolio weights, suggesting that it is a useful method of portfolio estimation for categorical time series data.

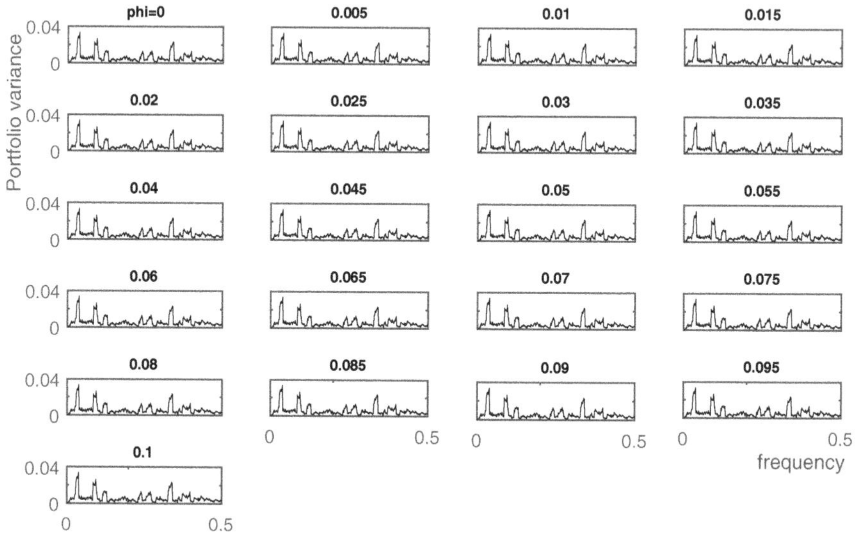

Figure 7.23: Estimated portfolio variance plots for the fourth 1000 sections

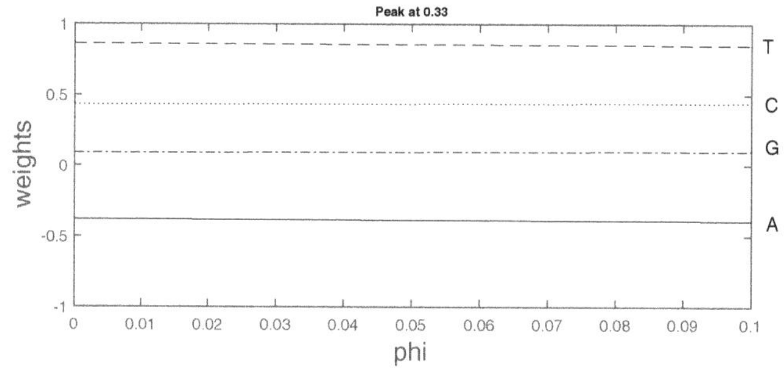

Figure 7.24: Estimated portfolio weights for the first 1000 sections and $\varphi = 0, \cdots, 0.1$

7.5 Application to Real-Value Time Series Data for Corticomuscular Functional Coupling for SpecEnv and the Portfolio Study

7.5.1 *Introduction*

Corticomuscular functional coupling (CMC) forms the functional connection between the brain and the peripheral organs. The evaluation of CMC is important for revealing pathophysiological mechanisms of many movement disorders, such as paresis and involuntary movements like dystonia, tremors, and myclonus (Mima and Hallett (1999)). The mutual relationship between electromyograms (EMGs) and electroencephalograms (EEG), which can be recorded easily from the scalp, has been studied in a number of cases (for review, see Mima and Hallett (1999)). In many of the previous studies on CMC, a linear model, such as a coherence (Ohara et al. (2000)) or multivariate autoregressive model (Honda et al. (2011)), has often been used to investigate the relationship between EEG and EMG time series.

The physiological natures of EEG and EMG are significantly different. Namely, EEG is the summation of postsynaptic potentials of neurons, which are characterized by relatively gradually

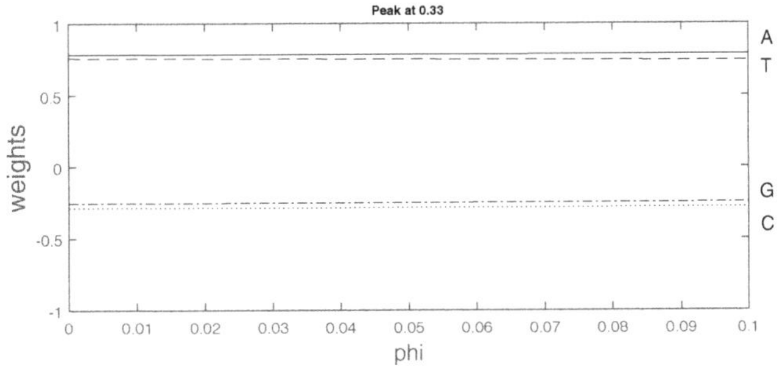

Figure 7.25: Estimated portfolio weights for the second 1000 sections and $\varphi = 0, \cdots, 0.1$

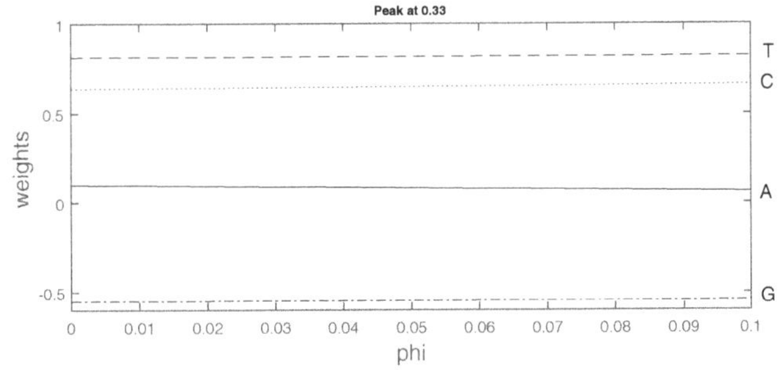

Figure 7.26: Estimated portfolio weights for the third 1000 sections and $\varphi = 0, \cdots, 0.1$

changing signal. By contrast, EMG is the summation of action potentials of the muscle spindle, which are impulse-like signals. Compared with EEG, the electrocorticogram (ECoG) more accurately reflects the electrical potential generated just below the recording electrode, thus making it possible to more directly observe the relationship between the cortex and other organs, such as muscles. Kato et al. (2006) proposed a multiplicatively modulated exponential autoregressive model (mmExpAR) to predict the dynamics of EMG modulating by ECoG. The mmExpAR was considered as a class of multiplicatively modulated nonlinear autoregressive model (mmNAR) and the local asymptotic normality (LAN, Le Cam (1986)) was also established. In this section, we employ the numerical example as a continuous time series data analysis.

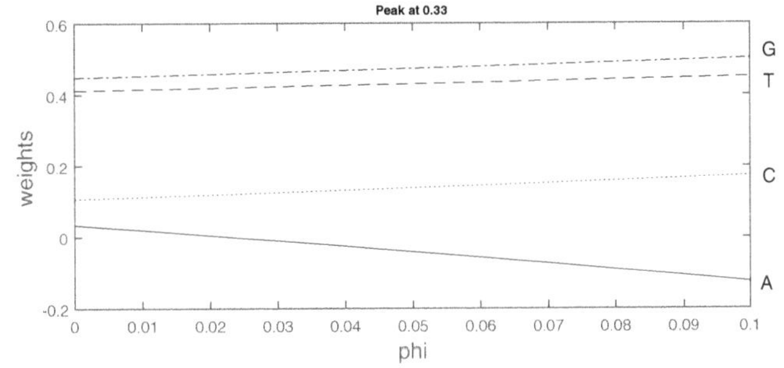

Figure 7.27: Estimated portfolio weights for the fourth 1000 sections and $\varphi = 0, \cdots, 0.1$

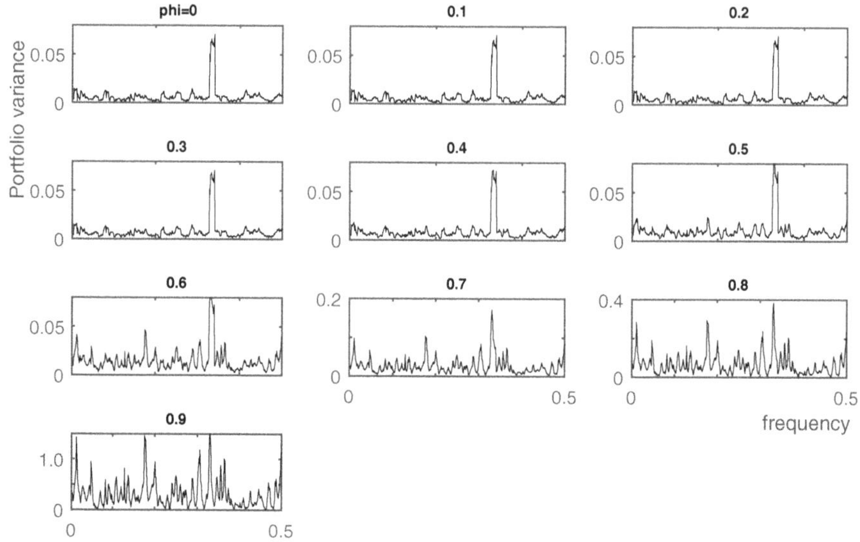

Figure 7.28: Outputs of portfolio variance for $0.1 < \varphi$

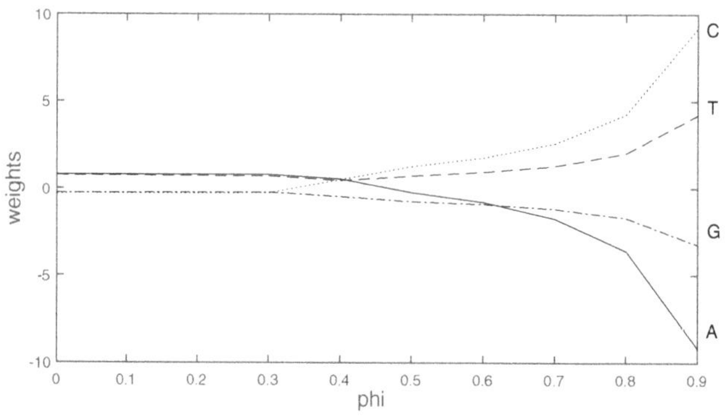

Figure 7.29: Outputs of portfolio weights for $0.1 < \varphi$

For the ECoG data, the subject was a 23-year-old female suffering from intractable seizures. Chronic subdural electrodes were implanted for surgical evaluation. ECoG data are from a monopolar recording obtained using platinum grid subdural electrodes (diameter 3 mm, interelectrode distance 1cm) placed on the hand area of the left primary motor cortex with the right mastoid as a reference. The location of the hand area was confirmed by electrical stimulation. Surface EMG data were obtained using a recorder on the right extensor carpi radialis muscle. The amplifier setting was 0.03–120Hz (−3dB) and the sampling frequency was 500 Hz. The motor task for the subject was isometric contraction of the right extensor carpi radialis muscle. The total number of data points was 20,000 sample points of 40-s duration. In light of the inhomogeneous characteristics of the entire dataset, we divided the data into stationary epochs with 1000 sample points (i.e., 20 epochs). The ECoG and EMG data for the second epoch is shown in Figure 7.30. To investigate the relationship between the cortex and other organs such as muscles, we used a dataset consisting of that from

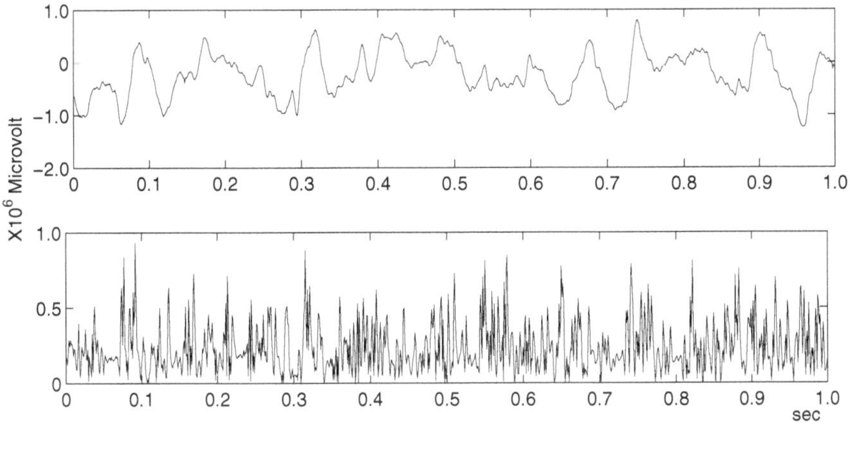

Figure 7.30: Original of ECoG and EMG data for one epoch

ECoG and an EMG. The nonlinear and non-Gaussian characteristics of the EMG data make it difficult to handle the data in a linear model. Kato et al. (2006) proposed a nonlinear model called the Multiplicatively Modulated Nonlinear Autoregressive Model (mmNAR) to predict the dynamics for generating ECoG. The model was presented by a radial basis function (RBF) of exogenous variables such as EMG.

Let $\{z_t : t = 1, \cdots, N\}$ be a time series of interest, and let $\{y_t : t = 1, \cdots, N\}$ be a series of exogenous variables. Suppose that $\{z_t\}$ is generated by

$$z_t = \mathbf{\Phi_0} + \sum_{p=1}^{P} \mathbf{\Phi_p z_{t-p}} + \varepsilon_t, \tag{7.5.1}$$

where the ε_t's are i.i.d. random variables, and

$$\begin{aligned} \mathbf{\Phi_0} &\equiv \Phi_0(\mathbf{Z_{t-1}}, \mathbf{C_{z,1}}, \cdots, \mathbf{C_{z,t}}) \\ &= \phi_{0,0} + \sum_{j=1}^{t_z} \phi_{0,j} \exp\{-\gamma_{z,j} \parallel \mathbf{Z_{t-1}} - \mathbf{C_{z,j}} \parallel^2\}, \end{aligned} \tag{7.5.2}$$

$$\begin{aligned} \mathbf{\Phi_p} &\equiv \Phi_p(\mathbf{Z_{t-1}}, \mathbf{Y_{t-l}}, \mathbf{C_{z,1}}, \cdots, \mathbf{C_{z,t_z}}, \mathbf{C_{y,1}}, \cdots, \mathbf{C_{y,t_y}}) \\ &= \phi_{p,0} + \sum_{j=1}^{t_z} \phi_{p,j} \exp\{-\gamma_{z,j} \parallel \mathbf{Z_{t-1}} - \mathbf{C_{z,j}} \parallel^2\} + \sum_{i=1}^{t_y} \phi_{p,i} \exp\{-\gamma_{y,i} \parallel \mathbf{Y_{t-l}} - \mathbf{C_{y,i}} \parallel^2\} \end{aligned} \tag{7.5.3}$$

with

$$\begin{aligned} \mathbf{Z_{t-1}} &= (z_{t-1}, \Delta z_{t-1}, \cdots, \Delta^{k_z-1} z_{t-1})^T \quad (k_z = 2, 3, \cdots, K_z), \\ \mathbf{Y_{t-l}} &= (y_{t-l}, \Delta y_{t-l}, \cdots, \Delta^{k_y-1} y_{t-l})^T \quad (k_y = 2, 3, \cdots, K_y) \end{aligned} \tag{7.5.4}$$

where t_z and t_y are the number of the centres in RBF of z_t and y_t, and $\phi_{i,j}$, $\{\gamma_{z,j} : \gamma_{z,j} > 0\}$ and $\{\gamma_{y,j} : \gamma_{y,j} > 0\}$ are unknown parameters. Also, l indicates a lag where the exogenous variables have significant time delays. Here, $c^*_{z,j}$ and $c^*_{y,j}$ mark the RBF's centre, and the difference orders are denoted by k_z and k_y.

$$\mathbf{C}_{z,j} \equiv (c_{z,j}, c_{z,j}^{(1)}, \cdots, c_{z,j}^{(k_z-1)})^T \tag{7.5.5}$$

$$\mathbf{C}_{y,j} \equiv (c_{y,j}, c_{y,j}^{(1)}, \cdots, c_{y,j}^{(k_y-1)})^T.$$

The mmNAR model is understood as a special form of vector conditional heteroscedastic autoregressive nonlinear (CHARN) models (Härdle et al. (1998)) introduced in Theorem 2.1.36.

In this section, we focus on the residual analysis. Stoffer et al. (2000) introduced the Spectral Envelope (SpecEnv) approach for real-valued time series data. In Stoffer et al. (2000), as an application, they performed a simple diagnostic procedure by using SpecEnv to aid in a residual analysis for fitting time series modeling. The method was motivated by a study by Tiao and Tsay (1994) indicating that the second-order moving average model (MA(2)) was applied to quarterly US real GNP data and the residuals from the model fit appeared to be uncorrelated. The SpecEnv explored that the residuals from the MA(2) fit were not i.i.d., and there was considerable power at the low frequencies. Based on that, we consider residual analysis from two methodological view points, SpecEnv and the mean-diversification efficient frontier defined in (7.4.4).

7.5.2 *Method*

We first obtain the residuals by fitting mmNAR and exponential AR (ExpAR) models (Haggan and Ozaki (1981)) to EMG data (ECoG data is used as the exogenous variable in the mmNAR model). The ExpAR model is given by

$$\text{ExpAR model} : z_t = \sum_{p=1}^{P} \left\{ \phi_p + \pi_p \exp(-\gamma z_{p-1}^2) \right\} z_{t-p} + \epsilon_t$$

where P is the AR order, ϕ_p, π_p indicate AR parameters, γ is a scaling parameter, and $\epsilon_t s$ are i.i.d. random variables. We consider some transformations for the residuals and place them as a vector.

To the vector, we secondarily calculate the SpecEnv $s(\lambda) = \sup_{\beta} \left\{ \frac{\beta^t f_X^{\mathrm{Re}}(\lambda)\beta}{\beta^t \mathbf{V} \beta} \right\}$ shown in Section 7.4.2.1. We investigate the maximum SpecEnv estimates and the scaling parameters at the frequency for the maximum SpecEnv peak. Furthermore, we apply the portfolio study to the vector by the procedure introduced in Section 7.4.2.3. The portfolio weights, returns and variances are estimated for each φ. Finally, we consider the correspondence of the portfolio weights to the scaling parameter and the interpretation to the residuals.

7.5.3 *Results and Discussion*

In this study, we use a section for epoch = 2 in the whole time series data shown in Kato et al. (2006). The ExpAR and mmNAR models were applied to the data, and the best fit models were identified by the optimum AR orders of ExpAR and mmNAR and time lags of mmNAR. For the best fit model, the obtained log-likelihood and AIC were -1.25×10^4 and 2.49×10^4 for ExpAR, and -1.05×10^4 and 2.10×10^4 for mmNAR. The minimum AIC selected 5 as the optimum AR order for ExpAR and 69 as the optimum time lag for the exogenous variable (ECoG) in mmNAR, following Kato et al. (2006). The autocorrelation of the residuals was assessed by a modified Q statistic (Ljung and Box (1978)). Q values indicated 26.2 for ExpAR and 26.9 for mmNAR, which were not larger than a given significant level of the chi-squared distribution ($46.9 = \chi^2_{0.05}(25)$, Kato et al. (2006)).

Next, we applied the residual diagnostic to the obtained residual series shown in Figure 7.31. For the residuals, a k-dimensional vector of continuous real-valued transformations should be set. We fixed the transformed set (vector) $\mathfrak{G} = \{x, |x|, x^2\}$ to the residual series x. The SpecEnv for the transformed set was calculated (smoothed using Daniell kernel $m = 7$), shown in Figure 7.32. In both cases for the ExpAR and mmNAR models, there were considerable peaks in the lower frequency domain. This suggests that the presence of spectral power at low frequency is associated with long-range dependencies. The peak for mmNAR was more than the peak for ExpAR; however, the SpecEnv around the middle and higher frequency domains for mmNAR were comparative lower than the SpecEnv on the same domain for ExpAR. This suggests a better fitting by mmNAR to the

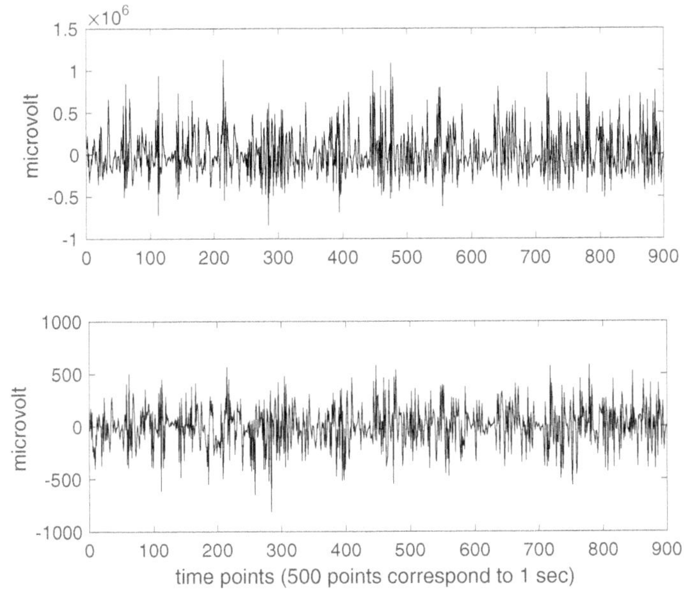

(a) Residuals series for applying the ExpAR model (upper) and the mmNAR model (lower).

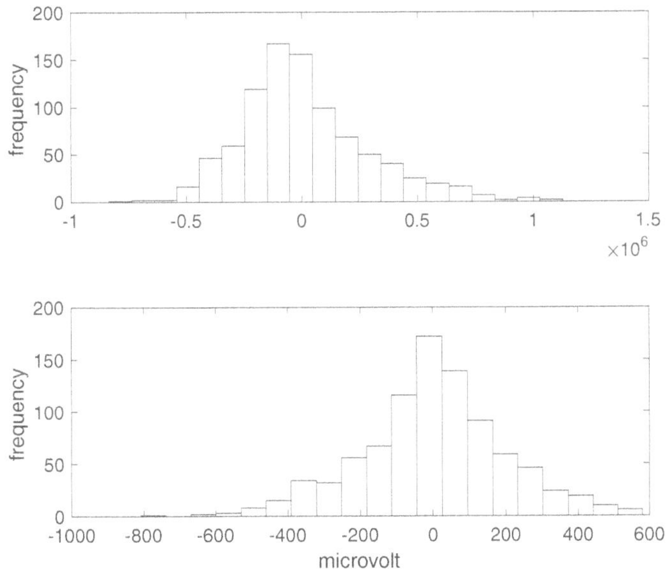

(b) Histogram of the residuals for ExpAR (upper) and mmNAR (lower).

Figure 7.31: Plots and the histogram for residuals

data. The scaling parameters at the maximum peaks for ExpAR are $\beta(0.011) = (1, 17.5, -3.0)$ (the second peak $\beta(0.04) = (1, 5.4, 0.086)$), which leads to an optimum transformation $y = x + 17.5|x| - 3.0x^2$ using the same notation as in Stoffer et al. (2000). Also, for mmNAR, the estimated maximum at the maximum peak was $\beta(0.011) = (1, 38.5, -8.7)$, which leads to an optimum transformation $y = x + 38.5|x| - 8.7x^2$.

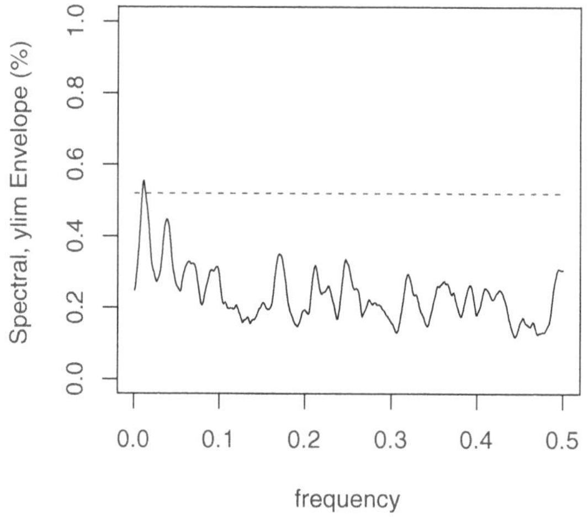

(a) ExpAR

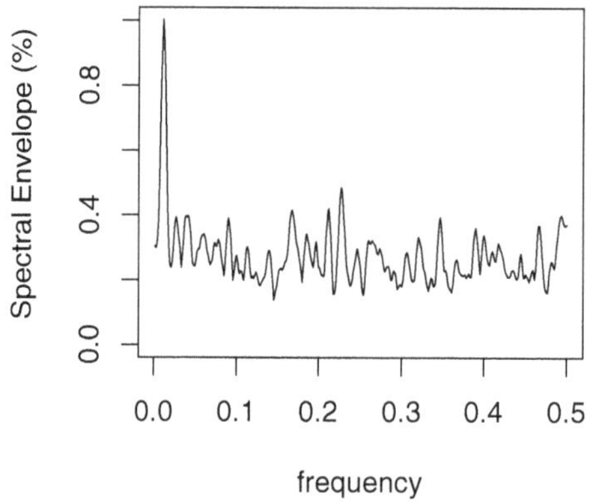

(b) mmNAR

Figure 7.32: SpecEnv for x, $|x|$, x^2 transformed residuals for ExpAR and mmNAR

Finally, we calculated the portfolio weights, variance and returns to the transformed residuals set. Figure 7.33 shows the estimates for $\varphi = 0, \ldots, 0.8$. The configurations of those plots indicated higher variance at lower frequencies, that is, long-range dependencies still remained in the prediction error. We saw the consistent tendency in the SpecEnv results. In Figure 7.33, the plot for each φ indicated that the frequency indicating the maximum peak was different from the frequency indicating the maximum SpecEnv. We reasoned that in the calculation procedure, the periodograms for SpecEnv were the same as the periodograms for the portfolio calculation. When calculating the mean and variance of the periodogram, the portfolio procedure required to aggregating periodograms vectors at some sequential frequency points around λ; however, in the case of SpecEnv, we didn't summarize the periodograms within a range and the eigenvalue of the periodogram was directly calculated for each λ. The SpecEnv and the variance of the portfolio for $\varphi = 0$ should be equivalent theoretically; however, this issue was caused by different numerical algorithms. On the other hand, the estimates for $\varphi = 0, \ldots, 0.5$ did not indicate clear distinct, but the estimates for $\varphi \geqq 0.6$ for ExpAR and the estimates for $\varphi = 0.65$ of mmNAR become more complicated. The estimated portfolio weights are illustrated in Figure 7.34. The weights for ExpAR model were calculated based on the maximum portfolio variance at the frequency, which was close to the frequency

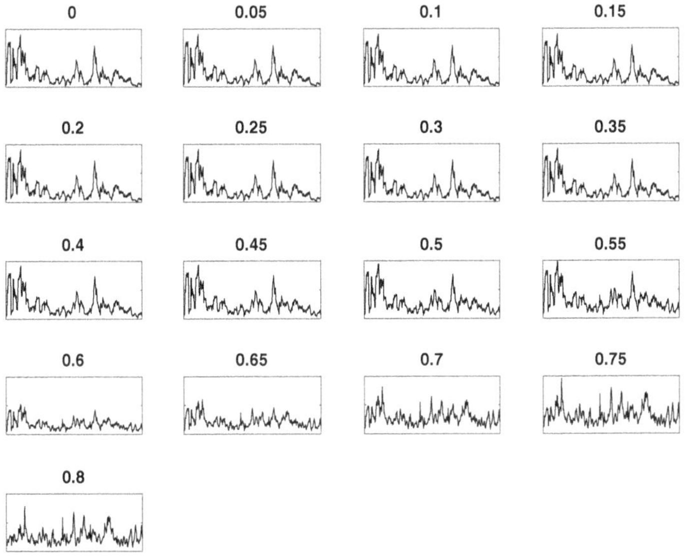

(a) Residuals by the ExpAR model.

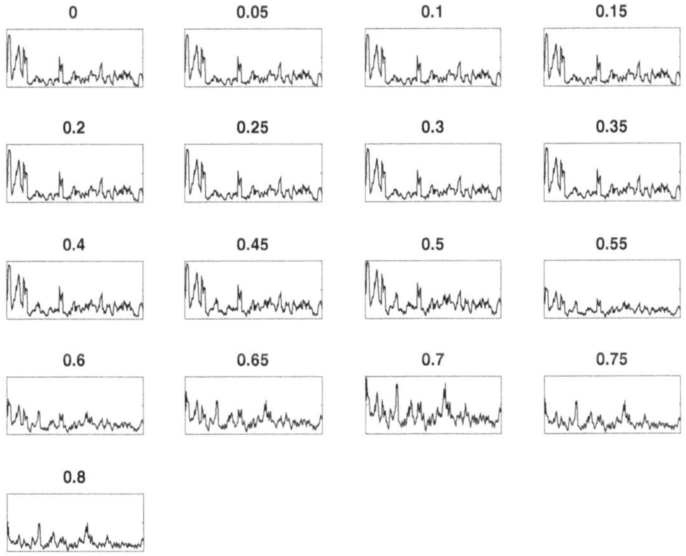

(b) Residuals by the mmNAR model.

Figure 7.33: The estimated portfolio variance for $\varphi = 0, \ldots, 0.8$

indicating the maximum SpecEnv peak (see Figure 7.32). In the case of mmNAR, the weights were calculated by the maximum portfolio variance for each φ because the maximum peak of the portfolio variance stands on a similar frequency to the frequency for the maximum SpecEnv. The degree of weights indicated a tendency $|w_{|x|}| > w_x > w_{x^2}$ in both models, where $w_{|x|}$, w_x and w_{x^2} indicated the portfolio weights for $|x|$, x and x^2. These results were consistent for the estimated scaling parameters

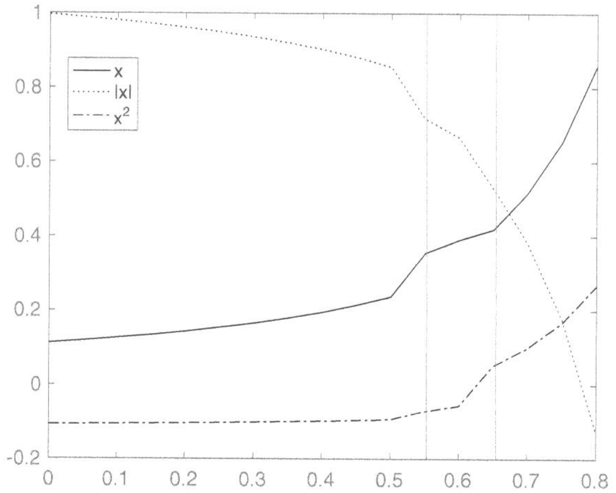

(a) Residuals by the ExpAR model.

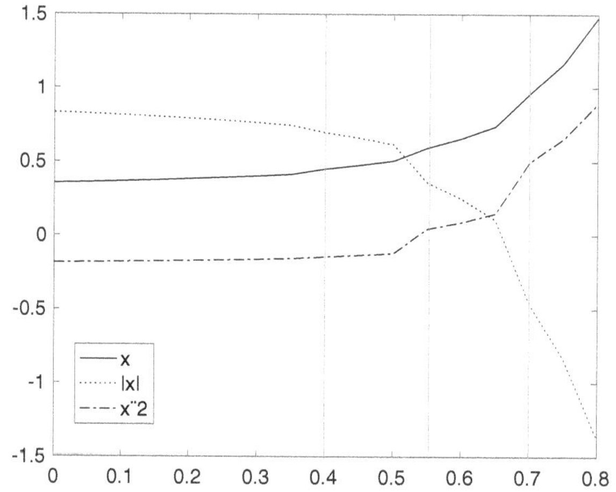

(b) Residuals by the mmNAR model.

Figure 7.34: The estimated portfolio weights for transformed residuals

of SpecEnv. In Figure 7.34, the vertical line shows the point that the frequency indicating the maximum portfolio variance was changed. The weights were dramatically changed for $\varphi > 0.5$. Finally, we shows the estimated portfolio return in Figure 7.35. The estimates were wider for $\varphi > 0.2$ in the case of ExpAR and $\varphi > 0.3$ in the case of mmNAR. The portfolio returns were interpreted as the mean that summarized the portfolio effect. On the other hand, the increase of prediction error is not good for time series modeling. For $\varphi < 0.2$ (enough small), the portfolio weights would correspond to the scaling parameter for SpecEnv.

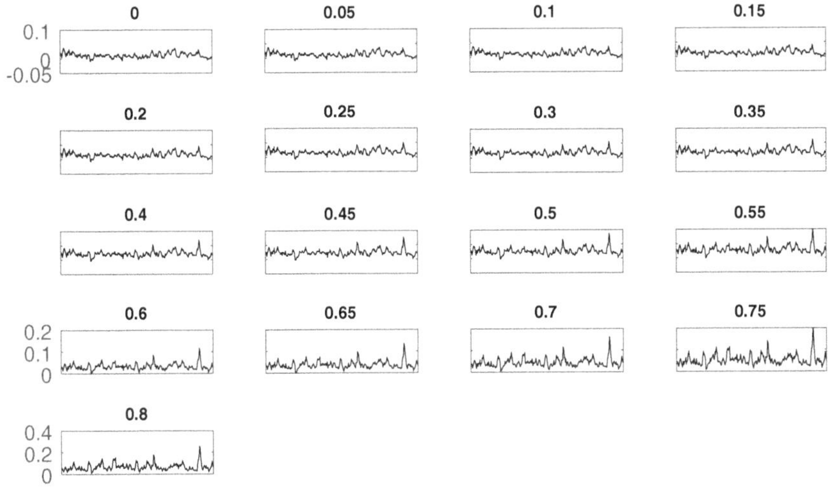

(a) Residuals by the ExpAR model.

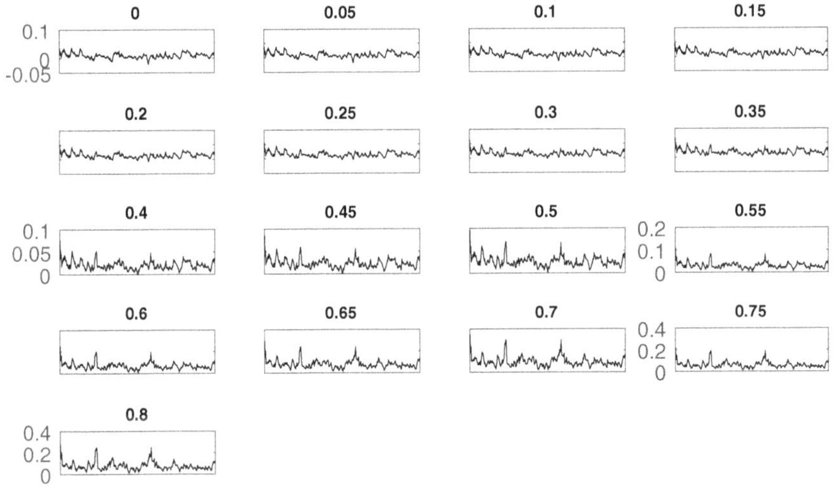

(b) Residuals by the mmNAR model.

Figure 7.35: The estimated portfolio returns for $\varphi = 0, \ldots, 0.8$

Problems

7.1 Assume that the kernel function $K(\cdot)$ in (7.1.9) is continuously differentiable, and satisfies $K(x) = K(-x)$, $\int K(x)dx = 1$ and $K(x) = 0$ for $x \notin [-1/2, 1/2]$ and the bandwidth b_n is of order $b_n = O\left(n^{-1/5}\right)$. Then, under the twice differentiablity of μ, show (see e.g., Shiraishi and Taniguchi (2007))

$$E\left\{\hat{\mu}\left(\frac{t}{n}\right)\right\} = \mu\left(\frac{t}{n}\right) + O\left(b_n^2\right) + O\left(\frac{1}{b_n n}\right).$$

7.2 Verify (7.2.1)-(7.2.4).

7.3 Generate artificial indicator data including '1' for 50 samples and '0' for 50 samples. For the two-groups samples, generate 1,000-array 100 samples, using normal distributions and ARCH models with different settings. Show the histograms of two-group samples.

7.4 To the same data generated in 7.3, apply rank order statistics to compare the two-group for each array. Plot the empirical distributions for the statistics using the statistics obtained for 1000 array.

7.5 For the same data generated in 7.3, fit an ARCH model and get the empirical residual for each array. Calculate the rank order statistics to compare two-group for each array. Show the empirical distributions of the statistics.

7.6 Take the first raw data obtained in 7.5 and reproduce the array data by shuffling samples. To the reproduced data, re-apply the rank order statistics for two-group (1-50 and 51-100 samples). Repeat the procedure 1000 times and plot the empirical distribution of the statistics. Calculate the p-value of the statistics of the original data obtained on the same row in 7.5.

7.7 Download the five (Des, Loi, Mil, Min and Chin) cohorts in Affy947 expression and the clinical data from the R code in Zhao et al. (2014).

a. To the GARCH residuals, apply other statistics for two-group comparisons, e.g. t statistics, Fold Change, and B statistics, instead of rank order statistics and identify the significant genes.
b. Show the common genes list among significant genes obtained by those statistics and check the common biological terms by GO analysis by the list.

7.8 What is the benefit of considering portfolio diversification?

7.9 What is the relationship between diversification measurement and information theory?

7.10 What are principal portfolios?

7.11 For the entropy

$$N_{\text{ENT}} \equiv \exp\left(-\sum_{i=1}^{N} p_i \ln p_i\right),$$

if it is a low number, what does it mean for the risk factor of the portfolio? If it achieves maximum value, what the portfolio contsructs?

7.12 Verify (7.4.2)

7.13 Verify (7.4.3).

7.14 Verify (7.4.5).

7.15 Generate simulation data using the R code shown in Section 7.4.3.1 by changing the parameter 'by' for 131 and 132.

a. Apply SpecEnv to the data and see the difference between parameter settings.

b. Apply (7.4.3) - (7.4.5) and see the difference for scaling parameters and the portfolio weights between parameer settings.

7.16 Download another DNA sequence data, called bnrf1hvs for "tsa3.rda" of the library astsa (Applied Statistical Time Series Analysis, the software by R.H. Shummway and D.S. Stoffer Time Series Analysis and Its Application: With R Examples, Springer) in R (Comprehensive R Archive Network, http://cran.r-project.org) and re-apply SpecEnv and the proposed portfolio estimation given by (7.4.3) – (7.4.5) to the data and show the scaling parameters and the portfolio weights.

Chapter 8

Theoretical Foundations and Technicalities

This chapter provides theoretical fundations of stochastic procsses and optimal statistical inference.

8.1 Limit Theorems for Stochastic Processes

Let $\{\boldsymbol{X}(t) = (X_1(t), \dots, X_m(t))' : t \in \mathbb{Z}\}$ be generated by

$$\boldsymbol{X}(t) = \sum_{j=0}^{\infty} A(j)\boldsymbol{U}(t-j), \quad t \in \mathbb{Z}, \tag{8.1.1}$$

where $\{\boldsymbol{U}(t) = (u_1(t), \dots, u_m(t))'\}$ is a sequence of m-vector random variables with

$$\mathrm{E}\{\boldsymbol{U}(t)\} = \boldsymbol{0}, \qquad E\{\boldsymbol{U}(t)\boldsymbol{U}(s)'\} = \boldsymbol{0} \quad (t \neq s),$$

and

$$\mathrm{Var}\{\boldsymbol{U}(t)\} = K = \{K_{ab} : a, b = 1, \dots, m\} \quad \text{for all } t \in \mathbb{Z},$$

and $A(j) = \{A_{ab}(j) : a, b = 1, \dots, m\}$, $j \in \mathbb{Z}$, are $m \times m$ matrices with $A(0) = \boldsymbol{I}_m$. We set down the following assumption.

(HT0)

$$\sum_{j=0}^{\infty} \mathrm{tr}\, A(j)KA(j)' < \infty.$$

Then, the process $\{\boldsymbol{X}_t\}$ is a second-order stationary process with spectral density matrix

$$\boldsymbol{f}(\lambda) = \{f_{ab}(\lambda) : a, b = 1, \dots, m\} \equiv (2\pi)^{-1}A(\lambda)KA(\lambda)^*,$$

where $A(\lambda) = \sum_{j=0}^{\infty} A(j)e^{ij\lambda}$.

For an observed stretch $\{\boldsymbol{X}(1), \dots, \boldsymbol{X}(n)\}$, let

$$\boldsymbol{I}_n(\lambda) = \frac{1}{2\pi n}\left\{\sum_{t=1}^{n} \boldsymbol{X}(t)e^{it\lambda}\right\}\left\{\sum_{t=1}^{n} \boldsymbol{X}(t)e^{it\lambda}\right\}^*, \tag{8.1.2}$$

which is called the *periodogram matrix*. In time series analysis, we often encounter a lot of statistics based on

$$F_n = \int_{-\pi}^{\pi} \mathrm{tr}\, \{\phi(\lambda)\boldsymbol{I}_n(\lambda)\}d\lambda, \text{ and } F \equiv \int_{-\pi}^{\pi} \mathrm{tr}\, \{\phi(\lambda)\boldsymbol{f}(\lambda)\}d\lambda, \tag{8.1.3}$$

where $\phi(\lambda)$ is an appropriate matrix-valued function. Hence, we will elucidate the asymptotics of

F_n. Here we mention two types of approaches. The first one is due to Brillinger (2001b). Assume that the process $\{\boldsymbol{X}(t)\}$ is strictly stationary and that all moments of it exist. Write

$$c^X_{a_1\dots a_k}(t_1,\dots,t_{k-1}) \equiv \mathrm{cum}\{X_{a_1}(0), X_{a_2}(t_1),\dots,X_{a_k}(t_{k-1})\}, \tag{8.1.4}$$

for $a_1,\dots,a_k = 1,\dots,m,\ k = 1,2,\dots$. Throughout this book, we denote by $\mathrm{cum}(Y_1,\dots,Y_k)$ the joint cumulant of random variables $Y_1,\dots,Y_k$ that is defined by the coefficients of $(it_1)\cdots(it_k)$ in the Taylor expansion of

$$\log[\mathrm{E}\{\exp(\boldsymbol{i}\sum_{j=1}^{k} t_jY_j)\}] \tag{8.1.5}$$

(for the fundamental properties, see Theorem 2.3.1 of Brillinger (2001b)). The following condition is introduced.

(B1) For each $j = 1,2,\dots,k-1$ and any k-tuple $a_1,a_2,\dots,a_k$ we have

$$\sum_{t_1,\dots,t_{k-1}=-\infty}^{\infty} (1+|t_j|)|c^X_{a_1\cdots a_k}(t_1,\dots,t_{k-1})| < \infty, \quad k = 2,3,\dots. \tag{8.1.6}$$

Under **(B1)** Brillinger (2001b) evaluated the Jth-order cumulant $c_n(J)$ of $\sqrt{n}(F_n - F)$, and showed that $c_n(J) \to 0$ as $n \to \infty$ for all $J \geq 3$, which leads to the asymptotic normality of $\sqrt{n}(F_n - F)$.

The second approach is due to Hosoya and Taniguchi (1982). Recall the linear process (8.1.1). Let $\mathcal{F}_t = \sigma\{\boldsymbol{U}(s); s \leq t\}$. Assuming that $\{\boldsymbol{U}(t)\}$ is fourth-order stationary, write $c^U_{a_1\cdots a_4}(t_1,t_2,t_3) = \mathrm{cum}\{u_{a_1}(0), u_{a_2}(t_1), u_{a_3}(t_2), u_{a_4}(t_3)\}$. Hosoya and Taniguchi (1982) introduced the following conditions.

(HT1) For each a_1, a_2 and s, there exists $\epsilon > 0$ such that

$$\mathrm{Var}\,[\mathrm{E}\{u_{a_1}(t)u_{a_2}(t+s)|\mathcal{F}_{t-\tau}\} - \delta(s,0)K_{a_1a_2}] = O(\tau^{-2-\epsilon}),$$

uniformly in t.

(HT2) For each a_j and t_j, $j = 1,\dots,4$ ($t_1 \leq t_2 \leq t_3 \leq t_4$), there exists $\eta > 0$ such that

$$\mathrm{E}\left|\mathrm{E}\left\{u_{a_1}(t_1)\,u_{a_2}(t_2)\,u_{a_3}(t_3)\,u_{a_4}(t_4)|\mathcal{F}_{t_1-\tau}\right\} - \mathrm{E}\{u_{a_1}(t_1)\,u_{a_2}(t_2)\,u_{a_3}(t_3)\,u_{a_4}(t_4)\}\right| = O(\tau^{-1-\eta}),$$

uniformly in t_1.

(HT3) For any $\rho > 0$ and for any integer $L \geq 0$, there exists $B_\rho > 0$ such that

$$\mathrm{E}\left[T(n,s)^2\chi\left\{T(n,s) > B_\rho\right\}\right] < \rho,$$

uniformly in n and s, where

$$T(n,s) = \left[\sum_{a_1,a_2=1}^{m}\sum_{r=0}^{L}\left\{\sum_{t=1}^{n}\frac{u_{a_1}(t+s)u_{a_2}(t+s+r) - K_{a_1a_2}\delta(0,r)}{\sqrt{n}}\right\}^2\right]^{1/2}.$$

(HT4) The spectral densities $f_{aa}(\lambda)$, $a = 1,\dots,m$, are square integrable.

(**HT5**) $\boldsymbol{f}(\lambda) \in Lip(\alpha)$, the Lipschitz class of degree α, $\alpha > 1/2$.

(**HT6**) For each a_j, $j = 1, \ldots, m$,

$$\sum_{t_1,t_2,t_3=-\infty}^{\infty} |c^U_{a_1\cdots a_4}(t_1, t_2, t_3)| < \infty.$$

In much literature a "higher-order martingale difference condition" for the innovation process is assumed, i.e.,

$$\begin{aligned} &\mathrm{E}\{u_{a_1}(t)|\mathcal{F}_{t-1}\} = 0 \text{ a.s.}, \quad \mathrm{E}\{u_{a_1}(t)u_{a_2}(t)|\mathcal{F}_{t-1}\} = K_{a_1a_2}, \text{ a.s.}, \\ &\mathrm{E}\{u_{a_1}(t)u_{a_2}(t)u_{a_3}(t)|\mathcal{F}_{t-1}\} = \gamma_{a_1a_2a_3} \text{ a.s., and so on} \end{aligned} \tag{8.1.7}$$

(e.g., Dunsmuir (1979), Dunsmuir and Hannan (1976)). What (**HT1**) and (**HT2**) mean amounts to a kind of "asymptotically" higher-order martingale condition, which seems more natural than (8.1.7).

It is seen that $\sqrt{n}(F_n - F)$ can be approximated by a finite linear combination of

$$\sqrt{n}\left\{\frac{1}{n}\sum_{t=1}^{n-s} X_{a_1}(t) X_{a_2}(t+s) - c^X_{a_1\,a_2}(s)\right\}, \tag{8.1.8}$$

where $s = 0, 1, \ldots, n-1$, $a_1, a_2 = 1, \ldots, m$. Recalling that $\{\boldsymbol{X}(t)\}$ is generated by (8.1.1), we see that (8.1.8) can be approximated by a finite linear combination of

$$\frac{1}{\sqrt{n}}\left\{\sum_{t=1}^{n} u_{a_1}(t)u_{a_2}(t+l) - \delta(l,0)K_{a_1a_2}\right\}, \quad l \leq s. \tag{8.1.9}$$

Applying Brown's central limit theorem (Theorem 2.1.29) to (8.1.9), Hosoya and Taniguchi (1982) elucidate the asymptotics of F_n. The two approaches are summarized as follows.

Theorem 8.1.1. *(Brillinger (2001b), Hosoya and Taniguchi (1982)) Suppose that one of the following conditions (B) and (HT) holds.*

(**B**) *$\{\boldsymbol{X}(t) : t \in \mathbb{Z}\}$ is strictly stationary, and satisfies the condition (B1).*

(**HT**) *$\{\boldsymbol{X}(t) : t \in \mathbb{Z}\}$ is generated by (8.1.1), and satisfies the conditions (HT0)–(HT6).*

Let $\phi_j(\lambda)$, $j = 1, \ldots, q$, be $m \times m$ matrix-valued continuous functions on $[-\pi, \pi]$ such that $\phi_j(\lambda) = \phi_j(\lambda)^*$ and $\phi_j(-\lambda) = \phi_j(\lambda)'$. Then

(**i**) *for each $j = 1, \ldots, q$,*

$$\int_{-\pi}^{\pi} \mathrm{tr}\{\phi_j(\lambda)\boldsymbol{I}_n(\lambda)\}d\lambda \xrightarrow{p} \int_{-\pi}^{\pi} \mathrm{tr}\{\phi_j(\lambda)\boldsymbol{f}(\lambda)\}d\lambda, \tag{8.1.10}$$

as $n \to \infty$,

(**ii**) *the quantities*

$$\sqrt{n}\int_{-\pi}^{\pi} \mathrm{tr}[\phi_j(\lambda)\{\boldsymbol{I}_n(\lambda) - \boldsymbol{f}(\lambda)\}]d\lambda, \quad j = 1, \ldots, q$$

have, asymptotically, a normal distribution with zero mean vector and covariance matrix V whose (j, l)th element is

$$\begin{aligned} &4\pi\int_{-\pi}^{\pi} \mathrm{tr}\{\boldsymbol{f}(\lambda)\phi_j(\lambda)\boldsymbol{f}(\lambda)\phi_l(\lambda)\}d\lambda \\ &+ 2\pi\sum_{r,t,u,v=1}^{m} \int\int_{-\pi}^{\pi} \phi^{(j)}_{rt}(\lambda_1)\phi^{(l)}_{uv}(\lambda_2)Q^X_{rtuv}(-\lambda_1, \lambda_2, -\lambda_2)d\lambda_1 d\lambda_2, \end{aligned}$$

where $\phi_{rt}^{(j)}(\lambda)$ *is the* (r, t)*th element of* $\phi_j(\lambda)$, *and*

$$Q_{rtuv}^X(\lambda_1, \lambda_2, \lambda_3) = (2\pi)^{-3} \sum_{t_1, t_2, t_3 = -\infty}^{\infty} \exp\{-i(\lambda_1 t_1 + \lambda_2 t_2 + \lambda_3 t_3)\} c_{rtuv}^X(t_1, t_2, t_3).$$

Since the results above are very general, we can apply them to various problems in time series analysis.

Let $X = \{X_t\}$ be a square integrable martingale, $X^{(c)}$ the continuous martingale component of X and $\Delta X_t = X_t - X_{t-}$. The *quadratic variation process* of X is defined by

$$[X]_t \equiv \left\langle X^{(c)} \right\rangle_t + \sum_{s \le t} \Delta X_s^2. \tag{8.1.11}$$

For another square integrable martingale $Y = \{Y_t\}$, define

$$[X, Y]_t \equiv \frac{1}{4} \{[X + Y, X + Y]_t - [X - Y, X - Y]_t\}. \tag{8.1.12}$$

If $\boldsymbol{X} = \{\boldsymbol{X}_t = (X_{1,t}, \ldots, X_{m,t})'\}$ is an m-dimensional vector-valued square integrable martingale, the quadratic variation $[\boldsymbol{X}] \equiv [\boldsymbol{X}_t, \boldsymbol{X}_t]$ is defined as the $\mathbb{R}^m \times \mathbb{R}^m$-valued process whose (j, k)th component is $[X_{j,t}, X_{k,t}]$.

We now mention the concept of stable and mixing convergence. Let $\{X_n\}$ be a sequence of random variables on a probability space $(\Omega, \mathcal{F}, P)$ converging in distribution to a random variable X. We say that the convergence is *stable* if for all the continuity points x of the distribution function of X and all $A \in \mathcal{F}$

$$\lim_{n \to \infty} P[\{X_n \le x\} \cap A] = Q_x(A) \tag{8.1.13}$$

exists and if $Q_x(A) \to P(A)$ as $x \to \infty$. We denote this convergence by

$$X_n \xrightarrow{d} X \quad \text{(stably)}. \tag{8.1.14}$$

If

$$\lim_{n \to \infty} P[\{X_n \le x\} \cap A] = P(X \le x)P(A), \tag{8.1.15}$$

for all $A \in F$ and for all the points of continuity of the distribution function of X, the limit is said to be *mixing*, and we write

$$X_n \xrightarrow{d} X \quad \text{(mixing)}. \tag{8.1.16}$$

The following is due to Sørensen (1991).

Theorem 8.1.2. *Let* $\boldsymbol{X} = \{\boldsymbol{X}_t = (X_1(t), \ldots, X_m(t))'\}$ *be an m-dimensional square integrable* $\mathcal{F}_t$*-martingale with the quadratic variation matrix* $[\boldsymbol{X}]$, *and set* $\boldsymbol{H}_t \equiv E(\boldsymbol{X}_t \boldsymbol{X}_t')$. *Suppose that there exist nonrandom variables* $k_{it}, i = 1, \ldots, m$, *with* $k_{it} > 0$ *and* $k_{it} \to \infty$ *as* $t \to \infty$. *Assume that the following conditions hold:*

(i) $k_{it}^{-1} E\left[\sup_{s \le t} |\Delta X_i(s)|\right] \to 0, \quad 1 \le i \le n,$ (8.1.17)

(ii) $\boldsymbol{K}_t^{-1} [\boldsymbol{X}]_t \boldsymbol{K}_t^{-1} \xrightarrow{p} \eta^2$,

where $\boldsymbol{K}_t \equiv diag(k_{1t}, \ldots, k_{mt})$ *and* η^2 *is a random nonnegative definite matrix, and*

(iii) $\boldsymbol{K}_t^{-1} \boldsymbol{H}_t \boldsymbol{K}_t^{-1} \to \Sigma$, *where $\boldsymbol{\Sigma}$ is a positive definite matrix. Then,*

$$\boldsymbol{K}_t^{-1}\boldsymbol{X}_t \to \boldsymbol{Z} \quad (stably), \tag{8.1.18}$$

where the distribution of $\boldsymbol{Z}$ is the normal variance mixture with characteristic function

$$Q(\boldsymbol{u}) = E\left[\exp\left(-\frac{1}{2}\boldsymbol{u}'\boldsymbol{\eta}^2\boldsymbol{u}\right)\right], \qquad \boldsymbol{u} = (u_1, \ldots, u_m)'. \tag{8.1.19}$$

Also, conditionally on $[\det \boldsymbol{\eta}^2 > 0]$,

$$(\boldsymbol{K}_t[\boldsymbol{X}]_t^{-1}\boldsymbol{K}_t)^{1/2}\boldsymbol{K}_t^{-1}\boldsymbol{X}_t \xrightarrow{d} N(\boldsymbol{0}, \boldsymbol{I}_m) \quad (mixing), \tag{8.1.20}$$

and

$$\boldsymbol{X}_t'[\boldsymbol{X}]_t^{-1}\boldsymbol{X}_t \xrightarrow{d} \chi_m^2 \quad (mixing). \tag{8.1.21}$$

8.2 Statistical Asymptotic Theory

This section discusses asymptotic properties of fundamental estimators for general stochastic processes.

Recall the vector-valued linear process

$$\boldsymbol{X}(t) = \sum_{j=0}^{\infty} A(j)\boldsymbol{U}(t-j), \tag{8.2.1}$$

defined by (8.1.1) satisfying (**HT0**)–(**HT6**) with spectral density matrix $\boldsymbol{f}(\lambda)$. Suppose that $\boldsymbol{f}(\lambda)$ is parameterized as $\boldsymbol{f}_{\boldsymbol{\theta}}(\lambda)$, $\boldsymbol{\theta} = (\theta_1, \ldots, \theta_r)' \in \Theta$, where Θ is a compact set of $\mathbb{R}^r$. We are now interested in estimation of $\boldsymbol{\theta}$ based on $\{\boldsymbol{X}(1), \ldots, \boldsymbol{X}(n)\}$. For this we introduce

$$D(\boldsymbol{f}_{\boldsymbol{\theta}}, \boldsymbol{I}_n) \equiv \int_{-\pi}^{\pi} [\log\det\{\boldsymbol{f}_{\boldsymbol{\theta}}(\lambda)\} + \mathrm{tr}\,\{\boldsymbol{I}_n(\lambda)\boldsymbol{f}_{\boldsymbol{\theta}}(\lambda)^{-1}\}]d\lambda. \tag{8.2.2}$$

If we assume that $\{\boldsymbol{X}(t)\}$ is Gaussian, it is known (c.f. Section 7.2 of Taniguchi and Kakizawa (2000)) that $-\frac{n}{4\pi}D(\boldsymbol{f}_{\boldsymbol{\theta}}, \boldsymbol{I}_n)$ is an approximation of the log-likelihood. However, henceforth, we do not assume Gaussianity of $\{\boldsymbol{X}(t)\}$; we call $\hat{\boldsymbol{\theta}} \equiv \arg\min_{\boldsymbol{\theta}\in\Theta} D(\boldsymbol{f}_{\boldsymbol{\theta}}, \boldsymbol{I}_n)$ a quasi-Gaussian maximum likelihood estimator, and write it as $\hat{\boldsymbol{\theta}}_{\mathrm{QGML}}$.

Assumption 8.2.1. **(i)** *The true value $\boldsymbol{\theta}_0$ of $\boldsymbol{\theta}$ belongs to Int Θ.*

(ii) *$D(\boldsymbol{f}_{\boldsymbol{\theta}}, \boldsymbol{f}_{\boldsymbol{\theta}_0}) \geq D(\boldsymbol{f}_{\boldsymbol{\theta}_0}, \boldsymbol{f}_{\boldsymbol{\theta}_0})$ for all $\boldsymbol{\theta} \in \Theta$, and the equality holds if and only if $\boldsymbol{\theta} = \boldsymbol{\theta}_0$.*

(iii) *$\boldsymbol{f}_{\boldsymbol{\theta}}(\lambda)$ is continuously twice differentiable with respect to $\boldsymbol{\theta}$.*

(iv) *The $r \times r$ matrix*

$$\Gamma(\boldsymbol{\theta}_0) = \left[\frac{1}{4\pi}\int_{-\pi}^{\pi} \mathrm{tr}\left\{\boldsymbol{f}_{\boldsymbol{\theta}_0}(\lambda)^{-1}\left(\frac{\partial}{\partial\theta_j}\boldsymbol{f}_{\boldsymbol{\theta}_0}(\lambda)\right)\boldsymbol{f}_{\boldsymbol{\theta}_0}(\lambda)^{-1}\left(\frac{\partial}{\partial\theta_l}\boldsymbol{f}_{\boldsymbol{\theta}_0}(\lambda)\right)\right\}d\lambda;\ j,l = 1, \ldots, r\right]$$

is nonsingular.

In view of Theorem 8.1.1 we can show the aymptotics of $\hat{\boldsymbol{\theta}}_{\mathrm{QGML}}$. We briefly mention the outline of derivation. Since $\boldsymbol{f}_{\boldsymbol{\theta}}$ is smooth with respect to $\boldsymbol{\theta}$, and $\boldsymbol{\theta}_0$ is unique, the consistency of $\hat{\boldsymbol{\theta}}_{\mathrm{QGML}}$ follows from (i) of Theorem 8.1.1. Expanding the right-hand side of

$$\boldsymbol{0} = \frac{\partial}{\partial\boldsymbol{\theta}}D(\boldsymbol{f}_{\hat{\boldsymbol{\theta}}_{\mathrm{QGML}}}, \boldsymbol{I}_n) \tag{8.2.3}$$

at $\boldsymbol{\theta}_0$, we observe that

$$\mathbf{0} = \frac{\partial}{\partial \boldsymbol{\theta}} D(\boldsymbol{f}_{\boldsymbol{\theta}_0}, \boldsymbol{I}_n) + \frac{\partial^2}{\partial \boldsymbol{\theta} \partial \boldsymbol{\theta}'} D(\boldsymbol{f}_{\boldsymbol{\theta}_0}, \boldsymbol{I}_n)(\hat{\boldsymbol{\theta}}_{\mathrm{QGML}} - \boldsymbol{\theta}_0) + \text{lower order terms}, \tag{8.2.4}$$

leading to

$$\sqrt{n}(\hat{\boldsymbol{\theta}}_{\mathrm{QGML}} - \boldsymbol{\theta}_0) = -\left[\frac{\partial^2}{\partial \boldsymbol{\theta} \partial \boldsymbol{\theta}'} D(\boldsymbol{f}_{\boldsymbol{\theta}_0}, \boldsymbol{I}_n)\right]^{-1} \sqrt{n}\left[\frac{\partial}{\partial \boldsymbol{\theta}} D(\boldsymbol{f}_{\boldsymbol{\theta}_0}, \boldsymbol{I}_n)\right] + \text{lower order terms}. \tag{8.2.5}$$

Application of (i) and (ii) in Theorem 8.1.1 to $\frac{\partial^2}{\partial \boldsymbol{\theta} \partial \boldsymbol{\theta}'} D(\boldsymbol{f}_{\boldsymbol{\theta}_0}, \boldsymbol{I}_n)$ and $\sqrt{n}\frac{\partial}{\partial \boldsymbol{\theta}} D(\boldsymbol{f}_{\boldsymbol{\theta}_0}, \boldsymbol{I}_n)$, respectively, yields

Theorem 8.2.2. *(Hosoya and Taniguchi (1982)) Under Assumption 8.2.1 and (HT0)–(HT6), the following assertions hold true.*

(i) $\hat{\boldsymbol{\theta}}_{QGML} \xrightarrow{p} \boldsymbol{\theta}_0 \quad as \quad n \to \infty.$ (8.2.6)

(ii) The distribution of $\sqrt{n}(\hat{\boldsymbol{\theta}}_{QGML} - \boldsymbol{\theta}_0)$ tends to the normal distribution with zero mean vector and covariance matrix

$$\Gamma(\boldsymbol{\theta}_0)^{-1} + \Gamma(\boldsymbol{\theta}_0)^{-1}\Pi(\boldsymbol{\theta}_0)\Gamma(\boldsymbol{\theta}_0)^{-1}, \tag{8.2.7}$$

where $\Pi(\boldsymbol{\theta}) = \{\Pi_{jl}(\boldsymbol{\theta})\}$ is an $r \times r$ matrix whose (j, l)th element is

$$\Pi_{jl}(\boldsymbol{\theta}) = \frac{1}{8\pi} \sum_{q,s,u,v=1}^{m} \int\int_{-\pi}^{\pi} \frac{\partial}{\partial \theta_j} \boldsymbol{f}_{\boldsymbol{\theta}}^{(q,s)}(\lambda_1) \frac{\partial}{\partial \theta_l} \boldsymbol{f}_{\boldsymbol{\theta}}^{(u,v)}(\lambda_2) \times Q_{qsuv}^X(-\lambda_1, \lambda_2, -\lambda_2) d\lambda_1 d\lambda_2.$$

Here $\boldsymbol{f}_{\boldsymbol{\theta}}^{(q,s)}(\lambda)$ is the (q, s)th element of $\boldsymbol{f}_{\boldsymbol{\theta}}(\lambda)^{-1}$.

The results above are very general, and can be applied to a variety of non-Gaussian vector processes (not only VAR, VMA and VARMA, but also second-order stationary "nonlinear processes"). It may be noted that the Wold decomposition theorem (e.g., Hannan (1970, p.137)) enables us to represent second-order stationary nonlinear processes as linear processes as in (8.1.1).

Although the model (8.1.1) is very general, the estimation criterion $D(\boldsymbol{f}_{\boldsymbol{\theta}_0}, \boldsymbol{I}_n)$ is a quadratic form motivated by Gaussian log-likelihood. In what follows we introduce an essentially non-Gaussian likelihood for linear processes.

Let $\{\boldsymbol{X}(t)\}$ be generated by

$$\boldsymbol{X}(t) = \sum_{j=0}^{\infty} A_{\boldsymbol{\theta}}(j)\boldsymbol{U}(t-j), \tag{8.2.8}$$

where $\boldsymbol{U}(t)$'s are i.i.d. with probability density function $p(\boldsymbol{u})$, and $\boldsymbol{\theta} \in \Theta$ (an open set) $\subset \mathbb{R}^r$.

Assumption 8.2.3. **(i)** $\lim_{\|\boldsymbol{u}\| \to \infty} p(\boldsymbol{u}) = 0$, $\int \boldsymbol{u} p(\boldsymbol{u}) d\boldsymbol{u} = \mathbf{0}$, $\int \boldsymbol{u}\boldsymbol{u}' p(\boldsymbol{u}) d\boldsymbol{u} = I_m$ *and* $\int \|\boldsymbol{u}\|^4 p(\boldsymbol{u}) d\boldsymbol{u} < \infty$.

(ii) *The derivatives Dp and $Dp^2 \equiv D(Dp)$ exist on $\mathbb{R}^m$, and every component of D^2p satisfies the Lipschitz condition.*

(iii) $\int |\phi(\boldsymbol{u})|^4 p(\boldsymbol{u}) d\boldsymbol{u} < \infty$ *and* $\int D^2 p(\boldsymbol{u}) d\boldsymbol{u} = \mathbf{0}$, *where* $\phi(\boldsymbol{u}) = p^{-1}Dp$.

We impose the following conditions on $A_{\boldsymbol{\theta}}(j)$'s.

Assumption 8.2.4. **(i)** *There exists* $0 < \rho_A < 1$ *so that*

$$|A_{\boldsymbol{\theta}}(j)| = O(\rho_A^j), \quad j \in \mathbb{N}.$$

(ii) *Every* $A_{\boldsymbol{\theta}}(j) = \{A_{\boldsymbol{\theta},ab}\}$ *is continuously two times differentiable with respect to* $\boldsymbol{\theta}$, *and the derivatives satisfy*

$$|\partial_{i_1} \cdots \partial_{i_k} A_{\boldsymbol{\theta},ab}(j)| = O(\gamma_A^j), \quad k = 1, 2, \ j \in \mathbb{N},$$

for some $0 < \gamma_A < 1$ *and for* $a, b = 1, \ldots, m$.

(iii) *Every* $\partial_{i_1}\partial_{i_2} A_{\boldsymbol{\theta},ab}(j)$ *satisfies the Lipschitz condition for all* $i_1, i_2 = 1, \ldots, m$ *and* $j \in \mathbb{N}$.

(iv) $\det\left\{\sum_{j=0}^{\infty} A_{\boldsymbol{\theta}}(j) z^j\right\} \neq 0$ *for* $|z| \leq 1$ *and*

$$\left\{\sum_{j=0}^{\infty} A_{\boldsymbol{\theta}}(j) z^j\right\}^{-1} = I_m + B_{\boldsymbol{\theta}}(1)z + B_{\boldsymbol{\theta}}(2)z^2 + \cdots, \quad |z| \leq 1,$$

where $|B_{\boldsymbol{\theta}}(j)| = O(\rho_B^j)$, $j \in \mathbb{N}$, *for some* $0 < \rho_B < 1$.

(v) *Every* $B_{\boldsymbol{\theta}}(j) = \{B_{\boldsymbol{\theta},ab}(j)\}$ *is continuously two times differentiable with respect to* $\boldsymbol{\theta}$, *and the derivatives satisfy*

$$|\partial_{i_1} \cdots \partial_{i_k} B_{\boldsymbol{\theta},ab}(j)| = O(\gamma_B^j), \quad k = 1, 2, \ j \in \mathbb{N},$$

for some $0 < \gamma_B < 1$ *and for* $a, b = 1, \ldots, m$.

(vi) *Every* $\partial_{i_1}\partial_{i_2} B_{\boldsymbol{\theta},ab}(j)$ *satisfies the Lipschitz condition for all* $i_1, i_2 = 1, \ldots, q$, $a, b = 1, \ldots, m$ *and* $j \in \mathbb{N}$.

If $\{\boldsymbol{X}(t)\}$ is a regular VARMA model, then the assumptions above are satisfied. Hence, Assumptions 8.2.3 and 8.2.4 are natural ones.

The likelihood function of $\{\boldsymbol{U}(s), s \leq 0, \boldsymbol{X}(1), \ldots, \boldsymbol{X}(n)\}$ is given in the form of

$$dQ_{n,\boldsymbol{\theta}} = \prod_{t=1}^{n} p\left\{\sum_{j=0}^{t-1} B_{\boldsymbol{\theta}} \boldsymbol{X}(t-j) + \sum_{r=0}^{\infty} c_{\boldsymbol{\theta}}(r,t)\boldsymbol{U}(-r)\right\} dQ_{\boldsymbol{u}}, \tag{8.2.9}$$

where $Q_{\boldsymbol{u}}$ is the probability distribution of $\{\boldsymbol{U}(s), \ s \leq 0\}$. Since $\{\boldsymbol{U}(s), s \leq 0\}$ are unobservable, we use the "quasi-likelihood"

$$L_n(\boldsymbol{\theta}) = \prod_{t=1}^{n} p\left\{\sum_{j=0}^{t-1} B_{\boldsymbol{\theta}}(j)\boldsymbol{X}(t-j)\right\} \tag{8.2.10}$$

for estimation of $\boldsymbol{\theta}$. A quasi-maximum likelihood estimator $\hat{\boldsymbol{\theta}}_{\text{QML}}$ of $\boldsymbol{\theta}$ is defined as a solution of the equation

$$\frac{\partial}{\partial \boldsymbol{\theta}}\left[\sum_{t=1}^{n} \log p\left\{\sum_{j=0}^{t-1} B_{\boldsymbol{\theta}}(j)\boldsymbol{X}(t-j)\right\}\right] = \boldsymbol{0} \tag{8.2.11}$$

with respect to $\boldsymbol{\theta}$. Henceforth we denote the true value of $\boldsymbol{\theta}$ by $\boldsymbol{\theta}_0$.

Theorem 8.2.5. *(Taniguchi and Kakizawa (2000, p. 72)) Under Assumptions 8.2.3 and 8.2.4, the following statements hold.*

(i) *There exists a statistic* $\hat{\boldsymbol{\theta}}_{\text{QML}}$ *that solves (8.2.11) such that for some* $c_1 > 0$,

$$P_{n,\boldsymbol{\theta}_0}[\sqrt{n}\|\hat{\boldsymbol{\theta}}_{\text{QML}} - \boldsymbol{\theta}_0\| \leq c_1 \log n] = 1 - o(1), \tag{8.2.12}$$

uniformly for $\boldsymbol{\theta}_0 \in \boldsymbol{C}$ *where* $\boldsymbol{C}$ *is a compact subset of* Θ.

(ii)

$$\sqrt{n}(\hat{\boldsymbol{\theta}}_{QML} - \boldsymbol{\theta}_0) \xrightarrow{d} N(\mathbf{0}, \Gamma(\boldsymbol{\theta}_0)^{-1}), \tag{8.2.13}$$

where $\Gamma(\boldsymbol{\theta})$ *is an* $r \times r$ *matrix whose* (i_1, i_2)*th component is given by*

$$\Gamma_{i_1 i_2}(\boldsymbol{\theta}) = \mathrm{tr}\left[\mathcal{F}(p) \sum_{j_1=1}^{\infty} \sum_{j_2=1}^{\infty} \{\partial_{i_1} B_{\boldsymbol{\theta}}(j_1)\} R(j_1 - j_2) \{\partial_{i_2} B_{\boldsymbol{\theta}}(j_2)'\}\right].$$

Here $R(j) = E\{\boldsymbol{X}(s)\boldsymbol{X}(s+j)'\}$ *and* $\mathcal{F}(p) = \int \phi(\boldsymbol{u})\phi(\boldsymbol{u})' p(\boldsymbol{u}) d\boldsymbol{u}$.

Next we return to the CHARN model (2.1.36):

$$\boldsymbol{X}_t = \boldsymbol{F}_{\boldsymbol{\theta}}(\boldsymbol{X}_{t-1}, \ldots, \boldsymbol{X}_{t-p}) + \boldsymbol{H}_{\boldsymbol{\theta}}(\boldsymbol{X}_{t-1}, \ldots, \boldsymbol{X}_{t-p})\boldsymbol{U}_t, \tag{8.2.14}$$

where $\boldsymbol{U}_t$'s are i.i.d. with probability density function $p(\boldsymbol{u})$.

Assumption 8.2.6. **(i)** *All the conditions (i)–(iv) for stationarity in Theorem 2.1.20 hold.*

(ii) $E_{\boldsymbol{\theta}}\|\boldsymbol{F}_{\boldsymbol{\theta}}(\boldsymbol{X}_{t-1}, \ldots, \boldsymbol{X}_{t-p})\|^2 < \infty$, $E_{\boldsymbol{\theta}}\|\boldsymbol{H}_{\boldsymbol{\theta}}(\boldsymbol{X}_{t-1}, \ldots, \boldsymbol{X}_{t-p})\|^2 < \infty$, *for all* $\boldsymbol{\theta} \in \Theta$.

(iii) *There exists* $c > 0$ *such that*

$$c \leq \|\boldsymbol{H}_{\boldsymbol{\theta}'}^{-1/2}(\boldsymbol{x})\boldsymbol{H}_{\boldsymbol{\theta}}(\boldsymbol{x})\boldsymbol{H}_{\boldsymbol{\theta}'}^{-1/2}(\boldsymbol{x})\| < \infty,$$

for all $\boldsymbol{\theta}, \boldsymbol{\theta}' \in \Theta$ *and for all* $\boldsymbol{x} \in \mathbb{R}^{mp}$.

(iv) $\boldsymbol{H}_{\boldsymbol{\theta}}$ *and* $\boldsymbol{F}_{\boldsymbol{\theta}}$ *are continuously differentiable with respect to* $\boldsymbol{\theta}$, *and their derivatives* $\partial_j \boldsymbol{H}_{\boldsymbol{\theta}}$ *and* $\partial_j \boldsymbol{F}_{\boldsymbol{\theta}}$ $(\partial_j = \partial/\partial\theta_j, j = 1, \ldots, r)$ *satisfy the condition that there exist square-integrable functions* A_j *and* B_j *such that*

$$\|\partial_j \boldsymbol{H}_{\boldsymbol{\theta}}\| \leq A_j \text{ and } \|\partial_j \boldsymbol{F}_{\boldsymbol{\theta}}\| \leq B_j,\ j = 1, \ldots, r, \text{ for all } \boldsymbol{\theta} \in \Theta.$$

(v) *The innovation density* $p(\cdot)$ *satisfies* $\lim_{\|\boldsymbol{u}\| \to \infty} \|\boldsymbol{u}\| p(\boldsymbol{u}) = 0$, $\int \boldsymbol{u}\boldsymbol{u}' p(\boldsymbol{u}) d\boldsymbol{u} = I_m$, *where* I_m *is the* $m \times m$ *identity matrix.*

(vi) *The continuous derivative* Dp *of* $p(\cdot)$ *exists on* $\mathbb{R}^m$, *and* $\int \|p^{-1}Dp\|^4 p(\boldsymbol{u}) d\boldsymbol{u} < \infty$, $\int \|\boldsymbol{u}\|^2 \|p^{-1}Dp\|^2 p(\boldsymbol{u}) d\boldsymbol{u} < \infty$.

Write $\eta_t(\boldsymbol{\theta}) \equiv \log p\{\boldsymbol{H}_{\boldsymbol{\theta}}^{-1}(\boldsymbol{X}_t - \boldsymbol{F}_{\boldsymbol{\theta}})\}\{\det \boldsymbol{H}_{\boldsymbol{\theta}}\}^{-1}$. We further impose the following assumption.

Assumption 8.2.7. **(i)** *The true value* $\boldsymbol{\theta}_0$ *of* $\boldsymbol{\theta}$ *is the unique maximum of*

$$\lim_{n\to\infty} n^{-1} \sum_{t=p}^{n} E\{\eta_t(\boldsymbol{\theta})\}$$

with respect to $\boldsymbol{\theta} \in \Theta$.

(ii) $\eta_t(\boldsymbol{\theta})$*'s are three times differentiable with respect to* $\boldsymbol{\theta}$, *and there exist functions* $Q_{ij}^t = Q_{ij}^t(\boldsymbol{X}_1, \ldots, \boldsymbol{X}_t)$ *and* $T_{ijk}^t = T_{ijk}^t(\boldsymbol{X}_1, \ldots, \boldsymbol{X}_t)$ *such that*

$$|\partial_i \partial_j \eta_t(\boldsymbol{\theta})| \leq Q_{ij}^t, \quad EQ_{ij}^t < \infty,$$

and

$$|\partial_i \partial_j \partial_k \eta_t(\boldsymbol{\theta})| \leq T_{ijk}^t, \quad ET_{ijk}^t < \infty.$$

Let $\boldsymbol{\Phi}_t \equiv \boldsymbol{H}_{\boldsymbol{\theta}}^{-1}(\boldsymbol{X}_t - \boldsymbol{F}_{\boldsymbol{\theta}})$, *and write*

$$\boldsymbol{W}_t = \begin{bmatrix} -vec'(\boldsymbol{H}_{\boldsymbol{\theta}}^{-1}\partial_1 \boldsymbol{H}_{\boldsymbol{\theta}}), & \partial_1 \boldsymbol{F}_{\boldsymbol{\theta}}' \cdot \boldsymbol{H}_{\boldsymbol{\theta}}^{-1} \\ \vdots & \\ -vec'(\boldsymbol{H}_{\boldsymbol{\theta}}^{-1}\partial_r \boldsymbol{H}_{\boldsymbol{\theta}}), & \partial_r \boldsymbol{F}_{\boldsymbol{\theta}}' \cdot \boldsymbol{H}_{\boldsymbol{\theta}}^{-1} \end{bmatrix} \quad \left(r \times \left(m^2 + m\right)\text{-matrix}\right)$$

$$\Psi_t = \begin{bmatrix} \{\Phi_t \otimes I_m\} p^{-1}(\Phi_t) D p(\Phi_t) + vec I_m \\ p^{-1}(\Phi_t) D p(\Phi_t) \end{bmatrix} \quad ((m^2+m) \times 1\text{-}vector),$$

$$\mathcal{F}(p) = E\{\Psi_t \Psi_t'\}, \quad ((m^2+m) \times (m^2+m)\text{-}matrix),$$

$$\Gamma(p, \theta) = E\{W_t \mathcal{F}(p) W'\}, \quad (r \times r\text{-}matrix).$$

For two hypothetical values $\theta, \theta' \in \Theta$,

$$\Lambda_n(\theta, \theta') \equiv \sum_{t=p}^{n} \log \frac{p\{H_{\theta'}^{-1}(X_t - F_{\theta'})\} \det H_{\theta}}{p\{H_{\theta}^{-1}(X_t - F_{\theta})\} \det H_{\theta'}}. \tag{8.2.15}$$

The maximum likelihood estimator of θ is given by

$$\hat{\theta}_{\mathrm{ML}} \equiv \arg\max_{\theta} \Lambda_n(\underline{\theta}, \theta), \tag{8.2.16}$$

where $\underline{\theta} \in \Theta$ is some fixed value.

Theorem 8.2.8. *(Kato et al. (2006)) Under Assumptions 8.2.6 and 8.2.7, the following statements hold.*

(i) $\hat{\theta}_{ML} \xrightarrow{p} \theta_0$ *as* $n \to \infty$,

(ii) $\sqrt{n}(\hat{\theta}_{ML} - \theta_0) \xrightarrow{d} N(\mathbf{0}, \Gamma(p, \theta_0)^{-1})$, *as* $n \to \infty$.

Let $X = \{X_t, \mathcal{F}_t, t \geq 0\}$ be an m-dimensional semimartingale with the canonical representation

$$X_t = X_0 + \alpha_t + X_t^{(c)} + \int_0^t \int_{\|x\|>1} x d\mu + \int_0^t \int_{\|x\|\leq 1} x d(\mu - \nu), \tag{8.2.17}$$

where $X^{(c)} \equiv \{X_t^{(c)}\}$ is the continuous martingale component of X_t, the components of $\alpha \equiv \{\alpha_t\}$ are processes of bounded variation (Definition 2.1.17), the random measure μ is defined by

$$\mu(\omega, dx, dt) = \sum_s \chi\{\Delta X_s(\omega) \neq 0\} \epsilon_{(x, \Delta X_s(\omega))}(dt, dx)$$

where ϵ_a is the Dirac measure at a and ν is the compensator of μ (see Jacod and Shiryaev (1987), Prakasa Rao (1999)). Letting $\beta \equiv \langle X^{(c)}, X^{(c)} \rangle$, the triple (α, β, ν) is called the triplet of P-local characteristics of the semimartingale X.

Because notations and regularity conditions for the asymptotic likelihood theory of semimartingales are unnecessarily complicated, in what follows, we state the essential line of discussions. For detail, see Prakasa Rao (1999) and Sørensen (1991), etc.

Suppose that the triple (α, β, ν) of (8.2.17) depends on unknown parameter $\theta = (\theta_1, \ldots, \theta_p)' \in \Theta$ (open set) $\subset \mathbb{R}^p$, i.e., $(\alpha_\theta, \beta_\theta, \nu_\theta)$. Then, based on the observation $\{X_s : 0 \leq s \leq t\}$ from (8.2.17), the log-likelihood function is given by

$$\begin{aligned} l_t(\theta) = & \int_0^t G_s(\theta) dX_s^{(c)} + \int_0^t A_s(\theta) ds \\ & + \int_0^t \int_{\mathbb{R}^m} B_s(\theta, X_{s-}, x)(\mu - \nu_\theta)(dx, ds) \\ & + \int_0^t \int_{\mathbb{R}^m} C_s(\theta, X_{s-}, x\mu(dx, ds)) \end{aligned} \tag{8.2.18}$$

where $G_s(\cdot)$, $A_s(\cdot)$, $B_s(\ ,\ ,\ )$ and $C_s(\ ,\ ,\ )$ are smooth with respect to θ, and $\mathcal{F}_s$-measurable with respect to the other arguments. Let

$$I_t(\theta) \equiv \left\langle \frac{\partial}{\partial \theta} l_t(\theta) \right\rangle_t, \quad J_t(\theta) \equiv \left[\frac{\partial}{\partial \theta} l_t(\theta) \right]_t, \quad j_t(\theta) \equiv -\frac{\partial^2}{\partial \theta \partial \theta'} l_t(\theta). \tag{8.2.19}$$

It is seen that

$$E_{\boldsymbol{\theta}}\{\boldsymbol{j}_t(\boldsymbol{\theta})\} = E_{\boldsymbol{\theta}}\{I_t(\boldsymbol{\theta})\} = E_{\boldsymbol{\theta}}\{J_t(\boldsymbol{\theta})\} = \boldsymbol{i}_t(\boldsymbol{\theta}), \text{ (say)}. \tag{8.2.20}$$

Let $D_t(\boldsymbol{\theta}) = \text{diag}(d_t(\boldsymbol{\theta})_{11}, \dots, d_t(\boldsymbol{\theta})_{pp})$, where $d_t(\boldsymbol{\theta})_{jj}$ is the jth diagonal element of $\boldsymbol{i}_t(\boldsymbol{\theta})$. Based on the law of large numbers (Theorem 2.1.27) and the central limit theorem (Theorem 8.1.2), the following results can be proved under appropriate regularity conditions (see Prakasa Rao (1999)).

Theorem 8.2.9. **(1)** *Under $P = P_{\boldsymbol{\theta}}$, almost surely on the set {the minimum eigenvalue of $I_t(\boldsymbol{\theta}) \to \infty$}, it holds that the likelihood equation*

$$\frac{\partial}{\partial \boldsymbol{\theta}} l_t(\boldsymbol{\theta}) = \mathbf{0}, \tag{8.2.21}$$

has a solution $\hat{\boldsymbol{\theta}}_t$ satisfying $\hat{\boldsymbol{\theta}}_t \xrightarrow{p} \boldsymbol{\theta}$ as $t \to \infty$.

(2) *Assume that*

$$D_t(\boldsymbol{\theta})^{-1/2} \boldsymbol{j}_t(\tilde{\boldsymbol{\theta}}_t) D_t(\boldsymbol{\theta})^{-1/2} \xrightarrow{p} \eta^2(\boldsymbol{\theta}) \text{ under } P_{\boldsymbol{\theta}} \text{ as } t \to \infty, \tag{8.2.22}$$

for any $\tilde{\boldsymbol{\theta}}_t$ lying on the line segment connecting $\boldsymbol{\theta}$ and $\hat{\boldsymbol{\theta}}$. Then, conditionally on the event $\{\det(\eta^2(\boldsymbol{\theta})) > 0\}$,

$$D_t(\boldsymbol{\theta})^{1/2}(\hat{\boldsymbol{\theta}}_t - \boldsymbol{\theta}) \xrightarrow{d} \eta^{-2}(\boldsymbol{\theta})\boldsymbol{Z}, \qquad \text{(stably)}, \tag{8.2.23}$$

$$[D_t(\boldsymbol{\theta})^{-1/2} \boldsymbol{j}_t(\hat{\boldsymbol{\theta}}_t) D_t(\boldsymbol{\theta})^{-1/2}]^{1/2} D_t(\boldsymbol{\theta})^{1/2}(\hat{\boldsymbol{\theta}}_t - \boldsymbol{\theta}) \xrightarrow{d} N(\mathbf{0}, \boldsymbol{I}_p), \text{ (mixing)},$$

and

$$2[l_t(\hat{\boldsymbol{\theta}}_t) - l_t(\boldsymbol{\theta})] \xrightarrow{d} \chi_p^2, \text{ (mixing)}, \tag{8.2.24}$$

under $P_{\boldsymbol{\theta}}$ as $t \to \infty$. Here the random vector $\boldsymbol{Z}$ has the characteristic function

$$\phi(\boldsymbol{u}) = E_{\boldsymbol{\theta}}\left[\exp\left\{-\frac{1}{2}\boldsymbol{u}'\eta^2(\boldsymbol{\theta})\boldsymbol{u}\right\}\right], \quad \boldsymbol{u} = (u_1, \dots, u_p)'. \tag{8.2.25}$$

Example 8.2.10. *Let $\{X_t\}$ be generated by*

$$dX_t = \theta X_t dt + \sigma dW_t + dN_t, \ t \ge 0, \ X_0 = x_0 \tag{8.2.26}$$

where $\{W_t\}$ is a standard Wiener process and $\{N_t\}$ is a Poisson process with intensity λ. Assuming $\sigma > 0$ is known, we estimate (θ, λ) by the maximum likelihood estimator (MLE) from the observation $\{X_s : 0 \le s \le t\}$. The MLE are

$$\hat{\theta}_t = \frac{\int_0^t X_{s-} dX_s^{(c)}}{\int_0^t X_s^2 ds}, \tag{8.2.27}$$

and

$$\hat{\lambda}_t = \frac{N_t}{t}, \tag{8.2.28}$$

where $X_t^{(c)} = X_t - \sum_{s \le t} \Delta X_s$ and N_t is the number of jumps before t.

8.3 Statistical Optimal Theory

Portfolio Estimation Influenced by Non-Gaussian Innovations and Exogenous Variables

In the previous section we discussed the asymptotics of MLE for very general stochastic processes. Here we show the optimality of MLE, etc.

Lucien Le Cam introduced the concept of *local asymptotic normality* (LAN) for the likelihood ratio of regular statistical models. Once LAN is proved, the asymptotic optimality of estimators and tests is described in terms of the LAN property. In what follows we state general LAN theorems due to Le Cam (1986) and Swensen (1985).

Let $P_{0,n}$ and $P_{1,n}$ be two sequences of probability measures on certain measurable spaces $(\Omega_n, \mathcal{F}_n)$. Suppose that there is a sequence $\mathcal{F}_{n,k}$, $k = 1, \ldots, k_n$, of sub σ-algebras of $\mathcal{F}_n$ such that $\mathcal{F}_{n,k} \subset \mathcal{F}_{n,k+1}$ and such that $\mathcal{F}_{n,k_n} = \mathcal{F}_n$. Let $P_{i,n,k}$ be the restriction of $P_{i,n}$ to $\mathcal{F}_{n,k}$, and let $\gamma_{n,k}$ be the Radon–Nikodym density taken on $\mathcal{F}_{n,k}$ of the part of $P_{1,n,k}$ that is dominated by $P_{0,n,k}$. Let $Y_{n,k} = (\gamma_{n,k}/\gamma_{n,k-1})^{1/2} - 1$, where $\gamma_{n,0} = 1$ and $n = 1, 2, \ldots$. Then the logarithm of likelihood ratio

$$\Lambda_n = \log \frac{dP_{1,n}}{dP_{0,n}} \tag{8.3.1}$$

taken on $\mathcal{F}_n$ is written as $\Lambda_n = 2 \sum_k \log(Y_{n,k} + 1)$.

Theorem 8.3.1. *(Le Cam (1986)) Suppose that under $P_{0,n}$ the following conditions are satisfied:*

(L1) $\max_k |Y_{n,k}| \xrightarrow{p} 0$,

(L2) $\sum_k Y_{n,k}^2 \xrightarrow{p} \tau^2/4$,

(L3) $\sum_k E(Y_{n,k}^2 + 2Y_{n,k}|\mathcal{F}_{n,k-1}) \xrightarrow{p} 0$, *and*

(L4) $\sum_k E\{Y_{n,k}^2 \chi(|Y_{n,k}| > \delta)|\mathcal{F}_{n,k-1}\} \xrightarrow{p} 0$ *for some* $\delta > 0$.

Then, $\Lambda_n \xrightarrow{d} N(-\tau^2/2, \tau^2)$ *as* $n \to \infty$.

Remark 8.3.2. *If both $P_{0,n,k}$ and $P_{1,n,k}$ are absolutely continuous with respect to Lebesgue measure μ_k on $\mathbb{R}^k$, the condition* **(L3)** *above is always satisfied (Problem 8.1).*

A version of the theorem is given with replacing the $Y_{n,k}$ by martingale differences.

Theorem 8.3.3. *(Swensen (1985)) Suppose that there exists an adapted stochastic process $\{W_{n,k}, \mathcal{F}_{n,k}\}$ which satisfies*

(S1) $E(W_{n,k}|\mathcal{F}_{n,k}) = 0$ *a.s.*,

(S2) $E\{\sum_k (Y_{n,k} - W_{n,k})^2\} \to 0$,

(S3) $\sup_n E(\sum_k W_{n,k}^2) < \infty$,

(S4) $\max_k |W_{n,k}| \xrightarrow{p} 0$,

(S5) $\sum_k W_{n,k}^2 \xrightarrow{p} \tau^2/4$, *and*

(S6) $\sum_k E\{W_{n,k}^2 \chi(|W_{n,k}| > \delta)|\mathcal{F}_{n,k-1}\} \xrightarrow{p} 0$ *for some* $\delta > 0$,

where $\xrightarrow{p}$ *in* **(S4)**–**(S6)** *are under $P_{0,n}$. Then the conditions* **(L1)**, **(L2)** *and* **(L4)** *of Theorem 8.3.1 hold. Further, if* **(L3)** *of Theorem 8.3.1 is satisfied, then*

$$\Lambda_n \xrightarrow{d} N(-\tau^2/2, \tau^2)$$

as $n \to \infty$.

Remark 8.3.4. *(Kreiss (1990)) The conditions* **(S5)** *and* **(S6)** *can be replaced by*

(S5)′ $\sum_k E(W_{n,k}^2|\mathcal{F}_{n,k-1}) \xrightarrow{p} \tau^2/4$,

(S6)′ $\sum_k E\{W_{n,k}^2 \chi(|W_{n,k}| > \epsilon)|\mathcal{F}_{n,k-1}\} \xrightarrow{p} 0$ *for every* $\epsilon > 0$, *under $P_{0,n}$ (Problem 8.2).*

Using Theorem 8.3.3, we shall prove the LAN theorem for the m-vector non-Gaussian linear process defined by (8.2.8):

$$\text{(LP)} \qquad \boldsymbol{X}(t) = \sum_{j=0}^{\infty} A_{\boldsymbol{\theta}}(j)\boldsymbol{U}(t-j)$$

satisfying Assumption 8.2.3. First, we mention a brief review of LAN results for stochastic precesses. Swensen (1985) showed that the likelihood ratio of an autoregressive process of finite order with a trend is LAN. For ARMA processes, Hallin et al. (1985a) and Kreiss (1987) proved the LAN property, and applied the results to test and estimation theory. Kreiss (1990) showed the LAN for a class of autoregressive processes with infinite order. Also, Garel and Hallin (1995) established the LAN property for multivariate ARMA models with a linear trend.

Returning to (LP), let

$$A_{\boldsymbol{\theta}}(z) = \sum_{j=0}^{\infty} A_{\boldsymbol{\theta}}(j)z^j \quad (A_{\boldsymbol{\theta}}(0) = \boldsymbol{I}_m), \quad |z| \le 1, \tag{8.3.2}$$

where $A_{\boldsymbol{\theta}}(j) = \{A_{\boldsymbol{\theta},ab}(j); a, b = 1, \dots, m\}$, $j \in \mathbb{N}$. We make the following assumptions.

Assumption 8.3.5. **(i)** *For some d $(0 < d < 1/2)$, the coefficient matrices $A_{\boldsymbol{\theta}}(j)$ satisfy*

$$|A_{\boldsymbol{\theta}}(j)| = O(j^{-1+d}), \, j \in \mathbb{N} \tag{8.3.3}$$

where $|A_{\boldsymbol{\theta}}(j)|$ denotes the sum of the absolute values of the entries of $A_{\boldsymbol{\theta}}(j)$.

(ii) *Every $A_{\boldsymbol{\theta}}(j)$ is continuously two times differentiable with respect to $\boldsymbol{\theta}$, and the derivatives satisfy*

$$|\partial_{i_1}\partial_{i_2}\cdots\partial_{i_k}A_{\boldsymbol{\theta},ab}(j)| = O\{j^{-1+d}(\log j)^k\}, \; k = 0, 1, 2 \tag{8.3.4}$$

for $a, b = 1, \dots m$, where $\partial_i = \partial/\partial\theta_i$.

(iii) $\det A_{\boldsymbol{\theta}}(z) \neq 0$ *for all $|z| \le 1$ and $A_{\boldsymbol{\theta}}(z)^{-1}$ can be expanded as*

$$A_{\boldsymbol{\theta}}(z)^{-1} = \boldsymbol{I}_m + B_{\boldsymbol{\theta}}(1)z + B_{\boldsymbol{\theta}}(2)z^2 + \cdots, \tag{8.3.5}$$

where $B_{\boldsymbol{\theta}}(j) = \{B_{\boldsymbol{\theta},ab}(j); a, b = 1, \dots, m\}$, $j = 1, 2, \dots,$ satisfy

$$|B_{\boldsymbol{\theta}}(j)| = O(j^{-1-d}). \tag{8.3.6}$$

(iv) *Every $B_{\boldsymbol{\theta}}(j)$ is continuously two times differentiable with respect to $\boldsymbol{\theta}$, and the derivatives satisfy*

$$|\partial_{i_1}\partial_{i_2}\cdots\partial_{i_k}B_{\boldsymbol{\theta},ab}(j)| = O\{j^{-1-d}(\log j)^k\}, k = 0, 1, 2 \tag{8.3.7}$$

for $a, b = 1, \dots, m$.

Under Assumption 8.3.5, our model (LP) is representable as

$$\sum_{j=0}^{\infty} B_{\boldsymbol{\theta}}(j)\boldsymbol{X}(t-j) = \boldsymbol{U}(t), \;\; B_{\boldsymbol{\theta}}(0) = \boldsymbol{I}_m. \tag{8.3.8}$$

Then it follows that

$$\boldsymbol{U}(t) = \sum_{j=0}^{\infty} B_{\boldsymbol{\theta}}(j)\boldsymbol{X}(t-j) + \sum_{r=0}^{\infty} C_{\boldsymbol{\theta}}(r,t)\boldsymbol{U}(-r), \tag{8.3.9}$$

where

$$C_{\boldsymbol{\theta}}(r,t) = \sum_{r'=0}^{\infty} B_{\boldsymbol{\theta}}(r'+t)\boldsymbol{A}(r-r').$$

From Assumption 8.3.5, we can show

$$C_{\boldsymbol{\theta}}(r,t) = O(t^{-d/2})O(r^{-1+d}) \tag{8.3.10}$$

(Problem 8.3).

Let $Q_{n,\boldsymbol{\theta}}$ and $Q_{\boldsymbol{u}}$ be the probability distributions of $\{\boldsymbol{U}(s), s \leq 0, \boldsymbol{X}(1), \dots, \boldsymbol{X}(n)\}$ and $\{\boldsymbol{U}(s), s \leq 0\}$, respectively. Then, from (8.3.9) it follows that

$$dQ_{n,\boldsymbol{\theta}} = \prod_{t=1}^{n} p\left\{\sum_{j=0}^{t-1} B_{\boldsymbol{\theta}}(j)\boldsymbol{X}(t-j) + \sum_{r=0}^{\infty} C_{\boldsymbol{\theta}}(r,t)\boldsymbol{U}(-r)\right\} dQ_{\boldsymbol{u}}. \tag{8.3.11}$$

For two hypothetical values $\boldsymbol{\theta}, \boldsymbol{\theta}' \in \boldsymbol{\Theta}$, the log-likelihood ratio is

$$\begin{aligned}\Lambda_n(\boldsymbol{\theta},\boldsymbol{\theta}') &\equiv \log \frac{dQ_{n,\boldsymbol{\theta}'}}{dQ_{n,\boldsymbol{\theta}}} \\ &= \sum_{k=1}^{n} \log\Big[p\Big\{\boldsymbol{U}(k) + \sum_{j=0}^{k-1}(B_{\boldsymbol{\theta}'}(j) - B_{\boldsymbol{\theta}}(j))\boldsymbol{X}(k-j) \\ &\quad + \sum_{r=0}^{\infty}(C_{\boldsymbol{\theta}'}(r,k) - C_{\boldsymbol{\theta}}(r,k))\boldsymbol{U}(-r)\Big\}/p\{\boldsymbol{U}(k)\}\Big].\end{aligned} \tag{8.3.12}$$

We denote by $H(p;\boldsymbol{\theta})$ the hypothesis under which the underlying parameter is $\boldsymbol{\theta} \in \boldsymbol{\Theta}$ and the probability density of $\boldsymbol{U}_t$ is $p = (\cdot)$. Introduce the contiguous sequence

$$\boldsymbol{\theta}_n = \boldsymbol{\theta} + \frac{1}{\sqrt{n}}\boldsymbol{h}, \ \boldsymbol{h} = (h_1, \dots, h_p)' \in \mathcal{H} \subset \mathbb{R}^p, \tag{8.3.13}$$

and let

$$\mathcal{F}(p) = \int \phi(\boldsymbol{u})\phi(\boldsymbol{u})'p(\boldsymbol{u})d\boldsymbol{u} \quad \text{(Fisher information matrix)},$$

and

$$R(t) = E\{\boldsymbol{X}(s)\boldsymbol{X}(t+s)'\}, \ t \in \mathbb{Z},$$

where $\phi(\boldsymbol{u}) = p^{-1}\frac{d}{d\boldsymbol{u}}p(\boldsymbol{u})$.

Now we can state the local asymptotic normality (LAN) for the vector process (LP) ((8.2.8)). (For a proof, see Taniguchi and Kakizawa (2000)).

Theorem 8.3.6. *For the vector-valued non-Gaussian linear process, suppose that Assumptions 8.2.3 and 8.3.5 hold. Then the sequence of experiments*

$$\mathcal{E}_n = \{\mathbb{R}^{\mathbb{Z}}, \mathcal{B}^{\mathbb{Z}}, \{Q_{n,\boldsymbol{\theta}} : \boldsymbol{\theta} \in \boldsymbol{\Theta} \subset \mathbb{R}^p\}\}, \ n \in \mathbb{N},$$

is locally asymptotically normal and equicontinuous on compact subset C of $\mathcal{H}$. That is,

(1) *For all $\boldsymbol{\theta} \in \boldsymbol{\Theta}$, the log-likelihood ratio $\Lambda_n(\boldsymbol{\theta}, \boldsymbol{\theta}_n)$ has, under $H(p;\boldsymbol{\theta})$ as $n \to \infty$, the asymptotic expansion*

$$\Lambda(\boldsymbol{\theta},\boldsymbol{\theta}_n) = \Delta_n(\boldsymbol{h};\boldsymbol{\theta}) - \frac{1}{2}\Gamma_{\boldsymbol{h}}(\boldsymbol{\theta}) + o_p(1), \tag{8.3.14}$$

where

$$\Delta_{\boldsymbol{h}}(\boldsymbol{h};\boldsymbol{\theta}) = \frac{1}{\sqrt{n}} \sum_{k=1}^{n} \phi\{\boldsymbol{U}(k)\}' \sum_{j=1}^{k-1} B_{\boldsymbol{h}'\partial\boldsymbol{\theta}} \boldsymbol{X}(t-j), \tag{8.3.15}$$

and

$$\Gamma(\boldsymbol{\theta}) = \operatorname{tr}\left\{ \mathcal{F}(p) \sum_{j_1=1}^{\infty} \sum_{j_2=1}^{\infty} B_{\boldsymbol{h}'\partial\boldsymbol{\theta}}(j_1) R(j_1 - j_2) B_{\boldsymbol{h}'\partial\boldsymbol{\theta}}(j_2)' \right\}$$

with

$$B_{\boldsymbol{h}'\partial\boldsymbol{\theta}}(j) = \sum_{l=1}^{p} h_l \partial_l B_{\boldsymbol{\theta}}(j).$$

(2) *Under $H(p;\boldsymbol{\theta})$*

$$\Delta_{\boldsymbol{h}}(\boldsymbol{h};\boldsymbol{\theta}) \xrightarrow{d} N(\boldsymbol{0}, \Gamma_{\boldsymbol{h}}(\boldsymbol{\theta})). \tag{8.3.16}$$

(3) *For all $n \in \mathbb{N}$ and all $\boldsymbol{h} \in \mathcal{H}$, the mapping*

$$\boldsymbol{h} \to Q_{n,\boldsymbol{\theta}_n}$$

is continuous with respect to the variational distance

$$\|P - Q\| = \sup\{|P(A) - Q(A)| : A \in \mathcal{B}^{\mathbb{Z}}\}.$$

Henceforth we denote the distribution law of random vector $\boldsymbol{Y}_n$ under $Q_{n,\boldsymbol{\theta}}$ by $\mathcal{L}(\boldsymbol{Y}_n|Q_{n,\boldsymbol{\theta}})$, and its weak convergence to L by $\mathcal{L}(\boldsymbol{Y}_n|Q_{n,\boldsymbol{\theta}}) \xrightarrow{d} L$. Define the class $\mathcal{A}$ of sequences of estimators of $\boldsymbol{\theta}$, $\{T_n\}$ as follows:

$$\mathcal{A} = [\{T_n\} : \mathcal{L}\{\sqrt{n}(T_n - \boldsymbol{\theta}_n)|Q_{n,\boldsymbol{\theta}_n}\} \xrightarrow{d} L_{\boldsymbol{\theta}}(\cdot), \text{ a probability distribution}], \tag{8.3.17}$$

where $L_{\boldsymbol{\theta}}(\cdot)$ in general depends on $\{T_n\}$.

Let L be the class of all loss functions $l : \mathbb{R}^q \to [0, \infty)$ of the form $l(\boldsymbol{x}) = \tau(|\boldsymbol{x}|)$ which satisfies $\tau(0) = 0$ and $\tau(a) \leq \tau(b)$ if $a \leq b$. Typical examples are $l(\boldsymbol{x}) = \chi(|\boldsymbol{x}| > a)$ and $l(\boldsymbol{x}) = |\boldsymbol{x}|^p$, $p \geq 1$.

Definition 8.3.7. *Assume that the LAN property (8.3.14) holds. Then a sequence $\{\hat{\boldsymbol{\theta}}_n\}$ of estimators of $\boldsymbol{\theta}$ is said to be a sequence of asymptotically centering estimators if*

$$\sqrt{n}(\hat{\boldsymbol{\theta}}_n - \boldsymbol{\theta}) - \Gamma(\boldsymbol{\theta})^{-1}\Delta_n = o_p(1) \quad \text{in } Q_{n,\boldsymbol{\theta}}. \tag{8.3.18}$$

The following theorem can be proved along the lines of Jeganathan (1995).

Theorem 8.3.8. *Assume that LAN property (8.3.14) holds, and $\Delta \equiv N(\boldsymbol{0}, \Gamma(\boldsymbol{\theta}))$. Suppose that $\{T_n\} \in \mathcal{A}$. Then, the following statements hold.*

(i) *For any $l \in L$ satisfying $E\{l(\Delta)\} < \infty$*

$$\liminf_{n\to\infty} E[l\{\sqrt{n}(T_n - \boldsymbol{\theta})\}|Q_{n,\boldsymbol{\theta}}] \geq E\{l(\Gamma^{-1}\Delta)\}. \tag{8.3.19}$$

(ii) *If, for a nonconstant $l \in L$ satisfying $E\{l(\Delta)\} < \infty$,*

$$\limsup_{n\to\infty} E[l\{\sqrt{n}(T_n - \boldsymbol{\theta})\}|Q_{n,\boldsymbol{\theta}}] \leq E\{l(\Gamma^{-1}\Delta)\}, \tag{8.3.20}$$

then $\{T_n\}$ is a sequence of asymptotically centering estimators.

From Theorem 8.3.8, we will define the asymptotic efficiency as follows.

Definition 8.3.9. *A sequence of estimators* $\{\hat{\theta}_n\} \in \mathcal{A}$ *of* θ *is said to be asymptotically efficient if it is a sequence of asymptotically centering estimators.*

Recall Theorem 8.2.5. Under Assumptions 8.2.3 and 8.2.4, it is shown that

$$\sqrt{n}(\hat{\theta}_{\mathrm{QML}} - \theta_0) - \Gamma(\theta_0)^{-1}\Delta_n = o_p(1) \quad \text{in } Q_{n,\theta_0}. \tag{8.3.21}$$

Let S_n be a p-dimensional statistic based on $\{X(1), \ldots, X(n)\}$. The following lemma is called Le Cam's third lemma (e.g., Hájek and Šidák (1967)).

Lemma 8.3.10. *Assume that the LAN property (8.3.14) holds. If* $(S_n', \Lambda_n(\theta_0, \theta_n))' \xrightarrow{d} N_{p+1}(m, \Sigma)$ *under* Q_{n,θ_0}*, where* $m' = (\mu', -\sigma_{22}/2)$ *and*

$$\Sigma = \begin{pmatrix} \Sigma_{11} & \sigma_{12} \\ \sigma_{12}' & \sigma_{22} \end{pmatrix},$$

then

$$S_n \xrightarrow{d} N_p(\mu + \sigma_{12}, \Sigma_{11}) \quad \text{under } Q_{n,\theta_n}. \tag{8.3.22}$$

From (8.3.21) we observe that

$$\{\sqrt{n}(\hat{\theta}_{\mathrm{QML}} - \theta_0)', \Lambda_n(\theta_0, \theta_n)\}' \xrightarrow{d} N_{p+1}(m, \Sigma), \text{ under } Q_{n,\theta_n},$$

where $m' = (0', h'\Gamma(\theta_0)h)$ and

$$\Sigma = \begin{pmatrix} \Gamma(\theta_0)^{-1} & h \\ h' & h'\Gamma(\theta_0)h \end{pmatrix}.$$

By Le Cam's third lemma, it is seen that

$$\mathcal{L}\{\sqrt{n}(\hat{\theta}_{\mathrm{QML}} - \theta_n)|Q_{n,\theta_n}\} \xrightarrow{d} N_p(0, \Gamma(\theta_0)^{-1}),$$

leading to $\hat{\theta}_{\mathrm{QML}} \in \mathcal{A}$. Hence $\hat{\theta}_{\mathrm{QML}}$ is asymptotically efficient.

In what follows we denote by $E_{n,h}(\cdot)$ the expectation with respect to Q_{n,θ_n}. Let L_0 be a linear subspace of $\mathbb{R}^p$. Consider the problem of testing (H, A):

$$H : h \in L_0 \text{ against } A : h \in \mathbb{R}^p - L_0. \tag{8.3.23}$$

Definition 8.3.11. *A sequence of tests* φ_n, $n \in \mathbb{N}$, *is asymptotically unbiased of level* $\alpha \in [0, 1]$ *for* (H, A) *if*

$$\limsup_{n\to\infty} E_{n,h}(\varphi_n) \le \alpha \quad \text{if } h \in L_0,$$
$$\liminf_{n\to\infty} E_{n,h}(\varphi_n) \ge \alpha \quad \text{if } h \in \mathbb{R}^p - L_0.$$

The following results are essentially due to Strasser (1985).

Theorem 8.3.12. *Suppose that the LAN property (8.3.14) holds for the non-Gaussian vector linear process (8.2.8). Let* π_{L_0} *be the orthogonal projection of* $\mathbb{R}^p$ *onto* L_0, *id the identity map and* $\tilde{\Delta}_n \equiv \Gamma^{-1/2}\Delta_n$. *Then the following assertions hold true.*

(i) *For* $\alpha \in [0, 1]$, *we can choose* k_α *so that the sequence of tests*

$$\varphi_n^* = \begin{cases} 1 & \text{if } \|(id - \pi_{L_0}) \circ \tilde{\Delta}_n\| > k_\alpha, \\ 0 & \text{if } \|(id - \pi_{L_0}) \circ \tilde{\Delta}_n\| < k_\alpha, \end{cases}$$

is asymptotically unbiased of level α *for* (H, A).

(ii) *If $\{\varphi_n\}$ is another sequence that is asymptotically unbiased of level α for (H, A), then*

$$\limsup_{n\to\infty} \inf_{\boldsymbol{h}\in B_c} E_{n,\boldsymbol{h}}(\varphi_n) \le \lim_{n\to\infty} \inf_{\boldsymbol{h}\in B_c} E_{n,\boldsymbol{h}}(\varphi_n^*),$$

where $B_c = \{\boldsymbol{h} \in \mathbb{R}^p : \|\boldsymbol{h} - \pi_{L_0}(\boldsymbol{h})\| = c\}$. If the equality holds, we say that $\{\varphi_n\}$ is locally asymptotically optimal.

(iii) *If $\{\varphi_n\}$ is any sequence of tests satisfying for at least one $c > 0$,*

$$\limsup_{n\to\infty} E_{n,\mathbf{0}}(\varphi_n) \le \alpha,$$

and

$$\liminf_{n\to\infty} \inf_{\boldsymbol{h}\in B_c} E_{n,\boldsymbol{h}}(\varphi_n) \ge \lim_{n\to\infty} E_{n,\boldsymbol{h}}(\varphi_n^*)$$

whenever $\boldsymbol{h} \in B_c$ and $\boldsymbol{h} \in L_0^{\perp}$, then $\{\varphi_n\}$ converges in distribution to the distribution limit of φ_n^* under $Q_{n,\boldsymbol{\theta}}$.

Let $\mathcal{M}(B)$ be the linear space spanned by the columns of a matrix B. Consider the problem of testing

$$H : \sqrt{n}(\boldsymbol{\theta} - \boldsymbol{\theta}_0) \in \mathcal{M}(B) \tag{8.3.24}$$

for some given $p \times (p - l)$ matrix B of full rank, and given vector $\boldsymbol{\theta}_0 \in \mathbb{R}^p$. For this we introduce the test statistic

$$\psi_n^* = n(\hat{\boldsymbol{\theta}}_{\mathrm{QML}} - \boldsymbol{\theta}_0)'[\Gamma(\hat{\boldsymbol{\theta}}_{\mathrm{QML}}) - \Gamma(\hat{\boldsymbol{\theta}}_{\mathrm{QML}})B\{B'\Gamma(\hat{\boldsymbol{\theta}}_{\mathrm{QML}})B\}^{-1}B'\Gamma(\hat{\boldsymbol{\theta}}_{\mathrm{QML}})](\hat{\boldsymbol{\theta}}_{\mathrm{QML}} - \boldsymbol{\theta}_0). \tag{8.3.25}$$

Then, it can be shown that ψ_n^* is asymptotically equivalent to φ_n^* in Theorem 8.3.12; hence, the test rejecting H whenever ψ_n^* exceeds the α-quantile $\chi^2_{l:\alpha}$ of a chi-square distribution with l degrees of freedom has asymptotic level α, and is locally asymptotically optimal.

For the non-Gaussian vector linear process (8.2.8), we established the LAN theorem (Theorem 8.3.6) and the description of the optimal estimator (Theorem 8.3.8) and optimal test (Theorem 8.3.12). Hereafter we call the integration of these results LAN-OPT for short. LAN-OPT has been developed for various stochastic processes. We summarize them as follows.

<u>LAN-OPT results</u>

(1) Time series regression model with long-memory disturbances:

$$X_t = \boldsymbol{z}_t'\boldsymbol{\beta} + u_t, \tag{8.3.26}$$

where $\boldsymbol{z}_t$'s are observable nonrandom regressors, $\{u_t\}$ an unobservable FARIMA disturbance with unknown parameter $\boldsymbol{\eta}$, and $\boldsymbol{\theta} \equiv (\boldsymbol{\beta}', \boldsymbol{\eta}')'$ (Hallin et al. (1999)).

(2) ARCH(∞)-SM model $\{X_t\}$:

$$\begin{cases} X_t - \boldsymbol{\beta}'\boldsymbol{z}_t = s_t u_t, \\ s_t^2 = b_0 + \sum_{j=1}^{\infty} b_j(X_{t-j} - \boldsymbol{\beta}'\boldsymbol{z}_{t-j})^2, \end{cases} \tag{8.3.27}$$

where $\{u_t\} \sim$ i.i.d. $(0, 1)$, and $\boldsymbol{z}_t$ is $\sigma\{u_{t-1}, u_{t-2}, \dots\}$-measurable. Here $b_j = b_j(\boldsymbol{\eta})$, $j = 0, 1, \cdots$, and $\boldsymbol{\beta}$ and $\boldsymbol{\eta}$ are unknown parameter vectors, and $\boldsymbol{\theta} = (\boldsymbol{\beta}', \boldsymbol{\eta}')'$ (Lee and Taniguchi (2005)).

(3) Vector CHARN model $\{\boldsymbol{X}_t\}$ (defined by (8.2.14)):

$$\boldsymbol{X}_t = \boldsymbol{F}_{\boldsymbol{\theta}}(\boldsymbol{X}_{t-1}, \dots, \boldsymbol{X}_{t-p}) + \boldsymbol{H}_{\boldsymbol{\theta}}(\boldsymbol{X}_{t-1}, \dots, \boldsymbol{X}_{t-p})\boldsymbol{U}_t, \tag{8.3.28}$$

(Kato et al. (2006)).

(4) Non-Gaussian locally stationary process $\{X_{t,n}\}$:

$$X_{t,n} = \int_{-\pi}^{\pi} \exp(i\lambda t) A_{t,n}^{\theta}(\lambda) d\xi(\lambda), \tag{8.3.29}$$

where $A_{t,n}^{\theta}(\lambda)$ is the response function and $\{\xi(\lambda)\}$ is an orthogonal increment process (Hirukawa and Taniguchi (2006)).

(5) Vector non-Gaussian locally stationary process $\{\boldsymbol{X}_t\}$:

$$\boldsymbol{X}_{t,n} = \boldsymbol{\mu}\left(\boldsymbol{\theta}, \frac{t}{n}\right) + \int_{-\pi}^{\pi} e^{i\lambda t} \boldsymbol{A}_{\boldsymbol{\theta},t,n}(\lambda) d\boldsymbol{\xi}(\lambda), \tag{8.3.30}$$

where $\boldsymbol{\mu}(\boldsymbol{\theta}, t/n)$ is the mean function , and $\boldsymbol{A}_{\boldsymbol{\theta},t,n}(\lambda)$ is the matrix-valued response function. Here $\{\boldsymbol{\xi}(\lambda)\}$ is a vector-valued orthogonal increment process.

Next we will discuss an optimality of estimators for semimartingales. Let $\boldsymbol{X} = \{\boldsymbol{X}_s : s \leq t\}$ be generated by the semimartingale (8.2.17) with the triple $(\alpha_{\boldsymbol{\theta}}, \beta_{\boldsymbol{\theta}}, \nu_{\boldsymbol{\theta}})$, $\boldsymbol{\theta} \in \Theta \subset \mathbb{R}^p$. The log-likelihood function based on $\boldsymbol{X}$ is given by $l_t(\boldsymbol{\theta})$ in (8.2.18). Introduce the contiguous sequence

$$\boldsymbol{\theta}_t = \boldsymbol{\theta} + \boldsymbol{\delta}_t \boldsymbol{h}, \ \boldsymbol{h} = (h_1, \dots, h_p)' \in \mathcal{H} \subset \mathbb{R}^p, \tag{8.3.31}$$

where $\boldsymbol{\delta}_t = \text{diag}(\delta_{1,t}, \dots, \delta_{p,t})$, ($p \times p$-diagonal matrix), and $\delta_{j,t}$'s tend to 0 as $t \to \infty$. Let $\Lambda_t(\boldsymbol{\theta}, \boldsymbol{\theta}_t) \equiv l_t(\boldsymbol{\theta}_t) - l_t(\boldsymbol{\theta})$. Under appropriate regularity conditions, we get the following theorem (see Prakasa Rao (1999)).

Theorem 8.3.13. *For every* $\boldsymbol{h} \in \mathbb{R}^p$,

$$\Lambda(\boldsymbol{\theta}, \boldsymbol{\theta}_t) = \boldsymbol{h}' \Delta_t(\boldsymbol{\theta}) - \frac{1}{2} \boldsymbol{h}' G_t(\boldsymbol{\theta}) \boldsymbol{h} + o_{P_{\boldsymbol{\theta}}}(1), \tag{8.3.32}$$

and

$$(\Delta_t(\boldsymbol{\theta}), G_t(\boldsymbol{\theta})) \xrightarrow{d} (G^{1/2} \boldsymbol{W}, G(\boldsymbol{\theta})) \ as \ t \to \infty, \tag{8.3.33}$$

where

$$\Delta_t(\boldsymbol{\theta}) = \boldsymbol{\delta}_t \frac{\partial}{\partial \boldsymbol{\theta}} l_t(\boldsymbol{\theta}), \quad G_t(\boldsymbol{\theta}) = \boldsymbol{\delta}_t' \left[\frac{\partial}{\partial \boldsymbol{\theta}} l_t(\boldsymbol{\theta}) \right]_t \boldsymbol{\delta}_t.$$

Here $G(\boldsymbol{\theta})$ *and* $\boldsymbol{W}$ *are independent, and* $\mathcal{L}\{\boldsymbol{W}\} = N(\boldsymbol{0}, \boldsymbol{I}_p)$.

If (8.3.32) and (8.3.33) hold, the experiment $(\Omega, \mathcal{F}, \{\mathcal{F}_t : t \geq 0\}, \{P_{\boldsymbol{\theta}} : \boldsymbol{\theta} \in \Theta\})$ is said to be *locally asymptotic mixed normal* (LAMN).

Definition 8.3.14. *An estimator* $\tilde{\boldsymbol{\theta}}_t$ *of* $\boldsymbol{\theta}$ *is said to be regular at* $\boldsymbol{\theta}$ *if*

$$\mathcal{L}(\boldsymbol{\delta}_t^{-1}(\tilde{\boldsymbol{\theta}}_t - \boldsymbol{\theta}_t), G_t(\boldsymbol{\theta}) | P_{\boldsymbol{\theta}+\boldsymbol{\delta}_t \boldsymbol{h}}) \xrightarrow{d} (\boldsymbol{S}, G), \tag{8.3.34}$$

as $t \to \infty$, *where* $G = G(\boldsymbol{\theta})$ *and* $\boldsymbol{S}$ *is a random vector.*

Let L be the class of loss functions used in Theorem 8.3.8. The following theorem is due to Basawa and Scott (1983).

Theorem 8.3.15. *Assume* $l \in L$. *Then,*

$$\liminf_{t \to \infty} E_{\boldsymbol{\theta}}[l\{\boldsymbol{\delta}_t^{-1}(\tilde{\boldsymbol{\theta}}_t - \boldsymbol{\theta})\}] \geq E[l\{G^{-1/2} \boldsymbol{W}\}], \tag{8.3.35}$$

for any regular estimator $\tilde{\boldsymbol{\theta}}_t$ *of* $\boldsymbol{\theta}$.

From Theorem 8.3.15, it is natural to introduce the asymptotic efficiency of regular estimators as follows.

Definition 8.3.16. *A regular estimator $\tilde{\theta}_t$ of θ is asymptotically efficient if*

$$\mathcal{L}\{\delta_t^{-1}(\tilde{\theta}_t - \theta)|P_\theta\} \xrightarrow{d} G^{-1/2}W \text{ as } t \to \infty. \tag{8.3.36}$$

Recalling Theorem 8.2.9, we have,

Theorem 8.3.17. *Let $\hat{\theta}_t$ be the maximum likelihood estimator defined by (8.2.21). Then $\hat{\theta}_t$ is asymptotically efficient.*

8.4 Statistical Model Selection

So far we have introduced many parametric models for stochastic processes which are specified by unknown parameter $\theta \in \Theta \subset \mathbb{R}^p$. Assuming that $p = \dim\theta$ is known, we discussed the optimal estimation of θ. However, in practice, the order p of parametric models must be inferred from the data. The best known rule for determining the true value p_0 of p is probably Akaike's information criterion (AIC) (e.g., Akaike (1974)). A lot of the alternative criteria have been proposed by Akaike (1977, 1978), Schwarz (1978) and Hannan and Quinn (1979), among others. For fitting AR(p) models, Shibata (1976) derived the asymptotic distribution of the selected order $\hat{p}$ by AIC, and showed that AIC has a tendency to overestimate p_0. There are other criteria which attempt to correct the overfitting nature of AIC. Akaike (1977, 1978) and Schwarz (1978) proposed the Bayesian information criterion (BIC), and Hannan and Quinn (1979) proposed the Hannan and Quinn (HQ) criterion. If $\hat{p}$ is the selected order by BIC or HQ, then it is shown that $\hat{p} \xrightarrow{a.s.} p_0$. We say that BIC and HQ are consistent. This property is not shared by AIC, i.e., AIC is inconsistent. However, this does not mean to detract from AIC. For autoregressive models, Shibata (1980) showed that order selection by AIC is asymptotically efficient in a certain sense, while order selection by BIC or HQ is not so.

In this section we propose a generalized AIC (GAIC) for general stochastic models, which includes the original AIC as a special case. The general stochastic models concerned are characterized by a structure g. As examples of g we can take the probability distribution function for the i.i.d. case, the trend function for the regression model, the spectral density function for the stationary process, and the dynamic system function for the nonlinear time series model. Thus, our setting is very general. We will fit a class of parametric models $\{f_\theta : \dim\theta = p\}$ to g by use of a measure of disparity $D(f_\theta, g)$. Concrete examples of $D(f_\theta, g)$ were given in Taniguchi (1991, p. 2–3). Based on $D(f_\theta, g)$ we propose a generalized AIC (GAIC(p)) which determines the order p.

In the case of the Gaussian AR(p_0) model, Shibata (1976) derived the asymptotic distribution of selected order $\hat{p}$ by the usual AIC. The results essentially rely on the structure of Gaussian AR(p_0) and the AIC. Without Gaussianity, Quinn (1988) gave formulae for the limiting probability of overestimating the order of a multivariate AR model by use of AIC.

In this subsection we derive the asymptotic distribution of selected order $\hat{p}$ by GAIC(p). The results are applicable to very general statistical models and various disparities $D(f_\theta, g)$.

We recognize that lack of information, data contamination, and other factors beyond our control make it virtually certain that the true model is not strictly correct. In view of this, we may suppose that the true model g would be incompletely specified by uncertain prior information (e.g., Bancroft (1964)) and be contiguous to a fundamental parametric model f_{θ_0} with $\dim\theta_0 = p_0$ (see Koul and Saleh (1993), Claeskens and Hjort (2003), Saleh (2006)). One plausible parametric description for g is

$$g = f_{(\theta_0, h/\sqrt{n})}, \qquad h = (h_1, \cdots, h_{K-p_0})', \tag{8.4.1}$$

where n is the sample size, and the true order is K. Then we may write $g = f_{\theta_0} + O(n^{-1/2})$. In this setting, Claeskens and Hjort (2003) proposed a focused information criterion (FIC) based on the

MSE of the deviation of a class of submodels, and discussed properties of FIC. In what follows, we will derive the asymptotic distribution of $\hat{p}$ by GAIC under (8.4.1).

We now introduce the GAIC as follows. Let $\boldsymbol{X}_n = (X_1, \cdots, X_n)'$ be a collection of m-dimensional random vectors X_t which are not necessarily i.i.d. Hence, our results can be applied to various regression models, multivariate models and time series models. Let g be a function which characterizes $\{X_t\}$. We will fit a class of parametric models $\mathcal{P} = \{f_{\boldsymbol{\theta}} : \boldsymbol{\theta} \in \Theta \subset \mathbb{R}^p\}$ to g by use of a measure of disparity $D(f_{\boldsymbol{\theta}}, g)$, which takes the minimum if and only if $f_{\boldsymbol{\theta}} = g$. Based on the observed stretch $\boldsymbol{X}_n$ we estimate $\boldsymbol{\theta}$ by the value $\hat{\boldsymbol{\theta}}_n = \hat{\boldsymbol{\theta}}_n(\boldsymbol{X}_n)$ which minimizes $D(f_{\boldsymbol{\theta}}, \hat{g}_n)$ with respect to $\boldsymbol{\theta}$, where $\hat{g}_n = \hat{g}_n(\boldsymbol{X}_n)$ is an appropriate estimator of g. Nearness between $f_{\hat{\boldsymbol{\theta}}_n}$ and g is measured by

$$E_{\hat{\boldsymbol{\theta}}_n}\{D(f_{\hat{\boldsymbol{\theta}}_n}, g)\}, \tag{8.4.2}$$

where $E_{\hat{\boldsymbol{\theta}}_n}(\cdot)$ is the expectation of $(\cdot)$ with respect to the asymptotic distribution of $\hat{\boldsymbol{\theta}}_n$. In what follows we give the derivation of GAIC. For this, all the appropriate regularity conditions for the ordinary asymptotics of $\hat{\boldsymbol{\theta}}_n$ and $\hat{g}_n$ are assumed (e.g., smoothness of the model and the central limit theorem for $\hat{\boldsymbol{\theta}}_n$). Define the pseudo true value $\boldsymbol{\theta}_0$ of $\boldsymbol{\theta}$ by the requirement

$$D(f_{\boldsymbol{\theta}_0}, g) = \min_{\boldsymbol{\theta} \in \Theta} D(f_{\boldsymbol{\theta}}, g). \tag{8.4.3}$$

Expanding $D(f_{\hat{\boldsymbol{\theta}}_n}, g)$ around $\boldsymbol{\theta}_0$ we have,

$$\begin{aligned} E_{\hat{\boldsymbol{\theta}}_n}\{D(f_{\hat{\boldsymbol{\theta}}_n}, g)\} &\approx D(f_{\boldsymbol{\theta}_0}, g) + (\hat{\boldsymbol{\theta}}_n - \boldsymbol{\theta}_0)' \frac{\partial}{\partial \boldsymbol{\theta}} D(f_{\boldsymbol{\theta}_0}, g) \\ &\quad + \frac{1}{2}(\hat{\boldsymbol{\theta}}_n - \boldsymbol{\theta}_0)' \frac{\partial^2}{\partial \boldsymbol{\theta} \partial \boldsymbol{\theta}'} D(f_{\boldsymbol{\theta}_0}, g)(\hat{\boldsymbol{\theta}}_n - \boldsymbol{\theta}_0), \end{aligned} \tag{8.4.4}$$

where $\approx$ means that the left-hand side is approximated by the asymptotic expectation of the right-hand side. From (8.4.3) it follows that

$$\frac{\partial}{\partial \boldsymbol{\theta}} D(f_{\boldsymbol{\theta}_0}, g) = \mathbf{0}, \tag{8.4.5}$$

which leads to

$$E_{\hat{\boldsymbol{\theta}}_n}\{D(f_{\hat{\boldsymbol{\theta}}_n}, g)\} \approx D(f_{\boldsymbol{\theta}_0}, g) + \frac{1}{2}(\hat{\boldsymbol{\theta}}_n - \boldsymbol{\theta}_0)' \frac{\partial^2}{\partial \boldsymbol{\theta} \partial \boldsymbol{\theta}'} D(f_{\boldsymbol{\theta}_0}, g)(\hat{\boldsymbol{\theta}}_n - \boldsymbol{\theta}_0). \tag{8.4.6}$$

On the other hand,

$$\begin{aligned} D(f_{\boldsymbol{\theta}_0}, g) &= D(f_{\hat{\boldsymbol{\theta}}_n}, \hat{g}_n) + \{D(f_{\boldsymbol{\theta}_0}, \hat{g}_n) - D(f_{\hat{\boldsymbol{\theta}}_n}, \hat{g}_n)\} + \{D(f_{\boldsymbol{\theta}_0}, g) - D(f_{\boldsymbol{\theta}_0}, \hat{g}_n)\} \\ &\approx D(f_{\hat{\boldsymbol{\theta}}_n}, \hat{g}_n) + \frac{1}{2}(\hat{\boldsymbol{\theta}}_n - \boldsymbol{\theta}_0)' \frac{\partial^2}{\partial \boldsymbol{\theta} \partial \boldsymbol{\theta}'} D(f_{\hat{\boldsymbol{\theta}}_n}, \hat{g}_n)(\hat{\boldsymbol{\theta}}_n - \boldsymbol{\theta}_0) + M, \end{aligned} \tag{8.4.7}$$

where $M = D(f_{\boldsymbol{\theta}_0}, g) - D(f_{\boldsymbol{\theta}_0}, \hat{g}_n)$. From (8.4.6) and (8.4.7) it holds that

$$E_{\hat{\boldsymbol{\theta}}_n}\{D(f_{\hat{\boldsymbol{\theta}}_n}, g)\} \approx D(f_{\hat{\boldsymbol{\theta}}_n}, \hat{g}_n) + \frac{1}{2}(\hat{\boldsymbol{\theta}}_n - \boldsymbol{\theta}_0)' J(\boldsymbol{\theta}_0)(\hat{\boldsymbol{\theta}}_n - \boldsymbol{\theta}_0) + M, \tag{8.4.8}$$

where $J(\boldsymbol{\theta}_0) = \frac{\partial^2}{\partial \boldsymbol{\theta} \partial \boldsymbol{\theta}'} D(f_{\boldsymbol{\theta}_0}, g)$. Under natural regularity conditions, we can show

$$\sqrt{n}(\hat{\boldsymbol{\theta}}_n - \boldsymbol{\theta}_0) \xrightarrow{d} N(\mathbf{0}, J(\boldsymbol{\theta}_0)^{-1} I(\boldsymbol{\theta}_0) J(\boldsymbol{\theta}_0)^{-1}), \tag{8.4.9}$$

where $I(\boldsymbol{\theta}_0) = \lim_{n \to \infty} \mathrm{Var}[\sqrt{n} \frac{\partial}{\partial \boldsymbol{\theta}} D(f_{\boldsymbol{\theta}_0}, \hat{g}_n)]$. Since the asymptotic mean of M is zero, (8.4.8) and (8.4.9) validate that

$$T_n(p) = D(f_{\hat{\boldsymbol{\theta}}_n}, \hat{g}_n) + \frac{1}{n} \mathrm{tr}\{J(\hat{\boldsymbol{\theta}}_n)^{-1} I(\hat{\boldsymbol{\theta}}_n)\} \tag{8.4.10}$$

is an "asymptotically unbiased estimator" of $E_{\hat{\theta}_n}\{D(f_{\hat{\theta}_n}, g)\}$. If $g \in \{f_\theta : \theta \in \Theta\}$, i.e., the true model belongs to the family of parametric models, it often holds that $I(\theta_0) = J(\theta_0)$. Hence $T_n'(p) = D(f_{\hat{\theta}_n}, \hat{g}_n) + n^{-1}p$ is an "asymptotically unbiased estimator" of $E_{\hat{\theta}_n}\{D(f_{\hat{\theta}_n}, g)\}$. Multiplying $T_n'(p)$ by n we call

$$\mathrm{GAIC}(p) = nD(f_{\hat{\theta}_n}, \hat{g}_n) + p \tag{8.4.11}$$

a generalized Akaike information criterion (GAIC). The essential philosophy of AIC is an "asymptotically unbiased estimator" for disparity between the true structure g and estimated model $f_{\hat{\theta}_n}$. GAIC is very general, and can be applied to a variety of statistical models.

Example 8.4.1. *[AIC in i.i.d. case] Let $X_1, \cdots, X_n$ be a sequence of i.i.d. random variables with probability density $g(\cdot)$. Assume that $g \in \mathcal{P} = \{f_\theta : \theta \in \Theta \subset \mathbb{R}^p\}$. We set $D(f_\theta, g) = -\int g(x) \log f_\theta(x)dx$ and $\int \hat{g}_n(x)dx$ = the empirical distribution function. Then $D(f_\theta, \hat{g}_n) = -n^{-1}\sum_{t=1}^n \log f_\theta(X_t)$. Let $\hat{\theta}_n$ be the maximum likelihood estimator (MLE) of θ. The GAIC is given by*

$$GAIC(p) = -\log \prod_{t=1}^{n} f_{\hat{\theta}_n}(X_t) + p, \tag{8.4.12}$$

which is equivalent to the original AIC

$$AIC(p) = -2\log \textit{(maximum likelihood)} + 2p, \tag{8.4.13}$$

(see Akaike (1974)).

Example 8.4.2. *[AIC for ARMA models] Suppose that $\{X_1, \cdots, X_n\}$ is an observed stretch from a Gaussian causal ARMA(p_1, p_2) process with spectral density $g(\lambda)$. Let*

$$\mathcal{P} = \left\{ f_\theta : f_\theta(\lambda) = \frac{\sigma^2}{2\pi} \frac{|1 + \sum_{j=1}^{p_2} a_j e^{ij\lambda}|^2}{|1 + \sum_{j=1}^{p_1} b_j e^{ij\lambda}|^2} \right\} \tag{8.4.14}$$

be a class of parametric causal ARMA (p_1, p_2) spectral models, where $\theta = (a_1, \ldots, a_{p_2}, b_1, \ldots, b_{p_1})'$. In this case we set

$$D(f_\theta, g) = \frac{1}{4\pi}\int_{-\pi}^{\pi} \left\{ \log f_\theta(\lambda) + \frac{g(\lambda)}{f_\theta(\lambda)} \right\} d\lambda, \tag{8.4.15}$$

and

$$\hat{g}_n(\lambda) = \frac{1}{2\pi n}\left| \sum_{t=1}^{n} X_t e^{it\lambda} \right|^2 \qquad \text{(periodogram).}$$

Write $f_\theta(\lambda) = \frac{\sigma^2}{2\pi} h_\theta(\lambda)$. For $h_\theta(\lambda)$, it is shown that

$$\int_{-\pi}^{\pi} \log h_\theta(\lambda) d\lambda = 0, \tag{8.4.16}$$

(e.g., Brockwell and Davis (2006, p. 191)). Then

$$\begin{aligned} \hat{\theta}_n &= \arg\min_{\theta \in \Theta} D(f_\theta, \hat{g}_n) \\ &= \arg\min_{\theta \in \Theta} \int_{-\pi}^{\pi} \frac{\hat{g}_n(\lambda)}{h_\theta(\lambda)} d\lambda, \end{aligned} \tag{8.4.17}$$

which is called a quasi-Gaussian MLE of $\boldsymbol{\theta}$. For σ^2, its quasi-Gaussian MLE $\hat{\sigma}_n{}^2(p)$, $(p = p_1 + p_2)$ is given by

$$\frac{\partial}{\partial\sigma^2}\int_{-\pi}^{\pi}\left\{\log\sigma^2 + \frac{\hat{g}_n(\lambda)}{\frac{\sigma^2}{2\pi}h_{\hat{\boldsymbol{\theta}}_n}(\lambda)}\right\}d\lambda\Bigg|_{\sigma^2=\hat{\sigma}_n^2(p)} = 0, \tag{8.4.18}$$

which leads to

$$\hat{\sigma}_n^2(p) = \int_{-\pi}^{\pi}\frac{\hat{g}_n(\lambda)}{h_{\hat{\boldsymbol{\theta}}_n}(\lambda)}d\lambda. \tag{8.4.19}$$

From (8.4.15) and (8.4.19) it is seen that the GAIC is equivalent to

$$GAIC(p) = \frac{n}{2}\log\hat{\sigma}_n^2(p) + p. \tag{8.4.20}$$

For simplicity, we assumed Gaussianity of the process; hence, it is possible to use (8.4.20) for non-Gaussian ARMA processes.

Example 8.4.3. *[AIC for Regression Model] Let $X_1, \cdots, X_n$ be a sequence of random variables generated by*

$$X_t = \beta_1 + \beta_2 z_{2t} + \cdots + \beta_p z_{pt} + u_t, \tag{8.4.21}$$

where u_t's are i.i.d. with probability density $g(\cdot)$, and $\{z_{jt}\}$, $j = 2, \cdots, p$, are nonrandom variables satisfying $\sum_{t=1}^{n} z_{jt}^2 = O(n)$. As in Example 8.4.2, we set

$$D(f_{\boldsymbol{\theta}}, g) = -\int g(x)\log f_{\boldsymbol{\theta}}(x)dx,$$

$$\int \hat{g}_n(x)dx = \text{the empirical distribution function},$$

which lead to

$$D(f_{\boldsymbol{\theta}}, \hat{g}_n) = -\frac{1}{n}\sum_{t=1}^{n}\log f_{\boldsymbol{\theta}}(X_t - \beta_1 - \beta_2 z_{2t} - \cdots - \beta_p z_{pt}). \tag{8.4.22}$$

If we take $f_{\boldsymbol{\theta}}(u_t) = (2\pi\sigma^2)^{-1/2}exp(-u_t^2/2\sigma^2)$ and $\boldsymbol{\theta} = (\beta_1, \cdots, \beta_p, \sigma^2)'$, then

$$GAIC(p) = \frac{n}{2}\log\hat{\sigma}^2(p) + p, \tag{8.4.23}$$

where

$$\hat{\sigma}^2(p) = \frac{1}{n}\sum_{t=1}^{n}(X_t - \hat{\beta}_1 - \hat{\beta}_2 z_{2t} - \cdots - \hat{\beta}_p z_{pt})^2,$$

and $(\hat{\beta}_1, \cdots, \hat{\beta}_p)$ is the least squares estimator of $(\beta_1, \cdots, \beta_p)$.

Example 8.4.4. *[GAIC for CHARN Model] Let $\{X_1, \cdots, X_n\}$ be generated by the following CHARN model:*

$$X_t = \psi_{\boldsymbol{\theta}}(X_{t-1}, \cdots, X_{t-q}) + h_{\boldsymbol{\theta}}(X_{t-1}, \ldots, X_{t-r})u_t, \tag{8.4.24}$$

where $\psi_{\boldsymbol{\theta}} : \mathbb{R}^q \to \mathbb{R}$ is a measurable function, $h_{\boldsymbol{\theta}} : \mathbb{R}^r \to \mathbb{R}$ is a positive measurable function, and $\{u_t\}$ is a sequence of i.i.d. random variables with known probability density $p(\cdot)$, and u_t is *independent of* $\{X_s : s < t\}$, *and* $\boldsymbol{\theta} = (\theta_1, \cdots, \theta_p)'$. *For the asymptotic theory,* $\psi_{\boldsymbol{\theta}}(\cdot)$ *and* $h_{\boldsymbol{\theta}}(\cdot)$ *are assumed to satisfy appropriate regularity conditions (see Kato et al. (2006)). Then we set*

$$f_{\boldsymbol{\theta}}(x_1, \cdots, x_n) = \text{the joint pdf model for } (X_1, \cdots, X_n) \text{ under (8.4.24)},$$

and

$$g(x_1, \cdots, x_n) = \text{the true joint pdf of } (X_1, \cdots, X_n).$$

If we take

$$D(f_{\boldsymbol{\theta}}, g) = -\int \cdots \int g(x_1, \cdots, x_n) \log f_{\boldsymbol{\theta}}(x_1, \cdots, x_n) dx_1 \cdots x_n,$$

$$\int \cdots \int \hat{g}_n(x_1, \cdots, x_n) dx_1 \cdots dx_n = \text{the empirical distribution function.}$$

Then,

$$D(f_{\boldsymbol{\theta}}, \hat{g}_n) = -\frac{1}{n} \sum_{t=\max(q,r)}^{n} \log p\{h_{\boldsymbol{\theta}}^{-1}(X_t - \psi_{\boldsymbol{\theta}})\} h_{\boldsymbol{\theta}}^{-1},$$

leading to

$$GAIC(p) = -2 \sum_{t=max(q,r)}^{n} \log p\{h_{\hat{\boldsymbol{\theta}}_{ML}}^{-1}(X_t - \psi_{\hat{\boldsymbol{\theta}}_{ML}})\} h_{\hat{\boldsymbol{\theta}}_{ML}}^{-1} + 2p, \tag{8.4.25}$$

where $\hat{\boldsymbol{\theta}}_{ML}$ *is the MLE of* $\boldsymbol{\theta}$.

As we saw above, GAIC is a very general information criterion which determines the order of model. By scanning p successively from zero to some upper limit L, the order of model is chosen by the p that gives the minimum of GAIC(p), $p = 0, 1, \cdots, L$. We denote the selected order by $\hat{p}$.

To describe the asymptotics of GAIC, we introduce natural regularity conditions for $f_{\boldsymbol{\theta}}$ and $\hat{\boldsymbol{\theta}}_n$, which a wide class of statistical models satisfy. In what follows we write $\boldsymbol{\theta}$ as $\boldsymbol{\theta}(p)$ if we need to emphasize the dimension.

Assumption 8.4.5. **(i)** $f_{\boldsymbol{\theta}}(\cdot)$ *is continuously twice differentiable with respect to* $\boldsymbol{\theta}$.

(ii) *The fitted model is nested, i.e.,* $\boldsymbol{\theta}(p+1) = (\boldsymbol{\theta}(p)', \theta_{p+1})'$.

(iii) *As* $n \to \infty$, *the estimator* $\hat{\boldsymbol{\theta}}_n$ *satisfies (8.4.9), and* $I(\boldsymbol{\theta}) = J(\boldsymbol{\theta})$ *if* $g = f_{\boldsymbol{\theta}}$.

(iv) *If* $g = f_{\boldsymbol{\theta}(p_0)}$, *then* $D(f_{\boldsymbol{\theta}(p)}, g)$ *is uniquely minimized at* $p = p_0$.

The assumption (iii) is not restrictive. For Examples 8.4.1 and 8.4.3, the estimators of ML type are known to satisfy it. For Example 8.4.2, the q-MLE satisfies it (see Taniguchi and Kakizawa (2000, p. 58)). For Example 8.4.4, by use of the asymptotic theory for CHARN models (see Kato et al. (2006)), we can see that (iii) is satisfied.

In the case of the Gaussian AR(p_0) model, Shibata (1976) derived the asymptotic distribution of selected order $\hat{p}$ by the usual AIC. The results essentially rely on the structure of Gaussian AR(p) and AIC. In what follows we derive the asymptotic distribution of selected order $\hat{p}$ by GAIC(p) for very general statistical models including Examples 8.4.1–8.4.4. Let $\alpha_i = P(\chi_i^2 > 2i)$, and let

$$\rho(l, \{\alpha_i\}) = \sum\nolimits_l^* \left\{ \prod_{i=1}^{l} \frac{1}{r_i!} \left(\frac{\alpha_i}{i}\right)^{r_i} \right\}, \tag{8.4.26}$$

where $\sum_l^*$ extends over all l-tuples $(r_1, \cdots, r_l)$ of nonnegative integers satisfying $r_1 + 2r_2 + \cdots + lr_l = l$.

Theorem 8.4.6. *(Taniguchi and Hirukawa (2012)) Suppose* $g = f_{\theta(p_0)}$. *Then, under Assumption 8.4.5, the asymptotic distribution of* $\hat{p}$ *selected by GAIC(p) is given by*

$$\lim_{n\to\infty} P(\hat{p} = p) = \begin{cases} 0, & 0 \le p < p_0, \\ \rho(p - p_0, \{\alpha_i\})\rho(L - p, \{1 - \alpha_i\}), & p_0 \le p \le L, \end{cases} \tag{8.4.27}$$

where $\rho(0, \{\alpha_i\}) = \rho(0, \{1 - \alpha_i\}) = 1$.

Proof. We write $\hat{\theta}_n$ and θ_0 as $\hat{\theta}_n(p)$ and $\theta_0(p)$ to emphasize their dimension. For $0 \le p < p_0$,

$$\begin{aligned} P\{\hat{p} = p\} &\le P\{\text{GAIC}(p) < \text{GAIC}(p_0)\} \\ &= P\{nD(f_{\hat{\theta}_n(p)}, \hat{g}_n) - nD(f_{\hat{\theta}_n(p_0)}, \hat{g}_n) < p_0 - p\}. \end{aligned} \tag{8.4.28}$$

As $n \to \infty$, $n\{D(f_{\hat{\theta}_n(p)}, \hat{g}_n) - D(f_{\hat{\theta}_n(p_0)}, \hat{g}_n)\}$ tends to $n\{D(f_{\theta_0(p)}, g) - D(f_{\theta_0(p_0)}, g)\}$ in probability. From (iv) of Assumption 8.4.5, it follows that $n\{D(f_{\theta_0(p)}, g) - D(f_{\theta_0(p_0)}, g)\}$ diverges to infinity, which implies that the right-hand side of (8.4.28) converges to zero. Hence,

$$\lim_{n\to\infty} P\{\hat{p} = p\} = 0, \quad \text{for } 0 \le p < p_0. \tag{8.4.29}$$

Next, we examine the case of $p_0 \le p \le L$. First,

$$\begin{aligned} \text{GAIC}(p) &= nD(f_{\hat{\theta}_n(p)}, \hat{g}_n) + p \\ &= n\Big[D(f_{\theta_0(p)}, \hat{g}_n) - (\theta_0(p) - \hat{\theta}_n(p))' \frac{\partial}{\partial \theta} D(f_{\hat{\theta}_n(p)}, \hat{g}_n) \\ &\quad - \frac{1}{2}(\theta_0(p) - \hat{\theta}_n(p))' \frac{\partial^2}{\partial\theta\partial\theta'} D(f_{\hat{\theta}_n(p)}, \hat{g}_n)(\theta_0(p) - \hat{\theta}_n(p))\Big] + p \\ &\quad + \text{lower order terms} \\ &= nD(f_{\theta_0(p)}, \hat{g}_n) \\ &\quad - \frac{1}{2}\sqrt{n}(\hat{\theta}_n(p) - \theta_0(p))' \frac{\partial^2}{\partial\theta\partial\theta'} D(f_{\hat{\theta}_n(p)}, g)\sqrt{n}(\hat{\theta}_n(p) - \theta_0(p)) + p \\ &\quad + \text{lower order terms.} \end{aligned} \tag{8.4.30}$$

Since $D(f_{\theta_0(m)}, \hat{g}_n) = D(f_{\theta_0(l)}, \hat{g}_n)$ for $p_0 \le m < l \le L$, we investigate the asymptotics of

$$Q_n(p) = \frac{1}{2}\sqrt{n}(\hat{\theta}_n(p) - \theta_0(p))' \frac{\partial^2}{\partial\theta\partial\theta'} D(f_{\theta_0(p)}, g)\sqrt{n}(\hat{\theta}_n(p) - \theta_0(p)), \quad (p_0 \le p \le L). \tag{8.4.31}$$

For $p_0 \le m < l \le L$, write the $l \times l$ Fisher information matrix $\mathcal{I}\{\theta_0(l)\}$, and divide it as a 2×2 block matrix:

$$\mathcal{I}\{\theta_0(l)\} = \begin{pmatrix} \overbrace{F_{11}}^{m} & \overbrace{F_{12}}^{l-m} \\ F_{21} & F_{22} \end{pmatrix} \begin{matrix} \}m \\ \}l-m \end{matrix} \tag{8.4.32}$$

where $F_{11} = \mathcal{I}\{\theta_0(m)\}$ is the $m \times m$ Fisher information matrix. Also we write

$$Z_1 = \begin{pmatrix} Z_{11} \\ Z_{21} \end{pmatrix} = \sqrt{n} \begin{pmatrix} \frac{\partial}{\partial\theta(m)} D(f_{\theta_0(m)}, \hat{g}_n) \\ \frac{\partial}{\partial\theta_{l-m}(l)} D(f_{\theta_0(m)}, \hat{g}_n) \end{pmatrix} \begin{matrix} \}m \\ \}l-m \end{matrix} \tag{8.4.33}$$

where $\boldsymbol{\theta}_{l-m}(l)$ is defined by $\boldsymbol{\theta}(l) = (\boldsymbol{\theta}(m)', \boldsymbol{\theta}_{l-m}(l)')'$. The matrix formula leads to

$$\mathcal{I}\{\boldsymbol{\theta}_0(l)\}^{-1} = \begin{pmatrix} \boldsymbol{F}_{11}^{-1} + \boldsymbol{F}_{11}^{-1}\boldsymbol{F}_{12}\boldsymbol{F}_{22\cdot 1}^{-1}\boldsymbol{F}_{21}\boldsymbol{F}_{11}^{-1} & -\boldsymbol{F}_{11}^{-1}\boldsymbol{F}_{12}\boldsymbol{F}_{22\cdot 1}^{-1} \\ -\boldsymbol{F}_{22\cdot 1}^{-1}\boldsymbol{F}_{21}\boldsymbol{F}_{11}^{-1} & \boldsymbol{F}_{22\cdot 1}^{-1} \end{pmatrix} \tag{8.4.34}$$

where $\boldsymbol{F}_{22\cdot 1} = \boldsymbol{F}_{22} - \boldsymbol{F}_{21}\boldsymbol{F}_{11}^{-1}\boldsymbol{F}_{12}$. From the general asymptotic theory,

$$\sqrt{n}(\hat{\boldsymbol{\theta}}_n(l) - \boldsymbol{\theta}_0(l)) = \mathcal{I}\{\boldsymbol{\theta}_0(l)\}^{-1}\boldsymbol{Z}_1 + o_p(1), \tag{8.4.35}$$

$$\sqrt{n}(\hat{\boldsymbol{\theta}}_n(m) - \boldsymbol{\theta}_0(m)) = \boldsymbol{F}_{11}^{-1}\boldsymbol{Z}_{11} + o_p(1). \tag{8.4.36}$$

From (8.4.34)–(8.4.36) it follows that

$$Q_n(l) - Q_n(m) = \frac{1}{2}(\boldsymbol{Z}_{21} - \boldsymbol{F}_{21}\boldsymbol{F}_{11}^{-1}\boldsymbol{Z}_{11})'\boldsymbol{F}_{22\cdot 1}^{-1}(\boldsymbol{Z}_{21} - \boldsymbol{F}_{21}\boldsymbol{F}_{11}^{-1}\boldsymbol{Z}_{11}) \tag{8.4.37}$$

for $p_0 \le m < l \le L$. Here it is seen that

$$\boldsymbol{Z}_{21} - \boldsymbol{F}_{21}\boldsymbol{F}_{11}^{-1}\boldsymbol{Z}_{11} \xrightarrow{d} N(\boldsymbol{0}, \boldsymbol{F}_{22\cdot 1}) \tag{8.4.38}$$

and $\boldsymbol{Z}_{21} - \boldsymbol{F}_{21}\boldsymbol{F}_{11}^{-1}\boldsymbol{Z}_{11}$ is asymptotically independent of $\boldsymbol{Z}_{11}$, i.e., it is asymptotically independent of $\sqrt{n}(\hat{\boldsymbol{\theta}}_n(m) - \boldsymbol{\theta}_0(m))$ (recall (8.4.36)). Therefore, for any m and l satisfying $p_0 \le m < l \le L$, it holds that, as $n \to \infty$,

$$Q_n(l) - Q_n(m) \xrightarrow{d} \frac{1}{2}\chi^2_{l-m}, \tag{8.4.39}$$

$$Q_n(l) - Q_n(m) \text{ is asymptotically independent of } \sqrt{n}(\hat{\boldsymbol{\theta}}_n(m) - \boldsymbol{\theta}_0(m)). \tag{8.4.40}$$

Now we turn to evaluate $P(\hat{p} = p)$ in the case when $p_0 \le p \le L$. For $p_0 \le p \le L$, from (8.4.30) and (8.4.33) it follows that

$$\begin{aligned} P(\hat{p} = p) &= P\{\text{GAIC}(p) \le \text{GAIC}(m),\ p_0 \le m \le L\} + o(1) \\ &= P\{-Q_n(p) + p \le -Q_n(m) + m,\ p_0 \le m \le L\} + o(1) \\ &= P\{-Q_n(p) + p \le -Q_n(m) + m,\ p_0 \le m < p, \text{ and} \\ &\quad - Q_n(p) + p \le -Q_n(m) + m,\ p < m \le L\} + o(1) \\ &= P\{p - m \le Q_n(p) - Q_n(m),\ p_0 \le m < p, \text{ and} \\ &\quad Q_n(m) - Q_n(p) \le m - p,\ p < m \le L\} + o(1) \\ &= (A) + o(1), \quad \text{(say)}. \end{aligned} \tag{8.4.41}$$

From (8.4.39) and (8.4.40) we can express (8.4.41) as

$$(A) = P[S_1 > 0, \cdots, S_{p-p_0} > 0] \cdot P[S_1 < 0, \cdots, S_{L-p} < 0] + o(1), \tag{8.4.42}$$

where $S_i = (Y_1 - 2) + \cdots + (Y_i - 2)$ with $Y_i \sim$ i.i.d. χ_1^2. Applying Shibata (1976) and Spitzer (1956) to (8.4.42) we can see that, for $p_0 \le p \le L$,

$$\lim_{n\to\infty} P(\hat{p} = p) = \rho(p - p_0, \{\alpha_i\})\rho(L - p, \{1 - \alpha_i\}).$$

□

Next we derive the asymptotic distribution of selected order under contiguous models. We recognize that lack of information, data contamination, and other factors beyond our control make it virtually certain that the true model g is not strictly correct and incompletely specified by certain prior information (e.g., Bancroft (1964)), and is contiguous to a fundamental parametric model $f_{\boldsymbol{\theta}_0}$. We may suppose that the true model is of the form

$$g(x) = f_{\boldsymbol{\theta}_0}(x) + \frac{1}{\sqrt{n}}h(x), \tag{8.4.43}$$

where $\dim \boldsymbol{\theta}_0 = p_0$ and n is the sample size, and $h(x)$ is a perturbation function. To be more specific, if we make a parametric version of (8.4.43), it is

$$g(x) = f_{(\boldsymbol{\theta}_0, \boldsymbol{h}/\sqrt{n})}(x), \tag{8.4.44}$$

where $\boldsymbol{h} = (h_1, \cdots, h_{K-p_0})'$. In practice, it is natural that the true model is (8.4.44), and that the true order of (8.4.44) is K. In the analysis of preliminary test estimation, models of the form (8.4.44) are often used (e.g., Koul and Saleh (1993) and Saleh (2006)).

In what follows, we derive the asymptotic distribution of the selected order $\hat{p}$ by GAIC under the models (8.4.44). Suppose that the observed stretch $\boldsymbol{X}_n = (X_1, \cdots, X_n)'$ specified by the true structure $f_{(\boldsymbol{\theta}_0, \boldsymbol{h}/\sqrt{n})}$ has the probability distribution $\boldsymbol{P}^{(n)}_{(\boldsymbol{\theta}_0, \boldsymbol{\eta})}$ with $\boldsymbol{\eta} = \boldsymbol{h}/\sqrt{n}$.

Assumption 8.4.7. **(i)** *If* $\boldsymbol{h} = \boldsymbol{0}$, *the model* $f_{\boldsymbol{\theta}} = f_{(\boldsymbol{\theta}, \boldsymbol{0})}$ *satisfies Assumption 8.4.5.*

(ii) $\{\boldsymbol{P}^{(n)}_{(\boldsymbol{\theta}_0, \boldsymbol{\eta})}\}$ *has the LAN property, i.e., it satisfies*

$$\log \frac{d\boldsymbol{P}^{(n)}_{(\boldsymbol{\theta}_0, \boldsymbol{h}/\sqrt{n})}}{d\boldsymbol{P}^{(n)}_{(\boldsymbol{\theta}_0, \boldsymbol{0})}} = \boldsymbol{h}'\Delta_n - \frac{1}{2}\boldsymbol{h}'\Gamma(\boldsymbol{\theta}_0)\boldsymbol{h} + o_p(1), \tag{8.4.45}$$

$$\Delta_n \xrightarrow{d} N(\boldsymbol{0}, \Gamma(\boldsymbol{\theta}_0)), \qquad \text{under } \boldsymbol{P}^{(n)}_{(\boldsymbol{\theta}_0, \boldsymbol{0})},$$

where $\Gamma(\boldsymbol{\theta}_0)$ *is the Fisher information matrix.*

We assume $p_0 < K \leq L$ and fit the model $f_{\boldsymbol{\theta}(p)}$, $0 \leq p \leq L$ to the data from (8.4.44). We write $(\boldsymbol{\theta}_0, \boldsymbol{0})$ as $\boldsymbol{\theta}_0(p)$, $p \geq p_0$, if the dimension of the fitted model is p. When $p_0 < l \leq L$, let

$$\boldsymbol{Z}_1^l = \begin{pmatrix} \boldsymbol{Z}_{11}^{(l-1)} \\ Z_{21}^{(l)} \end{pmatrix} = \sqrt{n}\begin{pmatrix} \frac{\partial}{\partial \boldsymbol{\theta}(l-1)} D(f_{\boldsymbol{\theta}_0(l-1)}, \hat{g}_n) \\ \frac{\partial}{\partial \theta_l} D(f_{\boldsymbol{\theta}_0(l)}, \hat{g}_n) \end{pmatrix} \begin{matrix} \}l-1 \\ \}1. \end{matrix} \tag{8.4.46}$$

Assumption 8.4.8. *Under* $\boldsymbol{P}^{(n)}_{(\boldsymbol{\theta}_0, \boldsymbol{0})}$, *for* $p_0 < l \leq L$, *the random vectors* $(\boldsymbol{Z}_1^{(l)'}, \Delta_n')'$ *are asymptotically normal with*

$$\lim_{n\to\infty} \text{Cov}(\boldsymbol{Z}_1^{(l)}, \Delta_n) = \{\sigma_{j,k}(l); j = 1, \cdots, l, k = 1, \cdots, K - p_0\}, (say)$$
$$(l \times (K - p_0)\text{-}matrix), \tag{8.4.47}$$

and

$$\lim_{n\to\infty} \text{Var}(\boldsymbol{Z}_1^{(l)}) = \begin{pmatrix} \overbrace{\boldsymbol{F}_{11}^{(l-1)}}^{l-1} & \overbrace{\boldsymbol{F}_{12}^{(l-1)}}^{1} \\ \boldsymbol{F}_{21}^{(l-1)} & F_{22}^{(l)} \end{pmatrix} \begin{matrix} \}l-1 \\ \}1. \end{matrix} \tag{8.4.48}$$

Write

$$\Sigma^{(l-1)} = \begin{pmatrix} \sigma_{1,1}(l) & \cdots & \sigma_{1,K-p_0}(l) \\ \vdots & & \vdots \\ \sigma_{l-1,1}(l) & \cdots & \sigma_{l-1,K-p_0}(l) \end{pmatrix},$$

$$\sigma^{(l)} = (\sigma_{l,1}(l), \cdots, \sigma_{l,K-p_0}(l))',$$
$$F_{22\cdot 1}^{(l)} = F_{22}^{(l)} - \boldsymbol{F}_{21}^{(l-1)} \boldsymbol{F}_{11}^{(l-1)^{-1}} \boldsymbol{F}_{12}^{(l-1)}, \quad l = p_0 + 1, \cdots, L$$

and

$$\mu_l = (\sigma^{(l)'} \boldsymbol{h} - \boldsymbol{F}_{21}^{(l-1)} \boldsymbol{F}_{11}^{(l-1)^{-1}} \Sigma^{(l-1)} \boldsymbol{h}) \sqrt{F_{22\cdot 1}^{(l)}}.$$

Letting $W_1, \cdots, W_{L-p_0} \sim$ i.i.d. $N(0, 1)$, define

$$S_{j,m} = (W_m - \mu_{m+p_0})^2 + \cdots + (W_{m-j+1} - \mu_{m+p_0-j+1})^2 - 2j,$$
$$(1 \leq m \leq L - p_0, j = 1, \cdots, m),$$
$$T_{j,k} = (W_{k+1} - \mu_{k+p_0+1})^2 + \cdots + (W_{k+j} - \mu_{k+p_0+j})^2 - 2j,$$
$$(0 \leq k \leq L - p_0 - 1, j = 1, \cdots, L - p_0 - k).$$

If $0 \leq p < p_0$, we can prove

$$\lim_{n\to\infty} \boldsymbol{P}_{(\boldsymbol{\theta}_0, \boldsymbol{h}/\sqrt{n})}^{(n)} \{\hat{p} = p\} = 0, \tag{8.4.49}$$

as in the proof of Theorem 8.4.6 . Next, we discuss the case of $p_0 \leq p \leq L$. Under $P = \boldsymbol{P}_{(\boldsymbol{\theta}_0, \boldsymbol{0})}^{(n)}$, the probability $P(\hat{p} = p)$ is evaluated in (8.4.41). Recall (8.4.37) and (8.4.38). From Assumptions 8.4.7 and 8.4.8, applying Le Cam's third lemma (Lemma 8.3.10) to (8.4.38), it is seen that

$$Z_{21}^{(l)} - \boldsymbol{F}_{21}^{(l-1)} \boldsymbol{F}_{11}^{(l-1)^{-1}} \boldsymbol{Z}_{11}^{(l-1)} \xrightarrow{d} N(\sigma^{(l)'} \boldsymbol{h} - \boldsymbol{F}_{21}^{(l-1)} \boldsymbol{F}_{11}^{(l-1)^{-1}} \Sigma^{(l-1)} \boldsymbol{h}, F_{22\cdot 1}) \tag{8.4.50}$$

under $\boldsymbol{P}_{(\boldsymbol{\theta}_0, \boldsymbol{h}/\sqrt{n})}^{(n)}$. Then, returning to (8.4.41), we obtain the following theorem.

Theorem 8.4.9. *(Taniguchi and Hirukawa (2012)) Assume Assumptions 8.4.7 and 8.4.8. Let $\hat{p}$ be a selected order by GAIC(p). Then, under $\boldsymbol{P}_{(\boldsymbol{\theta}_0, \boldsymbol{h}/\sqrt{n})}^{(n)}$, the asymptotic distribution of $\hat{p}$ is given by*

$$\lim_{n\to\infty} \boldsymbol{P}_{(\boldsymbol{\theta}_0, \boldsymbol{h}/\sqrt{n})}^{(n)} \{\hat{p} = p\} = \begin{cases} 0, & 0 \leq p < p_0 \\ \beta(p - p_0)\gamma(p - p_0), & p_0 \leq p \leq L, \end{cases} \tag{8.4.51}$$

where

$$\beta(m) = P\left\{\bigcap_{j=1}^{m} (S_{j,m} > 0)\right\}, \ \gamma(k) = P\left\{\bigcap_{j=1}^{L-p_0-k} (T_{j,k} \leq 0)\right\},$$
$$\beta(0) = \gamma(L) = 1.$$

Let $\{X_t : t = 1, \cdots, n\}$ be a stationary process with spectral density $f_{\boldsymbol{\theta}}(\lambda) = f_{(\boldsymbol{\theta}_0, \boldsymbol{h}/\sqrt{n})}(\lambda)$. Here $\dim \boldsymbol{\theta}_0 = p_0$ and $\dim \boldsymbol{h} = K - p_0$. We use the criterion (8.4.20) based on (8.4.15). Assume Assumptions 8.4.7 and 8.4.8. Then it can be shown that

$$\sigma_{j,k}(l) = \frac{1}{4\pi} \int_{-\pi}^{\pi} \frac{\partial}{\partial\theta_j} \log f_{(\boldsymbol{\theta}_0, \boldsymbol{0}_{l-p_0})}(\lambda) \frac{\partial}{\partial\theta_{p_0+k}} \log f_{(\boldsymbol{\theta}_0, \boldsymbol{0}_{K-p_0})}(\lambda) d\lambda, \tag{8.4.52}$$

$l = p_0 + 1, \cdots, L, \ \ j = 1, \cdots, l, \ \ k = 1, \cdots, K - p_0$, and $\boldsymbol{0}_r = (\overbrace{0, \cdots, 0}^{r})'$. Write the (j, k)th element of the right-hand side of (8.4.48) as $F_{jk}(l)$. Then,

$$F_{j,k}(l) = \frac{1}{4\pi} \int_{-\pi}^{\pi} \frac{\partial}{\partial\theta_j} \log f_{(\boldsymbol{\theta}_0, \boldsymbol{0}_{l-p_0})}(\lambda) \frac{\partial}{\partial\theta_k} \log f_{(\boldsymbol{\theta}_0, \boldsymbol{0}_{l-p_0})}(\lambda) d\lambda, \tag{8.4.53}$$

where $l = p_0 + 1, \cdots, L,\ \ j, k = 1, \cdots, l$. In principle we can express the key quantities μ_l in terms of spectral density $f_{\boldsymbol{\theta}}$. However, even for simple ARMA models, it is very difficult to express μ_l by $\boldsymbol{\theta}$ explicitly. To avoid this problem, we consider the following exponential spectral density model:

$$\begin{aligned} f_{\boldsymbol{\theta}(L)}(\lambda) &= \sigma^2 \exp\left[\sum_{j=1}^{L} \theta_j \cos(j\lambda)\right] \\ &= \sigma^2 \exp\left[\sum_{j=1}^{p_0} \theta_j \cos(j\lambda) + \frac{1}{\sqrt{n}} \sum_{j=1}^{L-p_0} h_j \cos\{(p_0 + j)\lambda\}\right]. \end{aligned} \tag{8.4.54}$$

The form of the true model is

$$\begin{aligned} f_{\boldsymbol{\theta}}(\lambda) &= f_{\boldsymbol{\theta}_0, \boldsymbol{h}/\sqrt{n}}(\lambda) \\ &\equiv \sigma^2 \exp\left[\sum_{j=1}^{p_0} \theta_{0,j} \cos(j\lambda) + \frac{1}{\sqrt{n}} \sum_{j=1}^{K-p_0} h_j \cos\{(p_0 + j)\lambda\}\right], \end{aligned} \tag{8.4.55}$$

where $\boldsymbol{\theta}_0 = (\theta_{0,1}, \cdots, \theta_{0,p_0})'$ and $\boldsymbol{h} = (h_1, \cdots, h_{K-p_0})'$. Then it is seen that

$$\mu_l = \begin{cases} \frac{1}{8} h_{l-p_0}, & l = p_0 + 1, \cdots, K, \\ 0, & l = K + 1, \cdots, L. \end{cases} \tag{8.4.56}$$

From Theorem 8.4.9 and (8.4.56) we can see that

$$\lim_{n\to\infty} \boldsymbol{P}^{(n)}_{(\boldsymbol{\theta}_0, \boldsymbol{h}/\sqrt{n})}\{\hat{p} = p\} \text{ tends to 0 if } ||\boldsymbol{h}|| \text{ becomes large for } 0 \le p < K, \tag{8.4.57}$$

$$\lim_{n\to\infty} \boldsymbol{P}^{(n)}_{(\boldsymbol{\theta}_0, \boldsymbol{h}/\sqrt{n})}[\hat{p} \in \{K, K+1, \cdots, L\}] \text{ tends to 1 if } ||\boldsymbol{h}|| \text{ becomes large.} \tag{8.4.58}$$

Although, if $\boldsymbol{h} = \mathbf{0}$, we saw that GAIC is not a consistent order selection criterion in Theorem 8.4.6, the results (8.4.57) and (8.4.58) imply

(i) GAIC is admissible in the sense of (8.4.57) and (8.4.58).

Many other criteria have been proposed. We mention two typical ones. Akaike (1977, 1978) and Schwarz (1978) proposed

$$\text{BIC} = -2\log(\text{maximum likelihood}) + (\log n)(\text{number of parameters}). \tag{8.4.59}$$

In the setting of Example 8.4.2, Hannan and Quinn (1979) suggested the criterion

$$HQ(p) = \frac{n}{2} \log \hat{\sigma}_n^2(p) + cp \cdot \log\log n, \quad c > 1. \tag{8.4.60}$$

When $\boldsymbol{h} = \mathbf{0}$, it is known that the criteria BIC and HQ are consistent for "the true order p_0." Here we summarize the results following the proofs of Theorems 8.4.6 and 8.4.9.

(ii) BIC still retains consistency to "pretended order p_0 (not K)," and

(iii) HQ still retains consistency to "pretended order p_0 (not K)."

The statements (i)–(iii) characterize the aims of GAIC(AIC), BIC and HQ. That is, GAIC aims to select contiguous models (8.4.55) with the orders $K, K+1, \cdots, L$ including the true one, if our model is "incompletely specified." On the other hand, BIC and HQ aim to select the true model $f_{\boldsymbol{\theta}_0}$ if our model is "completely specified."

Now, we evaluate the asymptotic distribution of the selected order in (8.4.27) and (8.4.51). The asymptotic probabilities $\lim_{n\to\infty} P(\hat{p} = p)$ under the null hypothesis $\boldsymbol{\theta} = \boldsymbol{\theta}_0$ for $L = 10$ are listed in Table 8.1. In this table the elements for $p = p_0$ correspond to correctly estimated cases while those for $p > p_0$ correspond to overestimated cases. It is seen that the asymptotic probabilities of overestimations are not equal to zero, hence GAIC is not a consistent order selection criterion in contrast with BIC and HQ, which are consistent order selection criteria. However, they tend to zero as $p - p_0$ becomes large. Moreover, the asymptotic probabilities of correct selections, $p = p_0$, $1 \leq p_0 \leq L$ for $L = 5, 8, 10$ are plotted in Figure 8.1. We see that the correct selection probability becomes larger as L becomes smaller for each p_0. However, in the first place, the setting of $L < p_0$ causes the meaningless situation where we always underestimate the order.

Table 8.1: The asymptotic distribution of selected order under the null hypothesis, $\boldsymbol{\theta} = \boldsymbol{\theta}_0$

$p \backslash p_0$	1	2	3	4	5	6	7	8	9	10
1	0.719	0	0	0	0	0	0	0	0	0
2	0.113	0.721	0	0	0	0	0	0	0	0
3	0.058	0.114	0.724	0	0	0	0	0	0	0
4	0.035	0.058	0.115	0.729	0	0	0	0	0	0
5	0.023	0.036	0.059	0.116	0.735	0	0	0	0	0
6	0.016	0.024	0.036	0.060	0.117	0.745	0	0	0	0
7	0.012	0.017	0.024	0.037	0.061	0.120	0.760	0	0	0
8	0.009	0.012	0.017	0.025	0.038	0.063	0.124	0.787	0	0
9	0.007	0.010	0.013	0.019	0.027	0.041	0.067	0.133	0.843	0
10	0.006	0.009	0.011	0.016	0.022	0.032	0.048	0.080	0.157	1

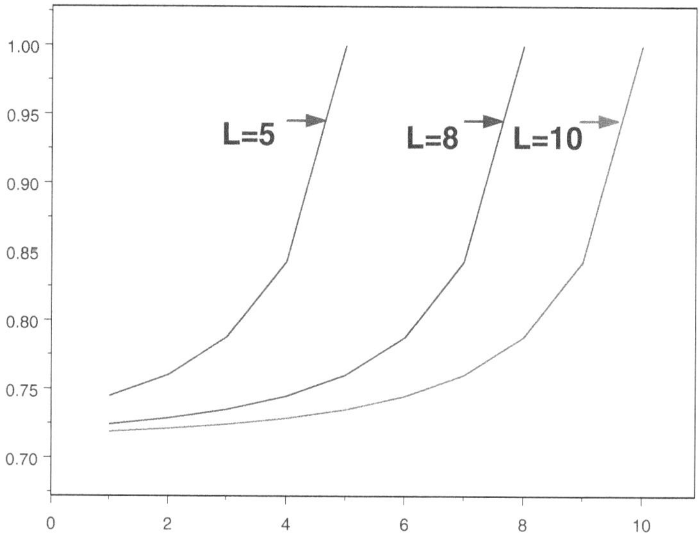

Figure 8.1: The asymptotic probabilities of correct selections

Next, we evaluate the asymptotic probabilities $\lim_{n\to\infty} \boldsymbol{P}^{(n)}_{(\boldsymbol{\theta}_0, \boldsymbol{h}/\sqrt{n})}\{\hat{p} = p\}$ under the local alternative hypothesis $\boldsymbol{\theta} = (\boldsymbol{\theta}_0', \boldsymbol{h}'/\sqrt{n})'$. We consider the exponential spectral density model (8.4.54) where the local alternative (true) model is given by (8.4.55). Here we assume that $h_l \equiv h$ are constant for $1 \leq l \leq K - p_0$ and $h_l \equiv 0$ for $K - p_0 + 1 \leq l \leq L - p_0$. Since the exact probabilities of $\beta(p - p_0)$ and $\gamma(p - p_0)$ in (8.4.51) are complicated, we replace them by the empirical relative frequencies of events $\{\cap_{j=1}^{p-p_0}(S_{j,p-p_0} > 0)\}$ and $\{\cap_{j=1}^{L-p}(T_{j,p-p_0} \leq 0)\}$ in the 1000 times iteration. To simplify the

Table 8.2: The asymptotic distribution of selected order under the local alternative hypothesis $\boldsymbol{\theta} = (\boldsymbol{\theta}_0', \boldsymbol{h}'/\sqrt{n})'$ for $h = 1$

$u \backslash R$	1	2	3	4	5	6	7	8	9	10
0	0.748	0.714	0.727	0.73	0.689	0.704	0.709	0.717	0.705	0.724
1	0.101	0.123	0.117	0.101	0.128	0.114	0.126	0.109	0.122	0.106
2	0.054	0.054	0.051	0.057	0.072	0.054	0.066	0.066	0.051	0.059
3	0.028	0.032	0.038	0.037	0.029	0.044	0.043	0.026	0.033	0.031
4	0.023	0.026	0.019	0.026	0.016	0.022	0.018	0.024	0.023	0.028
5	0.015	0.015	0.012	0.011	0.019	0.012	0.013	0.014	0.023	0.018
6	0.01	0.009	0.009	0.015	0.015	0.018	0.009	0.014	0.015	0.009
7	0.008	0.01	0.013	0.008	0.007	0.011	0.007	0.008	0.01	0.008
8	0.006	0.005	0.005	0.004	0.013	0.005	0.003	0.009	0.005	0.006
9	0.003	0.007	0.006	0.006	0.008	0.009	0.002	0.005	0.006	0.007
10	0.004	0.005	0.003	0.005	0.004	0.007	0.004	0.008	0.007	0.004

Table 8.3: The asymptotic distribution of selected order under the local alternative hypothesis $\boldsymbol{\theta} = (\boldsymbol{\theta}_0', \boldsymbol{h}'/\sqrt{n})'$ for $h = 10$

$u \backslash R$	1	2	3	4	5	6	7	8	9	10
0	0.464	0.366	0.262	0.233	0.213	0.205	0.163	0.154	0.124	0.111
1	0.332	0.208	0.173	0.137	0.094	0.088	0.071	0.056	0.065	0.052
2	0.067	0.236	0.166	0.149	0.103	0.085	0.06	0.07	0.05	0.051
3	0.041	0.067	0.242	0.149	0.107	0.094	0.078	0.058	0.056	0.041
4	0.027	0.048	0.053	0.182	0.133	0.105	0.099	0.077	0.058	0.048
5	0.028	0.023	0.036	0.051	0.201	0.127	0.107	0.08	0.062	0.066
6	0.011	0.019	0.022	0.035	0.056	0.167	0.13	0.092	0.081	0.06
7	0.015	0.007	0.017	0.022	0.04	0.05	0.192	0.137	0.108	0.087
8	0.007	0.009	0.014	0.017	0.032	0.033	0.054	0.179	0.135	0.11
9	0.004	0.01	0.007	0.012	0.006	0.025	0.025	0.054	0.202	0.114
10	0.004	0.007	0.008	0.013	0.015	0.021	0.021	0.043	0.059	0.26

notations we write $Q \equiv L - p_0$, $R \equiv K - p_0$ and $u = p - p_0$; then $\beta(p - p_0)$ and $\gamma(p - p_0)$ in (8.4.51) depend on (p, p_0, L) through u and Q. The asymptotic probabilities $\lim_{n\to\infty} \boldsymbol{P}^{(n)}_{(\boldsymbol{\theta}_0, \boldsymbol{h}/\sqrt{n})}\{\hat{p} = u + p_0\}$ under the local alternative hypothesis $\boldsymbol{\theta} = (\boldsymbol{\theta}_0', \boldsymbol{h}'/\sqrt{n})'$ for $h = 1, 10, 100$ are listed in Tables 8.2–8.4. Here we fix $Q = 10$ and evaluate the asymptotic probabilities for each $u = 0, \ldots, Q$ and $R = 1, \ldots, Q$. In these tables the elements for $u = R$ correspond to correctly estimated cases, while those for $u > R$ and $u < R$ correspond to overestimated and underestimated cases, respectively. Furthermore, the cases of $u = 0$ correspond to the acceptance of the null hypothesis under the local alternative hypothesis. From these tables we see that GAIC tends to select null hypothesis $u = 0$ for small h (see Table 8.2). On the other hand, GAIC tends to select local alternative hypothesis $u = R$ (true model) and does not select $u < R$ (smaller orders than the true one) for large h (see Table 8.4).

At first sight, GAIC may seem to have a bad performance since it is unstable and inconsistent in contrast with BIC and HQ, which are consistent order selection criteria of p_0. If $\|\boldsymbol{h}\|$ is small, the parameters in the local alternative part are negligible, so we expect to select the null hypothesis p_0. In such cases GAIC tends to select p_0, although the asymptotic probabilities of overestimations are not equal to zero. Therefore, GAIC is not so bad if we restrict ourselves to the negligible alternative and null hypotheses where BIC and HQ are surely desirable. However, $\|\boldsymbol{h}\|$ is clearly not always small.

Table 8.4: The asymptotic distribution of selected order under the local alternative hypothesis $\boldsymbol{\theta} = (\boldsymbol{\theta}_0', \boldsymbol{h}'/\sqrt{n})'$ for $h = 100$

$u \backslash R$	1	2	3	4	5	6	7	8	9	10
0	0	0	0	0	0	0	0	0	0	0
1	0.745	0	0	0	0	0	0	0	0	0
2	0.105	0.715	0	0	0	0	0	0	0	0
3	0.045	0.109	0.718	0	0	0	0	0	0	0
4	0.041	0.062	0.127	0.731	0	0	0	0	0	0
5	0.018	0.034	0.052	0.108	0.743	0	0	0	0	0
6	0.014	0.025	0.044	0.054	0.127	0.755	0	0	0	0
7	0.012	0.011	0.025	0.041	0.052	0.124	0.777	0	0	0
8	0.008	0.018	0.022	0.033	0.029	0.053	0.106	0.778	0	0
9	0.005	0.013	0.008	0.018	0.026	0.039	0.064	0.136	0.86	0
10	0.007	0.013	0.004	0.015	0.023	0.029	0.053	0.086	0.14	1

If $\|\boldsymbol{h}\|$ is large, the parameters in the local alternative part are evidently not ignorable, so we should select the larger order than the true order K to avoid misspecification. In such cases GAIC tends to select the true order K and does not underestimate, so we are guarded from misspecification. On the other hand, BIC and HQ still retain the consistency to "pretended order p_0" for any bounded $\boldsymbol{h}$. Note that such a situation may happen not only for the small sample but also for the large sample since $\boldsymbol{h}$ does not depend on n. Therefore, we can conclude that GAIC balances the order of the model with the magnitude of local alternative. Furthermore, GAIC protects us from misspecification under the unignorable alternative, while BIC and HQ always mislead us to underestimation.

8.5 Efficient Estimation for Portfolios

In this section, we discuss the asymptotic efficiency of estimators for optimal portfolio weights. In Section 3.2, the asymptotic distribution of portfolio weight estimators $\hat{g}$ for the processes are given. Shiraishi and Taniguchi (2008) showed that the usual portfolio weight estimators are not asymptotically efficient if the returns are dependent. Furthermore, they propose using maximum likelihood type estimators for optimal portfolio weight g, which are shown to be asymptotically efficient.
Although Shiraishi and Taniguchi (2008) dropped Gaussianity and independence for return processes, many empirical studies show that real time series data are generally nonstationary. To describe this phenomenon, Dahlhaus (1996a) introduced an important class of locally stationary processes, and discussed kernel methods for local estimators of the covariance structure. Regarding the parametric approach, Dahlhaus (1996b) developed asymptotic theory for a Gaussian locally stationary process, and derived the LAN property. Furthermore Hirukawa and Taniguchi (2006) dropped the Gaussian assumption, i.e., they proved the LAN theorem for a non-Gaussian locally stationary process with mean vector $\boldsymbol{\mu} = \mathbf{0}$, and applied the results to the asymptotic estimation. Regarding the application to the portfolio selection problem, Shiraishi and Taniguchi (2007) discussed the asymptotic property of estimators $\hat{g}$ when the returns are vector-valued locally stationary processes with time varying mean vector. Since the returns are nonstationary, the quantities $\boldsymbol{\mu}, \Sigma, \hat{\boldsymbol{\mu}}, \hat{\Sigma}$ depend on the time when the portfolio is constructed. Assuming parametric structure $\boldsymbol{\mu}_{\boldsymbol{\theta}}$ and $\Sigma_{\boldsymbol{\theta}}$, they proved the LAN theorem for non-Gaussian locally stationary processess with nonzero mean vector. In this setting they propose a parametric estimator $\hat{g} = g(\boldsymbol{\mu}_{\hat{\boldsymbol{\theta}}}, \Sigma_{\hat{\boldsymbol{\theta}}})$ where $\hat{\boldsymbol{\theta}}$ is a quasi-maximum likelihood estimator of $\boldsymbol{\theta}$. Then it is shown that $\hat{g}$ is asymptotically efficient.

8.5.1 *Traditional mean variance portfolio estimators*

Suppose that a sequence of random returns $\{\boldsymbol{X}(t) = (X_1(t), \dots, X_m(t))'; t \in \mathbb{Z}\}$ is a stationary process. Writing $\boldsymbol{\mu} = E\{\boldsymbol{X}(t)\}$ and $\Sigma = \mathrm{Var}\{\boldsymbol{X}(t)\} = E[\{\boldsymbol{X}(t) - \boldsymbol{\mu}\}\{\boldsymbol{X}(t) - \boldsymbol{\mu}\}']$, we define $(m + r)$-vector parameter $\theta = (\theta_1, \dots, \theta_{m+r})'$ by

$$\theta = (\boldsymbol{\mu}', vech(\Sigma)')'$$

where $r = m(m + 1)/2$. We consider estimating a general function $g(\theta)$ of θ, which expresses an optimal portfolio weight for a mean variance model such as (3.1.2), (3.1.3), (3.1.5) and (3.1.7). Here it should be noted that the portfolio weight $\boldsymbol{w} = (w_1, \dots, w_m)'$ satisfies the restriction $\boldsymbol{w}'\boldsymbol{e} = 1$. Then we have only to estimate the subvector $(w_1, \dots, w_{m-1})'$. Hence we assume that the function $g(\cdot)$ is $(m - 1)$-dimensional, i.e.,

$$g : \theta \to \mathbb{R}^{m-1}. \tag{8.5.1}$$

In this setting we address the problem of statistical estimation for $g(\theta)$, which describes various mean variance optimal portfolio weights.
Let the return process $\{\boldsymbol{X}(t) = (X_1(t), \dots, X_m(t))'; t \in \mathbb{Z}\}$ be an m-vector linear process

$$\boldsymbol{X}(t) = \sum_{j=0}^{\infty} A(j)\boldsymbol{U}(t - j) + \boldsymbol{\mu} \tag{8.5.2}$$

where $\{\boldsymbol{U}(t) = (u_1(t), \dots, u_m(t))'\}$ is a sequence of independent and identically distributed (i.i.d.) m-vector random variables with $E\{\boldsymbol{U}(t)\} = \boldsymbol{0}$, $\mathrm{Var}\{\boldsymbol{U}(t)\} = K$ (for short $\{\boldsymbol{U}(t) \sim$ i.i.d. $(\boldsymbol{0}, K)\}$) and fourth-order cumulants. Here

$$A(j) = (A_{ab}(j))_{a,b=1,\dots,m},\ A(0) = I_m,\ \boldsymbol{\mu} = (\mu_a)_{a=1,\dots,m},\ K = (K_{ab})_{a,b=1,\dots,m}.$$

The class of $\{\boldsymbol{X}(t)\}$ includes that of non-Gaussian vector-valued causal ARMA models. Hence it is sufficiently rich. The process $\{\boldsymbol{X}(t)\}$ is a second-order stationary process with spectral density matrix

$$f(\lambda) = (f_{ab}(\lambda))_{a,b=1,\dots,m} = \frac{1}{2\pi} A(\lambda) K A(\lambda)^*$$

where $A(\lambda) = \sum_{j=0}^{\infty} A(j) e^{ij\lambda}$.
From the partial realization $\{\boldsymbol{X}(1), \dots, \boldsymbol{X}(n)\}$, we introduce

$$\hat{\theta} = (\hat{\boldsymbol{\mu}}', vech(\hat{\Sigma})')', \quad \tilde{\theta} = (\hat{\boldsymbol{\mu}}', vech(\tilde{\Sigma})')'$$

where

$$\hat{\boldsymbol{\mu}} = \frac{1}{n}\sum_{t=1}^{n} \boldsymbol{X}(t),\ \hat{\Sigma} = \frac{1}{n}\sum_{t=1}^{n}\{\boldsymbol{X}(t) - \hat{\boldsymbol{\mu}}\}\{\boldsymbol{X}(t) - \hat{\boldsymbol{\mu}}\}',\ \tilde{\Sigma} = \frac{1}{n}\sum_{t=1}^{n}\{\boldsymbol{X}(t) - \boldsymbol{\mu}\}\{\boldsymbol{X}(t) - \boldsymbol{\mu}\}'.$$

Since we have already discussed the asymptotic normality in Section 3.2, we discuss the asymptotic Gaussian efficiency for $g(\hat{\theta})$. Fundamental results concerning the asymptotic efficiency of sample autocovariance matrices of vector Gaussian processes were obtained by Kakizawa (1999). He compared the asymptotic variance (AV) of sample autocovariance matrices with the inverse of the corresponding Fisher information matrix ($\mathcal{F}^{-1}$), and gave the condition for the asymptotic efficiency (i.e., condition for $AV = \mathcal{F}^{-1}$). Based on this we will discuss the asymptotic efficiency of $g(\hat{\theta})$.
Suppose that $\{\boldsymbol{X}(t)\}$ is a zero-mean Gaussian m-vector stationary process with spectral density matrix $f(\lambda)$, and satisfies the following assumptions.

Assumption 8.5.1. *(i) $f(\lambda)$ is parameterized by $\boldsymbol{\eta} = (\eta_1, \dots, \eta_q)' \in \mathcal{H} \subset \mathbb{R}^q$ (i.e. $f(\lambda) = f_{\eta}(\lambda)$). (ii) For $A^{(j)}(l) \equiv \int_{-\pi}^{\pi} \partial f_{\eta}(\lambda)/\partial \eta_j d\lambda,\ j = 1, \dots, q, l \in \mathbb{Z}$, it holds that $\sum_{l=-\infty}^{\infty} \|A^{(j)}(l)\| < \infty$. (iii) $q \geq m(m+1)/2$.*

Assumption 8.5.2. *There exists a positive constant c (independent of λ) such that $f_{\eta}(\lambda) - cI_m$ is positive semi-definite, where I_m is the $m \times m$ identity matrix.*

The limit of the averaged Fisher information matrix is given by

$$\mathcal{F}(\eta) = \frac{1}{4\pi} \int_{-\pi}^{\pi} \Delta(\lambda)^* [\{f_{\eta}(\lambda)^{-1}\}' \otimes f_{\eta}(\lambda)^{-1}] \Delta(\lambda) d\lambda$$

where

$$\Delta(\lambda) = (vec\{\partial f_{\eta}(\lambda)/\partial \eta_1\}, \dots, vec\{\partial f_{\eta}(\lambda)/\partial \eta_q\}) \qquad (m^2 \times q) - matrix.$$

Assumption 8.5.3. *The matrix $\mathcal{F}(\eta)$ is positive definite.*

We introduce an $m^2 \times m(m+1)/2$ matrix

$$\Phi = (vec\{\psi_{11}\}, \dots, vec\{\psi_{m1}\}, vec\{\psi_{22}\}, \dots, vec\{\psi_{mm}\}),$$

where $\psi_{ab} = \frac{1}{2}\{E_{ab} + E_{ba}\}$ and E_{ab} is a matrix of which the (a, b)th element is 1 and others are 0. Then we have the following theorem.

Theorem 8.5.4. *Under Assumptions 8.5.1–8.5.3 and the Gaussianity of $\{\boldsymbol{X}(t)\}$, $g(\hat{\theta})$ is asymptotically efficient if and only if there exists a matrix C (independent of λ) such that*

$$\{f_{\eta}(\lambda)' \otimes f_{\eta}(\lambda)\}\Phi = (vec\{\partial f_{\eta}(\lambda)/\partial \eta_1\}, \dots, vec\{\partial f_{\eta}(\lambda)/\partial \eta_q\})C. \tag{8.5.3}$$

Proof of Theorem 8.5.4. Recall Theorem 3.2.5. In the asymptotic variance matrix of $\sqrt{n}\{g(\hat{\theta}) - g(\theta)\}$, the transformation matrix $(\partial g/\partial \theta')$ does not depend on the goodness of estimation. Thus, if the matrix $\Omega = \begin{pmatrix} \Omega_1 & \mathbf{0} \\ \mathbf{0} & \Omega_2^G \end{pmatrix}$ is minimized in the sense of matrix, the estimator $g(\hat{\theta})$ becomes asymptotically efficient. Regarding the Ω_2^G part, Kakizawa (1999) gave a necessarily and sufficient condition for Ω_2^G to attain the lower bound matrix $\mathcal{F}(\eta)^{-1}$, which is given by the condition (8.5.3). Regarding the Ω_1 part, it is known that $\Omega_1 = 2\pi f(0)$ is equal to the asymptotic variance of the BLUE estimator of $\boldsymbol{\mu}$ (e.g., Hannan (1970)). Hence, Ω_1 attains the lower bound matrix, which completes the proof.□

This theorem implies that if (8.5.3) is not satisfied, the estimator $g(\hat{\theta})$ is not asymptotically efficient. This is a strong warning to use the traditional portfolio estimators even for Gaussian dependent returns. The interpretation of (8.5.3) is difficult. But Kakizawa (1999) showed that (8.5.3) is satisfied by VAR(p) models. The following are examples which do not satisfy (8.5.3).

Example 8.5.5. *(VARMA(p_1, p_2) process) Consider the $m \times m$ spectral density matrix of an m-vector ARMA(p_1, p_2) process,*

$$f(\lambda) = \frac{1}{2\pi} \Theta\{\exp(i\lambda)\}^{-1} \Psi\{\exp(i\lambda)\} \Sigma \Psi\{\exp(i\lambda)\}^* \Theta\{\exp(i\lambda)\}^{-1*} \tag{8.5.4}$$

where $\Psi(z) = I_m - \Psi_1 z - \cdots - \Psi_{p_2} z^{p_2}$ and $\Theta(z) = I_m - \Theta_1 z - \cdots - \Theta_{p_1} z^{p_1}$ satisfy $\det \Psi(z) \neq 0$, $\det \Theta(z) \neq 0$ for all $|z| \leq 1$. From Kakizawa (1999) it follows that (8.5.4) does not satisfy (8.5.3), if $p_1 < p_2$, hence $g(\hat{\theta})$ is not asymptotically efficient if $p_1 < p_2$.

Example 8.5.6. *(An exponential model) Consider the $m \times m$ spectral density matrix of exponential type,*

$$f(\lambda) = \exp\left\{\sum_{j \neq 0} A(j) \cos(j\lambda)\right\} \tag{8.5.5}$$

where $A(j)$'s are $m \times m$ matrices, and $\exp\{\cdot\}$ is the matrix exponential (for the definition, see Bellman (1960)). Since (8.5.5) does not satisfy (8.5.3), $g(\hat{\theta})$ is not asymptotically efficient.

8.5.2 *Efficient mean variance portfolio estimators*

In the previous subsection, we showed that the traditional mean variance portfolio weight estimator $g(\hat{\theta})$ is not asymptotically efficient generally even if $\{\boldsymbol{X}(t)\}$ is a Gaussian stationary process. Therefore, we propose maximum likelihood type estimators for $g(\theta)$, which are asymptotically efficient.

Gaussian Efficient Estimator Let $\{\boldsymbol{X}(t)\}$ be the linear Gaussian process defined by (8.5.2) with spectral density matrix $f_{\eta}(\lambda), \eta \in \mathcal{H}$. We assume $\eta = vech(\Sigma)$. Denote by $\mathrm{I}_n(\lambda)$, the periodogram matrix constructed from a partial realization $\{\boldsymbol{X}(1), \dots, \boldsymbol{X}(n)\}$:

$$\mathrm{I}_n(\lambda) = \frac{1}{2\pi n} \sum_{t=1}^{n} [\{\boldsymbol{X}(t) - \hat{\mu}\}e^{it\lambda}\}][\{\boldsymbol{X}(t) - \hat{\mu}\}e^{it\lambda}\}]^*, \quad for \ -\pi \le \lambda \le \pi.$$

To estimate η we introduce

$$D(f_{\eta}, \mathrm{I}_n) = \int_{-\pi}^{\pi} \left[\log \det f_{\eta}(\lambda) + \mathrm{tr}\{f_{\eta}^{-1}(\lambda)\mathrm{I}_n(\lambda)\}\right] d\lambda,$$

(e.g., Hosoya and Taniguchi (1982)). A quasi-Gaussian maximum likelihood estimator $\hat{\eta}$ of η is defined by

$$\hat{\eta} = \arg\min_{\eta \in \mathcal{H}} D(f_{\eta}, \mathrm{I}_n).$$

Let $\{\boldsymbol{X}(t)\}$ be the Gaussian m-vector linear process defined by (8.5.2). For the process $\{\boldsymbol{X}(t)\}$ and the true spectral density matrix $f(\lambda) = (f_{ab}(\lambda))_{a,b=1,\dots,m}$, we impose the following.

Assumption 8.5.7. *(i) For each a, b and l,*

$$\mathrm{Var}[E\{u_a(t)u_b(t+l)|\mathcal{F}_{t-\tau}\} - \delta(l,0)K_{ab}] = O\left(\tau^{-2-\alpha}\right), \quad \alpha > 0,$$

uniformly in t.

$$(ii)\ E[E\{u_{a_1}(t_1)u_{a_2}(t_2)u_{a_3}(t_3)u_{a_4}(t_4)|\mathcal{F}_{t_1-\tau}) - E\{u_{a_1}(t_1)u_{a_2}(t_2)u_{a_3}(t_3)u_{a_4}(t_4)\}] = O\left(\tau^{-1-\alpha}\right)$$

uniformly in t_1, where $t_1 \le t_2 \le t_3 \le t_4$ and $\alpha > 0$.

(iii) the spectral densities $f_{aa}(\lambda)$ $(a = 1, \dots, m)$ are square integrable.

(iv) $\sum_{j_1,j_2,j_3=-\infty}^{\infty} |c^{\boldsymbol{U}}_{a_1\cdots a_4}(j_1, j_2, j_3)| < \infty$ *where $c^{\boldsymbol{U}}_{a_1\cdots a_4}(j_1, j_2, j_3)$'s are the joint fourth-order cumulants of $u_{a_1}(t), u_{a_2}(t+j_1), u_{a_3}(t+j_3)$ and $u_{a_4}(t+j_3)$.*

(v) $f(\lambda) \in Lip(\alpha)$, the Lipschitz class of degree $\alpha, \alpha > 1/2$.

Then, Hosoya and Taniguchi (1982) showed the following.

Lemma 8.5.8. *(Theorem 3.1 of Hosoya and Taniguchi (1982)) Suppose that η exists uniquely and lies in $\mathcal{H} \subset \mathbb{R}^q$. Under Assumption 8.5.7, we have*

(i) $p\text{-}\lim_{n\to\infty} \hat{\eta} = \eta$

(ii) $\sqrt{n}(\hat{\eta} - \eta) \xrightarrow{\mathcal{L}} N(0, V(\eta))$

where

$$V(\eta) = \int_{-\pi}^{\pi} \left[\frac{\partial^2}{\partial\eta\partial\eta'}\mathrm{tr}(f_{\eta}(\lambda)^{-1}f(\lambda)) + \frac{\partial^2}{\partial\eta\partial\eta'}\log\det(f_{\eta}(\lambda))\right] d\lambda.$$

If $\{\boldsymbol{X}(t)\}$ is Gaussian, then $V(\eta) = \mathcal{F}(\eta)^{-1}$. Therefore, $\hat{\eta}$ is Gaussian asymptotically efficient, hence $g(\tilde{\theta})$, with $\tilde{\theta} = (\hat{\mu}', \hat{\eta}')'$, is Gaussian asymptotically efficient. Since the solution of $\partial D(f_\eta, \mathrm{I}_n)/\partial\eta = 0$ is generally nonlinear with respect to η, we use the Newton–Raphson iteration procedure. A feasible procedure is

$$\hat{\eta}^{(1)} = vech(\hat{\Sigma})$$

$$\hat{\eta}^{(k)} = \hat{\eta}^{(k-1)} - \left[\frac{\partial^2 D(f_\eta, \mathrm{I}_n)}{\partial\eta\partial\eta'}\right]^{-1} \frac{\partial D(f_\eta, \mathrm{I}_n)}{\partial\eta}\bigg|_{\eta=\hat{\eta}^{(k-1)}} \quad (k \geq 2). \tag{8.5.6}$$

From Theorem 5.1 of Hosoya and Taniguchi (1982) $\sqrt{n}(\hat{\eta}^{(2)} - \eta)$ has the same limiting distribution as $\sqrt{n}(\hat{\eta} - \eta)$, which implies that $\hat{\eta}^{(2)}$ is asymptotically efficient. Therefore, $g(\theta^{(2)}), \theta^{(2)} = (\hat{\mu}', \hat{\eta}^{(2)\prime})'$ becomes Gaussian asymptotically efficient. In (8.5.6), we can express

$$\frac{\partial D(f_\eta, \mathrm{I}_n)}{\partial\eta_{ij}} = \frac{1}{2\pi}\int_{-\pi}^{\pi} \mathrm{tr}\left[f_\eta(\lambda)^{-1}E_{ij}(I_m - f_\eta(\lambda)^{-1}\mathrm{I}_n(\lambda))\right]d\lambda, \tag{8.5.7}$$

$$\begin{aligned}\frac{\partial^2 D(f_\eta, \mathrm{I}_n)}{\partial\eta_{ij}\partial\eta_{kl}} &= \left(\frac{1}{2\pi}\right)^2 \int_{-\pi}^{\pi} \mathrm{tr}[-f_\eta(\lambda)^{-1}E_{kl}f_\eta(\lambda)^{-1}E_{ij}\left(I_m - f_\eta(\lambda)^{-1}\mathrm{I}_n(\lambda)\right) \\ &\quad + f_\eta(\lambda)^{-1}E_{ij}f_\eta(\lambda)^{-1}E_{kl}f_\eta(\lambda)^{-1}\mathrm{I}_n(\lambda)]d\lambda\end{aligned} \tag{8.5.8}$$

where η_{ij} is the $(i, j)th$ element of Σ. To make the step (8.5.6) feasible, we may replace f_η in (8.5.7) and (8.5.8) by a nonparametric spectral estimator

$$\hat{f}_\eta(\lambda) = \int_{-\pi}^{\pi} W_n(\lambda - \mu)\mathrm{I}_n(\mu)d\mu \tag{8.5.9}$$

where $W_n(\lambda)$ is a nonnegative, even, periodic function with period 2π whose mass becomes, as n increases, more and more concentrated around $\lambda = 0 \pmod{2\pi}$ (details see, e.g., p. 390 of Taniguchi and Kakizawa (2000)). Then appropriate regularity conditions (6.1.17) of Taniguchi and Kakizawa (2000) showed

$$\max_{\lambda\in[-\pi,\pi]} \|\hat{f}_\eta(\lambda) - f_\eta(\lambda)\|_E \xrightarrow{p} 0$$

where $\|A\|_E$ is the Euclidean norm of the matrix A; $\|A\|_E = \{tr(AA^*)\}^{1/2}$. Therefore, $g(\hat{\theta}^{(2)})$ with $\hat{\theta}^{(2)} = (\hat{\mu}', \hat{\hat{\eta}}^{(2)\prime})$ (where $\hat{\hat{\eta}}^{(2)}$ equivalents $\hat{\eta}^{(2)}$ applied $\hat{f}_\eta$ instead of f_η) is also Gaussian asymptotically efficient from this result, the asymptotic property of BLUE estimators and the δ-method.

Non-Gaussian Efficient Estimator Next we discuss construction of an efficient estimator when $\{\boldsymbol{X}(t)\}$ is non-Gaussian. Fundamental results concerning the asymptotic efficiency of estimators for unknown parameter θ of vector linear processes were obtained by Taniguchi and Kakizawa (2000). They showed the quasi-maximum likelihood estimator $\hat{\theta}_{QML}$ is asymptotically efficient.

Let $\{\boldsymbol{X}(t)\}$ be a non-Gaussian m-vector stationary process defined by (8.5.2), where $\boldsymbol{U}(t)$'s are i.i.d. m-vector random variables with probability density $p(\boldsymbol{u}) > 0$ on $\mathbb{R}^m$, and $A(j) \equiv A_\theta(j) = (A_{\theta,ab}(j))_{a,b=1,\ldots,m}$ for $j \in \mathbb{N}$ and $\boldsymbol{\mu} \equiv \boldsymbol{\mu}_\theta$, are $m \times m$ matrices and m-vector depending on a parameter vector $\theta = (\theta_1, \ldots, \theta_q)' \in \Theta \subset \mathbb{R}^q$, respectively. Here $A_\theta(0) = I_m$, the $m \times m$ identity matrix. We impose the following assumption.

Assumption 8.5.9. *(A1)(i) There exists* $0 < \rho_A < 1$ *so that* $|A_\theta(j)| = O(\rho_A^j)$, $j \in \mathbb{N}$ *where* $|A|$ *is the sum of the absolute values of the entries of A.*

(ii) Every $A_\theta(j) = \{A_{\theta,ab}(j)\}$ *is continuously two times differentiable with respect to* θ, *and the derivatives satisfy*

$$|\partial_{i_1}\cdots\partial_{i_k}A_{\theta,ab}(j)| = O(\gamma_A^j), \quad k = 1, 2, j \in \mathbb{N}$$

for some $0 < \gamma_A < 1$ *and for* $a, b = 1, \ldots, m$.

(iii) Every $\partial_{i_1}\partial_{i_2}A_{\theta,ab}(j)$ *satisfies the Lipschitz condition for all* $i_1, i_2 = 1, \ldots, q, a, b = 1, \ldots, m$ *and* $j \in \mathbb{N}$.

(iv) $\det\{\sum_{j=0}^{\infty} A_\theta(j)z^j\} \neq 0$ *for* $|z| \leq 1$ *and*

$$\det\{\sum_{j=0}^{\infty} A_\theta(j)z^j\}^{-1} = I_m + B_\theta(1)z + B_\theta(2)z^2 + \cdots, \quad |z| \leq 1$$

where $|B_\theta(j)| = O(\rho_B^j)$, $j \in \mathbb{N}$, *for some* $0 < \rho_B < 1$.

(v) Every $B_\theta(j) = \{B_{\theta,ab}(j)\}$ *is continuously two times differentiable with respect to* θ, *and the derivatives satisfy*

$$|\partial_{i_1} \cdots \partial_{i_k} B_{\theta,ab}(j)| = O(\gamma_B^j), \quad k = 1, 2, j \in \mathbb{N}$$

for some $0 < \gamma_B < 1$ *and for* $a, b = 1, \ldots, m$.

(vi) Every $\partial_{i_1}\partial_{i_2}B_{\theta,ab}(j)$ *satisfies the Lipschitz condition for all* $i_1, i_2 = 1, \ldots, q, a, b = 1, \ldots, m$ *and* $j \in \mathbb{N}$.

(A2)(i)

$$\lim_{\|\boldsymbol{u}\| \to \infty} p(\boldsymbol{u})| = 0, \quad \int \boldsymbol{u} p(\boldsymbol{u}) d\boldsymbol{u} = \boldsymbol{0}, \quad \int \boldsymbol{u}\boldsymbol{u}' p(\boldsymbol{u}) d\boldsymbol{u} = I_m \quad \text{and} \quad \int \|\mathrm{u}\|^4 \mathrm{p}(\mathrm{u})\mathrm{du} < \infty.$$

(ii) The derivatives Dp *and* $D^2p \equiv D(Dp)$ *exist on* $\mathbb{R}^n$, *and every component of* D^2p *satisfies the Lipschitz condition.*

(iii)

$$\int |\phi(\boldsymbol{u})|^4 \, p(\boldsymbol{u}) d\boldsymbol{u} < \infty \quad \text{and} \quad \int \mathrm{D}^2\mathrm{p}(\mathrm{u})\mathrm{du} = 0$$

where $\phi(\boldsymbol{u}) = p^{-1}Dp$.

A quasi-maximum likelihood estimator $\hat{\theta}_{QML}$ is defined as a solution of the equation

$$\frac{\partial}{\partial \theta}\left[\sum_{t=1}^{n} \log p\left\{\sum_{j=0}^{t-1} B_\theta(j)(\boldsymbol{X}(t-j) - \boldsymbol{\mu}_\theta)\right\}\right] = \boldsymbol{0}$$

with respect to θ. Then Taniguchi and Kakizawa (2000) showed the following result.

Lemma 8.5.10. *(Theorem 3.1.12 of Taniguchi and Kakizawa (2000)) Under Assumption 8.5.9,* $\hat{\theta}_{QML}$ *is asymptotically efficient.*

As we saw in the previous paragraph, the theory of linear time series is well established. However, it is not sufficient for linear models such as ARMA models to describe the real world. Using ARCH models, Hafner (1998) gave an extensive analysis for finance data. The results by Hafner (1998) reveal that many relationships in real data are "nonlinear." Thus the analysis of nonlinear time series is becoming an important component of time series analysis. Härdle and Gasser (1984) introduced vector conditional heteroscedastic autoregressive nonlinear (CHARN) models, which include the usual AR and ARCH models. Kato et al. (2006) established the LAN property for the

CHARN model.
Suppose that $\{\boldsymbol{X}(t)\}$ is generated by

$$\boldsymbol{X}(t) = \boldsymbol{F}_\theta(\boldsymbol{X}(t-1),\ldots,\boldsymbol{X}(t-p_1),\boldsymbol{\mu}) + \boldsymbol{H}_\theta(\boldsymbol{X}(t-1),\ldots,\boldsymbol{X}(t-p_2))\boldsymbol{U}(t), \tag{8.5.10}$$

where $\boldsymbol{F}_\theta : \mathbb{R}^{m(p_1+1)} \to \mathbb{R}^m$ is a vector-valued measurable function, $\boldsymbol{H}_\theta : \mathbb{R}^{mp_2} \to \mathbb{R}^m \times \mathbb{R}^m$ is a positive definite matrix-valued measurable function, and $\{\boldsymbol{U}(t) = (u_1(t),\ldots,u_m(t))'\}$ is a sequence of i.i.d. random variables with $E[\boldsymbol{U}(t)] = \boldsymbol{0}$, $E|\boldsymbol{U}(t)| < \infty$ and $\boldsymbol{U}(t)$ is independent of $\{\boldsymbol{X}(s), s < t\}$. Henceforth, without loss of generality we assume $p_1 + 1 = p_2(= p)$, and make the following assumption.

Assumption 8.5.11. *(A-i) $\boldsymbol{U}(t)$ has a density $p(\boldsymbol{u}) > 0$ a.e. on $\mathbb{R}^m$.*

(A-ii) There exist constants $a_{ij} \geq 0, b_{ij} \geq 0, 1 \leq i \leq m, 1 \leq j \leq p$ such that, as $|\boldsymbol{x}| \to \infty$,

$$|\boldsymbol{F}_\theta(\boldsymbol{x})| \leq \sum_{i=1}^m \sum_{j=1}^p a_{ij}|x_{ij}| + o(|\boldsymbol{x}|) \quad \text{and} \quad |\mathrm{H}_\theta(\mathrm{x})| \leq \sum_{i=1}^m \sum_{j=1}^p \mathrm{b}_{ij}|\mathrm{x}_{ij}| + \mathrm{o}(|\mathrm{x}|).$$

(A-iii) $\boldsymbol{H}_\theta(\boldsymbol{x})$ is continuous and symmetric on $\mathbb{R}^{mp}$, and there exists a positive constant λ such that

$$\lambda_{min}\{\boldsymbol{H}_\theta(\boldsymbol{x})\} \geq \lambda \quad \textit{for all } \boldsymbol{x} \in \mathbb{R}^{mp}$$

where $\lambda_{min}\{(\cdot)\}$ is the minimum eigenvalue of $(\cdot)$.

(A-iv) $\max_{1\leq i\leq m}\{\sum_{j=1}^p a_{ij} + E|U_1|\sum_{j=1}^p b_{ij}\} < 1.$

(B-i) For all $\theta \in \Theta$

$$E_\theta\|\boldsymbol{F}_\theta(\boldsymbol{X}(t-1),\ldots,\boldsymbol{X}(t-p),\boldsymbol{\mu})\|^2 < \infty \quad \text{and} \quad \mathrm{E}_\theta\|\mathrm{H}_\theta(\mathrm{X}(t-1),\ldots,\mathrm{X}(t-p))\|^2 < \infty.$$

(B-ii) There exists $c > 0$ such that

$$c \leq \|\boldsymbol{H}_{\theta'}^{-1/2}(\boldsymbol{x})\boldsymbol{H}_\theta(\boldsymbol{x})\boldsymbol{H}_{\theta'}^{-1/2}(\boldsymbol{x})\| < \infty$$

for all $\theta, \theta' \in \Theta$ and for all $\boldsymbol{x} \in \mathbb{R}^{mp}$.

(B-iii) $\boldsymbol{H}_\theta$ and $\boldsymbol{F}_\theta$ are continuously differentiable with respect to θ, and their derivatives $\partial_j\boldsymbol{H}_\theta$ and $\partial_j\boldsymbol{F}_\theta(\partial_j = \partial/\partial\theta_j)$, $j = 1,\ldots,q$, satisfy the condition that there exist square-integrable functions $A(j)$ and $B(j)$ such that

$$\|\partial_j\boldsymbol{H}_\theta\| \leq A(j) \quad \text{and} \quad \|\partial_j\mathrm{F}_\theta\| \leq \mathrm{B}(j) \quad \text{for } j = 1,\ldots,q \text{ and for all } \theta \in \Theta.$$

(B-iv) $p(\cdot)$ satisfies

$$\lim_{\|\boldsymbol{u}\|\to\infty} \|\boldsymbol{u}\|p(\boldsymbol{u}) = 0 \quad \text{and} \quad \int \mathrm{uu}'\mathrm{p}(\mathrm{u})\mathrm{du} = \mathrm{I}_m.$$

(B-v) The continuous derivative Dp of $p(\cdot)$ exists on $\mathbb{R}^m$, and

$$\int \|p^{-1}Dp\|^4 p(\boldsymbol{u})d\boldsymbol{u} < \infty \quad \text{and} \quad \int \|\mathrm{u}\|^2\|\mathrm{p}^{-1}\mathrm{Dp}\|^2\mathrm{p}(\mathrm{u})\mathrm{du} < \infty.$$

Suppose that an observed stretch $\boldsymbol{X}^{(n)} = \{\boldsymbol{X}(1),\ldots,\boldsymbol{X}(n)\}$ from (8.5.10) is available. We denote the probability distribution of $\boldsymbol{X}^{(n)}$ by $P_{n,\theta}$. For two hypothetical values $\theta, \theta' \in \Theta$, the log-likelihood ratio based on $\boldsymbol{X}^{(n)}$ is

$$\Lambda_n(\theta,\theta') = \log\frac{dP_{n,\theta'}}{dP_{n,\theta}} = \sum_{t=p}^n \log\frac{p\{\boldsymbol{H}_{\theta'}^{-1}(\boldsymbol{X}(t) - \boldsymbol{F}_{\theta'})\}\det\boldsymbol{H}_\theta}{p\{\boldsymbol{H}_\theta^{-1}(\boldsymbol{X}(t) - \boldsymbol{F}_\theta)\}\det\boldsymbol{H}_{\theta'}}.$$

As an estimator of θ, we use the following maximum likelihood estimator (MLE):

$$\hat{\theta}_{ML} \equiv \arg\max_{\theta} \Lambda_n(\theta_0, \theta)$$

where $\theta_0 \in \Theta$ is some fixed value. Write $\eta_t(\theta) \equiv \log p\{\boldsymbol{H}_\theta^{-1}(\boldsymbol{X}(t) - \boldsymbol{F}_\theta)\} \times \{\det \boldsymbol{H}_\theta\}^{-1}$. We further impose the following assumption.

Assumption 8.5.12. *For $i, j, k = 1, \ldots, q$, there exist functions $Q_{ij}^t = Q_{ij}^t(\boldsymbol{X}(1), \ldots, \boldsymbol{X}(t))$ and $T_{ijk}^t = T_{ijk}^t(\boldsymbol{X}(1), \ldots, \boldsymbol{X}(t))$ such that*

$$|\partial_i \partial_j \eta_t(\theta)| \leq Q_{ij}^t, \quad E[Q_{ij}^t] < \infty$$

and

$$|\partial_i \partial_j \partial_k \eta_t(\theta)| \leq T_{ijk}^t, \quad E[T_{ijk}^t] < \infty.$$

Then Kato et al. (2006) showed the following result.

Lemma 8.5.13. *(Kato et al. (2006)) Under Assumptions 8.5.11 and 8.5.12, $\hat{\theta}_{ML}$ is asymptotically efficient.*

From Lemma 8.5.10 or 8.5.13, the optimal portfolio weight estimator $g(\hat{\theta}_{QML})$ or $g(\hat{\theta}_{ML})$ is also asymptotically efficient by the δ-method.

Locally Stationary Process The definition of a locally stationary process is as follows:

Definition 8.5.14. *A sequence of stochastic processes $\boldsymbol{X}(t, T) = (X_1(t, T), \ldots, X_m(t, T))'(t = 1, \ldots, T; T \in \mathbb{N})$ is called an m-dimensional non-Gaussian locally stationary process with transfer function matrix A° and mean function vector $\boldsymbol{\mu} = \boldsymbol{\mu}(\cdot)$ if there exists a representation*

$$\boldsymbol{X}(t, T) = \boldsymbol{\mu}\left(\frac{t}{T}\right) + \int_{-\pi}^{\pi} \exp(i\lambda t) A^\circ(t, T, \lambda) d\boldsymbol{\xi}(\lambda) \tag{8.5.11}$$

with the following properties:

(i) $\boldsymbol{\xi}(\lambda)$ is a right-continuous complex-valued non-Gaussian vector process on $[-\pi, \pi]$ (see Brockwell and Davis (2006)) with $\overline{\xi_a(\lambda)} = \xi_a(-\lambda)$ and

$$cum\{d\xi_{a_1}(\lambda_1), \ldots, d\xi_{a_k}(\lambda_k)\} = \eta(\sum_{j=1}^{k} \lambda_j) h_k(a_1, \ldots, a_k) d\lambda_1 \ldots d\lambda_k$$

where $\eta(\sum_{j=1}^k \lambda_j) = \sum_{l=-\infty}^{\infty} \delta(\sum_{j=1}^k \lambda_j + 2\pi l)$ is the period 2π extension of the Dirac delta function, $cum\{\ldots\}$ denotes the kth-order cumulant, and $h_1 = 0, h_2(a_1, a_2) = \frac{1}{2\pi} c_{a_1 a_2}, h_k(a_1, \ldots, a_k) = \frac{1}{(2\pi)^{k-1}} c_{a_1 \ldots a_k}$ for all $k > 2$.

(ii) There exist a constant K and a 2π-periodic matrix-valued function $A(u, \lambda) = \{A_{ab}(u, \lambda) : a, b = 1, \ldots, m\} : [0, 1] \times \mathbb{R} \to \mathbb{C}^{m \times m}$ with $\overline{A(u, \lambda)} = A(u, -\lambda)$ and

$$\sup_{t, \lambda} \left| A_{ab}^\circ(t, T, \lambda) - A_{ab}\left(\frac{t}{T}, \lambda\right) \right| \leq KT^{-1}$$

for all $a, b = 1, \ldots, m$ and $T \in \mathbb{N}$. $A(u, \lambda)$ and $\boldsymbol{\mu}(u)$ are assumed to be continuous in u.

A typical example is the following ARMA process with time-varying coefficients:

$$\sum_{j=0}^{p} \alpha(j, u) X(t - j, T) = \sum_{j=0}^{q} \beta(j, u) u(t - j), \quad u = \frac{t}{T},$$

where $u(t) \equiv \int_{-\pi}^{\pi} e^{it\lambda} d\xi(\lambda)$ is a stationary non-Gaussian process.

Let $\boldsymbol{X}(1,T),\ldots,\boldsymbol{X}(T,T)$ be realizations of an m-dimensional locally stationary non-Gaussian process with mean function vector $\boldsymbol{\mu}$ and transfer function matrix A°. We assume that $\boldsymbol{\mu}\left(\frac{t}{T}\right)$ and $A^\circ(t,T,\lambda)$ are parameterized by an unknown vector $\theta = (\theta_1,\ldots,\theta_q) \in \Theta$, i.e., $\boldsymbol{\mu} = \boldsymbol{\mu}_\theta$ and $A^\circ(t,T,\lambda) = A^\circ_\theta(t,T,\lambda)$, where Θ is a compact subset of $\mathbb{R}^q$. The time-varying spectral density is given by $f_\theta(u,\lambda) := A_\theta(u,\lambda)\overline{A_\theta(u,\lambda)'}$. Introducing the notations $\nabla_i = \frac{\partial}{\partial\theta_i}, \nabla = (\nabla_1\ldots,\nabla_q)', \nabla_{ij} = \frac{\partial}{\partial\theta_i}\frac{\partial}{\partial\theta_j}, \nabla^2 = (\nabla_{ij})_{i,j=1,\ldots,q}$, we make the following assumption.

Assumption 8.5.15. *(i) $\boldsymbol{\mu}_\theta$ and A_θ are uniformly bounded from above and below.*

(ii) There exists a constant K with

$$\sup_{t,\lambda}\left|\nabla^s\left\{A^\circ_{\theta,ab}(t,T,\lambda) - A_{\theta,ab}\left(\frac{t}{T},\lambda\right)\right\}\right| \le KT^{-1}$$

for $s = 0,1,2$ and all $a,b = 1,\ldots,m$. The components of $A_\theta(u,\lambda)$, $\nabla A_\theta(u,\lambda)$ and $\nabla^2 A_\theta(u,\lambda)$ are differentiable in u and λ with uniformly continuous derivatives $\frac{\partial}{\partial u}\frac{\partial}{\partial\lambda}$.

Letting

$$\boldsymbol{U}(t) = (u_1(t),\ldots,u_m(t))' \equiv \int_{-\pi}^{\pi}\exp(i\lambda t)d\boldsymbol{\xi}(\lambda),$$

we further assume the following.

Assumption 8.5.16. *(i) $\boldsymbol{U}(t)$'s are i.i.d. m-vector random variables with mean vector $\mathbf{0}$ and variance matrix I_m, and the components u_a have finite fourth-order cumulant $c_{a_1a_2a_3a_4} \equiv cum(u_{a_1}(t),u_{a_2}(t),u_{a_3}(t),u_{a_4}(t)), a_1,\ldots,a_4 = 1,\ldots,m$. Furthermore $\boldsymbol{U}(t)$ has the probability density function $p(\boldsymbol{u}) > 0$.*

(ii) $p(\cdot)$ satisfies

$$\lim_{||\boldsymbol{u}||\to\infty} p(\boldsymbol{u}) = 0, \quad \text{and} \quad \lim_{||\mathrm{u}||\to\infty} \mathrm{up(u)} = 0.$$

(iii) The continuous derivatives Dp and $D^2p \equiv D(Dp)'$ of $p(\cdot)$ exist on $\mathbb{R}^m$. Furthermore each component of D^2p satisfies the Lipschitz condition.

(iv) All the components of

$$\mathcal{F}(p) = \int \phi(\boldsymbol{u})\phi(\boldsymbol{u})'p(\boldsymbol{u})d\boldsymbol{u},$$

$E\{\boldsymbol{U}(t)\boldsymbol{U}(t)'\phi(\boldsymbol{U}(t))\phi(\boldsymbol{U}(t))'\}$, $E\{\phi(\boldsymbol{U}(t))\phi(\boldsymbol{U}(t))'\boldsymbol{U}(t)\}$ and $E\{\phi(\boldsymbol{u})\phi(\boldsymbol{u})'\phi(\boldsymbol{u})\phi(\boldsymbol{u})'\}$ are bounded, and

$$\int D^2p(\boldsymbol{u})d\boldsymbol{u} = \mathbf{0}, \quad \lim_{||\boldsymbol{u}||\to\infty} ||\boldsymbol{u}||^2 Dp(\boldsymbol{u}) = \mathbf{0},$$

where $\phi(\boldsymbol{u}) = p^{-1}Dp(\boldsymbol{u})$.

(v) $\boldsymbol{X}(t,T)$ has the MA(∞) and AR(∞) representations

$$\boldsymbol{X}(t,T) = \boldsymbol{\mu}_\theta\left(\frac{t}{T}\right) + \sum_{j=0}^{\infty} a^\circ_\theta(t,T,j)\boldsymbol{U}(t-j),$$

$$a_\theta^\circ(t,T,j)\boldsymbol{U}(t) = \sum_{k=0}^{\infty} b_\theta^\circ(t,T,k)\left\{\boldsymbol{X}(t-k,T) - \boldsymbol{\mu}_\theta\left(\frac{t-k}{T}\right)\right\},$$

where $a_\theta^\circ(t,T,j), b_\theta^\circ(t,T,k)$ are $m \times m$ matrices with $b_\theta^\circ(t,T,0) \equiv I_m$ and $a_\theta^\circ(t,T,j) = a_\theta^\circ(0,T,j) = a_\theta^\circ(j)$ for $t \leq 0$.

(vi) Every $a_\theta^\circ(t,T,j)$ is continuously three times differentiable with respect to θ, and the derivatives satisfy

$$\sup_{t,T}\left\{\sum_{j=0}^{\infty}(1+j)|\nabla_{i_1}\cdots\nabla_{i_s}a_\theta^\circ(t,T,j)|\right\} < \infty \quad \text{for } s = 0,1,2,3.$$

(vii) Every $b_\theta^\circ(t,T,k)$ is continuously three times differentiable with respect to θ, and the derivatives satisfy

$$\sup_{t,T}\left\{\sum_{k=0}^{\infty}(1+k)|\nabla_{i_1}\cdots\nabla_{i_s}b_\theta^\circ(t,T,k)|\right\} < \infty \quad \text{for } s = 0,1,2,3. \tag{8.5.12}$$

(viii)

$$a_\theta^\circ(t,T,0)^{-1}\nabla a_\theta^\circ(t,T,0) = \frac{1}{4\pi}\int_{-\pi}^{\pi} f_\theta^\circ(t,T\lambda)^{-1}\nabla f_\theta^\circ(t,T,\lambda)d\lambda,$$

where $f_\theta^\circ(t,T,\lambda) = A_\theta^\circ(t,T,\lambda)\overline{A_\theta^\circ(t,T,\lambda)}'$.

(ix) The components of $\boldsymbol{\mu}_\theta(u)$, $\nabla\boldsymbol{\mu}_\theta(u)$ and $\nabla^2\boldsymbol{\mu}_\theta(u)$ are differentiable in u with uniformly continuous derivatives.

By Assumption 8.5.16.(i) we have

$$a_\theta^\circ(t,T,0)\boldsymbol{U}(t) = \sum_{k=0}^{t-1} b_\theta^\circ(t,T,k)\left\{\boldsymbol{X}(t-k,T) - \boldsymbol{\mu}_\theta\left(\frac{t-k}{T}\right)\right\} + \sum_{r=0}^{\infty} c_\theta^\circ(t,T,r)\boldsymbol{U}(-r), \tag{8.5.13}$$

where

$$c_\theta^\circ(t,T,r) = \sum_{s=0}^{r} b_\theta^\circ(t,T,t+s)a_\theta^\circ(r-s).$$

From Assumption 8.5.16. (v)–(ix), we can see that

$$\sum_{r=0}^{\infty}|c_\theta^\circ(t,T,r)| = O(t^{-1}). \tag{8.5.14}$$

For the special case of $m = 1$ and $\boldsymbol{\mu}_\theta = \boldsymbol{0}$, Hirukawa and Taniguchi (2006) showed the local asymptotic normal (LAN) property for $\{\boldsymbol{X}(t,T)\}$. In what follows we show the LAN property for the vector case with $\boldsymbol{\mu}_\theta \neq \boldsymbol{0}$. This extension is not straightforward, and is important for applications.
Let $\mathcal{P}_{\theta,T}$ and $\mathcal{P}_U$ be the probability distributions of $\{\boldsymbol{U}(s), s \leq 0, \boldsymbol{X}(1,T),\ldots,\boldsymbol{X}(T,T)\}$ and $\{\boldsymbol{U}(s), s \leq 0\}$, respectively. It is easy to see that a linear transformation L_θ exists (whose Jacobian is $\prod_{t=1}^{T}|a_\theta^\circ(t,T,0)|^{-1}$), which maps $\{\boldsymbol{U}(s), s \leq 0, \boldsymbol{X}(1,T),\ldots,\boldsymbol{X}(T,T)\}$ into $\{\boldsymbol{U}(s), s \leq 0\}$. Then recalling (8.5.14), we obtain

$$d\mathcal{P}_{\theta,T} = \prod_{t=1}^{T}|a_\theta^\circ(t,T,0)|^{-1}p\Bigg(a_\theta^\circ(t,T,0)^{-1}\Bigg[\sum_{k=0}^{t-1} b_\theta^\circ(t,T,k)\Big\{\boldsymbol{X}(t-k,T)$$

$$- \mu_\theta\left(\frac{t-k}{T}\right)\Big\} + \sum_{r=0}^{\infty} c_\theta^\circ(t,T,r)\boldsymbol{U}(-r)\Big]\Big) d\mathcal{P}_{\boldsymbol{U}}.$$

Let $a_\theta^\circ(t,T,0)\boldsymbol{U}(t) \equiv z_{\theta,t,T}$. Then $z_{\theta,t,T}$ has the probability density function

$$g_{\theta,t,T}(\cdot) = |a_\theta^\circ(t,T,0)|^{-1} p\left\{a_\theta^\circ(t,T,0)^{-1}(\cdot)\right\}.$$

Denote by $H(p;\theta)$ the hypothesis under which the underlying parameter is $\theta \in \Theta$ and the probability density of $\boldsymbol{U}(t)$ is $p = p(\cdot)$. We define

$$\theta_T = \theta + \frac{1}{\sqrt{T}} h, \quad h = (h_1, \dots, h_q)' \in \mathcal{H} \subset \mathbb{R}^q.$$

For two hypothetical values $\theta, \theta_T \in \Theta$, the log-likelihood ratio is

$$\Lambda_T(\theta,\theta_T) \equiv \log \frac{d\mathcal{P}_{\theta_T,T}}{d\mathcal{P}_{\theta,T}} = 2\sum_{t=1}^{T} \log \Phi_{t,T}(\theta,\theta_T),$$

where

$$\begin{aligned}\Phi_{t,T}^2(\theta,\theta_T) &= \frac{g_{\theta_T,t,T}\left\{\sum_{k=0}^{t-1} b_{\theta_T}^\circ(t,T,k)\boldsymbol{X}(t-k,T) + \sum_{r=0}^{\infty} c_{\theta_T}^\circ(t,T,r)\boldsymbol{U}(-r)\right\}}{g_{\theta,t,T}\left\{\sum_{k=0}^{t-1} b_\theta^\circ(t,T,k)\boldsymbol{X}(t-k,T) + \sum_{r=0}^{\infty} c_\theta^\circ(t,T,r)\boldsymbol{U}(-r)\right\}} \\ &= \frac{g_{\theta_T,t,T}\left(z_{\theta,t,T} + q_{t,T}^{(1)} + q_{t,T}^{(2)}\right)}{g_{\theta,t,T}\left(z_{\theta,t,T}\right)}\end{aligned}$$

with

$$\begin{aligned}q_{t,T}^{(1)} &= \sum_{k=1}^{t-1}\left\{\frac{1}{\sqrt{T}}\nabla_h b_\theta^\circ(t,T,k) + \frac{1}{2T}\nabla_h^2 b_{\theta^*}^\circ(t,T,k)\right\}\left\{\boldsymbol{X}(t-k,T) - \mu_\theta\left(\frac{t-k}{T}\right)\right\} \\ &\quad + \sum_{r=0}^{\infty}\frac{1}{\sqrt{T}}\nabla_h c_{\theta^{**}}^\circ(t,T,r)\boldsymbol{U}(-r) \qquad (8.5.15)\end{aligned}$$

$$q_{t,T}^{(2)} = -\sum_{k=0}^{t-1} b_{\theta_T}^\circ(t,T,k)\left\{\mu_{\theta_T}\left(\frac{t-k}{T}\right) - \mu_\theta\left(\frac{t-k}{T}\right)\right\}. \qquad (8.5.16)$$

Here the operators ∇_h and ∇_h^2 imply

$$\nabla_h a_\theta^\circ(t,T,0) = \sum_{l=1}^{q} h_l \nabla_l a_\theta^\circ(t,T,0), \quad \nabla_h^2 b_\theta^\circ(t,T,k) = \sum_{l_1=1}^{q}\sum_{l_2=1}^{q} h_{l_1}h_{l_2}\nabla_{l_1,l_2} b_\theta^\circ(t,T,k),$$

respectively, and θ^* and θ^{**} are points on the segment between θ and $\theta + h/\sqrt{T}$ and $q_{t,T}^{(1)}$ corresponds to $q_{t,T}$ in Hirukawa and Taniguchi (2006).

Now we can state the local asymptotic normality (LAN) for our locally stationary processes.

Theorem 8.5.17. *[LAN theorem] Suppose that Assumptions 8.5.15 and 8.5.16 hold. Then the sequence of experiments*

$$\mathcal{E}_T = \{\mathbb{R}^{\mathbb{Z}}, \mathcal{B}^{\mathbb{Z}}, \{\mathcal{P}_{\theta,T} : \theta \in \Theta \subset \mathbb{R}^q\}\}, \quad T \in \mathbb{N},$$

where $\mathcal{B}^{\mathbb{Z}}$ denotes the Borel σ-field on $\mathbb{R}^{\mathbb{Z}}$, is locally asymptotically normal and equicontinuous on a compact subset C of $\mathcal{H}$. That is

(i) *For all $\theta \in \Theta$, the log-likelihood ratio $\Lambda_T(\theta, \theta_T)$ admits, under $H(p;\theta)$, as $T \to \infty$, the asymptotic representation*

$$\Lambda_T(\theta, \theta_T) = \Delta_T(h;\theta) - \frac{1}{2}\Gamma_h(\theta) + o_p(1),$$

where

$$\begin{aligned}\Delta_T(h;\theta) = & \frac{1}{\sqrt{T}}\sum_{t=1}^{T}\phi(\boldsymbol{U}(t))'\sum_{k=1}^{t-1}a_\theta^\circ(t,T,0)^{-1}\nabla_h b_\theta^\circ(t,T,k)\left\{\boldsymbol{X}(t-k,T)-\boldsymbol{\mu}_\theta\left(\frac{t-k}{T}\right)\right\}\\ & -\frac{1}{\sqrt{T}}\sum_{t=1}^{T} tr\left[a_\theta^\circ(t,T,0)^{-1}\nabla_h a_\theta^\circ(t,T,0)\{I_m+\boldsymbol{U}(t)\phi(\boldsymbol{U}(t))'\}\right]\\ & -\frac{1}{\sqrt{T}}\sum_{t=1}^{T}\phi(\boldsymbol{U}(t))'\sum_{k=0}^{t-1}a_\theta^\circ(t,T,0)^{-1}b_\theta^\circ(t,T,k)\nabla_h\boldsymbol{\mu}_\theta\left(\frac{t-k}{T}\right)\end{aligned}$$

and

$$\begin{aligned}\Gamma_h(\theta) = & \frac{1}{4\pi}\int_0^1\int_{-\pi}^{\pi} tr\Big[\mathcal{F}(p)f_\theta(u,\lambda)^{-1}\nabla_h f_\theta(u,\lambda)\{f_\theta(u,\lambda)^{-1}\nabla_h f_\theta(u,\lambda)\}'\Big]d\lambda du\\ & +\frac{1}{16\pi^2}\int_0^1 E\Big[\phi(\boldsymbol{U}(0))'\Big\{\int_{-\pi}^{\pi} f_\theta(u,\lambda)^{-1}\nabla_h f_\theta(u,\lambda)d\lambda\Big\}\\ & \times\{\boldsymbol{U}(0)\boldsymbol{U}(0)'-2I_m\}\Big\{\int_{-\pi}^{\pi} f_\theta(u,\lambda)^{-1}\nabla_h f_\theta(u,\lambda)\,d\lambda\Big\}\phi(\boldsymbol{U}(0))\Big]du\\ & -\frac{1}{16\pi^2}\int_0^1 tr\Big\{\int_{-\pi}^{\pi} f_\theta(u,\lambda)^{-1}\nabla_h f_\theta(u,\lambda)d\lambda\Big\}^2 du\\ & +\int_0^1\nabla_h\boldsymbol{\mu}_\theta'(u)\{A_\theta(u,0)^{-1}\}'\mathcal{F}(p)A_\theta(u,0)^{-1}\nabla_h\boldsymbol{\mu}_\theta(u)du\\ & +\frac{1}{2\pi}\int_0^1\int_{-\pi}^{\pi}E\Big\{\phi(\boldsymbol{U}(0))'f_\theta(u,\lambda)^{-1}\nabla_h f_\theta(u,\lambda)\boldsymbol{U}(0)\phi(\boldsymbol{U}(0))'\\ & \times A_\theta(u,0)^{-1}\nabla_h\boldsymbol{\mu}_\theta(u)\Big\}d\lambda du.\end{aligned}$$

(ii) *Under $H(p;\theta)$,*

$$\Delta_T(h;\theta)\xrightarrow{\mathcal{L}} N(\boldsymbol{0},\Gamma_h(\theta)).$$

(iii) *For all $T \in \mathbb{N}$ and all $h \in \mathcal{H}$, the mapping $h \to \mathcal{P}_{\theta_T,T}$ is continuous with respect to the variational distance*

$$||\mathcal{P}-\mathcal{Q}|| = sup\{|\mathcal{P}(A)-\mathcal{Q}(A)| : A \in \mathcal{B}^{\mathbb{Z}}\}.$$

Proof of Theorem 8.5.17. See Appendix.□

Note that the third term of $\Delta_T(h;\theta)$ and the fourth and fifth terms of $\Gamma_h(\theta)$ vanish if $\boldsymbol{\mu}_\theta \equiv 0$. Next we estimate the true value θ_0 of unknown parameter θ of the model by the quasi-likelihood estimator

$$\hat{\theta}_{QML} := \arg\min_{\theta\in\Theta}\mathcal{L}_T(\theta),$$

where

$$\mathcal{L}_T(\theta) = \sum_{t=1}^{T}\left(\log p\left[a_\theta^\circ(t,T,0)^{-1}\sum_{k=0}^{t-1}b_\theta^\circ(t,T,k)\left\{\boldsymbol{X}(t-k,T)-\boldsymbol{\mu}\left(\frac{t-k}{T}\right)\right\}\right]-\log|a_\theta^\circ(t,T,0)|\right).$$

Theorem 8.5.18. *Under Assumptions 8.5.15 and 8.5.16,*

$$\sqrt{T}\left(\hat{\theta}_{QML} - \theta\right) \xrightarrow{\mathcal{L}} N\left(\mathbf{0}, V^{-1}\right)$$

where

$$\begin{aligned}
V_{ij} = &\frac{1}{4\pi}\int_0^1 \int_{-\pi}^{\pi} tr\left[f_\theta(u,\lambda)^{-1}\nabla_i f_\theta(u,\lambda)\mathcal{F}(p)\{f_\theta(u,\lambda)^{-1}\nabla_j f_\theta(u,\lambda)\}'\right] d\lambda du \\
&+ \frac{1}{16\pi^2}\int_0^1 E\Big[\phi(\boldsymbol{U}(0))'\Big\{\int_{-\pi}^{\pi} f_\theta(u,\lambda)^{-1}\nabla_i f_\theta(u,\lambda) d\lambda\Big\}\{\boldsymbol{U}(0)\boldsymbol{U}(0)' - 2I_m\} \\
&\times \Big\{\int_{-\pi}^{\pi} f_\theta(u,\lambda)^{-1}\nabla_j f_\theta(u,\lambda)\, d\lambda\Big\}' \phi(\boldsymbol{U}(0))\Big] du \\
&- \frac{1}{16\pi^2}\int_0^1 tr\Big\{\int_{-\pi}^{\pi} f_\theta(u,\lambda)^{-1}\nabla_i f_\theta(u,\lambda) d\lambda\Big\} tr\Big\{\int_{-\pi}^{\pi} f_\theta(u,\lambda)^{-1}\nabla_j f_\theta(u,\lambda) d\lambda\Big\} du \\
&+ \int_0^1 \nabla_i \boldsymbol{\mu}_\theta(u)'\{A_\theta(u,0)^{-1}\}'\mathcal{F}(p)A_\theta(u,0)^{-1}\nabla_j \boldsymbol{\mu}_\theta(u) du \\
&+ \frac{1}{2\pi}\int_0^1 \int_{-\pi}^{\pi} E\Big\{\phi(\boldsymbol{U}(0))' f_\theta(u,\lambda)^{-1}\nabla_i f_\theta(u,\lambda)\boldsymbol{U}(0)\phi(\boldsymbol{U}(0))' A_\theta(u,0)^{-1}\nabla_j \boldsymbol{\mu}_\theta(u)\Big\} d\lambda du.
\end{aligned}$$

Proof of Theorem 8.5.18. See Appendix.□

Write the optimal portfolio weight function $g(\theta)$ as

$$g(\theta) \equiv g\left[\boldsymbol{\mu}_\theta\left(\frac{T}{T}\right)', vech\left\{\Sigma_\theta\left(\frac{T}{T}\right)\right\}'\right]$$

where

$$\begin{aligned}
\Sigma_\theta\left(\frac{T}{T}\right) &= E\left[\left\{\boldsymbol{X}(T,T) - \boldsymbol{\mu}_\theta\left(\frac{T}{T}\right)\right\}\left\{\boldsymbol{X}(T,T) - \boldsymbol{\mu}_\theta\left(\frac{T}{T}\right)\right\}'\right] \\
&= \sum_{j=0}^{\infty} a_\theta^\circ(T,T,j) a_\theta^\circ(T,T,j)'.
\end{aligned}$$

Then we have the following theorem.

Theorem 8.5.19. *Under Assumptions 8.5.15, 8.5.16 and 3.2.4,*

$$\sqrt{T}\left\{g(\hat{\theta}_{QML}) - g(\theta)\right\} \xrightarrow{\mathcal{L}} N\left(\mathbf{0}, \left(\frac{\partial g}{\partial \theta'}\right) V^{-1} \left(\frac{\partial g}{\partial \theta'}\right)'\right).$$

Proof of Theorem 8.5.19. The proof follows from Theorem 8.5.18 and the δ-method.□

From this theorem and general results by Hirukawa and Taniguchi (2006) on the asymptotic efficiency of $\hat{\theta}_{QML}$, we can see that our estimator $g(\hat{\theta}_{QML})$ for optimal portfolio weight is $\sqrt{T}$-consistent and asymptotically efficient. Hence, we get a parametric optimal estimator for portfolio coefficients.

8.5.3 Appendix

Proof of Theorem 8.5.17. To prove this theorem we check sufficient conditions (S1)–(S6) (below) for the LAN given by Swensen (1985). Let $\mathcal{F}_t$ be the σ-algebra generated by $\{\boldsymbol{U}(t), s \le 0, X(1,T), \ldots, X(t,T)\}, t \le T$. Since $X(t,T)$ fulfills difference equations (8.5.13) we obtain

$$a_\theta^\circ(t,T,0) = \sum_{k=0}^{t-1} b_\theta^\circ(t,T,k)A_\theta^\circ(t-k,T,\lambda)\exp(-i\lambda k) + \sum_{r=0}^{\infty} c_\theta^\circ(t,T,r)\exp\{-i\lambda(r+t)\},\ a.e. \quad (8.5.17)$$

Hence we get

$$\sum_{k=1}^{t-1} \nabla b_\theta^\circ(t,T,k) A_\theta^\circ(t-k,T,\lambda)\exp(-i\lambda k)$$
$$= \nabla a_\theta^\circ(t,T,0) - \sum_{k=0}^{t-1} b_\theta^\circ(t,T,k)\nabla A_\theta^\circ(t-k,T,\lambda)\exp(-i\lambda k)$$
$$- \sum_{r=0}^{\infty} \nabla c_\theta^\circ(t,T,r)\exp\{-i\lambda(r+t)\},\ a.e.$$

We set down

$$\begin{aligned} Y_{t,T} &= \Phi_{t,T}(\theta,\theta_T) - 1 \\ &= \frac{g^{\frac{1}{2}}_{\theta_T,t,T}(z_{\theta,t,T} + q^{(1)}_{t,T}) - g^{\frac{1}{2}}_{\theta,t,T}(z_{\theta,t,T})}{g^{\frac{1}{2}}_{\theta,t,T}(z_{\theta,t,T})} \\ &\quad + \frac{g^{\frac{1}{2}}_{\theta_T,t,T}(z_{\theta,t,T} + q^{(1)}_{t,T} + q^{(2)}_{t,T}) - g^{\frac{1}{2}}_{\theta_T,t,T}(z_{\theta,t,T} + q^{(1)}_{t,T})}{g^{\frac{1}{2}}_{\theta,t,T}(z_{\theta,t,T})} \\ &= Y^{(1)}_{t,T} + Y^{(2)}_{t,T} \quad (say), \end{aligned} \tag{8.5.18}$$

and

$$\begin{aligned} W_{t,T} = &\frac{1}{2\sqrt{T}}\Bigg(\phi(\boldsymbol{U}(t))' a_\theta^\circ(t,T,0)^{-1} \sum_{k=1}^{t-1} \nabla_h b_{\theta_T}^\circ(t,T,k)\left\{\boldsymbol{X}(t-k,T) - \boldsymbol{\mu}_\theta\left(\frac{t-k}{T}\right)\right\} \\ &- tr\left[a_\theta^\circ(t,T,0)^{-1}\nabla_h a_\theta^\circ(t,T,0)\left\{I_m + \boldsymbol{U}(t)\phi(\boldsymbol{U}(t))'\right\}\right]\Bigg) \\ &- \frac{1}{2\sqrt{T}}\phi(\boldsymbol{U}(t))' a_\theta^\circ(t,T,0)^{-1} \sum_{k=0}^{t-1} b_{\theta_T}^\circ(t,T,k)\nabla_h \boldsymbol{\mu}_\theta\left(\frac{t-k}{T}\right) \\ = &W^{(1)}_{t,T} + W^{(2)}_{t,T} \quad (say). \end{aligned} \tag{8.5.19}$$

(S-1). $E(W_{t,T}|\mathcal{F}_{t-1}) = 0\ a.e.\ (t = 1,\ldots,T)$.

Since $E(W^{(1)}_{t,T}|\mathcal{F}_{t-1}) = 0$, we have

$$\begin{aligned} E(W_{t,T}|\mathcal{F}_{t-1}) &= E(W^{(1)}_{t,T} + W^{(2)}_{t,T}|\mathcal{F}_{t-1}) \\ &= -\frac{1}{2\sqrt{T}}E\left\{\phi(\boldsymbol{U}(t))|\mathcal{F}_{t-1}\right\}' a_\theta^\circ(t,T,0)^{-1}\sum_{k=0}^{t-1} b_{\theta_T}^\circ(t,T,k)\nabla_h\boldsymbol{\mu}_\theta\left(\frac{t-k}{T}\right) \\ &= 0 \quad (by\ Assumption\ 8.5.16), \end{aligned}$$

hence (S-1) holds.

(S-2). $E\left\{\sum_{t=1}^{T}(Y_{t,T} - W_{t,T})^2\right\} \to 0.$

First, recalling (8.5.18) and (8.5.19) we have

$$E\left\{(Y_{t,T} - W_{t,T})^2\right\} = E\left\{\left(Y^{(1)}_{t,T} - W^{(1)}_{t,T} + Y^{(2)}_{t,T} - W^{(2)}_{t,T}\right)^2\right\}$$

$$\leq 2E\left\{\left(Y_{t,T}^{(1)} - W_{t,T}^{(1)}\right)^2\right\} + 2E\left\{\left(Y_{t,T}^{(2)} - W_{t,T}^{(2)}\right)^2\right\},$$

and from Hirukawa and Taniguchi (2006)

$$E\left\{\sum_{t=1}^{T}\left(Y_{t,T}^{(1)} - W_{t,T}^{(1)}\right)^2\right\} \to 0. \tag{8.5.20}$$

Therefore if we can show

$$E\left\{\sum_{t=1}^{T}\left(Y_{t,T}^{(2)} - W_{t,T}^{(2)}\right)^2\right\} \to 0,$$

then it completes the proof of (S-2). From Taylor expansion we can see that

$$\begin{aligned}
Y_{t,T}^{(2)} = \frac{1}{g_{\theta,t,T}^{\frac{1}{2}}(z_{\theta,t,T})}&\left(\left[g_{\theta_T,t,T}^{\frac{1}{2}}(z_{\theta,t,T}) + Dg_{\theta_T,t,T}^{\frac{1}{2}}(z_{\theta,t,T})'(q_{t,T}^{(1)} + q_{t,T}^{(2)})\right.\right.\\
&+ \frac{1}{2}(q_{t,T}^{(1)} + q_{t,T}^{(2)})'D^2 g_{\theta_T,t,T}^{\frac{1}{2}}\left\{z_{\theta,t,T} + \alpha_1(q_{t,T}^{(1)} + q_{t,T}^{(2)})\right\}(q_{t,T}^{(1)} + q_{t,T}^{(2)})\Big]\\
&\left.- \left\{g_{\theta_T,t,T}^{\frac{1}{2}}(z_{\theta,t,T}) + Dg_{\theta_T,t,T}^{\frac{1}{2}}(z_{\theta,t,T})'q_{t,T}^{(1)} + \frac{1}{2}(q_{t,T}^{(1)})'D^2 g_{\theta_T,t,T}^{\frac{1}{2}}(z_{\theta,t,T} + \alpha_2 q_{t,T}^{(1)})q_{t,T}^{(1)}\right\}\right)\\
= \frac{Dg_{\theta_T,t,T}^{\frac{1}{2}}(z_{\theta,t,T})'q_{t,T}^{(2)}}{g_{\theta,t,T}^{\frac{1}{2}}(z_{\theta,t,T})}& + \frac{1}{2}R_1\{\boldsymbol{U}(t)\}\ (say)
\end{aligned}$$

where $0 < \alpha_1, \alpha_2 < 1$. Since

$$\begin{aligned}
Dg_{\theta_T,t,T}^{\frac{1}{2}}(z_{\theta,t,T}) &= Dg_{\theta,t,T}^{\frac{1}{2}}(z_{\theta,t,T}) + \frac{1}{\sqrt{T}}\tilde{\nabla}Dg_{\tilde{\theta},t,T}^{\frac{1}{2}}(z_{\theta,t,T})|_{\tilde{\theta}=\tilde{\theta}^*}' h\\
Dg_{\theta,t,T}^{\frac{1}{2}}(z_{\theta,t,T}) &= \frac{a_\theta^\circ(t,T,0)^{-1}Dp\left\{a_\theta^\circ(t,T,0)^{-1}z_{\theta,t,T}\right\}}{2|a_\theta^\circ(t,T,0)|g_{\theta_T,t,T}^{\frac{1}{2}}(z_{\theta,t,T})},
\end{aligned}$$

we can see that

$$\begin{aligned}
\frac{Dg_{\theta_T,t,T}^{\frac{1}{2}}(z_{\theta,t,T})'q_{t,T}^{(2)}}{g_{\theta,t,T}^{\frac{1}{2}}(z_{\theta,t,T})} &= \frac{Dg_{\theta,t,T}^{\frac{1}{2}}(z_{\theta,t,T})'q_{t,T}^{(2)}}{g_{\theta,t,T}^{\frac{1}{2}}(z_{\theta,t,T})} + \frac{h'\tilde{\nabla}Dg_{\tilde{\theta},t,T}^{\frac{1}{2}}(z_{\theta,t,T})|_{\tilde{\theta}=\tilde{\theta}^*}q_{t,T}^{(2)}}{\sqrt{T}g_{\theta,t,T}^{\frac{1}{2}}(z_{\theta,t,T})}\\
&= \frac{Dg_{\theta,t,T}^{\frac{1}{2}}(z_{\theta,t,T})'q_{t,T}^{(2)}}{g_{\theta,t,T}^{\frac{1}{2}}(z_{\theta,t,T})} + R_2\{\boldsymbol{U}(t)\}\ (say)\\
&= W_{t,T}^{(2)} + R_2\{\boldsymbol{U}(t)\} + O_p\left(\frac{1}{T}\right)
\end{aligned}$$

where $\tilde{\nabla} = \frac{\partial}{\partial\tilde{\theta}'}$ and $\tilde{\theta}^*$ is a point on the segment between θ and $\theta + h/\sqrt{T}$. Hence,

$$\begin{aligned}
E\left\{\sum_{t=1}^{T}\left(Y_{t,T}^{(2)} - W_{t,T}^{(2)}\right)^2\right\} &= E\left(\sum_{t=1}^{T}\left[\frac{1}{2}R_1\{\boldsymbol{U}(t)\} + R_2\{\boldsymbol{U}(t)\} + O_p\left(\frac{1}{T}\right)\right]^2\right)\\
&\leq E\left(\sum_{t=1}^{T}\left[R_1\{\boldsymbol{U}(t)\}^2 + 4R_2\{\boldsymbol{U}(t)\}^2 + O_p\left(\frac{1}{T^2}\right)\right]\right)
\end{aligned}$$

$$= \sum_{t=1}^{T} E\left[R_1\{\boldsymbol{U}(t)\}^2\right] + 4\sum_{t=1}^{T} E\left[R_2\{\boldsymbol{U}(t)\}^2\right] + O\left(\frac{1}{T}\right).$$

From (8.5.12), (8.5.14) and (8.5.15),

$$E\left(q_{t,T}^{(1)}\right)^2 = O\left(\frac{1}{T}\right),$$

and from Assumption 8.5.16 (ix), (8.5.12) and (8.5.16),

$$E\left(q_{t,T}^{(2)}\right)^2 = O\left(\frac{1}{T}\right).$$

Hence,

$$E\left\{\sum_{t=1}^{T}\left(Y_{t,T}^{(2)} - W_{t,T}^{(2)}\right)^2\right\} = o(1).$$

This, together with (8.5.20), completes the proof of (S-2).

(S-3). $\sup_T E\left(\sum_{t=1}^{T} W_{t,T}^2\right) < \infty.$

We have

$$\begin{aligned} E(W_{t,T})^2 &= E\left\{\left(W_{t,T}^{(1)} + W_{t,T}^{(2)}\right)^2\right\} \\ &= E\left\{\left(W_{t,T}^{(1)}\right)^2\right\} + E\left\{\left(W_{t,T}^{(2)}\right)^2\right\} + 2E\left(W_{t,T}^{(1)}W_{t,T}^{(2)}\right) \\ &= J_{t,T}^{(1)} + J_{t,T}^{(2)} + J_{t,T}^{(3)} \quad (say). \end{aligned}$$

From Hirukawa and Taniguchi (2006), it follows that

$$\begin{aligned} \sum_{t=1}^{T} J_{t,T}^{(1)} = &\frac{1}{16\pi}\int_0^1\int_{-\pi}^{\pi} tr\left\{\mathcal{F}(p)f_\theta(u,\lambda)^{-1}\nabla_h f_\theta(u,\lambda)\{f_\theta(u,\lambda)^{-1}\nabla_h f_\theta(u,\lambda)\}'\right\}d\lambda du \\ &+ \frac{1}{64\pi}\int_0^1 E\Big[\phi(\boldsymbol{U}(0))'\Big\{\int_{-\pi}^{\pi} f_\theta(u,\lambda)^{-1}\nabla_h f_\theta(u,\lambda)d\lambda\Big\}\{\boldsymbol{U}(0)\boldsymbol{U}(0)' - 2I_m\} \\ &\times\Big\{\int_{-\pi}^{\pi} f_\theta(u,\lambda)^{-1}\nabla_h f_\theta(u,\lambda)\, d\lambda\Big\}\phi(\boldsymbol{U}(0))\Big]du \\ &- \frac{1}{64\pi}\int_0^1 tr\Big\{\int_{-\pi}^{\pi} f_\theta(u,\lambda)^{-1}\nabla_h f_\theta(u,\lambda)d\lambda\Big\}^2 du + o(1). \end{aligned}$$

From (8.5.19),

$$\begin{aligned} J_{t,T}^{(2)} &= E\left(W_{t,T}^{(2)}\right)^2 \\ &= E\left\{-\frac{1}{2\sqrt{T}}\phi(\boldsymbol{U}(t))'a_\theta^\circ(t,T,0)^{-1}\sum_{k=0}^{t-1} b_\theta^\circ(t,T,k)\nabla_h\boldsymbol{\mu}_\theta\left(\frac{t-k}{T}\right)\right\}^2. \end{aligned}$$

From Taylor expansion and Assumptions 8.5.15 (i) and 8.5.16 (ix),

$$\nabla_h\boldsymbol{\mu}_\theta\left(\frac{t-k}{T}\right) = \nabla_h\boldsymbol{\mu}_\theta\left(\frac{t}{T}\right) + O\left(\frac{t}{T}\right)$$

$$A_\theta\left(\frac{t-k}{T},\lambda\right) = A_\theta\left(\frac{t}{T},\lambda\right) + O\left(\frac{t}{T}\right) \tag{8.5.21}$$

for $k = 0, 1, \ldots, t-1$. Hence from Assumption 8.5.15 (i), (8.5.14), (8.5.17) and (8.5.21), we obtain

$$a_\theta^\circ(t,T,0) = \sum_{k=0}^{t-1} b_\theta^\circ(t,T,k) A_\theta\left(\frac{t}{T},0\right) + O\left(\frac{t}{T}\right)$$

and

$$a_\theta^\circ(t,T,0)^{-1} \sum_{k=0}^{t-1} b_\theta^\circ(t,T,k) \nabla_h \boldsymbol{\mu}_\theta\left(\frac{t-k}{T}\right) = A_\theta\left(\frac{t}{T},0\right)^{-1} \nabla_h \boldsymbol{\mu}_\theta\left(\frac{t}{T}\right) + O\left(\frac{t}{T}\right).$$

Therefore, we have

$$\begin{aligned} J_{t,T}^{(2)} &= E\left[\left\{-\frac{1}{2\sqrt{T}}\phi(\boldsymbol{U}(t))' A_\theta\left(\frac{t}{T},0\right)^{-1} \nabla_h \boldsymbol{\mu}_\theta\left(\frac{t}{T}\right) + O\left(\frac{t}{T}\right)\right\}^2\right] \\ &= \frac{1}{4T}\left[\nabla_h \boldsymbol{\mu}_\theta'\left(\frac{t}{T}\right)\left\{A_\theta\left(\frac{t}{T},0\right)^{-1}\right\}' \mathcal{F}(p) A_\theta\left(\frac{t}{T},0\right)^{-1} \nabla_h \boldsymbol{\mu}_\theta\left(\frac{t}{T}\right)\right] + O\left(\frac{t^2}{T^3}\right) \end{aligned}$$

and

$$\sum_{t=1}^{T} J_{t,T}^{(2)} = \frac{1}{4}\left\{\int_0^1 \nabla_h \boldsymbol{\mu}_\theta'(u)\{A_\theta(u,0)^{-1}\}' \mathcal{F}(p) A_\theta(u,0)^{-1} \nabla_h \boldsymbol{\mu}_\theta(u) du\right\} + o(1).$$

Next it is seen that

$$\begin{aligned} J_{t,T}^{(3)} &= 2E\left(W_{t,T}^{(1)} W_{t,T}^{(2)}\right) \\ &= -\frac{1}{8\pi T}\int_{-\pi}^{\pi}\left\{\phi(\boldsymbol{U}(t))' f_\theta\left(\frac{t}{T},\lambda\right)^{-1} \nabla_h f_\theta\left(\frac{t}{T},\lambda\right) \boldsymbol{U}(t)\phi(\boldsymbol{U}(t))' A_\theta\left(\frac{t}{T},0\right)^{-1}\right. \\ &\quad \left. \times \nabla_h \boldsymbol{\mu}_\theta\left(\frac{t}{T}\right)\right\} d\lambda + O\left(\frac{t}{T^2}\right) \end{aligned}$$

which implies

$$\begin{aligned} \sum_{t=1}^{T} J_{t,T}^{(3)} = -\frac{1}{8\pi}\int_0^1 \int_{-\pi}^{\pi} & \left\{\phi(\boldsymbol{U}(t))' f_\theta(u,\lambda)^{-1} \nabla_h f_\theta(u,\lambda) \boldsymbol{U}(t) \phi(\boldsymbol{U}(t))' A_\theta(u,0)^{-1}\right. \\ & \left. \times \nabla_h \boldsymbol{\mu}_\theta(u)\right\} d\lambda du + o(1). \end{aligned}$$

(S-5). $\sum_{t=1}^{T} E\left(W_{t,T}^2 | \mathcal{F}_{t-1}\right) \xrightarrow{\mathcal{P}} \frac{\Gamma(h,\theta)}{4}.$

We write

$$W_{t,T}^{(1)} = I_{t,T}^{(1)} + I_{t,T}^{(2)}$$

where

$$\begin{aligned} I_{t,T}^{(1)} &= \frac{1}{2\sqrt{T}}\phi(\boldsymbol{U}(t))' a_\theta^\circ(t,T,0)^{-1} \sum_{k=1}^{t-1} \nabla_h b_\theta^\circ(t,T,k)\left\{\boldsymbol{X}(t-k,T) - \boldsymbol{\mu}_\theta\left(\frac{t-k}{T}\right)\right\} \\ I_{t,T}^{(2)} &= -\frac{1}{2\sqrt{T}} tr\left[a_\theta^\circ(t,T,0)^{-1} \nabla_h a_\theta^\circ(t,T,0)\left\{I_m + \boldsymbol{U}(t)\phi(\boldsymbol{U}(t))'\right\}\right]. \end{aligned}$$

From the definition of $\{W_{t,T}\}$ and Assumption 8.5.15 (ii), we can see that

$$\begin{aligned}
&E\left\{W_{t,T}^2|\mathcal{F}_{t-1}\right\}\\
&= E\left\{\left(W_{t,T}^{(1)}+W_{t,T}^{(2)}\right)^2|\mathcal{F}_{t-1}\right\}\\
&= E\left\{\left(I_{t,T}^{(1)}\right)^2|\mathcal{F}_{t-1}\right\}+2E\left\{I_{t,T}^{(1)}\left(I_{t,T}^{(2)}+W_{t,T}^{(2)}\right)|\mathcal{F}_{t-1}\right\}+E\left\{\left(I_{t,T}^{(2)}+W_{t,T}^{(2)}\right)^2\right\}\\
&\equiv L_{t,T}^{(1)}+L_{t,T}^{(2)}+E\left\{\left(I_{t,T}^{(2)}+W_{t,T}^{(2)}\right)^2\right\} \quad (say).
\end{aligned}$$

Hence we get

$$\begin{aligned}
&\mathrm{Var}\left\{\sum_{t=1}^{T}E\left(W_{t,T}^2|\mathcal{F}_{t-1}\right)\right\}\\
&= \sum_{t_1,t_2=1}^{T}\left\{cum\left(L_{t_1,T}^{(1)},L_{t_2,T}^{(1)}\right)+cum\left(L_{t_1,T}^{(2)},L_{t_2,T}^{(2)}\right)+2cum\left(L_{t_1,T}^{(1)},L_{t_2,T}^{(2)}\right)\right\}\\
&\equiv L^{(1)}+L^{(2)}+L^{(3)}, \quad (say).
\end{aligned}$$

According to Hirukawa and Taniguchi (2006), it follows that

$$L^{(1)} = O\left(\frac{1}{T}\right)+o(1).$$

Writing

$$\tilde{W}_{t,T} = a_\theta^\circ(t,T,0)^{-1}\sum_{k=1}^{t-1}\nabla_h b_\theta^\circ(t,T,k)\left\{\boldsymbol{X}(t-k,T)-\boldsymbol{\mu}_\theta\left(\frac{t-k}{T}\right)\right\}$$

we observe that

$$\begin{aligned}
L_{t,T}^{(2)} &= E\left\{I_{t,T}^{(1)}\left(I_{t_1,T}^{(2)}+W_{t,T}^{(2)}\right)|\mathcal{F}_{t_1-1}\right\}\\
&= -\frac{1}{4T}\tilde{W}_{t,T}'E\Bigg(\phi(\boldsymbol{U}(t))\phi(\boldsymbol{U}(t))'a_\theta^\circ(t,T,0)^{-1}\Big\{\nabla_h a_\theta^\circ(t,T,0)\boldsymbol{U}(t)\\
&\quad+\sum_{k=0}^{t-1}b_\theta^\circ(t,T,k)\nabla_h\boldsymbol{\mu}_\theta\left(\frac{t-k}{T}\right)\Big\}\Bigg)\\
&= \tilde{W}_{t,T}'\tilde{L}_{t,T}^{(2)} \quad (say).
\end{aligned}$$

Hence

$$\begin{aligned}
L^{(2)} &= \sum_{t_1,t_2=1}^{T}cum\left(L_{t_1,T}^{(2)},L_{t_2,T}^{(2)}\right)\\
&= \sum_{t_1,t_2=1}^{T}(\tilde{L}_{t_1,T}^{(2)})'cum\left(\tilde{W}_{t_1,T},\tilde{W}_{t_2,T}\right)\tilde{L}_{t_2,T}^{(2)}\\
&= o(1),
\end{aligned}$$

and similarly $L^{(3)} = o(1)$. Therefore, we can see that

$$\mathrm{Var}\left\{\sum_{t=1}^{T}E\left(W_{t,T}^2|\mathcal{F}_{t-1}\right)\right\}\to 0 \quad as \quad T\to\infty,$$

hence

$$\sum_{t=1}^{T} E\left(W_{t,T}^2|\mathcal{F}_{t-1}\right) \overset{\mathcal{P}}{\to} \frac{\Gamma(h,\theta)}{4}.$$

(S-4) and (S-6) can be proved similarly as in Hirukawa and Taniguchi (2006). Hence the proofs of (i) and (ii) of the theorem are completed.

Finally, (iii) of the theorem follows from Scheffe's theorem and the continuity of $p(\cdot)$. □

Proof of Theorem 8.5.18. We can see

$$\mathbf{0} = \frac{1}{\sqrt{T}}\nabla\mathcal{L}_T(\theta_0) + \left[\frac{1}{T}\left\{\nabla^2\mathcal{L}_T(\theta_0)\right\}' + R_T(\theta^*)\right]\sqrt{T}\left(\hat{\theta}_{QML} - \theta_0\right),$$

where $R_T(\theta^*) = \frac{1}{T}\left[\left\{\nabla^2\mathcal{L}_T(\theta^*)\right\}' - \left\{\nabla^2\mathcal{L}_T(\theta_0)\right\}'\right]$ with $|\theta^* - \theta_0| \leq |\hat{\theta}_{QML} - \theta_0|$.

Here it should be noted that the consistency of $\hat{\theta}_{QML}$ is proved in Theorem 2 of Hirukawa and Taniguchi (2006). Hence $\theta^* = \theta_0 + o_p(1)$. Let the ith component of $\frac{1}{\sqrt{T}}\nabla\mathcal{L}_T(\theta_0)$ and the (i, j)th component of $\frac{1}{T}\left\{\nabla^2\mathcal{L}_T(\theta_0)\right\}$ be $g_{\theta_0}^{(T)}(i)$ and $h_{\theta_0}^{(T)}(i, j)$, respectively. Then

$$\begin{aligned}
g_{\theta}^{(T)}(i) &= \frac{1}{\sqrt{T}}\nabla_i\mathcal{L}_T(\theta)\\
&= \frac{1}{\sqrt{T}}\sum_{t=1}^{T}\Bigg(\phi(\tilde{W}_{t,T})'\Big[-a_\theta^\circ(t,T,0)^{-1}\nabla_i a_\theta^\circ(t,T,0)a_\theta^\circ(t,T,0)^{-1}\\
&\qquad\times\sum_{k=0}^{t-1} b_\theta^\circ(t,T,k)\left\{\boldsymbol{X}(t-k,T) - \boldsymbol{\mu}_\theta\left(\frac{t-k}{T}\right)\right\}\\
&\qquad + a_\theta^\circ(t,T,0)^{-1}\sum_{k=1}^{t-1}\nabla_i b_\theta^\circ(t,T,k)\left\{\boldsymbol{X}(t-k,T) - \boldsymbol{\mu}_\theta\left(\frac{t-k}{T}\right)\right\}\\
&\qquad - a_\theta^\circ(t,T,0)^{-1}\sum_{k=0}^{t-1} b_\theta^\circ(t,T,k)\nabla_i\boldsymbol{\mu}_\theta\left(\frac{t-k}{T}\right)\Big]\\
&\qquad - tr\{a_\theta^\circ(t,T,0)^{-1}\nabla_i a_\theta^\circ(t,T,0)\}\Bigg).
\end{aligned} \tag{8.5.22}$$

Denote the ith component of the $\Delta_T(e;\theta)$ and the (i, j)th component of $\Gamma_e(\theta)$ by $\Delta_T(e;\theta)^{(i)}$ and $\Gamma_e(\theta)^{(i,j)}$ where $e = (1, \ldots, 1)'$ ($q \times 1$-vector), respectively. Then, under Assumptions 8.5.15 and 8.5.16, it can be shown that

$$\begin{aligned}
(8.5.22) = &\frac{1}{\sqrt{T}}\sum_{t=1}^{T}\phi\left(\boldsymbol{U}(t) + O_p\left(t^{-1}\right)\right)'\sum_{k=1}^{t-1} a_\theta^\circ(t,T,0)^{-1}\nabla_i b_\theta^\circ(t,T,k)\\
&\times\left\{\boldsymbol{X}(t-k,T) - \boldsymbol{\mu}_\theta\left(\frac{t-k}{T}\right)\right\}\\
&-\frac{1}{\sqrt{T}}\sum_{t=1}^{T}\phi\left(\boldsymbol{U}(t) + O_p\left(t^{-1}\right)\right)' a_\theta^\circ(t,T,0)^{-1}\nabla_i a_\theta^\circ(t,T,0)\left\{\boldsymbol{U}(t) + O_p\left(t^{-1}\right)\right\}\\
&-\sqrt{T}tr\left\{a_\theta^\circ(t,T,0)^{-1}\nabla_i a_\theta^\circ(t,T,0)\right\}\\
&-\frac{1}{\sqrt{T}}\sum_{t=1}^{T}\phi\left(\boldsymbol{U}(t) + O_p(t^{-1})\right)'\sum_{k=1}^{t-1} a_\theta^\circ(t,T,0)^{-1}b_\theta^\circ(t,T,k)\nabla_i\boldsymbol{\mu}_\theta\left(\frac{t-k}{T}\right)
\end{aligned}$$

$$= \Delta_T(e;\theta)^{(i)} + O_p\left(\frac{\log T}{\sqrt{T}}\right).$$

Similarly,

$$h_\theta^{(T)}(i,j) = -\Gamma_e(\theta)^{(i,j)} + O_p\left(\frac{\log T}{\sqrt{T}}\right).$$

Hence we can see that

$$\sqrt{T}\left(\hat{\theta}_{QML} - \theta_0\right) = -\left[\frac{1}{T}\left\{\nabla^2 \mathcal{L}_T(\theta_0)\right\}' + R_T(\theta^*)\right]^{-1} \frac{1}{\sqrt{T}}\nabla\mathcal{L}_T(\theta_0)$$

$$\xrightarrow{\mathcal{L}} \Gamma_e(\theta_0)^{-1}\mathbf{Y},$$

where $\boldsymbol{Y} \sim N(\mathbf{0}, \Gamma_e(\theta_0))$. Hence,

$$\sqrt{T}\left(\hat{\theta}_{QML} - \theta_0\right) \xrightarrow{\mathcal{L}} N(\mathbf{0}, V^{-1})$$

where $V = \Gamma_e(\theta_0)$. □

8.6 Shrinkage Estimation

This section develops a comprehensive study of shrinkage estimation for dependent observations. We discuss the shrinkage estimation of the mean and autocovariance functions for second-order stationary processes. This approach is extended to the introduction of a shrink predictor for stationary processes. Then, we show that the shrinkage estimators and predictors improve the mean square error of nonshrunken fundamental estimators and predictors.

Let $\boldsymbol{X}(1), \cdots, \boldsymbol{X}(n)$ be a sequence of independent and identically distributed random vectors distributed as $\mathbb{N}_k(\boldsymbol{\theta}, \boldsymbol{I_k})$, where $\boldsymbol{I_k}$ is the k × k identity matrix. The sample mean $\bar{\boldsymbol{X}}_n = n^{-1}\sum_{t=1}^{n}\boldsymbol{X}(n)$ seems the most fundamental and natural estimator of $\boldsymbol{\theta}$. However, if k≥3, Stein (1956) showed that $\bar{\boldsymbol{X}}_n$ is not admissible with respect to the mean squared error loss function. Furthermore James and Stein (1961) proposed a shrinkage estimator $\widehat{\boldsymbol{\theta}}_n$ defined by

$$\widehat{\boldsymbol{\theta}}_n = \left(1 - \frac{k-2}{n\|\bar{\boldsymbol{X}}_n\|^2}\right)\bar{\boldsymbol{X}}_n, \tag{8.6.1}$$

which improves on $\bar{\mathrm{X}}_n$ with respect to mean squared error when k≥3.

In what follows we investigate the mean squared error of $\widehat{\boldsymbol{\theta}}_n$ and $\bar{\boldsymbol{X}}_n$ when $\boldsymbol{X}(1), \cdots, \boldsymbol{X}(n)$ are from a k-dimensional vector stochastic process.

Suppose that $\{\boldsymbol{X}(t) : t \in \mathbb{Z}\}$ is a k-dimensional Gaussian stationary process with mean vector $E\{\boldsymbol{X}(t)\} = \boldsymbol{\theta}$ and autocovariance matrix $\boldsymbol{S}(\ell) = E[\{\boldsymbol{X}(t) - \boldsymbol{\theta}\}\{\boldsymbol{X}(t+\ell) - \boldsymbol{\theta}\}']$, such that

$$\sum_{\ell=-\infty}^{\infty} \|\boldsymbol{S}(\ell)\| < \infty, \tag{8.6.2}$$

where $\|\cdot\|$ is the Euclidean norm. The process $\{\boldsymbol{X}(t)\}$ is therefore assumed to be short memory. It has the spectral density matrix

$$\boldsymbol{f}(\lambda) = \frac{1}{2\pi}\sum_{\ell=-\infty}^{\infty} \boldsymbol{S}(\ell)e^{-i\ell\lambda}.$$

Next we investigate the comparison of $\bar{\boldsymbol{X}}_n$ and the Stein–James estimator $\widehat{\boldsymbol{\theta}}_n$ in terms of mean squared error. For this we introduce the Cesaro sum approximation for the spectral density, which is defined by

$$\boldsymbol{f}_n^c(\lambda) = \frac{1}{2\pi}\sum_{\ell=-n+1}^{n-1}\left(1 - \frac{|\ell|}{n}\right)\boldsymbol{S}(\ell)e^{-i\ell\lambda},$$

and which for short we call the C-spectral density matrix. Then

$$\mathrm{Var}(\bar{\boldsymbol{X}}_n) = \frac{2\pi}{n}\boldsymbol{f}_n^c(0). \tag{8.6.3}$$

Since $\boldsymbol{f}(\lambda)$ is evidently continuous with respect to λ, the C-spectral density matrix $\boldsymbol{f}_n^c(\lambda)$ converges uniformly to $\boldsymbol{f}(\lambda)$ on $[-\pi, \pi]$ as $n \to \infty$ (Hannan (1970, p. 507)). Hence

$$\lim_{n\to\infty} \mathrm{Var}(\sqrt{n}\bar{\boldsymbol{X}}_n) = 2\pi\boldsymbol{f}(0). \tag{8.6.4}$$

Let $\nu_1, \cdots, \nu_k(\nu_1 < \cdots < \nu_k)$ and $\nu_{1,n}, \cdots, \nu_{k,n}(\nu_{1,n} \leq \cdots \leq \nu_{k,n})$ be the eigenvalues of $2\pi\boldsymbol{f}(0)$ and $2\pi\boldsymbol{f}_n^c(0)$, respectively. From Magnus and Neudecker (1999, p. 163)) it follows that $\nu_{j,n} \to \nu_j$ as $n \to \infty$ for $j = 1, \cdots, k$. We will evaluate

$$\mathrm{DMSE}_n \equiv E[n\|\bar{\boldsymbol{X}}_n - \boldsymbol{\theta}\|^2] - E[n\|\widehat{\boldsymbol{\theta}}_n - \boldsymbol{\theta}\|^2].$$

Since the behaviour of DMSE_n when $\boldsymbol{\theta} = \boldsymbol{0}$ is very different from that when $\boldsymbol{\theta} \neq \boldsymbol{0}$, we first give the result for $\boldsymbol{\theta} = \boldsymbol{0}$. The following four theorems are due to Taniguchi and Hirukawa (2005).

Theorem 8.6.1. *Suppose that $\boldsymbol{\theta} = \boldsymbol{0}$ and (8.6.2) holds. Then we have the following.*
(i) $(k-2)\{2 - (\frac{\nu_{k,n}}{\nu_{1,n}})^{k/2}\nu_{k,n}^{-1}\} \leq \mathrm{DMSE}_n \leq (k-2)\{2 - (\frac{\nu_{1,n}}{\nu_{k,n}})^{k/2}\nu_{1,n}^{-1}\}$,
which implies that $\widehat{\boldsymbol{\theta}}_n$ improves upon $\bar{\boldsymbol{X}}_n$ if $2 > (\frac{\nu_{k,n}}{\nu_{1,n}})^{k/2}\nu_{k,n}{}^{-1}$.
(ii) $(k-2)\{2 - (\frac{\nu_k}{\nu_1})^{k/2}\nu_k^{-1}\} \leq \lim_{n\to\infty} \mathrm{DMSE}_n \leq (k-2)\{2 - (\frac{\nu_1}{\nu_k})^{k/2}\nu_1^{-1}\}$,
which implies that $\widehat{\boldsymbol{\theta}}_n$ improves upon $\bar{\boldsymbol{X}}_n$ asymptotically if $2 > (\frac{\nu_k}{\nu_1})^{k/2}\nu_k{}^{-1}$.

Proof. From the definition of DMSE_n, we can see that

$$\begin{aligned}\mathrm{DMSE}_n &= 2(k-2)E\Big(\Big\langle\frac{1}{\|\bar{\boldsymbol{X}}_n\|^2}\bar{\boldsymbol{X}}_n, \bar{\boldsymbol{X}}_n\Big\rangle\Big) - \frac{(k-2)^2}{n}E\Big(\Big\langle\frac{1}{\|\bar{\boldsymbol{X}}_n\|^2}\bar{\boldsymbol{X}}_n, \frac{1}{\|\bar{\boldsymbol{X}}_n\|^2}\bar{\boldsymbol{X}}_n\Big\rangle\Big)\\ &= (k-2)\Big\{2 - \Big(\frac{k-2}{n}\Big)E\Big(\frac{1}{\|\bar{\boldsymbol{X}}_n\|^2}\Big)\Big\},\end{aligned} \tag{8.6.5}$$

where $< A, B >$ is the inner product of the vectors A and B. Write $\boldsymbol{Z}_n = \sqrt{n}\bar{\boldsymbol{X}}_n$. From (8.6.3) and the Gaussianity of $\{\boldsymbol{X}(t)\}$, it follows that

$$\boldsymbol{Z}_n \sim \mathrm{N}(\boldsymbol{0}, \boldsymbol{P}\boldsymbol{V}_n\boldsymbol{P}'),$$

where $\boldsymbol{V}_n = \mathrm{diag}(\nu_{1,n}, \cdots, \nu_{k,n})$ and $\boldsymbol{P}$ is a k × k orthogonal matrix which diagonalizes $2\pi\boldsymbol{f}_n^c(0)$. Therefore

$$\mathrm{DMSE}_n = (k-2)\Big\{2 - (k-2)E\Big(\frac{1}{\|\boldsymbol{Z}_n\|^2}\Big)\Big\}. \tag{8.6.6}$$

Since (8.6.6) is invariant if $\boldsymbol{Z}_n$ is transformed in $\boldsymbol{P}'\boldsymbol{Z}_n$, we may assume that $\boldsymbol{Z}_n \sim \mathrm{N}(\boldsymbol{0}, \boldsymbol{V}_n)$. Next we evaluate

$$E\Big(\frac{1}{\|\boldsymbol{Z}_n\|^2}\Big) = \int\cdots\int_{\mathbb{R}^k} \frac{1}{\|z\|^2}\frac{1}{(2\pi)^{k/2}|\boldsymbol{V}_n|^{1/2}}\exp\Big(-\frac{1}{2}z'\boldsymbol{V}_n{}^{-1}z\Big)dz, \tag{8.6.7}$$

where $z = (z_1, \cdots, z_k)' \in \mathbb{R}^k$. First, the upper bound of (8.6.7) is evaluated as

$$\mathrm{UP} \equiv \int\cdots\int_{\mathbb{R}^k} \frac{1}{\|z\|^2}\frac{1}{(2\pi)^{k/2}\nu_{1,n}^{k/2}}\exp\Big(-\frac{1}{2}z'z/\nu_{k,n}\Big)dz.$$

Consider the polar transformation

$$z_1 = r\sin\rho_1$$

$$
\begin{aligned}
z_2 &= r\cos\rho_1 \sin\rho_2 \\
&\ \vdots \\
z_{k-1} &= r\cos\rho_1 \cdots \cos\rho_{k-2} \sin\rho_{k-1} \\
z_k &= r\cos\rho_1 \cdots \cos\rho_{k-2} \cos\rho_{k-1}
\end{aligned}
$$

where $-\pi/2 < \rho_i \le \pi/2$, for $i = 1, \cdots, k-2$, and $-\pi < \rho_{k-1} \le \pi$. The Jacobian is

$$r^{k-1} \cos^{k-2}\rho_1 \cos^{k-3}\rho_2 \cdots \cos\rho_{k-2}$$

(Anderson (1984, Chapter7)). By this transformation we observe that

$$\mathrm{UP} = v_{1,n}^{-k/2} v_{k,n}^{k/2-1} \Big(\frac{1}{k-2}\Big),$$

which, together with (8.6.6), implies that

$$\mathrm{DMSE}_n \ge (k-2)(2 - v_{1,n}^{-k/2} v_{k,n}^{k/2-1}). \tag{8.6.8}$$

Similarly we can show that

$$\mathrm{DMSE}_n \le (k-2)(2 - v_{k,n}^{-k/2} v_{1,n}^{k/2-1}). \tag{8.6.9}$$

Statement *(i)* follows (8.6.8) and (8.6.9). Taking the limit as $n \to \infty$ in (*i*) leads to (*ii*). □

We examine the upper and lower bounds for $\lim_{n\to\infty} \mathrm{DMSE}_n$ numerically. Let

$$\mathrm{L} = (k-2)\Big\{2 - \Big(\frac{\nu_k}{\nu_1}\Big)^{k/2} \frac{1}{\nu_k}\Big\}, \quad \mathrm{U} = (k-2)\Big\{2 - \Big(\frac{\nu_1}{\nu_k}\Big)^{k/2} \frac{1}{\nu_1}\Big\}. \tag{8.6.10}$$

Example 8.6.2.
(i) (Vector autoregression) Suppose that $\{\boldsymbol{X}(t)\}$ has the spectral density matrix

$$f(\lambda) = \begin{pmatrix} (2\pi)^{-1}|1-\eta_1 e^{i\lambda}|^{-2} & 0 & 0 \\ 0 & (2\pi)^{-1}|1-\eta_2 e^{i\lambda}|^{-2} & 0 \\ 0 & 0 & (2\pi)^{-1}|1-\eta_3 e^{i\lambda}|^{-2} \end{pmatrix},$$

where $0 \le \eta_1 < \eta_2 < \eta_3 < 1$. In this case,

$$\nu_1 = |1-\eta_1|^{-2}, \quad \nu_2 = |1-\eta_2|^{-2}, \quad \nu_3 = |1-\eta_3|^{-2},$$

so that

$$\mathrm{L} = \Big\{2 - \Big(\frac{|1-\eta_1|}{|1-\eta_3|}\Big)^3 |1-\eta_3|^2\Big\}, \mathrm{U} = \Big\{2 - \Big(\frac{|1-\eta_3|}{|1-\eta_1|}\Big)^3 |1-\eta_1|^2\Big\}.$$

(ii) (Vector moving average) Suppose that $\{\boldsymbol{X}(t)\}$ has the spectral density matrix

$$f(\lambda) = \begin{pmatrix} (2\pi)^{-1}|1-\eta_1 e^{i\lambda}|^{2} & 0 & 0 \\ 0 & (2\pi)^{-1}|1-\eta_2 e^{i\lambda}|^{2} & 0 \\ 0 & 0 & (2\pi)^{-1}|1-\eta_3 e^{i\lambda}|^{2} \end{pmatrix},$$

where $0 \le \eta_1 < \eta_2 < \eta_3 < 1$. In this case,

$$\nu_1 = |1-\eta_3|^{2}, \quad \nu_2 = |1-\eta_2|^{2}, \quad \nu_3 = |1-\eta_1|^{2},$$

so that

$$\mathrm{L} = \Big\{2 - \Big(\frac{|1-\eta_1|}{|1-\eta_3|}\Big)^3 \frac{1}{|1-\eta_1|^2}\Big\}, \mathrm{U} = \Big\{2 - \Big(\frac{|1-\eta_3|}{|1-\eta_1|}\Big)^3 \frac{1}{|1-\eta_3|^2}\Big\}.$$

Figures 8.2 (a) and (b) show L and U, respectively.

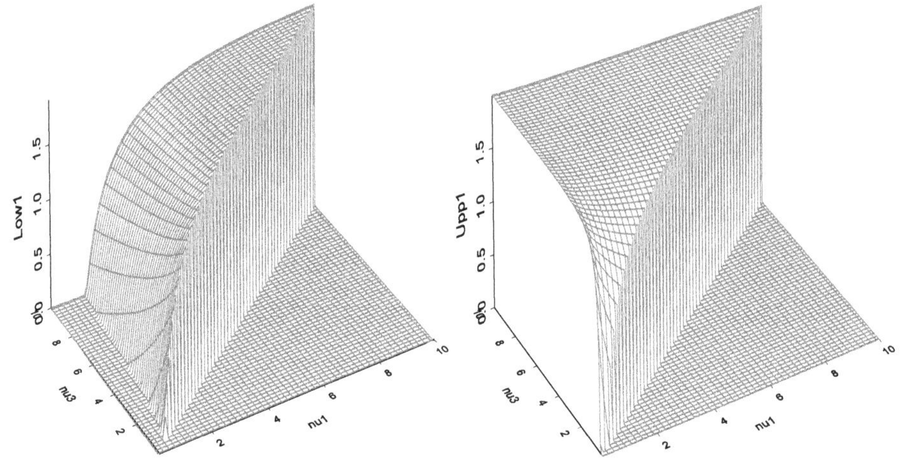

Figure 8.2: (a) L (left-hand side), (b) U (right-hand side)

Figure 8.2 shows that $\widehat{\boldsymbol{\theta}}_n$ *improves upon* $\bar{\boldsymbol{X}}_n$ *if* $\nu_1 \approx \nu_3$ *and* ν_1 *and* ν_3 *are not close to 0.*

Next we discuss the case of $\boldsymbol{\theta} \neq \mathbf{0}$.

Theorem 8.6.3. *Suppose that* $\boldsymbol{\theta} \neq \mathbf{0}$ *and that (8.6.2) holds.*
(i) Then

$$\mathrm{DMSE}_n = \frac{1}{n}\triangle_n(k, \boldsymbol{\theta}) + o\Big(\frac{1}{n}\Big), \tag{8.6.11}$$

where

$$\triangle_n(k, \boldsymbol{\theta}) = \frac{2(k-2)}{||\boldsymbol{\theta}||^2}\Big[\mathrm{tr}\{2\pi \boldsymbol{f}_{\boldsymbol{n}}^c(0)\} - \frac{2\boldsymbol{\theta}'\{2\pi \boldsymbol{f}_{\boldsymbol{n}}^c(0)\}\boldsymbol{\theta}}{||\boldsymbol{\theta}||^2} - \frac{k-2}{2}\Big]. \tag{8.6.12}$$

Hence,

$$\triangle_n(k, \boldsymbol{\theta}) \geq \frac{2(k-2)}{||\boldsymbol{\theta}||^2}\Big(\sum_{j=1}^{k-1} \nu_{j,n} - \nu_{k,n} - \frac{k-2}{2}\Big), \tag{8.6.13}$$

which implies that $\widehat{\boldsymbol{\theta}}_n$ *improves upon* $\bar{\boldsymbol{X}}_n$ *up to* $O(n^{-1})$ *if*

$$\sum_{j=1}^{k-1} \nu_{j,n} - \nu_{k,n} - \frac{k-2}{2} > 0. \tag{8.6.14}$$

(ii) Taking the limit of (8.6.13), we have

$$\lim_{n\to\infty} \triangle_n(k, \boldsymbol{\theta}) \geq \frac{2(k-2)}{||\boldsymbol{\theta}||^2}\Big(\sum_{j=1}^{k-1} \nu_j - \nu_k - \frac{k-2}{2}\Big), \tag{8.6.15}$$

which implies that $\widehat{\boldsymbol{\theta}}_n$ *improves upon* $\bar{\boldsymbol{X}}_n$ *asymptotically up to* $O(n^{-1})$ *if*

$$\sum_{j=1}^{k-1} \nu_j - \nu_k - \frac{k-2}{2} > 0.$$

Proof. First, we obtain

$$\mathrm{DMSE}_n = n\Big\{E\Big(\big\langle \bar{X}_n - \theta, \bar{X}_n - \theta\big\rangle\Big) - E\Big(\big\langle \bar{X}_n - \theta - \frac{k-2}{n\|\bar{X}_n\|^2}\bar{X}_n, \bar{X}_n - \theta - \frac{k-2}{n\|\bar{X}_n\|^2}\bar{X}_n\big\rangle\Big)$$

$$= n\Big\{2E\Big(\big\langle \bar{X}_n - \theta, \frac{k-2}{n\|\bar{X}_n\|^2}\bar{X}_n\big\rangle\Big) - \frac{(k-2)^2}{n}E\Big(\frac{1}{\|\bar{X}_n\|^2}\Big)\Big\}.$$

We can write $\bar{X}_n - \theta = (1/\sqrt{n})Z_n$, where $Z_n \sim \mathbf{N}(0, 2\pi f_n^c(0))$.
Hence,

$$\mathrm{DMSE}_n = \frac{2}{\sqrt{n}}E\Big\{\big\langle Z_n, \frac{k-2}{\|\theta + \frac{1}{\sqrt{n}}Z_n\|^2}(\theta + \frac{1}{\sqrt{n}}Z_n)\big\rangle\Big\} - \frac{(k-2)^2}{n}E\Big(\frac{1}{\|\bar{X}_n\|^2}\Big)\Big\}$$

$$= (\mathbf{I}) + (\mathbf{II}), (say).$$

From Theorem 4.5.1 of Brillinger (1981, p. 98), there exists a constaut C > 0 such that

$$\limsup_{n\to\infty} \|Z_n\| \le C\sqrt{\log n},$$

almost everywhere, which shows that, almost everywhere

$$\frac{1}{\|\theta + \frac{1}{\sqrt{n}}Z_n\|^2} = \frac{1}{\|\theta\|^2 + \frac{2}{\sqrt{n}} < Z_n, \theta > + O(\frac{\log n}{n})} = \frac{1}{\|\theta\|^2}\frac{1}{\{1 + \frac{2}{\sqrt{n}\|\theta\|^2} < Z_n, \theta > + O(\frac{\log n}{n})\}}$$

$$= \frac{1}{\|\theta\|^2}\Big\{1 - \frac{2}{\sqrt{n}\|\theta\|^2} < Z_n, \theta > + O(\frac{\log n}{n})\Big\}.$$

From this and by Fatou's lemma it is seen that

$$(\mathbf{I}) = \frac{2(k-2)}{\sqrt{n}}E\Big\{\big\langle Z_n, (\theta + \frac{1}{\sqrt{n}}Z_n)(\|\theta\|^{-2} - \frac{2}{\sqrt{n}}\|\theta\|^{-4} < Z_n, \theta >)\big\rangle\Big\} + o(n^{-1})$$

$$= \frac{2(k-2)}{\sqrt{n}}\Big\{E(< Z_n, \frac{1}{\sqrt{n}}Z_n > \|\theta\|^{-2}) - E(< Z_n, \theta >^2 \frac{2}{\sqrt{n}}\|\theta\|^{-4})\Big\} + o(n^{-1})$$

$$= \frac{2(k-2)}{n\|\theta\|^2}\Big[\mathbf{tr}\{2\pi f_n^c(0)\} - \frac{2}{\|\theta\|^2}E(< Z_n, \theta >^2)\Big] + o(n^{-1})$$

$$= \frac{2(k-2)}{n\|\theta\|^2}\Big[\mathbf{tr}\{2\pi f_n^c(0)\} - \frac{2\theta'\{2\pi f_n^c(0)\}\theta}{\|\theta\|^2}\Big] + o(n^{-1}).$$

By a similar argument we have that

$$(\mathbf{II}) = -\frac{(k-2)^2}{n}\|\theta\|^{-2} + o(n^{-1}).$$

Then, the assertions (8.6.11) and (8.6.12) follow. Inequality (8.6.13) follows from the fact that

$$\nu_{k,n} = \max_{\theta} \frac{\theta'\{2\pi f_n^c(0)\}\theta}{\|\theta\|^2}.$$

□

Remark 8.6.4. *If $\{X(t)\}$ is a sequence of i.i.d. Gaussian random vectors with mean θ and identity variance matrix, then the right-hand side of (8.6.13) becomes $(k-2)^2/\|\theta\|^2 > 0$, which implies that $\widehat{\theta}_n$ always improves upon $\bar{X}_n$.*

Remark 8.6.5. *Although in the case of $\theta \neq 0$, DMSE_n is of other $O(n^{-1})$, if $\|\theta\|$ is very near to 0, or if all the eigenvalues $\nu_{j,n}$ or ν_j are very large so that (8.6.13) and (8.6.15) become large, Theorem 8.6.3 implies that $\widehat{\theta}_n$ improves upon $\bar{X}_n$ substantially even if n is fairly large.*

Remark 8.6.6. *If k = 3, then the lower bound (8.6.15) is*

$$\frac{2}{\|\boldsymbol{\theta}\|^2}(\nu_1 + \nu_2 - \nu_3 - 1/2). \tag{8.6.16}$$

If, in Example 8.6.2(i), the autoregression coefficients η_1, η_2 and η_3 are near to 1 with $0 \le \eta_1 < \eta_2 < \eta_3 < 1$, then (8.6.16) $\to \infty$ as $\eta_1 \to 1$. Hence $\widehat{\boldsymbol{\theta}}_n$ improves upon $\bar{\boldsymbol{X}}_n$ greatly in such a case. In financial time series analysis, many cases show "near unit root" behaviour, and therefore it is important to consider shrinkage estimators.

In what follows we introduce Stein–James estimators for long-memory processes. Suppose that $\{\boldsymbol{X} : t \in \mathbb{Z}\}$ is a k-dimensional Gaussian stationary process with autocovariance matrix $\boldsymbol{S}(\ell)$, but where the $\boldsymbol{S}(\ell)$'s do not satisfy (8.6.2). Instead of that we assume that

$$\boldsymbol{S}(\ell) \sim \ell^{2d-1}, \ell \in \mathbb{Z}, \tag{8.6.17}$$

when $0 < d < \frac{1}{2}$, and $\sim$ means that $\boldsymbol{S}(\ell)/\ell^{2d-1}$ tends to some limit as $|\ell| \to \infty$. Recalling (8.6.3), we see that

$$\text{var}(n^{1/2-d}\bar{\boldsymbol{X}}_n) = \frac{1}{n^{2d}}\sum_{-n+1}^{n-1}\Big(1 - \frac{|\ell|}{n}\Big)\boldsymbol{S}(\ell) = \boldsymbol{H}_n \quad (say), \tag{8.6.18}$$

and from (8.6.17) that there exists a matrix such that $\lim_{n\to\infty}\boldsymbol{H}_n = \boldsymbol{H}$. Let $u_1, \cdots, u_k (u_1 < \cdots < u_k)$ and $u_{1,n}, \cdots, u_{k,n} (u_{1,n} \le \cdots \le u_{k,n})$ be the eigenvalues of $\boldsymbol{H}$ and $\boldsymbol{H}_n$, respectively. Write

$$\text{DMSE}_n^L \equiv E(n^{1-2d}\|\bar{\boldsymbol{X}}_n - \boldsymbol{\theta}\|^2) - E(n^{1-2d}\|\tilde{\boldsymbol{\theta}}_n - \boldsymbol{\theta}\|^2), \tag{8.6.19}$$

where $\tilde{\boldsymbol{\theta}}_n = \{1 - (k-2)/(n^{1-2d}\|\bar{\boldsymbol{X}}_n\|^2)\}\bar{\boldsymbol{X}}_n$. For $\boldsymbol{\theta} = \boldsymbol{0}$, we can prove, similary to Theorem 8.6.1, that

$$\begin{aligned}(k-2)\Big\{2-\Big(\frac{u_{k,n}}{u_{1,n}}\Big)^{k/2}\frac{1}{u_{k,n}}\Big\} &\le \text{DMSE}_n^L \le (k-2)\Big\{2-\Big(\frac{u_{1,n}}{u_{k,n}}\Big)^{k/2}\frac{1}{u_{k,n}}\Big\}\\ (k-2)\Big\{2-\Big(\frac{u_k}{u_1}\Big)^{k/2}\frac{1}{u_k}\Big\} &\le \lim_{n\to\infty}\text{DMSE}_n^L \le (k-2)\Big\{2-\Big(\frac{u_1}{u_k}\Big)^{k/2}\frac{1}{u_k}\Big\}\end{aligned} \tag{8.6.20}$$

if $k \ge 3$.

Next we evaluate the lower and upper bounds of (8.6.20) more explicitly. We assume that the spectral density matrix of $\{\boldsymbol{X}(t)\}$ is diagonalized by an orthogonal matrix to become

$$\begin{pmatrix}\frac{1}{2\pi}\frac{1}{|1-e^{i\lambda}|^{2d}}g_{11}(\lambda) & & \mathbf{0}\\ & \ddots & \\ \mathbf{0} & & \frac{1}{2\pi}\frac{1}{|1-e^{i\lambda}|^{2d}}g_{kk}(\lambda)\end{pmatrix}, \tag{8.6.21}$$

where the $g_{jj}(\lambda)$'s are bounded away from zero and are continuous at $\lambda = 0$, with

$$g_{11}(0) < \cdots < g_{kk}(0).$$

In this case, from Samarov and Taqqu (1988) it follows that the matrix H becomes

$$\begin{pmatrix}g_{11}(0)\frac{\Gamma(1-2d)}{\Gamma(d)\Gamma(1-d)d(1+2d)} & & \mathbf{0}\\ & \ddots & \\ \mathbf{0} & & g_{kk}(0)\frac{\Gamma(1-2d)}{\Gamma(d)\Gamma(1-d)d(1+2d)}\end{pmatrix}. \tag{8.6.22}$$

This is summarized in Theorem 8.6.7.

Theorem 8.6.7. *Suppose that $\boldsymbol{\theta} = \mathbf{0}$ and that (8.6.21) holds. Then,*

$$
\begin{aligned}
(k-2)\Big[2-\Big\{\frac{g_{kk}(0)}{g_{11}(0)}\Big\}^{k/2}\frac{\Gamma(d)\Gamma(1-d)d(1+2d)}{g_{kk}(0)\Gamma(1-2d)}\Big]
&\le \lim_{n\to\infty} \mathrm{DMSE}_n^L \\
&\le (k-2)\Big[2-\Big\{\frac{g_{11}(0)}{g_{kk}(0)}\Big\}^{k/2}\frac{\Gamma(d)\Gamma(1-d)d(1+2d)}{g_{11}(0)\Gamma(1-2d)}\Big]. \qquad (8.6.23)
\end{aligned}
$$

Let us examine the upper and lower bounds U and L of (8.6.23).

Example 8.6.8. *Let $k = 3$ and $g_{11}(\lambda) = 1, g_{22}(\lambda) = |1-\eta_1 e^{i\lambda}|^{-2}, g_{33}(\lambda) = |1-\eta_2 e^{i\lambda}|^{-2}$ with $0 < \eta_1 < \eta_2 < 1$. Then*

$$
L = \Big\{2 - \frac{1}{|1-\eta_2|}\frac{\Gamma(d)\Gamma(1-d)d(1+2d)}{\Gamma(1-2d)}\Big\}, \quad U = \Big\{2 - |1-\eta_2|^3\frac{\Gamma(d)\Gamma(1-d)d(1+2d)}{\Gamma(1-2d)}\Big\}. \qquad (8.6.24)
$$

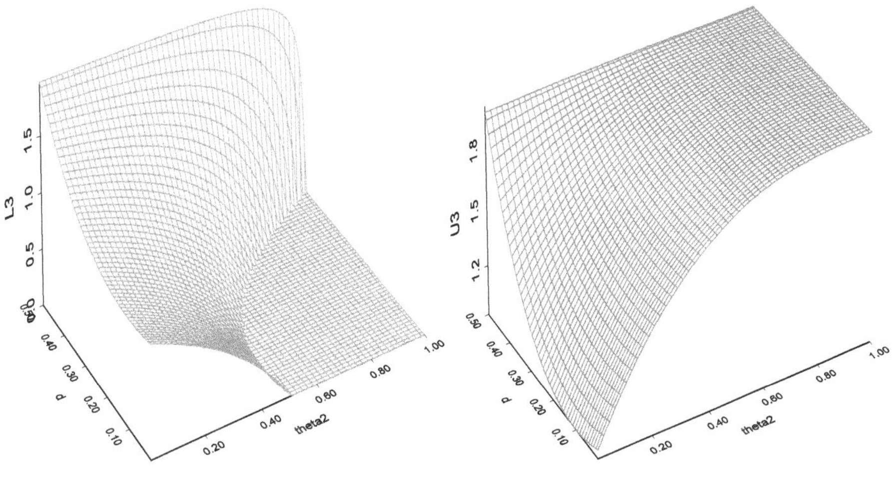

Figure 8.3: (a) L (left-hand side), (b) U (right-hand side)

We plotted L and U in Figure 8.3. We observe that, if d increases toward 1/2 and if η_2 is not large, $\tilde{\boldsymbol{\theta}}_n$ improves upon $\bar{\boldsymbol{X}}_n$. Next we discuss the case of $\boldsymbol{\theta} \neq \mathbf{0}$.

Theorem 8.6.9. *Suppose that $\boldsymbol{\theta} \neq \mathbf{0}$ and that (8.6.21) holds. Then*

$$
\mathrm{DMSE}_n^L = \frac{1}{n^{1-2d}}\Delta_n^L(k,\boldsymbol{\theta}) + o\Big(\frac{1}{n^{1-2d}}\Big),
$$

where

$$
\Delta_n^L(k,\boldsymbol{\theta}) = \frac{2(k-2)}{\|\boldsymbol{\theta}\|^2}\Big\{\mathrm{tr}(H_n) - \frac{2\boldsymbol{\theta}' H_n \boldsymbol{\theta}}{\|\boldsymbol{\theta}\|^2} - \frac{k-2}{2}\Big\}.
$$

Hence,

$$
\Delta_n^L(k,\boldsymbol{\theta}) \ge \frac{2(k-2)}{\|\boldsymbol{\theta}\|^2}\Big(\sum_{j=1}^{k-1} u_{j,n} - u_{k,n} - \frac{k-2}{2}\Big). \qquad (8.6.25)
$$

Taking the limit of (8.6.25), we obtain

$$
\lim_{n\to\infty} \Delta_n^L(k,\boldsymbol{\theta}) \ge \frac{2(k-2)}{\|\boldsymbol{\theta}\|^2}\Big(\sum_{j=1}^{k-1} u_j - u_k - \frac{k-2}{2}\Big). \qquad (8.6.26)
$$

If the right-hand sides of (8.6.25) and (8.6.26) are positive, $\tilde{\boldsymbol{\theta}}_n$ improves upon $\bar{\boldsymbol{X}}_n$.

Proof. Write $\bar{X}_n - \theta = (1/n^{1/2-d})Z_n$, and replace $2\pi f_n^c(0)$ by H_n, in the proof of Theorem 8.6.3. From Theorem 2 of Taqqu (1977) it follows that there exists $C > 0$ such that

$$\|Z_n\| \leq C\sqrt{\log\log n}$$

almost surely; this is the law of the iterated logarithm for the sum of long-memory processes. Then, along similar lines to Theorem 2, the results follow. □

In the context of Example 8.6.8, we observe that $\tilde{\theta}_n$ improves upon $\bar{X}_n$ when $\|\theta\|$ decreases towards zero or d increases toward 0.5.

So far we assumed that the mean of the concerned process is a constant vector. However, it is more natural that the mean structure is expressed as a regression form on other variables. In what follows, we consider the linear regression model of the form

$$y(t) = B'x(t) + \varepsilon(t),\ t = 1, \cdots, n, \tag{8.6.27}$$

where $y(t) = (y_1(t), \cdots, y_p(t))'$, $B \equiv (b_ij)$ is a $q \times p$ matrix of unknown coefficients and $\{\varepsilon(t)\}$ is a p-dimensional Gaussian stationary process with mean vector $E[\varepsilon(t)] = \mathbf{0}$ and spectral density matrix $f(\lambda)$. Here $x(t) = (x_1, \cdots, x_q(t))'$ is a $q \times 1$ vector of regressors satisfying the assumptions below. Let $d_i^2(n) \equiv \sum_{t=1}^n \{x_i(t)\}^2$, $i = 1, \cdots, q$.

Assumption 8.6.10. $\lim_{n\to\infty} d_i^2(n) = \infty$, $i = 1, \cdots, q$.

Assumption 8.6.11. $\lim_{n\to\infty} \dfrac{\{x_i(m)\}^2}{d_i^2(n)} = 0$, $i = 1, \cdots, q$, *and, for some* $\kappa_i > 0$,

$$\frac{x_i(n)}{n^{\kappa_i}} = O(1),\ i = 1, \cdots, q.$$

Assumption 8.6.12. *For each* $i, j = 1, \cdots, q$, *there exists the limit*

$$\rho_{ij}(h) \equiv \lim_{n\to\infty} \frac{\sum_{t=1}^n x_i(t)x_j(t+h)}{d_i(n)d_j(n)}.$$

Let $R(h) \equiv (\rho_{ij}(h))$.

Assumption 8.6.13. $R(0)$ *is nonsingular.*

From Assumptions 8.6.10–8.6.13, we can express $R(h)$ as

$$R(h) = \int_{-\pi}^{\pi} e^{ih\lambda} dM(\lambda), \tag{8.6.28}$$

where $M(\lambda)$ is a matrix function whose increments are Hermitian nonnegative, and which is uniquely defined if it is required to be continuous from the right and null at $\lambda = -\pi$. Assumptions 8.6.10–8.6.13 are a slight modification of Grenander's condition. Now we make an additional assumption.

Assumption 8.6.14. $\{\varepsilon(t)\}$ *has an absolutely continuous spectrum which is piecewise continuous with no discontinuities at the jumps of* $M(\lambda)$.

We rewrite (8.6.27) in the tensor notation

$$y = (I_p \otimes X)\beta + \varepsilon = U\beta + \varepsilon, \tag{8.6.29}$$

wherein $y_i(t)$ is in row $(i-1)p+t$ of y, $\varepsilon_i(t)$ is in row $(i-1)p+t$ of ε, X has $x_j(t)$ in row t column j, β has b_{ij} in row $(j-1)q+i$ and $U \equiv I_p \otimes X$.

If we are interested in estimation of β based on y, the most fundamental candidate is the least squares estimator $\hat{\beta}_{LS} = (U'U)^{-1}U'y$. When $\varepsilon(t)$'s are i.i.d., it is known that $\hat{\beta}_{LS}$ is not admissible if $pq \geq 3$. Stein (1956) proposed an alternative estimator,

$$\hat{\beta}_{JS} \equiv \left(1 - \frac{c}{\|(I_p \otimes D_n)(\hat{\beta}_{LS} - b)\|^2}\right)(\hat{\beta}_{LS} - b) + b, \tag{8.6.30}$$

which is called the James–Stein estimator for β. Here $c > 0$ and $D_n \equiv \mathrm{diag}(d_1(n), \cdots, d_q(n))$, b is a preassigned $pq \times 1$ vector toward which we shrink $\hat{\beta}_{LS}$. Stein showed that the risk of $\hat{\beta}_{JS}$ is everywhere smaller than that of $\hat{\beta}_{LS}$ under the MSE loss function

$$\mathrm{MSE}_n(\hat{\beta}) \equiv E[\|(I_p \otimes D_n)(\hat{\beta} - \beta)\|^2]$$

if $pq \geq 3$ and $c = pq - 2$.

In what follows, we compare the risk of $\hat{\beta}_{JS}$ with that of $\hat{\beta}_{LS}$ when $\{\varepsilon(t)\}$ is a Gaussian stationary process. From Gaussianity of $\{\varepsilon(t)\}$, it is seen that

$$(I_p \otimes D_n)(\hat{\beta}_{LS} - b) \sim N((I_p \otimes D_n)(\beta - b), C_n), \tag{8.6.31}$$

where

$$C_n \equiv (I_p \otimes D_n)(U'U)^{-1}U'cov(\varepsilon)U(U'U)^{-1}(I_p \otimes D_n).$$

From Theorem 8 of Hannan (1970, p. 216), if Assumptions 8.6.10–8.6.14 hold, it follows that

$$C \equiv \lim_{n\to\infty} C_n = (I_p \otimes R(0))^{-1} \int_{-\pi}^{\pi} 2\pi f(\lambda) \otimes M'(d\lambda)(I_p \otimes R(0))^{-1}. \tag{8.6.32}$$

Let $\nu_{1,n}, \cdots, \nu_{pq,n}$ $(\nu_{1,n} \leq \cdots \leq \nu_{pq,n})$ and $\nu_1, \cdots, \nu_{pq}$ $(\nu_1 \leq \cdots \leq \nu_{pq})$ be the eigenvalues of C_n and C, respectively. We evaluate

$$\begin{aligned}\mathrm{DMSE}_n &\equiv E[\|(I_p \otimes D_n)(\hat{\beta}_{LS} - \beta)\|^2] - E[\|(I_p \otimes D_n)(\hat{\beta}_{JS} - \beta)\|^2] \\ &= -c^2 E\left[\frac{1}{\|(I_p \otimes D_n)(\hat{b}_{LS} - \beta)\|^2}\right] \\ &\quad + 2c\left(1 - E\left[\frac{< (I_p \otimes D_n)(\beta - b), (I_p \otimes D_n)(\hat{\beta}_{LS} - b) >}{\|(I_p \otimes D_n)(\hat{\beta}_{LS} - b)\|^2}\right]\right).\end{aligned} \tag{8.6.33}$$

Because the behaviour of DMSE_n in the case of $\beta - b = \mathbf{0}$ is very different from that in the case of $\beta - b \neq \mathbf{0}$, first, we give the result for $\beta - b = \mathbf{0}$. Theorems 8.6.15 and 8.6.16 are generalizations of Theorems 8.6.1 and 8.6.3, respectively (for the proofs, see Senda and Taniguchi (2006)).

Theorem 8.6.15. *In the case of $\boldsymbol{\beta} - \boldsymbol{b} = \mathbf{0}$, suppose that Assumptions 8.6.10–8.6.14 hold and that $pq \geq 3$. Then,*
(i)

$$c\left\{2 - \frac{c}{pq-2}\left(\frac{\nu_{pq,n}}{\nu_{1,n}}\right)^{\frac{pq}{2}}\frac{1}{\nu_{pq,n}}\right\} \leq DMSE_n \leq c\left\{2 - \frac{c}{pq-2}\left(\frac{\nu_{1,n}}{\nu_{pq,n}}\right)^{\frac{pq}{2}}\frac{1}{\nu_{1,n}}\right\}, \tag{8.6.34}$$

which implies that $\hat{\boldsymbol{\beta}}_{JS}$ improves $\hat{\boldsymbol{\beta}}_{LS}$ if the left-hand side of (8.6.34) is positive.
(ii)

$$c\left\{2 - \frac{c}{pq-2}\left(\frac{\nu_{pq}}{\nu_{1}}\right)^{\frac{pq}{2}}\frac{1}{\nu_{pq}}\right\} \leq \lim_{n\to\infty} DMSE_n \leq c\left\{2 - \frac{c}{pq-2}\left(\frac{\nu_{1}}{\nu_{pq}}\right)^{\frac{pq}{2}}\frac{1}{\nu_{1}}\right\}, \tag{8.6.35}$$

which implies that $\hat{\boldsymbol{\beta}}_{JS}$ improves $\hat{\boldsymbol{\beta}}_{LS}$ asymptotically if the left-hand side of (8.6.35) is positive.

From this theorem we observe that, if $\nu_{1,n} \approx \cdots \approx \nu_{pq,n} \nearrow \infty$ ($\nu_1 \approx \cdots \approx \nu_{pq} \nearrow \infty$), then $\hat{\boldsymbol{\beta}}_{JS}$ improves $\hat{\boldsymbol{\beta}}_{LS}$ enormously.

Next, we discuss the case of $\boldsymbol{\beta} - \boldsymbol{b} \neq \mathbf{0}$. Letting $d(n)^2 \equiv \sum_{i=1}^{q} d_i^2(n)$, we note that $\|(\boldsymbol{I}_p \otimes \boldsymbol{D}_n)(\boldsymbol{\beta} - \boldsymbol{b})\|^2 = O(d^2(n))$.

Theorem 8.6.16. *In the case of $\boldsymbol{\beta} - \boldsymbol{b} \neq \mathbf{0}$, suppose that Assumptions 8.6.10–8.6.14 , and that $pq \geq 3$. Then,*
(i)

$$DMSE_n = \frac{2c}{\|(\boldsymbol{I}_p \otimes \boldsymbol{D}_n)(\boldsymbol{\beta} - \boldsymbol{b})\|^2}\{\Delta_n(c, \boldsymbol{\beta} - \boldsymbol{b}) + o(1)\}, \tag{8.6.36}$$

where

$$\Delta_n(c, \boldsymbol{\beta} - \boldsymbol{b}) = \mathrm{tr}\boldsymbol{C}_n - \frac{2(\boldsymbol{\beta} - \boldsymbol{b})'(\boldsymbol{I}_p \otimes \boldsymbol{D}_n)\boldsymbol{C}_n(\boldsymbol{I}_p \otimes \boldsymbol{D}_n)(\boldsymbol{\beta} - \boldsymbol{b})}{\|(\boldsymbol{I}_p \otimes \boldsymbol{D}_n)(\boldsymbol{\beta} - \boldsymbol{b})\|^2} - \frac{c}{2}.$$

Here we also have the inequality

$$\Delta_n(c, \boldsymbol{\beta} - \boldsymbol{b}) \geq \sum_{j=1}^{pq-1} \nu_{j,n} - \nu_{pq,n} - \frac{c}{2}, \tag{8.6.37}$$

which implies that $\hat{\boldsymbol{\beta}}_{JS}$ improves $\hat{\boldsymbol{\beta}}_{LS}$ up to $1/d^2(n)$ order if the right-hand side of (8.6.37) is positive.
(ii) Taking the limit of (8.6.37), we have

$$\lim_{n\to\infty} \Delta_n(c, \boldsymbol{\beta} - \boldsymbol{b}) \geq \sum_{j=1}^{pq-1} \nu_j - \nu_{pq} - \frac{c}{2}, \tag{8.6.38}$$

which implies that $\hat{\boldsymbol{\beta}}_{JS}$ improves $\hat{\boldsymbol{\beta}}_{LS}$ asymptotically up to $1/d^2(n)$ order if the right-hand side of (8.6.38) is positive.

From (8.6.38) we observe that if $\nu_1 \approx \nu_2 \approx \cdots \approx \nu_{pq}$ and $\nu_1 \nearrow \infty$, then, $\hat{\boldsymbol{\beta}}_{JS}$ improves $\hat{\boldsymbol{\beta}}_{LS}$ enormously.

Example 8.6.17. *Let $p = 1$ and $x_i(t) = \cos\lambda_i t$, $i = 1, \cdots, q$ with $0 \le \lambda_1 < \cdots < \lambda_q \le \pi$ (time series regression model with cyclic trend). Then $d_i^2(n) = O(n)$. We find that*
$\boldsymbol{R}(h) = \mathrm{diag}(\cos\lambda_1 h, \cdots, \cos\lambda_q h)$, *because*

$$\rho_{ij}(h) = \begin{cases} \cos\lambda_i h & \text{if } i = j, \\ 0 & \text{if } i \ne j. \end{cases}$$

Hence $\boldsymbol{M}(\lambda)$ has jumps at $\lambda = \pm\lambda_i$, $i = 1, \cdots, q$. The jump at $\lambda = \pm\lambda_i$ is $1/2$ at the ith diagonal and 0's elsewhere. Suppose that $\{\boldsymbol{\varepsilon}(t)\}$ is a Gaussian stationary process with spectral density $f(\lambda)$ satisfying Assumption 8.6.14. Then it follows that

$$\boldsymbol{C} = \mathrm{diag}(2\pi f(\lambda_1), \cdots, 2\pi f(\lambda_q))$$

.

Let $f(\lambda_{(j)})$ be the jth largest value of $f(\lambda_1), \cdots, f(\lambda_q)$. Then

$$\nu_j = 2\pi f(\lambda_{(j)}),\ j = 1, \cdots, q.$$

Thus the right-hand side of (8.6.38) is

$$2\pi\left\{\sum_{j=1}^{q-1} f(\lambda_{(j)}) - f(\lambda_{(q)})\right\} - \frac{c}{2}. \tag{8.6.39}$$

If all the values of $f(\lambda_{(j)})$'s are near and large, then (8.6.39) becomes large, which means that $\hat{\boldsymbol{\beta}}_{JS}$ improves $\hat{\boldsymbol{\beta}}_{LS}$ enormously. Because our model is very fundamental in econometrics, etc., the results seem important.

Next we give numerical comparisons between $\hat{\boldsymbol{\beta}}_{LS}$ and $\hat{\boldsymbol{\beta}}_{JS}$ by Monte Carlo simulations in the models of Example 8.6.17. We approximate

$$\mathrm{MSE}_n(\hat{\boldsymbol{\beta}}) = E[\|(\boldsymbol{I}_p \otimes \boldsymbol{D}_n)(\hat{\boldsymbol{\beta}} - \boldsymbol{\beta})\|^2]$$

by

$$\frac{1}{\mathrm{rep}-1}\sum_{j=1}^{\mathrm{rep}}\left\|(\boldsymbol{I}_p \otimes \boldsymbol{D}_n)\left(\hat{\boldsymbol{\beta}}_j - \frac{1}{\mathrm{rep}}\sum_{k=1}^{\mathrm{rep}}\hat{\boldsymbol{\beta}}_k\right)\right\|^2 + \left\|(\boldsymbol{I}_p \otimes \boldsymbol{D}_n)'\left(\frac{1}{\mathrm{rep}}\sum_{k=1}^{\mathrm{rep}}\hat{\boldsymbol{\beta}}_k - \boldsymbol{\beta}\right)\right\|^2, \tag{8.6.40}$$

where $\hat{\boldsymbol{\beta}}_k$ is the estimate in the kth simulation. For $n = 100$, $q = 10$ and $\lambda_j = 10^{-1}\pi(j-1)$, and rep $= 1000$, we calculate (8.6.40).

Table 8.5: MSE for $\hat{\boldsymbol{\beta}}_{LS}$ and $\hat{\boldsymbol{\beta}}_{JS}$

	$\theta = 0.1$	$\theta = 0.9$
MSE of $\hat{\boldsymbol{\beta}}_{LS}$	10.234	107.623
MSE of $\hat{\boldsymbol{\beta}}_{JS}$	5.102	94.770

Table 8.5 shows the values of MSE for the case when $\boldsymbol{\beta} - \boldsymbol{b} = (0.1, \cdots, 0.1)'$ and the model is AR(1) with coefficient $\theta = 0.1$ and 0.9. Then we observe that $\hat{\boldsymbol{\beta}}_{JS}$ improves $\hat{\boldsymbol{\beta}}_{LS}$ in both cases.

8.7 Shrinkage Interpolation for Stationary Processes

Interpolation is an important issue for statistics, e.g., missing data analysis. So we can apply the results to estimation of portfolio coefficients when there is a missing asset value. This section discusses the interpolation problem. When the spectral density of the time series is known, the best interpolators for missing data are discussed in Grenander and Rosenblatt (1957) and Hannan (1970). However, the spectral density is usually unknown, so in order to get the interpolator, we may use pseudo spectral density. In this case, Taniguchi (1981) discussed the misspecified interpolation problem and showed the robustness. The pseudo interpolator seems good, but similarly to the prediction problem, it is worth considering whether the better pseudo interpolator with respect to mean squared interpolation error (MSIE) exists. In this section, we show that the pseudo interpolator is not optimal with respect to MSIE by proposing a pseudo shrinkage interpolator motivated by James and Stein (1961). Under the appropriate conditions, it improves the usual pseudo interpolator in the sense of MSIE. This section is based on the results by Suto and Taniguchi (2016).

Let $\{\boldsymbol{X}(t); t \in \mathbb{Z}\}$ be a q-dimensional stationary process with mean $\mathbf{0}$ and true spectral density matrix $\boldsymbol{g}(\lambda)$. We discuss the interpolation problem, that when we observed the data $\{\boldsymbol{X}(t); t \in \mathbb{Z}\}$ except $t = 0$, we estimate $\boldsymbol{X}(0)$. According to Hannan (1970), the optimal response function of the interpolating filter is given as

$$\boldsymbol{h}(\lambda) = \boldsymbol{I}_q - \left(\frac{1}{2\pi}\int_{-\pi}^{\pi} \boldsymbol{g}(\mu)^{-1} d\mu\right)^{-1} \boldsymbol{g}(\lambda)^{-1}, \tag{8.7.1}$$

where $\boldsymbol{I}_q$ is $q \times q$ identity matrix, and the interpolation error matrix is

$$\Sigma = \left(\frac{1}{2\pi}\int_{-\pi}^{\pi} \{2\pi \boldsymbol{g}(\lambda)\}^{-1} d\lambda\right)^{-1}. \tag{8.7.2}$$

In actual analysis, it is natural that the true spectral density $\boldsymbol{g}(\lambda)$ is unknown and we often use pseudo spectral density $\boldsymbol{f}(\lambda)$. In this case, we obtain the pseudo response function of the interpolating filter

$$\boldsymbol{h}^f(\lambda) = \boldsymbol{I}_q - \left(\frac{1}{2\pi}\int_{-\pi}^{\pi} \boldsymbol{f}(\mu)^{-1} d\mu\right)^{-1} \boldsymbol{f}(\lambda)^{-1}, \tag{8.7.3}$$

and the pseudo interpolation error matrix

$$\Sigma^f = \left(\frac{1}{2\pi}\int_{-\pi}^{\pi} \boldsymbol{f}(\lambda)^{-1} d\lambda\right)^{-1} \left(\int_{-\pi}^{\pi} \boldsymbol{f}(\lambda)^{-1}\boldsymbol{g}(\lambda)\boldsymbol{f}(\lambda)^{-1} d\lambda\right) \left(\frac{1}{2\pi}\int_{-\pi}^{\pi} \boldsymbol{f}(\lambda)^{-1} d\lambda\right)^{-1}, \tag{8.7.4}$$

(cf. Taniguchi (1981)). In what follows, we propose shrinking the interpolator, which is motivated by James and Stein (1961). The usual pseudo interpolator is given by

$$\hat{\boldsymbol{X}}^f(0) = \int_{-\pi}^{\pi} \boldsymbol{h}^f(\lambda) d\boldsymbol{Z}(\lambda),$$

where $\boldsymbol{Z}(\lambda)$ is the spectral measure of $\boldsymbol{X}(t)$. We introduce the shrinkage version of $\hat{\boldsymbol{X}}^f(0)$,

$$\tilde{\boldsymbol{X}}^f(0) = \left(1 - \frac{c}{E[\|\hat{\boldsymbol{X}}^f(0)\|^2]}\right) \hat{\boldsymbol{X}}^f(0), \tag{8.7.5}$$

where c is a constant and we call it shrinkage constant. Then we can evaluate the mean squared interpolation error of (8.7.5) as follows.

Theorem 8.7.1.

(i)

$$E[\|\boldsymbol{X}(0) - \tilde{\boldsymbol{X}}^{\boldsymbol{f}}(0)\|^2] = E[\|\boldsymbol{X}(0) - \hat{\boldsymbol{X}}^{\boldsymbol{f}}(0)\|^2] + \frac{2c}{E[\|\hat{\boldsymbol{X}}^{\boldsymbol{f}}(0)\|^2]}B(\boldsymbol{f},\boldsymbol{g}) + \frac{c^2}{E[\|\hat{\boldsymbol{X}}^{\boldsymbol{f}}(0)\|^2]}, \tag{8.7.6}$$

where

$$\begin{aligned} &B(\boldsymbol{f},\boldsymbol{g}) \\ &= \mathrm{tr}\int_{-\pi}^{\pi}(\boldsymbol{I}_q - \boldsymbol{h}^{\boldsymbol{f}}(\lambda))\boldsymbol{g}(\lambda)\boldsymbol{h}^{\boldsymbol{f}}(\lambda)^* d\lambda \\ &= \mathrm{tr}\Bigg[\left(\frac{1}{2\pi}\int_{-\pi}^{\pi}\boldsymbol{f}(\lambda)^{-1}d\lambda\right)^{-1}\left(\int_{-\pi}^{\pi}\boldsymbol{f}(\lambda)^{-1}\boldsymbol{g}(\lambda)d\lambda\right) \\ &\quad - \left(\frac{1}{2\pi}\int_{-\pi}^{\pi}\boldsymbol{f}(\lambda)^{-1}d\lambda\right)^{-1}\left(\int_{-\pi}^{\pi}\boldsymbol{f}(\lambda)^{-1}\boldsymbol{g}(\lambda)\boldsymbol{f}(\lambda)^{-1}d\lambda\right)\left(\frac{1}{2\pi}\int_{-\pi}^{\pi}\boldsymbol{f}(\lambda)^{-1}d\lambda\right)^{-1}\Bigg]. \end{aligned} \tag{8.7.7}$$

(ii) If $B(\boldsymbol{f},\boldsymbol{g}) \neq 0$, the optimal shrinkage constant is $c = -B(\boldsymbol{f},\boldsymbol{g})$. Then

$$E[\|\boldsymbol{X}(0) - \hat{\boldsymbol{X}}^{\boldsymbol{f}}(0)\|^2] = E[\|\boldsymbol{X}(0) - \hat{\boldsymbol{X}}^{\boldsymbol{f}}(0)\|^2] - \frac{B(\boldsymbol{f},\boldsymbol{g})^2}{E[\|\hat{\boldsymbol{X}}^{\boldsymbol{f}}(0)\|^2]}, \tag{8.7.8}$$

which implies that the shrinkage interpolator improves the usual pseudo interpolator in the sense of MSIE.

Proof. (i) For the vectors $\boldsymbol{X}$ and $\boldsymbol{Y}$, define the inner product as $\langle \boldsymbol{X}, \boldsymbol{Y}\rangle = \mathrm{tr}(\boldsymbol{X}\boldsymbol{Y}^*)$. Then

$$\begin{aligned} &E[\|\boldsymbol{X}(0) - \tilde{\boldsymbol{X}}^{\boldsymbol{f}}(0)\|^2] \\ &= E\left[\left\|\boldsymbol{X}(0) - \hat{\boldsymbol{X}}^{\boldsymbol{f}}(0) + \frac{c}{E[\|\boldsymbol{X}^{\boldsymbol{f}}(0)\|^2]}\hat{\boldsymbol{X}}^{\boldsymbol{f}}(0)\right\|^2\right] \\ &= E[\|\boldsymbol{X}(0) - \hat{\boldsymbol{X}}^{\boldsymbol{f}}(0)\|^2] + 2E\left[\left\langle \boldsymbol{X}(0) - \hat{\boldsymbol{X}}^{\boldsymbol{f}}(0), \frac{c}{E[\|\boldsymbol{X}^{\boldsymbol{f}}(0)\|^2]}\hat{\boldsymbol{X}}^{\boldsymbol{f}}(0)\right\rangle\right] \\ &\quad + \frac{c^2}{(E[\|\boldsymbol{X}^{\boldsymbol{f}}(0))\|^2])^2}E[\|\hat{\boldsymbol{X}}^{\boldsymbol{f}}(0)\|^2] \\ &= E[\|\boldsymbol{X}(0) - \hat{\boldsymbol{X}}^{\boldsymbol{f}}(0)\|^2] + \frac{2c}{E[\|\boldsymbol{X}^{\boldsymbol{f}}(0)\|^2]}E\left[\mathrm{tr}\left\{\left(\int_{-\pi}^{\pi}(\boldsymbol{I}_q - \boldsymbol{h}^{\boldsymbol{f}}(\lambda))d\boldsymbol{Z}(\lambda)\right)\left(\int_{-\pi}^{\pi}\boldsymbol{h}^{\boldsymbol{f}}(\lambda)d\boldsymbol{Z}(\lambda)\right)^*\right\}\right] \\ &\quad + \frac{c^2}{E[\|\boldsymbol{X}^{\boldsymbol{f}}(0)\|^2]} \\ &= (\text{RHS of } (8.7.6)). \end{aligned}$$

(ii) RHS of (8.7.6) is the quadratic form of c. Thus the shrinkage constant c, which minimizes RHS of (8.7.6), is $c = -B(\boldsymbol{f},\boldsymbol{g})$, and we can get the result. □

Remark 8.7.2. *$B(\boldsymbol{f},\boldsymbol{g})$ can be regarded as a disparity between $\boldsymbol{f}(\lambda)$ and $\boldsymbol{g}(\lambda)$ (see Suto et al. (2016)).*

Because the above shrinkage interpolator (8.7.5) includes the unknown value $E[\|\hat{\boldsymbol{X}}^{\boldsymbol{f}}(0)\|^2]$, we have to estimate $E[\|\hat{\boldsymbol{X}}^{\boldsymbol{f}}(0)\|^2]$. Since we can see that

$$E[\|\hat{\boldsymbol{X}}^{\boldsymbol{f}}(0)\|^2] = E\left[\mathrm{tr}\left\{\left(\int_{-\pi}^{\pi}\boldsymbol{h}^{\boldsymbol{f}}(\lambda)d\boldsymbol{Z}(\lambda)\right)\left(\int_{-\pi}^{\pi}\boldsymbol{h}^{\boldsymbol{f}}(\lambda)d\boldsymbol{Z}(\lambda)\right)^*\right\}\right]$$

$$= \text{tr} \int_{-\pi}^{\pi} \boldsymbol{h}^{\boldsymbol{f}}(\lambda)\boldsymbol{g}(\lambda)\boldsymbol{h}^{\boldsymbol{f}}(\lambda)^* d\lambda,$$

we propose an estimator of $E[\|\hat{\boldsymbol{X}}^{\boldsymbol{f}}(0)\|^2]$ as

$$\hat{E}[\|\hat{\boldsymbol{X}}^{\boldsymbol{f}}(0)\|^2] = \text{tr} \int_{-\pi}^{\pi} \boldsymbol{h}^{\boldsymbol{f}}(\lambda) I_{n,\boldsymbol{X}}(\lambda)\boldsymbol{h}^{\boldsymbol{f}}(\lambda)^* d\lambda, \tag{8.7.9}$$

where $I_{n,\boldsymbol{X}}(\lambda)$ is the periodogram defined by

$$I_{n,\boldsymbol{X}}(\lambda) = \frac{1}{2\pi n} \left\{ \sum_{t=1}^{n} \boldsymbol{X}(t) e^{it\lambda} \right\} \left\{ \sum_{t=1}^{n} \boldsymbol{X}(t) e^{it\lambda} \right\}^* .$$

We see the goodness of the estimator (8.7.9) as follows. For this, we need the following assumption.

Assumption 8.7.3. *Let* $\{\boldsymbol{X}(t)\}$ *be generated by*

$$\boldsymbol{X}(t) = \sum_{j=0}^{\infty} G(j) e(t-j),$$

where $e(j)$*'s are q-dimensional random vectors with* $E(e(j)) = \boldsymbol{0}$*,* $E(e(t)e(s)') = \delta(t,s)K$*,* $\det K > 0$ *and* $G(j)$*'s satisfy* $\text{tr} \sum_{j=0}^{\infty} G(j)KG(j)' < \infty$*. Suppose that Hosoya and Taniguchi (1982) conditions (HT1)–(HT6) are satisfied (see Section 8.1).*

Under this assumption, using Theorem 8.1.1, we can see the estimator (8.7.9) is a $\sqrt{n}$-consistent estimator of $E[\|\hat{\boldsymbol{X}}^{\boldsymbol{f}}(0)\|^2]$. Then we can consider the sample version of (8.7.5) such as

$$\tilde{\boldsymbol{X}}_S^{\boldsymbol{f}}(0) = \left(1 - \frac{c}{\hat{E}[\|\hat{\boldsymbol{X}}^{\boldsymbol{f}}(0)\|^2]}\right) \hat{\boldsymbol{X}}^{\boldsymbol{f}}(0), \tag{8.7.10}$$

and we can evaluate the MSIE as follows.

Theorem 8.7.4. *Under Assumption 8.7.3, if* $B(\boldsymbol{f},\boldsymbol{g}) \neq 0$*, the optimal shrinkage constant is* $c = -B(\boldsymbol{f},\boldsymbol{g})$*. Then*

$$E[\|\boldsymbol{X}(0) - \tilde{\boldsymbol{X}}_S^{\boldsymbol{f}}(0)\|^2] = E[\|\boldsymbol{X}(0) - \hat{\boldsymbol{X}}^{\boldsymbol{f}}(0)\|^2] - \frac{B(\boldsymbol{f},\boldsymbol{g})^2}{E[\|\hat{\boldsymbol{X}}^{\boldsymbol{f}}(0)\|^2]} + O\left(\frac{1}{\sqrt{n}}\right), \tag{8.7.11}$$

which implies that the shrinkage interpolator improves the usual pseudo interpolator asymptotically in the sense of MSIE.

Proof. Let

$$\begin{aligned} s_n &= \sqrt{n}\{\hat{E}[\|\hat{\boldsymbol{X}}^{\boldsymbol{f}}(0)\|^2] - E[\|\hat{\boldsymbol{X}}^{\boldsymbol{f}}(0)\|^2]\} \\ &= \sqrt{n}\, \text{tr} \left[\int_{-\pi}^{\pi} \boldsymbol{h}^{\boldsymbol{f}}(\lambda) I_{n,\boldsymbol{X}}(\lambda) \boldsymbol{h}^{\boldsymbol{f}}(\lambda)^* d\lambda - \int_{-\pi}^{\pi} \boldsymbol{h}^{\boldsymbol{f}}(\lambda)\boldsymbol{g}(\lambda)\boldsymbol{h}^{\boldsymbol{f}}(\lambda)^* d\lambda \right]. \end{aligned}$$

Then, by Lemma A.2.2 of Hosoya and Taniguchi (1982), we can see

$$s_n^2 < \infty \quad a.s.$$

From this, we obtain

$$\text{tr} \int_{-\pi}^{\pi} \boldsymbol{h}^{\boldsymbol{f}}(\lambda) I_{n,\boldsymbol{X}}(\lambda) \boldsymbol{h}^{\boldsymbol{f}}(\lambda)^* d\lambda$$

$$= \mathrm{tr} \int_{-\pi}^{\pi} \boldsymbol{h}^{\boldsymbol{f}}(\lambda)\boldsymbol{g}(\lambda)\boldsymbol{h}^{\boldsymbol{f}}(\lambda)^* d\lambda + \frac{d}{\sqrt{n}} \quad a.s., \tag{8.7.12}$$

where d is a positive constant. Thus

$$\begin{aligned}
\frac{1}{\hat{E}[\|\hat{\boldsymbol{X}}^{\boldsymbol{f}}(0)\|^2]} &= \frac{1}{\mathrm{tr} \int_{-\pi}^{\pi} \boldsymbol{h}^{\boldsymbol{f}}(\lambda)\boldsymbol{g}(\lambda)\boldsymbol{h}^{\boldsymbol{f}}(\lambda)^* d\lambda + \frac{d}{\sqrt{n}}} \quad a.s. \\
&= \frac{1}{\mathrm{tr} \int_{-\pi}^{\pi} \boldsymbol{h}^{\boldsymbol{f}}(\lambda)\boldsymbol{g}(\lambda)\boldsymbol{h}^{\boldsymbol{f}}(\lambda)^* d\lambda \left\{ 1 + \frac{d}{\sqrt{n}} \frac{1}{\mathrm{tr} \int_{-\pi}^{\pi} \boldsymbol{h}^{\boldsymbol{f}}(\lambda)\boldsymbol{g}(\lambda)\boldsymbol{h}^{\boldsymbol{f}}(\lambda)^* d\lambda} \right\}} \quad a.s. \\
&= \frac{1}{\mathrm{tr} \int_{-\pi}^{\pi} \boldsymbol{h}^{\boldsymbol{f}}(\lambda)\boldsymbol{g}(\lambda)\boldsymbol{h}^{\boldsymbol{f}}(\lambda)^* d\lambda} \left\{ 1 - \frac{d}{\sqrt{n}} \frac{1}{\mathrm{tr} \int_{-\pi}^{\pi} \boldsymbol{h}^{\boldsymbol{f}}(\lambda)\boldsymbol{g}(\lambda)\boldsymbol{h}^{\boldsymbol{f}}(\lambda)^* d\lambda} + O\left(\frac{1}{n}\right) \right\} \\
&\quad a.s.
\end{aligned} \tag{8.7.13}$$

Substituting (8.7.13) into $\tilde{\boldsymbol{X}}^{\boldsymbol{f}}_S(0)$, we can evaluate the interpolation error of $\tilde{\boldsymbol{X}}^{\boldsymbol{f}}_S(0)$ as

$$\begin{aligned}
&E[\|\boldsymbol{X}(0) - \tilde{\boldsymbol{X}}^{\boldsymbol{f}}_S(0)\|^2] \\
&= E\left[\left\| \boldsymbol{X}(0) - \hat{\boldsymbol{X}}^{\boldsymbol{f}}(0) + \frac{c}{E[\|\hat{\boldsymbol{X}}^{\boldsymbol{f}}(0)\|^2]} \hat{\boldsymbol{X}}^{\boldsymbol{f}}(0) \right\|^2 \right] + \frac{d^2}{n} \frac{c^2}{E[\|\hat{\boldsymbol{X}}^{\boldsymbol{f}}(0)\|^2]} \\
&\quad - \frac{2d}{\sqrt{n}} \frac{c}{(E[\|\hat{\boldsymbol{X}}^{\boldsymbol{f}}(0)\|^2])^2} \mathrm{tr} \int_{-\pi}^{\pi} \boldsymbol{g}(\lambda)\boldsymbol{h}^{\boldsymbol{f}}(\lambda)^* d\lambda + \frac{2d}{\sqrt{n}} \frac{c}{E[\|\hat{\boldsymbol{X}}^{\boldsymbol{f}}(0)\|^2]} \\
&\quad - \frac{2d}{\sqrt{n}} \frac{c^2}{(E[\|\hat{\boldsymbol{X}}^{\boldsymbol{f}}(0)\|^2])^2} + O\left(\frac{1}{n}\right) \\
&= E[\|\boldsymbol{X}(0) - \hat{\boldsymbol{X}}^{\boldsymbol{f}}(0)\|^2] + O\left(\frac{1}{\sqrt{n}}\right),
\end{aligned}$$

which completes the proof. □

Here $B(\boldsymbol{f}, \boldsymbol{g})$ depends on the unknown true spectral density $\boldsymbol{g}(\lambda)$. To use the shrinkage interpolator practically, we must estimate $B(\boldsymbol{f}, \boldsymbol{g})$. Because $B(\boldsymbol{f}, \boldsymbol{g})$ is written as (8.7.7), which is the integral functional of $\boldsymbol{g}(\lambda)$, we propose

$$\widehat{B(\boldsymbol{f}, \boldsymbol{g})} = B(\boldsymbol{f}, I_{n,\boldsymbol{X}}) \tag{8.7.14}$$

as an estimator of $B(\boldsymbol{f}, \boldsymbol{g})$. Under Assumption 8.7.3, we can also see the estimator (8.7.14) is a $\sqrt{n}$-consistent estimator due to Theorem 8.1.1. Therefore we can evaluate the MSIE of the practical shrinkage interpolator

$$\tilde{\tilde{\boldsymbol{X}}}^{\boldsymbol{f}}_S(0) = \left(1 - \frac{-\widehat{B(\boldsymbol{f}, \boldsymbol{g})}}{\hat{E}[\|\hat{\boldsymbol{X}}^{\boldsymbol{f}}(0)\|^2]} \right) \hat{\boldsymbol{X}}^{\boldsymbol{f}}(0). \tag{8.7.15}$$

Theorem 8.7.5. *Under Assumption 8.7.3, the MSIE of the practical shrinkage estimator* $\tilde{\tilde{\boldsymbol{X}}}^{\boldsymbol{f}}_S(0)$ *is evaluated as*

$$E[\|\boldsymbol{X}(0) - \tilde{\tilde{\boldsymbol{X}}}^{\boldsymbol{f}}_S(0)\|^2] = E[\|\boldsymbol{X}(0) - \hat{\boldsymbol{X}}^{\boldsymbol{f}}(0)\|^2] - \frac{B(\boldsymbol{f}, \boldsymbol{g})^2}{E[\|\hat{\boldsymbol{X}}^{\boldsymbol{f}}(0)\|^2]} + O\left(\frac{1}{\sqrt{n}}\right), \tag{8.7.16}$$

which implies that the shrinkage interpolator improves the usual pseudo interpolator asymptotically in the sense of MSIE.

Proof. Similarly to the previous proof, we can see

$$\widehat{B(f,g)} = B(f,g) + \frac{\tilde{d}}{\sqrt{n}}, \quad (\tilde{d} > 0), \quad a.s.$$

Then we can evaluate the interpolation error of the practical shrinkage interpolator $\tilde{\hat{X}}_S^f(0)$ as (8.7.16). □

Remark 8.7.6. *For the problem of prediction, we can construct a shrinkage predictor as in (8.7.15), and show the result as in Theorem 8.7.5 (see Hamada and Taniguchi (2014)).*

In what follows, we see that the pseudo shrinkage interpolator is useful under the misspecification of spectral density. Assume the scalar-valued data $\{X(t)\}$ are generated by the AR(1) model

$$X(t) = 0.3X(t-1) + \varepsilon(t), \qquad \varepsilon(t) \sim \text{i.i.d. } t(10), \tag{8.7.17}$$

where $t(10)$ is Student's t-distribution with degree of freedom 10. The true spectral density is

$$g(\lambda) = \frac{1.25}{2\pi} \frac{1}{|1 - 0.3e^{i\lambda}|^2}.$$

Here we consider that we fit a pseudo parametric spectral density which is

$$f(\lambda, \theta) = \frac{1.25}{2\pi} \frac{1}{|1 - (0.3 + \theta)e^{i\lambda} + 0.3\theta e^{2i\lambda}|^2}.$$

Figure 8.4 shows the behaviours of $g(\lambda)$ and $f(\lambda, \theta)$, $(\theta = 0.1, 0.3, 0.9)$.

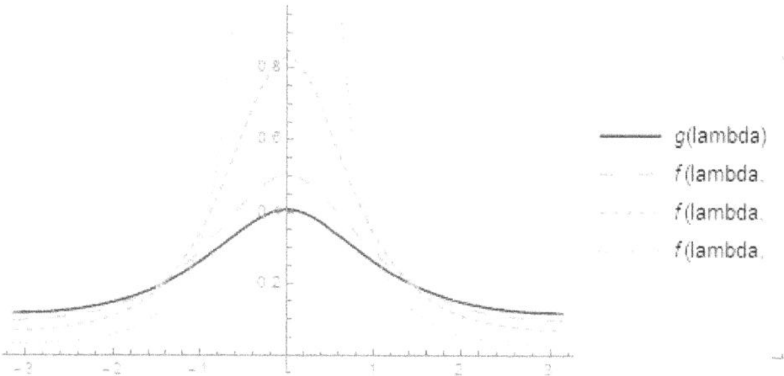

Figure 8.4: Plots of $g(\lambda)$ and $f(\lambda, \theta)$, $(\theta = 0.1, 0.3, 0.9)$

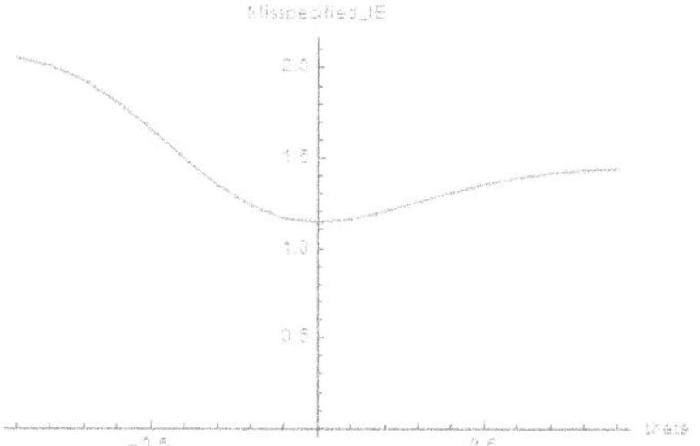

Figure 8.5: Misspecified interpolation error with θ

From Figure 8.4, when $|\theta|$ becomes close to 1, the disparity between $f(\lambda, \theta)$ and $g(\lambda)$ becomes large. Figure 8.5 also shows that the misspecified interpolation error becomes large when $|\theta|$ is near to 1. Since θ indicates the disparity between $f(\lambda, \theta)$ and $g(\lambda)$, we call θ the disparity parameter. As we saw, the misspecification of spectral density is troublesome; thus we should note that influence.

Under the situation, we see that the practical shrinkage interpolator defined by (8.7.15) is more useful than the usual interpolator. Furthermore, we observe that that shrinkage interpolator is effective when the misspecification of the spectral density is large.

First, we generate the data $\{X(-n), \ldots, X(-1), X(0), X(1), \ldots, X(n)\}$ from the model (8.7.17). Next, by using $\{X(-n), \ldots, X(-1), X(1), \ldots, X(n)\}$ without $X(0)$ and pseudo spectral density $f(\lambda, \theta)$, we calculate the usual interpolator $\hat{X}^f(0)$. Then, we calculate the shrinkage interpolator $\tilde{\hat{X}}_S^f(0)$ defined by (8.7.15). We make 500 replications, and plot the values of the sample MSIE such that

$$\widehat{MSUIE} = \frac{1}{500}(X(0) - \hat{X}^f(0))^2$$

and

$$\widehat{MSSIE} = \frac{1}{500}(X(0) - \tilde{\tilde{X}}_S^f(0))^2$$

with θ. Figure 8.6 shows that the shrinkage interpolator usually improves the usual interpolator in the sense of MSIE, and when the misspecification of the spectral density is large, the improvement of the shrinkage interpolator becomes large. We can say that the shrinkage interpolator is useful under the misspecification of the spectra.

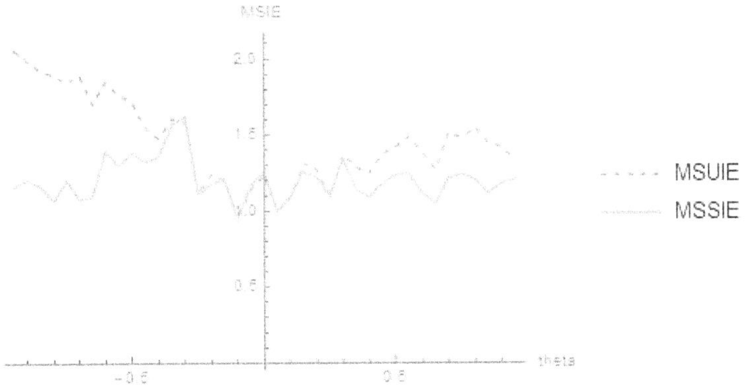

Figure 8.6: Plots of $\widehat{MSUIE}$ and $\widehat{MSSIE}$

Problems

8.1. Show the statement of Remark (8.3.2).

8.2. Show the statement of Remark (8.3.4).

8.3. Verify (8.3.10).

8.4. Derive $GAIC(p)$ in (8.4.20).

8.5. Verify 8.4.23.

8.6. Show $GAIC(p)$ for CHARN given by (8.4.25).

8.7. Show that 8.5.4 does not satisfy (8.5.3).

8.8. Show that 8.5.5 does not satisfy (8.5.3).

8.9. Show the formula 8.6.32.

8.10. Generate $\boldsymbol{X}(1), ..., \boldsymbol{X}(100)$ from $N_3(\boldsymbol{0}, \boldsymbol{I}_3)$, and repeat this procedure 1000 times. Calculating $\hat{\boldsymbol{X}}_n$ and $\hat{\boldsymbol{\theta}}_n$ in 8.6.1 1000 times, compare their empirical MSEs.

8.11. Generate $\mathbf{X}(1), ..., \mathbf{X}(100)$ from a Gaussian stationary process with the spectral density matrix given in (i) of Example 8.6.2 for the case when $\eta_1 = 0.1, \eta_2 = 0.5$ and $\eta_3 = 0.8$. Then, repeat this procedure 1000 times. Calculating $\bar{\mathbf{X}}_n$ and $\hat{\theta}_n$ in (8.6.1), compare their empirical MSEs.

8.12. Verify the formula (8.6.39).

Bibliography

C. Acerbi. *Coherent Representations of Subjective Risk-Aversion*, In: G. Szegö (ed.). John Wiley & Sons Inc., New York, 2004.

Carlo Acerbi and Dirk Tasche. On the coherence of expected shortfall. (cond-mat/0104295), April 2001.

C. J. Adcock and K. Shutes. An analysis of skewness and skewness persistence in three emerging markets. *Emerging Markets Review*, 6(4):396–418, 2005.

Yacine Aït-Sahalia. Disentangling diffusion from jumps. *Journal of Financial Economics*, 74(3): 487–528, 2004.

Yacine Aït-Sahalia and Michael W. Brandt. Variable selection for portfolio choice. *The Journal of Finance*, 56(4):1297–1351, 2001.

Yacine Aït-Sahalia and Lars Hansen, editors. *Handbook of Financial Econometrics, Volume 1: Tools and Techniques (Handbooks in Finance)*. North Holland, 2009.

Yacine Aït-Sahalia, Julio Cacho-Diaz, and T.R. Hurd. Portfolio choice with jumps: A closed-form solution. *Ann. Appl. Probab.*, 19(2):556–584, 2009.

Yacine Aït-Sahalia, Per A. Mykland, and Lan Zhang. Ultra high frequency volatility estimation with dependent microstructure noise. *J. Econometrics*, 160(1):160–175, 2011.

H. Akaike. Information theory and an extension of the maximum likelihood principle. In B. N. Petrov and F. Csadki, (eds.) *2nd International Symposium on Information Theory*, pages 267–281. Budapest: Akademiai Kiado, 1973.

Hirotugu Akaike. A new look at the statistical model identification. *IEEE Trans. Automatic Control*, AC-19:716–723, 1974. System identification and time-series analysis.

Hirotugu Akaike. On entropy maximization principle. In *Applications of Statistics (Proc. Sympos., Wright State Univ., Dayton, Ohio, 1976)*, pages 27–41. North-Holland, Amsterdam, 1977.

Hirotugu Akaike. A Bayesian analysis of the minimum AIC procedure. *Ann. Inst. Statist. Math.*, 30(1):9–14, 1978.

Gordon J. Alexander and Alexandre M. Baptista. Economic implications of using a mean-var model for portfolio selection: A comparison with mean-variance analysis. *Journal of Economic Dynamics and Control*, 26(7):1159–1193, 2002.

M. Allais. Le comportement de l'homme rationnel rationnel devant le risque: critique des postulats et axiomes de l'école américaine. *Econometrica*, 21:503–546, 1953.

Michael Allen and Somnath Datta. A note on bootstrapping M-estimators in ARMA models. *J. Time Ser. Anal.*, 20(4):365–379, 1999.

T. W. Anderson. *The Statistical Analysis of Time Series*. John Wiley & Sons, Inc., New York, 1971.

T. W. Anderson. *An Introduction to Multivariate Statistical Analysis*. Wiley Series in Probability and Mathematical Statistics: Probability and Mathematical Statistics. John Wiley & Sons Inc., New York, second edition, 1984.

T. W. Anderson. *An Introduction to Multivariate Statistical Analysis.* Wiley Series in Probability and Statistics. Wiley-Interscience [John Wiley & Sons], Hoboken, NJ, third edition, 2003.

Fred C. Andrews. Asymptotic behavior of some rank tests for analysis of variance. *Ann. Math. Statistics*, 25:724–736, 1954.

Philippe Artzner, Freddy Delbaen, Jean-Marc Eber, and David Heath. Coherent measures of risk. *Mathematical Finance*, 9(3):203–228, 1999.

Alexander Aue, Siegfried Hörmann, Lajos Horváth, Marie Hušková, and Josef G. Steinebach. Sequential testing for the stability of high-frequency portfolio betas. *Econometric Theory*, 28 (4):804–837, 2012.

Zhidong Bai, Huixia Liu, and Wing-Keung Wong. Enhancement of the applicability of Markowitz's portfolio optimization by utilizing random matrix theory. *J. Math. Finance*, 19 (4):639–667, 2009.

T. A. Bancroft. Analysis and inference for incompletely specified models involving the use of preliminary test(s) of significance. *Biometrics*, 20:427–442, 1964.

Gopal Basak, Ravi Jagannathan, and Guoqiang Sun. A direct test for the mean variance efficiency of a portfolio. *J. Econom. Dynam. Control*, 26(7-8):1195–1215, 2002.

Ishwar V. Basawa and David John Scott. *Asymptotic Optimal Inference for Nonergodic Models*, volume 17 of *Lecture Notes in Statistics.* Springer-Verlag, New York, 1983.

G. W. Bassett Jr., R. Koenker, and G. Kordas. Pessimistic portfolio allocation and Choquet expected utility. *Journal of Financial Econometrics*, 2(4):477–492, 2004.

Richard Bellman. *Introduction to Matrix Analysis.* McGraw–Hill Book Co., Inc., New York, 1960.

Richard Bellman. *Dynamic Programming (Princeton Landmarks in Mathematics).* Princeton Univ Pr, 2010.

Yoav Benjamini and Yosef Hochberg. Controlling the false discovery rate: A practical and powerful approach to multiple testing. *Journal of the Royal Statistical Society. Series B (Methodological)*, pages 289–300, 1995.

Jan Beran. *Statistics for Long-Memory Processes*, volume 61 of *Monographs on Statistics and Applied Probability.* Chapman & Hall, New York, 1994.

P. J. Bickel, C. A. J. Klaassen, Y. Ritov, and J. A. Wellner. *Efficient and Adaptive Estimation for Semiparametric Models.* Johns Hopkins series in the mathematical sciences. Springer, New York, 1998.

F. Black and R. Litterman. Global portfolio optimization. *Financial Analysts Journal*, 48(5): 28–43, 1992.

Avrim Blum and Adam Kalai. Universal portfolios with and without transaction costs. *Machine Learning*, 35(3):193–205, 1999.

Taras Bodnar, Nestor Parolya, and Wolfgang Schmid. Estimation of the global minimum variance portfolio in high dimensions. *arXiv preprint arXiv:1406.0437*, 2014.

Tim Bollerslev. Generalized autoregressive conditional heteroskedasticity. *J. Econometrics*, 31(3): 307–327, 1986.

Tim Bollerslev, Robert F. Engle, and Jeffrey M. Wooldridge. A capital asset pricing model with time-varying covariances. *Journal of Political Economy*, 96(1):116–131, 1988.

B. M. Bolstad, F. Collin, J. Brettschneider, K. Simpson, L. Cope, R. A. Irizarry, and T. P. Speed. Quality assessment of affymetrix genechip data. In *Bioinformatics and Computational Biology Solutions Using R and Bioconductor*, pages 33–47. Springer, 2005.

Arup Bose. Edgeworth correction by bootstrap in autoregressions. *Ann. Statist.*, 16(4):1709–1722, 1988.

Michael W. Brandt, Amit Goyal, Pedro Santa-Clara, and Jonathan R. Stroud. A simulation approach to dynamic portfolio choice with an application to learning about return predictability. *Review of Financial Studies*, 18(3):831–873, 2005.

Douglas T. Breeden. An intertemporal asset pricing model with stochastic consumption and investment opportunities. *Journal of Financial Economics*, 7(3):265–296, 1979.

David R. Brillinger. *Time Series: Data Analysis and Theory*, volume 36. SIAM, 2001a.

David R. Brillinger. *Time Series*, volume 36 of *Classics in Applied Mathematics*. Society for Industrial and Applied Mathematics (SIAM), Philadelphia, 2001b. Data analysis and theory, Reprint of the 1981 edition.

D. R. Brillinger. *Time Series: Data Analysis and Theory*. Classics in Applied Mathematics. Society for Industrial and Applied Mathematics, 2001c.

Peter J. Brockwell and Richard A. Davis. *Time Series: Theory and Methods*. Springer Series in Statistics. Springer, New York, 2006. Reprint of the second (1991) edition.

B. M. Brown. Martingale central limit theorems. *Ann. Math. Statist.*, 42:59–66, 1971.

Peter Bühlmann. Sieve bootstrap for time series. *Bernoulli*, 3(2):123–148, 1997.

Anna Campain and Yee Hwa Yang. Comparison study of microarray meta-analysis methods. *BMC Bioinformatics*, 11(1):1, 2010.

John Y. Campbell, Andrew W. Lo, and Archie Craig Mackinlay. *The Econometrics of Financial Markets*. Princeton Univ Pr, 1996.

Gary Chamberlain. Funds, factors, and diversification in arbitrage pricing models. *Econometrica*, 51(5):1305–1323, 1983.

Gary Chamberlain and Michael Rothschild. Arbitrage, factor structure, and mean-variance analysis on large asset markets. *Econometrica*, 51(5):1281–1304, 1983.

S. Ajay Chandra and Masanobu Taniguchi. Asymptotics of rank order statistics for arch residual empirical processes. *Stochastic Processes and Their Applications*, 104(2):301–324, 2003.

C.W.S. Chen, Yi-Tung Hsu, and M. Taniguchi. Discriminant analysis for quantile regression. *Subumitted for publication*, 2016.

Min Chen and Hong Zhi An. A note on the stationarity and the existence of moments of the garch model. *Statistica Sinica*, pages 505–510, 1998.

Herman Chernoff and I. Richard Savage. Asymptotic normality and efficiency of certain nonparametric test statistics. *Ann. Math. Statist.*, 29:972–994, 1958.

S. H. Chew and K. R. MacCrimmmon. Alpha-nu choice theory: A generalization of expected utility theory. *Working Paper No. 669, University of British Columbia, Vancouver*, 1979.

Koei Chin, Sandy DeVries, Jane Fridlyand, Paul T. Spellman, Ritu Roydasgupta, Wen-Lin Kuo, Anna Lapuk, Richard M. Neve, Zuwei Qian, and Tom Ryder. Genomic and transcriptional aberrations linked to breast cancer pathophysiologies. *Cancer Cell*, 10(6):529–541, 2006.

Ondřej Chochola, Marie Hušková, Zuzana Prášková, and Josef G. Steinebach. Robust monitoring of CAPM portfolio betas. *J. Multivariate Anal.*, 115:374–395, 2013.

Ondřej Chochola, Marie Hušková, Zuzana Prášková, and Josef G. Steinebach. Robust monitoring of CAPM portfolio betas II. *J. Multivariate Anal.*, 132:58–81, 2014.

G. Choquet. Théorie des capacitiés. *Annals Institut Fourier*, 5:131–295, 1953.

Gerda Claeskens and Nils Lid Hjort. The focused information criterion. *Journal of the American Statistical Association*, 98(464):900–916, 2003.

G. Connor. The three types of factor models: A comparison of their explanatory power. *Financial Analysts Journal*, pages 42–46, 1995.

Thomas M. Cover. Universal portfolios. *Math. Finance*, 1(1):1–29, 1991.

Thomas M. Cover and Erik Ordentlich. Universal portfolios with side information. *IEEE Trans. Inform. Theory*, 42(2):348–363, 1996.

Thomas M. Cover and Joy A. Thomas. *Elements of Information Theory*. Wiley-Interscience [John Wiley & Sons], Hoboken, NJ, second edition, 2006.

Jaksa Cvitanic, Ali Lazrak, and Tan Wang. Implications of the Sharpe ratio as a performance measure in multi-period settings. *Journal of Economic Dynamics and Control*, 32(5):1622–1649, May 2008.

R. Dahlhaus. On the Kullback-Leibler information divergence of locally stationary processes. *Stochastic Process. Appl.*, 62(1):139–168, 1996a.

R. Dahlhaus. Maximum likelihood estimation and model selection for locally stationary processes. *J. Nonparametr. Statist.*, 6(2-3):171–191, 1996b.

R. Dahlhaus. *Asymptotic Statistical Inference for Nonstationary Processes with Evolutionary Spectra*, volume 115 of *Lecture Notes in Statistics* Springer, New York, 1996c.

Somnath Datta. Limit theory and bootstrap for explosive and partially explosive autoregression. *Stochastic Process. Appl.*, 57(2):285–304, 1995.

Somnath Datta. On asymptotic properties of bootstrap for AR(1) processes. *J. Statist. Plann. Inference*, 53(3):361–374, 1996.

Giovanni De Luca, Marc G. Genton, and Nicola Loperfido. A multivariate skew-GARCH model. In *Econometric Analysis of Financial and Economic Time Series. Part A*, volume 20 of *Adv. Econom.*, pages 33–57. Emerald/JAI, Bingley, 2008.

A. P. Dempster, N. M. Laird, and D. B. Rubin. Maximum likelihood from incomplete data via the EM algorithm. *J. Roy. Statist. Soc. Ser. B*, 39(1):1–38, 1977. With discussion.

M. A. H. Dempster, Gautam Mitra, and Georg Pflug, editors. *Quantitative Fund Management*. Chapman & Hall/CRC Financial Mathematics Series. CRC Press, Boca Raton, FL, 2009. Reprinted from Quant. Finance **7** (2007), no. 1, no. 2 and no. 4.

Christine Desmedt, Fanny Piette, Sherene Loi, Yixin Wang, Françoise Lallemand, Benjamin Haibe-Kains, Giuseppe Viale, Mauro Delorenzi, Yi Zhang, and Mahasti Saghatchian d'Assignies. Strong time dependence of the 76-gene prognostic signature for node-negative breast cancer patients in the transbig multicenter independent validation series. *Clinical Cancer Research*, 13(11):3207–3214, 2007.

Maximilian Diehn, Gavin Sherlock, Gail Binkley, Heng Jin, John C Matese, Tina Hernandez-Boussard, Christian A. Rees, J. Michael Cherry, David Botstein, and Patrick O. Brown. Source: A unified genomic resource of functional annotations, ontologies, and gene expression data. *Nucleic Acids Research*, 31(1):219–223, 2003.

Feike C. Drost, Chris A. J. Klaassen, and Bas J. M. Werker. Adaptive estimation in time-series models. *Ann. Statist.*, 25(2):786–817, 04 1997.

W. Dunsmuir. A central limit theorem for parameter estimation in stationary vector time series and its application to models for a signal observed with noise. *Ann. Statist.*, 7(3):490–506, 1979.

W. Dunsmuir and E. J. Hannan. Vector linear time series models. *Advances in Appl. Probability*, 8(2):339–364, 1976.

Eran Eden, Roy Navon, Israel Steinfeld, Doron Lipson, and Zohar Yakhini. GOrilla: A tool for discovery and visualization of enriched GO terms in ranked gene lists. *BMC Bioinformatics*, 10 (1):1, 2009.

Ron Edgar, Michael Domrachev, and Alex E. Lash. Gene expression omnibus: Ncbi gene expression and hybridization array data repository. *Nucleic Acids Research*, 30(1):207–210, 2002.

B. Efron. Bootstrap methods: Another look at the jackknife. *Ann. Statist.*, 7(1):1–26, 1979.

Noureddine El Karoui. High-dimensionality effects in the Markowitz problem and other quadratic programs with linear constraints: Risk underestimation. *Ann. Statist.*, 38(6):3487–3566, 2010.

E. J. Elton, M. J. Gruber, S. J. Brown, and W. N. Goetzmann. *Mordern Portfolio Theory and Investment Analysis*. John Wiley & Sons Inc., New York, 7th edition, 2007.

Robert F. Engle. Autoregressive conditional heteroscedasticity with estimates of the variance of United Kingdom inflation. *Econometrica: Journal of the Econometric Society*, pages 987–1007, 1982.

Robert F. Engle. *Multivarite ARCH with Factor Structures — Cointegration in Variance*. University of California Press, 1987.

Robert F. Engle and Kenneth F. Kroner. Multivariate simultaneous generalized arch. *Econometric Theory*, 11(1):122–150, 1995.

E. F. Fama and K. R. French. The cross-section of expected stock returns. *The Journal of Finance*, 47(2):427–465, 1992.

Eugene F. Fama. Multiperiod consumption-investment decisions. *American Economic Review*, 60 (1):163–74, March 1970.

Jianqing Fan, Mingjin Wang, and Qiwei Yao. Modelling multivariate volatilities via conditionally uncorrelated components. *Journal of the Royal Statistical Society. Series B (Statistical Methodology)*, 70(4):679–702, 2008.

Hsing Fang and Tsong-Yue Lai. Co-kurtosis and capital asset pricing. *The Financial Review*, 32 (2):293, 1997.

Peter C. Fishburn. Mean-risk analysis with risk associated with below-target returns. *American Economic Review*, 67(2):116–26, March 1977.

Daniéle Florens-Zmirou. Approximate discrete-time schemes for statistics of diffusion processes. *Statistics*, 20(4):547–557, 1989.

Mario Forni, Marc Hallin, Marco Lippi, and Lucrezia Reichlin. The generalized dynamic-factor model: Identification and estimation. *Review of Economics and Statistics*, 82(4):540–554, 2000.

Mario Forni, Marc Hallin, Marco Lippi, and Lucrezia Reichlin. The generalized dynamic factor model: One-sided estimation and forecasting. *J. Amer. Statist. Assoc.*, 100(471):830–840, 2005.

Gabriel Frahm. Linear statistical inference for global and local minimum variance portfolios. *Statistical Papers*, 51(4):789–812, 2010.

P. A. Frost and J. E. Savarino. An empirical Bayes approach to efficient portfolio selection. *Journal of Financial and Quantitative Analysis*, 21(3):293–305, 1986.

Alexei A. Gaivoronski and Fabio Stella. Stochastic nonstationary optimization for finding universal portfolios. *Ann. Oper. Res.*, 100:165–188 (2001), 2000. Research in stochastic programming (Vancouver, BC, 1998).

Bernard Garel and Marc Hallin. Local asymptotic normality of multivariate ARMA processes with a linear trend. *Ann. Inst. Statist. Math.*, 47(3):551–579, 1995.

Joseph L. Gastwirth and Stephen S. Wolff. An elementary method for obtaining lower bounds on the asymptotic power of rank tests. *Ann. Math. Statist.*, 39:2128–2130, 1968.

John Geweke. Measurement of linear dependence and feedback between multiple time series. *Journal of the American Statistical Association*, 77(378):304–313, 1982.

R. Giacometti and S. Lozza. Risk measures for asset allocation models, ed. Szegö. *Risk Measures for the 21st Century*. John Wiley & Sons Inc., New York, 2004.

Liudas Giraitis, Piotr Kokoszka, and Remigijus Leipus. Stationary ARCH models: Dependence structure and central limit theorem. *Econometric Theory*, 16(1):3–22, 2000.

Kimberly Glass and Michelle Girvan. Annotation enrichment analysis: An alternative method for evaluating the functional properties of gene sets. *arXiv preprint arXiv:1208.4127*, 2012.

Konstantin Glombek. *High-Dimensionality in Statistics and Portfolio Optimization*. BoD – Books on Demand, 2012.

Sílvia Gonçalves and Lutz Kilian. Bootstrapping autoregressions with conditional heteroskedasticity of unknown form. *J. Econometrics*, 123(1):89–120, 2004.

Christian Gouriéroux and Joann Jasiak. *Financial Econometrics: Problems, Models, and Methods (Princeton Series in Finance)*. Princeton Univ Pr, illustrated edition, 2001.

Clive W. J. Granger. Investigating causal relations by econometric models and cross-spectral methods. *Econometrica: Journal of the Econometric Society*, pages 424–438, 1969.

Ulf Grenander and Murray Rosenblatt. *Statistical Analysis of Stationary Time Series*. John Wiley & Sons, New York, 1957.

Christian M. Hafner. *Nonlinear Time Series Analysis with Applications to Foreign Exchange Rate Volatility*. Contributions to Economics. Physica-Verlag, Heidelberg, 1998.

Valérie Haggan and Tohru Ozaki. Modelling nonlinear random vibrations using an amplitude-dependent autoregressive time series model. *Biometrika*, 68(1):189–196, 1981.

Jaroslav Hájek. Some extensions of the {W}ald-{W}olfowitz-{N}oether theorem. *Ann. Math. Statist.*, 32:506–523, 1961.

Jaroslav Hájek. Asymptotically most powerful rank-order tests. *Ann. Math. Statist.*, 33:1124–1147, 1962.

Jaroslav Hájek and Zbyněk Šidák. *Theory of Rank Tests*. Academic Press, New York, 1967.

P. Hall and C. C. Heyde. *Martingale Limit Theory and Its Application*. Academic Press Inc. [Harcourt Brace Jovanovich Publishers], New York, 1980. Probability and Mathematical Statistics.

Peter Hall. *The Bootstrap and Edgeworth Expansion*. Springer, New York, pages xiv+352, 1992.

Marc Hallin and Davy Paindaveine. Optimal tests for multivariate location based on interdirections and pseudo-Mahalanobis ranks. *Ann. Statist.*, 30(4):1103–1133, 2002.

Marc Hallin and Davy Paindaveine. Affine-invariant aligned rank tests for the multivariate general linear model with {VARMA} errors. *Journal of Multivariate Analysis*, 93(1):122–163, 2005.

Marc Hallin and Davy Paindaveine. Semiparametrically efficient rank-based inference for shape. I. Optimal rank-based tests for sphericity. *Ann Statistics*, 34(6):2707–2756, 2006.

Marc Hallin and Madan L. Puri. Optimal rank-based procedures for time series analysis: Testing an {ARMA} model against other {ARMA} models. *Ann. Statist.*, 16(1):402–432, 1988.

Marc Hallin and Madan L. Puri. Rank tests for time series analysis: A survey. In *New Directions in Time Series Analysis, {P}art {I}*, volume 45 of *IMA Vol. Math. Appl.*, pages 111–153. Springer, New York, 1992.

Marc Hallin and Bas J. M. Werker. Semi-parametric efficiency, distribution-freeness and invariance. *Bernoulli*, 9(1):137–165, 2003.

Marc Hallin, Jean-François Ingenbleek, and Madan L. Puri. Linear serial rank tests for randomness against {ARMA} alternatives. *Ann. Statist.*, 13(3):1156–1181, 1985a.

Marc Hallin, Masanobu Taniguchi, Abdeslam Serroukh, and Kokyo Choy. Local asymptotic normality for regression models with long-memory disturbance. *Ann. Statist.*, 27(6):2054–2080, 1999.

Kenta Hamada and Masanobu Taniguchi. Statistical portfolio estimation for non-Gaussian return processes. *Advances in Science, Technology and Environmentology*, B10:69–83, 2014.

Kenta Hamada, Dong Wei Ye, and Masanobu Taniguchi. Statistical portfolio estimation under the utility function depending on exogenous variables. *Adv. Decis. Sci.*, pages Art. ID 127571, 15, 2012.

E. J. Hannan. *Multiple Time Series*. John Wiley and Sons, Inc., New York, 1970.

E. J. Hannan and B. G. Quinn. The determination of the order of an autoregression. *J. Roy. Statist. Soc. Ser. B*, 41(2):190–195, 1979.

W. Härdle and T. Gasser. Robust non-parametric function fitting. *Journal of the Royal Statistical Society. Series B (Methodological)*, 46(1):42–51, 1984.

W. Härdle, A. Tsybakov, and L. Yang. Nonparametric vector autoregression. *J. Statist. Plann. Inference*, 68(2):221–245, 1998.

Campbell R. Harvey and Akhtar Siddique. Conditional skewness in asset pricing tests. *Journal of Finance*, pages 1263–1295, 2000.

Trevor Hastie, Robert Tibshirani, and Jerome Friedman. *The Elements of Statistical Learning*. Springer Series in Statistics. Springer, New York, second edition, 2009. Data mining, inference, and prediction.

Edwin Hewitt and Karl Stromberg. *Real and Abstract Analysis. A Modern Treatment of the Theory of Functions of a Real Variable*. Springer-Verlag, New York, 1965.

C. C. Heyde. Fixed sample and asymptotic optimality for classes of estimating functions. *Contemporary Mathematics*. 80:241–247, 1988.

Christopher C. Heyde. *Quasi-Likelihood and Its Application*. Springer Series in Statistics. Springer-Verlag, New York, 1997. A general approach to optimal parameter estimation.

Junichi Hirukawa. Cluster analysis for non-Gaussian locally stationary processes. *Int. J. Theor. Appl. Finance*, 9(1):113–132, 2006.

Junichi Hirukawa and Masanobu Taniguchi. LAN theorem for non-Gaussian locally stationary processes and its applications. *J. Statist. Plann. Inference*, 136(3):640–688, 2006.

Junichi Hirukawa, Hiroyuki Taniai, Marc Hallin, and Masanobu Taniguchi. Rank-based inference for multivariate nonlinear and long-memory time series models. *Journal of the Japan Statistical Society*, 40(1):167–187, 2010.

J. L. Hodges and Erich Leo Lehmann. The efficiency of some nonparametric competitors of the {t}-test. *Ann. Math. Statist.*, 27:324–335, 1956.

M. Honda, H. Kato, S. Ohara, A. Ikeda, Y. Inoue, T. Mihara, K. Baba, and H. Shibasaki. Stochastic time-series analysis of intercortical functional coupling during sustained muscle contraction. *31st Annual Meeting Society for Neuroscience Abstracts*, San Diego, Nov 10–15, 2001.

Lajos Horváth, Piotr Kokoszka, and Gilles Teyssiere. Empirical process of the squared residuals of an arch sequence. *Annals of Statistics*, pages 445–469, 2001.

Yuzo Hosoya. The decomposition and measurement of the interdependency between second-order stationary processes. *Probability Theory and Related Fields*, 88(4):429–444, 1991.

Yuzo Hosoya and Masanobu Taniguchi. A central limit theorem for stationary processes and the parameter estimation of linear processes. *Ann. Statist.*, 10(1):132–153, 1982.

Soosung Hwang and Stephen E. Satchell. Modelling emerging market risk premia using higher moments. *Return Distributions in Finance*, page 75, 1999.

I. A. Ibragimov and V. N. Solev. A certain condition for the regularity of {G}aussian stationary sequence. *Zap. Naučn. Sem. Leningrad. Otdel. Mat. Inst. Steklov. (LOMI)*, 12:113–125, 1969.

I.A. Ibragimov. A central limit theorem for a class of dependent random variables. *Theory of Probability & Its Applications*, 8(1):83–89, 1963.

Il\cprimedar Abdulovich Ibragimov and Y. A. Rozanov. *Gaussian Random Processes*, volume 9 of *Applications of Mathematics*. Springer-Verlag, New York, 1978.

Rafael A. Irizarry, Benjamin M. Bolstad, Francois Collin, Leslie M. Cope, Bridget Hobbs, and Terence P. Speed. Summaries of affymetrix genechip probe level data. *Nucleic Acids Research*, 31(4):e15–e15, 2003.

Jean Jacod and Albert N. Shiryaev. *Limit Theorems for Stochastic Processes*, volume 288. Springer-Verlag, Berlin, 1987.

W. James and Charles Stein. Estimation with quadratic loss. In *Proc. 4th Berkeley Sympos. Math. Statist. and Prob., Vol. I*, pages 361–379, Berkeley, Calif., 1961. Univ. California Press.

Harold Jeffreys. *Theory of Probability*. Third edition. Clarendon Press, Oxford, 1961.

P. Jeganathan. Some aspects of asymptotic theory with applications to time series models. *Econometric Theory*, 11(5):818–887, 1995. Trending multiple time series (New Haven, CT, 1993).

Kiho Jeong, Wolfgang K. Härdle, and Song Song. A consistent nonparametric test for causality in quantile. *Econometric Theory*, 28(04):861–887, 2012.

J. David Jobson and Bob Korkie. Estimation for Markowitz efficient portfolios. *Journal of the American Statistical Association*, 75(371):544–554, 1980.

J. D. Jobson and R. M. Korkie. Putting Markowitz theory to work. *The Journal of Portfolio Management*, 7(4):70–74, 1981.

J. D. Jobson, B. Korkie, and V. Ratti. Improved estimation for Markowitz portfolios using James-Stein type estimators. 41:279–284, 1979.

Harry Joe. *Dependence Modeling with Copulas*, volume 134 of *Monographs on Statistics and Applied Probability*. CRC Press, Boca Raton, FL, 2015.

Geweke John. The Dynamic Factor Analysis of Economic Time Series, *Latent Variables in Socio-Economic Models*, 1977.

P. Jorion. Bayes-Stein estimation for portfolio analysis. *Journal of Financial and Quantitative Analysis*, 21(3):279–292, 1986.

Yoshihide Kakizawa. Note on the asymptotic efficiency of sample covariances in Gaussian vector stationary processes. *J. Time Ser. Anal.*, 20(5):551–558, 1999.

Iakovos Kakouris and Berç Rustem. Robust portfolio optimization with copulas. *European J. Oper. Res.*, 235(1):28–37, 2014.

Shinji Kataoka. A stochastic programming model. *Econometrica*, 31:181–196, 1963.

H. Kato, M. Taniguchi, and M. Honda. Statistical analysis for multiplicatively modulated nonlinear autoregressive model and its applications to electrophysiological signal analysis in humans. *Signal Processing, IEEE Transactions on*, 54(9):3414–3425, 2006.

Daniel MacRae Keenan. Limiting behavior of functionals of higher-order sample cumulant spectra. *Ann. Statist.*, 15(1):134–151, 1987.

J. L. Kelly. A new interpretation of information rate. *Bell. System Tech. J.*, 35:917–926, 1956.

Mathieu Kessler. Estimation of an ergodic diffusion from discrete observations. *Scand. J. Statist.*, 24(2):211–229, 1997.

Keith Knight. Limiting distributions for L_1 regression estimators under general conditions. *The Annals of Statistics*, 26(2):755–770, 1998.

I. Kobayashi, T. Yamashita, and M. Taniguchi. Portfolio estimation under causal variables. *Special Issue on the Financial and Pension Mathematical Science. ASTE*, B9:85–100, 2013.

Roger Koenker. *Quantile Regression (Econometric Society Monographs)*. Cambridge University Press, 2005.

Roger Koenker and Gilbert Bassett, Jr. Regression quantiles. *Econometrica*, 46(1):33–50, 1978.

Roger Koenker and Zhijie Xiao. Quantile autoregression. *J. Amer. Statist. Assoc.*, 101(475):980–990, 2006.

Hiroshi Konno and Hiroaki Yamazaki. Mean-absolute deviation portfolio optimization model and its applications to Tokyo stock market. *Management Science*, 37(5):519–531, May 1991.

Amnon Koren, Itay Tirosh, and Naama Barkai. Autocorrelation analysis reveals widespread spatial biases in microarray experiments. *BMC Genomics*, 8(1):164, 2007.

Hira L. Koul and A. K. Md. E. Saleh. R-estimation of the parameters of autoregressive [AR(p)] models. *Ann. Statist.*, 21(1):534–551, 1993.

Alan Kraus and Robert H. Litzenberger. Skewness preference and the valuation of risk assets*. *The Journal of Finance*, 31(4):1085–1100, 1976.

Jens-Peter Kreiss. On adaptive estimation in stationary ARMA processes. *Ann. Statist.*, 15(1): 112–133, 1987.

Jens-Peter Kreiss. Local asymptotic normality for autoregression with infinite order. *J. Statist. Plann. Inference*, 26(2):185–219, 1990.

Jens-Peter Kreiss and Jürgen Franke. Bootstrapping stationary autoregressive moving-average models. *J. Time Ser. Anal.*, 13(4):297–317, 1992.

William H. Kruskal and W. Allen Wallis. Use of ranks in {one-criterion} variance analysis. *Journal of the American Statistical Association*, 47(260):583–621, 1952.

S. N. Lahiri. *Resampling Methods for Dependent Data*. Springer Series in Statistics. Springer-Verlag, New York, 2003.

G. J. Lauprete, A. M. Samarov, and R. E. Welsch. Robust portfolio optimization. *Metrika*, 55(1): 139–149, 2002.

Lucien Le Cam. *Asymptotic Methods in Statistical Decision Theory*. Springer Series in Statistics. Springer-Verlag, New York, 1986.

O. Ledoit and M. Wolf. Improved estimation of the covariance matrix of stock returns with an application to portfolio selection. *Journal of Empirical Finance*, 10(5):603–621, 2003.

O. Ledoit and M. Wolf. Honey, I shrunk the sample covariance matrix. *The Journal of Portfolio Management*, 30(4):110–119, 2004.

Sangyeol Lee and Masanobu Taniguchi. Asymptotic theory for ARCH-SM models: LAN and residual empirical processes. *Statist. Sinica*, 15(1):215–234, 2005.

Erich Leo Lehmann and J. P. Romano. *Testing Statistical Hypotheses*. Springer Texts in Statistics. Springer, 2005.

Markus Leippold, Fabio Trojani, and Paolo Vanini. A geometric approach to multiperiod mean variance optimization of assets and liabilities. *Journal of Economic Dynamics and Control*, 28:1079–1113, 2004.

Duan Li and Wan-Lung Ng. Optimal dynamic portfolio selection: Multiperiod mean-variance formulation. *Mathematical Finance*, 10(3):387–406, 2000.

Kian-Guan Lim. A new test of the three-moment capital asset pricing model. *Journal of Financial and Quantitative Analysis*, 24(02):205–216, 1989.

Wen-Ling Lin. Alternative estimators for factor GARCH models–A Monte Carlo comparison. *Journal of Applied Econometrics*, 7(3):259–279, 1992.

John Lintner. The valuation of risk assets and the selection of risky investments in stock portfolios and capital budgets: A reply. *The Review of Economics and Statistics*, 51(2):222–224, May 1969.

Greta M. Ljung and George E. P. Box. On a measure of lack of fit in time series models. *Biometrika*, 65(2):297–303, 1978.

Sherene Loi, Benjamin Haibe-Kains, Christine Desmedt, Françoise Lallemand, Andrew M. Tutt, Cheryl Gillet, Paul Ellis, Adrian Harris, Jonas Bergh, and John A. Foekens. Definition of clinically distinct molecular subtypes in estrogen receptor–positive breast carcinomas through genomic grade. *Journal of Clinical Oncology*, 25(10):1239–1246, 2007.

Zudi Lu and Zhenyu Jiang. L_1 geometric ergodicity of a multivariate nonlinear AR model with an ARCH term. *Statist. Probab. Lett.*, 51(2):121–130, 2001.

Zudi Lu, Dag Johan Steinskog, Dag Tjøstheim, and Qiwei Yao. Adaptively varying-coefficient spatiotemporal models. *Journal of the Royal Statistical Society: Series B (Statistical Methodology)*, 71(4):859–880, 2009.

David G. Luenberger. *Investment Science*. Oxford Univ Pr (Txt), illustrated edition edition, 1997.

Helmut Lütkepohl. General-to-specific or specific-to-general modelling? An opinion on current econometric terminology. *Journal of Econometrics*, 136(1):319–324, 2007.

Jan R. Magnus and Heinz Neudecker. *Matrix Differential Calculus with Applications in Statistics and Econometrics*. Wiley Series in Probability and Statistics. John Wiley & Sons Ltd., Chichester, 1999. Revised reprint of the 1988 original.

Yannick Malevergne and Didier Sornette. *Extreme Financial Risks: From Dependence to Risk Management (Springer Finance)*. Springer, 2006 edition, 2005.

Cecilia Mancini. Disentangling the jumps of the diffusion in a geometric jumping Brownian motion. *Giornale dell' Istituto Italiano degli Attuari*, 64(44):19–47, 2001.

Vladimir A. Marčenko and Leonid Andreevich Pastur. Distribution of eigenvalues for some sets of random matrices. *Sbornik: Mathematics*, 1(4):457–483, 1967.

H. Markowitz. Portfolio selection. *The Journal of Finance*, 7:77–91, 1952.

Harry M. Markowitz. *Portfolio Selection: Efficient Diversification of Investments*. Cowles Foundation for Research in Economics at Yale University, Monograph 16. John Wiley & Sons Inc., New York, 1959.

David S. Matteson and Ruey S. Tsay. Dynamic orthogonal components for multivariate time series. *Journal of the American Statistical Association*, 106(496), 2011.

Alexander J. McNeil, Ruediger Frey, and Paul Embrechts. *Quantitative Risk Management: Concepts, Techniques and Tools (Princeton Series in Finance)*. Princeton Univ Pr, revised edition, 2015.

Robert C. Merton. Lifetime portfolio selection under uncertainty: The continuous-time case. *The Review of Economics and Statistics*, 51(3):247–57, August 1969.

Robert C. Merton. An intertemporal capital asset pricing model. *Econometrica*, 41:867–887, 1973.

Robert C. Merton. *Continuous-Time Finance (Macroeconomics and Finance)*. Wiley-Blackwell, 1992.

Attilio Meucci. Managing diversification. https://papers.ssrn.com/sol3/papers.cfm?abstract_id=1358533 2010.

Anders Milhøj. The moment structure of arch processes. *Scandinavian Journal of Statistics*, pages 281–292, 1985.

Lance D. Miller, Johanna Smeds, Joshy George, Vinsensius B. Vega, Liza Vergara, Alexander Ploner, Yudi Pawitan, Per Hall, Sigrid Klaar, and Edison T. Liu. An expression signature for p53 status in human breast cancer predicts mutation status, transcriptional effects, and patient survival. *Proceedings of the National Academy of Sciences of the United States of America*, 102(38):13550–13555, 2005.

Tatsuya Mima and Mark Hallett. Corticomuscular coherence: A review. *Journal of Clinical Neurophysiology*, 16(6):501, 1999.

Andy J. Minn, Gaorav P. Gupta, Peter M. Siegel, Paula D. Bos, Weiping Shu, Dilip D. Giri, Agnes Viale, Adam B. Olshen, William L. Gerald, and Joan Massagué. Genes that mediate breast cancer metastasis to lung. *Nature*, 436(7050):518–524, 2005.

J. Mossin. Equilibrium in capital asset market. *Econometrica*, 35:768–783, 1966.

J. Mossin. Optimal multiperiod portfolio policies. *The Journal of Business*, 41(2):215–229, 1968.

Kai Wang Ng, Guo-Liang Tian, and Man-Lai Tang. *Dirichlet and Related Distributions*. Wiley Series in Probability and Statistics. John Wiley & Sons Ltd., Chichester, 2011. Theory, methods and applications.

Chapados Nicolas, editor. *Portfolio Choice Problems: An Introductory Survey of Single and Multiperiod Models (Springer Briefs in Electrical and Computer Engineering)*. Springer, 2011.

Gottfried E. Noether. On a theorem of Pitman. *The Annals of Mathematical Statistics*, 26(1): 64–68, 1955.

Shinji Ohara, Takashi Nagamine, Akio Ikeda, Takeharu Kunieda, Riki Matsumoto, Waro Taki, Nobuo Hashimoto, Koichi Baba, Tadahiro Mihara, Stephan Salenius et al. Electrocorticogram–electromyogram coherence during isometric contraction of hand muscle in human. *Clinical Neurophysiology*, 111(11):2014–2024, 2000.

Annamaria Olivieri and Ermanno Pitacco. Solvency requirements for life annuities. In *Proceedings of the AFIR 2000 Colloquium, Tromso, Norway*, pages 547–571, 2000.

Tohru Ozaki and Mitsunori Iino. An innovation approach to non-Gaussian time series analysis. *Journal of Applied Probability*, pages 78–92, 2001.

Helder P. Palaro and Luiz Koodi Hotta. Using conditional copula to estimate value at risk. *Journal of Data Science*, 4:93–115, 2006.

Charles M. Perou, Therese Sørlie, Michael B. Eisen, Matt van de Rijn, Stefanie S. Jeffrey, Christian A. Rees, Jonathan R. Pollack, Douglas T. Ross, Hilde Johnsen, and Lars A. Akslen. Molecular portraits of human breast tumours. *Nature*, 406(6797):747–752, 2000.

Georg Ch. Pflug. Some remarks on the value-at-risk and the conditional value-at-risk. In *Probabilistic Constrained Optimization*, pages 272–281. Springer, 2000.

Dinh Tuan Pham and Philippe Garat. Blind separation of mixture of independent sources through a quasi-maximum likelihood approach. *IEEE T Signal Proces*, 45(7):1712–1725, 1997.

E. J. G. Pitman. *Lecture Notes on Nonparametric Statistical Inference: Lectures Given for the University of North Carolina*. Mathematisch Centrum, 1948.

Stanley R. Pliska. *Introduction to Mathematical Finance: Discrete Time Models*. Wiley, first edition, 1997.

Vassilis Polimenis. The distributional capm: Connecting risk premia to return distributions. *UC Riverside Anderson School Working Paper*, (02-04), 2002.

B. L. S. Prakasa Rao. Asymptotic theory for nonlinear least squares estimator for diffusion processes. *Math. Operationsforsch. Statist. Ser. Statist.*, 14(2):195–209, 1983.

B. L. S. Prakasa Rao. Statistical inference from sampled data for stochastic processes. In *Statistical Inference from Stochastic Processes (Ithaca, NY, 1987)*, volume 80 of *Contemp. Math.*, pages 249–284. Amer. Math. Soc., Providence, RI, 1988.

B. L. S. Prakasa Rao. *Semimartingales and Their Statistical Inference*, volume 83 of *Monographs on Statistics and Applied Probability*. Chapman & Hall/CRC, Boca Raton, FL, 1999.

J. W. Pratt. Risk aversion in the small and in the large. *Econometrica*, 32:122–136, 1964.

Jean-Luc Prigent. *Portfolio Optimization and Performance Analysis*. Chapman & Hall/CRC Financial Mathematics Series. Chapman & Hall/CRC, Boca Raton, FL, 2007.

Madan Lal Puri and Pranab Kumar Sen. *Nonparametric Methods in Multivariate Analysis (Wiley Series in Probability and Statistics).* Wiley, first edition, 1971.

John Quiggin. A theory of anticipated utility. *Journal of Economic Behavior & Organization*, 3 (4):323–343, December 1982.

B. G. Quinn. A note on AIC order determination for multivariate autoregressions. *J. Time Ser. Anal.*, 9(3):241–245, 1988.

Svetlozar T. Rachev, John S. J. Hsu, Biliana S. Bagasheva, and Frank J. Fabozzi CFA. *Bayesian Methods in Finance (Frank J. Fabozzi Series).* Wiley, first edition, 2008.

Karim Rezaul, Jay Kumar Thumar, Deborah H. Lundgren, Jimmy K. Eng, Kevin P. Claffey, Lori Wilson, and David K. Han. Differential protein expression profiles in estrogen receptor positive and negative breast cancer tissues using label-free quantitative proteomics. *Genes & Cancer*, 1 (3):251–271, 2010.

R. T. Rockafellar and S. P. Uryasev. Optimization of conditional value-at-risk. *Journal of Risk*, 2: 21–42, 2000.

B. Rosenberg. "Extra-Market Components of Covariance in Security Returns," *Journal of Financial and Quantitative Analysis*, 9, 263–274, 1974.

Stephen A. Ross. The arbitrage theory of capital asset pricing. *Journal of Economic Theory*, 13 (3):341–360, December 1976.

A. D. Roy. Safety-first and the holding of assets. *Econometrica*, 20:431–449, 1952.

Mark E. Rubinstein. The fundamental theorem of parameter-preference security valuation. *Journal of Financial and Quantitative Analysis*, 8(01):61–69, 1973.

Alexander M. Rudin and Jonathan S. Morgan. A portfolio diversification index. *The Journal of Portfolio Management*, 32(2):81–89, 2006.

David Ruppert. *Statistics and Finance.* Springer Texts in Statistics. Springer-Verlag, New York, 2004. An introduction.

A. K. Md. Ehsanes Saleh. *Theory of Preliminary Test and Stein-Type Estimation with Applications.* Wiley Series in Probability and Statistics. Wiley-Interscience [John Wiley & Sons], Hoboken, NJ, 2006.

Alexander Samarov and Murad S. Taqqu. On the efficiency of the sample mean in long-memory noise. *Journal of Time Series Analysis*, 9(2):191–200, 1988.

Paul A. Samuelson. Lifetime portfolio selection by dynamic stochastic programming. *Review of Economics and Statistics*, 51(3):239–46, August 1969.

Paul A. Samuelson. The fundamental approximation theorem of portfolio analysis in terms of means, variances, and higher moments. *Review of Economic Studies*, 37(4):537–42, October 1970.

Thomas J. Sargent and Christopher A. Sims. Business cycle modeling without pretending to have too much a priori economic theory. *New Methods in Business Cycle Research*, 1:145–168, 1977.

João R. Sato, André Fujita, Elisson F. Cardoso, Carlos E. Thomaz, Michael J. Brammer, and Edson Amaro. Analyzing the connectivity between regions of interest: An approach based on cluster Granger causality for fmri data analysis. *Neuroimage*, 52(4):1444–1455, 2010.

Kenichi Sato. *Lévy Processes and Infinitely Divisible Distributions*, volume 68 of *Cambridge Studies in Advanced Mathematics.* Cambridge University Press, Cambridge, 1999. Translated from the 1990 Japanese original, Revised by the author.

Bernd Michael Scherer and R. Douglas Martin. *Introduction to Modern Portfolio Optimization With NUOPT And S-PLUS.* Springer-Verlag, first ed. 2005. corr. 2nd. printing edition, 2005.

David Schmeidler. Subjective probability and expected utility without additivity. *Econometrica*, 57(3):571–87, May 1989.

Gideon Schwarz. Estimating the dimension of a model. *Ann. Statist.*, 6(2):461–464, 1978.

Motohiro Senda and Masanobu Taniguchi. James-Stein estimators for time series regression models. *J. Multivariate Anal.*, 97(9):1984–1996, 2006.

C. E. Shannon. A mathematical theory of communication. *Bell System Tech. J.*, 27:379–423, 623–656, 1948.

W. Sharpe. Capital asset prices: A theory of market equilibrium under conditions of risk. *The Journal of Finance*, 19:425–442, 1964.

Ritei Shibata. Selection of the order of an autoregressive model by Akaike's information criterion. *Biometrika*, 63(1):117–126, 1976.

Ritei Shibata. Asymptotically efficient selection of the order of the model for estimating parameters of a linear process. *Ann. Statist.*, 8(1):147–164, 1980.

Y. Shimizu. *Estimation of diffusion processes with jumps from discrete observations, Zenkin Tenkai, University of Tokyo, Dec 4, 2002.* PhD thesis, Master thesis 2003, Graduate School of Mathematical Sciences, The University of Tokyo, 2002.

Yasutaka Shimizu. Threshold estimation for jump-type stochastic processes from discrete observations (in Japanese). *Proc. Inst. Statist. Math.*, 57(1):97–118, 2009.

Yasutaka Shimizu and Nakahiro Yoshida. Estimation of parameters for diffusion processes with jumps from discrete observations. *Stat. Inference Stoch. Process.*, 9(3):227–277, 2006.

Hiroshi Shiraishi. A simulation approach to statistical estimation of multiperiod optimal portfolios. *Adv. Decis. Sci.*, pages Art. ID 341476, 13, 2012.

Hiroshi Shiraishi and Masanobu Taniguchi. Statistical estimation of optimal portfolios for locally stationary returns of assets. *Int. J. Theor. Appl. Finance*, 10(1):129–154, 2007.

Hiroshi Shiraishi and Masanobu Taniguchi. Statistical estimation of optimal portfolios for non-Gaussian dependent returns of assets. *J. Forecast.*, 27(3):193–215, 2008.

Hiroshi Shiraishi, Hiroaki Ogata, Tomoyuki Amano, Valentin Patilea, David Veredas, and Masanobu Taniguchi. Optimal portfolios with end-of-period target. *Adv. Decis. Sci.*, pages Art. ID 703465, 13, 2012.

Albert N. Shiryaev. *Essentials of Stochastic Finance*, volume 3 of *Advanced Series on Statistical Science & Applied Probability*. World Scientific Publishing Co. Inc., River Edge, NJ, 1999. Facts, models, theory, translated from the Russian manuscript by N. Kruzhilin.

Ali Shojaie and George Michailidis. Discovering graphical Granger causality using the truncating lasso penalty. *Bioinformatics*, 26(18):i517–i523, 2010.

Steven E. Shreve. *Stochastic Calculus for Finance. II.* Springer Finance. Springer-Verlag, New York, 2004. Continuous-time models.

R. H. Shumway and Z. A. Der. Deconvolution of multiple time series. *Technometrics*, 27(4):385, 1985.

Robert H. Shumway. Time-frequency clustering and discriminant analysis. *Statistics & Probability Letters*, 63(3):307–314, 2003.

Z. Sidak, P. K. Sen, and Jaroslav Hájek. *Theory of Rank Tests.* Probability and Mathematical Statistics. Elsevier Science, San Diego, 1999.

Andrew H. Sims, Graeme J. Smethurst, Yvonne Hey, Michal J. Okoniewski, Stuart D. Pepper, Anthony Howell, Crispin J. Miller, and Robert B. Clarke. The removal of multiplicative, systematic bias allows integration of breast cancer gene expression datasets–improving meta-analysis and prediction of prognosis. *BMC Medical Genomics*, 1(1):1, 2008.

Christopher A. Sims. Money, income, and causality. *The American Economic Review*, 62(4): 540–552, 1972.

Hiroko Kato Solvang and Masanobu Taniguchi. Microarray analysis using rank order statistics for arch residual empirical process. *Open Journal of Statistics*, 7(01):54–71, 2017a.

Hiroko Kato Solvang and Masanobu Taniguchi. Portfolio estimation for spectral density of categorical time series data. *Far East Journal of Theoretical Statistics*, 53(1):19–33, 2017b.

Michael Sørensen. Likelihood methods for diffusions with jumps. In *Statistical Inference in Stochastic Processes*, volume 6 of *Probab. Pure Appl.*, pages 67–105. Dekker, New York, 1991.

Frank Spitzer. A combinatorial lemma and its application to probability theory. *Trans. Amer. Math. Soc.*, 82:323–339, 1956.

Charles Stein. Inadmissibility of the usual estimator for the mean of a multivariate normal distribution. In *Proceedings of the Third Berkeley Symposium on Mathematical Statistics and Probability, 1954–1955, vol. I*, pages 197–206, Berkeley, 1956. University of California Press.

David S. Stoffer. Nonparametric frequency detection and optimal coding in molecular biology. In *Modeling Uncertainty*, pages 129–154. Springer, 2005.

David S. Stoffer, David E. Tyler, and Andrew J. McDougall. Spectral analysis for categorical time series: Scaling and the spectral envelope. *Biometrika*, 80(3):611–622, 1993.

David S. Stoffer, David E. Tyler, and David A. Wendt. The spectral envelope and its applications. *Statist. Sci.*, 15(3):224–253, 2000.

William F. Stout. *Almost Sure Convergence*. Academic Press [A subsidiary of Harcourt Brace Jovanovich, Publishers], New York, 1974. Probability and Mathematical Statistics, Vol. 24.

Helmut Strasser. *Mathematical Theory of Statistics*, volume 7 of *de Gruyter Studies in Mathematics*. Walter de Gruyter & Co., Berlin, 1985. Statistical experiments and asymptotic decision theory.

Frantisek Stulajter. Comparison of different copula assumptions and their application in portfolio construction. *Ekonomická Revue-Central European Review of Economic Issues*, 12, 2009.

Y. Suto and M. Taniguchi. Shrinkage interpolation for stationary processes. *ASTE, Research Institute for Science and Engineering, Waseda University Special Issue "Financial and Pension Mathematical Science*," 13:35–42, 2016.

Y. Suto, Y. Liu, and M. Taniguchi. Asymptotic theory of parameter estimation by a contrast function based on interpolation error. *Statist. Infer. Stochast. Process*, 19(1):93–110, 2016.

Anders Rygh Swensen. The asymptotic distribution of the likelihood ratio for autoregressive time series with a regression trend. *J. Multivariate Anal.*, 16(1):54–70, 1985.

Gábor J. Székely and Maria L. Rizzo. Brownian distance covariance. *Ann. Appl. Stat.*, 3(4): 1236–1265, 2009.

Gábor J. Székely, Maria L. Rizzo, and Nail K. Bakirov. Measuring and testing dependence by correlation of distances. *Ann. Statist.*, 35(6):2769–2794, 2007.

H. Taniai and T. Shiohama. Statistically efficient construction of α-risk-minimizing portfolio. *Advances in Decision Sciences*, 2012.

Masanobu Taniguchi. Robust regression and interpolation for time series. *Journal of Time Series Analysis*, 2(1):53–62, 1981.

Masanobu Taniguchi. *Higher Order Asymptotic Theory for Time Series Analysis*. Springer, 1991.

Masanobu Taniguchi and Junichi Hirukawa. The Stein-James estimator for short- and long-memory Gaussian processes. *Biometrika*, 92(3):737–746, 2005.

Masanobu Taniguchi and Junichi Hirukawa. Generalized information criterion. *J. Time Series Anal.*, 33(2):287–297, 2012.

Masanobu Taniguchi and Yoshihide Kakizawa. *Asymptotic Theory of Statistical Inference for Time Series*. Springer Series in Statistics. Springer-Verlag, New York, 2000.

Masanobu Taniguchi, Madan L. Puri, and Masao Kondo. Nonparametric approach for non-Gaussian vector stationary processes. *J. Multivariate Anal.*, 56(2):259–283, 1996.

Masanobu Taniguchi, Junichi Hirukawa, and Kenichiro Tamaki. *Optimal Statistical Inference in Financial Engineering*. Chapman & Hall/CRC, Boca Raton, FL, 2008.

Masanobu Taniguchi, Alexandre Petkovic, Takehiro Kase, Thomas DiCiccio, and Anna Clara Monti. Robust portfolio estimation under skew-normal return processes. *The European Journal of Finance*, 1–22, 2012.

Murad S. Taqqu. Law of the iterated logarithm for sums of non-linear functions of Gaussian variables that exhibit a long range dependence. *Probability Theory and Related Fields*, 40(3): 203–238, 1977.

L. Telser. Safety first and hedging. *Review of Economic and Statistics*, 23:1–16, 1955.

Andrew E. Teschendorff, Michel Journée, Pierre A. Absil, Rodolphe Sepulchre, and Carlos Caldas. Elucidating the altered transcriptional programs in breast cancer using independent component analysis. *PLoS Comput Biol*, 3(8):e161, 2007.

H. Theil and A. S. Goldberger. On pure and mixed statistical estimation in economics. *International Economic Review*, 2(1):65–78, 1961.

George C. Tiao and Ruey S. Tsay. Some advances in non-linear and adaptive modelling in time-series. *Journal of Forecasting*, 13(2):109–131, 1994.

Dag Tjøstheim. Estimation in nonlinear time series models. *Stochastic Process. Appl.*, 21(2): 251–273, 1986.

Howell Tong. *Non-linear time series: A dynamical system approach*. Oxford University Press, 1990.

Ruey S. Tsay. *Analysis of Financial Time Series*. Wiley Series in Probability and Statistics. John Wiley & Sons Inc., Hoboken, NJ, third edition, 2010.

Virginia Goss Tusher, Robert Tibshirani, and Gilbert Chu. Significance analysis of microarrays applied to the ionizing radiation response. *Proceedings of the National Academy of Sciences*, 98(9):5116–5121, 2001.

Amos Tversky and Daniel Kahneman. Prospect theory: An analysis of decision under risk. *Econometrica*, 47(2):263–292, 1979.

Amos Tversky and Daniel Kahneman. Advances in prospect theory: Cumulative representation of uncertainty. *Journal of Risk and Uncertainty*, 5(4):297–323, October 1992.

Roy van der Weide. GO-GARCH: a multivariate generalized orthogonal GARCH model. *Journal of Applied Econometrics*, 17(5):549–564, 2002.

Martin H. van Vliet, Fabien Reyal, Hugo M. Horlings, Marc J. van de Vijver, Marcel J. T. Reinders, and Lodewyk F. A. Wessels. Pooling breast cancer datasets has a synergetic effect on classification performance and improves signature stability. *BMC Genomics*, 9(1):375, 2008.

John von Neumann and Oskar Morgenstern. *Theory of Games and Economic Behavior*. Princeton University Press, Princeton, NJ, 1944.

I. D. Vrontos, P. Dellaportas, and D. N. Politis. A full-factor multivariate GARCH model. *The Econometrics Journal*, 6(2):312–334, 2003.

Shouyang Wang and Yusen Xia. *Portfolio Selection and Asset Pricing*, volume 514 of *Lecture Notes in Economics and Mathematical Systems*. Springer-Verlag, Berlin, 2002.

Frank Wilcoxon. Individual comparisons by ranking methods. *Biometrics Bulletin*, 1(6):80–83, 1945.

C.-F. J. Wu. Jackknife, bootstrap and other resampling methods in regression analysis. *Ann. Statist.*, 14(4):1261–1350, 1986. With discussion and a rejoinder by the author.

Yingying Xu, Zhuwu Wu, Long Jiang, and Xuefeng Song. A maximum entropy method for a robust portfolio problem. *Entropy*, 16(6):3401–3415, 2014.

Menahem E. Yaari. The dual theory of choice under risk. *Econometrica*, 55(1):95–115, January 1987.

Yee Hwa Yang, Yuanyuan Xiao, and Mark R. Segal. Identifying differentially expressed genes from microarray experiments via statistic synthesis. *Bioinformatics*, 21(7):1084–1093, 2005.

Nakahiro Yoshida. Estimation for diffusion processes from discrete observation. *J. Multivariate Anal.*, 41(2):220–242, 1992.

Ken-ichi Yoshihara. Limiting behavior of {U}-statistics for stationary, absolutely regular processes. *Z. Wahrscheinlichkeitstheorie und Verw. Gebiete*, 35(3):237–252, 1976.

Xi Zhao, Einar Andreas Rødland, Therese Sørlie, Bjørn Naume, Anita Langerød, Arnoldo Frigessi, Vessela N. Kristensen, Anne-Lise Børresen-Dale, and Ole Christian Lingjærde. Combining gene signatures improves prediction of breast cancer survival. *PLoS One*, 6(3):e17845, 2011.

Xi Zhao, Einar Andreas Rødland, Therese Sørlie, Hans Kristian Moen Vollan, Hege G. Russnes, Vessela N. Kristensen, Ole Christian Lingjærde, and Anne-Lise Børresen-Dale. Systematic assessment of prognostic gene signatures for breast cancer shows distinct influence of time and er status. *BMC Cancer*, 14(1):1, 2014.

Rongxi Zhou, Ru Cai, and Guanqun Tong. Applications of entropy in finance: A review. *Entropy*, 15(11):4909–4931, 2013.

Author Index

Subject Index

Milton Keynes UK
Ingram Content Group UK Ltd.
UKHW051931141024
449569UK00027B/1441